Low-Speed Aerodynamics, Second Edition

Low-speed aerodynamics is important in the design and operation of aircraft flying at low Mach number and of ground and marine vehicles. This book offers a modern treatment of the subject, both the theory of inviscid, incompressible, and irrotational aerodynamics and the computational techniques now available to solve complex problems.

A unique feature of the text is that the computational approach (from a single vortex element to a three-dimensional panel formulation) is interwoven throughout. Thus, the reader can learn about classical methods of the past, while also learning how to use numerical methods to solve real-world aerodynamic problems. This second edition, updates the first edition with a new chapter on the laminar boundary layer, the latest versions of computational techniques, and additional coverage of interaction problems. It includes a systematic treatment of two-dimensional panel methods and a detailed presentation of computational techniques for three-dimensional and unsteady flows. With extensive illustrations and examples, this book will be useful for senior and beginning graduate-level courses, as well as a helpful reference tool for practicing engineers.

Joseph Katz is Professor of Aerospace Engineering and Engineering Mechanics at San Diego State University.

Allen Plotkin is Professor of Aerospace Engineering and Engineering Mechanics at San Diego State University.

Cambridge Aerospace Series

Editors:
MICHAEL J. RYCROFT AND WEI SHYY

1. J. M. Rolfe and K. J. Staples (eds.): *Flight Simulation*
2. P. Berlin: *The Geostationary Applications Satellite*
3. M. J. T. Smith: *Aircraft Noise*
4. N. X. Vinh: *Flight Mechanics of High-Performance Aircraft*
5. W. A. Mair and D. L. Birdsall: *Aircraft Performance*
6. M. J. Abzug and E. E. Larrabee: *Airplane Stability and Control*
7. M. J. Sidi: *Spacecraft Dynamics and Control*
8. J. D. Anderson: *A History of Aerodynamics*
9. A. M. Cruise, J. A. Bowles, C. V. Goodall, and T. J. Patrick: *Principles of Space Instrument Design*
10. G. A. Khoury and J. D. Gillett (eds.): *Airship Technology*
11. J. Fielding: *Introduction to Aircraft Design*
12. J. G. Leishman: *Principles of Helicopter Aerodynamics*
13. J. Katz and A. Plotkin: *Low Speed Aerodynamics*, Second Edition

Low-Speed Aerodynamics
Second Edition

JOSEPH KATZ
San Diego State University

ALLEN PLOTKIN
San Diego State University

PUBLISHED BY THE PRESS SYNDICATE OF THE UNIVERSITY OF CAMBRIDGE
The Pitt Building, Trumpington Street, Cambridge, United Kingdom

CAMBRIDGE UNIVERSITY PRESS
The Edinburgh Building, Cambridge CB2 2RU, UK
40 West 20th Street, New York, NY 10011-4211, USA
10 Stamford Road, Oakleigh, VIC 3166, Australia
Ruiz de Alarcón 13, 28014 Madrid, Spain
Dock House, The Waterfront, Cape Town 8001, South Africa

http://www.cambridge.org

© Cambridge University Press 2001

This book is in copyright. Subject to statutory exception
and to the provisions of relevant collective licensing agreements,
no reproduction of any part may take place without
the written permission of Cambridge University Press.

First published 2001

Printed in the United States of America

Typeface Times New Roman 10/12 pt.   *System* LATEX 2_ε [TB]

A catalog record for this book is available from the British Library.

Library of Congress Cataloging in Publication Data
Katz, Joseph, 1947–
  Low speed aerodynamics / Joseph Katz, Allen Plotkin. – 2nd ed.
    p. cm. – (Cambridge aerospace series ; 13)
  ISBN 0-521-66219-2
  1. Aerodynamics.   I. Plotkin, Allen.   II. Title.   III. Series.
  TL570 .K34   2000
  629.132'3 – dc21   00-031270

ISBN 0 521 66219 2 hardback
ISBN 0 521 66552 3 paperback

Contents

Preface		*page* xiii
Preface to the First Edition		xv

1	Introduction and Background		1
	1.1	Description of Fluid Motion	1
	1.2	Choice of Coordinate System	2
	1.3	Pathlines, Streak Lines, and Streamlines	3
	1.4	Forces in a Fluid	4
	1.5	Integral Form of the Fluid Dynamic Equations	6
	1.6	Differential Form of the Fluid Dynamic Equations	8
	1.7	Dimensional Analysis of the Fluid Dynamic Equations	14
	1.8	Flow with High Reynolds Number	17
	1.9	Similarity of Flows	19

2	Fundamentals of Inviscid, Incompressible Flow		21
	2.1	Angular Velocity, Vorticity, and Circulation	21
	2.2	Rate of Change of Vorticity	24
	2.3	Rate of Change of Circulation: Kelvin's Theorem	25
	2.4	Irrotational Flow and the Velocity Potential	26
	2.5	Boundary and Infinity Conditions	27
	2.6	Bernoulli's Equation for the Pressure	28
	2.7	Simply and Multiply Connected Regions	29
	2.8	Uniqueness of the Solution	30
	2.9	Vortex Quantities	32
	2.10	Two-Dimensional Vortex	34
	2.11	The Biot–Savart Law	36
	2.12	The Velocity Induced by a Straight Vortex Segment	38
	2.13	The Stream Function	41

3	General Solution of the Incompressible, Potential Flow Equations		44
	3.1	Statement of the Potential Flow Problem	44
	3.2	The General Solution, Based on Green's Identity	44
	3.3	Summary: Methodology of Solution	48
	3.4	Basic Solution: Point Source	49
	3.5	Basic Solution: Point Doublet	51
	3.6	Basic Solution: Polynomials	54
	3.7	Two-Dimensional Version of the Basic Solutions	56
	3.8	Basic Solution: Vortex	58
	3.9	Principle of Superposition	60

3.10	Superposition of Sources and Free Stream: Rankine's Oval	60
3.11	Superposition of Doublet and Free Stream: Flow around a Cylinder	62
3.12	Superposition of a Three-Dimensional Doublet and Free Stream: Flow around a Sphere	67
3.13	Some Remarks about the Flow over the Cylinder and the Sphere	69
3.14	Surface Distribution of the Basic Solutions	70

4 Small-Disturbance Flow over Three-Dimensional Wings: Formulation of the Problem — 75

4.1	Definition of the Problem	75
4.2	The Boundary Condition on the Wing	76
4.3	Separation of the Thickness and the Lifting Problems	78
4.4	Symmetric Wing with Nonzero Thickness at Zero Angle of Attack	79
4.5	Zero-Thickness Cambered Wing at Angle of Attack–Lifting Surfaces	82
4.6	The Aerodynamic Loads	85
4.7	The Vortex Wake	88
4.8	Linearized Theory of Small-Disturbance Compressible Flow	90

5 Small-Disturbance Flow over Two-Dimensional Airfoils — 94

5.1	Symmetric Airfoil with Nonzero Thickness at Zero Angle of Attack	94
5.2	Zero-Thickness Airfoil at Angle of Attack	100
5.3	Classical Solution of the Lifting Problem	104
5.4	Aerodynamic Forces and Moments on a Thin Airfoil	106
5.5	The Lumped-Vortex Element	114
5.6	Summary and Conclusions from Thin Airfoil Theory	120

6 Exact Solutions with Complex Variables — 122

6.1	Summary of Complex Variable Theory	122
6.2	The Complex Potential	125
6.3	Simple Examples	126
6.3.1	Uniform Stream and Singular Solutions	126
6.3.2	Flow in a Corner	127
6.4	Blasius Formula, Kutta–Joukowski Theorem	128
6.5	Conformal Mapping and the Joukowski Transformation	128
6.5.1	Flat Plate Airfoil	130
6.5.2	Leading-Edge Suction	131
6.5.3	Flow Normal to a Flat Plate	133
6.5.4	Circular Arc Airfoil	134
6.5.5	Symmetric Joukowski Airfoil	135
6.6	Airfoil with Finite Trailing-Edge Angle	137
6.7	Summary of Pressure Distributions for Exact Airfoil Solutions	138
6.8	Method of Images	141
6.9	Generalized Kutta–Joukowski Theorem	146

7 Perturbation Methods — 151

7.1	Thin-Airfoil Problem	151
7.2	Second-Order Solution	154
7.3	Leading-Edge Solution	157

	7.4	Matched Asymptotic Expansions	160
	7.5	Thin Airfoil between Wind Tunnel Walls	163
8	Three-Dimensional Small-Disturbance Solutions		167
	8.1	Finite Wing: The Lifting Line Model	167
	8.1.1	Definition of the Problem	167
	8.1.2	The Lifting-Line Model	168
	8.1.3	The Aerodynamic Loads	172
	8.1.4	The Elliptic Lift Distribution	173
	8.1.5	General Spanwise Circulation Distribution	178
	8.1.6	Twisted Elliptic Wing	181
	8.1.7	Conclusions from Lifting-Line Theory	183
	8.2	Slender Wing Theory	184
	8.2.1	Definition of the Problem	184
	8.2.2	Solution of the Flow over Slender Pointed Wings	186
	8.2.3	The Method of R. T. Jones	192
	8.2.4	Conclusions from Slender Wing Theory	194
	8.3	Slender Body Theory	195
	8.3.1	Axisymmetric Longitudinal Flow Past a Slender Body of Revolution	196
	8.3.2	Transverse Flow Past a Slender Body of Revolution	198
	8.3.3	Pressure and Force Information	199
	8.3.4	Conclusions from Slender Body Theory	201
	8.4	Far Field Calculation of Induced Drag	201
9	Numerical (Panel) Methods		206
	9.1	Basic Formulation	206
	9.2	The Boundary Conditions	207
	9.3	Physical Considerations	209
	9.4	Reduction of the Problem to a Set of Linear Algebraic Equations	213
	9.5	Aerodynamic Loads	216
	9.6	Preliminary Considerations, Prior to Establishing Numerical Solutions	217
	9.7	Steps toward Constructing a Numerical Solution	220
	9.8	Example: Solution of Thin Airfoil with the Lumped-Vortex Element	222
	9.9	Accounting for Effects of Compressibility and Viscosity	226
10	Singularity Elements and Influence Coefficients		230
	10.1	Two-Dimensional Point Singularity Elements	230
	10.1.1	Two-Dimensional Point Source	230
	10.1.2	Two-Dimensional Point Doublet	231
	10.1.3	Two-Dimensional Point Vortex	231
	10.2	Two-Dimensional Constant-Strength Singularity Elements	232
	10.2.1	Constant-Strength Source Distribution	233
	10.2.2	Constant-Strength Doublet Distribution	235
	10.2.3	Constant-Strength Vortex Distribution	236
	10.3	Two-Dimensional Linear-Strength Singularity Elements	237
	10.3.1	Linear Source Distribution	238

10.3.2	Linear Doublet Distribution	239
10.3.3	Linear Vortex Distribution	241
10.3.4	Quadratic Doublet Distribution	242
10.4	Three-Dimensional Constant-Strength Singularity Elements	244
10.4.1	Quadrilateral Source	245
10.4.2	Quadrilateral Doublet	247
10.4.3	Constant Doublet Panel Equivalence to Vortex Ring	250
10.4.4	Comparison of Near and Far Field Formulas	251
10.4.5	Constant-Strength Vortex Line Segment	251
10.4.6	Vortex Ring	255
10.4.7	Horseshoe Vortex	256
10.5	Three-Dimensional Higher Order Elements	258

11 Two-Dimensional Numerical Solutions — 262

11.1	Point Singularity Solutions	262
11.1.1	Discrete Vortex Method	263
11.1.2	Discrete Source Method	272
11.2	Constant-Strength Singularity Solutions (Using the Neumann B.C.)	276
11.2.1	Constant Strength Source Method	276
11.2.2	Constant-Strength Doublet Method	280
11.2.3	Constant-Strength Vortex Method	284
11.3	Constant-Potential (Dirichlet Boundary Condition) Methods	288
11.3.1	Combined Source and Doublet Method	290
11.3.2	Constant-Strength Doublet Method	294
11.4	Linearly Varying Singularity Strength Methods (Using the Neumann B.C.)	298
11.4.1	Linear-Strength Source Method	299
11.4.2	Linear-Strength Vortex Method	303
11.5	Linearly Varying Singularity Strength Methods (Using the Dirichlet B.C.)	306
11.5.1	Linear Source/Doublet Method	306
11.5.2	Linear Doublet Method	312
11.6	Methods Based on Quadratic Doublet Distribution (Using the Dirichlet B.C.)	315
11.6.1	Linear Source/Quadratic Doublet Method	315
11.6.2	Quadratic Doublet Method	320
11.7	Some Conclusions about Panel Methods	323

12 Three-Dimensional Numerical Solutions — 331

12.1	Lifting-Line Solution by Horseshoe Elements	331
12.2	Modeling of Symmetry and Reflections from Solid Boundaries	338
12.3	Lifting-Surface Solution by Vortex Ring Elements	340
12.4	Introduction to Panel Codes: A Brief History	351
12.5	First-Order Potential-Based Panel Methods	353
12.6	Higher Order Panel Methods	358
12.7	Sample Solutions with Panel Codes	360

13	Unsteady Incompressible Potential Flow		369
	13.1	Formulation of the Problem and Choice of Coordinates	369
	13.2	Method of Solution	373
	13.3	Additional Physical Considerations	375
	13.4	Computation of Pressures	376
	13.5	Examples for the Unsteady Boundary Condition	377
	13.6	Summary of Solution Methodology	380
	13.7	Sudden Acceleration of a Flat Plate	381
	13.7.1	The Added Mass	385
	13.8	Unsteady Motion of a Two-Dimensional Thin Airfoil	387
	13.8.1	Kinematics	388
	13.8.2	Wake Model	389
	13.8.3	Solution by the Time-Stepping Method	391
	13.8.4	Fluid Dynamic Loads	394
	13.9	Unsteady Motion of a Slender Wing	400
	13.9.1	Kinematics	401
	13.9.2	Solution of the Flow over the Unsteady Slender Wing	401
	13.10	Algorithm for Unsteady Airfoil Using the Lumped-Vortex Element	407
	13.11	Some Remarks about the Unsteady Kutta Condition	416
	13.12	Unsteady Lifting-Surface Solution by Vortex Ring Elements	419
	13.13	Unsteady Panel Methods	433
14	The Laminar Boundary Layer		448
	14.1	The Concept of the Boundary Layer	448
	14.2	Boundary Layer on a Curved Surface	452
	14.3	Similar Solutions to the Boundary Layer Equations	457
	14.4	The von Karman Integral Momentum Equation	463
	14.5	Solutions Using the von Karman Integral Equation	467
	14.5.1	Approximate Polynomial Solution	468
	14.5.2	The Correlation Method of Thwaites	469
	14.6	Weak Interactions, the Goldstein Singularity, and Wakes	471
	14.7	Two-Equation Integral Boundary Layer Method	473
	14.8	Viscous–Inviscid Interaction Method	475
	14.9	Concluding Example: The Flow over a Symmetric Airfoil	479
15	Enhancement of the Potential Flow Model		483
	15.1	Wake Rollup	483
	15.2	Coupling between Potential Flow and Boundary Layer Solvers	487
	15.2.1	The Laminar/Turbulent Boundary Layer and Transition	487
	15.2.2	Viscous–Inviscid Coupling, Including Turbulent Boundary Layer	491
	15.3	Influence of Viscous Flow Effects on Airfoil Design	495
	15.3.1	Low Drag Considerations	498
	15.3.2	High Lift Considerations	499
	15.4	Flow over Wings at High Angles of Attack	505

	15.4.1 Flow Separation on Wings with Unswept Leading Edge – Experimental Observations	508
	15.4.2 Flow Separation on Wings with Unswept Leading Edge – Modeling	510
	15.4.3 Flow Separation on Wings with Highly Swept Leading Edge – Experimental Observations	516
	15.4.4 Modeling of Highly Swept Leading-Edge Separation	523
	15.5 Possible Additional Features of Panel Codes	528
A	Airfoil Integrals	537
B	Singularity Distribution Integrals	540
C	Principal Value of the Lifting Surface Integral I_L	545
D	Sample Computer Programs	546

Index   611

Preface

Our goal in writing this Second Edition of *Low-Speed Aerodynamics* remains the same, to present a comprehensive and up-to-date treatment of the subject of inviscid, incompressible, and irrotational aerodynamics. It is still true that for most practical aerodynamic and hydrodynamic problems, the classical model of a thin viscous boundary layer along a body's surface, surrounded by a mainly inviscid flowfield, has produced important engineering results. This approach requires first the solution of the inviscid flow to obtain the pressure field and consequently the forces such as lift and induced drag. Then, a solution of the viscous flow in the thin boundary layer allows for the calculation of the skin friction effects.

The First Edition provides the theory and related computational methods for the solution of the inviscid flow problem. This material is complemented in the Second Edition with a new Chapter 14, "The Laminar Boundary Layer," whose goal is to provide a modern discussion of the coupling of the inviscid outer flow with the viscous boundary layer. First, an introduction to the classical boundary-layer theory of Prandtl is presented. The need for an interactive approach (to replace the classical sequential one) to the coupling is discussed and a viscous–inviscid interaction method is presented. Examples for extending this approach, which include transition to turbulence, are provided in the final Chapter 15.

In addition, updated versions of the computational methods are presented and several topics are improved and updated throughout the text. For example, more coverage is given of aerodynamic interaction problems such as multiple wings, ground effect, wall corrections, and the presence of a free surface.

We would like to thank Turgut Sarpkaya of the Naval Postgraduate School and H. K. Cheng of USC for their input in Chapter 14 and particularly Mark Drela of MIT who provided a detailed description of his solution technique, which formed the basis for the material in Sections 14.7 and 14.8. Finally, we would like to acknowledge the continuing love and support of our wives, Hilda Katz and Selena Plotkin.

Preface to the First Edition

Our goal in writing this book is to present a comprehensive and up-to-date treatment of the subject of inviscid, incompressible, and irrotational aerodynamics. Over the last several years there has been a widespread use of computational (surface singularity) methods for the solution of problems of concern to the low-speed aerodynamicist and a need has developed for a text to provide the theoretical basis for these methods as well as to provide a smooth transition from the classical small-disturbance methods of the past to the computational methods of the present. This book was written in response to this need. A unique feature of this book is that the computational approach (from a single vortex element to a three-dimensional panel formulation) is interwoven throughout so that it serves as a teaching tool in the understanding of the classical methods as well as a vehicle for the reader to obtain solutions to complex problems that previously could not be dealt with in the context of a textbook. The reader will be introduced to different levels of complexity in the numerical modeling of an aerodynamic problem and will be able to assemble codes to implement a solution.

We have purposely limited our scope to inviscid, incompressible, and irrotational aerodynamics so that we can present a truly comprehensive coverage of the material. The book brings together topics currently scattered throughout the literature. It provides a detailed presentation of computational techniques for three-dimensional and unsteady flows. It includes a systematic and detailed (including computer programs) treatment of two-dimensional panel methods with variations in singularity type, order of singularity, Neumann or Dirichlet boundary conditions, and velocity or potential-based approaches.

This book is divided into three main parts. In the first, Chapters 1–3, the basic theory is developed. In the second part, Chapters 4–8, an analytical approach to the solution of the problem is taken. Chapters 4, 5, and 8 deal with the small-disturbance version of the problem and the classical methods of thin-airfoil theory, lifting line theory, slender wing theory, and slender body theory. In this part exact solutions via complex variable theory and perturbation methods for obtaining higher-order small disturbance approximations are also included. The third part, Chapters 9–14, presents a systematic treatment of the surface singularity distribution technique for obtaining numerical solutions for incompressible potential flows. A general methodology for assembling a numerical solution is developed and applied to a series of increasingly complex aerodynamic elements (two-dimensional, three-dimensional, and unsteady problems are treated).

The book is designed to be used as a textbook for a course in low-speed aerodynamics at either the advanced senior or first-year graduate levels. The complete text can be covered in a one-year course and a one-quarter or one-semester course can be constructed by choosing the topics that the instructor would like to emphasize. For example, a senior elective course which concentrated on two-dimensional steady aerodynamics might include Chapters 1–3, 4, 5, 9, 11, 8, 12, and 14. A traditional graduate course which emphasized an analytical treatment of the subject might include Chapters 1–3, 4, 5–7, 8, 9, and 13 and a course which emphasized a numerical approach (panel methods) might include Chapters 1–3 and 9–14 and a treatment of pre- and postprocessors. It has been assumed that the reader has taken a first course in fluid mechanics and has a mathematical background which includes an exposure to vector calculus, partial differential equations, and complex variables.

We believe that the topics covered by this text are needed by the fluid dynamicist because of the complex nature of the fluid dynamic equations which has led to a mainly experimental approach for dealing with most engineering research and development programs. In a wider sense, such an approach uses tools such as wind tunnels or large computer codes where the engineer/user is experimenting and testing ideas with some trial and error logic in mind. Therefore, even in the era of supercomputers and sophisticated experimental tools, there is a need for simplified models that allow for an easy grasp of the dominant physical effects (e.g., having a simple lifting vortex in mind, one can immediately tell that the first wing in a tandem formation has the larger lift).

For most practical aerodynamic and hydrodynamic problems, the classical model of a thin viscous boundary layer along a body's surface, surrounded by a mainly inviscid flowfield, has produced important engineering results. This approach requires first the solution of the inviscid flow to obtain the pressure field and consequently the forces such as lift and induced drag. Then, a solution of the viscous flow in the thin boundary layer allows for the calculation of the skin friction effects. This methodology has been used successfully throughout the twentieth century for most airplane and marine vessel designs. Recently, due to developments in computer capacity and speed, the inviscid flowfield over complex and detailed geometries (such as airplanes, cars, etc.) can be computed by this approach (panel methods). Thus, for the near future, since these methods are the main tools of low-speed aerodynamicists all over the world, a need exists for a clear and systematic explanation of how and why (and for which cases) these methods work. This book is one attempt to respond to this need.

We would like to thank graduate students Lindsey Browne and especially Steven Yon who developed the two-dimensional panel codes in Chapter 11 and checked the integrals in Chapter 10. Allen Plotkin would like to thank his teachers Richard Skalak, Krishnamurthy Karamcheti, Milton Van Dyke, and Irmgard Flugge-Lotz, his parents Claire and Oscar for their love and support, and his children Jennifer Anne and Samantha Rose and especially his wife Selena for their love, support, and patience. Joseph Katz would like to thank his parents Janka and Jeno, his children Shirley, Ronny, and Danny, and his wife Hilda for their love, support, and patience. The support of the Low-Speed Aerodynamic Branch at NASA Ames is acknowledged by Joseph Katz for their inspiration that initiated this project and for their help during past years in the various stages of developing the methods presented in this book.

CHAPTER 1

Introduction and Background

The differential equations that are generally used in the solution of problems relevant to low-speed aerodynamics are a simplified version of the governing equations of fluid dynamics. Also, most engineers when faced with finding a solution to a practical aerodynamic problem, find themselves operating large computer codes rather than developing simple analytical models to guide them in their analysis. For this reason, it is important to start with a brief development of the principles upon which the general fluid dynamic equations are based. Then we will be in a position to consider the physical reasoning behind the assumptions introduced to generate simplified versions of the equations that still correctly model the aerodynamic phenomena being studied. It is hoped that this approach will give the engineer the ability to appreciate both the power and the limitations of the techniques that will be presented in this text. In this chapter we will derive the conservation of mass and momentum balance equations and show how they are reduced to obtain the equations that will be used in the rest of the text to model flows of interest to the low-speed aerodynamicist.

1.1   Description of Fluid Motion

The fluid being studied here is modeled as a continuum, and infinitesimally small regions of the fluid (with a fixed mass) are called fluid elements or fluid particles. The motion of the fluid can be described by two different methods. One adopts the particle point of view and follows the motion of the individual particles. The other adopts the field point of view and provides the flow variables as functions of position in space and time.

The particle point of view, which uses the approach of classical mechanics, is called the *Lagrangian method*. To trace the motion of each fluid particle, it is convenient to introduce a Cartesian coordinate system with the coordinates x, y, and z. The position of any fluid particle P (see Fig. 1.1) is then given by

$$\begin{aligned} x &= x_P(x_0, y_0, z_0, t) \\ y &= y_P(x_0, y_0, z_0, t) \\ z &= z_P(x_0, y_0, z_0, t) \end{aligned} \tag{1.1}$$

where (x_0, y_0, z_0) is the position of P at some initial time $t = 0$. (Note that the quantity (x_0, y_0, z_0) represents the vector with components x_0, y_0, and z_0.) The components of the velocity of this particle are then given by

$$\begin{aligned} u &= \partial x/\partial t \\ v &= \partial y/\partial t \\ w &= \partial z/\partial t \end{aligned} \tag{1.2}$$

and those of the acceleration by

$$\begin{aligned} a_x &= \partial^2 x/\partial t^2 \\ a_y &= \partial^2 y/\partial t^2 \\ a_z &= \partial^2 z/\partial t^2 \end{aligned} \tag{1.3}$$

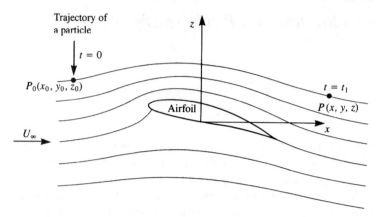

Figure 1.1 Particle trajectory lines in a steady-state flow over an airfoil as viewed from a body-fixed coordinate system.

The Lagrangian formulation requires the evaluation of the motion of each fluid particle. For most practical applications this abundance of information is neither necessary nor useful and the analysis is cumbersome.

The field point of view, called the *Eulerian method*, provides the spatial distribution of flow variables at each instant during the motion. For example, if a Cartesian coordinate system is used, the components of the fluid velocity are given by

$$
\begin{aligned}
u &= u(x, y, z, t) \\
v &= v(x, y, z, t) \\
w &= w(x, y, z, t)
\end{aligned}
\tag{1.4}
$$

The Eulerian approach provides information about the fluid variables that is consistent with the information supplied by most experimental techniques and that is in a form appropriate for most practical applications. For these reasons the Eulerian description of fluid motion is the most widely used.

1.2 Choice of Coordinate System

For the following chapters, when possible, primarily a Cartesian coordinate system will be used. Other coordinate systems such as curvilinear, cylindrical, spherical, etc. will be introduced and used if necessary, mainly to simplify the treatment of certain problems. Also, from the kinematic point of view, a careful choice of a coordinate system can considerably simplify the solution of a problem. As an example, consider the forward motion of an airfoil, with a constant speed U_∞, in a fluid that is otherwise at rest – as shown in Fig. 1.1. Here, the origin of the coordinate system is attached to the moving airfoil and the trajectory of a fluid particle inserted at point P_0 at $t = 0$ is shown in the figure. By following the trajectories of several particles a more complete description of the flowfield is obtained in the figure. It is important to observe that for a constant-velocity forward motion of the airfoil, in this frame of reference, these trajectory lines become independent of time. That is, if various particles are introduced at the same point in space, then they will follow the same trajectory.

Now let us examine the same flow, but from a coordinate system that is fixed relative to the undisturbed fluid. At $t = 0$, the airfoil was at the right side of Fig. 1.2 and as a result of its constant-velocity forward motion (with a speed U_∞ toward the left side of the page), later at $t = t_1$ it has moved to the new position indicated in the figure. A typical particle's

1.3 Pathlines, Streak Lines, and Streamlines

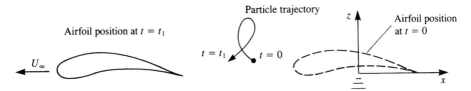

Figure 1.2 Particle trajectory line for the airfoil of Fig. 1.1 as viewed from a stationary inertial frame.

trajectory line between $t = 0$ and $t = t_1$, for this case, is shown in Fig. 1.2. The particle's motion now depends on time, and a new trajectory has to be established for each particle.

This simple example depicts the importance of good coordinate system selection. For many problems where a constant velocity and a fixed geometry (with time) are present, the use of a body-fixed frame of reference will result in a steady or time-independent flow.

1.3 Pathlines, Streak Lines, and Streamlines

Three sets of curves are normally associated with providing a pictorial description of a fluid motion: pathlines, streak lines, and streamlines.

Pathlines: A curve describing the trajectory of a fluid element is called a pathline or a particle path. Pathlines are obtained in the Lagrangian approach by an integration of the equations of dynamics for each fluid particle. If the velocity field of a fluid motion is given in the Eulerian framework by Eq. (1.4) in a body-fixed frame, the pathline for a particle at P_0 in Fig. 1.1 can be obtained by an integration of the velocity. For steady flows the pathlines in the body-fixed frame become independent of time and can be drawn as in the case of flow over the airfoil shown in Fig. 1.1.

Streak Lines: In many cases of experimental flow visualization, particles (e.g., dye or smoke) are introduced into the flow at a fixed point in space. The line connecting all of these particles is called a streak line. To construct streak lines using the Lagrangian approach, draw a series of pathlines for particles passing through a given point in space and, at a particular instant in time, connect the ends of these pathlines.

Streamlines: Another set of curves can be obtained (at a given time) by lines that are parallel to the local velocity vector. To express analytically the equation of a streamline at a certain instant of time, at any point P in the fluid, the velocity[1] $\mathbf{q}$ must be parallel to the streamline element $d\mathbf{l}$ (Fig. 1.3). Therefore, on a streamline:

$$\mathbf{q} \times d\mathbf{l} = 0 \qquad (1.5)$$

If the velocity vector is $\mathbf{q} = (u, v, w)$, then the vector equation (Eq. (1.5)) reduces to the following scalar equations:

$$\begin{aligned} w\,dy - v\,dz &= 0 \\ u\,dz - w\,dx &= 0 \\ v\,dx - u\,dy &= 0 \end{aligned} \qquad (1.6)$$

or in a differential equation form:

$$\frac{dx}{u} = \frac{dy}{v} = \frac{dz}{w} \qquad (1.6a)$$

---

[1] Bold letters in this book represent vectors.

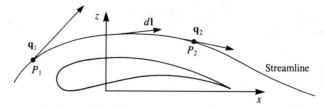

Figure 1.3 Description of a streamline.

In Eq. (1.6a), the velocity (u, v, w) is a function of the coordinates and of time. However, for steady flows the streamlines are independent of time and streamlines, pathlines, and streak lines become identical, as shown in Fig. 1.1.

1.4   Forces in a Fluid

Prior to discussing the dynamics of fluid motion, the types of forces that act on a fluid element should be identified. Here, we consider forces such as body forces per unit mass **f** and surface forces resulting from the stress vector **t**. The body forces are independent of any contact with the fluid, as in the case of gravitational or magnetic forces, and their magnitude is proportional to the local mass.

To define the stress vector **t** at a point, consider the force **F** acting on a planar area S (shown in Fig. 1.4) with **n** being an outward normal to S. Then

$$\mathbf{t} = \lim_{S \to 0} \left(\frac{\mathbf{F}}{S} \right)$$

To obtain the components of the stress vector consider the force equilibrium on an infinitesimal tetrahedral fluid element, shown in Fig. 1.5. According to Batchelor[1.1] (p. 10) this equilibrium yields the components in the x_1, x_2, and x_3 directions:

$$t_i = \sum_{j=1}^{3} \tau_{ij} n_j, \quad i = 1, 2, 3 \tag{1.7}$$

where the subscripts 1, 2, and 3 denote the three coordinate directions. A similar treatment of the moment equilibrium results in the symmetry of the stress vector components so that $\tau_{ij} = \tau_{ji}$.

These stress components τ_{ij} are shown schematically on a cubical element in Fig. 1.6. Note that τ_{ij} acts in the x_i direction on a surface whose outward normal points in the x_j direction. This indicial notation allows a simpler presentation of the equations, and the

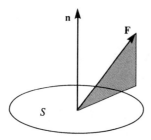

Figure 1.4 Force **F** acting on a surface S.

1.4 Forces in a Fluid

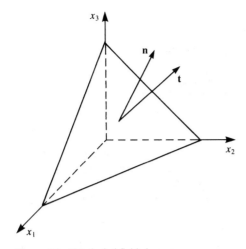

Figure 1.5 Tetrahedral fluid element.

subscripts 1, 2, and 3 denote the coordinate directions x, y, and z, respectively. For example,

$$x_1 = x, \qquad x_2 = y, \qquad x_3 = z$$

and

$$q_1 = u, \qquad q_2 = v, \qquad q_3 = w$$

The stress components shown on the cubical fluid element of Fig. 1.6 can be summarized in a matrix form or in an indicial form as follows:

$$\begin{pmatrix} \tau_{xx} & \tau_{xy} & \tau_{xz} \\ \tau_{yx} & \tau_{yy} & \tau_{yz} \\ \tau_{zx} & \tau_{zy} & \tau_{zz} \end{pmatrix} = \begin{pmatrix} \tau_{11} & \tau_{12} & \tau_{13} \\ \tau_{21} & \tau_{22} & \tau_{23} \\ \tau_{31} & \tau_{32} & \tau_{33} \end{pmatrix} = \tau_{ij} \tag{1.8}$$

Also, it is customary to sum over any index that is repeated such that

$$\sum_{j=1}^{3} \tau_{ij} n_j \equiv \tau_{ij} n_j \qquad \text{for } i = 1, 2, 3 \tag{1.9}$$

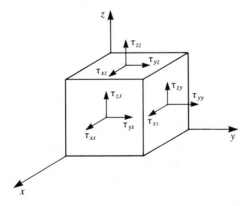

Figure 1.6 Stress components on a cubical fluid element.

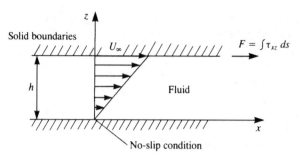

Figure 1.7 Flow between a stationary (lower) and a moving (upper) plate.

and to interpret an equation with a free index (as i in Eq. (1.9)) as being valid for all values of that index.

For a Newtonian fluid (where the stress components τ_{ij} are linear in the derivatives $\partial q_i/\partial x_j$), the stress components are related to the velocity field by (see, for example, Batchelor,[1.1] p. 147)

$$\tau_{ij} = \left(-p - \frac{2}{3}\mu\frac{\partial q_k}{\partial x_k}\right)\delta_{ij} + \mu\left(\frac{\partial q_i}{\partial x_j} + \frac{\partial q_j}{\partial x_i}\right) \quad (1.10)$$

where μ is the viscosity coefficient, p is the pressure, the dummy variable k is summed from 1 to 3, and δ_{ij} is the Kronecker delta function defined by

$$\delta_{ij} \equiv \begin{pmatrix} 1, & i = j \\ 0, & i \neq j \end{pmatrix}$$

When the fluid is at rest, the tangential stresses vanish and the normal stress component becomes simply the pressure. Thus the stress components become

$$\tau_{ij} = \begin{pmatrix} -p & 0 & 0 \\ 0 & -p & 0 \\ 0 & 0 & -p \end{pmatrix} \quad (1.11)$$

Another interesting case of Eq. (1.10) is the one-degree-of-freedom shear flow between a stationary and a moving infinite plate with a velocity U_∞ (shown in Fig. 1.7), without pressure gradients. This flow is called *Couette flow* (see, for example, Yuan,[1.2] p. 260) and the shear stress becomes

$$\tau_{xz} = \mu\frac{\partial u}{\partial z} = \frac{\mu U_\infty}{h} \quad (1.12)$$

Since there is no pressure gradient in the flow, the fluid motion in the x direction is entirely due to the action of the viscous forces. The force **F** on the plate can be found by integrating τ_{xz} on the upper moving surface.

1.5   Integral Form of the Fluid Dynamic Equations

To develop the governing integral and differential equations describing the fluid motion, the various properties of the fluid are investigated in an arbitrary control volume that is stationary and submerged in the fluid (Fig. 1.8). These properties can be density, momentum, energy, etc., and any change with time of one of them for the fluid flowing through the control volume is the sum of the accumulation of the property in the control volume and the transfer of this property out of the control volume through its boundaries.

1.5 Integral Form of the Fluid Dynamic Equations

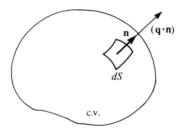

Figure 1.8 A control volume in the fluid.

For example, the conservation of mass can be analyzed by observing the changes in fluid density ρ for the control volume (c.v.). The mass $m_{c.v.}$ within the control volume is then

$$m_{c.v.} = \int_{c.v.} \rho \, dV \qquad (1.13)$$

where dV is the volume element. The accumulation of mass within the control volume is

$$\frac{\partial m_{c.v.}}{\partial t} = \frac{\partial}{\partial t} \int_{c.v.} \rho \, dV \qquad (1.13a)$$

The change in the mass within the control volume, due to the mass leaving (m_{out}) and to the mass entering (m_{in}) through the boundaries (c.s.), is

$$m_{\text{out}} - m_{\text{in}} = \int_{c.s.} \rho(\mathbf{q} \cdot \mathbf{n}) \, dS \qquad (1.14)$$

where $\mathbf{q}$ is the velocity vector (u, v, w) and $\rho \mathbf{q} \cdot \mathbf{n}$ is the rate of mass leaving across and normal to the surface element dS ($\mathbf{n}$ is the outward normal), as shown in Fig. 1.8. Since mass is conserved, and no new material is being produced, then the sum of Eq. (1.13a) and Eq. (1.14) must be equal to zero:

$$\frac{dm_{c.v.}}{dt} = \frac{\partial}{\partial t} \int_{c.v.} \rho \, dV + \int_{c.s.} \rho(\mathbf{q} \cdot \mathbf{n}) \, dS = 0 \qquad (1.15)$$

Equation (1.15) is the integral representation of the conservation of mass. It simply states that any change in the mass of the fluid in the control volume is equal to the rate of mass being transported across the control surface (c.s.) boundaries.

In a similar manner the rate of change in the momentum of the fluid flowing through the control volume at any instant $d(m\mathbf{q})_{c.v.}/dt$ is the sum of the accumulation of the momentum per unit volume $\rho \mathbf{q}$ within the control volume and of the change of the momentum across the control surface boundaries:

$$\frac{d(m\mathbf{q})_{c.v.}}{dt} = \frac{\partial}{\partial t} \int_{c.v.} \rho \mathbf{q} \, dV + \int_{c.s.} \rho \mathbf{q}(\mathbf{q} \cdot \mathbf{n}) \, dS \qquad (1.16)$$

This change in the momentum, as given in Eq. (1.16), according to Newton's second law must be equal to the forces $\Sigma \mathbf{F}$ applied to the fluid inside the control volume:

$$\frac{d(m\mathbf{q})_{c.v.}}{dt} = \Sigma \mathbf{F} \qquad (1.17)$$

The forces acting on the fluid in the control volume in the x_i direction are either body forces ρf_i per unit volume or surface forces $n_j \tau_{ij}$ per unit area, as discussed in Section 1.4:

$$(\Sigma \mathbf{F})_i = \int_{c.v.} \rho f_i \, dV + \int_{c.s.} n_j \tau_{ij} \, dS \qquad (1.18)$$

where $\mathbf{n}$ is the unit normal vector that points outward from the control volume.

By substituting Eqs. (1.16) and (1.18) into Eq. (1.17), the integral form of the momentum equation in the i direction is obtained:

$$\frac{\partial}{\partial t}\int_{c.v.} \rho q_i\, dV + \int_{c.s.} \rho q_i (\mathbf{q}\cdot\mathbf{n})\, dS = \int_{c.v.} \rho f_i\, dV + \int_{c.s.} n_j \tau_{ij}\, dS \qquad (1.19)$$

This approach can be used to develop additional governing equations, such as the energy equation. However, for the fluid dynamic cases that are being considered here, the mass and the momentum equations are sufficient to describe the fluid motion.

1.6 Differential Form of the Fluid Dynamic Equations

Equations (1.15) and (1.19) are the integral forms of the conservation of mass and momentum equations. In many cases, though, the differential representation is more useful. In order to derive the differential form of the conservation of mass equation, both integrals of Eq. (1.15) should be volume integrals. This can be accomplished by the use of the divergence theorem (see, Kellogg,[1.3] p. 39), which states that for a vector $\mathbf{q}$:

$$\int_{c.s.} \mathbf{n}\cdot\mathbf{q}\, dS = \int_{c.v.} \nabla\cdot\mathbf{q}\, dV \qquad (1.20)$$

If $\mathbf{q}$ is the flow velocity vector then this equation states that the fluid flux through the boundary of the control surface (left-hand side) is equal to the rate of expansion of the fluid (right-hand side) inside the control volume. In Eq. (1.20), ∇ is the gradient operator, which, in Cartesian coordinates, is

$$\nabla = \mathbf{i}\frac{\partial}{\partial x} + \mathbf{j}\frac{\partial}{\partial y} + \mathbf{k}\frac{\partial}{\partial z}$$

or in indicial form

$$\nabla = \mathbf{e}_j \frac{\partial}{\partial x_j}$$

where $\mathbf{e}_j$ is the unit vector ($\mathbf{i}$, $\mathbf{j}$, $\mathbf{k}$, for $j = 1, 2, 3$). Thus the indicial form of the divergence theorem becomes

$$\int_{c.s.} n_j q_j\, dS = \int_{c.v.} \frac{\partial q_j}{\partial x_j}\, dV \qquad (1.20a)$$

An application of Eq. (1.20) to the surface integral term in Eq. (1.15) transforms it to a volume integral:

$$\int_{c.s.} \rho(\mathbf{q}\cdot\mathbf{n})\, dS = \int_{c.v.} (\nabla\cdot\rho\mathbf{q})\, dV$$

This allows the two terms to be combined as one volume integral:

$$\int_{c.v.} \left(\frac{\partial \rho}{\partial t} + \nabla\cdot\rho\mathbf{q}\right) dV = 0$$

where the time derivative is taken inside the integral since the control volume is stationary. Because the equation must hold for an arbitrary control volume anywhere in the fluid, the integrand is also equal to zero. Thus, the following differential form of the conservation of mass or the continuity equation is obtained:

$$\frac{\partial \rho}{\partial t} + \nabla\cdot\rho\mathbf{q} = 0 \qquad (1.21)$$

1.6 Differential Form of the Fluid Dynamic Equations

Expansion of the second term of Eq. (1.21) yields

$$\frac{\partial \rho}{\partial t} + \mathbf{q} \cdot \nabla \rho + \rho \nabla \cdot \mathbf{q} = 0 \tag{1.21a}$$

and in Cartesian coordinates

$$\frac{\partial \rho}{\partial t} + u\frac{\partial \rho}{\partial x} + v\frac{\partial \rho}{\partial y} + w\frac{\partial \rho}{\partial z} + \rho\left(\frac{\partial u}{\partial x} + \frac{\partial v}{\partial y} + \frac{\partial w}{\partial z}\right) = 0 \tag{1.21b}$$

Use of the material derivative

$$\frac{D}{Dt} \equiv \frac{\partial}{\partial t} + \mathbf{q} \cdot \nabla = \frac{\partial}{\partial t} + u\frac{\partial}{\partial x} + v\frac{\partial}{\partial y} + w\frac{\partial}{\partial z}$$

transforms Eq. (1.21) into

$$\frac{D\rho}{Dt} + \rho \nabla \cdot \mathbf{q} = 0 \tag{1.21c}$$

The material derivative D/Dt represents the rate of change following a fluid particle. For example, the acceleration of a fluid particle is given by

$$\mathbf{a} = \frac{D\mathbf{q}}{Dt} = \frac{\partial \mathbf{q}}{\partial t} + \mathbf{q} \cdot \nabla \mathbf{q} \tag{1.22}$$

An incompressible fluid is a fluid whose elements cannot experience volume change. Since by definition the mass of a fluid element is constant, the fluid elements of an incompressible fluid must have constant density. (A homogeneous incompressible fluid is therefore a constant-density fluid.) The continuity equation (Eq. (1.21)) for an incompressible fluid reduces to

$$\nabla \cdot \mathbf{q} = \frac{\partial u}{\partial x} + \frac{\partial v}{\partial y} + \frac{\partial w}{\partial z} = 0 \tag{1.23}$$

Note that the incompressible continuity equation does not have time derivatives (but time dependency can be introduced via time-dependent boundary conditions).

To obtain the differential form of the momentum equation, the divergence theorem (Eq. (1.20a)) is applied to the surface integral terms of Eq. (1.19):

$$\int_{c.s.} \rho q_i (\mathbf{q} \cdot \mathbf{n}) \, dS = \int_{c.v.} \nabla \cdot \rho q_i \mathbf{q} \, dV$$

$$\int_{c.s.} n_j \tau_{ij} \, dS = \int_{c.v.} \frac{\partial \tau_{ij}}{\partial x_j} \, dV$$

Substitution of these results into Eq. (1.19) yields

$$\int_{c.v.} \left[\frac{\partial}{\partial t}(\rho q_i) + \nabla \cdot \rho q_i \mathbf{q} - \rho f_i - \frac{\partial \tau_{ij}}{\partial x_j}\right] dV = 0 \tag{1.24}$$

Since this integral holds for an arbitrary control volume, the integrand must be zero and therefore

$$\frac{\partial}{\partial t}(\rho q_i) + \nabla \cdot \rho q_i \mathbf{q} = \rho f_i + \frac{\partial \tau_{ij}}{\partial x_j} \quad (i = 1, 2, 3) \tag{1.25}$$

Expanding the left-hand side of Eq. (1.25) first, and then using the continuity equation, we can reduce the left-hand side to

$$\frac{\partial}{\partial t}(\rho q_i) + \nabla \cdot (\rho q_i \mathbf{q}) = q_i \left[\frac{\partial \rho}{\partial t} + \nabla \cdot \rho \mathbf{q}\right] + \rho \left[\frac{\partial q_i}{\partial t} + \mathbf{q} \cdot \nabla q_i\right] = \rho \frac{Dq_i}{Dt}$$

(Note that the fluid acceleration is

$$a_i = \frac{Dq_i}{Dt}$$

which, according to Newton's second law, when multiplied by the mass per volume must be equal to ΣF_i.)

So, after substituting this form of the acceleration term into Eq. (1.25), the differential form of the momentum equation becomes $\rho a_i = \Sigma F_i$ or

$$\rho \frac{Dq_i}{Dt} = \rho f_i + \frac{\partial \tau_{ij}}{\partial x_j} \quad (i = 1, 2, 3) \tag{1.26}$$

and in Cartesian coordinates

$$\rho \left(\frac{\partial u}{\partial t} + u \frac{\partial u}{\partial x} + v \frac{\partial u}{\partial y} + w \frac{\partial u}{\partial z} \right) = \Sigma F_x = \rho f_x + \frac{\partial \tau_{xx}}{\partial x} + \frac{\partial \tau_{xy}}{\partial y} + \frac{\partial \tau_{xz}}{\partial z} \tag{1.26a}$$

$$\rho \left(\frac{\partial v}{\partial t} + u \frac{\partial v}{\partial x} + v \frac{\partial v}{\partial y} + w \frac{\partial v}{\partial z} \right) = \Sigma F_y = \rho f_y + \frac{\partial \tau_{xy}}{\partial x} + \frac{\partial \tau_{yy}}{\partial y} + \frac{\partial \tau_{yz}}{\partial z} \tag{1.26b}$$

$$\rho \left(\frac{\partial w}{\partial t} + u \frac{\partial w}{\partial x} + v \frac{\partial w}{\partial y} + w \frac{\partial w}{\partial z} \right) = \Sigma F_z = \rho f_z + \frac{\partial \tau_{xz}}{\partial x} + \frac{\partial \tau_{yz}}{\partial y} + \frac{\partial \tau_{zz}}{\partial z} \tag{1.26c}$$

(Note that in Eqs. (1.26a–c) the symmetry of the stress vector has been enforced.) For a Newtonian fluid the stress components τ_{ij} are given by Eq. (1.10), and by substituting them into Eqs. (1.26a–c), the Navier–Stokes equations are obtained:

$$\rho \left(\frac{\partial q_i}{\partial t} + \mathbf{q} \cdot \nabla q_i \right) = \rho f_i - \frac{\partial}{\partial x_i} \left(p + \frac{2}{3} \mu \nabla \cdot \mathbf{q} \right) + \frac{\partial}{\partial x_j} \mu \left(\frac{\partial q_i}{\partial x_j} + \frac{\partial q_j}{\partial x_i} \right)$$
$$(i = 1, 2, 3) \tag{1.27}$$

which in Cartesian coordinates are

$$\rho \left(\frac{\partial u}{\partial t} + \mathbf{q} \cdot \nabla u \right) = \rho f_x - \frac{\partial p}{\partial x} + \frac{\partial}{\partial x} \left\{ \mu \left[2 \frac{\partial u}{\partial x} - \frac{2}{3} (\nabla \cdot \mathbf{q}) \right] \right\}$$
$$+ \frac{\partial}{\partial y} \left[\mu \left(\frac{\partial u}{\partial y} + \frac{\partial v}{\partial x} \right) \right] + \frac{\partial}{\partial z} \left[\mu \left(\frac{\partial w}{\partial x} + \frac{\partial u}{\partial z} \right) \right] \tag{1.27a}$$

$$\rho \left(\frac{\partial v}{\partial t} + \mathbf{q} \cdot \nabla v \right) = \rho f_y - \frac{\partial p}{\partial y} + \frac{\partial}{\partial y} \left\{ \mu \left[2 \frac{\partial v}{\partial y} - \frac{2}{3} (\nabla \cdot \mathbf{q}) \right] \right\}$$
$$+ \frac{\partial}{\partial z} \left[\mu \left(\frac{\partial v}{\partial z} + \frac{\partial w}{\partial y} \right) \right] + \frac{\partial}{\partial x} \left[\mu \left(\frac{\partial u}{\partial y} + \frac{\partial v}{\partial x} \right) \right] \tag{1.27b}$$

$$\rho \left(\frac{\partial w}{\partial t} + \mathbf{q} \cdot \nabla w \right) = \rho f_z - \frac{\partial p}{\partial z} + \frac{\partial}{\partial z} \left\{ \mu \left[2 \frac{\partial w}{\partial z} - \frac{2}{3} (\nabla \cdot \mathbf{q}) \right] \right\}$$
$$+ \frac{\partial}{\partial x} \left[\mu \left(\frac{\partial w}{\partial x} + \frac{\partial u}{\partial z} \right) \right] + \frac{\partial}{\partial y} \left[\mu \left(\frac{\partial v}{\partial z} + \frac{\partial w}{\partial y} \right) \right] \tag{1.27c}$$

1.6 Differential Form of the Fluid Dynamic Equations

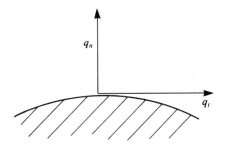

Figure 1.9 Direction of tangential and normal velocity components near a solid boundary.

Typical boundary conditions for this problem require that on stationary solid boundaries (Fig. 1.9) both the normal and tangential velocity components will reduce to zero:

$$q_n = 0 \quad \text{(on solid surface)} \tag{1.28a}$$

$$q_t = 0 \quad \text{(on solid surface)} \tag{1.28b}$$

The number of exact solutions to the Navier–Stokes equations is small because of the nonlinearity of the differential equations. However, in many situations some terms can be neglected so that simpler equations can be obtained. For example, by assuming constant viscosity coefficient μ, Eq. (1.27) becomes

$$\rho\left(\frac{\partial \mathbf{q}}{\partial t} + \mathbf{q} \cdot \nabla \mathbf{q}\right) = \rho \mathbf{f} - \nabla p + \mu \nabla^2 \mathbf{q} + \frac{\mu}{3}\nabla(\nabla \cdot \mathbf{q}) \tag{1.29}$$

Furthermore, by assuming an incompressible fluid (for which the continuity equation (Eq. (1.23)) becomes $\nabla \cdot \mathbf{q} = 0$), Eq. (1.27) reduces to

$$\rho\left(\frac{\partial \mathbf{q}}{\partial t} + \mathbf{q} \cdot \nabla \mathbf{q}\right) = \rho \mathbf{f} - \nabla p + \mu \nabla^2 \mathbf{q} \tag{1.30}$$

For an inviscid compressible fluid

$$\frac{\partial \mathbf{q}}{\partial t} + \mathbf{q} \cdot \nabla \mathbf{q} = \mathbf{f} - \frac{\nabla p}{\rho} \tag{1.31}$$

This equation is called the *Euler equation*.

In situations in which the problem has cylindrical or spherical symmetry, the use of appropriate coordinates can simplify the solution. As an example, we present the fundamental equations for an incompressible fluid with constant viscosity. The cylindrical coordinate system is described in Fig. 1.10, and for this example the r, θ coordinates are in a plane normal to the x coordinate. The operators ∇, ∇^2, and D/Dt in the r, θ, x system are (see Pai,[1.4] p. 38 or Yuan,[1.2] p. 132)

$$\nabla = \left(\mathbf{e}_r \frac{\partial}{\partial r}, \mathbf{e}_\theta \frac{1}{r}\frac{\partial}{\partial \theta}, \mathbf{e}_x \frac{\partial}{\partial x}\right) \tag{1.32}$$

$$\nabla^2 = \frac{\partial^2}{\partial r^2} + \frac{1}{r}\frac{\partial}{\partial r} + \frac{1}{r^2}\frac{\partial^2}{\partial \theta^2} + \frac{\partial^2}{\partial x^2} \tag{1.33}$$

$$\frac{D}{Dt} = \frac{\partial}{\partial t} + q_r \frac{\partial}{\partial r} + \frac{q_\theta}{r}\frac{\partial}{\partial \theta} + q_x \frac{\partial}{\partial x} \tag{1.34}$$

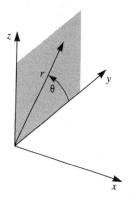

Figure 1.10 Cylindrical coordinate system.

The continuity equation in cylindrical coordinates for an incompressible fluid then becomes

$$\frac{\partial q_r}{\partial r} + \frac{1}{r}\frac{\partial q_\theta}{\partial \theta} + \frac{\partial q_x}{\partial x} + \frac{q_r}{r} = 0 \tag{1.35}$$

The momentum equations for an incompressible fluid are

r direction:

$$\rho\left(\frac{Dq_r}{Dt} - \frac{q_\theta^2}{r}\right) = \rho f_r - \frac{\partial p}{\partial r} + \mu\left(\nabla^2 q_r - \frac{q_r}{r^2} - \frac{2}{r^2}\frac{\partial q_\theta}{\partial \theta}\right) \tag{1.36}$$

θ direction:

$$\rho\left(\frac{Dq_\theta}{Dt} + \frac{q_r q_\theta}{r}\right) = \rho f_\theta - \frac{1}{r}\frac{\partial p}{\partial \theta} + \mu\left(\nabla^2 q_\theta + \frac{2}{r^2}\frac{\partial q_r}{\partial \theta} - \frac{q_\theta}{r^2}\right) \tag{1.37}$$

x direction:

$$\rho\frac{Dq_x}{Dt} = \rho f_x - \frac{\partial p}{\partial x} + \mu\nabla^2 q_x \tag{1.38}$$

A spherical coordinate system with the coordinates r, θ, φ is described in Fig. 1.11. The operators ∇, ∇^2, and D/Dt in the r, θ, φ system are (Karamcheti,[1.5] Chapter 2 or Yuan,[1.2] p. 132)

$$\nabla = \left(\mathbf{e}_r\frac{\partial}{\partial r}, \mathbf{e}_\theta\frac{1}{r}\frac{\partial}{\partial \theta}, \mathbf{e}_\varphi\frac{1}{r\sin\theta}\frac{\partial}{\partial \varphi}\right) \tag{1.39}$$

$$\nabla^2 = \frac{1}{r^2}\frac{\partial}{\partial r}\left(r^2\frac{\partial}{\partial r}\right) + \frac{1}{r^2\sin\theta}\frac{\partial}{\partial \theta}\left(\sin\theta\frac{\partial}{\partial \theta}\right) + \frac{1}{r^2\sin^2\theta}\frac{\partial^2}{\partial \varphi^2} \tag{1.40}$$

$$\frac{D}{Dt} = \frac{\partial}{\partial t} + q_r\frac{\partial}{\partial r} + \frac{q_\theta}{r}\frac{\partial}{\partial \theta} + \frac{q_\varphi}{r\sin\theta}\frac{\partial}{\partial \varphi} \tag{1.41}$$

The continuity equation in spherical coordinates for an incompressible fluid becomes (Pai,[1.4] p. 40)

$$\frac{1}{r}\frac{\partial(r^2 q_r)}{\partial r} + \frac{1}{\sin\theta}\frac{\partial(q_\theta \sin\theta)}{\partial \theta} + \frac{1}{\sin\theta}\frac{\partial q_\varphi}{\partial \varphi} = 0 \tag{1.42}$$

1.6 Differential Form of the Fluid Dynamic Equations

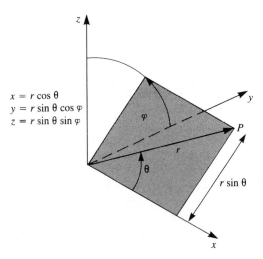

Figure 1.11 Spherical coordinate system.

The momentum equations for an incompressible fluid are (Pai,[1.4] p. 40)

r direction:

$$\rho\left(\frac{Dq_r}{Dt} - \frac{q_\varphi^2 + q_\theta^2}{r}\right)$$

$$= \rho f_r - \frac{\partial p}{\partial r} + \mu\left(\nabla^2 q_r - \frac{2q_r}{r^2} - \frac{2}{r^2}\frac{\partial q_\theta}{\partial \theta} - \frac{2q_\theta \cot\theta}{r^2} - \frac{2}{r^2 \sin\theta}\frac{\partial q_\varphi}{\partial \varphi}\right) \tag{1.43}$$

θ direction:

$$\rho\left(\frac{Dq_\theta}{Dt} + \frac{q_r q_\theta}{r} - \frac{q_\varphi^2 \cot\theta}{r}\right)$$

$$= \rho f_\theta - \frac{1}{r}\frac{\partial p}{\partial \theta} + \mu\left(\nabla^2 q_\theta + \frac{2}{r^2}\frac{\partial q_r}{\partial \theta} - \frac{q_\theta}{r^2 \sin^2\theta} - \frac{2\cos\theta}{r^2 \sin^2\theta}\frac{\partial q_\varphi}{\partial \varphi}\right) \tag{1.44}$$

φ direction:

$$\rho\left(\frac{Dq_\varphi}{Dt} + \frac{q_\varphi q_r}{r} + \frac{q_\theta q_\varphi \cot\theta}{r}\right)$$

$$= \rho f_\varphi - \frac{1}{r\sin\theta}\frac{\partial p}{\partial \varphi} + \mu\left(\nabla^2 q_\varphi - \frac{q_\varphi}{r^2 \sin^2\theta} + \frac{2}{r^2 \sin\theta}\frac{\partial q_r}{\partial \varphi} + \frac{2\cos\theta}{r^2 \sin^2\theta}\frac{\partial q_\theta}{\partial \varphi}\right) \tag{1.45}$$

When a two-dimensional flowfield is treated in this text, it will be described in either a Cartesian coordinate system with coordinates x and z or in a corresponding polar coordinate system with coordinates r and θ (see Fig. 1.12). In this polar coordinate system, the continuity equation for an incompressible fluid is obtained from Eq. (1.35) by eliminating $\partial q_x/\partial x$, and the r- and θ-momentum equations for an incompressible fluid are identical to Eqs. (1.36) and (1.37), respectively.

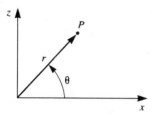

Figure 1.12 Two-dimensional polar coordinate system.

1.7 Dimensional Analysis of the Fluid Dynamic Equations

The governing equations developed in the previous section (e.g., Eq. (1.27)) are very complex and their solution, even by numerical methods, is difficult for many practical applications. If some of the terms causing this complexity can be neglected in certain regions of the flowfield, while the dominant physical features are still retained, then a set of simplified equations can be obtained (and probably solved with less effort). In this section, some of the conditions for simplifying the governing equations will be discussed.

To determine the relative magnitude of the various elements in the governing differential equations, the following dimensional analysis is performed. For simplicity, consider the fluid dynamic equations with constant properties (μ = constant, and ρ = constant):

$$\nabla \cdot \mathbf{q} = 0 \tag{1.23}$$

$$\rho \left(\frac{\partial \mathbf{q}}{\partial t} + \mathbf{q} \cdot \nabla \mathbf{q} \right) = \rho \mathbf{f} - \nabla p + \mu \nabla^2 \mathbf{q} \tag{1.30}$$

The first step is to define some characteristic or reference quantities, relevant to the physical problem to be studied:

L – reference length (e.g., wing's chord)
V – reference speed (e.g., the free-stream speed)
T – characteristic time (e.g., one cycle of a periodic process, or L/V)
p_0 – reference pressure (e.g., free-stream pressure, p_∞)
f_0 – body force (e.g., magnitude of earth's gravitation, g)

With the aid of these characteristic quantities we can define the following nondimensional variables:

$$\begin{aligned}
x^* &= \frac{x}{L}, \quad y^* = \frac{y}{L}, \quad z^* = \frac{z}{L} \\
u^* &= \frac{u}{V}, \quad v^* = \frac{v}{V}, \quad w^* = \frac{w}{V} \\
t^* &= \frac{t}{T} \\
p^* &= \frac{p}{p_0} \\
\mathbf{f}^* &= \frac{\mathbf{f}}{f_0}
\end{aligned} \tag{1.46}$$

If these characteristic magnitudes are properly selected, then all the nondimensional variables in Eq. (1.46) will be of the order of one. Next, the governing equations need to be rewritten using the quantities of Eq. (1.46). As an example, the first term of the continuity equation becomes

$$\frac{\partial u}{\partial x} = \frac{\partial u}{\partial u^*}\frac{\partial u^*}{\partial x^*}\frac{\partial x^*}{\partial x} = \frac{V}{L}\left(\frac{\partial u^*}{\partial x^*}\right)$$

and the transformed incompressible continuity equation is

$$\frac{V}{L}\left(\frac{\partial u^*}{\partial x^*} + \frac{\partial v^*}{\partial y^*} + \frac{\partial w^*}{\partial z^*}\right) = 0 \quad (1.47)$$

After a similar treatment, the momentum equation in the x direction becomes

$$\rho\left(\frac{V}{T}\frac{\partial u^*}{\partial t^*} + V\frac{V}{L}u^*\frac{\partial u^*}{\partial x^*} + V\frac{V}{L}v^*\frac{\partial u^*}{\partial y^*} + V\frac{V}{L}w^*\frac{\partial u^*}{\partial z^*}\right)$$

$$= \rho f_0 f_x^* - \frac{p_0}{L}\frac{\partial p^*}{\partial x^*} + \mu\frac{V}{L^2}\left(\frac{\partial^2 u^*}{\partial x^{*2}} + \frac{\partial^2 u^*}{\partial y^{*2}} + \frac{\partial^2 u^*}{\partial z^{*2}}\right) \quad (1.48)$$

The corresponding equations in the y and z directions can be obtained by the same procedure. Now, multiplying Eq. (1.47) by L/V and Eq. (1.48) by $L/\rho V^2$ we end up with

$$\frac{\partial u^*}{\partial x^*} + \frac{\partial v^*}{\partial y^*} + \frac{\partial w^*}{\partial z^*} = 0 \quad (1.49)$$

$$\left(\frac{L}{TV}\right)\frac{\partial u^*}{\partial t^*} + u^*\frac{\partial u^*}{\partial x^*} + v^*\frac{\partial u^*}{\partial y^*} + w^*\frac{\partial u^*}{\partial z^*}$$

$$= \left(\frac{Lf_0}{V^2}\right)f_x^* - \left(\frac{p_0}{\rho V^2}\right)\frac{\partial p^*}{\partial x^*} + \left(\frac{\mu}{\rho VL}\right)\left(\frac{\partial^2 u^*}{\partial x^{*2}} + \frac{\partial^2 u^*}{\partial y^{*2}} + \frac{\partial^2 u^*}{\partial z^{*2}}\right) \quad (1.50)$$

If all the nondimensional variables in Eq. (1.46) are of order one, then all terms appearing with an asterisk (*) will also be of order one, and the relative magnitude of each group in the equations is fixed by the nondimensional numbers appearing inside the parentheses. In the continuity equation (Eq. (1.49)), all terms have the same order of magnitude and for an arbitrary three-dimensional flow all terms are equally important. In the momentum equation the first nondimensional number is

$$\Omega = \frac{L}{TV} \quad (1.51)$$

which is a time constant and signifies the importance of time-dependent phenomena. A more frequently used form of this nondimensional number is the *Strouhal number*, where the characteristic time is the inverse of the frequency ω of a periodic occurrence (e.g., wake shedding frequency behind a separated airfoil):

$$St = \frac{L}{\left(\frac{1}{\omega}\right)V} = \frac{\omega L}{V} \quad (1.52)$$

If the Strouhal number is very small, perhaps due to very low frequencies, then the time-dependent first term in Eq. (1.50) can be neglected compared to the terms of order one.

The second group of nondimensional numbers (when gravity is the body force and f_0 is the gravitational acceleration g) is called the *Froude number*, which stands for the ratio of inertial force to gravitational force:

$$Fr = \frac{V}{\sqrt{Lg}} \quad (1.53)$$

Small values of Fr (note that Fr^{-2} appears in Eq. (1.50)) will mean that body forces such as gravity should be included in the equations, as in the case of free surface river flows, waterfalls, ship hydrodynamics, etc.

The third nondimensional number is the *Euler number*, which represents the ratio between the pressure and the inertia forces:

$$Eu = \frac{p_0}{\rho V^2} \quad (1.54)$$

A frequently used quantity related to the Euler number is the pressure coefficient C_p, which measures the nondimensional pressure difference, relative to a reference pressure p_0:

$$C_p \equiv \frac{p - p_0}{(1/2)\rho V^2} \quad (1.55)$$

The last nondimensional group in Eq. (1.50) represents the ratio between the inertial and viscous forces and is called the *Reynolds* number:

$$Re = \frac{\rho V L}{\mu} = \frac{V L}{\nu} \quad (1.56)$$

where ν is the kinematic viscosity given by

$$\nu = \frac{\mu}{\rho} \quad (1.57)$$

For the flow of gases, from the kinetic theory point of view (see Yuan,[1.2] p. 257) the viscosity can be connected to the characteristic velocity of the molecules c and to the mean distance λ that they travel between collisions (mean free path), by

$$\mu \approx \rho \frac{c\lambda}{3}$$

Substituting this into Eq. (1.56) yields

$$Re \approx \left(\frac{V}{c}\right)\left(\frac{L}{\lambda}\right)$$

This formulation shows that the Reynolds number represents the scaling of the velocity-times-length, compared to the molecular scale.

The conditions for neglecting the viscous terms when $Re \gg 1$ will be discussed in more detail in the next section.

For simplicity, at the beginning of this analysis an incompressible fluid was assumed. However, if compressibility is to be considered, an additional nondimensional number, called the *Mach number*, appears. It is the ratio of the velocity to the speed of sound a:

$$M = \frac{V}{a} \quad (1.58)$$

Note that the Euler number can be related to the Mach number since $p/\rho \sim a^2$ (see also Section 4.8).

Density changes caused by pressure changes are negligible if (see Karamcheti,[1.5] p. 23)

$$M \ll 1, \quad \frac{M^2}{Fr^2} \ll 1, \quad \frac{M^2}{Re} \ll 1 \quad (1.59)$$

and if these conditions are met, an incompressible fluid can be assumed.

1.8 Flow with High Reynolds Number

The most important outcome of the nondimensionalizing process of the governing equations is that now the relative magnitude of the terms appearing in the equations can be determined and compared. If desired, small terms can be neglected, resulting in simplified equations that are easier to solve but still contain the dominant physical effects.

In the case of the continuity equation all terms have the same magnitude and none is to be neglected. For the momentum equation the relative magnitude of the terms can be obtained by substituting Eqs. (1.51)–(1.56) into Eq. (1.50), and for the x direction we get

$$\Omega \frac{\partial u^*}{\partial t^*} + u^* \frac{\partial u^*}{\partial x^*} + v^* \frac{\partial u^*}{\partial y^*} + w^* \frac{\partial u^*}{\partial z^*}$$
$$= \left(\frac{1}{Fr^2}\right) f_x^* - Eu \frac{\partial p^*}{\partial x^*} + \left(\frac{1}{Re}\right)\left(\frac{\partial^2 u^*}{\partial x^{*2}} + \frac{\partial^2 u^*}{\partial y^{*2}} + \frac{\partial^2 u^*}{\partial z^{*2}}\right) \qquad (1.60)$$

Before proceeding further let us examine the range of Reynolds number and Mach number for some typical engineering problems. Since the viscosity of typical fluids such as air and water is very small, a wide variety of practical engineering problems (aircraft low-speed aerodynamics, hydrodynamics of naval vessels, etc.) fall within the $Re \gg 1$ range, as shown in Fig. 1.13. So for situations when the Reynolds number is high, the viscous terms become small compared to the other terms of order one in Eq. (1.60). But before neglecting these terms, a closer look at the high Reynolds number flow condition is needed. As an example, consider the flow over an airfoil, as shown in Fig. 1.14. In general, based on the assumption of high Reynolds number the viscous terms of Eq. (1.60) (or Eq. (1.30)) can be neglected in the outer flow regions (outside the immediate vicinity of a solid surface where $\nabla^2 \mathbf{q} \approx$ order 1). Therefore, in this outer flow region, the solution can be approximated by solving the incompressible continuity and the Euler equations:

$$\nabla \cdot \mathbf{q} = 0 \qquad (1.61)$$

$$\frac{\partial \mathbf{q}}{\partial t} + \mathbf{q} \cdot \nabla \mathbf{q} = \mathbf{f} - \frac{\nabla p}{\rho} \qquad (1.62)$$

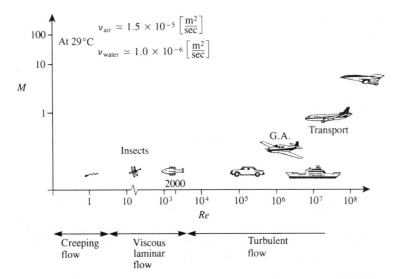

Figure 1.13 Range of Reynolds number and Mach number for some typical fluid flows.

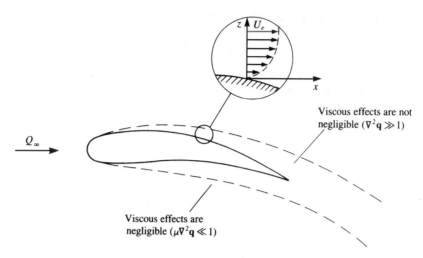

Figure 1.14 Flow regions in a high Reynolds number flow.

Equation (1.62) is a first-order partial differential equation and the solid surface boundary condition requires the specification of only one velocity component compared to all velocity components needed for Eq. (1.30) in the previous section. Since the flow is assumed to be inviscid, there is no physical reason for the tangential velocity component to be zero on a stationary solid surface and therefore what remains from the no-slip boundary condition (Eq. (1.28b)) is that the normal component of velocity must be zero:

$$q_n = 0 \quad \text{(on solid surface)} \tag{1.63}$$

However, a closer investigation of such flowfields reveals that near the solid boundaries in the fluid, shear flow derivatives such as $\nabla^2 \mathbf{q}$ become large and the viscous terms cannot be neglected even for high values of the Reynolds number (Fig. 1.14). For example, near the surface of a streamlined two-dimensional body submerged in a steady flow in the x direction (with no body forces) the Navier–Stokes equations can be reduced to the classical boundary layer equations (see Schlichting,[1.6] p. 131) where now x represents distance along the body surface and z is measured normal to the surface. The momentum equation in the x direction is

$$\rho \left(u \frac{\partial u}{\partial x} + w \frac{\partial u}{\partial z} \right) = -\frac{\partial p}{\partial x} + \mu \frac{\partial^2 u}{\partial z^2} \tag{1.64}$$

and that in the normal z direction is

$$0 = -\frac{\partial p}{\partial z} \tag{1.65}$$

So, in conclusion, for high Reynolds number flows there are two dominant regions in the flowfield:

1. The outer flow (away from the solid boundaries) where the viscous effects are negligible. A solution for the inviscid flow in this region provides information about the pressure distribution and the related forces.
2. The thin boundary layer (near the solid boundaries) where the viscous effects cannot be neglected. Solution of the boundary layer equations will provide information about the shear stress distribution and the related (friction) forces.

For the solution of the boundary layer equations, the no-slip boundary condition is applied on the solid boundary. The tangential velocity profile inside the boundary layer is shown in Fig. 1.14; we see that as the outer region is approached, the tangential velocity component becomes independent of z. The interface between the boundary layer region and the outer flow region is not precisely defined and occurs at a distance δ, the boundary layer thickness, from the wall. For large values of the Reynolds number the ratio of the boundary layer thickness to a characteristic length of the body (an airfoil's chord, for example) is proportional to $Re^{-1/2}$ (see Schlichting,[1.6] p. 129). Therefore, the normal extent of the boundary layer region is negligible when viewed on the length scale of the outer region.

A detailed solution for the complete flowfield of such a high Reynolds number flow proceeds as follows:

1. A solution is found for the inviscid flow past the body. For this solution the boundary condition of zero velocity normal to the solid surface is applied at the surface of the body (which is indistinguishable from the edge of the boundary layer on the scale of the chord). The tangential velocity component on the body surface U_e is then obtained as part of the inviscid solution and the pressure distribution along the solid surface is then determined.
2. Note that in the boundary layer equations (Eqs. (1.64) and (1.65)) the pressure does not vary across the boundary layer and is said to be impressed on the boundary layer. Therefore, the surface pressure distribution is taken from the inviscid solution in (1) and inserted into Eq. (1.64). Also, U_e is taken from the inviscid solution as the tangential component of the velocity at the edge of the boundary layer and is used as a boundary condition in the solution of the boundary layer equations.

Solving for a high Reynolds number flowfield with the assumption of an inviscid fluid is therefore the first step toward solving the complete physical problem. (Additional iterations between the inviscid outer flow and the boundary layer region in search of an improved solution are possible and are discussed in Chapters 9, 14, and 15.)

1.9 Similarity of Flows

Another interesting aspect of the process of nondimensionalizing the equations in the previous section is that two different flows are considered to be similar if the nondimensional numbers of Eq. (1.60) are the same. For most practical cases, where gravity and unsteady effects are negligible, only the Reynolds and the Mach numbers need to be matched. A possible implementation of this principle is in water or wind-tunnel testing, where the scale of the model differs from that of the actual flow conditions.

For example, many airplanes are tested in small scale first (e.g., 1/5-th scale). To keep the Reynolds number the same then either the airspeed or the air density must be increased (e.g., by a factor of 5). This is a typical conflict that test engineers face, since increasing the airspeed 5 times will bring the Mach number to an unreasonably high range. The second alternative of reducing the kinematic viscosity ν by compressing the air is possible in only a very few wind tunnels, and in most cases matching both of these nondimensional numbers is difficult.

Another possible way to apply the similarity principle is to exchange fluids between the actual and the test conditions (e.g., water with air where the ratio of kinematic viscosity is about 1:15). Thus a 1/15-scale model of a submarine can be tested in a wind tunnel at true speed conditions. Usually it is better to increase the speed in the wind tunnel and then even a smaller scale model can be tested (of course the Mach number is not always matched but for such low Mach number applications this is less critical).

References

[1.1] Batchelor, G. K., *An Introduction to Fluid Dynamics*, Cambridge University Press, 1967.
[1.2] Yuan, S. W., *Foundations of Fluid Mechanics*, Prentice-Hall, 1969.
[1.3] Kellogg, O. D., *Foundation of Potential Theory*, Dover, 1953.
[1.4] Pai, S.-I., *Viscous Flow Theory*, Van Nostrand, 1956.
[1.5] Karamcheti, K., *Principles of Ideal-Fluid Aerodynamics*, R. E. Krieger Publishing Co., 1980.
[1.6] Schlichting, H., *Boundary-Layer Theory*, McGraw-Hill, 1979.

Problems

1.1. The velocity components of a two-dimensional flowfield are given by

$$u(x,y) = k\left[\frac{x^2+y^2-1}{(x^2+y^2-1)^2+4y^2}\right]$$

$$v(x,y) = 2k\left[\frac{xy}{(x^2+y^2-1)^2+4y^2}\right]$$

where k is a constant. Does this flow satisfy the incompressible continuity equation?

1.2. The velocity components of a three-dimensional, incompressible flow are given by

$$u = 2x, \qquad v = -y, \qquad w = -z$$

Determine the equations of the streamlines passing through point (1,1,1).

1.3. The velocity components of a two-dimensional flow are given by

$$u = \frac{ky}{x^2+y^2}$$

$$v = \frac{-kx}{x^2+y^2}$$

where k is a constant.
a. Obtain the equations of the streamlines.
b. Does this flow satisfy the incompressible continuity equation?

1.4. The two-dimensional, incompressible, viscous, laminar flow between two parallel plates due to a constant pressure gradient dp/dx is called *Poiseuille flow* (shown in Fig. 1.15). Simplify the continuity and momentum equations for this case and specify the boundary conditions on the wall (at $z = \pm h/2$). Determine the velocity distribution $u(z)$ between the plates and the shearing stress $\tau_{zx}(z = h/2) = -\mu(\partial u/\partial z)|_{h/2}$ on the wall.

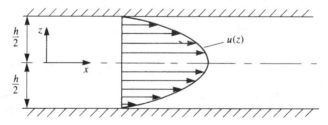

Figure 1.15 Two-dimensional viscous incompressible flow between two parallel plates.

CHAPTER 2

Fundamentals of Inviscid, Incompressible Flow

In Chapter 1 it was established that for flows at high Reynolds number the effects of viscosity are effectively confined to thin boundary layers and thin wakes. For this reason our study of low-speed aerodynamics will be limited to flows outside these limited regions where the flow is assumed to be inviscid and incompressible. To develop the mathematical equations that govern these flows and the tools that we will need to solve the equations it is necessary to study rotation in the fluid and to demonstrate its relationship to the effects of viscosity.

It is the goal of this chapter to define the mathematical problem (differential equation and boundary conditions) of low-speed aerodynamics whose solution will occupy us for the remainder of the book.

2.1 Angular Velocity, Vorticity, and Circulation

The arbitrary motion of a fluid element consists of translation, rotation, and deformation. To illustrate the rotation of a moving fluid element, consider at $t = t_0$ the control volume shown in Fig. 2.1. Here, for simplicity, we select an infinitesimal rectangular element that is being translated in the $z = 0$ plane by a velocity (u, v) of its corner no. 1. The lengths of the sides, parallel to the x and y directions, are Δx and Δy, respectively. Because of the velocity variations within the fluid the element may deform and rotate, and, for example, the x component of the velocity at the upper corner (no. 4) of the element will be $(u + (\partial u/\partial y)\Delta y)$, where higher order terms in the small quantities Δx and Δy are neglected. At a later time (e.g., $t = t_0 + \Delta t$), this will cause the deformation shown at the right-hand side of Fig. 2.1. The angular velocity component ω_z (note that positive direction in the figure follows the right-hand rule) of the fluid element can be obtained by averaging the instantaneous angular velocities of the segments 1–2 and 1–4 of the element. The instantaneous angular velocity of segment 1–2 is the difference in the linear velocities of the two edges of this segment, divided by the distance (Δx):

$$\text{angular velocity of segment 1–2} \approx \frac{\text{relative velocity}}{\text{radius}}$$
$$= \frac{v + (\partial v/\partial x)\Delta x - v}{\Delta x} = \frac{\partial v}{\partial x}$$

and the angular velocity of the 1–4 segment is

$$\frac{-[u + (\partial u/\partial y)\Delta y] + u}{\Delta y} = -\frac{\partial u}{\partial y}$$

The z component of the angular velocity of the fluid element is then the average of these two components:

$$\omega_z = \frac{1}{2}\left(\frac{\partial v}{\partial x} - \frac{\partial u}{\partial y}\right)$$

The two additional components of the angular velocity can be obtained similarly, and in

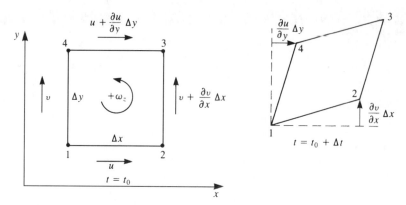

Figure 2.1 Angular velocity of a rectangular fluid element.

vector form the angular velocity becomes

$$\boldsymbol{\omega} = \frac{1}{2}\nabla \times \mathbf{q} \tag{2.1}$$

It is convenient to define the vorticity ζ as twice the angular velocity:

$$\boldsymbol{\zeta} \equiv 2\boldsymbol{\omega} = \nabla \times \mathbf{q} \tag{2.2}$$

In Cartesian coordinates the vorticity components are

$$\begin{aligned}\zeta_x &= 2\omega_x = \left(\frac{\partial w}{\partial y} - \frac{\partial v}{\partial z}\right) \\ \zeta_y &= 2\omega_y = \left(\frac{\partial u}{\partial z} - \frac{\partial w}{\partial x}\right) \\ \zeta_z &= 2\omega_z = \left(\frac{\partial v}{\partial x} - \frac{\partial u}{\partial y}\right)\end{aligned} \tag{2.2a}$$

Now consider an open surface S, shown in Fig. 2.2, which has the closed curve C as its boundary. With the use of Stokes's theorem (see Kellogg,[1.3] p. 73) the vorticity on the surface S can be related to the line integral around C:

$$\int_S \nabla \times \mathbf{q} \cdot \mathbf{n}\,dS = \int_S \boldsymbol{\zeta} \cdot \mathbf{n}\,dS = \oint_C \mathbf{q} \cdot d\mathbf{l}$$

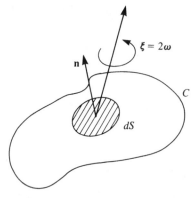

Figure 2.2 The relation between surface and line integrals.

2.1 Angular Velocity, Vorticity, and Circulation

where **n** is normal to S. The integral on the right-hand side is called the circulation and denoted by Γ:

$$\Gamma \equiv \oint_C \mathbf{q} \cdot d\mathbf{l} \tag{2.3}$$

This relation can be illustrated again with the simple fluid element of Fig. 2.1. The circulation $\Delta\Gamma$ is obtained by the evaluation of the closed line integral of the tangential velocity component around the fluid element. Note that the *positive* direction corresponds to the *positive* direction of ω:

$$\Delta\Gamma = \oint_C \mathbf{q} \cdot d\mathbf{l} = u\Delta x + \left(v + \frac{\partial v}{\partial x}\Delta x\right)\Delta y - \left(u + \frac{\partial u}{\partial y}\Delta y\right)\Delta x - v\Delta y$$

$$= \left(\frac{\partial v}{\partial x} - \frac{\partial u}{\partial y}\right)\Delta x \Delta y = \int_S \zeta_z \, dS$$

For the general three-dimensional case these conclusions can be summarized as

$$\Gamma \equiv \oint_C \mathbf{q} \cdot d\mathbf{l} = \int_S \nabla \times \mathbf{q} \cdot \mathbf{n}\, dS = \int_S \boldsymbol{\zeta} \cdot \mathbf{n}\, dS \tag{2.4}$$

The circulation is therefore somehow tied to the rotation in the fluid (e.g., to the angular velocity of a solid body type rotation). In Fig. 2.3 two examples are shown to illustrate the concept of circulation. The curve C (dashed lines) is taken to be a circle in each case. In Fig. 2.3a the flowfield consists of concentric circular streamlines in the counterclockwise direction. It is clear that along the circular integration path C (Fig. 2.3a) $\mathbf{q}$ and $d\mathbf{l}$ in Eq. (2.3) are positive for all $d\mathbf{l}$ and therefore C has a positive circulation. In Fig. 2.3b the flowfield is the symmetric flow of a uniform stream past a circular cylinder. It is clear from the symmetry that the circulation is zero for this case.

To illustrate the motion of a fluid with rotation consider the control volume shown in Fig. 2.4a, moving along the path l. Let us assume that the viscous forces are very large and the fluid will rotate as a rigid body, while following the path l. In this case $\nabla \times \mathbf{q} \neq 0$ and the flow is called *rotational*. For the fluid motion described in Fig. 2.4b, the shear forces in the fluid are negligible, and the fluid will not be rotated by the shear force of the neighboring fluid elements. In this case $\nabla \times \mathbf{q} = 0$ and the flow is considered to be *irrotational*.

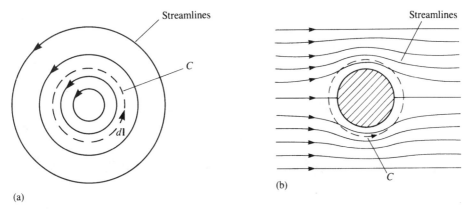

Figure 2.3 Flow fields with (a) and without (b) circulation.

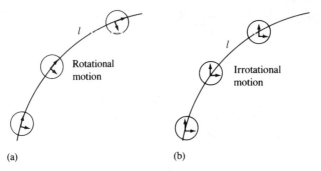

Figure 2.4 Rotational and irrotational motion of a fluid element.

2.2 Rate of Change of Vorticity

To obtain an equation that governs the rate of change of vorticity of a fluid element, we start with the incompressible Navier–Stokes equations in Cartesian coordinates (Eq. (1.30))

$$\frac{\partial \mathbf{q}}{\partial t} + \mathbf{q} \cdot \nabla \mathbf{q} = \mathbf{f} - \nabla \frac{p}{\rho} + \nu \nabla^2 \mathbf{q} \tag{1.30}$$

The convective acceleration term is rewritten using the vector identity

$$\mathbf{q} \cdot \nabla \mathbf{q} = \nabla \frac{q^2}{2} - \mathbf{q} \times \boldsymbol{\zeta} \tag{2.5}$$

Now take the curl of Eq. (1.30), with the second term on the left-hand side replaced by the right-hand side of Eq. (2.5). Note that for a scalar A, $\nabla \times \nabla A \equiv 0$ and therefore the pressure term vanishes:

$$\frac{\partial \boldsymbol{\zeta}}{\partial t} - \nabla \times (\mathbf{q} \times \boldsymbol{\zeta}) = \nabla \times \mathbf{f} + \nu \nabla^2 \boldsymbol{\zeta} \tag{2.6}$$

To simplify the result, we use the following vector identity:

$$\nabla \times (\mathbf{q} \times \boldsymbol{\zeta}) = \mathbf{q} \nabla \cdot \boldsymbol{\zeta} - \mathbf{q} \cdot \nabla \boldsymbol{\zeta} + \boldsymbol{\zeta} \cdot \nabla \mathbf{q} - \boldsymbol{\zeta} \nabla \cdot \mathbf{q} \tag{2.7}$$

along with the incompressible continuity equation and the fact that the vorticity is divergence free (note that for any vector $\mathbf{A}$, $\nabla \cdot \nabla \times \mathbf{A} \equiv 0$). If we also assume that the body force acting is conservative (irrotational, such as gravity) then

$$\nabla \times \mathbf{f} = 0$$

and the rate of change of vorticity equation becomes

$$\frac{D\boldsymbol{\zeta}}{Dt} = \frac{\partial \boldsymbol{\zeta}}{\partial t} + \mathbf{q} \cdot \nabla \boldsymbol{\zeta} = \boldsymbol{\zeta} \cdot \nabla \mathbf{q} + \nu \nabla^2 \boldsymbol{\zeta} \tag{2.8}$$

The inviscid incompressible version of the vorticity transport equation is then

$$\frac{D\boldsymbol{\zeta}}{Dt} = \boldsymbol{\zeta} \cdot \nabla \mathbf{q} \tag{2.9}$$

For a flow that is two-dimensional, the vorticity is perpendicular to the flow direction and Eq. (2.8) becomes

$$\frac{D\boldsymbol{\zeta}}{Dt} = \nu \nabla^2 \boldsymbol{\zeta} \tag{2.10}$$

and for the two-dimensional flow of an inviscid, incompressible fluid

$$\frac{D\zeta}{Dt} = 0 \tag{2.11}$$

and the vorticity of each fluid element is seen to remain constant.

The vorticity equation (Eq. (2.8)) strongly resembles the Navier–Stokes equation and for very high values of the Reynolds number we see that the vorticity that is created at the solid boundary is convected along with the flow at a much faster rate than it can be diffused out across the flow and so it remains in the confines of the boundary layer and trailing wake. The fluid in the outer portion of the flowfield (the part that we will study) is seen to be effectively rotation free (irrotational) as well as inviscid.

The above observation can be illustrated for the two-dimensional case using the nondimensional quantities defined in Eq. (1.46). Then, Eq. (2.10) can be rewritten in nondimensional form as

$$\frac{D\zeta_z^*}{Dt^*} = \frac{1}{Re}\nabla^{*2}\zeta_z^* \tag{2.10a}$$

where the Reynolds number, Re, is defined in Eq. (1.56). Here a two-dimensional flow in the x–y plane is assumed and therefore the vorticity points in the z direction. The left-hand side in this equation is the rate at which vorticity is accumulated, which is equal to the rate it is being generated (near the solid boundaries of solid surfaces). It is clear from Eq. (2.10a) that for high Reynolds number flows, vorticity generation is small and can be neglected outside the boundary layer. Thus for an irrotational fluid Eq. (2.2) reduces to

$$\frac{\partial w}{\partial y} = \frac{\partial v}{\partial z}$$
$$\frac{\partial u}{\partial z} = \frac{\partial w}{\partial x} \tag{2.12}$$
$$\frac{\partial v}{\partial x} = \frac{\partial u}{\partial y}$$

2.3   Rate of Change of Circulation: Kelvin's Theorem

Consider the circulation around a fluid curve (which always passes through the same fluid particles) in an incompressible inviscid flow with conservative body forces acting. The time rate of change of the circulation of this fluid curve C is given as

$$\frac{D\Gamma}{Dt} = \frac{D}{Dt}\oint_C \mathbf{q}\cdot d\mathbf{l} = \oint_C \frac{D\mathbf{q}}{Dt}\cdot d\mathbf{l} + \oint_C \mathbf{q}\cdot \frac{D}{Dt}d\mathbf{l} \tag{2.13}$$

Since C is a fluid curve, we have

$$\frac{D\mathbf{q}}{Dt} = \mathbf{a} \quad \text{and} \quad \frac{D}{Dt}d\mathbf{l} = d\mathbf{q}$$

and therefore

$$\frac{D\Gamma}{Dt} = \oint_C \mathbf{a}\cdot d\mathbf{l} \tag{2.14}$$

since the closed integral of an exact differential that is a function of the coordinates and time only is $\oint_C \mathbf{q}\cdot d\mathbf{q} = \oint_C d(q^2/2) = 0$. The acceleration $\mathbf{a}$ is obtained from the Euler equation (Eq. (1.62)) and is

$$\mathbf{a} = -\nabla\left(\frac{p}{\rho}\right) + \mathbf{f}$$

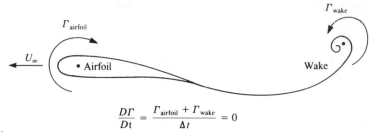

Figure 2.5 Circulation caused by an airfoil after it is suddenly set into motion.

Substitution into Eq. (2.14) yields the result that the circulation of a fluid curve remains constant:

$$\frac{D\Gamma}{Dt} = 0 = -\oint_C d\left(\frac{p}{\rho}\right) + \oint_C \mathbf{f} \cdot d\mathbf{l} \tag{2.15}$$

since the integral of a perfect differential around a closed path is zero and the work done by a conservative force around a closed path is also zero. The result in Eq. (2.15) is a form of angular momentum conservation known as *Kelvin's theorem* (after the British scientist who published his theorem in 1869), which states that: The time rate of change of circulation around a closed curve consisting of the same fluid elements is zero. For example, consider an airfoil as in Fig. 2.5, which prior to $t = 0$ was at rest and then at $t > 0$ was suddenly set into a constant forward motion. As the airfoil moves through the fluid a circulation Γ_{airfoil} develops around it. In order to comply with Kelvin's theorem a starting vortex Γ_{wake} must exist such that the total circulation around a line surrounding both the airfoil and the wake remains unchanged:

$$\frac{D\Gamma}{Dt} = \frac{1}{\Delta t}(\Gamma_{\text{airfoil}} + \Gamma_{\text{wake}}) = 0 \tag{2.16}$$

This is possible only if the starting vortex circulation equals the airfoil's circulation, but with rotation in the opposite direction.

2.4 Irrotational Flow and the Velocity Potential

It has been shown that the vorticity in the high Reynolds number flowfields being studied is confined to the boundary layer and wake regions where the influence of viscosity is not negligible and so it is appropriate to assume an irrotational as well as inviscid flow outside these confined regions. (The results of Sections 2.2 and 2.3 will be used when it is necessary to model regions of vorticity in the flowfield.)

Consider the following line integral in a simply connected region, along the line C:

$$\int_C \mathbf{q} \cdot d\mathbf{l} = \int_C u\,dx + v\,dy + w\,dz \tag{2.17}$$

If the flow is irrotational in this region then $u\,dx + v\,dy + w\,dz$ is an exact differential (see Kreyszig,[2.1] p. 475) of a potential Φ that is independent of the integration path C and is a function of the location of the point $P(x, y, z)$:

$$\Phi(x, y, z) = \int_{P_0}^{P} u\,dx + v\,dy + w\,dz \tag{2.18}$$

where P_0 is an arbitrary reference point. Φ is called the velocity potential, and the velocity at each point can be obtained as its gradient

$$\mathbf{q} = \nabla \Phi \tag{2.19}$$

In Cartesian coordinates the velocity components are given by

$$u = \frac{\partial \Phi}{\partial x}, \qquad v = \frac{\partial \Phi}{\partial y}, \qquad w = \frac{\partial \Phi}{\partial z} \tag{2.20}$$

The substitution of Eq. (2.19) into the continuity equation (Eq. (1.23)) leads to the following differential equation for the velocity potential:

$$\nabla \cdot \mathbf{q} = \nabla \cdot \nabla \Phi = \nabla^2 \Phi = 0 \tag{2.21}$$

which is *Laplace's equation* (named after the French mathematician Pierre S. De Laplace (1749–1827)). It is a statement of the incompressible continuity equation for an irrotational fluid. Note that Laplace's equation is a linear differential equation. Since the fluid's viscosity has been neglected, the no-slip boundary condition on a solid–fluid boundary cannot be enforced and only Eq. (1.28a) is required. In a more general form, the boundary condition states that the normal component of the relative velocity between the fluid and the solid surface (which may have a velocity $\mathbf{q}_B$) is zero on the boundary:

$$\mathbf{n} \cdot (\mathbf{q} - \mathbf{q}_B) = 0 \tag{2.22}$$

This boundary condition is physically reasonable and is consistent with the proper mathematical formulation of the problem as will be shown later in the chapter.

For an irrotational, inviscid, incompressible flow it now appears that the velocity field can be obtained from a solution of Laplace's equation for the velocity potential. Note that we have not yet used the Euler equation, which connects the velocity to the pressure. Once the velocity field is obtained it is necessary to also obtain the pressure distribution on the body surface to allow for a calculation of the aerodynamic forces and moments.

2.5 Boundary and Infinity Conditions

Laplace's equation for the velocity potential is the governing partial differential equation for the velocity for an inviscid, incompressible, and irrotational flow. It is an elliptic differential equation that results in a boundary-value problem. For aerodynamic problems the boundary conditions need to be specified on all solid surfaces and at infinity. One form of the boundary condition on a solid–fluid interface is given in Eq. (2.22). Another statement of this boundary condition, which will prove useful in applications, is obtained in the following way.

Let the solid surface be given by

$$F(x, y, z, t) = 0 \tag{2.23}$$

in Cartesian coordinates. Particles on the surface move with velocity $\mathbf{q}_B$ such that F remains zero. Therefore the derivative of F following the surface particles must be zero:

$$\left(\frac{D}{Dt}\right)_B F \equiv \frac{\partial F}{\partial t} + \mathbf{q}_B \cdot \nabla F = 0 \tag{2.24}$$

Equation (2.22) can be rewritten as

$$\mathbf{q} \cdot \nabla F = \mathbf{q}_B \cdot \nabla F \tag{2.25}$$

since the normal to the surface **n** is proportional to the gradient of F,

$$\mathbf{n} = \frac{\nabla F}{|\nabla F|} \qquad (2.26)$$

If Eq. (2.25) is now substituted into Eq. (2.24) the boundary condition becomes

$$\frac{\partial F}{\partial t} + \mathbf{q} \cdot \nabla F = \frac{DF}{Dt} = 0 \qquad (2.27)$$

At infinity, the disturbance **q**, due to the body moving through a fluid that was initially at rest, decays to zero. In a space-fixed frame of reference the velocity of such fluid (at rest) is therefore zero at infinity (far from the solid boundaries of the body):

$$\lim_{r \to \infty} \mathbf{q} = 0 \qquad (2.28)$$

2.6 Bernoulli's Equation for the Pressure

The incompressible Euler equation (Eq. (1.31)) can be rewritten with the use of Eq. (2.5) as

$$\frac{\partial \mathbf{q}}{\partial t} - \mathbf{q} \times \boldsymbol{\zeta} + \nabla \frac{q^2}{2} = \mathbf{f} - \nabla \frac{p}{\rho} \qquad (2.29)$$

For irrotational flow $\boldsymbol{\zeta} = 0$ and the time derivative of the velocity can be written as

$$\frac{\partial \mathbf{q}}{\partial t} = \frac{\partial}{\partial t} \nabla \Phi = \nabla \left(\frac{\partial \Phi}{\partial t} \right) \qquad (2.30)$$

Let us also assume that the body force is conservative with a potential E,

$$\mathbf{f} = -\nabla E \qquad (2.31)$$

If gravity is the body force acting and the z axis points upward, then $E = gz$.

The Euler equation for incompressible irrotational flow with a conservative body force (by substituting Eqs. (2.30) and (2.31) into Eq. (2.29)) then becomes

$$\nabla \left(E + \frac{p}{\rho} + \frac{q^2}{2} + \frac{\partial \Phi}{\partial t} \right) = 0 \qquad (2.32)$$

Equation (2.32) is true if the quantity in parentheses is a function of time only, that is,

$$E + \frac{p}{\rho} + \frac{q^2}{2} + \frac{\partial \Phi}{\partial t} = C(t) \qquad (2.33)$$

This is the *Bernoulli equation* (named after the Dutch/Swiss mathematician, Daniel Bernoulli (1700–1782)) for inviscid incompressible irrotational flow. A more useful form of the Bernoulli equation is obtained by comparing the quantities on the left-hand side of Eq. (2.33) at two points in the fluid; the first is an arbitrary point and the second is a reference point at infinity. The equation becomes

$$\left[E + \frac{p}{\rho} + \frac{q^2}{2} + \frac{\partial \Phi}{\partial t} \right] = \left[E + \frac{p}{\rho} + \frac{q^2}{2} + \frac{\partial \Phi}{\partial t} \right]\bigg|_{\infty} \qquad (2.34)$$

If the reference condition is chosen such that $E_\infty = 0$, $\Phi_\infty = $ const., and $\mathbf{q}_\infty = 0$ then the pressure p at any point in the fluid can be calculated from

$$\frac{p_\infty - p}{\rho} = \frac{\partial \Phi}{\partial t} + E + \frac{q^2}{2} \qquad (2.35)$$

2.7 Simply and Multiply Connected Regions

If the flow is steady, incompressible, but rotational the Bernoulli equation (Eq. (2.34)) is still valid with the time-derivative term set equal to zero if the constant on the right-hand side is now allowed to vary from streamline to streamline. (This is because the product $\mathbf{q} \times \boldsymbol{\zeta}$ is normal to the streamline $d\mathbf{l}$ and their dot product vanishes along the streamline. Consequently, Eq. (2.34) can be used in a rotational fluid between two points lying on the same streamline.)

2.7 Simply and Multiply Connected Regions

The region exterior to a two-dimensional airfoil and that exterior to a three-dimensional wing or body are fundamentally different in a mathematical sense and lead to velocity potentials with different properties. To point out the difference in these regions, we need to introduce a few basic definitions.

A *reducible* curve in a region can be contracted to a point without leaving the region. For example, in the region exterior to an airfoil, any curve surrounding the airfoil is not reducible and any curve not surrounding it is reducible. A *simply connected* region is one where all closed curves are reducible. (The region exterior to a finite three-dimensional body is simply connected. Any curve surrounding the body can be translated away from the body and then contracted.) A *barrier* is a curve that is inserted into a region but is not a part of the resulting modified region. The insertion of barriers into a region can change it from being multiply connected to being simply connected. The degree of connectivity of a region is $n + 1$, where n is the minimum number of barriers needed to make the remaining region simply connected. For example, consider the region in Fig. 2.6 exterior to an airfoil. Draw a barrier from the trailing edge to downstream infinity. The original region minus the barrier is now simply connected. (Note that curves in the region can no longer surround the airfoil.) Therefore $n = 1$ and the original region is doubly connected.

Consider irrotational motion in a simply connected region. The circulation around any curve is given by

$$\Gamma = \oint \mathbf{q} \cdot d\mathbf{l} = \oint \nabla \Phi \cdot d\mathbf{l} = \oint d\Phi \tag{2.36}$$

With the use of Eqs. (2.4) and with $\zeta = 0$ the circulation is seen to be zero. Also, since the integral of $d\Phi$ around any curve is zero (Eq. (2.36)), the velocity potential is single valued.

Now consider irrotational motion in the doubly connected region exterior to an airfoil as shown in Fig. 2.7. For any curve not surrounding the airfoil, the above results for the simply connected region apply and the circulation is zero. Now insert a barrier as shown in the figure. Consider the curve consisting of C_1 and C_2, which surround the airfoil, and the two sides of the barrier. Since the region excluding the barrier is simply connected, the circulation around this curve is zero. This leads to the following equation:

$$\oint_{C_1} \mathbf{q} \cdot d\mathbf{l} - \oint_{C_2} \mathbf{q} \cdot d\mathbf{l} + \int_A^B \mathbf{q} \cdot d\mathbf{l} + \int_B^A \mathbf{q} \cdot d\mathbf{l} = 0$$

Note that the first term is the circulation around C_1 and the second is minus the circulation around C_2. Also, the contributions from the barrier cancel for steady flow (since the barrier

Figure 2.6 Flow exterior to an airfoil in a doubly connected region.

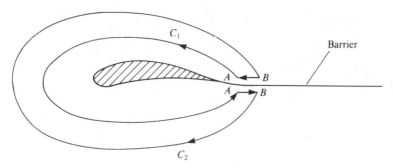

Figure 2.7 Integration lines along a simply connected region.

cannot be along a vortex sheet). The circulation around curves C_1 and C_2 (and any other curves surrounding the airfoil once) are the same and may be nonzero. From Eq. (2.36) the velocity potential is not single valued if there is a nonzero circulation.

2.8 Uniqueness of the Solution

The physical problem of finding the velocity field for the flow created, say, by the motion of an airfoil or wing has been reduced to the mathematical problem of solving Laplace's equation for the velocity potential with suitable boundary conditions for the velocity on the body and at infinity. In a space-fixed reference frame, this mathematical problem is

$$\nabla^2 \Phi = 0 \tag{2.37a}$$

$$\frac{\partial \Phi}{\partial n} = \mathbf{n} \cdot \mathbf{q}_B \quad \text{on body} \tag{2.37b}$$

$$\nabla \Phi \to 0 \quad \text{at} \quad r \to \infty \tag{2.37c}$$

Since the body boundary condition is on the normal derivative of the potential and since the flow is in the region exterior to the body, the mathematical problem of Eqs. (2.37a,b,c) is called the Neumann exterior problem. In what follows we will answer the question "Is there a unique solution to the Neumann exterior problem?" We will discover that the answer is different, depending on whether the region is simply or multiply connected.

Let us consider a simply connected region first. This will apply to the region outside of a three-dimensional body, but care must be taken in extending the results to wings since the flowfield is not irrotational everywhere (for instance in the wakes). Assume that there are two solutions Φ_1 and Φ_2 to the mathematical problem posed in Eqs. (2.37a,b,c). Then the difference

$$\Phi_1 - \Phi_2 \equiv \Phi_D$$

satisfies Laplace's equation, the homogeneous version of Eq. (2.37b), and Eq. (2.37c).

One form of Green's theorem (named after the English mathematician George Green (1793–1841)) (Ref. 1.5, p. 135) is obtained by applying the divergence theorem to the function $\Phi \nabla \Phi$, where Φ is a solution of Laplace's equation, R is the fluid region, and S is its boundary. The result is

$$\int_R \nabla \Phi \cdot \nabla \Phi \, dV = \int_S \Phi \frac{\partial \Phi}{\partial n} \, dS \tag{2.38}$$

2.8 Uniqueness of the Solution

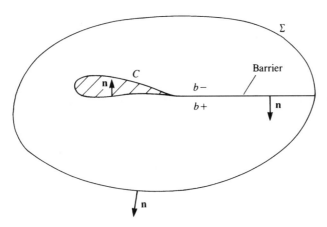

Figure 2.8 Double connected region exterior to an airfoil.

Now apply Eq. (2.38) to Φ_D for the region R between the body B and an arbitrary surface Σ surrounding B to get

$$\int_R \nabla \Phi_D \cdot \nabla \Phi_D \, dV = \int_B \Phi_D \frac{\partial \Phi_D}{\partial n} \, dS + \int_\Sigma \Phi_D \frac{\partial \Phi_D}{\partial n} \, dS \qquad (2.39)$$

If we let Σ go to infinity the integral over Σ vanishes and since $\partial \phi_D / \partial n = 0$ on B we are left with

$$\int_R \nabla \Phi_D \cdot \nabla \Phi_D \, dV = 0 \qquad (2.40)$$

Since the integrand is always greater than or equal to zero, it must be zero and consequently the difference $\Phi_1 - \Phi_2$ can at most be a constant. Therefore, the solution to the Neumann exterior problem in a simply connected region is unique to within a constant.

Consider now the doubly connected region exterior to the airfoil C in Fig. 2.8. Again let Φ_1 and Φ_2 be solutions and take

$$\Phi_1 - \Phi_2 = \Phi_D$$

Green's theorem is now applied to the function Φ_D in the region σ between the airfoil C and the curve Σ surrounding it. Note that the integrals are still volume and surface integrals and that the integrands do not vary normal to the plane of motion.

Insert a barrier b joining C and Σ and denote the two sides of the barrier as $b-$ and $b+$ as shown in the figure. Note that $\mathbf{n}$ is the outward normal to $b-$ and $-\mathbf{n}$ is the outward normal to $b+$. Equation (2.38) then becomes

$$\int_\sigma \nabla \Phi_D \cdot \nabla \Phi_D \, dV = \int_C \Phi_D \frac{\partial \Phi_D}{\partial n} \, dS + \int_\Sigma \Phi_D \frac{\partial \Phi_D}{\partial n} \, dS$$
$$+ \int_{b-} \Phi_D \frac{\partial \Phi_D}{\partial n} \, dS - \int_{b+} \Phi_D \frac{\partial \Phi_D}{\partial n} \, dS \qquad (2.41)$$

The integral around C is zero from the boundary condition and if we let Σ go to infinity the integral around Σ is zero also. Let Φ_D^- be Φ_D on $b-$ and Φ_D^+ be Φ_D on $b+$. Then Eq. (2.41) is

$$\int_\sigma \nabla \Phi_D \cdot \nabla \Phi_D \, dV = \int_{b-} \Phi_D^- \frac{\partial \Phi_D^-}{\partial n} \, dS - \int_{b+} \Phi_D^+ \frac{\partial \Phi_D^+}{\partial n} \, dS \qquad (2.42)$$

The normal derivative of Φ_D is continuous across the barrier and Eq. (2.42) can be written in terms of an integral over the barrier:

$$\int_\sigma \nabla\Phi_D \cdot \nabla\Phi_D \, dV = \int_{\text{barrier}} (\Phi_D^- - \Phi_D^+) \frac{\partial \Phi_D^-}{\partial n} \, dS \tag{2.43}$$

If we reintroduce the quantities Φ_1 and Φ_2 and rearrange the integrand we get

$$\int_\sigma \nabla\Phi_D \cdot \nabla\Phi_D \, dV = \int_{\text{barrier}} (\Phi_1^- - \Phi_1^+ + \Phi_2^+ - \Phi_2^-) \frac{\partial \Phi_D^-}{\partial n} \, dS \tag{2.44}$$

Note that the circulations associated with flows 1 and 2 are given by

$$\Gamma_1 = \Phi_1^+ - \Phi_1^-$$
$$\Gamma_2 = \Phi_2^+ - \Phi_2^-$$

and are constant, and finally

$$\int_\sigma \nabla\Phi_D \cdot \nabla\Phi_D \, dV = (\Gamma_2 - \Gamma_1) \int_{\text{barrier}} \frac{\partial \Phi_D^-}{\partial n} \, ds \tag{2.45}$$

Since in general we cannot require that the integral along the barrier be zero, the solution to the Neumann exterior problem is only uniquely determined to within a constant when $\Gamma_1 = \Gamma_2$ (when the circulation is specified as part of the problem statement). This result can be generalized for multiply connected regions in a similar manner. The value of the circulation cannot be specified on purely mathematical grounds but will be determined later on the basis of *physical* considerations.

2.9   Vortex Quantities

In conjunction with the velocity vector, we can define various quantities such as streamlines, stream tubes, and stream surfaces. Corresponding quantities can be defined for the vorticity vector and these will prove to be useful later on in the modeling of lifting flows.

The field lines (e.g., in Fig. 2.9) that are parallel to the vorticity vector are called *vortex lines* and these lines are described by

$$\boldsymbol{\zeta} \times d\mathbf{l} = 0 \tag{2.46}$$

where $d\mathbf{l}$ is a segment along the vortex line (as shown in Fig. 2.9). In Cartesian coordinates, this equation yields the differential equations for the vortex lines

$$\frac{dx}{\zeta_x} = \frac{dy}{\zeta_y} = \frac{dz}{\zeta_z} \tag{2.47}$$

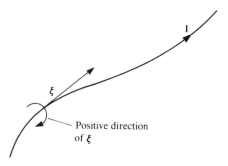

Figure 2.9   Vortex line.

2.9 Vortex Quantities

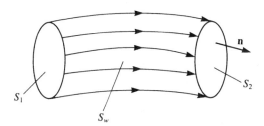

Figure 2.10 Vortex tube.

The vortex lines passing through an open curve in space form a *vortex surface* and the vortex lines passing through a closed curve in space form a *vortex tube*. A *vortex filament* is defined as a vortex tube of infinitesimal cross-sectional area.

The divergence of the vorticity is zero since the divergence of the curl of any vector is identically zero:

$$\nabla \cdot \boldsymbol{\zeta} = \nabla \cdot \nabla \times \mathbf{q} = 0 \tag{2.48}$$

Consider, at any instant, a region of space R enclosed by a surface S. An application of the divergence theorem yields

$$\int_S \boldsymbol{\zeta} \cdot \mathbf{n}\, dS = \int_R \nabla \cdot \boldsymbol{\zeta}\, dV = 0 \tag{2.49}$$

At some instant in time draw a vortex tube in the flow as shown in Fig. 2.10. Apply Eq. (2.49) to the region enclosed by the wall of the tube S_w and the surfaces S_1 and S_2 that cap the tube. Since on S_w the vorticity is parallel to the surface, the contribution of S_w vanishes and we are left with

$$\int_S \boldsymbol{\zeta} \cdot \mathbf{n}\, dS = \int_{S_1} \boldsymbol{\zeta} \cdot \mathbf{n}\, dS + \int_{S_2} \boldsymbol{\zeta} \cdot \mathbf{n}\, dS = 0 \tag{2.50}$$

Note that $\mathbf{n}$ is the outward normal and its direction is shown in the figure. If we denote $\mathbf{n}_v$ as being positive in the direction of the vorticity, then Eq. (2.50) becomes

$$\int_{S_1} \boldsymbol{\zeta} \cdot \mathbf{n}_v\, dS = \int_{S_2} \boldsymbol{\zeta} \cdot \mathbf{n}_v\, dS = \text{const.} \tag{2.51}$$

At each instant of time, the quantity in Eq. (2.51) is the same for any cross-sectional surface of the tube. Let C be any closed curve that surrounds the tube and lies on its wall. The circulation around C is given from Eq. (2.4) as

$$\Gamma_C = \int_S \boldsymbol{\zeta} \cdot \mathbf{n}_v\, dS = \text{const.} \tag{2.52}$$

and is seen to be constant along the tube. The results in Eqs. (2.51) and (2.52) express the spatial conservation of vorticity and are purely kinematical.

If Eq. (2.52) is applied to a vortex filament and $\mathbf{n}_v$ is chosen parallel to the vorticity vector, then

$$\Gamma_C = \zeta\, dS = \text{const.} \tag{2.53}$$

and the vorticity at any section of a vortex filament is seen to be inversely proportional to its cross-sectional area. A consequence of this result is that a vortex filament cannot end in the fluid since zero area would lead to an infinite value for the vorticity. This limiting case, however, is useful for the purposes of modeling and so it is convenient to define a vortex filament with a fixed circulation, zero cross-sectional area, and infinite vorticity as a vortex filament with concentrated vorticity.

Based on results similar to those of Section 2.3 and this section, the German scientist Hermann von Helmholtz (1821–1894) developed his vortex theorems for inviscid incompressible flows, which can be summarized as:

1. The strength of a vortex filament is constant along its length.
2. A vortex filament cannot start or end in a fluid (it must form a closed path or extend to infinity).
3. The fluid that forms a vortex tube continues to form a vortex tube and the strength of the vortex tube remains constant as the tube moves about (hence vortex elements, such as vortex lines, vortex tubes, vortex surfaces, etc., will remain vortex elements with time).

The first theorem is based on Eq. (2.53), while the second theorem follows from this. The third theorem is actually a combination of Helmholtz's third and fourth theorems and is a consequence of the inviscid vorticity transport equation (Eq. (2.9)).

2.10 Two-Dimensional Vortex

To illustrate a flowfield frequently called a two-dimensional vortex consider a two-dimensional rigid cylinder of radius R rotating in a viscous fluid at a constant angular velocity of ω_y, as shown in Fig. 2.11a. This motion results in a flow with circular streamlines

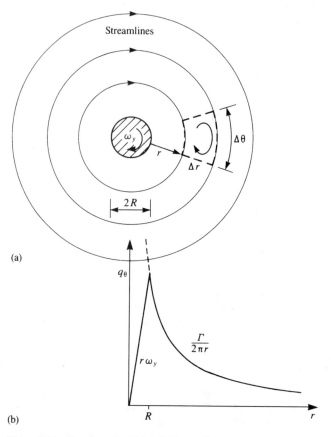

Figure 2.11 Two-dimensional flowfield around a cylindrical core rotating as a rigid body.

2.10 Two-Dimensional Vortex

and therefore the radial velocity component is zero. Consequently the continuity equation (Eq. (1.35)) in the r–θ plane becomes

$$\frac{\partial q_\theta}{\partial \theta} = 0 \tag{2.54}$$

Integrating this equation results in

$$q_\theta = q_\theta(r) \tag{2.55}$$

The Navier–Stokes equation in the r direction (Eq. (1.36)), after neglecting the body force terms, becomes

$$-\rho \frac{q_\theta^2}{r} = -\frac{\partial p}{\partial r} \tag{2.56}$$

Because q_θ is a function of r only, and because of the radial symmetry of the problem, the pressure must be either a function of r or a constant. Therefore, its derivative will not appear in the momentum equation in the θ direction (Eq. (1.37))

$$0 = \mu \left(\frac{\partial^2 q_\theta}{\partial r^2} + \frac{1}{r} \frac{\partial q_\theta}{\partial r} - \frac{q_\theta}{r^2} \right) \tag{2.57}$$

and since q_θ is a function of r only,

$$0 = \frac{d^2 q_\theta}{dr^2} + \frac{d}{dr}\left(\frac{q_\theta}{r} \right) \tag{2.58}$$

Integration with respect to r yields

$$\frac{dq_\theta}{dr} + \frac{q_\theta}{r} = C_1$$

where C_1 is the constant of integration. Rearranging this yields

$$\frac{1}{r} \frac{d}{dr}(r q_\theta) = C_1$$

and after an additional integration we get

$$q_\theta = \frac{C_1}{2} r + \frac{C_2}{r} \tag{2.59}$$

The boundary conditions are

$$\text{at} \quad r = R, \qquad q_\theta = -R\omega_y \tag{2.60a}$$
$$\text{at} \quad r = \infty, \qquad q_\theta = 0 \tag{2.60b}$$

The second boundary condition is satisfied only if $C_1 = 0$, and by using the first boundary condition, the velocity becomes

$$q_\theta = -\frac{R^2 \omega_y}{r} \tag{2.61}$$

From the vortex filament results (Eq. (2.53)), the circulation has the same sign as the vorticity, and it is therefore positive in the clockwise direction. The circulation around the circle of radius r concentric with (and larger than) the cylinder is found by using Eq. (2.3),

$$\Gamma = \int_{2\pi}^{0} q_\theta r \, d\theta = 2\omega_y \pi R^2 \tag{2.62}$$

and is constant. The tangential velocity can be rewritten as

$$q_\theta = -\frac{\Gamma}{2\pi r} \tag{2.63}$$

This velocity distribution is shown in Fig. 2.11b and is called *vortex flow*. If $r \to 0$ then the velocity becomes very large near the solid core, as shown by the dashed lines.

It has been demonstrated that Γ is the circulation generated by the rotating cylinder. However, to estimate the vorticity in the fluid, the integration line shown by the dashed lines in Fig. 2.11a is suggested. Integrating the velocity in a clockwise direction, and recalling that $q_r = 0$, we obtain

$$\oint \mathbf{q} \cdot d\mathbf{l} = 0 \cdot \Delta r + \frac{\Gamma}{2\pi(r + \Delta r)}(r + \Delta r)\Delta\theta - 0 \cdot \Delta r - \frac{\Gamma}{2\pi r}r\Delta\theta = 0$$

This indicates that this vortex flow is irrotational everywhere, excluding the rotating cylinder at the boundary of which *all* the vorticity is generated. When the core size approaches zero ($R \to 0$) then this flow is called an *irrotational vortex* (excluding the core point, where the velocity approaches infinity).

The three-dimensional velocity field induced by such an element is derived in the next section.

2.11 The Biot–Savart Law

At this point we have an incompressible fluid for which the continuity equation is

$$\nabla \cdot \mathbf{q} = 0 \tag{1.23}$$

and where vorticity ζ can exist; the problem is to determine the velocity field as a result of a known vorticity distribution. We may express the velocity field as the curl of a vector field $\mathbf{B}$, such that

$$\mathbf{q} = \nabla \times \mathbf{B} \tag{2.64}$$

Since the curl of a gradient vector is zero, $\mathbf{B}$ is indeterminate to within the gradient of a scalar function of position and time, and $\mathbf{B}$ can be selected such that

$$\nabla \cdot \mathbf{B} = 0 \tag{2.65}$$

The vorticity then becomes

$$\zeta = \nabla \times \mathbf{q} = \nabla \times (\nabla \times \mathbf{B}) = \nabla(\nabla \cdot \mathbf{B}) - \nabla^2 \mathbf{B}$$

Application of Eq. (2.65) reduces this to Poisson's equation for the vector potential $\mathbf{B}$:

$$\zeta = -\nabla^2 \mathbf{B} \tag{2.66}$$

The solution of this equation, using Green's theorem (see Karamcheti,[1.5] p. 533) is

$$\mathbf{B} = \frac{1}{4\pi} \int_V \frac{\zeta}{|\mathbf{r}_0 - \mathbf{r}_1|} dV$$

Here $\mathbf{B}$ is evaluated at point P (which is a distance $\mathbf{r}_0$ from the origin, shown in Fig. 2.12) and is a result of integrating the vorticity ζ (at point $\mathbf{r}_1$) within the volume V. The velocity field is then the curl of $\mathbf{B}$:

$$\mathbf{q} = \frac{1}{4\pi} \int_V \nabla \times \frac{\zeta}{|\mathbf{r}_0 - \mathbf{r}_1|} dV \tag{2.67}$$

Before proceeding with this integration, let us consider an infinitesimal piece of the vorticity filament ζ, as shown in Fig. 2.13. The cross-sectional area dS is selected such that it is normal

2.11 The Biot–Savart Law

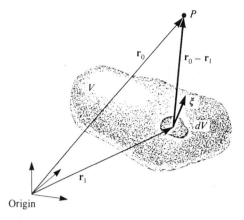

Figure 2.12 Velocity at point P due to a vortex distribution.

to ζ, and the direction $d\mathbf{l}$ on the filament is

$$d\mathbf{l} = \frac{\boldsymbol{\zeta}}{\zeta} dl$$

Also, the circulation Γ is

$$\Gamma = \zeta\, dS$$

and

$$dV = dS\, dl$$

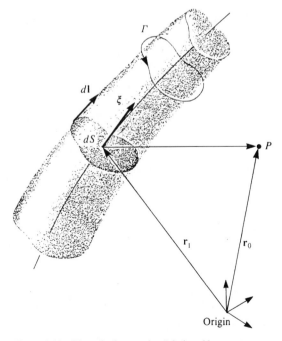

Figure 2.13 The velocity at point P induced by a vortex segment.

so that

$$\nabla \times \frac{\zeta}{|\mathbf{r}_0 - \mathbf{r}_1|} dV = \nabla \times \Gamma \frac{d\mathbf{l}}{|\mathbf{r}_0 - \mathbf{r}_1|}$$

Carrying out the curl operation while keeping $\mathbf{r}_1$ and $d\mathbf{l}$ fixed we get

$$\nabla \times \Gamma \frac{d\mathbf{l}}{|\mathbf{r}_0 - \mathbf{r}_1|} = \Gamma \frac{d\mathbf{l} \times (\mathbf{r}_0 - \mathbf{r}_1)}{|\mathbf{r}_0 - \mathbf{r}_1|^3}$$

Substitution of this result back into Eq. (2.67) results in the Biot–Savart law, which states

$$\mathbf{q} = \frac{\Gamma}{4\pi} \int \frac{d\mathbf{l} \times (\mathbf{r}_0 - \mathbf{r}_1)}{|\mathbf{r}_0 - \mathbf{r}_1|^3} \tag{2.68}$$

or in differential form

$$\Delta \mathbf{q} = \frac{\Gamma}{4\pi} \frac{d\mathbf{l} \times (\mathbf{r}_0 - \mathbf{r}_1)}{|\mathbf{r}_0 - \mathbf{r}_1|^3} \tag{2.68a}$$

A similar manipulation of Eq. (2.67) leads to the following result for the velocity due to a volume distribution of vorticity:

$$\mathbf{q} = \frac{1}{4\pi} \int_V \frac{\zeta \times (\mathbf{r}_0 - \mathbf{r}_1)}{|\mathbf{r}_0 - \mathbf{r}_1|^3} dV \tag{2.67a}$$

2.12 The Velocity Induced by a Straight Vortex Segment

In this section, the velocity induced by a straight vortex line segment is derived, based on the Biot–Savart law. It is clear that a vortex line cannot start or end in a fluid, and the following discussion is aimed at developing the contribution of a segment that is a section of a continuous vortex line. The vortex segment is placed at an arbitrary orientation in the (x, y, z) frame with constant circulation Γ, as shown in Fig. 2.14. The velocity induced by this vortex segment will have tangential components only as indicated in the figure. Also, the distance $\mathbf{r}_0 - \mathbf{r}_1$ between the vortex segment and the point P is $\mathbf{r}$. According to the Biot–Savart law (Eq. (2.68a)) the velocity induced by

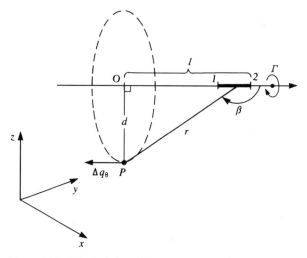

Figure 2.14 Velocity induced by a straight vortex segment.

2.12 The Velocity Induced by a Straight Vortex Segment

a segment $d\mathbf{l}$ on this line, at a point P, is

$$\Delta \mathbf{q} = \frac{\Gamma}{4\pi} \frac{d\mathbf{l} \times \mathbf{r}}{r^3} \tag{2.68b}$$

This may be rewritten in scalar form as

$$\Delta q_\theta = \frac{\Gamma}{4\pi} \frac{\sin \beta}{r^2} dl \tag{2.68c}$$

From the figure it is clear that

$$d = r \sin \beta \quad \text{and} \quad \tan(\pi - \beta) = \frac{d}{l}$$

and therefore

$$l = \frac{-d}{\tan \beta} \quad \text{and} \quad dl = \frac{d}{\sin^2 \beta} d\beta$$

Substituting these terms into Eq. (2.68c) we get

$$\Delta q_\theta = \frac{\Gamma}{4\pi} \frac{\sin^2 \beta}{d^2} \sin \beta \frac{d}{\sin^2 \beta} d\beta = \frac{\Gamma}{4\pi d} \sin \beta \, d\beta$$

This equation can be integrated over a section ($1 \to 2$) of the straight vortex segment of Fig. 2.15:

$$(q_\theta)_{1,2} = \frac{\Gamma}{4\pi d} \int_{\beta_1}^{\beta_2} \sin \beta \, d\beta = \frac{\Gamma}{4\pi d} (\cos \beta_1 - \cos \beta_2) \tag{2.69}$$

The results of this equation are shown schematically in Fig. 2.15. Thus, the velocity induced by a straight vortex segment is a function of its strength Γ, the distance d, and the two view angles β_1 and β_2.

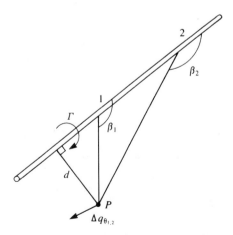

Figure 2.15 Definition of the view angles used for the vortex-induced velocity calculations.

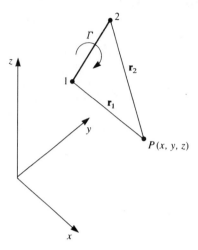

Figure 2.16 Nomenclature used for the velocity induced by a three-dimensional, straight vortex segment.

For the two-dimensional case (infinite vortex length) $\beta_1 = 0$, $\beta_2 = \pi$, and

$$q_\theta = \frac{\Gamma}{4\pi d} \int_0^\pi \sin\beta \, d\beta = \frac{\Gamma}{2\pi d} \qquad (2.70)$$

For the semi-infinite vortex line that starts at point O in Fig. 2.14, $\beta_1 = \pi/2$ and $\beta_2 = \pi$ and the induced velocity is

$$q_\theta = \frac{\Gamma}{4\pi d} \qquad (2.71)$$

which is exactly half of the previous value.

Equation (2.68b) can be modified to a form more convenient for numerical computations by using the definitions of Fig. 2.16. For the general three-dimensional case the two edges of the vortex segment will be located by $\mathbf{r}_1$ and $\mathbf{r}_2$ and the vector connecting the edges is

$$\mathbf{r}_0 = \mathbf{r}_1 - \mathbf{r}_2$$

as shown in Fig. 2.16. The distance d and the cosines of the angles β are then (Robinson and Laurman,[2.2] p. 33)

$$d = \frac{|\mathbf{r}_1 \times \mathbf{r}_2|}{|\mathbf{r}_0|}$$

$$\cos\beta_1 = \frac{\mathbf{r}_0 \cdot \mathbf{r}_1}{|\mathbf{r}_0||\mathbf{r}_1|}$$

$$\cos\beta_2 = \frac{\mathbf{r}_0 \cdot \mathbf{r}_2}{|\mathbf{r}_0||\mathbf{r}_2|}$$

The direction of the velocity $\mathbf{q}_{1,2}$ is normal to the plane created by the point P and the vortex edges 1, 2 and is given by

$$\frac{\mathbf{r}_1 \times \mathbf{r}_2}{|\mathbf{r}_1 \times \mathbf{r}_2|}$$

Substituting these quantities, and multiplying by this directional vector, we get an induced

2.13 The Stream Function

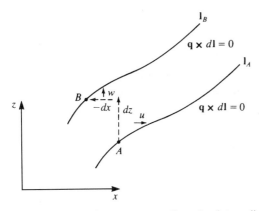

Figure 2.17 Flow between two two-dimensional streamlines.

velocity of

$$\mathbf{q}_{1,2} = \frac{\Gamma}{4\pi} \frac{\mathbf{r}_1 \times \mathbf{r}_2}{|\mathbf{r}_1 \times \mathbf{r}_2|^2} \mathbf{r}_0 \cdot \left(\frac{\mathbf{r}_1}{r_1} - \frac{\mathbf{r}_2}{r_2} \right) \qquad (2.72)$$

A more detailed procedure, including the numerical interpretation for using this formula, is provided in Section 10.4.5. The subroutine inputs are vortex strength Γ and the three (x, y, z) values of the points 1, 2, and P; the subroutine returns the three components of the induced velocity at point P.

2.13   The Stream Function

Consider two arbitrary streamlines in a two-dimensional steady flow, as shown in Fig. 2.17. The velocity $\mathbf{q}$ along these lines $\mathbf{l}$ is tangent to them so that

$$\mathbf{q} \times d\mathbf{l} = u\,dz - w\,dx = 0 \qquad (1.5)$$

and, therefore, the flux (volumetric flow rate) between two such lines is constant. This flow rate between these two curves is

$$\text{flux} = \int_A^B \mathbf{q} \cdot \mathbf{n}\,dl = \int_A^B u\,dz + w(-dx) \qquad (2.73)$$

where A and B are two arbitrary points on these lines. If a scalar function $\Psi(x, z)$ for this flux is to be introduced, such that its variation along a streamline will be zero (according to Eq. (1.5)), then based on these two equations (Eq. (1.6) and Eq. (2.73)) its relation to the velocity is

$$u = \frac{\partial \Psi}{\partial z}, \qquad w = -\frac{\partial \Psi}{\partial x} \qquad (2.74)$$

Substitution of this into Eq. (1.5) for the streamline results in

$$d\Psi = \frac{\partial \Psi}{\partial x} dx + \frac{\partial \Psi}{\partial z} dz = -w\,dx + u\,dz = 0 \qquad (2.75)$$

Therefore, $d\Psi$ along a streamline is zero, and between two different streamlines $d\Psi$ represents the volume flux (Eq. (2.73)). Integration of this equation results in

$$\Psi = \text{const.} \qquad \text{on streamlines} \qquad (2.76)$$

Substitution of Eqs. (2.74) into the continuity equation yields

$$\frac{\partial u}{\partial x} + \frac{\partial w}{\partial z} = \frac{\partial^2 \Psi}{\partial x \partial z} - \frac{\partial^2 \Psi}{\partial x \partial z} = 0 \qquad (2.77)$$

and therefore the continuity equation is automatically satisfied. Note that the stream function is valid for viscous flow, too, and if the irrotational flow requirement is added then $\zeta_y = 0$. Recall that the y component of the vorticity is

$$\zeta_y = \frac{\partial u}{\partial z} - \frac{\partial w}{\partial x} = \nabla^2 \Psi$$

and therefore for two-dimensional, incompressible, irrotational flow Ψ satisfies Laplace's equation

$$\nabla^2 \Psi = 0 \qquad (2.78)$$

It is possible to express the two-dimensional velocity in the x–z plane as

$$\mathbf{q} = \frac{\partial \Psi}{\partial z}\mathbf{i} - \frac{\partial \Psi}{\partial x}\mathbf{k} = \mathbf{j} \times \nabla \Psi$$

Thus

$$\mathbf{q} = \mathbf{j} \times \nabla \Psi \qquad (2.79)$$

Using this method, we can obtain the velocity in cylindrical coordinates (for the r–θ plane):

$$\mathbf{q} = \mathbf{j} \times \left(\frac{\partial \Psi}{\partial r}\mathbf{e}_r + \frac{1}{r}\frac{\partial \Psi}{\partial \theta}\mathbf{e}_\theta \right) = -\frac{\partial \Psi}{\partial r}\mathbf{e}_\theta + \frac{1}{r}\frac{\partial \Psi}{\partial \theta}\mathbf{e}_r$$

and the velocity components are

$$q_\theta = -\frac{\partial \Psi}{\partial r} \qquad (2.80a)$$

$$q_r = \frac{1}{r}\frac{\partial \Psi}{\partial \theta} \qquad (2.80b)$$

The relation between the stream function and the velocity potential can be found by equating the expressions for the velocity components (Eq. (2.20) and Eq. (2.74)); in Cartesian coordinates we have

$$\frac{\partial \Phi}{\partial x} = \frac{\partial \Psi}{\partial z}, \qquad \frac{\partial \Phi}{\partial z} = -\frac{\partial \Psi}{\partial x} \qquad (2.81)$$

and in cylindrical coordinates we have

$$\frac{\partial \Phi}{\partial r} = \frac{1}{r}\frac{\partial \Psi}{\partial \theta}, \qquad \frac{1}{r}\frac{\partial \Phi}{\partial \theta} = -\frac{\partial \Psi}{\partial r} \qquad (2.82)$$

These are the Cauchy–Riemann equations with which the complex flow potential will be defined in Chapter 6.

Laplace's equation in polar coordinates, expressed in terms of the stream function, is

$$\nabla^2 \Psi = \frac{\partial^2 \Psi}{\partial r^2} + \frac{1}{r}\frac{\partial \Psi}{\partial r} + \frac{1}{r^2}\frac{\partial^2 \Psi}{\partial \theta^2} = 0 \qquad (2.83)$$

To demonstrate the relation between the velocity potential and the stream function recall that along a streamline

$$d\Psi = u\,dz - w\,dx = 0 \qquad (2.84)$$

and similarly, along a constant potential line

$$d\Phi = u\,dx + w\,dz = 0 \tag{2.85}$$

Since the slopes of the streamlines and the potential lines are negative reciprocals, these lines are perpendicular to one another at any point in the flow.

Since constant stream function lines represent streamlines (Eq. (2.76)), the use of the stream function for two-dimensional flows is quite attractive (see Sections 3.7–3.11). However, the applicability of stream functions to three-dimensional flows, apart from the axisymmetric case, is more complicated (see Karamcheti,[1.5] Section 4.9). Therefore, in this book the velocity potential representation is preferred, except for a few two-dimensional examples that use the stream function representation.

References

[2.1] Kreyszig, E., *Advanced Engineering Mathematics*, 8th edition, John Wiley & Sons, 1999.
[2.2] Robinson, A., and Laurmann, J. A., *Wing Theory*, Cambridge University Press, 1956.

Problems

2.1. Write the scalar version of the inviscid, incompressible, vorticity transport equation in cylindrical coordinates for an axisymmetric flow.

2.2. Evaluate the boundary condition of Eq. (2.27) for a circle (and a sphere) whose radius is varying such that $r = a(t)$ in a fluid at rest at infinity.

2.3. a. Consider an incompressible potential flow in a fluid region V with boundary S. Find an equation for the kinetic energy in the region as an integral over S.
   b. Now consider the two-dimensional flow between concentric cylinders with radii a and b and velocity components $q_r = 0$ and $q_\theta = A/r$ (where A is constant). Calculate the kinetic energy in the fluid region using the result from (a).

2.4. a. Find the velocity induced at the center of a square vortex ring whose circulation is Γ and whose sides are of length a.
   b. Find the velocity along the z axis induced by a circular vortex ring that lies in the x–y plane, whose radius is a and circulation is Γ, and whose center is at the origin of coordinates.

2.5. Find the stream function for a two-dimensional flow whose velocity components are $u = 2Ax$ and $w = -2Az$.

CHAPTER 3

General Solution of the Incompressible, Potential Flow Equations

In the previous two chapters the fundamental fluid dynamic equations were formulated and the conditions leading to the simplified inviscid, incompressible, and irrotational flow problem were discussed. In this chapter, the basic methodology for obtaining the elementary solutions to this potential flow problem will be developed. Because of the linear nature of the potential flow problem, the differential equation does not have to be solved individually for flowfields having different geometry at their boundaries. Instead, the elementary solutions will be distributed in a manner that will satisfy each individual set of geometrical boundary conditions.

This approach, of distributing elementary solutions with unknown strength, allows a more systematic methodology for resolving the flowfield in both cases of the "classical" and the numerical methods.

3.1 Statement of the Potential Flow Problem

For most engineering applications the problem requires a solution in a fluid domain V that usually contains a solid body with additional boundaries that may define an outer flow boundary (e.g., a wing in a wind tunnel), as shown in Fig. 3.1. If the flow in the fluid region is considered to be incompressible and irrotational then the continuity equation reduces to

$$\nabla^2 \Phi = 0 \tag{3.1}$$

For a submerged body in the fluid, the velocity component normal to the body's surface and to other solid boundaries must be zero, and in a body-fixed coordinate system:

$$\nabla \Phi \cdot \mathbf{n} = 0 \tag{3.2}$$

Here $\mathbf{n}$ is a vector normal to the body's surface, and $\nabla \Phi$ is measured in a frame of reference attached to the body. Also, the disturbance created by the motion should decay far ($r \to \infty$) from the body:

$$\lim_{r \to \infty} (\nabla \Phi - \mathbf{v}) = 0 \tag{3.3}$$

where $\mathbf{r} = (x, y, z)$ and $\mathbf{v}$ is the relative velocity between the undisturbed fluid in V and the body (or the velocity at infinity seen by an observer moving with the body).

3.2 The General Solution, Based on Green's Identity

The mathematical problem of the previous section is described schematically by Fig. 3.1. Laplace's equation for the velocity potential must be solved for an arbitrary body with boundary S_B enclosed in a volume V, with the outer boundary S_∞. The boundary conditions in Eqs. (3.2) and (3.3) apply to S_B and S_∞, respectively. The normal $\mathbf{n}$ is defined such that it always points outside the region of interest V. Now, the vector appearing in the divergence theorem (e.g., $\mathbf{q}$ in Eq. (1.20)) is replaced by the vector $\Phi_1 \nabla \Phi_2 - \Phi_2 \nabla \Phi_1$,

3.2 The General Solution, Based on Green's Identity

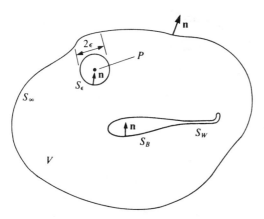

Figure 3.1 Nomenclature used to define the potential flow problem.

where Φ_1 and Φ_2 are two scalar functions of position. This results in

$$\int_S (\Phi_1 \nabla \Phi_2 - \Phi_2 \nabla \Phi_1) \cdot \mathbf{n}\, dS = \int_V (\Phi_1 \nabla^2 \Phi_2 - \Phi_2 \nabla^2 \Phi_1)\, dV \qquad (3.4)$$

This equation is one of Green's identities (Kellogg,[1.3] p. 215). Here the surface integral is taken over all the boundaries S, including a wake model S_W (which might model a surface across which a discontinuity in the velocity potential or the velocity may occur),

$$S = S_B + S_W + S_\infty$$

Also, let us set

$$\Phi_1 = \frac{1}{r} \quad \text{and} \quad \Phi_2 = \Phi \qquad (3.5)$$

where Φ is the potential of the flow of interest in V, and r is the distance from a point $P(x, y, z)$, as shown in the figure. As we shall see later, Φ_1 is the potential of a source (or sink) and is unbounded ($\frac{1}{r} \to \infty$) as P is approached and $r \to 0$. In the case where the point P is outside of V both Φ_1 and Φ_2 satisfy Laplace's equation and Eq. (3.4) becomes

$$\int_S \left(\frac{1}{r}\nabla\Phi - \Phi\nabla\frac{1}{r}\right) \cdot \mathbf{n}\, dS = 0 \qquad (3.6)$$

Of particular interest is the case when the point P is inside the region. The point P must now be excluded from the region of integration and it is surrounded by a small sphere of radius ϵ. Outside of the sphere and in the remaining region V the potential Φ_1 satisfies Laplace's equation $[\nabla^2(1/r) = 0]$. Similarly $\nabla^2 \Phi_2 = 0$ and Eq. (3.4) becomes

$$\int_{S+\text{sphere } \epsilon} \left(\frac{1}{r}\nabla\Phi - \Phi\nabla\frac{1}{r}\right) \cdot \mathbf{n}\, dS = 0 \qquad (3.6a)$$

To evaluate the integral over the sphere, introduce a spherical coordinate system at P and since the vector $\mathbf{n}$ points inside the small sphere, $\mathbf{n} = -\mathbf{e}_r$, $\mathbf{n} \cdot \nabla\Phi = -\partial\Phi/\partial r$, and $\nabla 1/r = -(1/r^2)\mathbf{e}_r$. Equation (3.6a) now becomes

$$-\int_{\text{sphere } \epsilon} \left(\frac{1}{r}\frac{\partial\Phi}{\partial r} + \frac{\Phi}{r^2}\right)dS + \int_S \left(\frac{1}{r}\nabla\Phi - \Phi\nabla\frac{1}{r}\right) \cdot \mathbf{n}\, dS = 0 \qquad (3.6b)$$

On the sphere surrounding P, $\int dS = 4\pi\epsilon^2$ (where $r = \epsilon$), and as $\epsilon \to 0$ (and assuming that the potential and its derivatives are well-behaved functions and therefore do not vary much in the small sphere) the first term in the first integral vanishes, while the second term yields

$$-\int_{\text{sphere }\epsilon} \left(\frac{\Phi}{r^2}\right) dS = -4\pi\,\Phi(P)$$

Equation (3.6b) then becomes

$$\Phi(P) = \frac{1}{4\pi} \int_S \left(\frac{1}{r}\nabla\Phi - \Phi\nabla\frac{1}{r}\right) \cdot \mathbf{n}\, dS \qquad (3.7)$$

This formula gives the value of $\Phi(P)$ at any point in the flow, within the region V, in terms of the values of Φ and $\partial\Phi/\partial n$ on the boundaries S.

If, for example, the point P lies on the boundary S_B then in order to exclude the point from V, the integration is carried out only around the surrounding hemisphere (submerged in V) with radius ϵ, and Eq. (3.7) becomes

$$\Phi(P) = \frac{1}{2\pi} \int_S \left(\frac{1}{r}\nabla\Phi - \Phi\nabla\frac{1}{r}\right) \cdot \mathbf{n}\, dS \qquad (3.7a)$$

Now consider a situation when the flow of interest occurs inside the boundary of S_B and the resulting "internal potential" is Φ_i. For this flow the point P (which is in the region V) is exterior to S_B, and applying Eq. (3.6) yields

$$0 = \frac{1}{4\pi} \int_{S_B} \left(\frac{1}{r}\nabla\Phi_i - \Phi_i\nabla\frac{1}{r}\right) \cdot \mathbf{n}\, dS \qquad (3.7b)$$

Here, $\mathbf{n}$ points outward from S_B. A form of Eq. (3.7) that includes the influence of the inner potential, as well, is obtained by adding Eq. (3.7) and Eq. (3.7b) (note that the minus sign is a result of the opposite direction of $\mathbf{n}$ for Φ_i):

$$\Phi(P) = \frac{1}{4\pi} \int_{S_B} \left[\frac{1}{r}\nabla(\Phi - \Phi_i) - (\Phi - \Phi_i)\nabla\frac{1}{r}\right] \cdot \mathbf{n}\, dS$$
$$+ \frac{1}{4\pi} \int_{S_W + S_\infty} \left(\frac{1}{r}\nabla\Phi - \Phi\nabla\frac{1}{r}\right) \cdot \mathbf{n}\, dS \qquad (3.8)$$

The contribution of the S_∞ integral in Eq. (3.8) (when S_∞ is considered to be far from S_B) can be defined as

$$\Phi_\infty(P) = \frac{1}{4\pi} \int_{S_\infty} \left(\frac{1}{r}\nabla\Phi - \Phi\nabla\frac{1}{r}\right) \cdot \mathbf{n}\, dS \qquad (3.9)$$

This potential, usually, depends on the selection of the coordinate system and, for example, in an inertial system where the body moves through an otherwise stationary fluid Φ_∞ can be selected as a constant in the region. Also, the wake surface is assumed to be thin, such that $\partial\Phi/\partial n$ is continuous across it (which means that no fluid-dynamic loads will be supported by the wake). With these assumptions Eq. (3.8) becomes

$$\Phi(P) = \frac{1}{4\pi} \int_{S_B} \left[\frac{1}{r}\nabla(\Phi - \Phi_i) - (\Phi - \Phi_i)\nabla\frac{1}{r}\right] \cdot \mathbf{n}\, dS$$
$$- \frac{1}{4\pi} \int_{S_W} \Phi\mathbf{n} \cdot \nabla\frac{1}{r}\, dS + \Phi_\infty(P) \qquad (3.10)$$

3.2 The General Solution, Based on Green's Identity

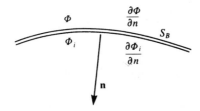

Figure 3.2 The velocity potential near a solid boundary S_B.

As was stated before, Eq. (3.7) (or Eq. (3.10)) provides the value of $\Phi(P)$ in terms of Φ and $\partial\Phi/\partial n$ on the boundaries. Therefore, the problem is reduced to determining the value of these quantities on the boundaries. For example, consider a segment of the boundary S_B as shown in Fig. 3.2; then the difference between the external and internal potentials can be defined as

$$-\mu = \Phi - \Phi_i \tag{3.11}$$

and the difference between the normal derivative of the external and internal potentials as

$$-\sigma = \frac{\partial \Phi}{\partial n} - \frac{\partial \Phi_i}{\partial n} \tag{3.12}$$

These elements are called *doublet* (μ) and *source* (σ) and the minus sign is a result of the normal vector **n** pointing into S_B. The properties of these elementary solutions will be investigated in the following sections. With the definitions of Eq. (3.11) and Eq. (3.12), Eq. (3.10) can be rewritten as

$$\Phi(P) = -\frac{1}{4\pi} \int_{S_B} \left[\sigma\left(\frac{1}{r}\right) - \mu \mathbf{n} \cdot \nabla\left(\frac{1}{r}\right) \right] dS$$
$$+ \frac{1}{4\pi} \int_{S_W} \left[\mu \mathbf{n} \cdot \nabla\left(\frac{1}{r}\right) \right] dS + \Phi_\infty(P) \tag{3.13}$$

and the doublet strength μ appearing in the second integral (over S_W) is the potential difference between the upper and lower wake surfaces (that is, if the wake thickness is zero, then $\mu = -\Delta\Phi$ on S_W). The vector **n** here is the local normal to the surface, which points in the doublet direction (as will be shown in Section 3.5). It is convenient to replace $\mathbf{n} \cdot \nabla$ by $\partial/\partial n$ in this equation, and it becomes

$$\Phi(P) = -\frac{1}{4\pi} \int_{S_B} \left[\sigma\left(\frac{1}{r}\right) - \mu \frac{\partial}{\partial n}\left(\frac{1}{r}\right) \right] dS + \frac{1}{4\pi} \int_{S_W} \left[\mu \frac{\partial}{\partial n}\left(\frac{1}{r}\right) \right] dS + \Phi_\infty(P)$$
$$\tag{3.13a}$$

Note that both source and doublet solutions decay as $r \to \infty$ and automatically fulfill the boundary condition of Eq. (3.3) (where **v** is the velocity due to Φ_∞).

To find the velocity potential in the region V, the strength of the distribution of doublets and sources on the surface must be determined. Also, Eq. (3.13) does not specify a unique combination of sources and doublets for a particular problem and a choice must be made in this matter (usually based on the physics of the problem).

It is possible to require that

$$\frac{\partial \Phi_i}{\partial n} = \frac{\partial \Phi}{\partial n} \quad \text{on} \quad S_B$$

and in this case the source term on S_B vanishes and only the doublet distribution remains. Alternatively, the potential can be defined such that

$$\Phi_i = \Phi \quad \text{on} \quad S_B$$

and in this case the doublet term on S_B vanishes and the problem will be modeled by a source distribution on the boundary.

In the two-dimensional case the source potential is $\Phi_1 = \ln r$ as will be shown in Section 3.7, and the two functions of Eq. (3.5) become

$$\Phi_1 = \ln r \quad \text{and} \quad \Phi_2 = \Phi \tag{3.14}$$

Also, at the point P, the integration is around a circle with radius ϵ and Eq. (3.6b) becomes

$$-\int_{\text{circle } \epsilon} \left(\ln r \frac{\partial \Phi}{\partial r} - \Phi \frac{1}{r} \right) dS + \int_S \left(\ln r \nabla \Phi - \Phi \nabla \ln r \right) \cdot \mathbf{n} \, dS = 0 \tag{3.15}$$

The circumference of the small circle around P is now $2\pi\epsilon$ (compared to $4\pi\epsilon^2$ in the three-dimensional case) and Eq. (3.7) in two dimensions is

$$\Phi(P) = -\frac{1}{2\pi} \int_S (\ln r \nabla \Phi - \Phi \nabla \ln r) \cdot \mathbf{n} \, dS \tag{3.16}$$

If the point P lies on the boundary S_B, then the integration is around a semicircle with radius ϵ and Eq. (3.16) becomes

$$\Phi(P) = -\frac{1}{\pi} \int_S (\ln r \nabla \Phi - \Phi \nabla \ln r) \cdot \mathbf{n} \, dS \tag{3.16a}$$

whereas if P is inside S_B the two-dimensional version of Eq. (3.7b) is

$$0 = -\frac{1}{2\pi} \int_{S_B} (\ln r \nabla \Phi_i - \Phi_i \nabla \ln r) \cdot \mathbf{n} \, dS \tag{3.16b}$$

With the definition of the far field potential Φ_∞ and the unit elements μ and σ being unchanged, Eq. (3.13a) for the two-dimensional case becomes

$$\Phi(P) = \frac{1}{2\pi} \int_{S_B} \left[\sigma \ln r - \mu \frac{\partial}{\partial n}(\ln r) \right] dS - \frac{1}{2\pi} \int_{S_W} \mu \frac{\partial}{\partial n}(\ln r) \, dS + \Phi_\infty(P) \tag{3.17}$$

Note that $\partial/\partial n$ is the orientation of the doublet as will be illustrated in Section 3.5 and that the wake model S_W in the steady, two-dimensional lifting case is needed to represent a discontinuity in the potential Φ.

3.3 Summary: Methodology of Solution

In view of Eq. (3.13) (Eq. (3.17) in two dimensions), it is possible to establish a fairly general approach to the solution of incompressible potential flow problems. The most important observation is that the solution of $\nabla^2 \Phi = 0$ can be obtained by distributing elementary solutions (sources and doublets) on the problem *boundaries* (S_B, S_W). These elementary solutions automatically fulfill the boundary condition of Eq. (3.3) by having velocity fields that decay as $r \to \infty$. However, at the point where $r = 0$, the velocity becomes singular, and therefore the basic elements are called *singular solutions*.

The general solution requires the integration of these basic solutions over any surface S containing these singularity elements because each element will have an effect on the whole fluid field.

The solution of a fluid dynamic problem is now reduced to finding the appropriate singularity element distribution over some known boundaries, so that the boundary condition (Eq. (3.2)) will be fulfilled. The main advantage of this formulation is its straightforward applicability to numerical methods. When the potential is specified on the problem boundaries then this type of mathematical problem is called the *Dirichlet problem* (Kellogg,[1.3] p. 286) and is frequently used in many numerical solutions (panel methods).

A more direct approach to the solution, from the physical point of view, is to specify the zero normal flow boundary condition (Eq. (3.2)) on the solid boundaries. This problem is known as the *Neumann problem* (Kellogg,[1.3] p. 286) and in order to evaluate the velocity field the potential is differentiated:

$$\nabla \Phi = -\frac{1}{4\pi} \int_{S_B} \sigma \nabla \left(\frac{1}{r}\right) dS + \frac{1}{4\pi} \int_{S_B+S_W} \mu \nabla \left[\frac{\partial}{\partial n}\left(\frac{1}{r}\right)\right] dS + \nabla \Phi_\infty \quad (3.18)$$

Again, the derivative $\partial/\partial n$ for the doublet indicates the orientation of the element as will be shown in Section 3.5. Substituting this equation into the boundary condition of Eq. (3.2) can serve as the basis of finding the unknown singularity distribution. (This can be done analytically or numerically.)

For a given set of boundary conditions, the above solution technique is not unique, and many problems can be solved by using a preferred type of singularity element or any linear combination of the two singularity types. Therefore, in many situations additional considerations are required (e.g., the method that will be presented in the next chapter to define the flow near sharp trailing edges of wings). Also, in a particular solution a mixed use of the above boundary conditions is possible for various regions in the flowfield (e.g., Neumann condition on one boundary and Dirichlet on another).

Prior to attempting to apply this methodology to the solution of particular problems, the features of the elementary solutions are analyzed in the next sections.

3.4 Basic Solution: Point Source

One of the two basic solutions presented in Eq. (3.13) is the source/sink. The potential of such a point source element (Fig. 3.3a), placed at the origin of a spherical coordinate system, is

$$\Phi = -\frac{\sigma}{4\pi r} \quad (3.19)$$

The velocity due to this element is obtained by using ∇ in spherical coordinates from Eq. (1.39). This will result in a velocity field with a radial component only

$$\mathbf{q} = -\frac{\sigma}{4\pi} \nabla \left(\frac{1}{r}\right) = \frac{\sigma}{4\pi} \frac{\mathbf{e}_r}{r^2} = \frac{\sigma}{4\pi} \frac{\mathbf{r}}{r^3} \quad (3.20)$$

which, in spherical coordinates, is

$$(q_r, q_\theta, q_\varphi) = \left(\frac{\partial \Phi}{\partial r}, 0, 0\right) = \left(\frac{\sigma}{4\pi r^2}, 0, 0\right) \quad (3.21)$$

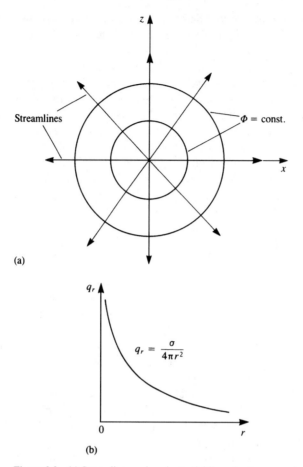

Figure 3.3 (a) Streamlines and equipotential lines due to source element at the origin, as viewed in the x–z plane. (b) Radial variation of the radial velocity component induced by a point source.

So the velocity in the radial direction decays with the rate of $1/r^2$ and is singular at $r = 0$, as shown in Fig. 3.3. Consider a source element of strength σ located at the origin as in Fig. 3.3. The volumetric flow rate through a spherical surface of radius r is

$$q_r 4\pi r^2 = \left(\frac{\sigma}{4\pi r^2}\right) \cdot 4\pi r^2 = \sigma$$

where $4\pi r^2$ is the surface area of the sphere. The positive σ, then, is the volumetric rate at which fluid is introduced at the source, whereas a negative σ is the rate at which flow is going into the sink. Note that this introduction of fluid at the source violates the conservation of mass; therefore, this point must be excluded from the region of solution.

If the point element is located at a point $\mathbf{r}_0$ and not at the origin, then the corresponding potential and velocity will be

$$\Phi = \frac{-\sigma}{4\pi |\mathbf{r} - \mathbf{r}_0|} \tag{3.22}$$

$$\mathbf{q} = \frac{\sigma}{4\pi} \frac{\mathbf{r} - \mathbf{r}_0}{|\mathbf{r} - \mathbf{r}_0|^3} \tag{3.23}$$

The Cartesian form of this equation, when the element is located at (x_0, y_0, z_0), is

$$\Phi(x, y, z) = \frac{-\sigma}{4\pi \sqrt{(x - x_0)^2 + (y - y_0)^2 + (z - z_0)^2}} \qquad (3.24)$$

The velocity components of this source element are

$$u(x, y, z) = \frac{\partial \Phi}{\partial x} = \frac{\sigma(x - x_0)}{4\pi [(x - x_0)^2 + (y - y_0)^2 + (z - z_0)^2]^{3/2}} \qquad (3.25a)$$

$$v(x, y, z) = \frac{\partial \Phi}{\partial y} = \frac{\sigma(y - y_0)}{4\pi [(x - x_0)^2 + (y - y_0)^2 + (z - z_0)^2]^{3/2}} \qquad (3.25b)$$

$$w(x, y, z) = \frac{\partial \Phi}{\partial z} = \frac{\sigma(z - z_0)}{4\pi [(x - x_0)^2 + (y - y_0)^2 + (z - z_0)^2]^{3/2}} \qquad (3.25c)$$

This basic point element can be integrated over a line l, a surface S, or a volume V to create corresponding singularity elements that can be used, for example, to construct panel elements. Consequently, these elements can be established by the following integrals:

$$\Phi(x, y, z) = \frac{-1}{4\pi} \int_l \frac{\sigma(x_0, y_0, z_0)\, dl}{\sqrt{(x - x_0)^2 + (y - y_0)^2 + (z - z_0)^2}} \qquad (3.26)$$

$$\Phi(x, y, z) = \frac{-1}{4\pi} \int_S \frac{\sigma(x_0, y_0, z_0)\, dS}{\sqrt{(x - x_0)^2 + (y - y_0)^2 + (z - z_0)^2}} \qquad (3.27)$$

$$\Phi(x, y, z) = \frac{-1}{4\pi} \int_V \frac{\sigma(x_0, y_0, z_0)\, dV}{\sqrt{(x - x_0)^2 + (y - y_0)^2 + (z - z_0)^2}} \qquad (3.28)$$

Note that σ in Eqs. (3.26), (3.27), and (3.28) represents the source strength per unit length, area, and volume, respectively. The velocity components induced by these distributions can be obtained by differentiating the corresponding potentials:

$$(u, v, w) = \left(\frac{\partial \Phi}{\partial x}, \frac{\partial \Phi}{\partial y}, \frac{\partial \Phi}{\partial z} \right)$$

3.5 Basic Solution: Point Doublet

The second basic solution, presented in Eq. (3.13), is the doublet

$$\Phi = \frac{\mu}{4\pi} \mathbf{n} \cdot \nabla \left(\frac{1}{r} \right) \qquad (3.29)$$

A closer observation reveals that $\Phi_{\text{doublet}} = -(\partial/\partial n)\Phi_{\text{source}}$ for elements of unit strength. This suggests that the doublet element can be developed from the source element. Consider a point sink at the origin and a point source at $\mathbf{l}$, as shown in Fig. 3.4. The potential at a point P, due to these two elements, is

$$\Phi = \frac{\sigma}{4\pi} \left(\frac{1}{|\mathbf{r}|} - \frac{1}{|\mathbf{r} - \mathbf{l}|} \right) \qquad (3.30)$$

Now, bringing the source and the sink together by letting $l \to 0$ and $\sigma \to \infty$ such that $l\sigma \to \mu$, where μ is finite, we obtain for the potential

$$\Phi = \lim_{\substack{l \to 0 \\ \sigma \to \infty \\ \sigma l \to \mu}} \frac{\sigma}{4\pi} \left(\frac{|\mathbf{r} - \mathbf{l}| - |\mathbf{r}|}{|\mathbf{r}||\mathbf{r} - \mathbf{l}|} \right)$$

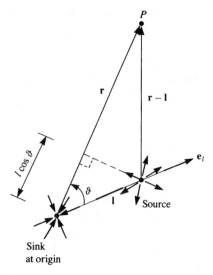

Figure 3.4 The influence of a point source and sink at point P.

As the distance l approaches zero

$$|\mathbf{r}||\mathbf{r} - \mathbf{l}| \to r^2$$

and the difference in length between $|\mathbf{r}|$ and $|\mathbf{r} - \mathbf{l}|$ becomes

$$(|\mathbf{r} - \mathbf{l}| - |\mathbf{r}|) \to -l \cos \vartheta$$

and the potential becomes

$$\Phi = \frac{-\mu}{4\pi} \frac{\cos \vartheta}{r^2} \tag{3.31}$$

The angle ϑ is between the unit vector $\mathbf{e}_l$ pointing in the sink-to-source direction (doublet axis) and the vector $\mathbf{r}$, as shown in the figure. Defining a vector doublet strength $\boldsymbol{\mu}$ that points in this direction $\boldsymbol{\mu} = \mu \, \mathbf{e}_l$ can further simplify this equation to

$$\Phi = \frac{-\boldsymbol{\mu} \cdot \mathbf{r}}{4\pi r^3} \tag{3.32}$$

Note that this doublet element is identical to the second term appearing in the general equation of the potential (Eq. (3.13) or Eq. (3.29)) if $\mathbf{e}_l$ is in the $\mathbf{n}$ direction; thus

$$\Phi_{\text{doublet}} = \frac{-\mathbf{e}_l \cdot \mathbf{r}}{4\pi r^3} = -\mathbf{e}_l \cdot \nabla\left(\frac{-1}{4\pi r}\right) = -\frac{\partial}{\partial n} \Phi_{\text{source}} \tag{3.33}$$

For example, for a doublet at the origin and the doublet strength vector $(\mu, 0, 0)$ aligned with the x axis ($\mathbf{e}_l = \mathbf{e}_x$ and $\vartheta = \theta$), the potential in spherical coordinates is

$$\Phi(r, \theta, \varphi) = \frac{-\mu \cos \theta}{4\pi r^2} \tag{3.34}$$

Furthermore, in Cartesian coordinates, the arbitrary orientation of $\boldsymbol{\mu}$ can be expressed in terms of three generic unit doublet elements whose axes are aligned with the coordinate directions:

$$(\mu, 0, 0), \quad (0, \mu, 0), \quad (0, 0, \mu)$$

3.5 Basic Solution: Point Doublet

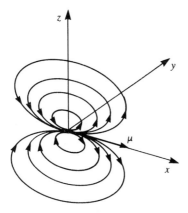

Figure 3.5 Sketch of the streamlines due to a doublet pointing in the x direction (e.g., a small jet engine blowing in the $\boldsymbol{\mu} = (\mu, 0, 0)$ direction).

The different elements can be derived for each of these three doublets by using Eq. (3.32) or by differentiating the corresponding term in Eq. (3.29) using $\partial/\partial n$ as the derivative in the direction of the three axes. The velocity potential due to such doublet elements, located at (x_0, y_0, z_0), is

$$\Phi(x, y, z) = \frac{\mu}{4\pi} \mathbf{n} \cdot \nabla \left(\frac{1}{|\mathbf{r} - \mathbf{r}_0|} \right) = \frac{\mu}{4\pi} \frac{\partial}{\partial n} \left(\frac{1}{|\mathbf{r} - \mathbf{r}_0|} \right) \quad (3.35)$$

Taking $\partial/\partial n$ in the x, y, and z directions yields

$$\Phi(x, y, z) = \frac{\mu}{4\pi} \begin{pmatrix} \frac{\partial}{\partial x} \\ \frac{\partial}{\partial y} \\ \frac{\partial}{\partial z} \end{pmatrix} \frac{1}{\sqrt{(x - x_0)^2 + (y - y_0)^2 + (z - z_0)^2}} \quad (3.36)$$

Equation (3.34) shows that the doublet element does not have a radial symmetry but rather has a directional property. Therefore, in Cartesian coordinates three elements are defined, one for each direction: x, y, or z (see for example the element pointing in the x direction in Fig. 3.5). After performing the differentiation in Eq. (3.36) in the x direction we obtain the velocity potential:

$$\Phi(x, y, z) = \frac{-\mu}{4\pi}(x - x_0)[(x - x_0)^2 + (y - y_0)^2 + (z - z_0)^2]^{-3/2} \quad (3.37)$$

The result of the differentiation in the y direction is

$$\Phi(x, y, z) = \frac{-\mu}{4\pi}(y - y_0)[(x - x_0)^2 + (y - y_0)^2 + (z - z_0)^2]^{-3/2} \quad (3.38)$$

and the result in the z direction is

$$\Phi(x, y, z) = \frac{-\mu}{4\pi}(z - z_0)[(x - x_0)^2 + (y - y_0)^2 + (z - z_0)^2]^{-3/2} \quad (3.39)$$

The velocity field, due to a x-directional point doublet $(\mu, 0, 0)$, is illustrated in Fig. 3.5. The velocity potential is given by Eq. (3.34) and the velocity components due to such an element at the origin are easily described in spherical coordinates:

$$q_r = \frac{\partial \Phi}{\partial r} = \frac{\mu \cos\theta}{2\pi r^3} \qquad (3.40)$$

$$q_\theta = \frac{1}{r}\frac{\partial \Phi}{\partial \theta} = \frac{\mu \sin\theta}{4\pi r^3} \qquad (3.41)$$

$$q_\varphi = \frac{1}{r\sin\theta}\frac{\partial \Phi}{\partial \varphi} = 0 \qquad (3.42)$$

The velocity components in Cartesian coordinates for this doublet at (x_0, y_0, z_0) can be obtained by differentiating the velocity potential in Eq. (3.37):

$$u = -\frac{\mu}{4\pi}\frac{(y-y_0)^2 + (z-z_0)^2 - 2(x-x_0)^2}{[(x-x_0)^2 + (y-y_0)^2 + (z-z_0)^2]^{5/2}} \qquad (3.43)$$

$$v = \frac{3\mu}{4\pi}\frac{(x-x_0)(y-y_0)}{[(x-x_0)^2 + (y-y_0)^2 + (z-z_0)^2]^{5/2}} \qquad (3.44)$$

$$w = \frac{3\mu}{4\pi}\frac{(x-x_0)(z-z_0)}{[(x-x_0)^2 + (y-y_0)^2 + (z-z_0)^2]^{5/2}} \qquad (3.45)$$

Again, this basic point element can be integrated over a line l, a surface S, or a volume V to create the corresponding singularity elements that can be used, for example, to construct panel elements. Consequently, these elements [e.g., for $(\mu, 0, 0)$] can be established by the following integrals:

$$\Phi(x, y, z) = \frac{-1}{4\pi}\int_l \frac{\mu(x_0, y_0, z_0)\cdot(x-x_0)\,dl}{[(x-x_0)^2 + (y-y_0)^2 + (z-z_0)^2]^{3/2}} \qquad (3.46)$$

$$\Phi(x, y, z) = \frac{-1}{4\pi}\int_S \frac{\mu(x_0, y_0, z_0)\cdot(x-x_0)\,dS}{[(x-x_0)^2 + (y-y_0)^2 + (z-z_0)^2]^{3/2}} \qquad (3.47)$$

$$\Phi(x, y, z) = \frac{-1}{4\pi}\int_V \frac{\mu(x_0, y_0, z_0)\cdot(x-x_0)\,dV}{[(x-x_0)^2 + (y-y_0)^2 + (z-z_0)^2]^{3/2}} \qquad (3.48)$$

3.6 Basic Solution: Polynomials

Since Laplace's equation is a second-order differential equation, a linear function of position will be a solution, too:

$$\Phi = Ax + By + Cz \qquad (3.49)$$

The velocity components due to such a potential are

$$u = \frac{\partial \Phi}{\partial x} = A \equiv U_\infty, \qquad v = \frac{\partial \Phi}{\partial y} = B \equiv V_\infty, \qquad w = \frac{\partial \Phi}{\partial z} = C \equiv W_\infty \qquad (3.50)$$

where U_∞, V_∞, and W_∞ are constant velocity components in the x, y, and z directions. Hence, the velocity potential due a constant free-stream flow in the x direction is

$$\Phi = U_\infty x \qquad (3.51)$$

and in general

$$\Phi = U_\infty x + V_\infty y + W_\infty z \qquad (3.52)$$

3.6 Basic Solution: Polynomials

Along the same lines, additional polynomial solutions can be sought and as an example let us consider the second-order polynomial with A, B, and C being constants:

$$\Phi = Ax^2 + By^2 + Cz^2 \tag{3.53}$$

To satisfy the continuity equation we must have

$$\nabla^2 \Phi = A + B + C = 0$$

There are numerous combinations of constants that will satisfy this condition. However, one combination where one of the constants is equal to zero (e.g., $B = 0$) describes an interesting flow condition. Consequently

$$A = -C$$

and by substituting this result into Eq. (3.53) we get a velocity potential of

$$\Phi = A(x^2 - z^2) \tag{3.54}$$

The velocity components for this two-dimensional flow in the x–z plane are

$$u = 2Ax, \quad v = 0, \quad w = -2Az \tag{3.55}$$

To visualize this flow, we need the streamline equation (Eq. (1.6a))

$$\frac{dx}{u} = \frac{dz}{w}$$

Substitution of the velocity components yields

$$\frac{dx}{2Ax} = \frac{dz}{-2Az}$$

Integration by separation of variables results in

$$xz = \text{const.} = D \tag{3.56}$$

The streamlines for different constant values of $D = 1, 2, 3\ldots$ are plotted in Fig. 3.6 and, for example, if only the first quadrant of the x–z plane is considered, then the potential

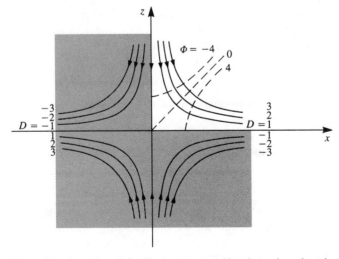

Figure 3.6 Streamlines defined by $xz = $ constant. Note that each quadrant describes a flow in a corner.

describes the flow around a corner. If the upper half of the x–z plane is considered then this flow describes a stagnation flow against a wall. Note that when $x = z = 0$, the velocity components $u = w = 0$ vanish too, which means that a stagnation point is present at the origin, and the coordinate axes x and z are also the stagnation streamline.

3.7 Two-Dimensional Version of the Basic Solutions

a. Source

We have seen in the three-dimensional case that a source element will have a radial velocity component only. Thus, in the two-dimensional r, θ coordinate system the tangential velocity component $q_\theta = 0$. The requirement that the flow be irrotational yields

$$\zeta_y = 2\omega_y = -\frac{1}{r}\left[\frac{\partial}{\partial r}(rq_\theta) - \frac{\partial}{\partial \theta}(q_r)\right] = \frac{1}{r}\frac{\partial}{\partial \theta}(q_r) = 0$$

and therefore the velocity component in the r direction is a function of r only [$q_r = q_r(r)$]. Also, the remaining radial velocity component must satisfy the continuity equation (Eq. (1.35))

$$\nabla \cdot \mathbf{q} = \frac{dq_r}{dr} + \frac{q_r}{r} = \frac{1}{r}\frac{d}{dr}(rq_r) = 0$$

This indicates that $rq_r = \text{const.} = \sigma/2\pi$, where σ is the area flow rate passing across a circle of radius r, and the resulting velocity components for a source element at the origin are

$$q_r = \frac{\partial \Phi}{\partial r} = \frac{\sigma}{2\pi r} \tag{3.57}$$

$$q_\theta = \frac{1}{r}\frac{\partial \Phi}{\partial \theta} = 0 \tag{3.58}$$

Integrating these equations we find the velocity potential

$$\Phi = \frac{\sigma}{2\pi}\ln r + C \tag{3.59}$$

and the constant C can be set to zero, as in the source potential used in Eq. (3.19).

The strength of the source is then σ, which represents the flux introduced by the source. This can be shown by observing the flux across a circle with a radius R. The velocity at that location, according to Eq. (3.57), is $\sigma/2\pi R$, and the flux is

$$q_r \, 2\pi R = \frac{\sigma}{2\pi R} 2\pi R = \sigma$$

So the velocity, as in the three-dimensional case, is in the radial direction only (Fig. 3.3a) and decays with a rate of $1/r$. At $r = 0$, the velocity is infinite and this singular point must be excluded from the region of the solution.

In Cartesian coordinates the corresponding equations for a source located at (x_0, z_0) are

$$\Phi(x, z) = \frac{\sigma}{2\pi}\ln\sqrt{(x-x_0)^2 + (z-z_0)^2} \tag{3.60}$$

$$u = \frac{\partial \Phi}{\partial x} = \frac{\sigma}{2\pi}\frac{x-x_0}{(x-x_0)^2 + (z-z_0)^2} \tag{3.61}$$

$$w = \frac{\partial \Phi}{\partial z} = \frac{\sigma}{2\pi}\frac{z-z_0}{(x-x_0)^2 + (z-z_0)^2} \tag{3.62}$$

3.7 Two-Dimensional Version of the Basic Solutions

In the two-dimensional case, the velocity components can be found as the derivatives of the stream function for a source at the origin. Recalling these formulas (Eqs. (2.80a,b)) and comparing with the velocity components we find

$$q_\theta = -\frac{\partial \Psi}{\partial r} = 0 \tag{3.63}$$

$$q_r = \frac{1}{r}\frac{\partial \Psi}{\partial \theta} = \frac{\sigma}{2\pi r} \tag{3.64}$$

Integrating Eqs. (3.63) and (3.64) and setting the constant of integration to zero yields

$$\Psi = \frac{\sigma}{2\pi}\theta \tag{3.65}$$

The streamlines (Eq. (3.65)) and the perpendicular constant potential lines (Eq. (3.59)) for the two-dimensional source resemble those of the three-dimensional case and are shown schematically in Fig. 3.3a.

b. Doublet

The two-dimensional doublet can be obtained by letting a point source and a point sink approach each other, such that their strength multiplied by their separation distance becomes the constant μ (as in Section 3.5). Because of the logarithmic dependence of the source potential, Eq. (3.32) becomes

$$\Phi(r) = \frac{-\boldsymbol{\mu} \cdot \mathbf{r}}{2\pi r^2} \tag{3.66}$$

which can be derived directly by using Eq. (3.33),

$$\Phi(r) = -\frac{\partial}{\partial n}\frac{\sigma}{2\pi}\ln r \tag{3.67}$$

and then replacing the source strength by μ. As an example, selecting $\mathbf{n}$ in the x direction yields

$$\boldsymbol{\mu} = (\mu, 0)$$

and Eq. (3.66) for a doublet at the origin becomes

$$\Phi(r, \theta) = \frac{-\mu \cos\theta}{2\pi r} \tag{3.68}$$

The velocity field due to this element can be obtained by differentiating the velocity potential:

$$q_r = \frac{\partial \Phi}{\partial r} = \frac{\mu \cos\theta}{2\pi r^2} \tag{3.69}$$

$$q_\theta = \frac{1}{r}\frac{\partial \Phi}{\partial \theta} = \frac{\mu \sin\theta}{2\pi r^2} \tag{3.70}$$

The velocity potential in Cartesian coordinates for such a doublet at the point (x_0, z_0) is

$$\Phi(x, z) = \frac{-\mu}{2\pi}\frac{x - x_0}{(x - x_0)^2 + (z - z_0)^2} \tag{3.71}$$

and the velocity components are

$$u = \frac{\mu}{2\pi}\frac{(x - x_0)^2 - (z - z_0)^2}{[(x - x_0)^2 + (z - z_0)^2]^2} \tag{3.72}$$

$$w = \frac{\mu}{2\pi}\frac{2(x - x_0)(z - z_0)}{[(x - x_0)^2 + (z - z_0)^2]^2} \tag{3.73}$$

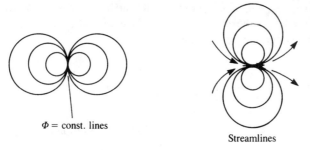

Figure 3.7 Streamlines and equipotential lines due to a two-dimensional doublet at the origin, pointing in the x direction.

To derive the stream function for this doublet element, located at the origin, write the above velocity components in terms of the stream function derivatives

$$q_\theta = -\frac{\partial \Psi}{\partial r} = \frac{\mu \sin\theta}{2\pi r^2} \tag{3.74}$$

$$q_r = \frac{1}{r}\frac{\partial \Psi}{\partial \theta} = \frac{\mu \cos\theta}{2\pi r^2} \tag{3.75}$$

Integrating Eqs. (3.74) and (3.75) and setting the constant of integration to zero (see streamlines in Fig. 3.7) we obtain

$$\Psi = \frac{\mu \sin\theta}{2\pi r} \tag{3.76}$$

Note that a similar doublet element where $\boldsymbol{\mu} = (0, \mu)$ can be derived by using Eq. (3.66) (or (3.67)).

3.8   Basic Solution: Vortex

The general solution to Laplace's equation as stated in Eqs. (3.13) and (3.17) consists of source and doublet distributions only. But, as indicated in Section 3.6, other solutions to Laplace's equation are possible, and based on the vortex flow of Section 2.10 we shall formulate the velocity potential and its derivatives for a point vortex (the three-dimensional velocity field is then given by the Biot–Savart law of Section 2.11). Therefore, it is desired to construct a singularity element with only a tangential velocity component, as shown in Fig. 3.8a, whose magnitude will decay in a manner similar to the decay of the radial velocity component of a two-dimensional source (e.g., will vary with $1/r$). The velocity components are then

$$q_r = 0$$
$$q_\theta = q_\theta(r, \theta)$$

Substitution of these velocity components into the continuity equation (Eq. (1.35)) results in q_θ being a function of r only,

$$q_\theta = q_\theta(r)$$

For irrotational flow, we substitute these relations into the vorticity expression to get

$$\zeta_y = 2\omega_y = -\frac{1}{r}\left[\frac{\partial}{\partial r}(rq_\theta) - \frac{\partial}{\partial \theta}(q_r)\right] = -\frac{1}{r}\frac{\partial}{\partial r}(rq_\theta) = 0$$

3.8 Basic Solution: Vortex

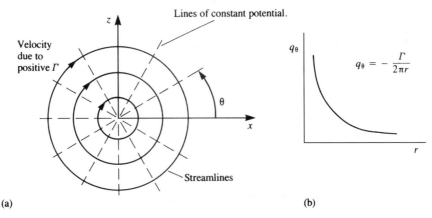

Figure 3.8 (a) Streamlines and equipotential lines for a two-dimensional vortex at the origin. (b) Radial variation of the tangential velocity component induced by a vortex.

Integrating with respect to r, we get

$$rq_\theta = \text{const.} = A$$

Thus the magnitude of the velocity varies with $1/r$, similarly to the radial velocity component of a source. The value of the constant A can be calculated by using the definition of the circulation Γ as in Eq. (2.36):

$$\Gamma = \oint \mathbf{q} \cdot d\mathbf{l} = \int_{2\pi}^{0} q_\theta \cdot r\, d\theta = -2\pi A$$

Note that positive Γ is defined according to the right-hand rule (positive clockwise); therefore, in the x–z plane as in Fig. 3.8 the line integral must be taken in the direction opposite to that of increasing θ. The constant A is then

$$A = -\frac{\Gamma}{2\pi}$$

and the velocity field is

$$q_r = 0 \tag{3.77}$$

$$q_\theta = -\frac{\Gamma}{2\pi r} \tag{3.78}$$

As expected, the tangential velocity component decays at a rate of $1/r$ as shown in Fig. 3.8b. The velocity potential for a vortex element at the origin can be obtained by integration of Eqs. (3.77) and (3.78):

$$\Phi = \int q_\theta r\, d\theta + C = -\frac{\Gamma}{2\pi}\theta + C \tag{3.79}$$

where C is an arbitrary constant that can be set to zero. Equation (3.79) indicates too that the velocity potential of a vortex is multivalued and depends on the number of revolutions around the vortex point. So when integrating around a vortex we do find vorticity concentrated at a zero-area point, but with finite circulation (see Sections 2.9 and 2.10). However, if we integrate $\mathbf{q} \cdot d\mathbf{l}$ around any closed curve in the field (not surrounding the vortex) the value of the integral will be zero (as shown at the end of Section 2.10 and in Fig. 2.11a). Thus, the vortex is a solution to the Laplace equation and results in an irrotational flow, excluding the vortex point itself.

Equations (3.77) to (3.79) in Cartesian coordinates for a vortex located at (x_0, z_0) are

$$\Phi = -\frac{\Gamma}{2\pi} \tan^{-1} \frac{z - z_0}{x - x_0} \tag{3.80}$$

$$u = \frac{\Gamma}{2\pi} \frac{z - z_0}{(z - z_0)^2 + (x - x_0)^2} \tag{3.81}$$

$$w = -\frac{\Gamma}{2\pi} \frac{x - x_0}{(z - z_0)^2 + (x - x_0)^2} \tag{3.82}$$

To derive the stream function for the two-dimensional vortex located at the origin, in the x–z (or r–θ) plane, consider the velocity components in terms of the stream function derivatives

$$q_\theta = -\frac{\partial \Psi}{\partial r} = -\frac{\Gamma}{2\pi r} \tag{3.83}$$

$$q_r = \frac{1}{r} \frac{\partial \Psi}{\partial \theta} = 0 \tag{3.84}$$

Integrating Eq. (3.83) and (3.84) and setting the constant of integration to zero we get

$$\Psi = \frac{\Gamma}{2\pi} \ln r \tag{3.85}$$

and the streamlines where $\Psi = \text{const.}$ are shown schematically in Fig. 3.8a.

3.9 Principle of Superposition

If $\Phi_1, \Phi_2, \ldots, \Phi_n$ are solutions of the Laplace equation (Eq. (3.1)), which is linear, then

$$\Phi = \sum_{k=1}^{n} c_k \Phi_k \tag{3.86}$$

is also a solution for that equation in that region. Here $c_1, c_2, \ldots, c_n$ are arbitrary constants and therefore

$$\nabla^2 \Phi = \sum_{k=1}^{n} c_k \nabla^2 \Phi_k = 0$$

This superposition principle is a very important property of the Laplace equation, paving the way for solutions of the flowfield near complex boundaries. In theory, by using a set of elementary solutions, the solution process (of satisfying a set of given boundary conditions) can be reduced to an algebraic search for the right linear combination of these elementary solutions.

3.10 Superposition of Sources and Free Stream: Rankine's Oval

As a first example for using the principle of superposition, consider the two-dimensional flow resulting from superimposing a source with a strength σ at $x = -x_0$, a sink with a strength $-\sigma$ at $x = +x_0$, both on the x axis, and a free stream flow with speed U_∞ in the x direction (Fig. 3.9). The velocity potential for this case will be

$$\Phi(x, z) = U_\infty x + \frac{\sigma}{2\pi} \ln(r_1) - \frac{\sigma}{2\pi} \ln(r_2) \tag{3.87}$$

3.10 Superposition of Sources and Free Stream: Rankine's Oval

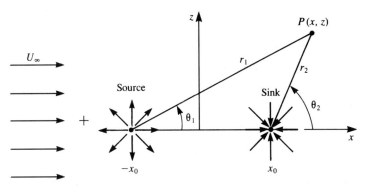

Figure 3.9 Combination of a free stream, a source, and a sink.

where $r_1 = [(x + x_0)^2 + z^2]^{1/2}$ and $r_2 = [(x - x_0)^2 + z^2]^{1/2}$. The stream function can be obtained by adding the stream functions of the individual elements:

$$\Psi(x, z) = U_\infty z + \frac{\sigma}{2\pi}\theta_1 - \frac{\sigma}{2\pi}\theta_2 \tag{3.88}$$

where $\theta_1 = \tan^{-1}[z/(x + x_0)]$ and $\theta_2 = \tan^{-1}[z/(x - x_0)]$. Substitution of r_1, r_2, θ_1, and θ_2 into the velocity potential and the stream function yields

$$\Phi(x, z) = U_\infty x + \frac{\sigma}{2\pi} \ln \sqrt{(x + x_0)^2 + z^2} - \frac{\sigma}{2\pi} \ln \sqrt{(x - x_0)^2 + z^2} \tag{3.87a}$$

$$\Psi(x, z) = U_\infty z + \frac{\sigma}{2\pi} \tan^{-1} \frac{z}{x + x_0} - \frac{\sigma}{2\pi} \tan^{-1} \frac{z}{x - x_0} \tag{3.88a}$$

The velocity field due to this potential is obtained by differentiating either the velocity potential or the stream function:

$$u = \frac{\partial \Phi}{\partial x} = U_\infty + \frac{\sigma}{2\pi} \frac{x + x_0}{(x + x_0)^2 + z^2} - \frac{\sigma}{2\pi} \frac{x - x_0}{(x - x_0)^2 + z^2} \tag{3.89}$$

$$w = \frac{\partial \Phi}{\partial z} = \frac{\sigma}{2\pi} \frac{z}{(x + x_0)^2 + z^2} - \frac{\sigma}{2\pi} \frac{z}{(x - x_0)^2 + z^2} \tag{3.90}$$

Because of the symmetry about the x axis the stagnation points are located along the x axis, at points further out than the location of the source and sink, say at $x = \pm a$ (Fig. 3.10a). The w component of the velocity at these points (and along the x axis) is automatically zero, too. The distance a is then found by setting the u component of the velocity to zero:

$$u(\pm a, 0) = U_\infty + \frac{\sigma}{2\pi} \frac{1}{(\pm a + x_0)} - \frac{\sigma}{2\pi} \frac{1}{(\pm a - x_0)} = U_\infty - \frac{\sigma}{\pi} \frac{x_0}{(a^2 - x_0^2)} = 0$$

and a is

$$a = \sqrt{\frac{\sigma x_0}{\pi U_\infty} + x_0^2} \tag{3.91}$$

Consider the stagnation streamline (which passes through the stagnation points). The value of Ψ for the stagnation streamline can be found by observing the value of Eq. (3.88) on the left-side stagnation point (where $\theta_1 = \theta_2 = \pi$ and $z = 0$). This results in $\Psi = 0$, which can be shown to be the same for the right-side stagnation point as well (where $\theta_1 = \theta_2 = 0$). The equation for the stagnation streamline is therefore

$$\Psi(x, z) = U_\infty z + \frac{\sigma}{2\pi} \tan^{-1} \frac{z}{x + x_0} - \frac{\sigma}{2\pi} \tan^{-1} \frac{z}{x - x_0} = 0 \tag{3.92}$$

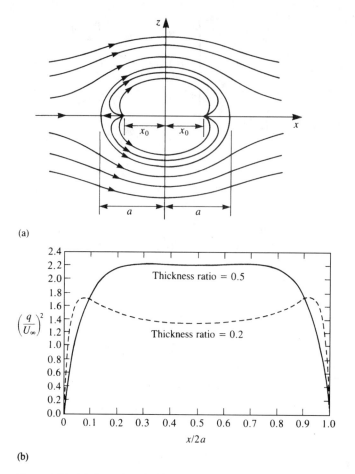

Figure 3.10 (a) Streamlines inside and outside of a Rankine oval. (b) Velocity distribution ($q^2 = u^2 + w^2$) on the surface of 20% and 50% thick Rankine ovals.

The streamlines of this flow, including the stagnation streamline, are sketched in Fig. 3.10a and the resulting velocity distribution is shown in Fig. 3.10b. Note that the stagnation streamline includes a closed oval shape (called Rankine's oval after W. J. M. Rankine, a Scottish engineer who lived in the nineteenth century) and the x axis (excluding the segment between $x = \pm a$). This flow (source and sink) can therefore be considered to model the flow past an oval of length $2a$. (For this application, the streamlines inside the oval have no physical significance.) The flow past a family of such ovals can be derived by varying the parameters σ and x_0 or a, and by plotting the corresponding streamlines.

3.11 Superposition of Doublet and Free Stream: Flow around a Cylinder

Consider the superposition of the free stream potential of Eq. (3.51), where $x = r\cos\theta$ in cylindrical coordinates, with the potential of a doublet (Eq. (3.68)) pointing in the negative x direction [$\boldsymbol{\mu} = (-\mu, 0)$]. The combined flow, as shown in Fig. 3.11, has the velocity potential

$$\Phi = U_\infty r \cos\theta + \frac{\mu}{2\pi}\frac{\cos\theta}{r} \tag{3.93}$$

3.11 Superposition of Doublet and Free Stream: Flow around a Cylinder

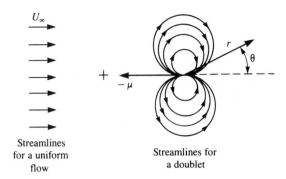

Figure 3.11 Addition of a uniform flow and a doublet to describe the flow around a cylinder.

The velocity field of this potential can be obtained by differentiating Eq. (3.93):

$$q_r = \frac{\partial \Phi}{\partial r} = \left(U_\infty - \frac{\mu}{2\pi r^2}\right) \cos\theta \qquad (3.94)$$

$$q_\theta = \frac{1}{r}\frac{\partial \Phi}{\partial \theta} = -\left(U_\infty + \frac{\mu}{2\pi r^2}\right) \sin\theta \qquad (3.95)$$

If this flow combination is thought of as a limiting case of the flow in Section 3.10 with the source and sink approaching each other, it is expected that the oval will approach a circle in this limit. To verify this, note that $q_r = 0$ for $r = [\mu/2\pi U_\infty]^{1/2}$ for all θ (from Eq. (3.94)) and the radial direction is normal to the circle. If we take $r = R$ as the radius of the circle, then the strength of the doublet is

$$\mu = U_\infty 2\pi R^2 \qquad (3.96)$$

Substitution of this value of μ into Eqs. (3.93), (3.94), and (3.95) results in the flowfield around a cylinder with a radius R:

$$\Phi = U_\infty \cos\theta \left(r + \frac{R^2}{r}\right) \qquad (3.97)$$

$$q_r = U_\infty \cos\theta \left(1 - \frac{R^2}{r^2}\right) \qquad (3.98)$$

$$q_\theta = -U_\infty \sin\theta \left(1 + \frac{R^2}{r^2}\right) \qquad (3.99)$$

For the two-dimensional case, evaluation of the stream function can readily provide the streamlines in the flow (by setting $\Psi = $ const). These results for the cylinder in a free stream can be obtained, too, by the superposition of the free stream and the doublet [with $(-\mu, 0)$ strength] stream functions:

$$\Psi = U_\infty r \sin\theta - \frac{\mu}{2\pi}\frac{\sin\theta}{r} \qquad (3.100)$$

The stagnation points on the circle are found by letting $q_\theta = 0$ in Eq. (3.99) and thus are at $\theta = 0$ and $\theta = \pi$. The value of Ψ at the stagnation points $\theta = 0$ and $\theta = \pi$ (and therefore along the stagnation streamline) is found from Eq. (3.100) to be $\Psi = 0$. This is equivalent to requiring that $q_r(R, \theta) = 0$, and the strength of μ again is given by Eq. (3.96). Substituting

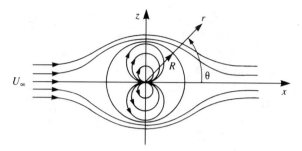

Figure 3.12 Streamlines due to the addition of a doublet and a uniform flow (flow around a cylinder).

μ in terms of the cylinder radius into Eq. (3.100) we obtain

$$\Psi = U_\infty \sin\theta \left(r - \frac{R^2}{r}\right) \tag{3.101}$$

This describes the streamlines of the flow around the cylinder with radius R (Fig. 3.12). These lines are perpendicular to the potential lines of Eq. (3.97).

To obtain the pressure distribution over the cylinder, the velocity components are evaluated at $r = R$:

$$q_r = 0, \qquad q_\theta = -2U_\infty \sin\theta \tag{3.102}$$

The pressure distribution at $r = R$ is obtained now with Bernoulli's equation

$$p_\infty + \frac{\rho}{2}U_\infty^2 = p + \frac{\rho}{2}q_\theta^2$$

Substitution of the value of q_θ at $r = R$ yields

$$p - p_\infty = \frac{1}{2}\rho U_\infty^2 (1 - 4\sin^2\theta) \tag{3.103}$$

and the pressure coefficient is

$$C_p = \frac{p - p_\infty}{(1/2)\rho U_\infty^2} = (1 - 4\sin^2\theta) \tag{3.104}$$

It can be easily observed that at the stagnation points $\theta = 0$ and π (where $q = 0$), $C_p = 1$. Also, the maximum speed occurs at the top and bottom of the cylinder ($\theta = \pi/2, 3\pi/2$) and the pressure coefficient there is -3.

To evaluate the components of the fluid dynamic force acting on the cylinder, the above pressure distribution must be integrated. Let L be the lift acting in the z direction and D the drag acting in the x direction. Integration of the components of the pressure force on an element of length $R\,d\theta$ leads to

$$L = \int_0^{2\pi} -pR\,d\theta\,\sin\theta = \int_0^{2\pi} -(p - p_\infty)R\,d\theta\,\sin\theta$$

$$= \frac{-1}{2}\rho U_\infty^2 \int_0^{2\pi} (1 - 4\sin^2\theta)R\sin\theta\,d\theta = 0 \tag{3.105}$$

$$D = \int_0^{2\pi} -pR\,d\theta\,\cos\theta = \int_0^{2\pi} -(p - p_\infty)R\,d\theta\,\cos\theta$$

$$= \frac{-1}{2}\rho U_\infty^2 \int_0^{2\pi} (1 - 4\sin^2\theta)R\cos\theta\,d\theta = 0 \tag{3.106}$$

3.11 Superposition of Doublet and Free Stream: Flow around a Cylinder

Figure 3.13 Hydrogen bubble visualization of the separated water flow around a cylinder at a Reynolds number of 0.2×10^6. (Courtesy of K. W. McAlister and L. W. Carr, U. S. Army Aeroflightdynamics Directorate, AVSCOM.)

Here the pressure was replaced by the pressure difference $p - p_\infty$ term of Eq. (3.103), and this has no effect on the results since the integral of a constant pressure p_∞ around a closed body is zero. A very interesting result of this potential flow is that the fore and aft symmetry leads to pressure loads that cancel out. In reality the flow separates and will not follow the cylinder's rear surface, as shown in Fig. 3.13. The pressure distribution due to this real flow, along with the results of Eq. (3.104), are plotted in Fig. 3.14. This shows that at the front section of the cylinder, where the flow is attached, the pressures are well predicted by this model. However, behind the cylinder, because of the flow separation, the pressure distribution is different.

In this example, because of the symmetry in the upper and the lower flows (about the x axis), no lift was generated. A lifting condition can be obtained by introducing an asymmetry, in the form of a clockwise vortex with strength Γ situated at the origin. The velocity potential for this case is

$$\Phi = U_\infty \cos\theta \left(r + \frac{R^2}{r} \right) - \frac{\Gamma}{2\pi}\theta \tag{3.107}$$

The velocity components are obtained by differentiating the velocity potential to get

$$q_r = U_\infty \cos\theta \left(1 - \frac{R^2}{r^2} \right) \tag{3.108}$$

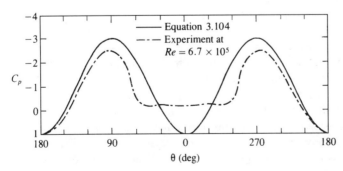

Figure 3.14 Theoretical pressure distribution (solid curve) around a cylinder compared with experimental data at Reynolds number of 6.7×10^5 (chain curve) from Ref. 1.6.

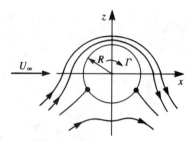

Figure 3.15 Streamlines for the flow around a cylinder with circulation Γ.

which is the same as for the cylinder without the circulation, and

$$q_\theta = -U_\infty \sin\theta \left(1 + \frac{R^2}{r^2}\right) - \frac{\Gamma}{2\pi r} \qquad (3.109)$$

This potential still describes the flow around a cylinder since at $r = R$ the radial velocity component becomes zero. The stagnation points can be obtained by finding the tangential velocity component at $r = R$,

$$q_\theta = -2U_\infty \sin\theta - \frac{\Gamma}{2\pi R} \qquad (3.110)$$

and by solving for $q_\theta = 0$,

$$\sin\theta_s = -\frac{\Gamma}{4\pi R U_\infty} \qquad (3.111)$$

These stagnation points (located at an angular position θ_s) are shown by the two dots in Fig. 3.15 and lie on the cylinder as long as $\Gamma \leq 4\pi R U_\infty$.

The lift and drag will be found by using Bernoulli's equation, but because of the fore and aft symmetry, no drag is expected from this calculation. For the lift, the tangential velocity component is substituted into the Bernoulli equation and

$$\begin{aligned}
L &= \int_0^{2\pi} -(p - p_\infty) R\, d\theta \sin\theta \\
&= -\int_0^{2\pi} \left[\frac{\rho U_\infty^2}{2} - \frac{\rho}{2}\left(2U_\infty \sin\theta + \frac{\Gamma}{2\pi R}\right)^2\right] \sin\theta R\, d\theta \\
&= \frac{\rho U_\infty \Gamma}{\pi} \int_0^{2\pi} \sin^2\theta\, d\theta = \underline{\rho U_\infty \Gamma}
\end{aligned} \qquad (3.112)$$

This very important result states that the force in this two-dimensional flow is directly proportional to the circulation and acts normal to the free stream. A generalization of this result was discovered independently by the German mathematician M. W. Kutta in 1902 and by the Russian physicist N. E. Joukowski in 1906. They observed that the lift per unit span on a lifting airfoil or cylinder is proportional to the circulation. Consequently the Kutta–Joukowski theorem (which will be derived in Chapter 6) states:

The resultant aerodynamic force in an incompressible, inviscid, irrotational flow in an unbounded fluid is of magnitude $\rho Q_\infty \Gamma$ per unit width and acts in a direction normal to the free stream. (Note that the speed of the free stream is taken to be Q_∞ since in the general case the stream may not be parallel to the x axis.)

3.12 Superposition of a Three-Dimensional Doublet and Free Stream

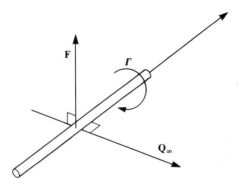

Figure 3.16 Notation used for the vector Kutta–Joukowski theorem.

Using vector notation, this can be expressed as

$$\mathbf{F} = \rho \mathbf{Q}_\infty \times \mathbf{\Gamma} \tag{3.113}$$

where **F** is the aerodynamic force per unit width and acts in the direction determined by the vector product, as shown schematically in Fig. 3.16. Note that positive $\mathbf{\Gamma}$ is defined according to the right-hand rule.

3.12 Superposition of a Three-Dimensional Doublet and Free Stream: Flow around a Sphere

The method of the previous section can be extended to study the case of the three-dimensional flow over a sphere. The velocity potential is obtained by the superposition of the free stream potential of Eq. (3.51) with a doublet pointing in the negative x direction (Eq. (3.34)). The combined velocity potential is

$$\Phi = U_\infty r \cos\theta + \frac{\mu}{4\pi} \frac{\cos\theta}{r^2} \tag{3.114}$$

The velocity field of this potential can be obtained by differentiating Eq. (3.114):

$$q_r = \frac{\partial \Phi}{\partial r} = \left(U_\infty - \frac{\mu}{2\pi r^3}\right) \cos\theta \tag{3.115}$$

$$q_\theta = \frac{1}{r}\frac{\partial \Phi}{\partial \theta} = -\left(U_\infty + \frac{\mu}{4\pi r^3}\right) \sin\theta \tag{3.116}$$

$$q_\varphi = \frac{1}{r \sin\theta}\frac{\partial \Phi}{\partial \varphi} = 0 \tag{3.117}$$

At the sphere surface, where $r = R$, the zero normal flow boundary condition is enforced ($q_r = 0$), so that

$$q_r = \left(U_\infty - \frac{\mu}{2\pi R^3}\right) \cos\theta = 0 \tag{3.118}$$

This condition is met at $\theta = \pi/2, 3\pi/2$ and, in general, when the quantity in the parentheses is zero. This second condition is used to determine the doublet strength

$$\mu = U_\infty 2\pi R^3 \tag{3.119}$$

which means that $q_r = 0$ at $r = R$, which is the radius of the sphere. Substituting the strength μ into the equations for the potential and the velocity components results in the flowfield around a sphere with a radius R:

$$\Phi = U_\infty \cos\theta \left(r + \frac{R^3}{2r^2} \right) \tag{3.120}$$

$$q_r = U_\infty \cos\theta \left(1 - \frac{R^3}{r^3} \right) \tag{3.121}$$

$$q_\theta = -U_\infty \sin\theta \left(1 + \frac{R^3}{2r^3} \right) \tag{3.122}$$

To obtain the pressure distribution over the sphere, the velocity components at $r = R$ are found:

$$q_r = 0, \qquad q_\theta = -\frac{3}{2} U_\infty \sin\theta \tag{3.123}$$

The stagnation points occur at $\theta = 0$ and $\theta = \pi$, and the maximum velocity occurs at $\theta = \pi/2$ or $\theta = 3\pi/2$. The value of the maximum velocity is $(3/2)U_\infty$, which is smaller than in the two-dimensional case.

The pressure distribution is obtained now with Bernoulli's equation

$$p - p_\infty = \frac{1}{2}\rho U_\infty^2 \left(1 - \frac{9}{4}\sin^2\theta \right) \tag{3.124}$$

and the pressure coefficient is

$$C_p = \frac{p - p_\infty}{(1/2)\rho U_\infty^2} = \left(1 - \frac{9}{4}\sin^2\theta \right) \tag{3.125}$$

It can be easily observed that at the stagnation points $\theta = 0$ and π (where $q = 0$), $C_p = 1$. Also, the maximum velocity occurs at the top and bottom of the sphere ($\theta = \pi/2, 3\pi/2$) and the pressure coefficient there is $-5/4$.

Because of symmetry, lift and drag will be zero, as in the case of the flow over the cylinder. However, the lift on a hemisphere is not zero (even without introducing circulation); this case is of particular interest in the field of road vehicle aerodynamics. The flow past a sphere can be interpreted to also represent the flow past a hemisphere on the ground since the x axis is a streamline and can be replaced by a solid surface.

The lift force acting on the hemisphere's upper surface is

$$L = -\int (p - p_\infty) \sin\theta \, \sin\varphi \, dS \tag{3.126}$$

and the surface element dS on the sphere is

$$dS = (R\sin\theta \, d\varphi)(R \, d\theta)$$

Substituting dS and the pressure from Eq. (3.124), we obtain the lift of the hemisphere:

$$\begin{aligned}
L &= -\int_0^\pi \int_0^\pi \frac{1}{2}\rho U_\infty^2 \left(1 - \frac{9}{4}\sin^2\theta \right) R^2 \sin^2\theta \sin\varphi \, d\theta \, d\varphi \\
&= -\frac{1}{2}\rho U_\infty^2 \int_0^\pi \left(1 - \frac{9}{4}\sin^2\theta \right) 2R^2 \sin^2\theta \, d\theta \\
&= -\rho R^2 U_\infty^2 \left(\frac{\pi}{2} - \frac{27\pi}{32} \right) = \frac{11}{32}\pi \rho R^2 U_\infty^2
\end{aligned} \tag{3.127}$$

The lift and drag coefficients due to the upper surface are then

$$C_L \equiv \frac{L}{(1/2)\rho U_\infty^2 (\pi/2) R^2} = \frac{11}{8} \tag{3.128}$$

$$C_D \equiv \frac{D}{(1/2)\rho U_\infty^2 (\pi/2) R^2} = 0 \tag{3.129}$$

For the complete configuration the forces due to the pressure distribution on the flat, lower surface of the hemisphere must be included, too, in this calculation.

3.13 Some Remarks about the Flow over the Cylinder and the Sphere

The examples of the flow over a cylinder and a sphere clearly demonstrate the principle of superposition as a tool for deriving particular solutions to Laplace's equation. From the physical point of view, these results fall in a range where potential flow based calculations are inaccurate owing to flow separation. The pressure distribution around the cylinder, as obtained from Eq. (3.104), is shown in Fig. 3.14 along with some typical experimental results. Clearly, at the frontal stagnation point ($\theta = \pi$) the results of Eq. (3.104) are close to the experimental data, whereas at the back the difference is large. This is a result of the streamlines not following the surface curvature and separating from the cylinder's surface as shown in Fig. 3.13; this is called *flow separation*.

The theoretical pressure distribution (Eq. (3.125)) for the sphere, along with the results for the cylinder, are shown in Fig. 3.17. Note that for the three-dimensional case the suction pressures are much smaller (relieving effect). Experimental data for the sphere show that the flow separates too but that the low pressure in the rear section is smaller. Consequently, the actual drag coefficient of a sphere is less than that of an equivalent cylinder, as shown in Fig. 3.18 (for $Re > 2,000$). These drag data are a result of the skin friction and flow separation pattern, which is strongly affected by the Reynolds number. Clearly, for the laminar flows ($Re < 2,000$) the drag is large owing to larger flow separation behind the body; both this separation region and resulting drag are reduced as the turbulent flow

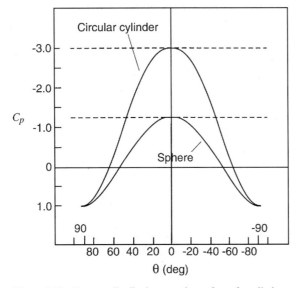

Figure 3.17 Pressure distribution over the surface of a cylinder and a sphere.

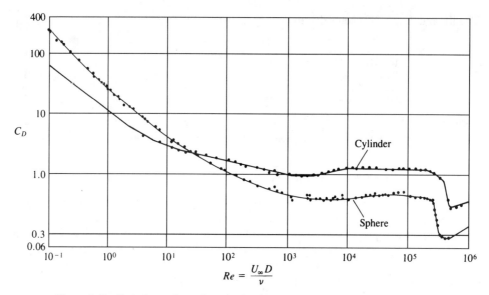

Figure 3.18 Typical experimental results for the drag coefficient for cylinders and spheres as a function of Reynolds number (from Ref. 1.6). Reproduced with permission of McGraw-Hill, Inc.

momentum transfer increases ($Re > 10^5$, see Schlichting,[1.6] p. 17). Note that the inviscid flow results do not account for flow separation and viscous friction near the body's surface and therefore the drag coefficient for both cylinder and sphere is *zero*. This fact disturbed the French mathematician d'Alembert, in the middle of the seventeenth century, who arrived at this conclusion that the drag of a closed body in two-dimensional, inviscid, incompressible flow is zero (even though he realized that experiments result in a finite drag). Ever since those early days of fluid dynamics this problem has been known as the *d'Alembert's paradox*.

3.14 Surface Distribution of the Basic Solutions

The results of Sections 3.2 and 3.3 indicate that a solution to the flow over arbitrary bodies can be obtained by distributing elementary singularity solutions over the modeled surfaces. Prior to applying this method to practical problems, the nature of each of the elementary solutions needs to be investigated. For simplicity, the two-dimensional point elements will be distributed continuously along the x axis in the region $x_1 \to x_2$.

a. *Source Distribution*

Consider the source distribution of strength per length $\sigma(x)$ along the x axis as shown in Fig. 3.19. The influence of this distribution at a point $P(x, z)$ is an integral of the influences of all the point elements:

$$\Phi(x, z) = \frac{1}{2\pi} \int_{x_1}^{x_2} \sigma(x_0) \ln \sqrt{(x - x_0)^2 + z^2} \, dx_0 \tag{3.130}$$

$$u(x, z) = \frac{1}{2\pi} \int_{x_1}^{x_2} \sigma(x_0) \frac{x - x_0}{(x - x_0)^2 + z^2} \, dx_0 \tag{3.131}$$

$$w(x, z) = \frac{1}{2\pi} \int_{x_1}^{x_2} \sigma(x_0) \frac{z}{(x - x_0)^2 + z^2} \, dx_0 \tag{3.132}$$

3.14 Surface Distribution of the Basic Solutions

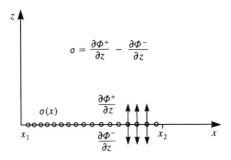

Figure 3.19 Source distribution along the x axis. [Note that $\frac{\partial \Phi^+}{\partial z} \equiv \frac{\partial \Phi}{\partial z}(x, 0+)$.]

To investigate the properties of such a distribution for future modeling purposes, the type of discontinuity across the surface needs to be examined. Since each source emits fluid in all directions, intuitively we can see that the resulting velocity direction will point away from the surface, as shown in Fig. 3.19. From the figure it is clear that there is a discontinuity in the w component at $z = 0$. Note that as $z \to 0$ the integrand in Eq. (3.132) is zero except when $x_0 = x$. Therefore, the value of the integral depends only on the contribution from this point. Consequently, $\sigma(x_0)$ can be moved out of the integral and replaced by $\sigma(x)$. This suggests that the limits of integration do not affect the value of the integral and for convenience can be replaced by $\pm\infty$. Also, from the z dependence of the integrand in Eq. (3.132), the velocity component when approaching $z = 0$ from above the x axis, w^+, is in the opposite direction to w^-, which is the component when approaching the axis from below. For the velocity component w^+, Eq. (3.132) becomes

$$w(x, 0+) = \lim_{z \to 0+} \frac{\sigma(x)}{2\pi} \int_{-\infty}^{\infty} \frac{z}{(x - x_0)^2 + z^2} \, dx_0 \qquad (3.133)$$

To evaluate this integral it is convenient to introduce a new integration variable λ:

$$\lambda = \frac{x - x_0}{z}$$

$$d\lambda = -\frac{dx_0}{z}$$

and the integration limits for $z \to 0+$ become $\pm\infty$. The transformed integral becomes

$$w(x, 0+) = \lim_{z \to 0+} \frac{\sigma(x)}{2\pi} \int_{-\infty}^{\infty} \frac{d\lambda}{1 + \lambda^2}$$

$$= \frac{\sigma(x)}{2\pi} \tan^{-1} \lambda \Big|_{-\infty}^{\infty}$$

$$= \frac{\sigma(x)}{2\pi} \left[\frac{\pi}{2} - \left(-\frac{\pi}{2}\right) \right] = \frac{\sigma(x)}{2} \qquad (3.134)$$

Therefore $w(x, 0\pm)$ become

$$w(x, 0\pm) = \frac{\partial \Phi}{\partial z}(x, 0\pm) = \pm\frac{\sigma(x)}{2} \qquad (3.135)$$

This element will be suitable to model flows that are symmetrical with respect to the x axis, and the total jump in the velocity component normal to the surface of the distribution is

$$w^+ - w^- = \sigma(x) \qquad (3.136)$$

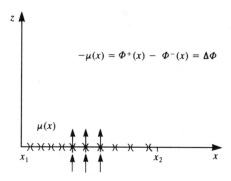

Figure 3.20 Doublet distribution along the x axis.

The u component is continuous across the x axis, and its evaluation needs additional considerations (e.g., as in Chapter 5).

b. *Doublet Distribution*

In a similar manner the influence of a two-dimensional doublet distribution, pointing in the z direction [$\mu = (0, \mu)$], at a point $P(x, z)$ is an integral of the influences of the point elements between $x_1 \to x_2$ (Fig. 3.20):

$$\Phi(x, z) = \frac{-1}{2\pi} \int_{x_1}^{x_2} \mu(x_0) \frac{z}{(x - x_0)^2 + z^2} \, dx_0 \qquad (3.137)$$

$$u(x, z) = \frac{1}{\pi} \int_{x_1}^{x_2} \mu(x_0) \frac{(x - x_0)z}{[(x - x_0)^2 + z^2]^2} \, dx_0 \qquad (3.138)$$

$$w(x, z) = \frac{-1}{2\pi} \int_{x_1}^{x_2} \mu(x_0) \frac{(x - x_0)^2 - z^2}{[(x - x_0)^2 + z^2]^2} \, dx_0 \qquad (3.139)$$

Note that the velocity potential in Eq. (3.137) is identical in form to the w component of the source (Eq. (3.132)). Approaching the surface, at $z = 0\pm$, this element creates a jump in the velocity potential. This analogy yields

$$\Phi(x, 0\pm) = \mp \frac{\mu(x)}{2} \qquad (3.140)$$

This leads to a discontinuous tangential velocity component given by

$$u(x, 0\pm) = \frac{\partial \Phi}{\partial x}(x, 0\pm) = \frac{\mp 1}{2} \frac{d\mu(x)}{dx} \qquad (3.141)$$

Since the doublet distribution begins at x_1 (e.g., $\mu(x \le x_1) = 0$), the circulation $\Gamma(x)$ around a path surrounding the segment $x_1 \to x$ is

$$\Gamma(x) = \int_{x_1}^{x} u(x_0, 0+) \, dx_0 + \int_{x}^{x_1} u(x_0, 0-) \, dx_0 = -\mu(x) \qquad (3.142)$$

which is equivalent to the jump in the potential

$$\Gamma(x) = \Phi(x, 0+) - \Phi(x, 0-) = -\mu(x) = \Delta\Phi(x) \qquad (3.143)$$

3.14 Surface Distribution of the Basic Solutions

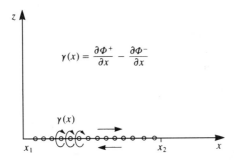

Figure 3.21 Vortex distribution along the x axis.

c. Vortex Distribution

In a similar manner the influence of a vortex distribution at a point $P(x, z)$ is an integral of the influences of the point elements between $x_1 \to x_2$ (Fig. 3.21):

$$\Phi(x, z) = -\frac{1}{2\pi} \int_{x_1}^{x_2} \gamma(x_0) \tan^{-1} \frac{z}{x - x_0} dx_0 \tag{3.144}$$

$$u(x, z) = \frac{1}{2\pi} \int_{x_1}^{x_2} \gamma(x_0) \frac{z}{(x - x_0)^2 + z^2} dx_0 \tag{3.145}$$

$$w(x, z) = -\frac{1}{2\pi} \int_{x_1}^{x_2} \gamma(x_0) \frac{x - x_0}{(x - x_0)^2 + z^2} dx_0 \tag{3.146}$$

Here the u component of the velocity is similar in form to Eqs. (3.132) and (3.137) and there is a jump in this component as $z = 0\pm$. The tangential velocity component is then

$$u(x, 0\pm) = \frac{\partial \Phi}{\partial x}(x, 0\pm) = \pm \frac{\gamma(x)}{2} \tag{3.147}$$

The contribution of this velocity jump to the potential jump, assuming that $\Phi = 0$ ahead of the vortex distribution, is

$$\Delta\Phi(x) = \Phi(x, 0+) - \Phi(x, 0-) = \int_{x_1}^{x} \frac{\gamma(x_0)}{2} dx_0 - \int_{x_1}^{x} \frac{-\gamma(x_0)}{2} dx_0$$

The circulation Γ is the closed integral of $u(x, 0)dx$, which is equivalent to that of Eq. (3.142). Therefore,

$$\Gamma(x) = \Phi(x, 0+) - \Phi(x, 0-) = \Delta\Phi(x) \tag{3.148}$$

Note that similar flow conditions can be modeled by either a vortex or a doublet distribution and the relation between these two distributions is

$$\Gamma = -\mu \tag{3.149}$$

A comparison of Eq. (3.141) with Eq. (3.147) indicates that a vortex distribution can be replaced by an equivalent doublet distribution such that

$$\gamma(x) = -\frac{d\mu(x)}{dx} \tag{3.150}$$

Problems

3.1. Consider a distribution of two-dimensional sources around a circle of radius R. The source strength is $f(\theta)$ per unit arc length. Find an analytic expression for the velocity potential of this source ring.

3.2. Consider the two-dimensional flow of a uniform stream of speed U_∞ past a source of strength Q. Find the stagnation point(s) and the equation of the stagnation streamline. Find the width of the generated semi-infinite body far downstream.

3.3. Consider the two-dimensional flow due to a uniform stream of speed U_∞ in the x direction, a clockwise vortex of circulation Γ at $(0, b)$, and an equal strength counterclockwise vortex at $(0, -b)$. Find the stream function for the limit $b \to 0$, $\Gamma \to \infty$, and where $2\Gamma b \to N$, a constant.

3.4. Consider the two-dimensional flow of a uniform stream of speed U_∞ along a wall with a semicircular bump of radius R. Find the lift on the bump.

3.5. Consider the two-dimensional flow of a uniform stream of speed U_∞ past a circle of radius R with circulation Γ. Find the lift force on the circle by an application of the integral momentum theorem for the fluid region in between the circle and a concentric circle at a large distance away.

CHAPTER 4

Small-Disturbance Flow over Three-Dimensional Wings: Formulation of the Problem

One of the first important applications of potential flow theory was the study of lifting surfaces (wings). Since the boundary conditions on a complex surface can considerably complicate the attempt to solve the problem by analytical means, some simplifying assumptions need to be introduced. In this chapter these assumptions will be applied to the formulation of the three-dimensional thin wing problem and the scene for the singularity solution technique will be set.

4.1   Definition of the Problem

Consider the finite wing shown in Fig. 4.1, which is moving at a constant speed in an otherwise undisturbed fluid. A Cartesian coordinate system is attached to the wing and the components of the free-stream velocity $\mathbf{Q}_\infty$ in the x, y, z frame of reference are U_∞, V_∞, and W_∞, respectively. The angle of attack α is defined as the angle between the free-stream velocity and the x axis

$$\alpha = \tan^{-1} \frac{W_\infty}{U_\infty}$$

and for the sake of simplicity the side slip condition is not included at this point ($V_\infty \equiv 0$).

If it is assumed that the fluid surrounding the wing and the wake is inviscid, incompressible, and irrotational, the resulting velocity field due to the motion of the wing can be obtained by solving the continuity equation

$$\nabla^2 \Phi^* = 0 \tag{4.1}$$

where Φ^* is the velocity potential, as defined in the wing frame of reference. The boundary conditions require that the disturbance induced by the wing will decay far from the wing:

$$\lim_{r \to \infty} \nabla \Phi^* = \mathbf{Q}_\infty \tag{4.2}$$

which is automatically fulfilled by the singular solutions such as for the source, doublet, or the vortex elements (derived in Chapter 3). Also, the normal component of velocity on the solid boundaries of the wing must be zero. Thus, in a frame of reference attached to the wing,

$$\nabla \Phi^* \cdot \mathbf{n} = 0 \tag{4.3}$$

where $\mathbf{n}$ is an outward normal to the surface (Fig. 4.1). So, basically, the problem reduces to finding a singularity distribution that will satisfy Eq. (4.3). Once this distribution is found, the velocity $\mathbf{q}$ at each point in the field is known and the corresponding pressure p will be calculated from the steady-state Bernoulli equation:

$$p_\infty + \frac{\rho}{2} Q_\infty^2 = p + \frac{\rho}{2} q^2 \tag{4.4}$$

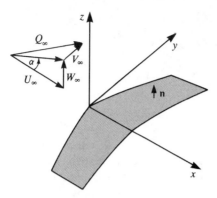

Figure 4.1 Nomenclature used for the definition of the finite wing problem.

The analytical solution of this problem, for an arbitrary wing shape, is complicated by the difficulty of specifying the boundary condition of Eq. (4.3) on a complex shape surface and by the shape of a wake. The need for a wake model follows immediately from the Helmholtz theorems (Section 2.9), which state that vorticity cannot end or start in the fluid. Consequently, if the wing is modeled by singularity elements that will introduce vorticity (as will be shown later in this chapter), these need to be "shed" into the flow in the form of a wake.

To overcome the difficulty of defining the zero normal flow boundary condition on an arbitrary wing shape some additional simplifying assumptions are made in the next section.

4.2 The Boundary Condition on the Wing

To satisfy the boundary condition of Eq. (4.3), on the wing, geometrical information about the shape of the solid boundaries is required. Let the wing solid surface be defined as

$$z = \eta(x, y) \tag{4.5}$$

and in the case of a wing with nonzero thickness two such functions will describe the upper (η_u) and the lower (η_l) surfaces (Fig. 4.2). To find the normal to the wing surface, a function $F(x, y, z)$ can be defined such that

$$F(x, y, z) \equiv z - \eta(x, y) = 0 \tag{4.6}$$

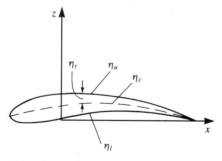

Figure 4.2 Definitions for wing thickness and upper, lower, and mean camberlines at an arbitrary spanwise location y.

4.2 The Boundary Condition on the Wing

and the outward normal on the wing upper surface is obtained by using Eq. (2.26):

$$\mathbf{n} = \frac{\nabla F}{|\nabla F|} = \frac{1}{|\nabla F|}\left(-\frac{\partial \eta}{\partial x}, -\frac{\partial \eta}{\partial y}, 1\right) \qquad (4.7)$$

whereas on the lower surface the outward normal is $-\mathbf{n}$.

The velocity potential due to the free-stream flow can be obtained by using the solution of Eq. (3.52):

$$\Phi_\infty = U_\infty x + W_\infty z \qquad (4.8)$$

and, since Eq. (4.1) is linear, its solution can be divided into two separate parts:

$$\Phi^* = \Phi + \Phi_\infty \qquad (4.9)$$

Substituting Eq. (4.7) and the derivatives of Eqs. (4.8) and (4.9) into the boundary condition (Eq. (4.3)) requiring no flow through the wing's solid boundaries results in

$$\nabla \Phi^* \cdot \mathbf{n} = \nabla \Phi^* \cdot \frac{\nabla F}{|\nabla F|}$$
$$= \left(\frac{\partial \Phi}{\partial x} + U_\infty, \frac{\partial \Phi}{\partial y}, \frac{\partial \Phi}{\partial z} + W_\infty\right) \cdot \frac{1}{|\nabla F|}\left(-\frac{\partial \eta}{\partial x}, -\frac{\partial \eta}{\partial y}, 1\right) = 0 \qquad (4.10)$$

The intermediate result of this brief investigation is that the unknown is the perturbation potential Φ, which represents the velocity induced by the motion of the wing in a stationary frame of reference. Consequently, the equation for the perturbation potential is

$$\nabla^2 \Phi = 0 \qquad (4.11)$$

and the boundary conditions on the wing surface are obtained by rearranging $\partial \Phi/\partial z$ in Eq. (4.10):

$$\frac{\partial \Phi}{\partial z} = \frac{\partial \eta}{\partial x}\left(U_\infty + \frac{\partial \Phi}{\partial x}\right) + \frac{\partial \eta}{\partial y}\left(\frac{\partial \Phi}{\partial y}\right) - W_\infty \qquad \text{on} \quad z = \eta \qquad (4.12)$$

Now, introducing the classical *small-disturbance approximation* will allow us to further simplify this boundary condition. Assume

$$\frac{|\partial \Phi/\partial x|}{Q_\infty}, \frac{|\partial \Phi/\partial y|}{Q_\infty}, \frac{|\partial \Phi/\partial z|}{Q_\infty} \ll 1 \qquad (4.13)$$

Then, from the boundary condition of Eq. (4.12), the following restrictions on the geometry will follow:

$$\left|\frac{\partial \eta}{\partial x}\right| \ll 1, \left|\frac{\partial \eta}{\partial y}\right| \ll 1, \quad \text{and} \quad \left|\frac{W_\infty}{U_\infty}\right| = \tan \alpha \approx \alpha \ll 1 \qquad (4.14)$$

This means that the wing must be thin compared to its chord. Also, near stagnation points and near the leading edge (where $\partial \eta/\partial x$ is not small), the small perturbation assumption is not valid.

Accounting for the above assumptions and recalling that for small α, $W_\infty \approx Q_\infty \alpha$ and $U_\infty \approx Q_\infty$, we can reduce the boundary condition of Eq. (4.12) to a much simpler form,

$$\frac{\partial \Phi}{\partial z}(x, y, \eta) = Q_\infty\left(\frac{\partial \eta}{\partial x} - \alpha\right) \qquad (4.15)$$

It is consistent with the above approximation to also transfer the boundary conditions from the wing surface to the x–y plane. This is accomplished by a Taylor series expansion of the dependent variable, that is,

$$\frac{\partial \Phi}{\partial z}(x, y, z = \eta) = \frac{\partial \Phi}{\partial z}(x, y, 0) + \eta \frac{\partial^2 \Phi}{\partial z^2}(x, y, 0) + O(\eta^2) \tag{4.16}$$

Along with the above small-disturbance approximation, only the first term from the expansion of Eq. (4.16) is used. Then the first-order approximation of the boundary condition, Eq. (4.12) (neglecting products of small quantities), becomes

$$\frac{\partial \Phi}{\partial z}(x, y, 0) = Q_\infty \left(\frac{\partial \eta}{\partial x} - \alpha\right) \tag{4.17}$$

A more precise treatment of the boundary conditions (for the two-dimensional airfoil problem) including proceeding to a higher order approximation will be considered in Chapter 7.

4.3 Separation of the Thickness and the Lifting Problems

At this point of the discussion, the boundary condition (Eq. (4.17)) is defined for a thin wing and is linear. The shape of the wing is then defined by the contours of the upper η_u and lower η_l surfaces as shown in Fig. 4.2,

$$z = \eta_u(x, y) \tag{4.18a}$$

$$z = \eta_l(x, y) \tag{4.18b}$$

This wing shape can also be expressed by using a thickness function η_t and a camber function η_c, such that

$$\eta_c = \frac{1}{2}(\eta_u + \eta_l) \tag{4.19a}$$

$$\eta_t = \frac{1}{2}(\eta_u - \eta_l) \tag{4.19b}$$

Therefore, the upper and the lower surfaces of the wing can be specified alternatively by using the local wing thickness and camberline (Fig. 4.2):

$$\eta_u = \eta_c + \eta_t \tag{4.20a}$$

$$\eta_l = \eta_c - \eta_t \tag{4.20b}$$

Now, the linear boundary condition (Eq. (4.17)) should be specified for both the upper and lower wing surfaces,

$$\frac{\partial \Phi}{\partial z}(x, y, 0+) = \left(\frac{\partial \eta_c}{\partial x} + \frac{\partial \eta_t}{\partial x}\right) Q_\infty - Q_\infty \alpha \tag{4.21a}$$

$$\frac{\partial \Phi}{\partial z}(x, y, 0-) = \left(\frac{\partial \eta_c}{\partial x} - \frac{\partial \eta_t}{\partial x}\right) Q_\infty - Q_\infty \alpha \tag{4.21b}$$

The boundary condition at infinity (Eq. (4.2)), for the perturbation potential Φ, now becomes

$$\lim_{r \to \infty} \nabla \Phi = 0 \tag{4.21c}$$

Since the continuity equation (Eq. (4.11)) as well as the boundary conditions (Eqs. (4.21a–c)) are linear, it is possible to solve three simpler problems and superimpose

4.4 Symmetric Wing with Nonzero Thickness at Zero Angle of Attack

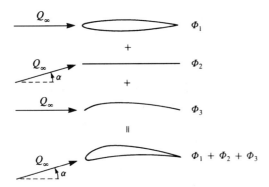

Figure 4.3 Decomposition of the thick cambered wing at an angle of attack into three simpler problems.

the three separate solutions according to Eqs. (4.21a) and (4.21b), as shown schematically in Fig. 4.3. Note that this decomposition of the solution is valid only if the small-disturbance approximation is applied to the wake model as well. These three subproblems are:

1. Symmetric wing with nonzero thickness at zero angle of attack (effect of thickness):

$$\nabla^2 \Phi_1 = 0 \quad (4.22)$$

with the boundary condition:

$$\frac{\partial \Phi_1}{\partial z}(x, y, 0\pm) = \pm \frac{\partial \eta_t}{\partial x} Q_\infty \quad (4.23)$$

where $+$ is for the upper and $-$ is for the lower surfaces.

2. Zero-thickness, uncambered wing at angle of attack (effect of angle of attack):

$$\nabla^2 \Phi_2 = 0 \quad (4.24)$$

$$\frac{\partial \Phi_2}{\partial z}(x, y, 0\pm) = -Q_\infty \alpha \quad (4.25)$$

3. Zero-thickness, cambered wing at zero angle of attack (effect of camber):

$$\nabla^2 \Phi_3 = 0 \quad (4.26)$$

$$\frac{\partial \Phi_3}{\partial z}(x, y, 0\pm) = \frac{\partial \eta_c}{\partial x} Q_\infty \quad (4.27)$$

The complete solution for the cambered wing with nonzero thickness at an angle of attack is then

$$\Phi = \Phi_1 + \Phi_2 + \Phi_3 \quad (4.28)$$

Of course, for Eq. (4.28) to be valid all three linear boundary conditions have to be fulfilled at the wing's projected area on the $z = 0$ plane.

4.4 Symmetric Wing with Nonzero Thickness at Zero Angle of Attack

Consider a symmetric wing with a thickness distribution of $\eta_t(x, y)$ at zero angle of attack, as shown in Fig. 4.4. The equation to be solved is

$$\nabla^2 \Phi = 0 \quad (4.29)$$

Here the subscript is dropped for simplicity. The approximate boundary condition to be

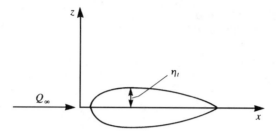

Figure 4.4 Definition of wing thickness η_t at an arbitrary spanwise location y.

fulfilled at the $z = 0$ plane is

$$\frac{\partial \Phi}{\partial z}(x, y, 0\pm) = \pm \frac{\partial \eta_t}{\partial x} Q_\infty \tag{4.30}$$

The solution of this problem can be obtained by distributing basic solution elements of Laplace's equation. Because of the symmetry, as explained in Chapter 3, a source/sink distribution placed at the wing section centerline, can be used to model the flow, as shown in Fig. 4.5.

Recall that the potential due to such a point source element σ (Eq. (3.19))

$$\Phi = \frac{-\sigma}{4\pi r} \tag{4.31}$$

where r is the distance from the point singularity located at (x_0, y_0, z_0) (see Section 3.4), that is,

$$r = \sqrt{(x - x_0)^2 + (y - y_0)^2 + (z - z_0)^2} \tag{4.32}$$

Now if these elements are distributed over the wing's projected area on the x–y plane ($z_0 = 0$), the velocity potential at an arbitrary point (x, y, z) will be

$$\Phi(x, y, z) = \frac{-1}{4\pi} \int_{\text{wing}} \frac{\sigma(x_0, y_0) \, dx_0 \, dy_0}{\sqrt{(x - x_0)^2 + (y - y_0)^2 + z^2}} \tag{4.33}$$

Note that the integration is done over the wing only (no wake). The normal velocity component $w(x, y, z)$ is obtained by differentiating Eq. (4.33) with respect to z:

$$w(x, y, z) = \frac{\partial \Phi}{\partial z} = \frac{z}{4\pi} \int_{\text{wing}} \frac{\sigma(x_0, y_0) \, dx_0 \, dy_0}{[(x - x_0)^2 + (y - y_0)^2 + z^2]^{3/2}} \tag{4.34}$$

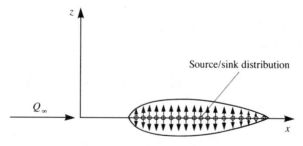

Figure 4.5 Method of modeling the thickness problem by a source/sink distribution.

4.4 Symmetric Wing with Nonzero Thickness at Zero Angle of Attack

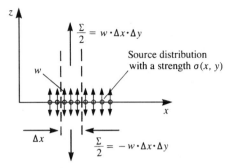

Figure 4.6 Segment of a source distribution on the $z = 0$ plane.

To find $w(x, y, 0)$, a limit process is required (see Section 3.14) and the result is

$$w(x, y, 0\pm) = \lim_{z \to 0\pm} w(x, y, z) = \pm \frac{\sigma(x, y)}{2} \qquad (4.35)$$

where $+$ is on the upper and $-$ is on the lower surface of the wing, respectively.

This result can be obtained by observing the volume flow rate due to a Δx long and Δy wide source element with a strength $\sigma(x, y)$. A two-dimensional section view is shown in Fig. 4.6. Following the definition of a source element (Section 3.4) the volumetric flow Σ produced by this element is then

$$\Sigma = \sigma(x, y)\Delta x \Delta y$$

But as $dz \to 0$ the flux from the sides becomes negligible (at $z = 0\pm$) and only the normal velocity component $w(x, y, 0\pm)$ contributes to the source flux. The above volume flow feeds the two sides (upper and lower) of the surface element and, therefore, $\Sigma = 2w(x, y, 0+)\Delta x \Delta y$. So by equating this flow rate with that produced by the source distribution

$$\Sigma = 2w(x, y, 0+)\Delta x \Delta y = \sigma(x, y)\Delta x \Delta y$$

we obtain again

$$w(x, y, 0\pm) = \pm \frac{\sigma(x, y)}{2} \qquad (4.35)$$

Substitution of Eq. (4.35) into the boundary condition results in

$$\frac{\partial \Phi}{\partial z}(x, y, 0\pm) = \pm \frac{\partial \eta_t}{\partial x} Q_\infty = \pm \frac{\sigma(x, y)}{2}$$

or

$$\sigma(x, y) = 2Q_\infty \frac{\partial \eta_t}{\partial x}(x, y) \qquad (4.36)$$

So in this case the solution for the source distribution is easily obtained after substituting Eq. (4.36) into Eq. (4.33) for the velocity potential and differentiating to obtain the velocity field:

$$\Phi(x, y, z) = \frac{-Q_\infty}{2\pi} \int_{\text{wing}} \frac{[\partial \eta_t(x_0, y_0)/\partial x]\, dx_0\, dy_0}{\sqrt{(x-x_0)^2 + (y-y_0)^2 + z^2}} \tag{4.37}$$

$$u(x, y, z) = \frac{Q_\infty}{2\pi} \int_{\text{wing}} \frac{[\partial \eta_t(x_0, y_0)/\partial x](x-x_0)\, dx_0\, dy_0}{[(x-x_0)^2 + (y-y_0) + z^2]^{3/2}} \tag{4.38}$$

$$v(x, y, z) = \frac{Q_\infty}{2\pi} \int_{\text{wing}} \frac{[\partial \eta_t(x_0, y_0)/\partial x](y-y_0)\, dx_0\, dy_0}{[(x-x_0)^2 + (y-y_0) + z^2]^{3/2}} \tag{4.39}$$

$$w(x, y, z) = \frac{Q_\infty}{2\pi} \int_{\text{wing}} \frac{[\partial \eta_t(x_0, y_0)/\partial x] z\, dx_0\, dy_0}{[(x-x_0)^2 + (y-y_0) + z^2]^{3/2}} \tag{4.40}$$

The pressure distribution due to this solution will be derived later, but it is easy to observe that since the pressure field is symmetric, there is no lift produced due to thickness.

4.5 Zero-Thickness Cambered Wing at Angle of Attack–Lifting Surfaces

Here we shall solve the two linear problems of angle of attack and camber together (Fig. 4.7). The problem to be solved is

$$\nabla^2 \Phi = 0 \tag{4.29}$$

with the boundary condition requiring no flow across the surface (evaluated at $z = 0$) as

$$\frac{\partial \Phi}{\partial z}(x, y, 0\pm) = Q_\infty \left(\frac{\partial \eta_c}{\partial x} - \alpha \right) \tag{4.41}$$

This problem is antisymmetric with respect to the z direction and can be solved by a doublet distribution or by a vortex distribution. These basic singularity elements are solutions to Eq. (4.29) and fulfill the boundary condition (Eq. (4.2)) at infinity. As mentioned in Section 2.9, vortex lines cannot begin and terminate in the fluid. This means that if the lifting problem is to be modeled with vortex elements they cannot be terminated at the wing and must be shed into the flow. So as not to generate force in the fluid, these free vortex elements must be parallel to the local flow direction, at any point on the wake. (This observation is based on the vector product $\mathbf{Q} \times \mathbf{\Gamma}$ in Eq. (3.113).)

In the following section two methods of representing lifting problems by a doublet or vortex distribution are presented. Also, as a consequence of the small-disturbance approximation, the wake is taken to be *planar* and placed on the $z = 0$ plane.

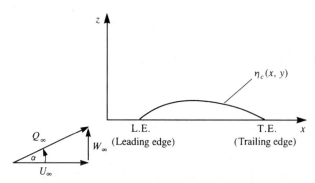

Figure 4.7 Nomenclature used for the definition of the thin, lifting-wing problem.

4.5 Zero-Thickness Cambered Wing at Angle of Attack–Lifting Surfaces

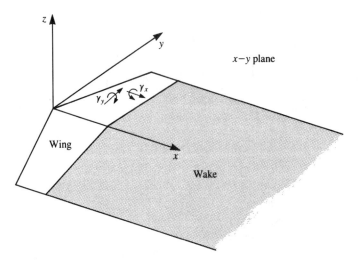

Figure 4.8 Lifting-surface model of a three-dimensional wing.

a. *Doublet Distribution*

To establish the lifting surface equation in terms of doublets the various directional derivatives of the term $1/r$ in the basic doublet solution have to be examined (see Section 3.5). The most suitable differentiation is with respect to z, which results in doublets pointing in the z direction that create a pressure jump in this direction. Consequently, this antisymmetric point element (Eq. (3.39)) placed at (x_0, y_0, z_0) will be used:

$$\Phi(x, y, z) = \frac{-\mu(x_0, y_0)(z - z_0)}{4\pi[(x - x_0)^2 + (y - y_0)^2 + (z - z_0)^2]^{3/2}} \quad (4.42)$$

The potential at an arbitrary point (x, y, z) due to these elements distributed over the wing and its wake, as shown in Fig. 4.8 ($z_0 = 0$), is

$$\Phi(x, y, z) = \frac{1}{4\pi} \int_{\text{wing+wake}} \frac{-\mu(x_0, y_0) z \, dx_0 \, dy_0}{[(x - x_0)^2 + (y - y_0)^2 + z^2]^{3/2}} \quad (4.43)$$

The velocity is obtained by differentiating Eq. (4.43) and letting $z \to 0$ on the wing. The limit for the tangential velocity components was derived in Section 3.14, whereas the limit process for the normal velocity component is more elaborate (see Ashley and Landahl,[4.1] p. 149). We obtain

$$u(x, y, 0\pm) = \frac{\partial \Phi}{\partial x} = \frac{\mp 1}{2} \frac{\partial \mu}{\partial x}$$

$$v(x, y, 0\pm) = \frac{\partial \Phi}{\partial y} = \frac{\mp 1}{2} \frac{\partial \mu}{\partial y}$$

$$w(x, y, 0\pm) = \frac{\partial \Phi}{\partial z}$$

$$= \frac{1}{4\pi} \int_{\text{wing+wake}} \frac{\mu(x_0, y_0)}{(y - y_0)^2} \left[1 + \frac{(x - x_0)}{\sqrt{(x - x_0)^2 + (y - y_0)^2}}\right] dx_0 \, dy_0$$

$$(4.44)$$

To construct the integral equation for the unknown $\mu(x, y)$, substitute Eq. (4.44) into the left-hand side of Eq. (4.41) to get

$$\frac{1}{4\pi} \int_{\text{wing+wake}} \frac{\mu(x_0, y_0)}{(y - y_0)^2} \left[1 + \frac{(x - x_0)}{\sqrt{(x - x_0)^2 + (y - y_0)^2}} \right] dx_0 \, dy_0$$

$$= Q_\infty \left(\frac{\partial \eta_c}{\partial x} - \alpha \right) \quad (4.45)$$

The strong singularity at $y = y_0$ in the integrals in Eqs. (4.44) and (4.45) is discussed in Appendix C.

b. *Vortex Distribution*

According to this model, vortex line distributions will be used over the wing and the wake, as in the case of the doublet distribution. This model is physically very easy to construct and the velocity $\Delta \mathbf{q}$ due a vortex line element $d\mathbf{l}$ with a strength of $\Delta \Gamma$ will be computed by the Biot–Savart law (r is defined by Eq. (4.32)):

$$\Delta \mathbf{q} = \frac{-1}{4\pi} \frac{\Delta \Gamma \mathbf{r} \times d\mathbf{l}}{r^3} \quad (2.68b)$$

Now if vortices are distributed over the wing and wake (Fig. 4.9), and if those elements that point in the y direction are denoted as γ_y, and in the x direction as γ_x, then the component of velocity normal to the wing (downwash), induced by these elements, is

$$w(x, y, z) = \frac{-1}{4\pi} \int_{\text{wing+wake}} \frac{\gamma_y(x - x_0) - \gamma_x(y - y_0)}{r^3} dx_0 \, dy_0 \quad (4.46)$$

It appears that in this formulation there are two unknown quantities per point (γ_x, γ_y) compared to one (μ) in the case of the doublet distribution. However, according to the Helmholtz vortex theorems (Section 2.9) vortex strength is constant along a vortex line, and if we consider the vortex distribution on the wing to consist of a large number of infinitesimal vortex lines then at any point on the wing $|\partial \gamma_x / \partial x| = |\partial \gamma_y / \partial y|$ and the final number of unknowns at a point is reduced to one.

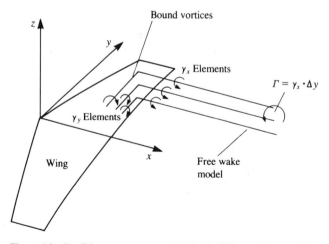

Figure 4.9 Possible vortex representation for the lifting-surface model.

As was shown earlier (in Section 3.14) for a vortex distribution,

$$u(x, y, 0\pm) = \frac{\partial \Phi}{\partial x} = \frac{\pm \gamma_y(x, y)}{2} \tag{4.47}$$

$$v(x, y, 0\pm) = \frac{\partial \Phi}{\partial y} = \frac{\mp \gamma_x(x, y)}{2} \tag{4.47a}$$

The velocity potential on the wing at any point x ($y = y_0 =$ const.) can be obtained by integrating the x component of the velocity along an x-wise line beginning at the leading edge (L.E.):

$$\Phi(x, y_0, 0\pm) = \int_{L.E.}^{x} u(x_1, y_0, 0\pm) \, dx_1 \tag{4.48}$$

and

$$\Delta\Phi(x, y_0) = \int_{L.E.}^{x} \gamma_y(x_1, y_0) \, dx_1 \tag{4.49}$$

To construct the lifting surface equation for the unknown γ, the wing-induced downwash of Eq. (4.46) must be equal and opposite in sign to the normal component of the free-stream velocity:

$$\frac{-1}{4\pi} \int_{\text{wing+wake}} \frac{\gamma_y(x - x_0) - \gamma_x(y - y_0)}{[(x - x_0)^2 + (y - y_0)^2]^{3/2}} \, dx_0 \, dy_0 = Q_\infty \left(\frac{\partial \eta_c}{\partial x} - \alpha \right) \tag{4.50}$$

Here again it is assumed that the boundary conditions and the vortices are placed on the $z = 0$ plane.

Solution for the unknown doublet or vortex strength in Eq. (4.45) or in Eq. (4.50) allows for the evaluation of the velocity distribution. The method of obtaining the corresponding pressure distribution is described in the next section.

4.6 The Aerodynamic Loads

Solution of the aforementioned problems (e.g., the thickness or lifting problems) results in the velocity field. To obtain the aerodynamic loads the pressures need to be resolved by using the Bernoulli equation (Eq. (4.4)). Also, the aerodynamic coefficients can be derived either in the wing or in the flow coordinate system. In this case of small-disturbance flow over wings, traditionally, the wing coordinates are selected as shown in Fig. 4.10.

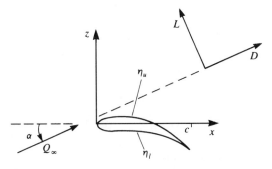

Figure 4.10 Wing-attached coordinate system.

The velocity at any point in the field is then a combination of the free-stream velocity and the perturbation velocity:

$$\mathbf{q} = \left(Q_\infty \cos\alpha + \frac{\partial \Phi}{\partial x}, \frac{\partial \Phi}{\partial y}, \frac{\partial \Phi}{\partial z} + Q_\infty \sin\alpha \right) \tag{4.51}$$

Substituting $\mathbf{q}$ into the Bernoulli equation (Eq. (4.4)) and taking into account the small-disturbance assumptions (Eqs. (4.13) and (4.14) and $\alpha \ll 1$) we obtain

$$p_\infty - p = \frac{\rho}{2}(q^2 - Q_\infty^2) = \frac{\rho}{2}\left[Q_\infty^2 \cos^2\alpha + 2Q_\infty \cos\alpha \frac{\partial \Phi}{\partial x} + \left(\frac{\partial \Phi}{\partial x}\right)^2 + \left(\frac{\partial \Phi}{\partial y}\right)^2 \right.$$
$$\left. + \left(Q_\infty \sin\alpha + \frac{\partial \Phi}{\partial z}\right)^2 - Q_\infty^2 \right] = \rho Q_\infty \frac{\partial \Phi}{\partial x} \tag{4.52}$$

The pressure coefficient C_p can be defined as

$$C_p \equiv \frac{p - p_\infty}{(1/2)\rho Q_\infty^2} = 1 - \left(\frac{q}{Q_\infty}\right)^2 = -2\frac{\partial \Phi/\partial x}{Q_\infty} \tag{4.53}$$

Note that at a stagnation point $q = 0$ and $C_p = 1$. In the undisturbed flow $q = Q_\infty$ and $C_p = 0$. The aerodynamic loads, then, can be calculated by integrating the pressures over the wing surface:

$$\mathbf{F} = -\int_{\text{wing}} p\mathbf{n}\, dS \tag{4.54}$$

When the surface shape is given as in Eq. (4.6) then the normal to the surface is given by Eq. (4.7), which with the small-disturbance approximation becomes

$$\mathbf{n} = \frac{1}{|\nabla F|}\left(-\frac{\partial \eta}{\partial x}, -\frac{\partial \eta}{\partial y}, 1\right) \approx \left(-\frac{\partial \eta}{\partial x}, -\frac{\partial \eta}{\partial y}, 1\right)$$

Consequently, the components of the force $\mathbf{F}$ can be defined as axial, side, and normal force,

$$F_x = \int_{\text{wing}} \left(p_u \frac{\partial \eta_u}{\partial x} - p_l \frac{\partial \eta_l}{\partial x} \right) dx\, dy \tag{4.55}$$

$$F_y = \int_{\text{wing}} \left(p_u \frac{\partial \eta_u}{\partial y} - p_l \frac{\partial \eta_l}{\partial y} \right) dx\, dy \tag{4.56}$$

$$F_z = \int_{\text{wing}} (p_l - p_u)\, dx\, dy \tag{4.57}$$

Here the subscripts u and l represent the upper and lower wing surfaces, respectively. Aerodynamicists frequently refer to the forces in the free-stream coordinates (Fig. 4.10), and therefore these forces must be transformed accordingly. For the small-disturbance case the angle of attack is small and therefore the lift and drag forces are

$$D = F_x \cos\alpha + F_z \sin\alpha$$
$$L = -F_x \sin\alpha + F_z \cos\alpha \approx F_z$$

Note that the evaluation of drag by integrating the pressure distribution is considered to be less accurate than the above formulation for the lift.

4.6 The Aerodynamic Loads

In the case when the wing is assumed to be thin, the pressure difference across the wing Δp is evaluated (positive Δp is in the $+z$ direction) as

$$\Delta p = p_l - p_u = p_\infty - \rho Q_\infty \frac{\partial \Phi}{\partial x}(x, y, 0-) - \left[p_\infty - \rho Q_\infty \frac{\partial \Phi}{\partial x}(x, y, 0+) \right]$$
$$= \rho Q_\infty \left[\frac{\partial \Phi}{\partial x}(x, y, 0+) - \frac{\partial \Phi}{\partial x}(x, y, 0-) \right] \quad (4.58)$$

If the singularity distribution is assumed to be placed on the x–y plane then the pressure differences become as follows:

1. Source distribution: Because of symmetry, $(\partial \Phi/\partial x)(x, y, 0+) = (\partial \Phi/\partial x)(x, y, 0-)$ and

$$\Delta p = \rho Q_\infty \left[\frac{\partial \Phi}{\partial x}(x, y, 0+) - \frac{\partial \Phi}{\partial x}(x, y, 0+) \right] = 0 \quad (4.59a)$$

2. Doublet distribution: In this case $(\partial \Phi/\partial x)(x, y, 0\pm) = (\mp 1/2)\partial \mu(x, y)/\partial x$ and the pressure difference becomes

$$\Delta p = \rho Q_\infty \frac{\partial}{\partial x} \Delta \Phi(x, y) = -\rho Q_\infty \frac{\partial \mu(x, y)}{\partial x} \quad (4.59b)$$

where $\Delta \Phi = \Phi_u - \Phi_l$.

3. Vortex distribution: For the vortex distribution on the x–y plane the pressure jump can be modeled with a vortex distribution $\gamma_y(x, y)$ that points in the y direction, such that $(\partial \Phi/\partial x)(x, y, 0\pm) = (\pm 1/2)\gamma_y(x, y)$. Therefore, the pressure difference becomes

$$\Delta p = \rho Q_\infty \frac{\partial}{\partial x} \Delta \Phi(x, y) = \rho Q_\infty \gamma_y(x, y) \quad (4.59c)$$

The aerodynamic moment can be derived in a similar manner and as an example the pitching moment about the y axis for a wing placed at the $z = 0$ surface is

$$M_{x=0} = -\int_{\text{wing}} \Delta p x \, dx \, dy \quad (4.60)$$

Usually, the aerodynamic loads are presented in a nondimensional form. In the case of the force coefficients where F is lift, drag, or side force the corresponding coefficients will have the form

$$C_F = \frac{F}{(1/2)\rho Q_\infty^2 S} \quad (4.61)$$

where S is a reference area (wing planform area for wings). Similarly the nondimensional moment coefficient becomes

$$C_M = \frac{M}{(1/2)\rho Q_\infty^2 Sb} \quad (4.62)$$

Here, again, M can be a moment about any arbitrary axis and b is a reference moment arm (e.g., wing span).

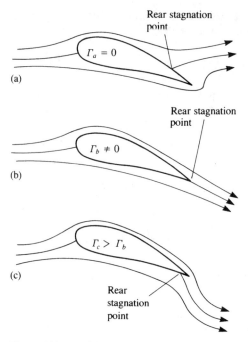

Figure 4.11 Possible solutions for the flow over an airfoil: (a) flow with zero circulation, (b) flow with circulation that will result in a smooth flow near the trailing edge, (c) flow with circulation larger than in case (b).

4.7   The Vortex Wake

The analysis followed up to this point suggests that by using distributions of the elementary solutions of Laplace's equation, the problem is reduced to finding a combination of these elements that will satisfy the zero normal flow boundary condition on solid surfaces. However, as in the case of the flow over a cylinder (Section 3.11), the solution is not unique and an arbitrary value can be selected for the circulation Γ. This problem is illustrated for the airfoil in Fig. 4.11, where in case (a) the circulation is zero. In case (b) the circulation is such that the flow at the trailing edge (T.E.) seems to be parallel at the edge. In case (c) the circulation is even larger and the flow turns downward near the trailing edge (this can be achieved, for example, by blowing). W. M. Kutta (the German mathematician who was the first to use this trailing-edge condition in a theoretical paper in 1902) suggested that from the physical point of view, case (b) seems to result in the right amount of circulation. The *Kutta condition* thus states that: *The flow leaves the sharp trailing edge of an airfoil smoothly and the velocity there is finite*. For the current modeling purposes this can be interpreted that the flow leaves the T.E. along the bisector line there. Also, since the trailing-edge angle is finite the normal component of the velocity, from both sides of the airfoil, must vanish. For a continuous velocity, this is possible only if this is a stagnation point. Therefore, it is useful to assume that the pressure difference there is also zero,

$$\Delta p_{T.E.} = 0 \tag{4.63}$$

Additionally if the circulation is modeled by a vortex distribution, then this can be expressed as

$$\gamma_{T.E.} = 0 \tag{4.63a}$$

4.7 The Vortex Wake

Figure 4.12 Flow near cusped trailing edge.

For a cusped trailing edge (where the angle is zero, as in Fig. 4.12), Eq. (4.63) must hold even though the trailing edge need not be a stagnation point.

Next, consider the lifting wing of Fig. 4.9. As was shown in the case of the cylinder, circulation is needed to generate lift. Assume that the vortex distribution used to model the lift is placed on the wing as the bound vortex $\gamma_y(x, y)$, where the subscript designates the direction of the circulation vector. But, according to Helmholtz's theorem, a vortex line cannot begin or end in the fluid and any change in $\gamma_y(x, y)$ must be followed by an equal change in $\gamma_x(x, y)$. Consequently, the wing will be modeled by constant-strength vortex lines, and if a change in the local strength of $\gamma_y(x, y)$ is needed then an additional vortex line will be added (or the vortex line is bent by $\pm 90°$) such that

$$\left| \frac{\partial \gamma_x(x, y)}{\partial x} \right| = \left| \frac{\partial \gamma_y(x, y)}{\partial y} \right| \tag{4.64}$$

This condition can also be obtained by requiring that the flow above the wing be vorticity free. Thus the vortex distribution induced velocity at a point slightly above ($z = 0+$) the wing is

$$u(x, y, 0+) = \frac{\gamma_y(x, y)}{2} \tag{4.65a}$$

$$v(x, y, 0+) = -\frac{\gamma_x(x, y)}{2} \tag{4.65b}$$

For the flow resulting from this vortex distribution to be vorticity free requires that

$$\omega_z = \frac{1}{2}\left(\frac{\partial v}{\partial x} - \frac{\partial u}{\partial y} \right) = \frac{1}{4}\left(\frac{-\partial \gamma_x(x, y)}{\partial x} - \frac{\partial \gamma_y(x, y)}{\partial y} \right) = 0$$

which is exactly the same result of Eq. (4.64).

Physically this means that any change in vorticity in one direction must be followed by a change in a normal direction (as shown in Fig. 4.13, where the wing and the vortex lines are in the x–y plane). Consequently, all vortex lines must be either infinitely long lines or closed vortex rings. In the case of the wing this means that the lifting vortices (bound vortices) cannot end at the wing (e.g., at the tip) and must be extended behind the wing into a wake. Furthermore, a lifting wing creates a starting vortex and this vortex may be located far downstream (see Fig. 2.5).

Next, the wake shape must be considered. If the wake is to be modeled by a vortex sheet (free vortex sheet) then from physical considerations it must be different from the bound circulation by not creating loads. The pressure difference across the sheet is obtained by a generalization (with vector notation) of Eq. (4.59c), and if there is no pressure difference across the vortex sheet then

$$\Delta p = \rho \mathbf{q} \times \boldsymbol{\gamma} = 0$$

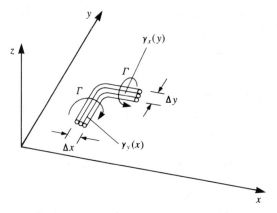

Figure 4.13 Since vortex lines cannot end in a fluid, the bound vortices are turned backward, parallel to the free stream.

or

$$\mathbf{q} \times \boldsymbol{\gamma} = 0 \quad (4.66)$$

where $\boldsymbol{\gamma} = (\gamma_x, \gamma_y, \gamma_z)$. This means that the velocity on the wake must be parallel to the wake vortices.

This consideration will be very helpful when proposing some simple models for the lifting wing problem in the following chapters.

A small-disturbance approximation applied to the wake model results in

$$\mathbf{Q}_\infty \times \boldsymbol{\gamma}_w = 0 \quad (4.66a)$$

implying that vortex lines in the wake are parallel to the free-stream, that is,

$$\mathbf{Q}_\infty \parallel \boldsymbol{\gamma}_w \quad (4.66b)$$

4.8 Linearized Theory of Small-Disturbance Compressible Flow

The potential flow model was based so far on the assumption of an incompressible fluid. In the case when the disturbance to the flow is small, it is possible to extend the methods of incompressible potential flow to cover cases with small effects of compressibility (e.g., low-speed subsonic flows). To investigate this possibility, the continuity equation (Eq. (1.21)) is rewritten in the form

$$\frac{-1}{\rho}\left(\frac{\partial \rho}{\partial t} + u\frac{\partial \rho}{\partial x} + v\frac{\partial \rho}{\partial y} + w\frac{\partial \rho}{\partial z}\right) = \frac{\partial u}{\partial x} + \frac{\partial v}{\partial y} + \frac{\partial w}{\partial z} \quad (4.67)$$

and the inviscid momentum equations (Eqs. (1.31)) are

$$\frac{\partial u}{\partial t} + u\frac{\partial u}{\partial x} + v\frac{\partial u}{\partial y} + w\frac{\partial u}{\partial z} = \frac{-1}{\rho}\frac{\partial p}{\partial x} \quad (4.68a)$$

$$\frac{\partial v}{\partial t} + u\frac{\partial v}{\partial x} + v\frac{\partial v}{\partial y} + w\frac{\partial v}{\partial z} = \frac{-1}{\rho}\frac{\partial p}{\partial y} \quad (4.68b)$$

$$\frac{\partial w}{\partial t} + u\frac{\partial w}{\partial x} + v\frac{\partial w}{\partial y} + w\frac{\partial w}{\partial z} = \frac{-1}{\rho}\frac{\partial p}{\partial z} \quad (4.68c)$$

4.8 Linearized Theory of Small-Disturbance Compressible Flow

For an isentropic fluid the propagation speed of the disturbance a (speed of sound) can be defined as

$$a^2 = \frac{\partial p}{\partial \rho} \tag{4.69}$$

and consequently the pressure terms in the momentum equation can be replaced (e.g., $\partial p/\partial x = a^2 \partial \rho/\partial x$, in the x direction). Multiplying the momentum equations by u, v, and w, respectively, and adding them together leads to

$$u\frac{\partial u}{\partial t} + v\frac{\partial v}{\partial t} + w\frac{\partial w}{\partial t} + u^2\frac{\partial u}{\partial x} + v^2\frac{\partial v}{\partial y} + w^2\frac{\partial w}{\partial z} + uv\frac{\partial u}{\partial y} + uv\frac{\partial v}{\partial x} + uw\frac{\partial u}{\partial z}$$

$$+ uw\frac{\partial w}{\partial x} + vw\frac{\partial v}{\partial z} + vw\frac{\partial w}{\partial y} = \frac{-a^2}{\rho}\left(u\frac{\partial \rho}{\partial x} + v\frac{\partial \rho}{\partial y} + w\frac{\partial \rho}{\partial z}\right)$$

Replacing the right-hand side with the continuity equation and recalling the irrotationality condition (Eq. (2.12), $\nabla \times \mathbf{q} = 0$) we obtain

$$\left(1 - \frac{u^2}{a^2}\right)\frac{\partial u}{\partial x} + \left(1 - \frac{v^2}{a^2}\right)\frac{\partial v}{\partial y} + \left(1 - \frac{w^2}{a^2}\right)\frac{\partial w}{\partial z} - 2\frac{uv}{a^2}\frac{\partial u}{\partial y} - 2\frac{vw}{a^2}\frac{\partial v}{\partial z} - 2\frac{uw}{a^2}\frac{\partial w}{\partial x}$$

$$+ \frac{1}{\rho}\frac{\partial \rho}{\partial t} - \frac{u}{a^2}\frac{\partial u}{\partial t} - \frac{v}{a^2}\frac{\partial v}{\partial t} - \frac{w}{a^2}\frac{\partial w}{\partial t} = 0 \tag{4.70}$$

Using the velocity potential Φ as defined in Eq. (2.19), and assuming that the free-stream velocity $\mathbf{Q}_\infty$ is parallel to the x axis (thus $\mathbf{Q}_\infty$ becomes $U_\infty \mathbf{i}$), and that the velocity perturbations caused by the motion of the body in the fluid are small, we get

$$\left|\frac{\partial \Phi}{\partial x}\right|, \left|\frac{\partial \Phi}{\partial y}\right|, \left|\frac{\partial \Phi}{\partial z}\right| \ll U_\infty \tag{4.71}$$

Based on these assumptions, the velocity components, in term of the perturbation velocity potential, are

$$u = U_\infty + \frac{\partial \Phi}{\partial x}$$

$$v = \frac{\partial \Phi}{\partial y} \tag{4.72}$$

$$w = \frac{\partial \Phi}{\partial z}$$

Assuming steady-state flow ($\partial/\partial t = 0$), and neglecting the smaller terms in Eq. (4.70), based on Eq. (4.71), we obtain

$$\left(1 - \frac{u^2}{a^2}\right)\frac{\partial u}{\partial x} + \frac{\partial v}{\partial y} + \frac{\partial w}{\partial z} = 0$$

Using the energy equation for an adiabatic flow, we can show that the local speed of sound can be replaced by its free-stream value and the small-disturbance equation becomes

$$(1 - M_\infty^2)\frac{\partial^2 \Phi}{\partial x^2} + \frac{\partial^2 \Phi}{\partial y^2} + \frac{\partial^2 \Phi}{\partial z^2} = 0 \tag{4.73}$$

For time-dependent flows the $\partial \rho/\partial t$ term in Eq. (4.70) needs to be evaluated by using the Bernoulli equation (Eq. (2.32)), but the result will introduce additional time-dependent terms. However, in the case of steady-state flows, the effect of compressibility is easily evaluated.

Using a simple coordinate transformation, called the *Prandtl–Glauert rule* (named after the German and British scientists, Ludwig Prandtl and Herman Glauert, circa 1922–27), we obtain for subsonic flow

$$x_M = \frac{x}{\sqrt{1-M_\infty^2}}$$
$$y_M = y \qquad (4.74)$$
$$z_M = z$$

Equation (4.73) can be reduced to Laplace's equation, and the results of incompressible flow can be applied (by using $\partial/\partial x_M = (1-M_\infty^2)^{-1/2}\partial/\partial x$). The subscript $(\ )_M$ represents here the flow for $M > 0$.

For example, the pressure coefficient of Eq. (4.53) becomes

$$C_p = -2\frac{\partial\Phi/\partial x_M}{Q_\infty} = -2\frac{\partial\Phi/\partial x}{Q_\infty}\frac{1}{\sqrt{1-M_\infty^2}} \qquad (4.75)$$

Similarly the lift and moment coefficients become

$$C_L(M > 0) = \frac{C_L(M=0)}{\sqrt{1-M_\infty^2}} \qquad (4.76)$$

$$C_M(M > 0) = \frac{C_M(M=0)}{\sqrt{1-M_\infty^2}} \qquad (4.77)$$

which indicates that at higher speeds the lift slope is increasing as shown by Fig. 4.14. Also, note that according to Eq. (4.74) the x coordinate is being stretched as the Mach number increases and therefore the results for $M = 0$ and $M > 0$ are for wings of different aspect ratio.

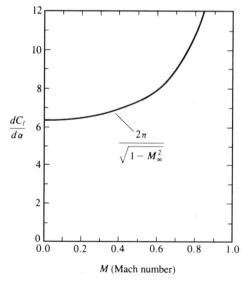

Figure 4.14 Variation of two-dimensional lift-curve slope with Mach number using Prandtl–Glauert formula.

Based on the results of Fig. 4.14 (for a two-dimensional airfoil), for small-disturbance flows the potential flow based models of this chapter are applicable at least up to $M_\infty = 0.5$.

Reference

[4.1] Ashley, H., and Landahl, M., *Aerodynamics of Wings and Bodies*, Addison-Wesley, 1965.

Problems

4.1. Consider a two-dimensional parabolic camberline with ϵ being its maximum height. The equation of the camberline is then
$$\eta_c(x) = 4\epsilon \frac{x}{c}\left[1 - \frac{x}{c}\right]$$
and the free-stream components in the airfoil frame of reference are (U_∞, W_∞). Derive the formula for the chordwise normal vector **n** and the exact boundary conditions on the camberline (by using Eq. (4.10)).

4.2. A two-dimensional distribution of doublets oriented in the vertical direction, with constant strength $\boldsymbol{\mu} = (0, \mu)$, is placed along the x axis ($0 < x < x_1$). Show that this doublet distribution is identical to a point vortex at the origin and at $x = x_1$. What is the strength of the point vortices?

4.3. Show that a vortex distribution of strength $\gamma(x)$ along the x axis ($x_1 < x < x_2$) is equivalent to a distribution of doublets oriented in the vertical direction (plus two vortices at $x = x_1$ and at $x = x_2$) and that the strength of this doublet distribution is
$$\mu(x) = \int_{x_1}^{x} \gamma(x_0)\,dx_0 \qquad x_1 < x < x_2$$
(Show that both singular distributions have the same velocity potential and velocity field.)

CHAPTER 5

Small-Disturbance Flow over Two-Dimensional Airfoils

The strategy presented in Chapter 3 postulates that a solution to the potential flow problem can be obtained by superimposing elementary solutions of Laplace's equation. Thus, the solution consists of finding the "right" combination of these elementary solutions that will fulfill the zero normal flow boundary condition. Using this approach, in the previous chapter the small-disturbance problem for a wing moving with a steady motion was established. This treatment allowed us to separate the problem into the solution of two linear subproblems, namely the thickness and lifting problems. In this chapter the simpler two-dimensional case of both the airfoil with nonzero thickness at zero angle of attack and the lifting zero-thickness airfoil will be solved, by using analytical techniques. These solutions can then be added to yield the complete small-disturbance solution for the flow past a thin airfoil.

5.1 Symmetric Airfoil with Nonzero Thickness at Zero Angle of Attack

Consider the two-dimensional symmetric airfoil, with a thickness distribution of $\eta_t(x)$, at zero angle of attack, as shown in Fig. 5.1. The velocity field will be obtained by solving the continuity equation

$$\nabla^2 \Phi = 0 \tag{5.1}$$

with the boundary condition requiring that the flow normal to the airfoil upper ($+\eta_t$) and lower surface ($-\eta_t$) be zero:

$$\frac{\partial \Phi}{\partial z}(x, 0\pm) = \pm \frac{d\eta_t}{dx} Q_\infty \tag{5.2}$$

This equation actually states that the sum of the free-stream and the airfoil-induced normal velocity components is zero on the surface $w(x, 0\pm) \mp (d\eta_t/dx)Q_\infty = 0$. Equation (5.2) is the two-dimensional version of the three-dimensional boundary condition (Eq. (4.30)) and Φ is the perturbation velocity potential. Recall that the boundary condition has been transferred to the $z = 0$ plane. Also, the boundary condition requiring that the disturbance due to the airfoil will decay far from it (Eq. (4.2)) is not stated because it is automatically fulfilled by the basic source, doublet, or vortex elements.

Because of the symmetry of the problem (relative to the $z = 0$ plane) we use a source distribution, which inherently has such a symmetric feature. These sources are placed on the x axis from $x = 0$ to $x = c$, as shown in Fig. 5.2. The potential of a source distribution can be obtained by observing the potential due to a single source element of strength σ_0, located at $(x_0, 0)$:

$$\Phi_{\sigma_0} = \frac{\sigma_0}{2\pi} \ln r = \frac{\sigma_0}{2\pi} \ln \sqrt{(x - x_0)^2 + z^2} = \frac{\sigma_0}{4\pi} \ln[(x - x_0)^2 + z^2] \tag{5.3}$$

5.1 Symmetric Airfoil with Nonzero Thickness at Zero Angle of Attack

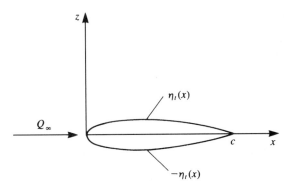

Figure 5.1 Two-dimensional thin symmetric airfoil at zero angle of attack.

The local radial velocity component q_r due to this element at an arbitrary point (x, z) is

$$q_r = \frac{\sigma_0}{2\pi r} \tag{5.4}$$

(Note that the tangential component is zero.)

In Cartesian coordinates this can be resolved into the x and z directions as $(u, w) = q_r(\cos\theta, \sin\theta)$. The same result can be obtained by differentiating Eq. (5.3):

$$u = \frac{\partial \Phi_{\sigma_0}}{\partial x} = \frac{\sigma_0}{2\pi} \frac{x - x_0}{(x - x_0)^2 + z^2} \tag{5.5}$$

$$w = \frac{\partial \Phi_{\sigma_0}}{\partial z} = \frac{\sigma_0}{2\pi} \frac{z}{(x - x_0)^2 + z^2} \tag{5.6}$$

As shown in Fig. 5.2, the airfoil thickness effect is modeled by a continuous $\sigma(x)$ distribution along the x axis. The velocity potential and the resulting velocity field can be obtained by integrating the contribution of the above point elements over the chord (from $x = 0$ to $x = c$); however, now $\sigma(x_0)$ is the source strength per unit length. We thus have

$$\Phi(x, z) = \frac{1}{2\pi} \int_0^c \sigma(x_0) \ln \sqrt{(x - x_0)^2 + z^2} \, dx_0 \tag{5.7}$$

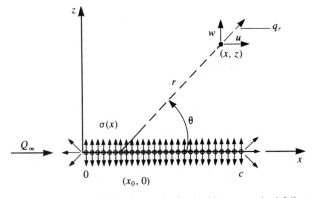

Figure 5.2 Source distribution model for the thin symmetric airfoil.

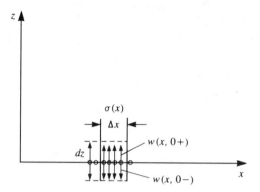

Figure 5.3 A Δx long segment of a source distribution along the x axis.

$$u(x, z) = \frac{1}{2\pi} \int_0^c \sigma(x_0) \frac{x - x_0}{(x - x_0)^2 + z^2} dx_0 \tag{5.8}$$

$$w(x, z) = \frac{1}{2\pi} \int_0^c \sigma(x_0) \frac{z}{(x - x_0)^2 + z^2} dx_0 \tag{5.9}$$

In order to substitute the velocity component $w(x, 0)$ into the boundary condition (Eq. (5.2)) the limit of Eq. (5.9) at $z = 0$ is needed. Following the results of Section 3.14, we obtain

$$w(x, 0\pm) = \lim_{z \to \pm 0} w(x, z) = \pm \frac{\sigma(x)}{2} \tag{5.10}$$

where $+$ is on the upper and $-$ is on the lower surface of the airfoil, respectively. Similarly to the three-dimensional case, this result can be obtained by observing the volume flow rate due to a Δx long element with a strength $\sigma(x)$, as shown in Fig. 5.3. As $dz \to 0$, the flux from the sides of the small element becomes negligible, compared to the flux due to the $w(x, 0\pm)$ component. The volumetric flow due to a Δx wide source element is $\sigma(x)\Delta x$, which must be equal to the flow rate fed by the two sides (upper and lower) of the surface, $2w(x, 0+)\Delta x$. Therefore,

$$2w(x, 0+)\Delta x = \sigma(x)\Delta x$$

and we obtain again

$$w(x, 0\pm) = \pm \frac{\sigma(x)}{2} \tag{5.10}$$

Substitution of Eq. (5.10) into the boundary condition results in

$$\frac{\partial \Phi}{\partial z}(x, 0\pm) = \pm \frac{d\eta_t}{dx} Q_\infty = \pm \frac{\sigma(x)}{2}$$

and therefore

$$\sigma(x) = 2Q_\infty \frac{d\eta_t}{dx} \tag{5.11}$$

Hence in this case the solution for the source distribution is easily obtained after substituting Eq. (5.11) into Eqs. (5.7)–(5.9):

$$\Phi(x, z) = \frac{Q_\infty}{\pi} \int_0^c \frac{d\eta_t(x_0)}{dx} \ln \sqrt{(x - x_0)^2 + z^2} \, dx_0 \tag{5.12}$$

5.1 Symmetric Airfoil with Nonzero Thickness at Zero Angle of Attack

$$u(x, z) = \frac{Q_\infty}{\pi} \int_0^c \frac{d\eta_t(x_0)}{dx} \frac{x - x_0}{(x - x_0)^2 + z^2} dx_0 \tag{5.13}$$

$$w(x, z) = \frac{Q_\infty}{\pi} \int_0^c \frac{d\eta_t(x_0)}{dx} \frac{z}{(x - x_0)^2 + z^2} dx_0 \tag{5.14}$$

It is clear from these equations that the u component of the velocity is symmetric, and the w component is antisymmetric (with respect to the x axis). Therefore, the pressure distribution is the same for the top and bottom surfaces and is evaluated at $z = 0$. The axial velocity component at $z = 0$ is then

$$u(x, 0) = \frac{Q_\infty}{\pi} \int_0^c \frac{d\eta_t(x_0)}{dx} \frac{1}{(x - x_0)} dx_0 \tag{5.15}$$

and the pressure is obtained by substituting this into the steady-state Bernoulli equation (Eq. (4.52)):

$$p - p_\infty = -\rho Q_\infty \frac{\partial \Phi}{\partial x} = -\rho Q_\infty u(x, 0) \tag{5.16}$$

and in terms of the pressure coefficient we have

$$C_p = \frac{p - p_\infty}{(1/2)\rho Q_\infty^2} = -2 \frac{u(x, 0)}{Q_\infty} \tag{5.17}$$

Evaluating the velocity at $z = 0\pm$, as in Eq. (5.15), we get the pressure coefficient as

$$C_p = \frac{-2}{\pi} \int_0^c \frac{d\eta_t(x_0)}{dx} \frac{1}{(x - x_0)} dx_0 \tag{5.18}$$

Since this pressure distribution is the same for the upper and for the lower surface the pressure difference between the upper and lower surface is zero:

$$\Delta p = p_l - p_u = 0 \tag{5.19}$$

and the aerodynamic lift is

$$L = \int_0^c \Delta p \, dx = 0 \tag{5.20}$$

For the drag force calculation the contribution of the upper and lower surfaces needs to be included using Eq. (4.55):

$$D = \int_0^c p_u \frac{d\eta_t}{dx} dx - \int_0^c p_l \frac{-d\eta_t}{dx} dx = 2 \int_0^c p_u \frac{d\eta_t}{dx} dx \tag{5.21}$$

Substituting the pressure from Eqs. (5.15) and (5.16) into Eq. (5.21) and observing that the integral of a constant pressure p_∞ over a closed body is zero we obtain

$$D = -2\rho \frac{Q_\infty^2}{\pi} \int_0^c \int_0^c \frac{[d\eta_t(x_0)/dx][d\eta_t(x)/dx]}{x - x_0} dx_0 \, dx \tag{5.21a}$$

It can be shown, using the symmetry properties of the integrand (see Moran,[5.1] pp. 87–88), that the drag is zero:

$$D = 0 \tag{5.21b}$$

This result can be obtained directly from the Kutta–Joukowski theorem (Section 3.11). Thus, the symmetrical airfoil at zero angle of attack does not generate lift, drag, or pitching moment. Evaluation of the velocity distribution needs to be done only to add this thickness effect to the lifting thin airfoil problem (as derived in the next section).

To obtain the velocity components from Eqs. (5.13) and (5.14) for points not lying on the strip ($0 < x < c, z = 0$), the integrals can be evaluated numerically or in closed form for certain simple geometries. However, when the axial component of the velocity or the pressure coefficient is to be determined on the airfoil surface using Eqs. (5.15) and (5.18) the integrands become infinite at $x = x_0$ and the integrals are not defined. It is noted that if the thickness is increasing at $x = x_0$, the integrand goes to $-\infty$ as x_0 is approached from the left and to $+\infty$ as x_0 is approached from the right (e.g., in Eq. (5.15)) and the integrand is antisymmetric in the neighborhood of $x = x_0$.

If the integral in Eq. (5.13) were evaluated at the actual airfoil surface the integrand would not be singular. It is the transfer of the boundary condition to the chordline and the subsequent result that the velocity components on the surface are equivalent to the components on the chordline that has led to the appearance of the improper integral for the surface pressure. We would expect from physical considerations that the surface pressure should be determinable from Eq. (5.18) and aerodynamicists generally agree that the *Cauchy principal value* of the integral is the appropriate one. The Cauchy principal value of the improper integral

$$\int_a^b f(x_0)\,dx_0$$

where

$$f(x_0) \to \infty \quad \text{at} \quad x_0 = x \quad \text{and} \quad a < x < b$$

is defined by

$$\int_a^b f(x_0)\,dx_0 = \lim_{\epsilon \to 0}\left[\int_a^{x-\epsilon} f(x_0)\,dx_0 + \int_{x+\epsilon}^b f(x_0)\,dx_0\right]$$

As an example, consider the following integral where the limits can be evaluated in closed form:

$$\int_0^c \frac{dx_0}{x - x_0} = \lim_{\epsilon \to 0}\left[\int_0^{x-\epsilon} \frac{dx_0}{x - x_0} - \int_{x+\epsilon}^c \frac{dx_0}{x_0 - x}\right]$$

$$= \lim_{\epsilon \to 0}\left[-\ln(x - x_0)\big|_0^{x-\epsilon} - \ln(x_0 - x)\big|_{x+\epsilon}^c\right]$$

$$= \lim_{\epsilon \to 0}[-\ln\epsilon + \ln x - \ln(c - x) + \ln\epsilon] = \ln\frac{x}{c - x}$$

Note that in the second integral the sign was changed to avoid obtaining the logarithm of a negative quantity.

In practice, if the integral can be evaluated in closed form the correct Cauchy principal value can be obtained by simply ignoring the limit process as long as the arguments of all logarithm terms are taken as their absolute values.

A frequently used principal value integral in many small-disturbance flow applications is the *Glauert integral* (see Glauert,[5.2] pp. 92–93), which has the form

$$\int_0^\pi \frac{\cos n\theta_0}{\cos\theta_0 - \cos\theta}\,d\theta_0 = \frac{\pi \sin n\theta}{\sin\theta}, \quad n = 0, 1, 2, \ldots \tag{5.22}$$

5.1 Symmetric Airfoil with Nonzero Thickness at Zero Angle of Attack

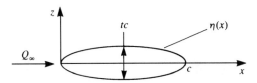

Figure 5.4 Thin ellipse in a uniform flow.

Example: Flow Past an Ellipse

To demonstrate the features of the pressure distribution obtained from this small-disturbance solution consider an ellipse with a thickness of $t \cdot c$ at zero angle of attack (Fig. 5.4). The equation for the surface is then

$$\frac{[x-(c/2)]^2}{(c/2)^2} + \frac{\eta^2}{(tc/2)^2} = 1$$

or

$$\eta = \pm t\sqrt{x(c-x)} \tag{5.23}$$

The derivative of the thickness function for the upper (+) and lower (−) surfaces is then

$$\frac{d\eta}{dx} = \pm\frac{t}{2}\frac{c-2x}{\sqrt{x(c-x)}}$$

The velocity distribution on the ellipse is obtained by substituting this into Eq. (5.15) (note that η here is $\pm\eta_t$):

$$u(x,0) = \frac{Q_\infty}{\pi}\int_0^c \frac{t}{2}\frac{c-2x_0}{\sqrt{x_0(c-x_0)}}\frac{1}{(x-x_0)}dx_0 \tag{5.24}$$

This integral needs to be evaluated in terms of its principal value. To enable use of Eq. (5.22) the following transformation is introduced:

$$x = \frac{c}{2}(1-\cos\theta) \tag{5.25}$$

and

$$dx = \frac{c}{2}\sin\theta\, d\theta \tag{5.25a}$$

which transforms the straight chord line into a semicircle. The leading edge of the ellipse ($x=0$) is now at $\theta=0$ and the trailing edge ($x=c$) is at $\theta=\pi$. With the aid of this transformation $d\eta_t/dx$ becomes

$$\frac{d\eta_t}{dx} = \frac{t}{2}\frac{c-c(1-\cos\theta)}{\sqrt{(c/2)(1-\cos\theta)[c-(c/2)(1-\cos\theta)]}} = t\frac{\cos\theta}{\sin\theta}$$

Substituting this into the u component of the velocity (Eq. (5.24)),

$$u(x,0) = \frac{tQ_\infty}{\pi}\int_0^\pi \frac{\cos\theta_0}{\cos\theta_0 - \cos\theta}d\theta_0$$

and with the aid of Glauert's integral (Eq. (5.22)) for $n=1$, we can reduce the axial velocity component to

$$u(x,0) = tQ_\infty \tag{5.26}$$

The pressure coefficient thus becomes

$$C_p = -2t \tag{5.27}$$

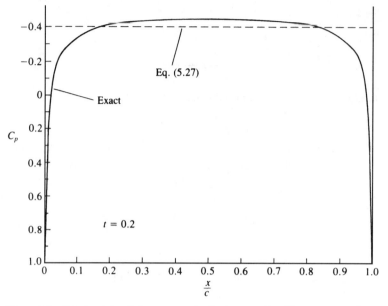

Figure 5.5 Calculated chordwise pressure distribution on a thin ellipse.

which indicates that the pressure coefficient is a constant. This result is plotted in Fig. 5.5 and compared with the exact solution obtained by complex variables (by Van Dyke,[5.3] p. 52). The maximum of $|-C_p|$ is well predicted but the solution near the front and rear stagnation points is incorrect. As the thickness ratio decreases the pressure distribution becomes more flat with a smaller stagnation region and therefore the accuracy of this solution improves.

5.2   Zero-Thickness Airfoil at Angle of Attack

It was demonstrated in Section 4.3 that the small-disturbance flow over thin airfoils can be divided into a thickness problem and a lifting problem due to angle of attack and chord camber. In this section the lifting problem will be addressed using the classical approach (Glauert,[5.2] pp. 87–93). To illustrate the problem, consider a thin cambered airfoil, at an angle of attack α, as shown schematically by Fig. 5.6. The flow is assumed to be inviscid,

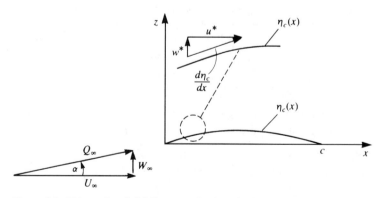

Figure 5.6   Thin cambered airfoil at an angle of attack.

5.2 Zero-Thickness Airfoil at Angle of Attack

incompressible, and irrotational and the continuity equation is

$$\nabla^2 \Phi = 0 \tag{5.28}$$

The airfoil camberline is placed close to the x axis with the leading edge at $x = 0$ and the trailing edge at $x = c$. The camberline of the airfoil is given by a known function $\eta_c = \eta_c(x)$. The boundary condition requiring no flow across the surface, as derived in Chapter 4 for the small-disturbance flow case, will be transferred to the $z = 0$ plane:

$$\frac{\partial \Phi}{\partial z}(x, 0\pm) = Q_\infty \left(\frac{d\eta_c}{dx} \cos \alpha - \sin \alpha \right) \approx Q_\infty \left(\frac{d\eta_c}{dx} - \alpha \right) \tag{5.29}$$

This equation actually states that the sum of the free-stream and the airfoil induced normal velocity components is zero on the surface $w(x, 0\pm) - Q_\infty(d\eta_c/dx - \alpha) = 0$. Also, note that this boundary condition can be obtained by requiring that the flow stay tangent to the camberline (see inset to Fig. 5.6). Thus, the slope of the local (total) velocity w^*/u^* must be equal to the camberline slope:

$$\frac{w^*}{u^*} = \frac{\partial \Phi^*/\partial z}{\partial \Phi^*/\partial x} = \frac{d\eta_c}{dx}$$

Recalling the definition of the total potential Φ^* (Eq. (4.9)), and enforcing the small-disturbance assumption (e.g., $W_\infty \approx Q_\infty \alpha$, $\partial \Phi/\partial x \ll U_\infty$, and $U_\infty = Q_\infty \cos \alpha \approx Q_\infty$) reduces this to the same boundary condition as in Eq. (5.29). We can also use Eq. (4.10) to get

$$\left(\frac{\partial \Phi}{\partial x} + U_\infty, \frac{\partial \Phi}{\partial z} + W_\infty \right) \cdot \frac{(-d\eta_c/dx, 1)}{\sqrt{(d\eta_c/dx)^2 + 1}} = \nabla \Phi^* \cdot \mathbf{n} = 0$$

where the normal vector $\mathbf{n}$ can be described in terms of the camberline η_c:

$$\mathbf{n} = \frac{(-d\eta_c/dx, 1)}{\sqrt{(d\eta_c/dx)^2 + 1}}$$

When considering a solution, based on a singularity element distribution, the antisymmetric nature of the problem (relative to the x axis, as in Fig. 5.6) needs to be observed. In Section 4.5, both doublet and vortex distributions were presented to model this antisymmetric lifting problem. Traditionally, however, the solution based on the vortex distribution is used, probably because of its easy derivation and physical descriptiveness. Also, the boundary condition requiring that the disturbance due to the airfoil will decay far from it (Eq. (4.2)) is not stated since it is automatically fulfilled by either the vortex or doublet elements. Consequently, a model based on the continuous vortex distribution (as shown in Fig. 5.7) is suggested for the solution of this problem. Furthermore, the vortex elements are transferred to the $z = 0$ plane, following the assumptions of small-disturbance flow where $\eta_c \ll c$.

To demonstrate the basic features of the proposed vortex distribution, consider a point vortex in the x–z plane, located at a point $(x_0, 0)$ with a strength of γ_0. Here $\gamma_0 = \gamma(x)\,dx$ at $x = x_0$ in Fig. 5.7. The velocity potential due to this element at a point (x, z) in the field is then

$$\Phi_{\gamma_0} = -\frac{\gamma_0}{2\pi} \theta = -\frac{\gamma_0}{2\pi} \tan^{-1}\left(\frac{z}{x - x_0} \right) \tag{5.30}$$

The velocity due to a vortex points only in the tangential direction; thus

$$q_\theta = -\frac{\gamma_0}{2\pi r}, \qquad q_r = 0 \tag{5.31}$$

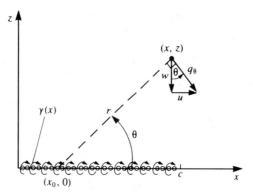

Figure 5.7 Vortex distribution based model for the thin lifting airfoil.

where $r = [(x - x_0)^2 + z^2]^{\frac{1}{2}}$. The minus sign in q_θ is a result of the angle θ being positive counterclockwise in Fig. 5.7. In Cartesian coordinates the components of the velocity will be $(u, w) = q_\theta(\sin\theta, -\cos\theta)$, or by simply differentiating Eq. (5.30),

$$u = \frac{\partial \Phi_{\gamma_0}}{\partial x} = \frac{\gamma_0}{2\pi} \frac{z}{(x - x_0)^2 + z^2} \tag{5.32a}$$

$$w = \frac{\partial \Phi_{\gamma_0}}{\partial z} = -\frac{\gamma_0}{2\pi} \frac{x - x_0}{(x - x_0)^2 + z^2} \tag{5.32b}$$

Note that if the field point is placed on the x axis, then the velocity due to the above element, normal to the x axis, is

$$w = \frac{-\gamma_0}{2\pi(x - x_0)} \qquad (x \neq x_0) \tag{5.33}$$

As shown in Fig. 5.7, this problem is being modeled by a vortex distribution that is placed on the x axis with the small-disturbance boundary conditions being fulfilled also on the x axis. The velocity potential and the resulting velocity field, due to such a vortex distribution (between the airfoil leading edge at $x = 0$ and its trailing edge at $x = c$) are

$$\Phi(x, z) = \frac{-1}{2\pi} \int_0^c \gamma(x_0) \tan^{-1}\left(\frac{z}{x - x_0}\right) dx_0 \tag{5.34}$$

$$u(x, z) = \frac{1}{2\pi} \int_0^c \gamma(x_0) \frac{z}{(x - x_0)^2 + z^2} dx_0 \tag{5.35}$$

$$w(x, z) = \frac{-1}{2\pi} \int_0^c \gamma(x_0) \frac{x - x_0}{(x - x_0)^2 + z^2} dx_0 \tag{5.36}$$

Here, $\gamma(x_0)$ is the vortex strength per unit length at x_0.

Since the boundary condition will be fulfilled at $z = 0$, it is useful to evaluate the velocity components there. The x component of the velocity above (+) and below (−) a vortex distribution was derived in Section 3.14:

$$u(x, 0\pm) = \lim_{z \to \pm 0} u(x, z) = \frac{\pm \gamma(x)}{2} \tag{5.37}$$

for $0 < x < c$, and this result is shown schematically in Fig. 5.8. The w component of the

5.2 Zero-Thickness Airfoil at Angle of Attack

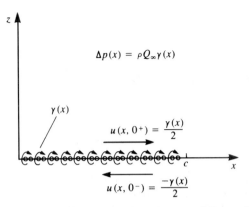

Figure 5.8 Tangential velocity and pressure difference due to a vortex distribution.

velocity at $z = 0$ can be obtained directly from Eq. (5.36) and is

$$w(x, 0) = \frac{-1}{2\pi} \int_0^c \gamma(x_0) \frac{dx_0}{x - x_0} \qquad (5.38)$$

The unknown vortex distribution $\gamma(x)$ has to satisfy the zero normal flow boundary condition on the airfoil. Therefore, substitution of the normal velocity component from Eq. (5.38) into the boundary condition (Eq. (5.29)) results in

$$\frac{\partial \Phi(x, 0)}{\partial z} = w(x, 0) = Q_\infty \left(\frac{d\eta_c}{dx} - \alpha \right)$$

or

$$\frac{-1}{2\pi} \int_0^c \gamma(x_0) \frac{dx_0}{x - x_0} = Q_\infty \left(\frac{d\eta_c}{dx} - \alpha \right), \qquad 0 < x < c \qquad (5.39)$$

This is the integral equation for $\gamma(x)$. However, the solution to this equation is not unique and an additional physical condition has to be added to obtain a unique solution. Such a condition will require that the flow leave the trailing edge smoothly and the velocity there be finite, that is,

$$\nabla \Phi < \infty \quad \text{(at trailing edges)} \qquad (5.40)$$

This is the Kutta condition discussed in Section 4.7. It can be interpreted now as a requirement for the pressure difference Δp [or $\gamma(x)$] to be equal to zero at the trailing edge:

$$\gamma(x = c) = 0 \qquad (5.41)$$

Once the velocity field is obtained, the pressure distribution can be calculated by the steady-state Bernoulli equation for small-disturbance flow over the airfoil (Eq. (5.16)):

$$p - p_\infty = -\rho Q_\infty u(x, 0\pm) = \mp \rho Q_\infty \frac{\gamma}{2} \qquad (5.42)$$

We can now calculate the pressure difference across the airfoil Δp (positive Δp is in the $+z$ direction), where above the airfoil $(\partial \Phi / \partial x)(x, 0+) = +\gamma/2$ and at the airfoil's lower surface $(\partial \Phi / \partial x)(x, 0-) = -\gamma/2$:

$$\Delta p = p_l - p_u = p_\infty - \rho Q_\infty \left(-\frac{\gamma}{2} \right) - \left[p_\infty - \rho Q_\infty \left(\frac{\gamma}{2} \right) \right] = \rho Q_\infty \gamma \qquad (5.43)$$

The pressure coefficient with the small-disturbance assumption then becomes

$$C_p = \frac{p - p_\infty}{\frac{1}{2}\rho Q_\infty^2} = \mp \frac{\gamma}{Q_\infty} \tag{5.44}$$

and the pressure difference coefficient between the lower and upper surfaces is

$$\Delta C_p = 2\frac{\gamma}{Q_\infty} \tag{5.44a}$$

5.3 Classical Solution of the Lifting Problem

The solution for the velocity distribution, pressure difference, and the aerodynamic loads on the thin, lifting airfoil requires the knowledge of the vortex distribution $\gamma(x)$ on the airfoil. This can be obtained by solving the integral equation (Eq. (5.39)), which is a form of the zero normal flow boundary condition. The classical approach (e.g., Glauert,[5.2] p. 88) is to approximate $\gamma(x)$ by a trigonometric expansion and then the problem reduces to finding the coefficient values of this expansion. Therefore, a transformation into trigonometric variables is needed. Such a transformation is described by Fig. 5.9 and is

$$x = \frac{c}{2}(1 - \cos\theta) \tag{5.45}$$

and

$$dx = \frac{c}{2}\sin\theta \, d\theta \tag{5.45a}$$

Note that the airfoil leading edge is at $x = 0$ ($\theta = 0$), and the trailing edge is at $x = c$ ($\theta = \pi$). Substitution of Eq. (5.45) into Eq. (5.39) results in the transformed integral equation

$$\frac{-1}{2\pi}\int_0^\pi \gamma(\theta_0)\frac{\sin\theta_0 \, d\theta_0}{\cos\theta_0 - \cos\theta} = Q_\infty\left[\frac{d\eta_c(\theta)}{dx} - \alpha\right], \quad 0 < \theta < \pi \tag{5.46}$$

This integration with θ_0 should hold for each point x (or θ) on the airfoil. The transformed Kutta condition now has the form

$$\gamma(\pi) = 0 \tag{5.47}$$

The next step is to find a vortex distribution that will satisfy these last two equations. A trigonometric expansion of the form

$$\sum_{n=1}^\infty A_n \sin(n\theta)$$

will satisfy the Kutta condition and is general enough that it can be used to represent the

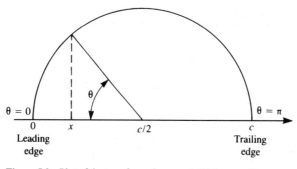

Figure 5.9 Plot of the transformation $x = (c/2)(1 - \cos\theta)$.

5.3 Classical Solution of the Lifting Problem

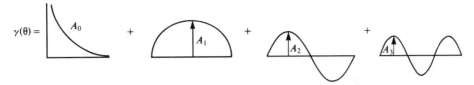

Figure 5.10 Schematic description of the first four terms in the series describing the circulation.

circulation distribution. However, experimental evidence shows a large suction peak at the airfoil's leading edge, which can be modeled by a function whose value is large at the leading edge and reduces to 0 at the trailing edge. Such a trigonometric expression is the cotangent function, which will be included, too, in the proposed vortex distribution:

$$A_0 \cot \frac{\theta}{2} = A_0 \frac{1 + \cos \theta}{\sin \theta}$$

The suggested solution for the circulation is shown graphically in Fig. 5.10. To cancel the $2Q_\infty$ term on the right-hand side of Eq. (5.46), the proposed function for the vortex distribution will be multiplied by this constant:

$$\gamma(\theta) = 2Q_\infty \left[A_0 \frac{1 + \cos \theta}{\sin \theta} + \sum_{n=1}^{\infty} A_n \sin(n\theta) \right] \tag{5.48}$$

An additional advantage of the first term is that it induces a constant downwash on the airfoil, as will be evident later on (see Eq. (5.53)). To determine the values of the A_n constants, Eq. (5.48) is substituted into Eq. (5.46) to give

$$\frac{-1}{2\pi} \int_0^\pi 2Q_\infty \left[A_0 \frac{1 + \cos \theta_0}{\sin \theta_0} + \sum_{n=1}^{\infty} A_n \sin(n\theta_0) \right] \frac{\sin \theta_0 \, d\theta_0}{\cos \theta_0 - \cos \theta}$$

$$= Q_\infty \left[\frac{d\eta_c(\theta)}{dx} - \alpha \right] \tag{5.49}$$

In this equation, each point θ is influenced by all the vortex elements of the airfoil – this requires the evaluation of the integral for each value of θ. Recalling Glauert's integral

$$\int_0^\pi \frac{\cos n\theta_0}{\cos \theta_0 - \cos \theta} d\theta_0 = \frac{\pi \sin n\theta}{\sin \theta}, \quad n = 0, 1, 2, \ldots \tag{5.22}$$

and replacing 1 by $\cos 0\theta$, the first term of the integral becomes

$$\frac{-1}{\pi} A_0 \int_0^\pi \frac{\cos 0\theta_0 + \cos \theta_0}{\sin \theta_0} \frac{\sin \theta_0 \, d\theta_0}{\cos \theta_0 - \cos \theta} = \frac{-1}{\pi} A_0(0 + \pi) = -A_0$$

whereas for the terms with the coefficients $A_1, A_2, \ldots$, the following trigonometric relation is used:

$$\sin n\theta_0 \sin \theta_0 = \frac{1}{2}[\cos(n-1)\theta_0 - \cos(n+1)\theta_0], \quad n = 1, 2, 3, \ldots$$

This allows the presentation of the nth term in the following form:

$$\frac{-1}{\pi} \int_0^\pi [A_n \sin(n\theta_0)] \frac{\sin \theta_0 \, d\theta_0}{\cos \theta_0 - \cos \theta}$$

$$= \frac{-A_n}{2\pi} \int_0^\pi [\cos(n-1)\theta_0 - \cos(n+1)\theta_0] \frac{d\theta_0}{\cos \theta_0 - \cos \theta}$$

Using Glauert's integral reduces this to

$$\frac{-A_n}{2\pi}\pi\left[\frac{\sin(n-1)\theta}{\sin\theta} - \frac{\sin(n+1)\theta}{\sin\theta}\right] = \frac{-A_n}{2}\left[-2\frac{\sin\theta\cos(n\theta)}{\sin\theta}\right] = A_n\cos(n\theta)$$

Substitution of this into Eq. (5.49) yields

$$-A_0 + \sum_{n=1}^{\infty} A_n \cos(n\theta) = \frac{d\eta_c(\theta)}{dx} - \alpha \qquad (5.50)$$

This is actually a Fourier expansion of the right-hand side of the equation that includes the information on the airfoil geometry. Multiplying both sides of the equation by $\cos m\theta$ and performing an integration from $0 \to \pi$, for each value of n, will result in the cancellation of all the nonorthogonal multipliers (where $m \neq n$). Consequently, for each value of n the value of the corresponding coefficient A_n is obtained:

$$A_0 = \alpha - \frac{1}{\pi}\int_0^\pi \frac{d\eta_c(\theta)}{dx} d\theta, \quad n = 0 \qquad (5.51)$$

$$A_n = \frac{2}{\pi}\int_0^\pi \frac{d\eta_c(\theta)}{dx} \cos n\theta \, d\theta, \quad n = 1, 2, 3, \ldots \qquad (5.52)$$

Note that Eq. (5.50) can be rewritten as an expansion of the downwash distribution $w = w(\theta)$ on the airfoil as

$$\frac{w}{Q_\infty} = -A_0 + \sum_{n=1}^{\infty} A_n \cos(n\theta) \qquad (5.53)$$

and it is clear that the downwash due to the first term (multiplied by A_0) of the vortex distribution is constant along the airfoil chord.

The slope $d\eta_c/dx$ can be expanded as a Fourier series such that

$$\frac{d\eta_c(\theta)}{dx} = \sum_{n=0}^{\infty} B_n \cos(n\theta) \qquad (5.50a)$$

and a comparison with Eq. (5.50) indicates that

$$B_0 = \alpha - A_0, \qquad B_n = A_n \quad n = 1, 2, \ldots, \infty$$

This allows the simplification of Eq. (5.53) such that the angle of attack and camber contributions to the downwash are explicitly displayed. A replacement of the A_n coefficients with the B_n coefficients in Eq. (5.53) results in

$$\frac{w}{Q_\infty} = -\alpha + \sum_{n=0}^{\infty} B_n \cos(n\theta) \qquad (5.53a)$$

5.4 Aerodynamic Forces and Moments on a Thin Airfoil

For a given airfoil geometry, the mean camberline $\eta_c(x)$ is a known function and the coefficients $A_0, A_1, A_2, \ldots$ can be computed by Eqs. (5.51) and (5.52). The pressure difference across the thin lifting surface $\Delta p(x)$ can be calculated by Eq. (5.43) and the aerodynamic coefficients can be evaluated. These aerodynamic coefficients are usually defined in the free-stream coordinate system such that the lift is normal and the drag force is parallel to the free-stream flow. To determine the aerodynamic lift and drag, consider the

5.4 Aerodynamic Forces and Moments on a Thin Airfoil

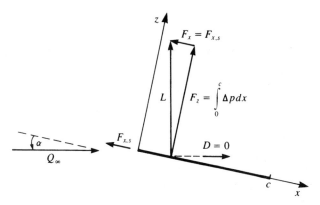

Figure 5.11 Fluid dynamic forces acting on a two-dimensional zero-thickness airfoil.

simple case shown in Fig. 5.11. The pressure difference can be evaluated by using Eq. (5.43)

$$\Delta p(x) = \rho Q_\infty \gamma(x)$$

and since the angle of attack is small Q_∞ is used instead of $Q_\infty \cos\alpha$. The normal force F_z is then

$$F_z = \int_0^c \Delta p(x)\,dx = \int_0^c \rho Q_\infty \gamma(x)\,dx = \rho Q_\infty \Gamma$$

where

$$\Gamma = \int_0^c \gamma(x)\,dx \tag{5.54}$$

Also, the flat plate of Fig. 5.11 is very thin and the x component of the force is zero:

$$F_x = 0$$

Based on this formulation, the lift and drag forces become

$$L = F_z, \qquad D = F_z \alpha$$

However, the Kutta–Joukowski theorem in Section 3.11 clearly states that the lift is perpendicular to the free-stream $\mathbf{Q}_\infty$. Thus, the aerodynamic lift is

$$L = \rho Q_\infty \Gamma \tag{5.55}$$

and the aerodynamic drag is

$$D = 0 \tag{5.56}$$

Therefore, an additional force must exist to balance these two calculations. This force is called the leading-edge suction force $F_{x.s}$ and is a result of the very high suction forces acting at the leading edge (where $q \to \infty$ and the local leading-edge radius is approaching zero). The strength of this leading-edge suction force is calculated in Section 6.5.2 using the exact solution near the leading edge of the flat plate (which is similar to the treatment of this problem by Lighthill[5.4]) and for the small angle of attack case is

$$F_{x.s} = -\rho Q_\infty \Gamma \alpha \tag{5.57}$$

This force cancels the drag component of the thin lifting airfoil obtained by integrating the pressure difference, so that the two-dimensional drag becomes zero. This result – that the aerodynamic drag in two-dimensional inviscid incompressible flow is zero – was obtained in 1744 by the French mathematician d'Alembert and hence is known as d'Alembert's paradox (since actual airfoils will have nonzero viscous drag). Exact solutions and numerical computations of the thick airfoil problem (where the velocity at the leading edge is finite) will verify this result in the following chapters.

To evaluate the lift of the thin airfoil, the circulation of Eq. (5.54) is calculated:

$$\Gamma = \int_0^c \gamma(x)\,dx = \int_0^\pi \gamma(\theta)\frac{c}{2}\sin\theta\,d\theta$$

$$= 2Q_\infty \int_0^\pi \left[A_0 \frac{1+\cos\theta}{\sin\theta} + \sum_{n=1}^\infty A_n \sin(n\theta) \right] \frac{c}{2} \sin\theta\,d\theta$$

If we recall that

$$\int_0^\pi (1+\cos\theta)\,d\theta = \pi$$

and that the integral of $\sin n\theta \sin\theta$ is nonzero only if $n = 1$

$$\int_0^\pi \sin n\theta \sin\theta\,d\theta = \begin{pmatrix} \frac{\pi}{2} & \text{when} & n = 1 \\ 0 & \text{when} & n \neq 1 \end{pmatrix}$$

the circulation becomes

$$\Gamma = Q_\infty c\pi \left(A_0 + \frac{A_1}{2} \right) \tag{5.58}$$

The lift per unit span, obtained from Eq. (5.55), is

$$L = \rho Q_\infty^2 c\pi \left(A_0 + \frac{A_1}{2} \right) \tag{5.59}$$

This equation indicates that only the first two terms of the circulation (shown in Fig. 5.10) will have an effect on the lift and the integration over the airfoil of the higher-order terms will cancel out. Since the pitching moment about the y axis is positive for a clockwise rotation, a minus sign needs to be included when calculating the moment M_0 relative to the airfoil's leading edge:

$$M_0 = -\int_0^c \Delta p x\,dx = -\rho Q_\infty \int_0^\pi \gamma(\theta)\frac{c}{2}(1-\cos\theta)\frac{c}{2}\sin\theta\,d\theta$$

$$= \rho Q_\infty \left[-\frac{c}{2}\Gamma + \frac{c^2}{4}\int_0^\pi \gamma(\theta)\sin\theta\cos\theta\,d\theta \right]$$

After some trigonometric manipulations this results in

$$M_0 = -\frac{c}{2}L + \rho\frac{c^2}{4}Q_\infty^2 \left(A_0\pi + A_2\frac{\pi}{2} \right) \tag{5.60}$$

and substituting the results for the lift, we get

$$M_0 = -\rho Q_\infty^2 \pi \frac{c^2}{4} \left(A_0 + A_1 - \frac{A_2}{2} \right) \tag{5.60a}$$

The moment M along the x axis can be described in terms of the lift and the moment at the leading edge as

$$M = M_0 + x \cdot F_z \approx M_0 + x \cdot L$$

5.4 Aerodynamic Forces and Moments on a Thin Airfoil

The center of pressure x_{cp} is defined as the point where the moment is zero (this can be considered to be the point where the resultant lift force acts):

$$M = M_0 + x_{cp} \cdot L = 0$$

which yields

$$x_{cp} = \frac{-M_0}{L} = \frac{c}{4} \frac{A_0 + A_1 - (A_2/2)}{A_0 + (A_1/2)} \tag{5.61}$$

Similarly the airfoil section aerodynamic coefficients can be derived:

$$C_l = \frac{L}{(1/2)\rho Q_\infty^2 c} = 2\pi\left(A_0 + \frac{A_1}{2}\right) \tag{5.62}$$

$$C_d = \frac{D}{(1/2)\rho Q_\infty^2 c} = 0 \tag{5.63}$$

$$C_{m_0} = \frac{M_0}{(1/2)\rho Q_\infty^2 c^2} = -\frac{\pi}{2}\left[A_0 + A_1 - \frac{A_2}{2}\right] \tag{5.64}$$

An observation of the coefficients of the circulation (Eqs. (5.51) and (5.52)) reveals that only the first term A_0 is a function of angle of attack α. Substitution of A_0 into the lift coefficient equation yields

$$C_l = 2\pi\left(\alpha - \frac{1}{\pi}\int_0^\pi \frac{d\eta_c(\theta)}{dx} d\theta + \frac{A_1}{2}\right) \tag{5.65}$$

Also, for a flat plate $d\eta_c/dx = 0$ and thus all terms except for $2\pi\alpha$, in Eq. (5.65) will vanish. Therefore, the terms including the effect of the camberline η_c are independent of angle of attack and are a constant for a particular chordline shape. This allows us to write the lift coefficient as

$$C_l = 2\pi(\alpha - \alpha_{L0}) \tag{5.66}$$

where α_{L0} is called the zero-lift angle and is a function of the camber. Further substitution of the value of A_1 from Eq. (5.52) yields

$$\alpha_{L0} = -\frac{1}{\pi}\int_0^\pi \frac{d\eta_c}{dx}(\cos\theta - 1)\,d\theta \tag{5.67}$$

By using the B_n coefficients of Eq. (5.50a) the lift coefficient becomes

$$C_l = 2\pi\left(\alpha - B_0 + \frac{B_1}{2}\right) \tag{5.62a}$$

Comparison with Eq. (5.66) indicates that the zero-lift angle can readily be obtained as

$$\alpha_{L0} = B_0 - \frac{B_1}{2} \tag{5.67a}$$

The lift slope can be defined as

$$C_{l_\alpha} \equiv \frac{\partial C_l}{\partial \alpha} = 2\pi \tag{5.68}$$

Equations (5.66)–(5.68) show that the lift slope of a two-dimensional airfoil is 2π and that the camber will have an effect similar to an angle of attack increment $\Delta\alpha$ but will not change the lift slope.

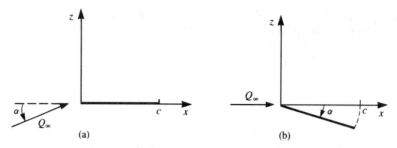

Figure 5.12 Free-stream and body coordinate systems for a flat plate at a small angle of attack.

Next, the pitching moment coefficient (Eq. (5.64)) can be rewritten, using the formula for the lift coefficient (Eq. (5.62)). Thus

$$C_{m_0} = -\frac{C_l}{4} + \frac{\pi}{4}(A_2 - A_1) \tag{5.69}$$

Since the coefficients A_1, A_2 are independent of angle of attack, only the first term in this equation depends on α. Therefore, if the moments are calculated relative to the airfoil quarter-chord point the first term in this equation disappears and the moment at this point becomes independent of angle of attack. This point is called the aerodynamic center x_{ac} and according to thin airfoil theory it is located at the quarter chord. Consequently, the pitching moment measured at this point is only due to the second term in Eq. (5.69):

$$C_{m_{c/4}} = \frac{\pi}{4}(A_2 - A_1) \tag{5.70}$$

The use of this formulation for some simple chordline shapes is demonstrated in the following examples.

Example 1: Flat Plate

As a first example, consider the thin, lifting model of a symmetric airfoil represented by a flat plate (shown in Fig. 5.12a). For this particular case there is no camber and $\eta_c(x) = 0$. Consequently, all terms having derivatives of the camberline will vanish, and the circulation coefficients become

$$A_0 = \alpha, \qquad A_1 = A_2 = \cdots = A_n = 0 \tag{5.71}$$

The circulation Γ for the flat plate airfoil is then

$$\Gamma = Q_\infty \pi c \alpha \tag{5.72}$$

and the lift and moment are obtained by substituting Eq. (5.71) into Eqs. (5.59) and (5.60):

$$L = \rho Q_\infty \Gamma = \rho Q_\infty^2 \pi c \alpha \tag{5.73}$$

$$M_0 = -\rho Q_\infty^2 \pi \frac{c^2}{4} \alpha \tag{5.74}$$

The lift and pitching moment coefficients are

$$C_l = 2\pi\alpha \tag{5.75}$$

$$C_{m_0} = -\frac{\pi}{2}\alpha \tag{5.76}$$

and the lift slope is again 2π as was shown in Eq. (5.66). The center of pressure is at

$$\frac{x_{cp}}{c} = \frac{-C_{m_0}}{C_l} = \frac{1}{4} \tag{5.77}$$

5.4 Aerodynamic Forces and Moments on a Thin Airfoil

Thus, for the symmetric thin airfoil, the center of pressure and the aerodynamic center are located at the quarter-chord location.

Because of the transfer of the boundary condition to the $z = 0$ plane, the airfoil trailing or leading edge can be at a certain small distance from this plane (as long as $\eta(x) \ll c$). As an example, let us solve this problem in the free-stream coordinate system, as shown in Fig. 5.12b. In this case the free-stream angle of attack is zero, but the chord can be expressed as

$$\eta(x) = -\alpha x \implies \frac{d\eta}{dx} = -\alpha$$

Substitution of this into Eqs. (5.51) and (5.52) yields

$$A_0 = \alpha \quad \text{and} \quad A_1 = A_2 = \cdots = A_n = 0$$

which is the same result as in Eq. (5.71). Thus both methods will lead to the same results.

For the symmetrical airfoil, the pressure coefficient difference ΔC_p can be found from Eq. (5.44a) by substituting A_0 and the corresponding circulation:

$$\Delta C_p = 2\frac{\gamma}{Q_\infty} = 4\frac{1+\cos\theta}{\sin\theta}\alpha \tag{5.78}$$

In terms of x (with Eq. (5.45)) this becomes

$$\Delta C_p = 4\sqrt{\frac{c-x}{x}}\alpha \tag{5.79}$$

The result of this formulation is plotted in Fig. 5.13a. In Fig. 5.13b a comparison is made with the results of a more accurate method (e.g., panel method) for a NACA 0012 symmetric airfoil. This indicates that the pressure difference is closely predicted over most of the airfoil. Near the leading edge, however, the flat plate solution is singular and the model is not accurate there.

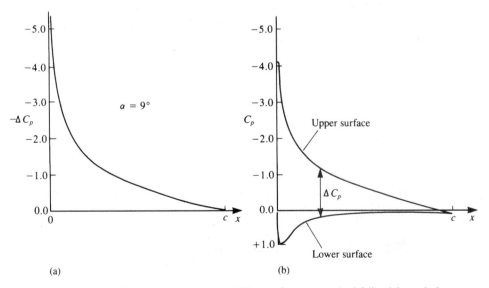

Figure 5.13 Typical chordwise pressure difference for a symmetric airfoil and the equivalent upper and lower surface pressures for a NACA 0012 airfoil.

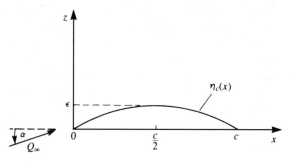

Figure 5.14 Parabolic arc airfoil.

Example 2: Thin Airfoil with a Parabolic Camber

As an example for a simple nonsymmetric chordline shape consider the parabolic camberline shown in Fig. 5.14, with ϵ being its maximum height. The equation of the camberline is then

$$\eta_c(x) = 4\epsilon \frac{x}{c}\left[1 - \frac{x}{c}\right] \tag{5.80}$$

and the camberline slope is

$$\frac{d\eta_c(x)}{dx} = 4\frac{\epsilon}{c}\left[1 - 2\frac{x}{c}\right] \tag{5.81}$$

Expressing this term by using the transformation $x = \frac{c}{2}(1 - \cos\theta)$ we obtain

$$\frac{d\eta_c(\theta)}{dx} = 4\frac{\epsilon}{c}\left[1 - \frac{2}{c}\frac{c}{2}(1 - \cos\theta)\right] = 4\frac{\epsilon}{c}\cos\theta \tag{5.82}$$

The coefficients A_n can be found by substituting this into Eq. (5.51) and (5.52). Because of the orthogonal nature of the integral $\int_0^\pi \cos n\theta \cos m\theta \, d\theta$ all terms where $m \neq n$ will vanish. So in this case, when $m = 1$,

$$A_0 = \alpha - 0$$

and only the first coefficient will be nonzero:

$$A_1 = 4\frac{\epsilon}{c}$$
$$A_2 = A_3 = \cdots = A_n = 0$$

This result can be found immediately by comparing Eq. (5.50a) with the camberline slope

$$\frac{d\eta_c(\theta)}{dx} = \sum_{n=0}^{\infty} B_n \cos(n\theta) = 4\frac{\epsilon}{c}\cos\theta$$

Therefore, clearly $B_1 = 4\frac{\epsilon}{c}$ and the other B_n coefficients are zero.

The lift and the moment of the parabolic camber airfoil can be obtained by substituting these results into Eqs. (5.59) and (5.60). The result is

$$L = \rho Q_\infty^2 \pi c \left(\alpha + 2\frac{\epsilon}{c} \right) \tag{5.83}$$

$$M_0 = -\rho Q_\infty^2 \pi \frac{c^2}{4} \left(\alpha + 4\frac{\epsilon}{c} \right) \tag{5.84}$$

and the corresponding aerodynamic coefficients are thus

$$C_l = 2\pi \left(\alpha + 2\frac{\epsilon}{c} \right) \tag{5.85}$$

$$C_{m_0} = -\frac{\pi}{2} \left(\alpha + 4\frac{\epsilon}{c} \right) \tag{5.86}$$

Comparing this result for the lift with Eq. (5.66) we find that the zero-lift angle is

$$\alpha_{L0} = -2\epsilon/c \tag{5.87}$$

This means that this airfoil will have zero lift when it is pitched to a negative angle of attack with a magnitude of $2\epsilon/c$.

The center of pressure is obtained by dividing the moment by the lift,

$$\frac{x_{cp}}{c} = \frac{1}{4} \frac{\alpha + 4\epsilon/c}{\alpha + 2\epsilon/c} \tag{5.88}$$

Note that at $\alpha = 0$ the center of pressure is at the midchord and as the angle of attack increases it moves toward the quarter chord.

Also, in this case the pitching moment about the aerodynamic center can be calculated using Eq. (5.70):

$$C_{m_{c/4}} = \frac{\pi}{4}(A_2 - A_1) = -\pi \frac{\epsilon}{c} \tag{5.89}$$

which indicates that the portion of the moment that is independent of angle of attack increases with increased curvature (as ϵ/c increases) of the camberline.

Example 3: Flapped Airfoil

One of the most frequently used control devices is the trailing-edge flap. The reason for mounting such a device at the trailing edge can be observed by examining the $(\cos\theta - 1)$ term in Eq. (5.67). This implies that the zero-lift angle is most influenced by the trailing-edge region where $\theta \to \pi$; therefore, relatively small deflections of the flap at the trailing edge will have noticeable effect.

To demonstrate the effect of the trailing-edge flap consider the following simple example. Here the main airfoil plane is placed on the x axis, and at a chordwise position $k \cdot c$ the flap is deflected by δ_f, as shown in Fig. 5.15 (for $\alpha = 0$). Although the trailing edge of the deflected airfoil is now not on the x axis, but because of the small-disturbance approximation of the boundary condition, the error introduced by using this coordinate system is within the accuracy of thin airfoil theory. It is assumed that the airfoil is continuous, and there is no gap at the flap hinge point. The slope of the camberline, for the case shown in the figure, is

$$\frac{d\eta_c}{dx} = 0 \quad \text{for} \quad 0 < x < kc \tag{5.90a}$$

$$\frac{d\eta_c}{dx} = -\delta_f \quad \text{for} \quad kc < x < c \tag{5.90b}$$

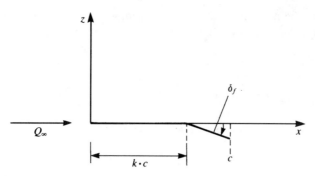

Figure 5.15 Thin flapped airfoil (without a gap at point $k \cdot c$).

Since the coefficients A_n are given as a function of the variable θ, the location of the hinge point θ_k can be found by using Eq. (5.45):

$$kc = \frac{c}{2}(1 - \cos\theta_k) \implies \cos\theta_k = 1 - 2k$$

The coefficients of Eqs. (5.51) and (5.52) are computed now only within the range θ_k to π, resulting in

$$A_0 = \alpha + \frac{1}{\pi}\int_{\theta_k}^{\pi} \delta_f \, d\theta = \alpha + \frac{\delta_f}{\pi}(\pi - \theta_k) \tag{5.91a}$$

$$A_n = -\frac{2}{\pi}\int_{\theta_k}^{\pi} \delta_f \cos n\theta \, d\theta = \frac{2\delta_f}{\pi}\frac{\sin n\theta_k}{n} \tag{5.91b}$$

Substituting the values of the first three A_n coefficients into Eq. (5.62) and Eq. (5.64) we obtain the lift and pitching moment coefficients:

$$C_l = 2\pi\left\{\alpha + \delta_f\left[\left(1 - \frac{\theta_k}{\pi}\right) + \frac{1}{\pi}\sin\theta_k\right]\right\} \tag{5.92}$$

$$C_{m_0} = -\frac{\pi}{2}\left[\alpha + \delta_f\left(1 - \frac{\theta_k}{\pi}\right) + \frac{2\delta_f}{\pi}\sin\theta_k - \frac{\delta_f}{2\pi}\sin 2\theta_k\right] \tag{5.93}$$

Setting $\alpha = 0$ allows the incremental effect of the flap to be obtained:

$$\Delta C_l = [2(\pi - \theta_k) + 2\sin\theta_k]\delta_f \tag{5.94}$$

$$\Delta C_{m_0} = -\frac{1}{2}\left[(\pi - \theta_k) + 2\sin\theta_k - \frac{1}{2}\sin 2\theta_k\right]\delta_f \tag{5.95}$$

The increment in the moment at the aerodynamic center, $c/4$, due to the flap deflection is obtained using Eq. (5.70) as

$$\Delta C_{m_{c/4}} = \left[\frac{1}{4}\sin 2\theta_k - \frac{1}{2}\sin\theta_k\right]\delta_f \tag{5.96}$$

5.5 The Lumped-Vortex Element

Based on the results for the lifting symmetrical airfoil (flat plate), it is possible to develop a simple "lifting element." The vortex distribution on such a flat plate airfoil can be obtained from Eq. (5.48) as

$$\gamma(\theta) = 2Q_\infty \alpha \frac{1 + \cos\theta}{\sin\theta} \tag{5.97}$$

5.5 The Lumped-Vortex Element

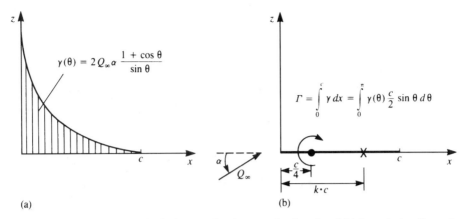

Figure 5.16 (a) Vortex distribution on a flat plate at angle of attack and (b) the equivalent "lumped-vortex" representation (the circulation Γ is the same for both models).

which is shown schematically in Fig. 5.16a. From a far field point of view, this can be replaced by a single vortex with the same strength $\Gamma = \int_0^c \gamma(x)\,dx$.

Since the lift of the symmetric airfoil

$$L = \rho Q_\infty \Gamma$$

acts at the center of pressure (at the quarter chord for the flat plate), the concentrated vortex is placed there.

If the lifting flat plate is to be represented by only one vortex Γ, then the boundary condition requiring zero normal flow at the surface can be specified at only one point too. Assuming that this point is at a distance $k \cdot c$ along the x axis (Fig. 5.16b) then we can specify the boundary condition of zero normal velocity as

$$\frac{-\Gamma}{2\pi[kc - (1/4)c]} + Q_\infty \alpha = 0 \tag{5.98}$$

For this model to simulate the results of the thin airfoil the corresponding value of the circulation for a flat plate (Eq. (5.72)) must be substituted:

$$\Gamma = \pi c Q_\infty \alpha$$

Thus

$$\frac{-\pi c Q_\infty \alpha}{2\pi[kc - (1/4)c]} + Q_\infty \alpha = 0$$

The solution of this equation provides the point at which the boundary condition needs to be specified. This *collocation point* is

$$k = \frac{3}{4} \tag{5.99}$$

Note that this representation is based on results that account for the Kutta condition at the trailing edge. This is the main reason for some of the good approximations that can be obtained when using this model. Some of the advantages of using this lifting element for the estimation of some aerodynamic effects are shown in the following examples. However, for the three interaction examples to follow, the lift force will be calculated with the use of the generalized Kutta–Joukowski theorem to be developed later in Section 6.9 (instead of

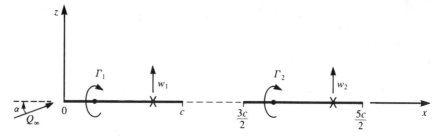

Figure 5.17 Lumped-vortex model for tandem airfoils.

the single-vortex case where $L = \rho Q_\infty \Gamma$). This is necessary since the airfoils studied are not in an unbounded fluid. The lift force on an airfoil is now (see Eq. (6.113))

$$L = \rho Q_\infty \Gamma \left(1 + \frac{\mathbf{Q}_\infty \cdot \mathbf{q}_I}{Q_\infty^2}\right) \tag{5.100}$$

where $\mathbf{q}_I$ is the velocity induced by other vortices at the airfoil vortex location.

Example 1: Tandem Airfoils

The useful application of this simple model can be demonstrated by investigating the lift of the two-airfoil system, shown in Fig. 5.17. The circulations of the two airfoils are represented by Γ_1 and Γ_2, and the two boundary conditions at the two collocation points require that the normal velocity component will be zero: $w_1 = w_2 = 0$. The normal velocity at each collocation point consists of the influence of the two vortices and the free-stream normal component and when specified at these points the two boundary conditions are

$$w_1 = \frac{-\Gamma_1}{2\pi c/2} + \frac{\Gamma_2}{2\pi c} + Q_\infty \alpha = 0 \tag{5.101a}$$

$$w_2 = \frac{-\Gamma_1}{2\pi 2c} + \frac{-\Gamma_2}{2\pi c/2} + Q_\infty \alpha = 0 \tag{5.101b}$$

The solution of this system is

$$\Gamma_1 = \frac{4}{3}\pi c Q_\infty \alpha, \qquad \Gamma_2 = \frac{2}{3}\pi c Q_\infty \alpha \tag{5.102}$$

The force on each airfoil can be obtained from Eq. (5.100). Note that for the small-angle approximation we are using here, the contribution of the streamwise velocity component of the other vortex is proportional to α and can be neglected.

Thus, clearly, the front airfoil has a larger lift owing to the upwash induced by the second airfoil, and because of the same but reversed interaction the second airfoil will have less lift. Also, this effect is stronger when the airfoils are closer and the interaction will disappear as the distance increases. The importance of this result is that the immediate effects of the tandem airfoil configuration could be estimated with minimum effort.

Example 2: Ground Effect

Another simple example is the airfoil near the ground, which is modeled by using the mirror-image method (Fig. 5.18). In order to create a straight streamline at the ground plane two symmetrically positioned airfoils are considered. Again, using the lumped-vortex element, the normal velocity component at the collocation point

5.5 The Lumped-Vortex Element

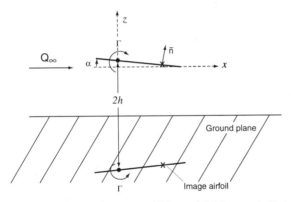

Figure 5.18 Lumped-vortex model for an airfoil in ground effect.

due to the bound vortex is $-\Gamma/2\pi(c/2)$. The influence of the image vortex, which has the same strength but in the opposite direction and is located at a distance $2h$ under the primary vortex, is calculated as follows. The velocity due to a vortex of circulation Γ at (x_0, z_0) at the point (x, z) is given by Eqs. (3.81) and (3.82) as

$$(u, w) = \frac{\Gamma}{2\pi} \frac{(z - z_0, x_0 - x)}{(x - x_0)^2 + (z - z_0)^2} \tag{5.103}$$

For the image vortex, $x_0 = 0$ and $z_0 = -2h$. For the collocation point $x = (c/2)\cos\alpha$ and $z = -(c/2)\sin\alpha$. The normal to the airfoil is

$$\mathbf{n} = \sin\alpha \mathbf{i} + \cos\alpha \mathbf{k} \tag{5.104}$$

The boundary condition at the collocation point is

$$-\frac{\Gamma}{\pi c} + \mathbf{q}_I \cdot \mathbf{n} + Q_\infty \sin\alpha = 0 \tag{5.105}$$

where the image vortex contribution is obtained using Eqs. (5.103) and (5.104). Note that the circulation of the image vortex is $(-\Gamma)$. After some manipulation, the circulation is found as

$$\Gamma = \pi Q_\infty c \sin\alpha \left(\frac{1 - (c/2h)\sin\alpha + c^2/16h^2}{1 - (c/4h)\sin\alpha} \right) \tag{5.106}$$

Note that $\Gamma = \pi Q_\infty c \sin\alpha$ is the exact solution for the flat plate in the absence of the ground plane (Eq. (6.36)). The lift force on the airfoil is given by Eq. (5.100) as

$$L = \rho Q_\infty \Gamma \left(1 - \frac{\Gamma}{4\pi Q_\infty h}\right) \tag{5.107}$$

To obtain the limit for the case when the airfoil is relatively far from the ground, substitute Eq. (5.106) into Eq. (5.107) and let c/h approach zero to get

$$L = \pi \rho Q_\infty^2 c \sin\alpha \left[1 - \frac{c}{2h}\sin\alpha + \frac{c^2}{16h^2}(1 + \sin^2\alpha) + O\left(\frac{c^3}{h^3}\right) \right] \tag{5.108}$$

Corresponding results for a parabolic arc airfoil (see Eq. (5.80)) at zero angle of attack in ground effect can be found in Coulliette and Plotkin.[5.5] The circulation

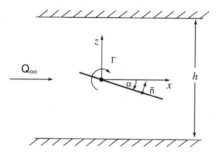

Figure 5.19 Lumped-vortex model for an airfoil between wind-tunnel walls.

for this case is

$$\Gamma = 2\pi Q_\infty \epsilon \left(\frac{1 + c^2/16h^2}{1 + \epsilon/2h} \right) \tag{5.109}$$

where ϵ is the maximum camber as used in Eq. (5.80) and h is measured from midchord. The lift force for large ground height is

$$L = 2\pi \rho Q_\infty^2 \epsilon \left[1 - \frac{\epsilon}{h} + \frac{c^2}{16h^2} + \frac{3}{2}\frac{\epsilon^2}{h^2} + O\left(\frac{1}{h^3}\right) \right] \tag{5.110}$$

Comparing the second terms in these two equations based on the single-vortex model (e.g., $\sim c\alpha/2h$ in Eq. (5.108) and $2\epsilon/2h$ in Eq. (5.110)) indicates a reduction in the free-stream speed (for the lifting case) due to the induced velocity $\sim \Gamma/4\pi 2h$ and the circulation (a trend that is reversed for inverted airfoils). The third term $(c^2/16h^2)$ increases the effect for either positive or negative lift airfoils, and it becomes more pronounced for smaller values of h, as shown by Katz and Plotkin,[5.6] (p. 137).

Example 3: Wind Tunnel Walls

To model the effect of wind-tunnel walls on the lift of a symmetrical airfoil, we place a flat plate of chord c at angle of attack α between two parallel walls a distance h apart. The quarter chord is at the center of the tunnel (see Fig. 5.19). We will seek a small-disturbance solution in the limit as c/h approaches zero. For small α, the collocation point is at $(c/2, 0)$ and the boundary condition there is

$$w\left(\frac{c}{2}, 0\right) + Q_\infty \alpha = 0 \tag{5.111}$$

where w is the z component of velocity for the airfoil vortex (at the origin) plus its image system. With the use of the complex potential for this configuration in Eq. (6.89) we get

$$w\left(\frac{c}{2}, 0\right) = \frac{\Gamma}{4h}\left(\tanh\frac{\pi c}{4h} - \coth\frac{\pi c}{4h} \right) \tag{5.112}$$

Now, for the hyperbolic functions, as A approaches zero, we have (see Gradshteyn and Ryzhik,[5.7] p. 35)

$$\tanh(A) = A - \frac{A^3}{3} + \cdots \tag{5.113a}$$

$$\coth(A) = \frac{1}{A} - \frac{A}{3} + \cdots \tag{5.113b}$$

5.5 The Lumped-Vortex Element

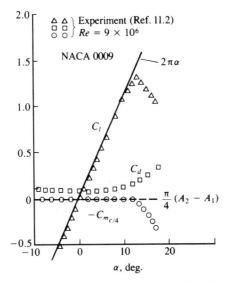

Figure 5.20 Lift and pitching moment of a NACA 0009 airfoil.

Substituting Eq. (5.112) in the boundary condition (Eq. (5.111)) and with the use of Eq. (5.113) we get

$$\frac{\Gamma}{4h}\left(\frac{2}{3}\frac{\pi c}{4h} - \frac{4h}{\pi c}\right) + Q_\infty \alpha = 0 \tag{5.114}$$

The circulation is then

$$\Gamma = \frac{\pi c Q_\infty \alpha}{1 - \pi^2 c^2/24h^2} \approx \pi c Q_\infty \alpha \left(1 + \frac{\pi^2}{24}\frac{c^2}{h^2}\right) \tag{5.115}$$

which is greater than the unbounded fluid result. For this small-disturbance approximation the Kutta–Joukowski theorem ($L = \rho Q_\infty \Gamma$) can be used for the lift, which is therefore

$$L = \pi \rho c Q_\infty^2 \alpha \left(1 + \frac{\pi^2}{24}\frac{c^2}{h^2}\right) \tag{5.116}$$

clearly showing the increase in lift as the wind-tunnel walls approach the airfoil.

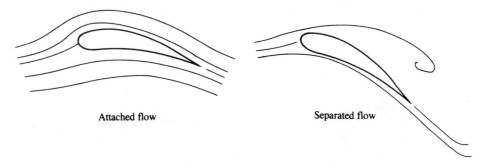

Figure 5.21 Streamlines of the (a) attached and (b) separated flow over an airfoil.

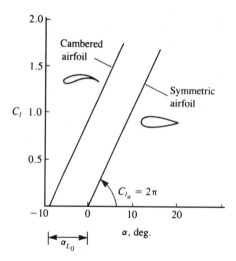

Figure 5.22 Schematic description of airfoil camber effect on the lift coefficient.

5.6 Summary and Conclusions from Thin Airfoil Theory

Up to this point, in order to be able to solve practical problems, the fluid dynamic equations were considerably simplified and even the boundary conditions were approximated. However, in spite of these simplifications, some very important results were obtained in this chapter:

1. The lift slope of a two-dimensional airfoil is 2π, as shown by Eq. (5.66).
2. The pitching moment at the aerodynamic center (at $c/4$) is independent of angle of attack (excluding airfoil's stalled conditions).

    These two very important results are very close to experimental data in the low angle of attack range, as shown in Fig. 5.20. When the angle of attack increases beyond the limits of the small angle of attack assumption, the streamlines do not follow the airfoil surface shape (Fig. 5.21) and the flow is considered to be *separated*. This results in loss of lift, as indicated by the experimental data in Fig. 5.20 (for $\alpha > 10°$) and this condition is called *airfoil stall*.
3. Airfoil camber does not change the lift slope and can be viewed as an additional angle of attack effect (α_{L0} in Eq. (5.66)). This is shown schematically by Fig. 5.22. The symmetric airfoil will have zero lift at $\alpha = 0$ while the airfoil with camber has an "effective" angle of attack that is larger by α_{L0}.
4. The trailing-edge section has a larger influence on the above camber effect. Therefore, if the lift of the airfoil needs to be changed without changing its angle of attack, then changing the chordline geometry (e.g., by flaps or slats) at the trailing-edge region is more effective than at the leading-edge region.
5. The effect of thickness on the airfoil lift is not treated in a satisfactory manner by the small-disturbance approach, but this will be calculated more accurately in the following two chapters.
6. The two-dimensional drag coefficient obtained by this model is zero and there is no drag associated with the generation of two-dimensional lift. Experimental airfoil data, however, include drag due to the viscous boundary layer on the airfoil, and this should be included in engineering calculations. The experimental drag coefficient values for the NACA 0009 airfoil are also plotted in Fig. 5.20 and for example the "zero-lift" drag coefficient is close to $C_d = 0.0055$.

References

[5.1] Moran, J., *An Introduction to Theoretical and Computational Aerodynamics*, John Wiley & Sons, 1984.

[5.2] Glauert, H., *The Elements of Aerofoil and Airscrew Theory*, 2nd edition, Cambridge University Press, 1959.

[5.3] Van Dyke, M., *Perturbation Methods in Fluid Mechanics*, The Parabolic Press, Stanford, CA, 1975.

[5.4] Lighthill, M. J., "A New Approach to Thin Airfoil Theory," *The Aeronautical Quarterly*, Vol. 3, Nov. 1951, pp. 193–210.

[5.5] Coulliette, C., and Plotkin, A., "Aerofoil Ground Effect Revisited," *The Aeronautical Journal*, Feb. 1996, pp. 65–74.

[5.6] Katz, J., and Plotkin, A., *Low Speed Aerodynamics – From Wing Theory to Panel Methods*, McGraw-Hill, 1991.

[5.7] Gradshteyn, I. S., and Ryzhik, I. M., *Table of Integrals, Series, and Products*, Academic Press, 1980.

Problems

5.1. Find the camberline shape that leads to a constant-pressure jump along the airfoil chordline for zero angle of attack.

5.2. Consider the flow of a uniform stream of speed Q_∞ at angle of attack α past a thin airfoil whose camberline is given by

$$\eta_c = h\left(1 - \frac{x}{c}\right)\left(1 - \frac{\lambda x}{c}\right)x$$

where $h \ll 1$ and λ is a constant. Show that

$$C_l = 2\pi(\alpha + \epsilon), \qquad C_{m_0} = 2\left(\mu - \frac{\pi\epsilon}{4}\right) - \frac{C_l}{4}$$

where $\epsilon = (h/8)(4 - 3\lambda)$ and $\mu = (\pi/64)h\lambda$.

Find the value of λ for the zero-lift angle to be zero.

5.3. Find the hinge moment for the flapped airfoil of Eq. (5.90).

5.4. Consider the flow of a uniform stream of speed Q_∞ at angle of attack α past a thin airfoil whose upper surface is given by the parabola in Eq. (5.80) and whose lower surface is $z = 0$. Find the lift coefficient and moment coefficient about the leading edge.

5.5. Consider the flow of a uniform stream of speed Q_∞ at angle of attack α past a biplane consisting of two flat plate airfoils of chord c located a distance h apart (no stagger). Find the lift coefficient for each airfoil using a single vortex to represent each one.

CHAPTER 6

Exact Solutions with Complex Variables

Approximate solutions to the exact potential flow problem are obtained in this book using both classical small-disturbance methods and numerical modeling. However, it is important to have exact solutions available to test the accuracy of the approximations and to assess their applicability. In this chapter complex variables will be used to obtain the solution to three model problems: the flat plate, the circular arc, and a symmetrical airfoil.

6.1 Summary of Complex Variable Theory

Prior to applying complex variable methods to potential flow problems, some of the principles are discussed briefly (for more details about the mathematics of complex variables see Churchill[6.1]). To begin, first define the imaginary unit i by

$$i^2 = -1 \tag{6.1}$$

Then any complex number Y can be written as

$$Y = a + ib \tag{6.2}$$

where a and b are real and are called the real and imaginary parts of Y, respectively. Every complex number therefore can be thought of as representing an ordered pair of real numbers (a, b) and as such may be represented geometrically by points in a plane. The complex number $Y = x + iz$ is shown in Fig. 6.1 in a Cartesian coordinate system with x and z axes. A polar coordinate version of Y with coordinates r and θ is also shown in the figure. Note that the absolute value of Y ($|Y|$) is defined as $[x^2 + z^2]^{\frac{1}{2}}$ and the argument of Y (arg $Y = \theta$) is defined as $\tan^{-1} z/x$. An exponential form of Y is expressed as

$$Y = re^{i\theta} \tag{6.3}$$

if the exponential term is defined as

$$e^{i\theta} \equiv \cos\theta + i\sin\theta \tag{6.4}$$

The complex conjugate of the complex number Y is defined as

$$\bar{Y} = x - iz$$

Otherwise the algebra of complex numbers is similar to the algebra of the term $(a + b)$, but note that $i^2 = -1$. As an example, the multiplication of a complex number by its conjugate is

$$Y\bar{Y} = (x + iz)(x - iz) = x^2 + z^2$$

A function f of the complex variable Y can be written in terms of its real and imaginary parts as

$$f(Y) = g(x, z) + ih(x, z) \tag{6.5}$$

Analytic functions of a complex variable are differentiable, which means that

$$\frac{df(Y)}{dY} = \lim_{\Delta Y \to 0} \frac{f(Y + \Delta Y) - f(Y)}{\Delta Y}$$

6.1 Summary of Complex Variable Theory

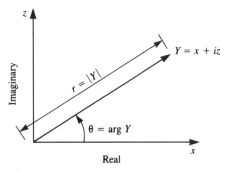

Figure 6.1 Complex plane.

exists for all possible paths ΔY. Now consider the derivative of $f(Y)$ along the x axis

$$\frac{df(Y)}{dY} = \lim_{\Delta x \to 0} \frac{\Delta g + i \Delta h}{\Delta x} = \frac{\partial g}{\partial x} + i \frac{\partial h}{\partial x}$$

Similarly, the derivative in the z direction is

$$\frac{df(Y)}{dY} = \lim_{i\Delta z \to 0} \frac{\Delta g + i \Delta h}{i \Delta z} = \frac{1}{i}\frac{\partial g}{\partial z} + \frac{\partial h}{\partial z} = \frac{\partial h}{\partial z} - i \frac{\partial g}{\partial z}$$

The derivatives must be independent of the direction of differentiation; therefore, equating the real and imaginary parts of these derivatives results in

$$\frac{\partial g}{\partial x} = \frac{\partial h}{\partial z}, \qquad \frac{\partial g}{\partial z} = -\frac{\partial h}{\partial x} \tag{6.6}$$

So, differentiability is guaranteed if the real and imaginary parts of f satisfy the above equations, which are called the *Cauchy–Riemann conditions*. Also, if a function of a complex variable is analytic, then the real and imaginary parts each satisfy Laplace's equation. Points in a region where $f(Y)$ is analytic are called *regular* points and points where $f(Y)$ is not analytic are called *singular* points.

Consider the integration of a complex function. If the function is analytic and the region is simply connected, then the integral

$$\int_A^B f(Y)\,dY$$

from point A to point B is independent of the path of integration and the integral around all closed paths is zero. The latter result is called the *Cauchy integral theorem*. Multiply connected regions are of interest since they include the region exterior to a two-dimensional airfoil as well as the region remaining once singular points are excluded by surrounding them with closed curves. Consider the region in Fig. 6.2 that is exterior to n curves $C_1, C_2, \ldots, C_n$ and consider a curve C that surrounds the n curves. An application of the Cauchy integral theorem in this region for a function f that is analytic inside C and outside the n curves yields the result that the integral around C is equal to the sum of the integrals around the n curves where all integrations are in the same direction:

$$\oint_C f(Y)\,dY = \oint_{C_1} f(Y)\,dY + \oint_{C_2} f(Y)\,dY + \cdots + \oint_{C_n} f(Y)\,dY \tag{6.7}$$

Consider the following results for power series expansions of the function $f(Y)$. If f is analytic at all points within a circle C_0 with center at Y_0, then at each point Y inside the

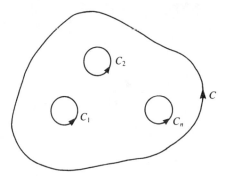

Figure 6.2 Integration in a multiply connected region.

circle f can be represented by the Taylor series expansion

$$f(Y) = f(Y_0) + f'(Y_0)(Y - Y_0) + \cdots + \frac{f^{(n)}(Y_0)}{n!}(Y - Y_0)^n + \cdots \tag{6.8}$$

Now consider the region exterior to the circle C_1 whose center is at Y_0 in Fig. 6.3. The function f is analytic in the annular region between C_1 and C_2. Then f can be represented by the Laurent series expansion

$$f(Y) = \sum_{-\infty}^{\infty} A_n (Y - Y_0)^n \tag{6.9}$$

Consider now the integration of a function with singularities. Let $f(Y)$ be analytic inside the curve C except at Y_0. Surround Y_0 by the circle C_0 (see Fig. 6.4) and represent f between C_0 and C by the Laurent series of Eq. (6.9). Then the integral around C becomes

$$\oint_C f(Y)\,dY = \sum_{-\infty}^{\infty} A_n \oint_{C_0} (Y - Y_0)^n \, dY = 2\pi i A_{-1} \tag{6.10}$$

where A_{-1} is the coefficient of the term $A_{-1}/(Y - Y_0)$ and is called the *residue* of $f(Y)$ at Y_0. If $f(Y)$ is analytic inside C except at a finite number of singularities (N), then a generalization of Eq. (6.10) leads to the residue theorem:

$$\oint_C f(Y)\,dY = 2\pi i \sum_{j=1}^{N} A_{-1}(Y_j) \tag{6.11}$$

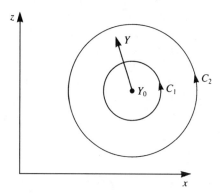

Figure 6.3 Region for Laurent series expansion.

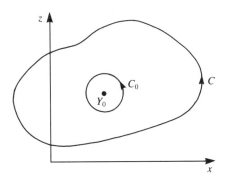

Figure 6.4 Integration of a function with singularities.

Complex variable theory is a powerful tool for the solution of two-dimensional incompressible potential flow problems through its mapping properties. Consider the function $f(Y)$ that generates the pair of values (g, h) for each pair of values (x, z). Each value of Y represents a point in the Y plane and each value of f can be thought of as representing a corresponding point in the f plane. The function $f(Y)$ therefore geometrically maps or transforms points (and also curves and regions) from the Y plane to the f plane (see Fig. 6.5).

When the mapping function $f(Y)$ is analytic, the mapping from the Y plane to the f plane is called *conformal* and has the following special property. Consider a curve C through the point Y_0 in the Y plane and the corresponding curve D through the corresponding point f_0 in the f plane (Fig. 6.6). If f is analytic at Y_0 and if $f'(Y_0) \neq 0$ then every curve through Y_0 in the Y plane is rotated by the amount $\arg f'(Y_0)$ when it is transformed into the f plane. This is illustrated in Fig. 6.6, which shows the two curves C_1 and C_2 that intersect at Y_0 in the Y plane and the corresponding curves D_1 and D_2 that intersect at f_0 in the f plane. For this conformal mapping, it is observed that the angle of intersection between the curves is preserved in the transformation. A point at which $f'(Y) = 0$ is called a *critical point* of the mapping and at a critical point the above intersection angle is not preserved.

6.2 The Complex Potential

Consider a steady, incompressible, inviscid, irrotational two-dimensional flow. The velocity potential and the stream function are related by the following equations (Eq. (2.81)):

$$\frac{\partial \Phi}{\partial x} = \frac{\partial \Psi}{\partial z}, \qquad \frac{\partial \Phi}{\partial z} = -\frac{\partial \Psi}{\partial x} \tag{6.12}$$

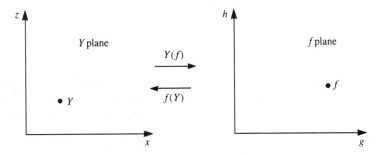

Figure 6.5 Mapping with a function of a complex variable.

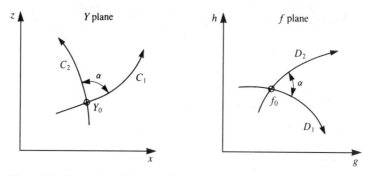

Figure 6.6 Preservation of the angle between intersecting curves for a conformal transformation.

and both satisfy Laplace's equation. Note that Eq. (6.12) yields the Cauchy Riemann conditions for Φ and Ψ to be the real and imaginary parts of an analytic function F of a complex variable. We define the complex potential as

$$F = \Phi + i\Psi \tag{6.13}$$

and note that its derivative

$$W(Y) = F' = \frac{dF}{dY} = u - iw \tag{6.14}$$

is the complex conjugate of the velocity and is called the complex velocity. Any analytic function of a complex variable can represent the complex potential of some flow.

6.3 Simple Examples

To evaluate the complex potential of two-dimensional flowfields, we shall apply Eq. (6.13) to the results of some basic flows that were treated in Chapter 3.

6.3.1 Uniform Stream and Singular Solutions

The complex potential for the flow of a uniform stream of speed Q_∞ in the x direction is obtained by substituting the results for the velocity potential and stream function into Eq. (6.13) to get

$$F = \Phi + i\Psi = Q_\infty(x + iz) = Q_\infty Y \tag{6.15}$$

Now, consider the stream to be at an angle α to the x axis and repeat the process. The complex potential becomes

$$F = Q_\infty(x\cos\alpha + z\sin\alpha) + iQ_\infty(-x\sin\alpha + z\cos\alpha)$$
$$= Q_\infty(\cos\alpha - i\sin\alpha)(x + iz) = Q_\infty Y e^{-i\alpha} \tag{6.15a}$$

This illustrates the general result that the complex potential for one flowfield can be made to represent the same flowfield rotated counterclockwise by α if Y is replaced by $Ye^{-i\alpha}$.

Consider a source of strength σ at the origin. Its complex potential can be obtained similarly, and using polar coordinates we get

$$F = \frac{\sigma}{2\pi}(\ln r + i\theta) = \frac{\sigma}{2\pi}\ln Y \tag{6.16}$$

6.3 Simple Examples

Note that it is easy to demonstrate that for a source at $Y = Y_0 = x_0 + iz_0$, the complex potential is

$$F = \frac{\sigma}{2\pi} \ln(Y - Y_0) \tag{6.16a}$$

and in general a flowfield can be translated by Y_0 by replacing Y by $Y - Y_0$ in the complex potential. The complex potential for a vortex with clockwise circulation Γ at $Y = Y_0$ is

$$F = \frac{i\Gamma}{2\pi} \ln(Y - Y_0) \tag{6.17}$$

The complex potential of a doublet at the origin whose axis is in the x direction is

$$F = -\frac{\mu}{2\pi} \frac{1}{Y} \tag{6.18}$$

Using the above rules, we find the complex potential for a doublet at $Y = Y_0$ with an axis at an angle α to the x direction is given by

$$F = -\frac{\mu}{2\pi(Y - Y_0)} e^{i\alpha} \tag{6.18a}$$

6.3.2 Flow in a Corner

A second approach (inverse) is where the flowfield shape is sought for a given complex potential F. For example, consider the complex potential

$$F = B Y^{\pi/\alpha}$$

where B is real. The stream function in polar coordinates is

$$\Psi = B r^{\pi/\alpha} \sin\left(\frac{\pi \theta}{\alpha}\right)$$

It can be seen that $\Psi = 0$ at $\theta = 0$ and $\theta = \alpha$, and therefore this potential represents flow in a corner as shown in Fig. 6.7. The complex velocity is

$$W = B \frac{\pi}{\alpha} Y^{\pi - \alpha/\alpha}$$

and at the corner $Y = 0$, the velocity is zero if $\alpha < \pi$ and infinite if $\alpha > \pi$. If $\alpha = \pi/2$, the flow can be considered to be either the flow in a right-angle corner or flow against a horizontal wall. This flow, called *stagnation point flow*, is shown in Fig. 3.6.

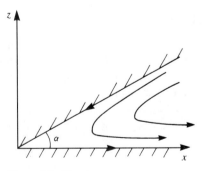

Figure 6.7 Flow in a corner.

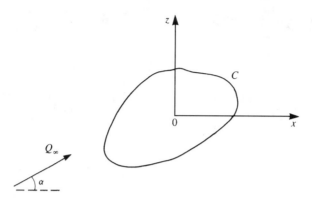

Figure 6.8 Coordinate system for use with Blasius formula.

6.4 Blasius Formula, Kutta–Joukowski Theorem

Consider the flow past a body whose contour is denoted by C (Fig. 6.8). Let the components of the aerodynamic force acting on the body be X and Z in the x and z directions, respectively. An integration of the pressure around the contour and an application of Bernoulli's equation then leads to the Blasius formula (see proof in Glauert,[5.2] pp. 80–81):

$$X - iZ = \frac{i\rho}{2} \int_C [W(Y)]^2 \, dY \tag{6.19}$$

Let the free-stream velocity be $Q_\infty e^{-i\alpha}$ and let the circulation around C be Γ (see Fig. 6.8). Then because the complex velocity is analytic outside of C (since the fluid is unbounded), we can write W in a Laurent series about $Y = 0$ (which is taken inside C):

$$W(Y) = Q_\infty e^{-i\alpha} + \frac{i\Gamma}{2\pi Y} + \frac{A_2}{Y^2} + \frac{A_3}{Y^3} + \frac{A_4}{Y^4} + \cdots \tag{6.20}$$

Now substituting into the Blasius formula and using the residue theorem we get

$$X - iZ = -i\rho Q_\infty \Gamma e^{-i\alpha} = \rho Q_\infty \Gamma e^{-i(\pi/2+\alpha)} \tag{6.21}$$

or

$$X + iZ = \rho Q_\infty \Gamma e^{i(\pi/2+\alpha)} \tag{6.21a}$$

The force is seen to act perpendicular to the stream $\mathbf{Q}_\infty$ and has the magnitude $D = 0$ and $L = \rho Q_\infty \Gamma$. This result is called the Kutta–Joukowski theorem.

6.5 Conformal Mapping and the Joukowski Transformation

The method of solution for our model airfoil problem is to map the airfoil (which is in the physical plane $Y = x + iz$) to a circular cylinder in the $f = g + ih$ plane through the conformal mapping $Y = Y(f)$. The solution in the circle plane has already been obtained (in Section 3.11). Let the complex potential in the circle plane be $F(f)$ and the complex velocity $W(f)$. Then the results in the physical plane are

$$F(Y) = F[f(Y)] \tag{6.22}$$

$$W(Y) = \frac{dF}{dY} = \frac{dF}{df}\frac{df}{dY} = W(f)\frac{1}{\frac{dY}{df}} \tag{6.23}$$

6.5 Conformal Mapping and the Joukowski Transformation

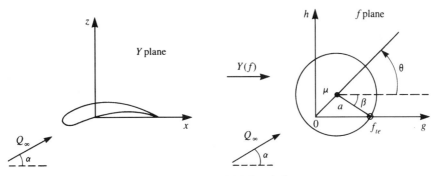

Figure 6.9 Joukowski transformation: mapping of airfoil to circle.

The complex velocity in the physical plane is given as a function of the transformation variable f. The following three model problems are all special cases of the Joukowski transformation

$$Y = f + \frac{C^2}{16f} \tag{6.24}$$

where C will be shown to be the chord for a flat plate, for a circular arc, and approximately for a symmetrical airfoil.

Consider the mapping from the airfoil to the circle shown in Fig. 6.9. The complex velocity at infinity in both planes is $Q_\infty e^{-i\alpha}$ and the transformation has two free parameters, the radius of the circle a and the center of the circle μ. The complex velocity in the circle plane is obtained with the aid of the results of the flow over a cylinder from Section 3.11:

$$W(f) = Q_\infty e^{-i\alpha} + \frac{i\Gamma}{2\pi} \frac{1}{f-\mu} - \frac{Q_\infty a^2 e^{i\alpha}}{(f-\mu)^2} \tag{6.25}$$

Since the airfoil has a sharp trailing edge and the circle has no corners, the transformation must have a critical point ($dY/df = 0$) at the point in the circle plane corresponding to the airfoil trailing edge. Denote this point by f_{te}. The Kutta condition requires the velocity at the airfoil trailing edge to be finite and therefore from Eq. (6.23) it can only be satisfied if

$$W(f_{te}) = 0 \tag{6.26}$$

In the circle plane $f_{te} = C/4$ and the coordinate system is shown in Fig. 6.9. (Note that $f = -C/4$ is also a critical point and must be placed inside the circle to avoid a velocity singularity in the flowfield. The critical points $f = \pm C/4$ transform to $Y = \pm C/2$.) From the figure, we see that

$$f_{te} - \mu = ae^{-i\beta} \tag{6.27}$$

If this is substituted into Eq. (6.25) for $W(f)$ and the Kutta condition is applied, we get

$$Q_\infty e^{-i\alpha} + \frac{i\Gamma}{2\pi a} e^{i\beta} - Q_\infty e^{i\alpha} e^{2i\beta} = 0$$

$$-2\pi a Q_\infty i e^{-i(\alpha+\beta)} + \Gamma + 2\pi a Q_\infty i e^{i(\alpha+\beta)} = 0$$

and the circulation is

$$\Gamma = 4\pi a Q_\infty \sin(\alpha + \beta) \tag{6.28}$$

The lift and lift coefficient are then given by

$$L = \rho Q_\infty \Gamma \tag{6.29}$$

$$C_l = \frac{L}{(1/2)\rho Q_\infty^2 c} = 8\pi \frac{a}{c} \sin(\alpha + \beta) \tag{6.29a}$$

Let the surface of the circle be given by

$$f = \mu + ae^{i\theta} \tag{6.30}$$

as shown in Fig. 6.9. The complex velocity on the circle is then obtained by substituting Eqs. (6.28) and (6.30) in Eq. (6.25), which gives

$$\begin{aligned}W(f) &= Q_\infty e^{-i\alpha} + 2i Q_\infty \sin(\alpha + \beta) e^{-i\theta} - Q_\infty e^{-i\alpha} e^{-2i\theta} \\ &= Q_\infty e^{-i\theta}\left[e^{-i(\alpha-\theta)} + 2i \sin(\alpha + \beta) - e^{i(\alpha-\theta)}\right] \\ &= 2i Q_\infty [\sin(\alpha + \beta) - \sin(\alpha - \theta)] e^{-i\theta}\end{aligned} \tag{6.31}$$

and the complex velocity on the airfoil surface is obtained from Eq. (6.23) as

$$W(Y) = \frac{W(f)}{1 - C^2/16 f^2} \tag{6.31a}$$

To find the complex velocity at the airfoil trailing edge, L'Hospital's rule must be applied since both $W(f)$ and dY/df are zero there. At the trailing edge $f = C/4$ and $\theta = 2\pi - \beta$, and the complex velocity is therefore

$$\begin{aligned}W\left(Y = \frac{C}{2}\right) &= \lim_{f \to f_{te}} \frac{dW/df}{d^2 Y/df^2} \\ &= \frac{[-2ia Q_\infty \sin(\alpha + \beta)/(f - \mu)^2] + [2a^2 Q_\infty e^{i\alpha}/(f - \mu)^3]}{2C^2/16 f^3}\end{aligned}$$

Using $f_{te} - \mu = ae^{i\theta}$, we get

$$W\left(Y = \frac{C}{2}\right) = \frac{Q_\infty C}{4a} e^{-2i\theta}\left[-i\sin(\alpha + \beta) + e^{i(\alpha-\theta)}\right] = Q_\infty \frac{C}{4a} e^{2i\beta} \cos(\alpha + \beta) \tag{6.32}$$

6.5.1 Flat Plate Airfoil

Choose the circle with its center at the origin and $a = C/4$. Then from Eq. (6.27),

$$\mu = \beta = 0 \tag{6.33}$$

The circle is given by $f = (C/4)e^{i\theta}$ and the corresponding airfoil is $Y = (C/2)\cos\theta$, which is seen to be a flat plate of chord $c = C$ (see Fig. 6.10). Note that $0 \le \theta \le \pi$ represents the top surface and $\pi \le \theta \le 2\pi$ represents the bottom. The complex velocity on the plate surface is obtained using Eqs. (6.31) and (6.31a) as

$$\begin{aligned}W &= \frac{2i Q_\infty[\sin\alpha - \sin(\alpha - \theta)]e^{-i\theta}}{1 - e^{-2i\theta}} = \frac{2i Q_\infty[\sin\alpha - \sin(\alpha - \theta)]e^{-i\theta}}{2i \sin\theta \, e^{-i\theta}} \\ &= Q_\infty \frac{[\sin\alpha - \sin(\alpha - \theta)]}{\sin\theta}\end{aligned} \tag{6.34}$$

and since $x = (c/2)\cos\theta$ then $\sin\theta = \pm[1 - (2x/c)^2]^{1/2}$, and we have

$$\frac{W}{Q_\infty} = \cos\alpha + \sin\alpha \frac{1 - \cos\theta}{\sin\theta} = \cos\alpha \pm \sin\alpha \sqrt{\frac{1 - 2x/c}{1 + 2x/c}} \tag{6.35}$$

6.5 Conformal Mapping and the Joukowski Transformation

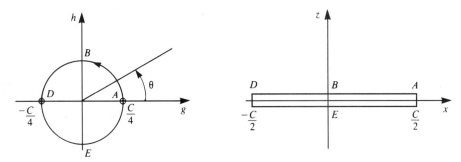

Figure 6.10 Flat plate airfoil mapping.

where the plus sign refers to the upper surface and the minus sign to the lower. Note that the trailing-edge velocity is $Q_\infty \cos\alpha$ and that the disturbance to the stream vanishes as the square root of distance from the trailing edge. Also, the velocity has a square root singularity at the plate's sharp leading edge.

For small α, Eq. (6.35) becomes

$$\frac{W}{Q_\infty} = 1 + \alpha \frac{1 - \cos\theta}{\sin\theta} \tag{6.35a}$$

Note that with the use of Eqs. (5.37), (5.48), and (5.71) (and considering the different definition of θ in Chapter 5), the solution is identical to the flat plate solution from thin airfoil theory.

The streamline patterns in the circle and plate planes are shown schematically in Fig. 6.11. Note that the forward stagnation point in the circle plane is at $\theta = \pi + 2\alpha$ and therefore the forward stagnation point on the plate is at $x = -(c/2)\sin 2\alpha$.

The circulation and lift force are given by Eqs. (6.28) and (6.29) as

$$\Gamma = \pi c Q_\infty \sin\alpha \tag{6.36}$$

$$L = \pi \rho c Q_\infty^2 \sin\alpha \tag{6.37}$$

and the lift coefficient is (Eq. (6.29a))

$$C_l = 2\pi \sin\alpha \tag{6.37a}$$

6.5.2 Leading-Edge Suction

In the previous section the force on a flat plate airfoil is obtained with the use of the Kutta–Joukowski theorem and is seen to be perpendicular to the free-stream direction.

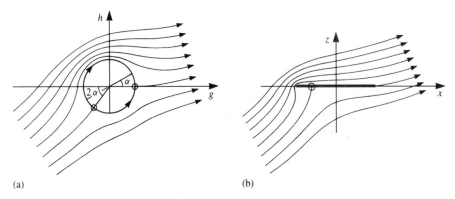

Figure 6.11 Schematic description of the streamlines in (a) circle and (b) flat plate airfoil planes.

An apparent problem arises if we attempt to find the force by an integration of the pressure distribution. On the surface of the plate the velocity is given in Eq. (6.35) and with the use of the Bernoulli equation the pressure difference across the plate is given as

$$\Delta p = 2\rho Q_\infty^2 \sin\alpha \cos\alpha \sqrt{\frac{1 - 2x/c}{1 + 2x/c}} \tag{6.38}$$

The force Z is perpendicular to the plate and is obtained by integrating the pressure difference along the plate to get

$$Z = \int_{-c/2}^{c/2} \Delta p \, dx = \pi \rho c Q_\infty^2 \sin\alpha \cos\alpha \tag{6.39}$$

The force obtained by these two different approaches is not the same in either magnitude or direction.

The difference can be explained by considering the flat plate as the limiting case of a thin airfoil as its thickness goes to zero. In this limit the pressure at the leading edge increases while the area upon which it acts decreases until in the flat plate limit the pressure is infinite and the area is zero. In this limit there is a finite contribution to the force that must be added to the result obtained by the pressure integration. To obtain this force we surround the plate leading edge by a small circle and calculate the force with the use of the Blasius formula.

The complex velocity on the plate is given in Eqs. (6.31) and (6.31a). The velocity on the circle at the leading edge is obtained by using Eq. (6.31) with $\beta = 0$ and $\theta = \pi$ and is

$$W(f) = -4i Q_\infty \sin\alpha \tag{6.40}$$

Near the leading edge f is approximately $-c/4$ and therefore we can take

$$W(Y) = \frac{f^2 W(f)}{f^2 - c^2/16} = \frac{ic}{2} Q_\infty \sin\alpha \frac{1}{f + c/4} \tag{6.41}$$

If the transformation in Eq. (6.24) is now inverted and Y is set approximately equal to $-c/2$, the transformation becomes

$$f = \frac{1}{2}(Y + \sqrt{Y^2 - c^2/4}) = -\frac{c}{4} + \frac{1}{2}i\sqrt{c}\sqrt{Y + c/2} \tag{6.42}$$

The complex velocity in the leading-edge region is therefore (from Eqs. (6.41) and (6.42))

$$W(Y) = \frac{Q_\infty \sqrt{c} \sin\alpha}{\sqrt{Y + c/2}} \tag{6.43}$$

This velocity is now substituted into the Blasius formula (Eq. (6.19)) to yield

$$X - iZ = \frac{i\rho}{2} \int W^2 \, dY = \frac{i\rho}{2} Q_\infty^2 c \sin^2\alpha \int \frac{dY}{Y + c/2} = -\pi \rho c Q_\infty^2 \sin^2\alpha \tag{6.44}$$

This leading-edge force acts along the plate in the upstream direction (Fig. 6.12) and is called the *leading-edge suction force*.

The total force is now obtained by the addition of the pressure force and the suction force (Eqs. (6.39) and (6.44)) and the resultant force is seen to be perpendicular to the stream and exactly equal to the result from the Kutta–Joukowski theorem (see Fig. 6.12).

A generalization of these results can be applied to the solution of the small-disturbance version of the thin-airfoil problem. Assume that this solution has the following complex

6.5 Conformal Mapping and the Joukowski Transformation

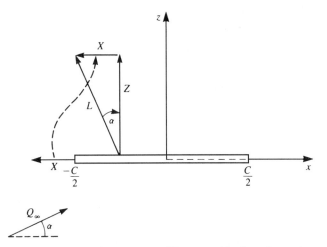

Figure 6.12 Forces due to pressure difference and leading-edge suction on the flat plate at angle of attack. Note that the resultant force (lift) is normal to the free stream Q_∞.

velocity in the neighborhood of the airfoil leading edge:

$$W(Y) = \frac{A}{\sqrt{Y + c/2}} \tag{6.45}$$

where A is a constant. Then the leading-edge suction force in this situation is given by the Blasius formula as

$$X = -\pi \rho A^2 \tag{6.46}$$

6.5.3 Flow Normal to a Flat Plate

Another interesting solution that can be obtained by this method is the solution for the flow normal to a flat plate. The complex potential for this flow in the circle plane is obtained by adding the potentials of a stream in the z direction and an opposing doublet (the flow is symmetric about midchord and has zero circulation) and is given by

$$F = -iQ_\infty \left[f - \frac{c^2}{16f} \right] \tag{6.47}$$

On the surface of the circle $f = \frac{c}{4}e^{i\theta}$ and the complex potential becomes

$$F = -iQ_\infty \left[\frac{c}{4}e^{i\theta} - \frac{c}{4}e^{-i\theta} \right] = \frac{Q_\infty c}{2}\sin\theta = \pm\frac{Q_\infty c}{2}\sqrt{1 - (2x/c)^2} \tag{6.48}$$

The complex potential on the surface is thus real and therefore it is equal to the velocity potential. The jump in potential across the plate is therefore given by

$$\Delta\Phi = Q_\infty c\sqrt{1 - (2x/c)^2} \tag{6.49}$$

Both an application of the Kutta–Joukowski theorem and a pressure integration yield the result that there is no force acting on the plate (recall that this is a potential flow solution without any flow separations!). Based on the results of the previous section, however, it is expected that symmetrically placed tip forces may be acting on the tips of the plate and these will be important in the slender wing application.

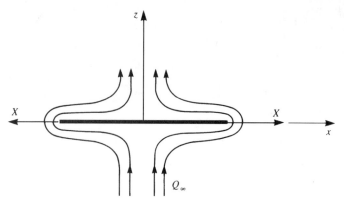

Figure 6.13 Suction force at the two tips of a flat plate in a normal flow (the two opposite forces cancel each other).

Consider the flow in the neighborhood of the left tip where f is approximately $-c/4$. The complex velocity at the corresponding point on the circle is obtained by a differentiation of the complex potential (Eq. (6.47)) as

$$W(f) = -2i Q_\infty \tag{6.50}$$

The analysis now proceeds in an identical fashion to the analysis in the previous section since the transformation is the same and the complex velocity in the neighborhood of the tip is

$$W(Y) = \frac{Q_\infty \sqrt{c}}{2} \frac{1}{\sqrt{Y + c/2}} \tag{6.51}$$

The tip force is then calculated to be

$$X = -\frac{\pi \rho c Q_\infty^2}{4} \tag{6.52}$$

The force acts to the left and from symmetry a tip force of equal magnitude acts on the right tip and points to the right (see Fig. 6.13).

6.5.4 Circular Arc Airfoil

The center of the circle is chosen on the imaginary axis in the f plane $\mu = im$ and from Eq. (6.27) $a = (C/4) \sec \beta$ and $m = a \sin \beta$. This choice results in the circular arc airfoil shown in Fig. 6.14a with chord $c = C$. Note that since the circle passes through both critical points A and D, the corresponding points on the airfoil are sharp. Also, points $B\,[f = i(a+m)]$ and $E\,[f = -i(a-m)]$ on the circle, at the top and bottom, both transform to the same point on the airfoil, $Y = 2im$. The schematic streamline pattern for the flow in both the physical and circle planes is shown in Fig. 6.14b. Note that the forward stagnation point on the circle occurs when $\theta = \pi + 2\alpha + \beta$ and therefore the forward stagnation point on the circular arc can be found from the transformation. The velocity at the trailing edge is given by Eq. (6.32) as

$$W\left(Y = \frac{c}{2}\right) = Q_\infty \cos \beta \cos(\alpha + \beta)\, e^{2i\beta} \tag{6.53}$$

The lift coefficient for the circular arc airfoil is given by Eq. (6.29a) as

$$C_l = \frac{2\pi \sin(\alpha + \beta)}{\cos \beta} \tag{6.54}$$

6.5 Conformal Mapping and the Joukowski Transformation

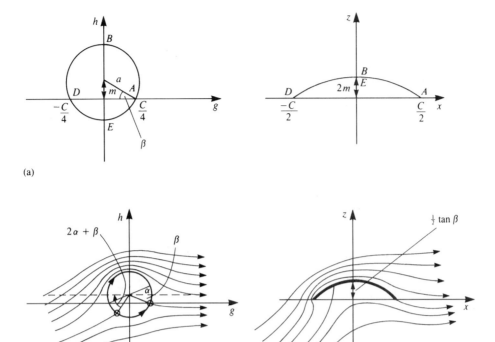

Figure 6.14 (a) Circular arc airfoil mapping. (b) Streamlines in the circle and the circular arc airfoil planes (at an angle of attack).

The zero-lift angle is seen to be equal to $-\beta$. The maximum camber ratio is defined as the ratio of the maximum ordinate $2m$ to the chord c and is $\frac{1}{2}\tan\beta$.

An interesting special case occurs when the circular arc is set at an angle of attack of zero. From above, it appears that the forward stagnation point is at the leading edge but since a critical point exists there, L'Hospital's rule must be used again and with $f = -c/4$ and $\theta = \pi + \beta$, the complex velocity at the leading edge is

$$W\left(Y = -\frac{c}{2}\right) = Q_\infty \cos^2\beta \, e^{-2i\beta} \tag{6.55}$$

This is equal in magnitude to the velocity at the trailing edge and the flow is seen to be symmetric with respect to the z axis. This is an example of a lifting flow with no stagnation points (see the streamline pattern in Fig. 6.15) and with a flow path of equal length for particles traveling along the upper and lower surfaces. The pressure coefficient is plotted in Fig. 6.16.

6.5.5 Symmetric Joukowski Airfoil

Let the center of the circle be taken on the real axis

$$\mu = -\epsilon C/4, \quad \epsilon > 0 \tag{6.56}$$

so that from Eq. (6.27)

$$\beta = 0, \quad a = \frac{C}{4}(1 + \epsilon) \tag{6.56a}$$

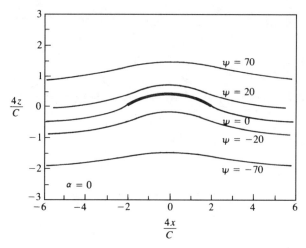

Figure 6.15 Streamlines for circular arc at zero angle of attack.

The circle is transformed into the airfoil shape shown in Fig. 6.17 (note that ϵ should be small). The surface of the airfoil is given by (Eq. (6.24))

$$Y = -\frac{\epsilon C}{4} + \frac{C}{4}(1+\epsilon)e^{i\theta} + \frac{C^2}{16[-\epsilon C/4 + (C/4)(1+\epsilon)e^{i\theta}]}$$

$$= \frac{C}{4}[-\epsilon + (1+\epsilon)\cos\theta + i(1+\epsilon)\sin\theta]$$

$$\times \left\{1 + \frac{1}{[-\epsilon + (1+\epsilon)\cos\theta]^2 + (1+\epsilon)^2\sin^2\theta}\right\} \tag{6.57}$$

Note that $Y(-\theta) = \bar{Y}$ and therefore the resulting airfoil is symmetric. The chord length c is given by

$$c = Y(\theta = 0) + |Y(\theta = \pi)| = \frac{C}{4}\left\{(1+2\epsilon)\left[1 + \frac{1}{(1+2\epsilon)^2}\right] + 2\right\}$$

$$= \frac{C}{4}\left\{3 + 2\epsilon + \frac{1}{(1+2\epsilon)}\right\} \tag{6.58}$$

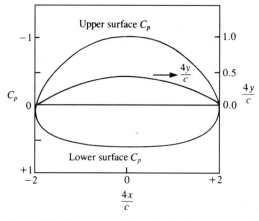

Figure 6.16 Pressure coefficient for circular arc at zero angle of attack.

6.6 Airfoil with Finite Trailing-Edge Angle

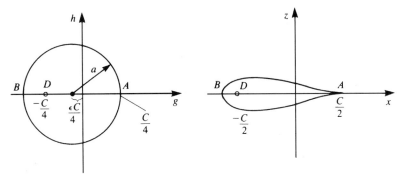

Figure 6.17 Symmetric Joukowski airfoil mapping.

For small ϵ, the chord length is approximated by

$$c \cong \frac{C}{4}\{3 + 2\epsilon + 1 - 2\epsilon + 4\epsilon^2 + \cdots\} = C\{1 + 4\epsilon^2 + \cdots\} \tag{6.59}$$

We can therefore take $c = C$. The velocity at the cusped trailing edge is given by Eq. (6.32) as

$$W(Y = c/2) = \frac{Q_\infty \cos\alpha}{1 + \epsilon} \tag{6.60}$$

and the lift coefficient is (when C is the chord) given by Eq. (6.29a) as

$$C_l = 2\pi(1 + \epsilon)\sin\alpha \tag{6.61}$$

The thickness ratio is approximately equal to 1.299ϵ.

6.6 Airfoil with Finite Trailing-Edge Angle

The Joukowski airfoils have cusped trailing edges as has been seen for the flat plate, circular arc, and symmetric examples. The cusped trailing edge presents some numerical difficulties for panel method solutions since in the neighborhood of the trailing edge the airfoil's upper and lower surfaces coincide. Therefore, for the purpose of providing exact solutions to test the results of the panel methods to be presented later, a mapping is introduced here that takes a symmetrical airfoil with a finite trailing-edge angle in the Y plane to a circle in the f plane. The transformation, which appeared in van de Vooren and de Jong,[6.2] is

$$Y = \frac{(f - a)^k}{(f - a\epsilon)^{k-1}} + \ell \tag{6.62}$$

The center of the circle is at the origin of coordinates in the f plane and the radius is a (see Fig. 6.18). Here ϵ is a thickness parameter, and k controls the trailing-edge angle while ℓ determines the chord length.

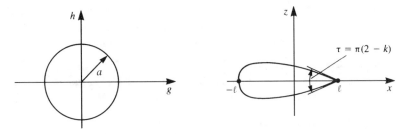

Figure 6.18 Mapping for airfoil with finite trailing-edge angle.

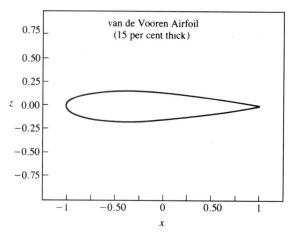

Figure 6.19 Finite trailing-edge angle airfoil with 15% thickness (x, z coordinates are normalized by airfoil semichord l).

For the circle $f = ae^{i\theta}$ the transformation becomes

$$Y = \frac{[a(\cos\theta - 1) + ia\sin\theta]^k}{[a(\cos\theta - \epsilon) + ia\sin\theta]^{k-1}} + \ell \tag{6.63}$$

Note that $\theta = 0$ corresponds to $Y = \ell$, the trailing edge. For $\theta = \pi$, the leading edge is given by

$$Y = \frac{-a2^k}{(1+\epsilon)^{k-1}} + \ell \tag{6.64}$$

For the chord length to be $c = 2\ell$, we set $Y = -\ell$ above to get

$$a = 2\ell(1+\epsilon)^{k-1} 2^{-k} \tag{6.65}$$

It can be shown that the trailing-edge angle (Fig. 6.18) is given by

$$\tau = \pi(2-k) \tag{6.66}$$

The airfoil with 15% thickness is shown in Fig. 6.19.

In Section 6.5 and this section mappings are presented that transform specific airfoil shapes into circles so that exact solutions to the incompressible potential flow problem are obtained. Theodorsen and Garrick[6.3] developed a numerical conformal mapping procedure to obtain solutions for arbitrary airfoil shapes and this procedure later became an integral part of more recent techniques. A review of modern methods for numerical conformal mapping can be found in Henrici.[6.4]

6.7   Summary of Pressure Distributions for Exact Airfoil Solutions

The exact solutions obtained in this chapter are very useful for the validation of various numerical methods. Therefore, the methods of calculation of the accurate analytical pressure distribution for several airfoil shapes are briefly summarized in this section.

a. *Circular Arc Thin Airfoil*

For an airfoil of chord c and camber ratio $2m/c = (1/2)\tan\beta$ the radius of the circle in the f plane is a (Fig. 6.14a), where

$$a = \sqrt{m^2 + \frac{c^2}{16}} \tag{6.67a}$$

6.7 Summary of Pressure Distributions for Exact Airfoil Solutions

and

$$m = a \sin \beta \tag{6.67b}$$

The x, z coordinates of the circular arc airfoil are then obtained from Eqs. (6.24) and (6.30) (where $\mu = im$):

$$x = a \cos \theta \left[1 + \frac{c^2}{16a^2(1 + 2 \sin \theta \sin \beta + \sin^2 \beta)} \right] \tag{6.68a}$$

$$z = a(\sin \theta + \sin \beta) \left[1 - \frac{c^2}{16a^2(1 + 2 \sin \theta \sin \beta + \sin^2 \beta)} \right] \tag{6.68b}$$

The velocity distribution is then calculated from Eqs. (6.30), (6.31), and (6.31a):

$$u = 2Q_\infty [\sin(\alpha + \beta) - \sin(\alpha - \theta)]$$
$$\times \left[\frac{a^4 \sin \theta (A^2 + B^2) + a^2(c^2/16)(B \cos \theta - A \sin \theta)}{(a^2 A - c^2/16)^2 + a^4 B^2} \right] \tag{6.69a}$$

$$w = -2Q_\infty [\sin(\alpha + \beta) - \sin(\alpha - \theta)]$$
$$\times \left[\frac{a^4 \cos \theta (A^2 + B^2) - a^2(c^2/16)(B \sin \theta + A \cos \theta)}{(a^2 A - c^2/16)^2 + a^4 B^2} \right] \tag{6.69b}$$

where

$$A = \cos^2 \theta - \sin^2 \theta - \sin^2 \beta - 2 \sin \beta \sin \theta$$
$$B = 2 \cos \theta \sin \theta + 2 \sin \beta \cos \theta$$

The pressure coefficient is then obtained directly from Bernoulli's equation as

$$C_p = 1 - \frac{u^2 + w^2}{Q_\infty^2} \tag{6.70}$$

Note that for $\beta = 0$ the equations for the velocity components reduce to the flat plate case, which is presented in Eq. (6.35).

b. Symmetric Joukowski Airfoil

For an airfoil of chord c and thickness parameter ϵ the radius of the circle in the f plane is a (Fig. 6.17), where

$$a = \frac{c}{4}(1 + \epsilon) \tag{6.56a}$$

where the airfoil chord c is

$$c = \frac{C}{4} \left(3 + 2\epsilon + \frac{1}{1 + 2\epsilon} \right) \tag{6.58}$$

The x, z coordinates of the airfoil are given in Eq. (6.57) as

$$x = \left(a \cos \theta - \frac{\epsilon C}{4} \right) \left[1 + \frac{C^2/16}{(a \cos \theta - \epsilon C/4)^2 + a^2 \cos^2 \theta} \right] \tag{6.71a}$$

$$z = a \sin \theta \left[1 - \frac{C^2/16}{(a \cos \theta - \epsilon C/4)^2 + a^2 \sin^2 \theta} \right] \tag{6.71b}$$

The velocity distribution is obtained from Eqs. (6.30), (6.31), and (6.31a) (where $\mu = -\epsilon C/4$):

$$u = 2Q_\infty[\sin\alpha - \sin(\alpha - \theta)]$$
$$\times \left[\frac{(A^2 - C^2/16)(A\sin\theta - B\cos\theta) + B(A\cos\theta + B\sin\theta)}{(A - C^2/16)^2 + B^2}\right] \quad (6.72a)$$

$$w = -2Q_\infty[\sin\alpha - \sin(\alpha - \theta)]$$
$$\times \left[\frac{(A^2 - C^2/16)(A\cos\theta + B\sin\theta) - B(A\sin\theta - B\cos\theta)}{(A - C^2/16)^2 + B^2}\right] \quad (6.72b)$$

where

$$A = \left(a\cos\theta - \frac{\epsilon C}{4}\right)^2 - a^2\sin^2\theta$$

$$B = 2a\sin\theta\left(a\cos\theta - \frac{\epsilon C}{4}\right)$$

and the pressure coefficient is calculated by using Eq. (6.70)

c. The van de Vooren Airfoil

The parameters for this airfoil are shown in Fig. 6.18 where the chord length is 2ℓ and is given from Eq. (6.65) as

$$2\ell = \frac{a2^k}{(1+\epsilon)^{k-1}} \quad (6.73)$$

Here ϵ is the thickness parameter, k is the trailing-edge angle parameter (see Eq. (6.66)) and a is the radius of the circle in the f plane.

The x, z coordinates of the Van de Vooren airfoil are then given in Eq. (6.63) as

$$x = \frac{r_1^k}{r_2^{k-1}}[\cos k\theta_1 \cos(k-1)\theta_2 + \sin k\theta_1 \sin(k-1)\theta_2] \quad (6.74a)$$

$$z = \frac{r_1^k}{r_2^{k-1}}[\sin k\theta_1 \cos(k-1)\theta_2 - \cos k\theta_1 \sin(k-1)\theta_2] \quad (6.74b)$$

where

$$r_1 = \sqrt{(a\cos\theta - a)^2 + a^2\sin^2\theta}$$

$$r_2 = \sqrt{(a\cos\theta - \epsilon a)^2 + a^2\sin^2\theta}$$

$$\theta_1 = \tan^{-1}\frac{a\sin\theta}{a\cos\theta - a} + \pi$$

$$\theta_2 = \tan^{-1}\frac{a\sin\theta}{a\cos\theta - \epsilon a} + n_1\pi$$

Here n_1 depends on the quadrant where θ_2 is being evaluated. (It has a value of 0 in the first quadrant, 1 in the second and third quadrants, and 2 in the fourth quadrant.)

6.8 Method of Images

The velocity distribution is then given from the solution in the circle plane plus the transformation (Eq. (6.62)) as

$$u = 2Q_\infty \frac{r_2^k}{r_1^{k-1}} \frac{\sin\alpha - \sin(\alpha - \theta)}{D_1^2 + D_2^2}(D_1 \sin\theta + D_2 \cos\theta) \tag{6.75a}$$

$$w = -2Q_\infty \frac{r_2^k}{r_1^{k-1}} \frac{\sin\alpha - \sin(\alpha - \theta)}{D_1^2 + D_2^2}(D_1 \cos\theta - D_2 \sin\theta) \tag{6.75b}$$

where

$$A = \cos(k-1)\theta_1 \cos k\theta_2 + \sin(k-1)\theta_1 \sin k\theta_2$$
$$B = \sin(k-1)\theta_1 \cos k\theta_2 - \cos(k-1)\theta_1 \sin k\theta_2$$
$$D_0 = a(1 - k + k\epsilon)$$
$$D_1 = A(a\cos\theta - D_0) - Ba\sin\theta$$
$$D_2 = A(a\sin\theta) + B(a\cos\theta - D_0)$$

The pressure distribution is again calculated by using Eq. (6.70).

6.8 Method of Images

Since the solution for the flow past bodies of aerodynamic interest can be represented by suitable distributions of singular solutions to Laplace's equation, it is important to study the representation of these singular solutions in the presence of additional boundaries, mainly straight ones, to be able to deal with ground planes and wind-tunnel walls, etc.

As an example, consider a two-dimensional source of strength σ located a distance h from a plane wall as shown in Fig. 6.20. Introduce a Cartesian coordinate system whose origin is at the source and whose x axis is parallel to the wall. In the absence of the wall, the velocity potential of the source is

$$\Phi = \frac{\sigma}{2\pi} \ln\sqrt{x^2 + z^2} \tag{6.76}$$

Since we would expect that the only singularity in the flowfield is due to the source, we look for a solution of the form

$$\Phi = \frac{\sigma}{2\pi} \ln\sqrt{x^2 + z^2} + \Phi_I \tag{6.77}$$

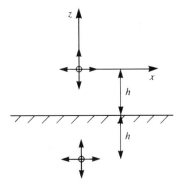

Figure 6.20 Image of source in plane wall.

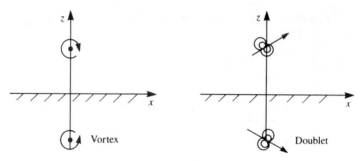

Figure 6.21 Image of vortex and doublet in plane wall.

where Φ_I satisfies Laplace's equation, has no singularities for $z > -h$, decays at infinity, and exactly cancels the normal component of velocity at the wall due to the source so that the wall boundary condition is satisfied. The boundary condition on Φ_I is therefore

$$\frac{\partial \Phi_I}{\partial z}(x, -h) = \frac{\sigma h}{2\pi} \frac{1}{x^2 + h^2} \tag{6.78}$$

From symmetry considerations, an "image" source at $(0, -2h)$ is investigated as a possible solution. Its velocity potential is

$$\Phi_I = \frac{\sigma}{2\pi} \ln \sqrt{x^2 + (z + 2h)^2} \tag{6.79}$$

and substitution into the boundary condition at the wall shows that it is satisfied. Similar image solutions for a doublet and a vortex are shown in Fig. 6.21. The complex potentials for the original singularities plus their images are

Source:
$$F(Y) = \frac{\sigma}{2\pi} \ln Y + \frac{\sigma}{2\pi} \ln(Y + 2ih) \tag{6.80}$$

Doublet:
$$F(Y) = -\frac{\mu}{2\pi Y} e^{i\alpha} - \frac{\mu}{2\pi} \frac{1}{(Y + 2ih)} e^{i(2\pi - \alpha)} \tag{6.81}$$

Vortex:
$$F(Y) = \frac{i\Gamma}{2\pi} \ln Y - \frac{i\Gamma}{2\pi} \ln(Y + 2ih) \tag{6.82}$$

Next consider a source placed midway between two parallel walls set a distance h apart. An image source at $(0, h)$ will satisfy the boundary condition on the upper wall but now both the original source and this image source must be canceled at the lower wall to satisfy the boundary condition there. Images at $(0, -h)$ and $(0, -2h)$ will take care of the lower wall but now two more images are needed for the upper wall and the process will continue until the complete image system plus the original source consists of an infinite stack of sources a distance h apart as shown in Fig. 6.22. The complex potential for this source stack is

$$F(Y) = \frac{\sigma}{2\pi}[\ln Y + \ln(Y - ih) + \ln(Y + ih) + \ln(Y - 2ih) + \ln(Y + 2ih) + \cdots] \tag{6.83}$$

6.8 Method of Images

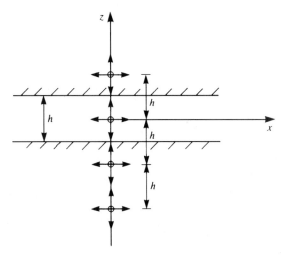

Figure 6.22 Image of source midway between parallel walls.

Each pair of images can be combined as

$$\ln(Y - inh) + \ln(Y + inh) = \ln n^2 h^2 + \ln\left(1 + \frac{Y^2}{n^2 h^2}\right)$$

and if the constant terms are neglected, the complex potential becomes

$$F(Y) = \frac{\sigma}{2\pi} \ln\left[Y \prod_{n=1}^{\infty}\left(1 + \frac{Y^2}{n^2 h^2}\right)\right] \tag{6.84}$$

The use of the following identity from Gradshteyn and Ryzhik (Ref. 5.7, p. 37),

$$\sinh A = A \prod_{k=1}^{\infty}\left(1 + \frac{A^2}{k^2 \pi^2}\right) \tag{6.85}$$

leads to the closed-form solution for the complex potential as

$$F(Y) = \frac{\sigma}{2\pi} \ln\left[\sinh \frac{\pi Y}{h}\right] \tag{6.86}$$

For a clockwise vortex of circulation Γ between parallel walls, an application of the iterative image procedure previously used for the source leads to the solution shown in Fig. 6.23, which consists of a stack of clockwise vortices at $Y = 0, \pm 2hi, \pm 4hi, \ldots$ and a stack of counterclockwise vortices at $\pm hi, \pm 3hi, \pm 5hi, \ldots$. From before, the complex potential for the clockwise stack is

$$F(Y) = \frac{\Gamma}{2\pi} \ln\left[\sinh \frac{\pi Y}{2h}\right] \tag{6.87}$$

The use of another identity from Gradshteyn and Ryzhik[5.7] (p. 37),

$$\cosh A = \prod_{k=0}^{\infty}\left(1 + \frac{4A^2}{(2k+1)^2 \pi^2}\right) \tag{6.88}$$

results in the following complex potential for the vortex between walls:

$$F(Y) = \frac{i\Gamma}{2\pi}\left[\ln\left(\sinh \frac{\pi Y}{2h}\right) - \ln\left(\cosh \frac{\pi Y}{2h}\right)\right] = \frac{i\Gamma}{2\pi} \ln\left(\tanh \frac{\pi Y}{2h}\right) \tag{6.89}$$

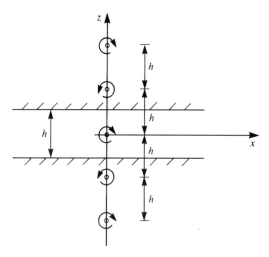

Figure 6.23 Image of clockwise vortex midway between parallel walls.

We have considered images of the singular solutions in a plane wall (for ground-effect applications) and between parallel walls (for wind-tunnel applications). Another possible application is the interaction of an airfoil with its wake or the wake of another airfoil (for unsteady motion) and since we have shown that an airfoil geometry can be transformed through conformal mapping into a circle, the image system for a singular solution in the presence of a circle will be studied.

The circle theorem due to Milne-Thomson[6.5] states that if the complex potential $F_1(Y)$ represents a flow without singularities for $|Y| < a$, then

$$F(Y) = F_1(Y) + \bar{F}_1\left(\frac{a^2}{Y}\right) \tag{6.90}$$

represents the same flow at infinity with a circular cylinder of radius a at the origin. The function $\bar{F}_1(Y)$ is defined in the following way: "If $F_1(x)$ is a function that takes complex values for real values of x, then $\bar{F}_1(x)$ is the function that takes the corresponding conjugate complex values for the same real values of x, and $\bar{F}_1(Y)$ is obtained by writing Y instead of x."

Consider the simple example where $F_1(Y) = UY$, a uniform stream in the x direction. $\bar{F}_1(Y)$ is seen to be also UY and therefore the flow of the uniform stream with a circle at the origin is given as

$$F(Y) = UY + U\frac{a^2}{Y} \tag{6.91}$$

which is simply the stream plus doublet solution previously derived. Now let

$$F_1(Y) = \frac{\sigma}{2\pi} \ln(Y - Y_0) \tag{6.92}$$

which is the complex potential for a source of strength σ at $Y = Y_0$. $\bar{F}_1(Y)$ is

$$\bar{F}_1(Y) = \frac{\sigma}{2\pi} \ln(Y - \bar{Y}_0) \tag{6.93}$$

and the complex potential for a source outside a circular cylinder becomes (Eq. (6.90))

$$F(Y) = \frac{\sigma}{2\pi}\left[\ln(Y - Y_0) + \ln\left(\frac{a^2}{Y} - \bar{Y}_0\right)\right] \tag{6.94}$$

6.8 Method of Images

The following manipulation will put the above result in the form of a recognizable image system:

$$F(Y) = \frac{\sigma}{2\pi}\left[\ln(Y - Y_0) + \ln\left(\frac{a^2}{Y} - \bar{Y}_0\right)\right]$$

$$= \frac{\sigma}{2\pi}\left[\ln(Y - Y_0) + \ln\frac{\bar{Y}_0}{Y} + \ln\left(Y - \frac{a^2}{\bar{Y}_0}\right)(-1)\right]$$

$$= \frac{\sigma}{2\pi}\left[\ln(Y - Y_0) - \ln Y + \ln\left(Y - \frac{a^2}{\bar{Y}_0}\right)\right] \quad (6.95)$$

where the constant terms have been neglected. It can be seen that the solution consists of the original source, an image source of the same strength at the image point, and a sink of the same strength at the origin. These three singularities line up along the same radial line from the origin as can be seen by writing the location of the image point as

$$\frac{a^2}{\bar{Y}_0} = \frac{a^2 Y_0}{\bar{Y}_0 Y_0} = \left(\frac{a^2}{|Y_0|^2}\right) Y_0$$

For a clockwise vortex of circulation Γ at $Y = Y_0$ outside a circle, the image system consists of a counterclockwise image at the same image point as for the source and a clockwise vortex at the origin. Both of these image systems are illustrated in Fig. 6.24.

As a final example, take

$$F_1(Y) = -\frac{\mu}{2\pi}\frac{e^{i\alpha}}{Y - Y_0} \quad (6.96)$$

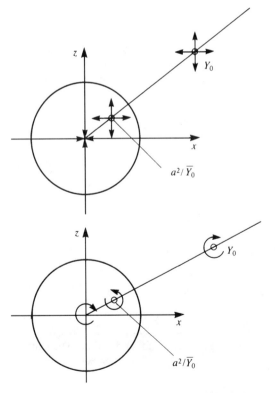

Figure 6.24 Image of source and vortex outside a circle.

which is the complex potential for a doublet of strength μ at $Y = Y_0$ whose axis is at an angle α to the x direction. With the use of the result that

$$\overline{\dfrac{A}{B}} = \dfrac{\bar{A}}{\bar{B}}$$

then

$$\bar{F}_1(Y) = -\dfrac{\mu}{2\pi} \dfrac{e^{-i\alpha}}{Y - \bar{Y}_0} \tag{6.97}$$

and the complex potential for the doublet outside of a circle becomes

$$F(Y) = -\dfrac{\mu}{2\pi}\left[\dfrac{e^{i\alpha}}{Y - Y_0} + \dfrac{e^{-i\alpha}}{a^2/Y - \bar{Y}_0}\right] \tag{6.98}$$

The following manipulation will put the result in the form of a recognizable image system:

$$\begin{aligned}F(Y) &= \dfrac{-\mu}{2\pi}\left[\dfrac{e^{i\alpha}}{Y - Y_0} + e^{-i\alpha}\left(\dfrac{-Y/\bar{Y}_0}{Y - a^2/\bar{Y}_0}\right)\right] \\ &= \dfrac{-\mu}{2\pi}\left[\dfrac{e^{i\alpha}}{Y - Y_0} + e^{-i\alpha}\left(\dfrac{-Y}{\bar{Y}_0(Y - a^2/\bar{Y}_0)} + \dfrac{a^2/\bar{Y}_0 - a^2/\bar{Y}_0}{\bar{Y}_0(Y - a^2/\bar{Y}_0)}\right)\right] \\ &= \dfrac{-\mu}{2\pi}\left[\dfrac{e^{i\alpha}}{Y - Y_0} - e^{-i\alpha}\dfrac{a^2}{\bar{Y}_0^2}\dfrac{1}{(Y - a^2/\bar{Y}_0)}\right] \\ &= \dfrac{-\mu}{2\pi}\left[\dfrac{e^{i\alpha}}{Y - Y_0} + \dfrac{a^2}{|Y_0|^2(Y - a^2/\bar{Y}_0)}e^{i(\pi - \alpha + 2\arg Y_0)}\right]\end{aligned} \tag{6.98a}$$

where the constant term has been neglected. The image of the doublet in a circle is therefore seen to be another doublet inside the circle at the image point previously derived for the source but with a reduced strength

$$\dfrac{\mu a^2}{|Y_0|^2}$$

For the special case of a doublet pointing outward along the radial line from the origin, $\arg Y_0 = \alpha$, and the complex potential becomes

$$F(Y) = \dfrac{-\mu}{2\pi}\left[\dfrac{e^{i\alpha}}{Y - \bar{Y}_0} + \dfrac{a^2}{|Y_0^2|}\dfrac{e^{i(\pi + \alpha)}}{Y - a^2/\bar{Y}_0}\right] \tag{6.99}$$

This doublet plus its image are shown in Fig. 6.25.

6.9 Generalized Kutta–Joukowski Theorem

For the study of interaction problems involving the flow past an airfoil in the presence of additional boundaries (ground effect, wind-tunnel walls, etc.) or additional airfoils (tandem, biplane, etc.), the airfoil is not in an unbounded fluid. The Kutta–Joukowski theorem (Eq. (6.21a)) therefore does not apply. In this section we will develop a generalized version of this theorem.

6.9 Generalized Kutta–Joukowski Theorem

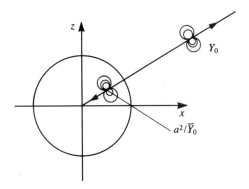

Figure 6.25 Image of radial doublet outside a circle.

For simplicity, let us model the airfoil as a single vortex of circulation Γ (lumped-vortex element). For interaction problems, the airfoil vortex will be in the presence of an image vortex of circulation Γ_I. Let us place the airfoil vortex at the origin of a Cartesian coordinate system and the image vortex at $Y_0 = a + ib$. Consider the flow of a uniform stream of speed Q_∞ in the x direction past the airfoil. The complex potential for the stream and the two vortices is

$$F = Q_\infty Y + \frac{i\Gamma}{2\pi} \ln Y + \frac{i\Gamma_I}{2\pi} \ln(Y - Y_0) \tag{6.100}$$

Let the curve C surround the airfoil vortex and the curve C_I surround the image vortex (see Fig. 6.26). The curve C_∞, where $Y \to \infty$, surrounds C and C_I. Now, the Blasius formula (Eq. (6.19)) provides the force on the airfoil

$$X - iZ = \frac{i\rho}{2} \int_C W^2 \, dY \tag{6.101}$$

where the complex velocity W is given by

$$W = Q_\infty + \frac{i\Gamma}{2\pi Y} + \frac{i\Gamma_I}{2\pi(Y - Y_0)} \tag{6.102}$$

With the use of Eq. (6.7),

$$\int_{C_\infty} W^2 \, dY = \int_C W^2 \, dY + \int_{C_I} W^2 \, dY \tag{6.103a}$$

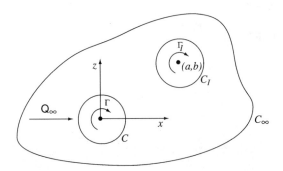

Figure 6.26 Nomenclature for the force on an airfoil in the presence of an additional vortex.

and therefore
$$\int_C W^2 \, dY = \int_{C_\infty} W^2 \, dY - \int_{C_I} W^2 \, dY \tag{6.103b}$$

Let us first evaluate the integral on C_∞. For $Y \to \infty$,
$$\frac{1}{Y - Y_0} = \frac{1}{Y(1 - \frac{Y_0}{Y})} = \frac{1}{Y}\left(1 + \frac{Y_0}{Y} + \frac{Y_0^2}{Y^2} + \cdots\right)$$

and
$$W = Q_\infty + \frac{i}{2\pi}\frac{(\Gamma + \Gamma_I)}{Y} + \cdots$$

and therefore
$$W^2 = Q_\infty^2 + \frac{iQ_\infty}{\pi}\frac{(\Gamma + \Gamma_I)}{Y} + \cdots \tag{6.104}$$

With the use of the residue theorem (Eq. (6.11)), we get
$$\frac{i\rho}{2}\int_{C_\infty} W^2 \, dY = -i\rho Q_\infty(\Gamma + \Gamma_I) \tag{6.105}$$

Now consider the integral on C_I. The complex velocity (Eq. (6.102)) is written as
$$W = f(Y) + \frac{i}{2\pi}\frac{\Gamma_I}{Y - Y_0} \tag{6.106a}$$

where $f(Y)$ is
$$f(Y) = Q_\infty + \frac{i\Gamma}{2\pi Y} \tag{6.106b}$$

Squaring the velocity yields
$$W^2 = f^2 + \frac{i\Gamma_I f}{\pi(Y - Y_0)} - \frac{\Gamma_I^2}{4\pi^2(Y - Y_0)^2} \tag{6.107}$$

With the use of the residue theorem we get
$$\frac{i\rho}{2}\int_{C_I} W^2 \, dY = -i\rho \Gamma_I f(Y_0) \tag{6.108}$$

and with the use of Eq. (6.106b) this becomes
$$\frac{i\rho}{2}\int_{C_I} W^2 \, dY = -i\rho \Gamma_I \left(Q_\infty + \frac{i\Gamma}{2\pi Y_0}\right) \tag{6.109}$$

If we use Eqs. (6.101), (6.103b), (6.105), and (6.109) and some manipulation the complex force is obtained as
$$X - iZ = -i\rho Q_\infty \Gamma\left(1 - \frac{i\Gamma_I}{2\pi Y_0 Q_\infty}\right) \tag{6.110}$$

The airfoil lift is therefore
$$Z = \rho Q_\infty \Gamma \, \text{Real}\left(1 - \frac{i\Gamma_I}{2\pi Y_0 Q_\infty}\right) \tag{6.111}$$

To get an interpretation of this result, consider the complex velocity of the image vortex evaluated at the location of the airfoil vortex ($Y = 0$):

$$W = -\frac{i\Gamma_I}{2\pi Y_0} \tag{6.112}$$

The real part (see Eq. (6.111)) of this velocity is its component in the direction of the uniform stream. We can therefore write the generalized Kutta–Joukowski theorem as

$$L = \rho Q_\infty \Gamma \left(1 + \frac{\mathbf{Q}_\infty \cdot \mathbf{q}_I}{Q_\infty^2}\right) \tag{6.113}$$

where $\mathbf{q}_I$ is the velocity of the image vortex evaluated at the location of the airfoil vortex.

References

[6.1] Churchill, R. V., *Complex Variables and Applications*, 2nd edition, McGraw-Hill, 1960.

[6.2] van de Vooren, A. I., and de Jong, L. S., "Calculation of Incompressible Flow About Aerofoils Using Source, Vortex or Doublet Distributions," Report TW-86 of the Math. Inst. of the University of Groningen, The Netherlands, 1969.

[6.3] Theodorsen, T., and Garrick, I. E., "General Potential Theory of Arbitrary Wing Sections," NACA TR 452, 1933.

[6.4] Henrici, P., *Applied and Computational Complex Analysis*, Vol. 3, Wiley, 1986.

[6.5] Milne-Thomson, L. M., "Hydrodynamical Images," *Proceedings of the Cambridge Philosophical Society*, Vol. 36, p. 246, 1940.

Problems

6.1. Consider the flow due to a doublet of strength μ at $(a, 0)$ whose axis is in the x direction and an equal doublet at $(-a, 0)$ whose axis lies in the opposite direction. Find the complex potential for the limiting case $a \to 0$, $\mu a \to M$. Sketch some streamlines.

6.2. Consider the flow of a uniform stream of speed Q_∞ in the x direction past two sources of strength σ at $(0, a)$ and $(0, -a)$. (a) Find the stagnation point(s) and discuss the significance of $\sigma/2\pi Q_\infty a$ (b) Sketch some streamlines (including the stagnation streamline) for the cases: $a \to 0$, $a \to \infty$, $\sigma/2\pi Q_\infty a < 1$, and $\sigma/2\pi Q_\infty a > 1$.

6.3. Consider the flow of a uniform stream of speed Q_∞ at an angle of attack α past an ellipse of semi-axes a and b

(a) Show that the transformation

$$Y = f + \frac{a^2 - b^2}{4f}$$

maps the ellipse into a circle of radius $(a + b)/2$ in the f plane.

(b) Sketch the streamlines of the flow. What are the values of θ for the stagnation points in the circle plane? Use these values plus the results in (a) to find the stagnation points on the ellipse.

(c) Plot the pressure distribution on the ellipse for $a = 1$, $b = .25$, and $\alpha = 30°$.

6.4. Consider the flow around a flat plate lying along the x axis from $-c/2 < x < c/2$ due to the presence of a clockwise vortex of circulation Γ one chord length downstream of the trailing edge. Find the complex potential and sketch some streamlines.

6.5. Consider the flow due to a source of strength σ between two parallel walls a distance h apart. The source is situated a distance ah from the bottom wall. Find the complex potential.

6.6. Consider the flow of a uniform stream of speed Q_∞ past a circular cylinder of radius a in the presence of a ground plane located a distance h from the center of the circle. For $a/h \ll 1$, use the method of images to find the complex potential (find the first few terms in the solution).

CHAPTER 7

Perturbation Methods

For the small-disturbance solution techniques that are treated in this book, approximations to the exact mathematical problem formulation are made to facilitate the determination of a solution. Since for incompressible and irrotational flow the governing partial differential equation is linear, the approximations are made to the body boundary condition. For example, for the three-dimensional wing in Chapter 4, only terms linear in thickness, camber, and angle of attack are kept and the boundary condition is transferred to the x–y plane. The solution technique is therefore a "first-order" thin wing theory.

The small-disturbance methods developed here can be thought of as providing the first term in a perturbation series expansion of the solution to the exact mathematical problem and terms that were neglected in determining the first term will come into play in the solution for the following terms. In this book we will follow the lead of Van Dyke[5.3] and use the thin-airfoil problem as the vehicle for the presentation of the ideas and some of the details of perturbation methods and their applicability to aerodynamics. First, the thin-airfoil solution will be introduced as the first term in a small-disturbance expansion and the mathematical problem for the next term will be derived. An example of a second-order solution will be presented and the failure of the expansion in the leading-edge region will be noted. A local solution applicable in the leading-edge region will be obtained and the method of matched asymptotic expansions will be used to provide a solution valid for the complete airfoil. Finally, the thin airfoil between wind-tunnel walls will be studied to illustrate an expansion within an expansion.

7.1   Thin-Airfoil Problem

Consider the two-dimensional airfoil problem as a special case of the three-dimensional wing problem of Chapter 4. The dependent variables are now functions of x and z and both the upper and lower airfoil surfaces are given by

$$f(x, z) = z - \eta(x) = 0, \qquad \frac{-c}{2} \leq x \leq \frac{c}{2} \tag{7.1}$$

Note that the origin is at midchord and that the airfoil chord is c (Fig. 7.1). This choice of the origin is made for convenience in the evaluation of the Cauchy principal value integrals that will appear in the example problems.

The perturbation velocity potential Φ is defined in Section 4.2 by

$$\Phi^* = \Phi + \Phi_\infty \tag{7.2}$$

where

$$\Phi_\infty = U_\infty x + W_\infty z = x Q_\infty \cos\alpha + z Q_\infty \sin\alpha \tag{7.3}$$

The exact airfoil boundary condition is the two-dimensional version of Eq. (4.12):

$$-\frac{d\eta}{dx}\left(\frac{\partial\Phi}{\partial x} + Q_\infty \cos\alpha\right) + \frac{\partial\Phi}{\partial z} + Q_\infty \sin\alpha = 0 \qquad \text{on} \quad z = \eta \tag{7.4}$$

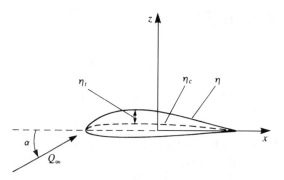

Figure 7.1 Coordinate system for airfoil problem.

with $U_\infty = Q_\infty \cos\alpha$ and $W_\infty = Q_\infty \sin\alpha$. The small-disturbance approximations and limitations on the geometry introduced in Chapter 4 apply and it is assumed that the order of magnitude of the airfoil thickness ratio, camber ratio, and angle of attack can all be represented by the small parameter ϵ.

Let us consider the following expansion for the perturbation velocity potential:

$$\Phi = \Phi_1 + \Phi_2 + \Phi_3 + \cdots \tag{7.5}$$

where

$$\Phi_i = O(\epsilon^i), \qquad i = 1, 2, 3, \ldots \tag{7.6}$$

and the order symbol $O(\epsilon)$ is defined by

$$g(\epsilon) = O(\epsilon) \quad \text{as} \quad \epsilon \to 0 \quad \text{if} \quad \lim_{\epsilon \to 0} \frac{g(\epsilon)}{\epsilon} < \infty \tag{7.7}$$

In this chapter we will carry the analysis through to second order to illustrate the method. Terms to $O(\epsilon^2)$ will be kept and therefore the components of the free-stream flow are written as

$$U_\infty = Q_\infty \cos\alpha = Q_\infty \left[1 - \frac{\alpha^2}{2} + O(\alpha^4)\right] \tag{7.8a}$$

$$W_\infty = Q_\infty \sin\alpha = Q_\infty [\alpha + O(\alpha^3)] \tag{7.8b}$$

The boundary condition will be transferred to the chord line as in Eq. (4.16) and the complete boundary condition with the above substitutions becomes

$$-\frac{d\eta}{dx}\left[Q_\infty + \frac{\partial \Phi_1}{\partial x}(x, 0\pm)\right] + Q_\infty \alpha + \frac{\partial \Phi_1}{\partial z}(x, 0\pm)$$

$$+ \eta \frac{\partial^2 \Phi_1}{\partial z^2}(x, 0\pm) + \frac{\partial \Phi_2}{\partial z}(x, 0\pm) = 0 \tag{7.9}$$

For this equation to be valid for all values of the perturbation parameter ϵ, the terms of the same order (ϵ, ϵ^2) must individually be zero. To show this, divide the equation by ϵ and take the limit as ϵ goes to zero. Then all of the terms of $O(\epsilon)$ must be zero. Now, subtract these terms from the original equation and repeat the process. This shows that all of the terms of $O(\epsilon^2)$ must be zero.

7.1 Thin-Airfoil Problem

The boundary conditions for the first- and second-order problems then become

$$O(\epsilon): \quad \frac{\partial \Phi_1}{\partial z}(x, 0\pm) = Q_\infty \frac{d\eta}{dx} - Q_\infty \alpha \tag{7.10}$$

$$O(\epsilon^2): \quad \frac{\partial \Phi_2}{\partial z}(x, 0\pm) = \frac{d\eta}{dx} \frac{\partial \Phi_1}{\partial x}(x, 0\pm) - \eta \frac{\partial^2 \Phi_1}{\partial z^2}(x, 0\pm) \tag{7.11}$$

If Laplace's equation for Φ_1 is used in the second-order condition, it becomes

$$\frac{\partial \Phi_2}{\partial z}(x, 0\pm) = \frac{d}{dx}\left[\eta \frac{\partial \Phi_1}{\partial x}(x, 0\pm)\right] \tag{7.12}$$

At this point it is noted that the first-order boundary condition is the one that was used in the thin-airfoil treatment in Chapter 5. Now let us separate the problems at each order into a nonlifting (symmetric or thickness) problem and a lifting (camber and angle of attack) problem and introduce the camber and thickness functions

$$\eta = \eta_c \pm \eta_t \tag{7.13a}$$

$$\Phi_1 = \Phi_{1L} + \Phi_{1T} \tag{7.13b}$$

$$\Phi_2 = \Phi_{2L} + \Phi_{2T} \tag{7.13c}$$

Note that the lifting potentials (Φ_{1L}, Φ_{2L}) are antisymmetric in z and the nonlifting potentials (Φ_{1T}, Φ_{2T}) are symmetric in z. Consequently, the z component of velocity w is continuous across the chord for the lifting problems and discontinuous for the nonlifting problems.

With the above definitions, Eqs. (7.10)–(7.12) become

$$\frac{\partial \Phi_{1L}}{\partial z}(x, 0\pm) = Q_\infty \frac{d\eta_c}{dx} - Q_\infty \alpha \tag{7.14a}$$

$$\frac{\partial \Phi_{1T}}{\partial z}(x, 0\pm) = \pm Q_\infty \frac{d\eta_t}{dx} \tag{7.14b}$$

$$\frac{\partial \Phi_{2L}}{\partial z}(x, 0\pm) = \frac{d}{dx}\left[\eta_c \frac{\partial \Phi_{1T}}{\partial x}(x, 0\pm) + \eta_t \frac{\partial \Phi_{1L}}{\partial x}(x, 0+)\right] \tag{7.15}$$

$$\frac{\partial \Phi_{2T}}{\partial z}(x, 0\pm) = \pm \frac{d}{dx}\left[\eta_t \frac{\partial \Phi_{1T}}{\partial x}(x, 0\pm) + \eta_c \frac{\partial \Phi_{1L}}{\partial x}(x, 0+)\right] \tag{7.16}$$

The complete mathematical problems that accompany the above boundary conditions (Eqs. (7.14a,b), (7.15), and (7.16)) include Laplace's equation for each velocity potential and a velocity field that decays to zero at infinity. A Kutta condition must be applied in the lifting problems and the nonlifting problems have zero circulation.

The solutions to the above mathematical problems can be obtained with the use of the theory of singular integral equations (see Newman,[7.1] Section 5.7). The first-order tangential velocity component is

$$u_{1T} \equiv \frac{\partial \Phi_{1T}}{\partial x}(x, 0\pm) = \frac{Q_\infty}{\pi} \int_{-c/2}^{c/2} \frac{d\eta_t}{dx}(x_0) \frac{dx_0}{x - x_0} \tag{7.17}$$

for the thickness problem and

$$u_{1L} \equiv \frac{\partial \Phi_{1L}}{\partial x}(x, 0+) = \frac{Q_\infty}{\pi} \frac{1}{\sqrt{c^2/4 - x^2}}$$

$$\times \left\{ \int_{-c/2}^{c/2} \frac{[(d\eta_c/dx)(x_0) - \alpha]\sqrt{c^2/4 - x_0^2}}{x - x_0} dx_0 + \frac{\Gamma}{2} \right\} \tag{7.18}$$

for the lifting problem. A source distribution for Φ_{1T} also leads to Eq. (7.17) (see Eq. (5.15)). A vortex distribution solution for Φ_{1L} leads to an integral equation for γ (Eq. (5.39)) and the solution to this integral equation is given in Eq. (7.18) where $\gamma = 2u_{1L}$.

The nonunique solution (of Eq. (5.39)) with arbitrary circulation is given because for another application the solution with zero circulation will be needed. If the Kutta condition is applied, then $u_{1L}(c/2) = 0$, and using Eq. (7.18) we get

$$\Gamma = -2 \int_{-c/2}^{c/2} \left[\frac{d\eta_c}{dx}(x_0) - \alpha \right] \frac{\sqrt{c^2/4 - x_0^2}}{c/2 - x_0} dx_0$$

$$= -2 \int_{-c/2}^{c/2} \left[\frac{d\eta_c}{dx}(x_0) - \alpha \right] \sqrt{\frac{c/2 + x_0}{c/2 - x_0}} dx_0$$

Substitution of this value of the circulation into Eq. (7.18) yields

$$u_{1L} = \frac{Q_\infty}{\pi} \frac{1}{\sqrt{c^2/4 - x^2}} \int_{-c/2}^{c/2} \left[\frac{d\eta_c}{dx}(x_0) - \alpha \right] \sqrt{\frac{c^2}{4} - x_0^2} \left[\frac{1}{x - x_0} - \frac{1}{c/2 - x_0} \right] dx_0$$

$$= \frac{Q_\infty}{\pi} \frac{1}{\sqrt{c^2/4 - x^2}} \int_{-c/2}^{c/2} \left[\frac{d\eta_c}{dx}(x_0) - \alpha \right] \sqrt{\frac{c^2}{4} - x_0^2} \frac{c/2 - x}{(x - x_0)(c/2 - x_0)} dx_0$$

$$= \frac{Q_\infty}{\pi} \sqrt{\frac{c/2 - x}{c/2 + x}} \int_{-c/2}^{c/2} \left[\frac{d\eta_c}{dx}(x_0) - \alpha \right] \sqrt{\frac{c/2 + x_0}{c/2 - x_0}} \frac{dx_0}{x - x_0}$$

and finally the lifting solution becomes

$$u_{1L} = Q_\infty \sqrt{\frac{c/2 - x}{c/2 + x}} \left\{ \alpha + \frac{1}{\pi} \int_{-c/2}^{c/2} \sqrt{\frac{c/2 + x_0}{c/2 - x_0}} \frac{d\eta_c}{dx}(x_0) \frac{dx_0}{x - x_0} \right\} \quad (7.19)$$

The x component of velocity on the airfoil surface is then given by $u = Q_\infty + u_{1T} \pm u_{1L}$ and the z component is obtained from the boundary conditions (Eq. (7.14a,b)).

7.2 Second-Order Solution

Consider the second-order solution. Define fictitious thickness and camber functions as

$$\eta_{c2} \equiv \eta_c \frac{u_{1T}}{Q_\infty} + \eta_t \frac{u_{1L}}{Q_\infty} \quad (7.20a)$$

$$\eta_{t2} \equiv \eta_t \frac{u_{1T}}{Q_\infty} + \eta_c \frac{u_{1L}}{Q_\infty} \quad (7.20b)$$

This puts the second-order problem in the same form as the first-order one at zero angle of attack (see boundary conditions in Eqs. (7.14)–(7.16)) and the solution can be written as

$$u_{2T} = \frac{Q_\infty}{\pi} \int_{-c/2}^{c/2} \frac{d\eta_{t2}}{dx}(x_0) \frac{dx_0}{x - x_0} \quad (7.21)$$

$$u_{2L} = \frac{Q_\infty}{\pi} \sqrt{\frac{c/2 - x}{c/2 + x}} \int_{-c/2}^{c/2} \sqrt{\frac{c/2 + x_0}{c/2 - x_0}} \frac{d\eta_{c2}}{dx}(x_0) \frac{dx_0}{x - x_0} \quad (7.22)$$

7.2 Second-Order Solution

The x component of velocity on the airfoil surface is then given as

$$u = Q_\infty - \frac{Q_\infty \alpha^2}{2} + u_{1T} + u_{2T} \pm u_{1L} \pm u_{2L} \tag{7.23}$$

and the z component is obtained from the boundary conditions (Eqs. (7.15) and (7.16)).

The surface speed on the airfoil is the magnitude of the velocity and to obtain its value at any order the velocity components at that order are substituted into

$$q = \sqrt{u^2 + w^2}$$

the expansion for the square root

$$(1+x)^{1/2} = 1 + \frac{x}{2} - \frac{x^2}{8} + \cdots \qquad \text{for } x < 1 \tag{7.24}$$

is used, the results are evaluated in terms of values on the chordline, and only terms up to the desired order are kept. The expressions for the surface speed correct to first and second order are derived as follows. On the surface, to second order,

$$q^2 = u^2 + w^2 = \left[Q_\infty\left(1 - \frac{\alpha^2}{2}\right) + \frac{\partial \Phi_1}{\partial x} + \frac{\partial \Phi_2}{\partial x}\right]^2 + \left[Q_\infty \alpha + \frac{\partial \Phi_1}{\partial z} + \frac{\partial \Phi_2}{\partial z}\right]^2$$

If the results are evaluated at $z = 0$ and terms up to second order are kept, we get

$$q_2^2 = Q_\infty^2 + 2Q_\infty(u_{1T} \pm u_{1L} + u_{2T} \pm u_{2L}) + 2Q_\infty \eta \frac{\partial}{\partial z}\frac{\partial \Phi_1}{\partial x}$$
$$+ (u_{1T} \pm u_{1L})^2 + 2Q_\infty \alpha \frac{\partial \Phi_1}{\partial z}(x, 0\pm) + \left[\frac{\partial \Phi_1}{\partial z}(x, 0\pm)\right]^2$$

Note that $(\partial^2 \Phi_1/\partial x \partial z)(x, 0\pm) = (\partial/\partial x)(\partial \Phi_1/\partial z)(x, 0\pm) = Q_\infty \frac{\partial}{\partial x}(\eta' - \alpha) = Q_\infty \eta''$. Then,

$$\frac{q_2^2}{Q_\infty^2} = 1 + \frac{2}{Q_\infty}(u_{1T} \pm u_{1L} + u_{2T} \pm u_{2L}) + \frac{(u_{1T} \pm u_{1L})^2}{Q_\infty^2}$$
$$+ 2\alpha(\eta' - \alpha) + 2\eta\eta'' + (\eta' - \alpha)^2$$
$$= 1 + \frac{2}{Q_\infty}(u_{1T} \pm u_{1L} + u_{2T} \pm u_{2L}) + \frac{(u_{1T} \pm u_{1L})^2}{Q_\infty^2} - \alpha^2 + (\eta')^2 + 2\eta\eta''$$

With the use of Eq. (7.24) we get

$$\frac{q_2}{Q_\infty} = 1 + \frac{1}{2}\left[\frac{2}{Q_\infty}(u_{1T} \pm u_{1L} + u_{2T} \pm u_{2L})\right.$$
$$\left. + \frac{(u_{1T} \pm u_{1L})^2}{Q_\infty^2} - \alpha^2 + (\eta')^2 + 2\eta\eta''\right] - \frac{1}{8}\frac{4}{Q_\infty^2}(u_{1T} \pm u_{1L})^2$$

Therefore,

$$\frac{q_1}{Q_\infty} = 1 + \frac{u_{1T}}{Q_\infty} \pm \frac{u_{1L}}{Q_\infty} \tag{7.25a}$$

$$\frac{q_2}{Q_\infty} = 1 + \frac{u_{1T}}{Q_\infty} \pm \frac{u_{1L}}{Q_\infty} + \frac{u_{2T}}{Q_\infty} \pm \frac{u_{2L}}{Q_\infty} + (\eta_c \pm \eta_t)(\eta_c'' \pm \eta_t'') + \frac{1}{2}(\eta_c' \pm \eta_t')^2 - \frac{\alpha^2}{2} \tag{7.25b}$$

The surface pressure coefficient (correct to second order) is obtained from Bernoulli's equation as follows:

$$C_p = 1 - \frac{q_2^2}{Q_\infty^2}$$

Compare the expression for q_2^2/Q_∞^2 above to the expression for q_2/Q_∞ in Eq. (7.25b) to observe that

$$\frac{q_2^2}{Q_\infty^2} = 1 + 2\left(\frac{q_2}{Q_\infty} - 1\right) + \frac{(u_{1T} \pm u_{1L})^2}{Q_\infty^2}$$

and therefore

$$C_p = -2\left(\frac{q_2}{Q_\infty} - 1\right) - \left(\frac{u_{1T}}{Q_\infty} \pm \frac{u_{1L}}{Q_\infty}\right)^2 \tag{7.26}$$

The airfoil lift coefficient can then be determined from

$$C_l = \frac{1}{c}\int_{-c/2}^{c/2} [C_p(x, 0-) - C_p(x, 0+)]\, dx \tag{7.27}$$

and with the use of Eqs. (7.25b) and (7.26) is

$$C_l = \frac{4}{c}\int_{-c/2}^{c/2} \left\{ \frac{u_{1L}}{Q_\infty} + \frac{u_{1T} u_{1L}}{Q_\infty^2} + \frac{u_{2L}}{Q_\infty} + \eta_t \eta_c'' + \eta_c \eta_t'' + \eta_c' \eta_t' \right\} dx \tag{7.28}$$

To illustrate the results of second-order thin-airfoil theory, consider a symmetric airfoil at angle of attack and the surface speed is to be calculated. The following thickness function represents a symmetric Joukowski airfoil to second order in thickness ratio (see Van Dyke,[5.3] p. 54):

$$\eta_t = \frac{c\tau_1}{2}\sqrt{1 - \frac{4x^2}{c^2}}\left(1 - \frac{2x}{c}\right) \tag{7.29}$$

where $\tau_1 = 4\tau/3\sqrt{3}$ and τ is the thickness ratio. To evaluate the Cauchy principal value integrals appearing in the equations for the x component of velocity at various orders, it will be advantageous to use Appendix A, which is reproduced from Ref. 7.2. Therefore, lengths must be scaled by half the chord length to obtain the limits of integration from -1 to $+1$. Introduce the nondimensional coordinate $\bar{x} = x/(c/2)$. The nondimensional thickness function for the Joukowski airfoil becomes

$$\bar{\eta}_t \equiv \frac{\eta_t}{c/2} = \tau_1(1 - \bar{x})\sqrt{1 - \bar{x}^2} \tag{7.30}$$

The nondimensional versions of Eqs. (7.17) and (7.19) for this symmetric airfoil become

$$\frac{u_{1T}}{Q_\infty} = \frac{1}{\pi}\int_{-1}^{1} \frac{d\bar{\eta}_t}{d\bar{x}}(\bar{x}_0)\frac{d\bar{x}_0}{\bar{x} - \bar{x}_0} \tag{7.31a}$$

$$\frac{u_{1L}}{Q_\infty} = \alpha\sqrt{\frac{1 - \bar{x}}{1 + \bar{x}}} \tag{7.31b}$$

The slope of the nondimensional thickness function is

$$\frac{d\bar{\eta}_t}{d\bar{x}} = \tau_1(1 - \bar{x}^2)^{-1/2}(-1 - \bar{x} + 2\bar{x}^2) \tag{7.32}$$

7.3 Leading-Edge Solution

and the first-order x component of velocity becomes

$$\frac{u}{Q_\infty} = 1 + \tau_1(1 - 2\bar{x}) \pm \alpha\sqrt{\frac{1-\bar{x}}{1+\bar{x}}} \tag{7.33}$$

The nondimensional thickness and camber functions at second order (Eqs. (7.20a,b)) are then

$$\bar{\eta}_{t2} = \tau_1^2\sqrt{1-\bar{x}^2}(1 - 3\bar{x} + 2\bar{x}^2) \tag{7.34a}$$

$$\bar{\eta}_{c2} = \tau_1\alpha(1-\bar{x})^2 \tag{7.34b}$$

and their nondimensional derivatives are

$$\frac{d\bar{\eta}_{t2}}{d\bar{x}} = \frac{3\tau_1^2}{\sqrt{1-\bar{x}^2}}(-1 + \bar{x} + 2\bar{x}^2 - 2\bar{x}^3) \tag{7.35a}$$

$$\frac{d\bar{\eta}_{c2}}{d\bar{x}} = -2\tau_1\alpha(1-\bar{x}) \tag{7.35b}$$

The second-order result for the x component of velocity is obtained from the nondimensional version of Eqs. (7.21), (7.22), and (7.23) and is

$$\frac{u}{Q_\infty} = 1 - \frac{\alpha^2}{2} + \tau_1(1 - 2\bar{x}) \pm \alpha\sqrt{\frac{1-\bar{x}}{1+\bar{x}}} - 6\tau_1^2\bar{x}\sqrt{\frac{1-\bar{x}}{1+\bar{x}}} \mp 2\tau_1\alpha\bar{x}\sqrt{\frac{1-\bar{x}}{1+\bar{x}}} \tag{7.36}$$

The second-order result for the surface speed from Eq. (7.25b) is obtained with some manipulation as

$$\frac{q_2}{Q_\infty} = 1 + \tau_1(1 - 2\bar{x}) \pm \alpha\sqrt{\frac{1-\bar{x}}{1+\bar{x}}}$$
$$- \frac{1}{2}\tau_1^2\frac{1-\bar{x}}{1+\bar{x}}(1+2\bar{x})^2 \mp 2\tau_1\alpha\bar{x}\sqrt{\frac{1-\bar{x}}{1+\bar{x}}} - \frac{\alpha^2}{2} \tag{7.37}$$

The first- and second-order surface speeds for a 10% thick Joukowski airfoil are shown in Fig. 7.2 and compared to the exact result from Chapter 6 for the case with zero angle of attack. It is noted that the first-order solution is not singular at the leading edge but that the leading-edge stagnation point and the acceleration region following it are not predicted by the theory. This is not surprising since the approximations of the theory are invalid in the neighborhood of a stagnation point and round edge. The deceleration region over the rear of the airfoil appears to be predicted well by the theory. The second-order surface speed improves the comparison with the exact results over most of the foil (including the maximum speed) but is now singular at the leading edge. If we were to continue to higher order, the solution would become more and more singular at the leading edge as the thin-airfoil theory is not able to predict the correct behavior in this region.

7.3 Leading-Edge Solution

Van Dyke[5.3] (pp. 50–52) shows that the perturbation expansion for the thin airfoil breaks down in the neighborhood of the round leading edge in a region whose extent is measured by the leading-edge or nose radius of the airfoil r (r is the radius of curvature

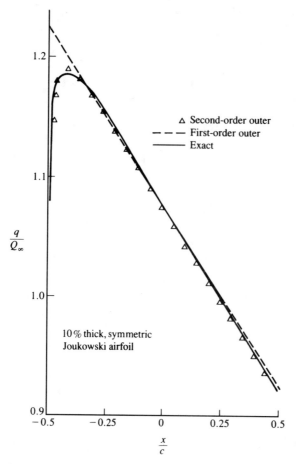

Figure 7.2 Surface speed results for 10% thick symmetric Joukowski airfoil.

at the leading edge). Also, r is $O(\epsilon^2)$. To illustrate the correct local solution in the neighborhood of the leading edge, let us consider a symmetric airfoil at zero angle of attack. Introduce the coordinate $s = x + c/2$, which is measured from the leading edge (Fig. 7.3). Many symmetric low-speed airfoil sections are analytic in the leading-edge region and their surfaces can be described by

$$z = \pm T_0 s^{1/2} \pm T_1 s^{3/2} \pm \cdots \qquad (7.38)$$

where $T_0, T_1, \ldots$ are constants. For small values of s (or for $s = O(\epsilon^2)$), the surface is given by the first term

$$z = \pm\sqrt{T_0^2 s} \qquad (7.39)$$

which is seen to be identical to the equation of a semi-infinite parabola, which can also be given by

$$z = \pm\sqrt{2rs} \qquad (7.40)$$

The local solution then is the symmetric flow past this parabola whose geometry is shown in Fig. 7.3 and since this solution is not valid in the far field, let us for the moment denote

7.3 Leading-Edge Solution

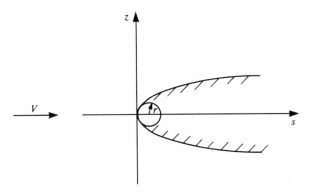

Figure 7.3 Symmetric flow past semi-infinite parabola.

the stream speed as V. The method of conformal mapping will be used to obtain the surface speed on the parabola. Consider the mapping

$$Y = -f^2 = \eta^2 - \xi^2 - 2i\xi\eta \tag{7.41}$$

where $Y = x + iz$ and $f = \xi + i\eta$. Then it can be seen that the curve $\xi = \xi_0$ in the f plane maps into the parabola

$$z = \pm\sqrt{4\xi_0^2(x + \xi_0^2)} \tag{7.42}$$

in the Y plane and the corresponding flowfields in the two planes are shown in Fig. 7.4.

The flow in the f plane is seen to be stagnation point flow against the wall $\xi = \xi_0$ and its complex potential is

$$F = -V(f - \xi_0)^2 \tag{7.43}$$

The constant V has been chosen to provide the correct far field solution in the parabola or physical plane. On the surface we have $f = i\eta$, $\xi = \xi_0$, and the complex velocity becomes

$$W(Y) = \frac{dF/df}{dY/df} = \frac{-2V(f - \xi_0)}{-2f} = \frac{Vi\eta}{(\xi_0 + i\eta)} \tag{7.44}$$

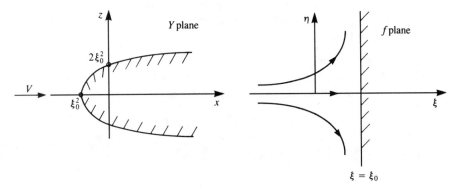

Figure 7.4 Mapping from parabola flowfield to stagnation flowfield.

Now, $\eta = [x + \xi_0^2]^{\frac{1}{2}}$ and if we introduce $s = x + \xi_0^2$, the surface speed on the parabola is

$$\frac{q}{V} = \sqrt{\frac{s}{s + \xi_0^2}} \tag{7.45}$$

Since the nose radius of the parabola is $r = 2\xi_0^2$ (Eq. (7.42) yields $z = \pm[4\xi_0^2 s]^{\frac{1}{2}}$), the desired local surface speed for the airfoil becomes

$$\frac{q}{V} = \sqrt{\frac{s}{s + r/2}} \tag{7.46}$$

The corresponding local solution for the airfoil problem with camber and angle of attack is given in Van Dyke.[7.2] We therefore have available two incomplete solutions to the problem we set out to solve at the beginning of the chapter. The thin-airfoil solution has been obtained correct to second order but it is not correct in the neighborhood of the leading edge. The local solution is exact in the neighborhood of the leading edge but does not describe the flow in the far field and it also contains an undetermined constant.

7.4 Matched Asymptotic Expansions

We will use the method of matched asymptotic expansions to obtain a solution that is uniformly valid over the airfoil surface. Our results will be presented essentially in outline form and further details are available in Van Dyke.[5.3] The success of the method is predicated on the observation that the two solutions complement each other and it is expected that in the limits of their applicability they approach each other (the limit of the thin-airfoil solution as the leading edge is approached should somehow be equivalent to the limit of the local solution as the distance from the leading edge is increased). The local solution is called the inner solution and the thin-airfoil solution is called the outer solution.

The formal task of matching the inner and outer solutions is achieved through the asymptotic matching principle (Van Dyke,[5.3] p. 90):

> The m-term inner expansion of the n-term outer expansion = the n-term outer expansion of the m-term inner expansion

Outer variables are scaled with the airfoil chord and inner variables are scaled with the nose radius, and m and n are integers, not necessarily equal. The definition of the m-term inner expansion of the n-term outer expansion is expressed in the following sequence of steps:

1. Rewrite the n-term outer expansion in inner variables.
2. Expand in an asymptotic series for small ϵ (or τ).
3. Keep m terms.

We will apply the above matching technique to the surface speed for the flow past a symmetric Joukowski airfoil at zero angle of attack. The three-term ($n = 3$) outer expansion in dimensionless coordinates is given in Eq. (7.37) as

$$\frac{q}{Q_\infty} = 1 + \tau_1(1 - 2\bar{x}) - \frac{\tau_1^2}{2}\frac{1 - \bar{x}}{1 + \bar{x}}(1 + 2\bar{x})^2 \tag{7.47}$$

In terms of $\bar{s} = 2s/c = \bar{x} + 1$, Eq. (7.47) becomes

$$q = Q_\infty\left[1 + \tau_1(3 - 2\bar{s}) - \frac{\tau_1^2}{2}\frac{2 - \bar{s}}{\bar{s}}(2\bar{s} - 1)^2\right] \tag{7.47a}$$

7.4 Matched Asymptotic Expansions

To second order in the thickness τ, the airfoil can be represented locally by the parabola so that the two-term ($m = 2$) inner expansion is given in Eq. (7.46) as

$$\frac{q}{V} = \sqrt{\frac{s}{s+r/2}} = \sqrt{\frac{\bar{s}}{\bar{s}+r/c}} \tag{7.48}$$

The Joukowski airfoil (Eq. (7.30)) becomes

$$\bar{z} = \pm \tau_1 (2 - \bar{s})\sqrt{2\bar{s} - \bar{s}^2} \tag{7.49}$$

and as $\bar{s} \to 0$ we get the parabola $\bar{z} = \pm[8\tau_1^2 \bar{s}]^{1/2}$. From Eq. (7.40), the parabola is $\bar{z} = \pm[4\bar{s}r/c]^{1/2}$. The nose radius is therefore $r = 2c\tau_1^2$.

Let us now do the matching for q.

For the *two-term inner expansion* (already in outer variables), we have

$$V\sqrt{\frac{\bar{s}}{\bar{s} + 2\tau_1^2}} = \frac{V}{\sqrt{1 + 2\tau_1^2/\bar{s}}} \tag{7.50a}$$

which when expanded for small τ_1 gives

$$V\left(1 - \frac{\tau_1^2}{\bar{s}}\right) + O(\tau_1^4) \tag{7.50b}$$

Its *three-term outer expansion* is

$$V\left(1 - \frac{\tau_1^2}{\bar{s}}\right) \tag{7.50c}$$

For the *three-term inner expansion*, we have

$$Q_\infty \left[1 + \tau_1(3 - 2\bar{s}) - \frac{\tau_1^2}{2}\frac{2-\bar{s}}{\bar{s}}(2\bar{s} - 1)^2\right] \tag{7.51a}$$

which, rewritten in inner variables (note that $\bar{S} = \bar{s}/\tau_1^2$), is

$$Q_\infty \left[1 + \tau_1(3 - 2\tau_1^2 \bar{S}) - \frac{1}{2}\frac{2 - \tau_1^2 \bar{S}}{\bar{S}}(2\tau_1^2 \bar{S} - 1)^2\right] \tag{7.51b}$$

and expanded for small τ_1 gives

$$Q_\infty \left[1 + 3\tau_1 - \frac{1}{\bar{S}}\right] + O(\tau_1^2) \tag{7.51c}$$

Its *two-term inner expansion* is

$$Q_\infty \left[1 - \frac{1}{\bar{S}} + 3\tau_1\right] = Q_\infty \left[1 + 3\tau_1 - \frac{\tau_1^2}{\bar{s}}\right] \tag{7.51d}$$

The matching is complete when we equate the results for q from Eqs. (7.50c) and (7.51d) to get $V = Q_\infty(1 + 3\tau_1)$. The local solution therefore experiences a free-stream speed that is larger than the actual one.

The final step in the analysis is to combine the inner and outer solutions to obtain a solution valid over the complete airfoil surface. At best the solution will be as accurate as either the inner or outer expansions in their regions of applicability. The combined solution is called a *composite expansion* (Van Dyke,[5.3] pp. 94–97) and we will use the additive composite

$$f_c^{(m,n)} = f_i^{(m)} + f_o^{(n)} - [f_o^{(n)}]_i^m, \qquad m = 2, n = 3 \tag{7.52}$$

The additive composite expansion is the sum of the inner and outer expansions minus the part they have in common (i refers to inner and o refers to outer). This common part, the last term in Eq. (7.52), is obtained during the matching process and is given in Eq. (7.50c). Our result is

$$\frac{q}{Q_\infty} = (1 + 3\tau_1)\sqrt{\frac{\bar{s}}{\bar{s} + 2\tau_1^2}} + 1 + \tau_1(3 - 2\bar{s})$$
$$- \frac{\tau_1^2}{2}\frac{2 - \bar{s}}{\bar{s}}(2\bar{s} - 1)^2 - \left[1 + 3\tau_1 - \frac{\tau_1^2}{\bar{s}}\right] \tag{7.53}$$

After some manipulations this result becomes

$$\frac{q}{Q_\infty} = (1 + 3\tau_1)\sqrt{\frac{\bar{s}}{\bar{s} + 2\tau_1^2}} - 2\tau_1\bar{s} + \frac{\tau_1^2}{2}(2\bar{s} - 3)^2 \tag{7.53a}$$

The important feature to note in the solution is that the singular part of the thin-airfoil result in the neighborhood of the leading edge has been removed. (Figure 7.5 compares the inner, outer, and composite expansions for a particular value of the thickness).

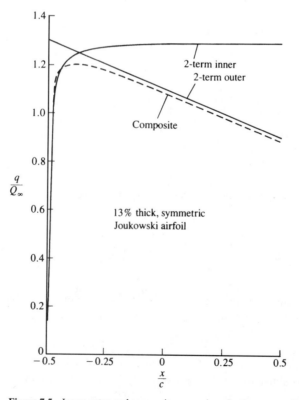

Figure 7.5 Inner, outer, and composite expansions for the symmetric Joukowski airfoil.

7.5 Thin Airfoil between Wind Tunnel Walls

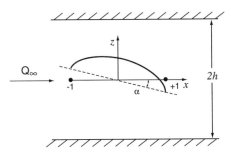

Figure 7.6 Coordinate system for thin airfoil between wind-tunnel walls.

7.5 Thin Airfoil between Wind Tunnel Walls

The zero-thickness airfoil between wind-tunnel walls problem will be studied as an example of a perturbation expansion for a case with two small parameters, the standard thin-airfoil parameter (camber or angle of attack), and the chord to tunnel height ratio. A thin airfoil is placed in a stream along the x axis (see Fig. 7.6) with its midchord in the center of a wind tunnel of height h chords. We will consider a solution linear in camber and angle of attack (first-order thin-airfoil theory) for $h \gg 1$. It is convenient to use dimensionless variables with lengths scaled by the semichord, speeds by the free stream speed, and the velocity potential by the product of the two. For simplicity, we will drop the bars on the dimensionless variables.

The airfoil boundary condition is transferred to the strip on the x axis with $-1 \leq x \leq 1$, and the mathematical problem for the perturbation velocity potential becomes

$$\nabla^2 \Phi = 0 \tag{7.54}$$

$$\frac{\partial \Phi}{\partial z}(x, 0\pm) = \frac{d\eta_c}{dx} - \alpha, \qquad -1 \leq x \leq 1 \tag{7.55}$$

$$\frac{\partial \Phi}{\partial z}(x, \pm h) = 0 \tag{7.56}$$

Equation (7.55) is the airfoil boundary condition from Eqs. (7.10) and (7.13a) and Eq. (7.56) is the wind-tunnel walls boundary condition. A Kutta condition must be applied at the airfoil trailing edge to complete the problem specification.

The solution is modeled by a distribution of vortices of circulation $\gamma(x)$ per unit length along the strip $-1 \leq x \leq 1$, $z = 0$ and corresponding image distributions are added (see Eq. (6.89)) to satisfy Eq. (7.56). The complex perturbation velocity potential for this flow is

$$f(Y) = \Phi + i\Psi = \frac{i}{2\pi} \int_{-1}^{1} \gamma(x_0) \ln \tanh \frac{\pi(Y - x_0)}{4h} dx_0 \tag{7.57}$$

Substitution of Eq. (7.57) in the airfoil boundary condition (Eq. (7.55)) results in an integral equation for the unknown circulation density $\gamma(x)$. This integral equation is written

$$\int_{-1}^{1} \gamma(x_0) K(x - x_0) dx_0 = 2\pi \left(\alpha - \frac{d\eta_c}{dx} \right) \tag{7.58}$$

where the kernel function is

$$K(x) = \frac{\pi}{4h} \left(\coth \frac{\pi x}{4h} - \tanh \frac{\pi x}{4h} \right) \tag{7.59}$$

To solve the integral equation we use an approach due to Keldysh and Lavrentiev[7.3] and seek an expansion in $1/h$ of the following form:

$$K(x) = \frac{1}{x} + h^{-1} \sum_{n=0}^{\infty} K_n \left(\frac{x}{h}\right)^n \tag{7.60a}$$

$$\gamma(x) = \sum_{n=0}^{\infty} h^{-n} \gamma_n(x) \tag{7.60b}$$

The expansion coefficients can be found with the use of Eqs. (5.113) and it is seen that only K_n with odd subscripts are nonzero and the first two are

$$K_1 = -\frac{\pi^2}{24}, \qquad K_3 = \frac{7\pi^4}{5760} \tag{7.61}$$

Equations (7.60) are substituted into the integral equation (Eq. (7.58)) and terms with like powers of $1/h$ are collected. The following system of thin-airfoil-like equations for the unknown $\gamma_n(x)$ is obtained:

$$\int_{-1}^{1} \frac{\gamma_0(x_0)}{x - x_0} \, dx_0 = 2\pi \left[\alpha - \frac{d\eta_c}{dx} \right] \equiv f_0(x) \tag{7.62a}$$

$$\int_{-1}^{1} \frac{\gamma_n(x_0)}{x - x_0} \, dx_0 = -\int_{-1}^{1} \sum_{m=0}^{n-1} K_m (x - x_0)^m \gamma_{n-m-1}(x_0) \, dx_0 \equiv f_n(x) \tag{7.62b}$$

The solution to Eqs. (7.62) that satisfies the Kutta condition is obtained with the help of Eq. (7.19) as

$$\gamma_n(x) = \frac{1}{\pi^2} \sqrt{\frac{1-x}{1+x}} \int_{-1}^{1} \sqrt{\frac{1+x_0}{1-x_0}} \frac{f_n(x_0)}{x_0 - x} \, dx_0 \tag{7.63}$$

The singular integrals are to be considered in the Cauchy principal value sense. For $n = 0$, the unbounded fluid result is recovered:

$$\gamma_0(x) = 2\sqrt{\frac{1-x}{1+x}} \left[\alpha + \frac{1}{\pi} \int_{-1}^{1} \sqrt{\frac{1+x_0}{1-x_0}} \frac{d\eta_c}{dx}(x_0) \frac{dx_0}{x - x_0} \right] \tag{7.64}$$

Let us find the first term in the expansions for the camber and angle of attack problems separately. Note that each expansion has all odd terms so that the terms we neglect are two orders smaller than the ones we keep. Since the camber problem requires the choice of a particular airfoil to proceed, let us begin with the angle of attack problem.

The expansion for the circulation density has terms for $n = 0, 2, 4, \ldots$, and for $n = 2$, the function on the right-hand side of the integral equation (Eq. (7.62b)) is

$$f_2 = -2K_1 \alpha \int_{-1}^{1} \sqrt{\frac{1-x_0}{1+x_0}} (x - x_0) \, dx_0$$

$$= \frac{\pi^2 \alpha}{12} \int_{-1}^{1} \frac{(1-x_0)(x-x_0)}{\sqrt{1-x_0^2}} \, dx_0 = \frac{\pi^3 \alpha}{12} (x + 1/2) \tag{7.65}$$

and the solution for the circulation density for $n = 2$ is found from Eq. (7.63) as

$$\gamma_2 = \frac{\pi \alpha}{12} \sqrt{\frac{1-x}{1+x}} \int_{-1}^{1} \sqrt{\frac{1+x_0}{1-x_0}} \frac{x_0 + 1/2}{x_0 - x} \, dx_0 = \frac{\pi^2 \alpha}{12} \sqrt{\frac{1-x}{1+x}} (x + 3/2) \tag{7.66}$$

(Note that $\int_{-1}^{1}[(1+x_0)/(1-x_0)]^{1/2}[f(x_0)/(x_0-x)]dx_0 = \int_{-1}^{1}[(1+x_0)/(1-x_0^2)^{1/2}]$ $[f(x_0)/(x_0-x)]dx_0$ is introduced so that the integrals in Appendix A can be used).

As an example to illustrate the camber effect, choose the parabolic arc camberline given by

$$\eta_c = \beta(1-x^2)$$

where $\beta = 2\epsilon/c$ (see Eq. (5.80)). The expansion for the circulation density has terms for $n = 0, 2, 4, \ldots$, and the unbounded fluid result ($n = 0$) is found from Eq. (7.64) as

$$\gamma_0(x) = -\frac{4\beta}{\pi}\sqrt{\frac{1-x}{1+x}}\int_{-1}^{1}x_0\sqrt{\frac{1+x_0}{1-x_0}}\frac{dx_0}{x-x_0} = 4\beta\sqrt{1-x^2} \quad (7.67)$$

For $n = 2$, the function on the right-hand side of the integral equation (Eq. (7.62b)) is

$$f_2 = \frac{\pi^2\beta}{6}\int_{-1}^{1}(x-x_0)\sqrt{1-x_0^2}\,dx_0 = \frac{\pi^3\beta x}{12} \quad (7.68)$$

and the solution for the circulation density for $n = 2$ is found from Eq. (7.63) as

$$\gamma_2 = \frac{\pi\beta}{12}\sqrt{\frac{1-x}{1+x}}\int_{-1}^{1}x_0\sqrt{\frac{1+x_0}{1-x_0}}\frac{dx_0}{x_0-x} = \frac{\pi^2\beta}{12}\sqrt{1-x^2} \quad (7.69)$$

The lift coefficient for the airfoil is found from the nondimensional circulation density as

$$C_l = \int_{-1}^{1}\gamma(x)\,dx \quad (7.70)$$

since the Kutta–Joukowski theorem can be used for this small-disturbance approximation. The lift coefficients for the separate angle of attack and camber effects that include the first term in the expansion are obtained by substituting Eqs. (7.66) and (7.69) into Eq. (7.70) to get

Angle of attack:

$$C_l = 2\pi\alpha\left[1 + \frac{\pi^2}{24}h^{-2} + O(h^{-4})\right] \quad (7.71a)$$

Parabolic arc camber:

$$C_l = 2\pi\beta\left[1 + \frac{\pi^2}{48}h^{-2} + O(h^{-4})\right] \quad (7.71b)$$

Note that the first term in Eq. (7.71a) is identical to the result obtained in Section 5.5 using a single-element lumped-vortex model. Additional terms for the angle of attack and parabolic arc solutions may be found in Plotkin.[7.4] It is seen that for these examples and for the assumptions connected with the expansions the wind-tunnel walls increase the lift due to angle of attack and camber.

References

[7.1] Newman, J. N., *Marine Hydrodynamics*, The MIT Press, 1977.
[7.2] Van Dyke, M., "Second-Order Subsonic Airfoil Theory Including Edge Effects," NACA Rep. 1274, 1956.

[7.3] Keldysh, M. V., and Lavrentiev, M. A., "On the Motion of a Wing under the Surface of a Heavy Fluid," CAHI, Moscow, 1935; translation in Science Translation Service, STS-75, Cambridge, MA, 1949.

[7.4] Plotkin, A., "Wind Tunnel Corrections for Lifting Thin Airfoils," *Journal of Applied Mechanics*, Vol. 49, June 1982, pp. 448–450.

Problems

7.1. Find the second-order surface speed for the flow past a thin ellipse at zero angle of attack.

7.2. Consider the flow of a uniform stream of speed Q_∞ past a wavy wall given by $z = \epsilon \sin \alpha x$, where ϵ is small compared to the wavelength $2\pi/\alpha$. Find the second-order perturbation velocity potential.

7.3. Consider the two-dimensional flow of a uniform stream of speed Q_∞ normal to the chordline of a thin symmetric body given by

$$z = \pm\epsilon f(x), \qquad -\frac{c}{2} < x < \frac{c}{2}, \qquad \epsilon \ll 1$$

The solution for the velocity potential can be written

$$\Phi = \Phi_0 + \epsilon \Phi_1 + \epsilon^2 \Phi_2 + \cdots$$

where Φ_0 represents the flow normal to a flat plate of length c and satisfies the mathematical problem

$$\nabla^2 \Phi = 0$$

$$\frac{\partial \Phi_0}{\partial z}(x, 0\pm) = 0$$

$$\Phi_0 = Q_\infty z \quad \text{as} \quad z \to \infty$$

(a) Write the mathematical problem for Φ_1 (in terms of Φ_0).
(b) Find Φ_1.

CHAPTER 8

Three-Dimensional Small-Disturbance Solutions

In this chapter, three-dimensional small-disturbance solutions will be derived for some simple cases such as the large aspect ratio wing, the slender pointed wing, and the slender cylindrical body. Flow problems requiring more detailed geometries will be treated in the forthcoming chapters.

8.1 Finite Wing: The Lifting Line Model

The three-dimensional lifting wing problem was formulated in Chapter 4, and it is clear that obtaining an analytic solution of the integral equations is difficult. However, it is possible to approximate the lifting properties of a wing by a single lifting line, an approximation that will allow a closed-form solution. In spite of the considerable simplifications in this model it captures the basic features of three-dimensional lifting flows and predicts the reduction of lift slope and the increase in induced drag with decreasing aspect ratio.

8.1.1 Definition of the Problem

Consider a lifting, thin, finite wing (described in Section 4.5 and shown in Fig. 8.1) that is moving at a constant speed in an otherwise undisturbed fluid. The free stream of speed Q_∞ has a small angle of attack α, relative to the coordinate system attached to the wing.

The velocity field for this potential flow problem can be obtained by solving Laplace's equation for the perturbation potential Φ:

$$\nabla^2 \Phi = 0 \tag{8.1}$$

Following Section 4.5, the boundary condition requiring no flow across the wing solid surface will be approximated at $z = 0$, for the case of small angle of attack, by

$$\frac{\partial \Phi}{\partial z}(x, y, 0\pm) = Q_\infty \left(\frac{\partial \eta}{\partial x} - \alpha \right) \tag{8.2}$$

where $\eta = \eta_c(x, y)$ is the camber surface placed on the x–y plane and for simplicity the subscript c is omitted in this chapter. For modeling the lifting surface, a vortex distribution is selected (as formulated in Section 4.5). The unknown vortex distribution $\gamma_x(x, y)$ and $\gamma_y(x, y)$ (shown in Fig. 4.9) is placed on the wing's projected area at the $z = 0$ plane. The resulting integral equation is

$$\frac{-1}{4\pi} \int_{\text{wing+wake}} \frac{\gamma_y(x-x_0) - \gamma_x(y-y_0)}{[(x-x_0)^2 + (y-y_0)^2]^{3/2}} \, dx_0 \, dy_0 = Q_\infty \left(\frac{\partial \eta}{\partial x} - \alpha \right) \tag{8.3}$$

which should hold for any point on the wing. A proper (and unique) solution for the vortex distribution will have to fulfill the Kutta condition along the trailing edge, such that the vorticity component parallel to the trailing edge ($\gamma_{\text{T.E.}}$) is zero:

$$\gamma_{\text{T.E.}} = 0 \tag{8.4}$$

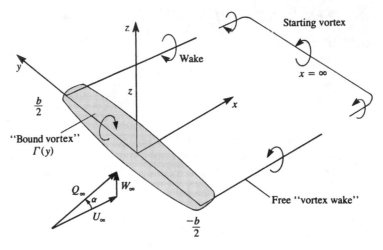

Figure 8.1 Far field horseshoe model of a finite wing.

Also, since vortex lines do not begin or end in a fluid (Eq. (4.64)), the solution must comply with

$$\left|\frac{\partial \gamma_x}{\partial x}\right| = \left|\frac{\partial \gamma_y}{\partial y}\right| \tag{8.5}$$

8.1.2 The Lifting-Line Model

The simplest model that can be suggested to solve this problem is where the chordwise circulation, at any spanwise station, is replaced by a single concentrated vortex. Also, these local vortices of circulation $\Gamma(y)$ will be placed along a single spanwise line. Based on the results of Section 5.5 for the two-dimensional lumped-vortex element, this vortex line will be placed at the wing's quarter-chord line along the span, $-b/2 < y < b/2$ (this bound vortex line is assumed to be straight and parallel to the y axis, as shown in Fig. 8.1). The above positioning of the vortex line at the wing's quarter-chord line effectively satisfies the Kutta condition of Eq. (8.4) as was shown in Section 5.5.

At this point, attention needs to be focused on the vortex theorems requiring that a vortex line cannot start or end abruptly in a fluid (or Eq. (8.5)). Therefore, if any change of the vortex line strength $d\Gamma(y)/dy$ is introduced, it must be followed by introducing a similar vorticity component in the other direction γ_x. In other words, the vortex line does not terminate at this point but it changes direction, and its strength remains constant.

The most physical application of these principles is to "shed" these trailing vortices into the flow and create a "wake" such that there will be no force acting on these free vortices. Following Section 4.7, this requirement reduces to the condition that the flow along this segment must be parallel to Γ (where positive Γ is according to the right-hand rule):

$$\mathbf{q} \times \boldsymbol{\Gamma}_{\text{wake}} = 0 \tag{8.6}$$

The most basic element that will fit these requirements will have the shape of a "horseshoe" vortex (Fig. 8.1), which will have constant "bound vorticity" Γ along its quarter-chord line, will turn backward at the wing tips and will continue far behind the wing, and eventually will be closed by the starting vortex. It is assumed here that the flow is steady and therefore the starting vortex is far downstream and its influence can be neglected.

8.1 Finite Wing: The Lifting Line Model

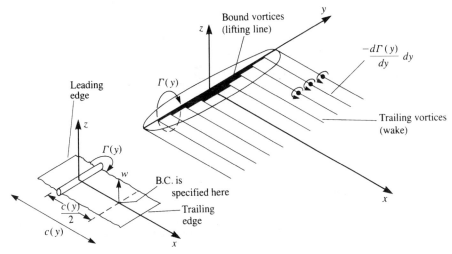

Figure 8.2 Lifting-line model consisting of horseshoe vortices. The bound vortex segment of all vortices is placed on the y axis.

A more refined model of the finite wing was first proposed by the German scientist Ludwig Prandtl (see Ref. 5.2) during World War 1 and it uses a large number of such spanwise horseshoe vortices, as illustrated by Fig. 8.2 (the following analysis is in the spirit of this early model). The straight bound vortex $\Gamma(y)$ in this case is placed along the y axis and at each spanwise station the leading edge is one-quarter chord ahead of this line and the local trailing edge is three-quarter chord behind the vortex line. Now, let us examine the integral equation (Eq. (8.3)) for the case of the flat lifting surface, where $\partial\eta/\partial x = 0$. The equation now simply states the boundary condition of Eq. (8.2):

$$\frac{\partial \Phi_{\text{wing}}}{\partial z} + \frac{\partial \Phi_{\text{wake}}}{\partial z} + Q_\infty \alpha = 0 \tag{8.2a}$$

That is, the sum of the normal velocity components induced by the wing $w_b = \partial\Phi_{\text{wing}}/\partial z$ and wake vortices $w_i = \partial\Phi_{\text{wake}}/\partial z$, plus the normal velocity component of the free-stream flow $Q_\infty \alpha$, must be zero on the wing's solid boundary:

$$w_b + w_i + Q_\infty \alpha = 0 \tag{8.7}$$

where w is considered to be positive in the $+z$ direction. The subscripts $(\ )_b$ and $(\ )_i$ stand for bound (on wing) and induced (by wake) influences, respectively.

The velocity component w_b induced by the lifting line on the section with a chord $c(y)$ can be estimated by using the lumped-vortex model with the downwash calculated at the collocation point located at the 3/4 chord. Consider the spanwise component ($-y_0 \leq y \leq y_0$) of a typical horseshoe vortex in Fig. 8.3 with strength $\Delta\Gamma(y_0)$ (where $\Delta\Gamma(y_0) = -(d\Gamma(y_0)/dy)dy_0$). The downwash Δw_b at the collocation point $(c/2, y)$ due to this segment is given by Eq. (2.69) (see Fig. 2.15, which defines the angles β in this formula) as

$$\Delta w_b = \frac{-\Delta\Gamma}{4\pi d}(\cos\beta_1 - \cos\beta_2)$$

$$= \frac{-\Delta\Gamma}{4\pi [c(y)/2]} \left[\frac{y + y_0}{\sqrt{(c/2)^2 + (y + y_0)^2}} + \frac{y_0 - y}{\sqrt{(c/2)^2 + (y_0 - y)^2}} \right]$$

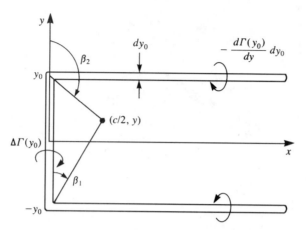

Figure 8.3 Velocity induced by the segments of a typical horseshoe element.

For a wing of large aspect ratio, we can neglect $(c/2)^2$ in the square root terms to get

$$\Delta w_b = \frac{-\Delta\Gamma(y_0)}{4\pi[c(y)/2]}[1+1]$$

The result for the complete lifting line (evaluated at y) is obtained by summing the results for all the horseshoe vortices and is

$$w_b = \frac{-\Gamma(y)}{2\pi[c(y)/2]} \qquad (8.8)$$

Note that this is identical to the result obtained by applying a locally two-dimensional lumped-vortex model at each spanwise station, where the downwash w_b is measured at the 3/4 chord due to a vortex $\Gamma(y)$ placed at the 1/4 chord position (see inset in the left-hand side of Fig. 8.2).

Next, the downwash due to the wing trailing vortices must be evaluated. Since a change in the spanwise circulation $\Gamma(y)$ is allowed, and since no vortex can begin or end in the flow, the local change in this circulation is shed into the wake. Thus, the wake is now constructed from semi-infinite vortex lines with the strength of $(d\Gamma/dy)dy$ (Fig. 8.2). Before we proceed with the solution, the velocity induced by a single, semi-infinite trailing vortex line with a strength of $\Delta\Gamma = -(d\Gamma(y_0)/dy)dy_0$ is evaluated (note that for positive $\Delta\Gamma$ on the $+y$ side of the wing, negative $d\Gamma/dy$ is needed). The right-hand-side wake vortex line is located at a spanwise location y_0, as shown in Fig. 8.3, and the downwash induced by this vortex at the collocation point $(c/2, y)$ is given by the result for a semi-infinite vortex line from Eq. (2.71). Since for a large aspect ratio wing $\beta_1 \approx \pi/2$, $\beta_2 \approx \pi$ (in Fig. 2.15) and therefore

$$w(y) = \frac{\Delta\Gamma(y_0)}{4\pi}\frac{1}{(y-y_0)} \qquad (8.9)$$

which is exactly one half of the velocity induced by an infinite (two-dimensional) vortex of strength $\Delta\Gamma(y_0)$. With the aid of this equation the normal velocity component induced by

8.1 Finite Wing: The Lifting Line Model

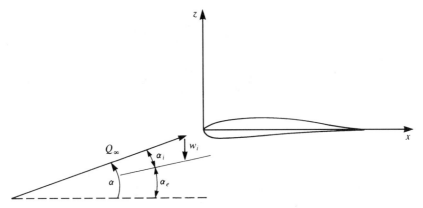

Figure 8.4 Two-dimensional section (in the $y = $ const. plane) of a three-dimensional wing. The angle of attack α is reduced by the induced downwash of the trailing vortices by α_i.

the trailing vortices of the wing becomes

$$w_i = \frac{1}{4\pi} \int_{-b/2}^{b/2} \frac{[-d\Gamma(y_0)/dy]\, dy_0}{y - y_0} \qquad (8.10)$$

(Note that since the trailing vortices are assumed to lie in the $z = 0$ plane, their induced spanwise velocity component is zero from the Biot–Savart law, Eq. (2.68b).) Assuming that the wing aspect ratio is large ($b/c(y) \gg 1$) has allowed us to treat a spanwise station as a *two-dimensional* section and to transfer the boundary condition to the local three-quarter chord. Substituting Eqs. (8.8) and (8.10) into Eq. (8.7) yields

$$\frac{-\Gamma(y)}{2\pi[c(y)/2]} - \frac{1}{4\pi} \int_{-b/2}^{b/2} \frac{[d\Gamma(y_0)/dy]\, dy_0}{y - y_0} + Q_\infty \alpha = 0 \qquad (8.11)$$

Dividing Eq. (8.11) by the free-stream speed Q_∞ results in

$$\frac{-\Gamma(y)}{\pi c(y) Q_\infty} - \frac{1}{4\pi Q_\infty} \int_{-b/2}^{b/2} \frac{[d\Gamma(y_0)/dy]\, dy_0}{y - y_0} + \alpha = 0 \qquad (8.11a)$$

This is the Prandtl lifting-line integrodifferential equation for the spanwise load distribution $\Gamma(y)$. This equation can be viewed as a combination of the angles (as shown in Fig. 8.4):

$$-\alpha_e - \alpha_i + \alpha = 0 \qquad (8.12)$$

where the induced downwash angle is (note that w is positive in the positive z direction)

$$\alpha_i \approx \frac{-w_i}{Q_\infty} \qquad (8.13)$$

Equation (8.12) can be rearranged as

$$\alpha_e = \alpha - \alpha_i \qquad (8.12a)$$

This means that in the case of the finite wing the effective angle of attack of a wing section α_e (the angle between the modified free-stream velocity **q** in Fig. 8.4 and the chord) is smaller than the actual geometric angle of attack α by α_i, which is a result of the downwash induced by the wake.

It is possible to generalize the result of this equation by assuming that the two-dimensional section has a local lift slope of m_0 and its local effective angle of attack is α_e. Now, if camber effects are to be accounted for too, then this angle is measured from the zero-lift angle of the section, such that

$$C_l(y) = \frac{\rho Q_\infty \Gamma(y)}{(1/2)\rho Q_\infty^2 c(y)} = m_0(y)\alpha_e(y) \tag{8.14}$$

Consequently, Eq. (8.12a) becomes

$$\alpha_e = \alpha - \alpha_i - \alpha_{L0} \tag{8.15}$$

where α_{L0} is the angle of zero lift due to the section camber (for cambered airfoils, usually α_{L0} is a negative number). A more general form of Eq. (8.11a) that allows for section camber and wing twist $\alpha(y)$ is now

$$\frac{-2\Gamma(y)}{m_0(y)c(y)Q_\infty} - \frac{1}{4\pi Q_\infty}\int_{-b/2}^{b/2}\frac{[d\Gamma(y_0)/dy]\,dy_0}{y - y_0} + \alpha(y) - \alpha_{L0}(y) = 0 \tag{8.16}$$

In this equation $\alpha(y)$ is the local angle of attack relative to $\mathbf{Q}_\infty$ and $\alpha_{L0}(y)$ is the airfoil section zero-lift angle. If it is assumed that these geometrical quantities are known, then $\Gamma(y)$ becomes the unknown in this equation. Also, at the wing tips the pressure difference [or the lift $\rho Q_\infty \Gamma(y = \pm b/2)$] must reduce to zero, that is,

$$\Gamma\left(y = \pm\frac{b}{2}\right) = 0 \tag{8.17}$$

8.1.3 The Aerodynamic Loads

The solution of Eq. (8.16) will provide the spanwise bound circulation distribution $\Gamma(y)$. To obtain the aerodynamic forces, the two-dimensional Kutta–Joukowski theorem will be applied (in the $y = $ const. plane). However, because of the wake-induced velocity, the free-stream vector will be rotated by $\alpha_i(y)$, as shown in Fig. 8.5. This angle can be calculated for a known $\Gamma(y)$ by using Eqs. (8.10) and (8.13):

$$\alpha_i = \frac{1}{4\pi Q_\infty}\int_{-b/2}^{b/2}\frac{[d\Gamma(y_0)/dy]\,dy_0}{y - y_0} \tag{8.18}$$

If we assume that α_i is small, then $\cos\alpha_i \approx 1$ and $\sin\alpha_i \approx \alpha_i$, and the lift of the wing is given by an integration of the local two-dimensional lift (see Eq. (3.113)) as

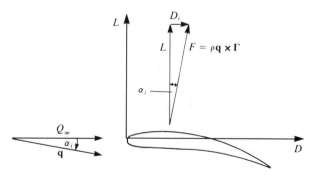

Figure 8.5 Tilting of the local (of section $y = $ const.) lift vector by the angle α_i induced by the trailing vortices.

8.1 Finite Wing: The Lifting Line Model

$$L = \rho Q_\infty \int_{-b/2}^{b/2} \Gamma(y)\, dy \tag{8.19}$$

while the drag force, which is created by turning the two-dimensional lift vector by the wake-induced flow, becomes

$$D_i = \rho Q_\infty \int_{-b/2}^{b/2} \alpha_i(y)\Gamma(y)\, dy \tag{8.20}$$

This drag is called *induced drag* because it is induced by the trailing vortices. This finite wing's drag is directly related to the lift and will diminish as the wingspan increases ($b \to \infty$). Equation (8.20) can also be rewritten in terms of the wake-induced downwash w_i:

$$D_i = -\rho \int_{-b/2}^{b/2} w_i(y)\Gamma(y)\, dy \tag{8.20a}$$

From the engineering point of view, the total drag D of a wing includes the induced drag D_i and the viscous drag D_0:

$$D = D_i + D_0$$

For example, the two-dimensional viscous drag of a NACA 0009 section is presented in Fig. 5.20.

8.1.4 The Elliptic Lift Distribution

The spanwise circulation distribution $\Gamma(y)$ for a given planform shape can be obtained by solving Eq. (8.16). In the particular case of an elliptic distribution of the circulation, the solution becomes rather simple since the downwash w_i becomes constant along the wing span. Also, as will be shown later, wings having such a spanwise distribution will have minimum induced drag. The proposed distribution of $\Gamma(y)$ is shown in Fig. 8.6 and is

$$\Gamma(y) = \Gamma_{\max}\left[1 - \left(\frac{y}{b/2}\right)^2\right]^{1/2} \tag{8.21}$$

This must be substituted into Eq. (8.16) so that the constant $\Gamma_{\max}$ can be evaluated. For simplicity, let us first calculate the downwash integral (second term in Eq. (8.16)). The term

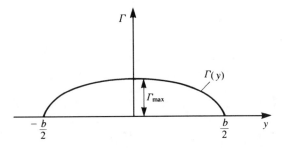

Figure 8.6 Elliptic spanwise distribution of the circulation $\Gamma(y)$.

$d\Gamma(y)/dy$ is evaluated by differentiating Eq. (8.21):

$$\frac{d\Gamma(y)}{dy} = \frac{\Gamma_{\max}}{2}\left[1 - \left(\frac{y}{b/2}\right)^2\right]^{-1/2}\left(-2\frac{4}{b^2}y\right)$$

and the downwash w_i is obtained by substituting this result into Eq. (8.10):

$$w_i(y) = \frac{\Gamma_{\max}}{\pi b^2}\int_{-b/2}^{b/2}\left[1 - \left(\frac{y_0}{b/2}\right)^2\right]^{-1/2}\frac{y_0}{y - y_0}\,dy_0 \tag{8.22}$$

Note that when $y = y_0$, this integral is singular and therefore must be evaluated based on Cauchy's principal value. It is possible to arrive at Glauert's integral (Eq. (5.22)) by the transformation

$$y = \frac{b}{2}\cos\theta \tag{8.23}$$

$$dy = -\frac{b}{2}\sin\theta\,d\theta \tag{8.23a}$$

and at the wingtips $y = -b/2$, $\theta = \pi$ and at $y = b/2$, $\theta = 0$. This reduces Eq. (8.21) to

$$\Gamma(\theta) = \Gamma_{\max}[1 - \cos^2\theta]^{1/2} = \Gamma_{\max}\sin\theta \tag{8.21a}$$

Substituting Eq. (8.23) into Eq. (8.22) yields

$$w_i = \frac{\Gamma_{\max}}{\pi b^2}\int_{\pi}^{0}[1 - \cos^2\theta_0]^{-1/2}\frac{(b/2)\cos\theta_0[(-b/2)\sin\theta_0]d\theta_0}{(b/2)(\cos\theta - \cos\theta_0)}$$

The principal value of this integral can be obtained by using the Glauert integral, Eq. (5.22):

$$w_i = \frac{-\Gamma_{\max}}{2\pi b}\int_0^{\pi}\frac{\cos\theta_0\,d\theta_0}{(\cos\theta_0 - \cos\theta)} = \frac{-\Gamma_{\max}}{2\pi b}\frac{\pi\sin\theta}{\sin\theta}$$

Consequently, w_i and α_i become

$$w_i = -\frac{\Gamma_{\max}}{2b} \tag{8.24}$$

$$\alpha_i = \frac{\Gamma_{\max}}{2bQ_\infty} \tag{8.24a}$$

and are constant along the wing span.

Another feature of the elliptic distribution is that the spanwise integral is simply half the area of an ellipse (with semi-axes $\Gamma_{\max}$ and $b/2$):

$$\int_{-b/2}^{b/2}\Gamma(y)\,dy = \frac{\pi}{2}\Gamma_{\max}\frac{b}{2} = \frac{\pi b}{4}\Gamma_{\max} \tag{8.25}$$

Consequently, the lift and the drag of the wing can be evaluated as

$$L = \rho Q_\infty\int_{-b/2}^{b/2}\Gamma(y)\,dy = \frac{\pi b}{4}\rho Q_\infty\Gamma_{\max} \tag{8.26}$$

$$D_i = \rho Q_\infty\int_{-b/2}^{b/2}\alpha_i\Gamma(y)\,dy = \alpha_i L = \left(\frac{\Gamma_{\max}}{2bQ_\infty}\right)\frac{\pi b}{4}\rho Q_\infty\Gamma_{\max} = \frac{\pi}{8}\rho\Gamma_{\max}^2 \tag{8.27}$$

8.1 Finite Wing: The Lifting Line Model

The lift and drag coefficients become

$$C_L \equiv \frac{L}{(1/2)\rho Q_\infty^2 S} = \frac{\pi}{2}\frac{b}{S}\frac{\Gamma_{max}}{Q_\infty} \tag{8.28}$$

$$C_{D_i} \equiv \frac{D_i}{(1/2)\rho Q_\infty^2 S} = \frac{\pi}{4S}\frac{\Gamma_{max}^2}{Q_\infty^2} = \frac{1}{\pi}\frac{S}{b^2}C_L^2 \tag{8.29}$$

Substituting the spanwise downwash (Eq. (8.24)) and the elliptic circulation distribution (Eq. (8.21)) into Eq. (8.16) yields

$$\frac{-2\Gamma_{max}}{m_0(y)c(y)Q_\infty}\left[1-\left(\frac{y}{b/2}\right)^2\right]^{1/2} - \frac{\Gamma_{max}}{2bQ_\infty} + \alpha(y) - \alpha_{L0}(y) = 0 \tag{8.30}$$

This equation provides the relation between the local chord $c(y)$ and the local angle of attack $\alpha(y)$ for the wing with the elliptic circulation distribution. If the chord $c(y)$ has an elliptic form, too, the constant Γ_{max} can be easily evaluated. Thus, assume

$$c(y) = c_0\left[1-\left(\frac{y}{b/2}\right)^2\right]^{1/2} \tag{8.31}$$

where c_0 is the root chord. Substituting Eq. (8.31) into Eq. (8.30) cancels the elliptic variation and we have

$$\frac{-2\Gamma_{max}}{m_0(y)c_0 Q_\infty} - \frac{\Gamma_{max}}{2bQ_\infty} + \alpha(y) - \alpha_{L0}(y) = 0 \tag{8.32}$$

For an elliptic planform with constant airfoil shape, all terms but $\alpha(y)$ in this equation are constant, and therefore this wing with an elliptic planform and load distribution is untwisted ($\alpha(y) = \alpha = $ const.). The value of Γ_{max} is then

$$\Gamma_{max} = \frac{2bQ_\infty(\alpha - \alpha_{L0})}{1 + 4b/m_0 c_0} \tag{8.33}$$

The area S of the elliptic wing is

$$S = \pi \frac{c_0}{2}\frac{b}{2} \tag{8.34}$$

Also, it is common to define the wing aspect ratio $\!R$ as

$$\!R \equiv \frac{b^2}{S} \tag{8.35}$$

Using the $\!R$ and the area S for the elliptic wing and substituting into Eq. (8.33), we obtain

$$\Gamma_{max} = \frac{2bQ_\infty(\alpha - \alpha_{L0})}{1 + \pi\!R/m_0} \tag{8.33a}$$

With this expression for Γ_{max} and using $m_0 = 2\pi$, the lift coefficient (Eq. (8.28)) becomes

$$C_L = \frac{2\pi}{1+2/\!R}(\alpha - \alpha_{L0}) \equiv C_{L_\alpha}(\alpha - \alpha_{L0}) \tag{8.36}$$

Here C_{L_α} is the three-dimensional wing lift slope and the most important conclusion of this analysis is that this slope becomes less as the wingspan becomes smaller due to the induced downwash. This is illustrated by Fig. 8.7, where for a wing with given α_{L0} the effective

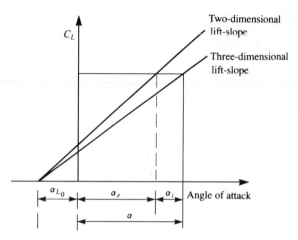

Figure 8.7 The reduction of lift slope for three-dimensional wings.

angle of attack is reduced by α_i according to Eq. (8.15). Consequently, for finite span wings, more incidence is needed to achieve the same lift coefficient as the wingspan decreases.

The induced drag coefficient is obtained by substituting Eq. (8.35) into Eq. (8.29):

$$C_{D_i} = \frac{1}{\pi \mathcal{R}} C_L^2 \qquad (8.37)$$

which indicates that as the wing aspect ratio increases the induced drag becomes smaller. Also, the induced drag for the finite elliptic wing will increase with a rate of C_L^2 as shown in Fig. 8.8.

The lift slope C_{L_α} versus $\mathcal{R}$ for the elliptic wing (Eq. (8.36)) is shown in Fig. 8.9. The lift slope of a two-dimensional wing is the largest (2π) and as the wingspan becomes smaller C_{L_α} decreases too.

The spanwise loading $L'(y)$ (lift per unit span) of the elliptic wing is obtained by using the Kutta–Joukowski theorem:

$$L'(y) = \rho Q_\infty \Gamma(y) = \rho Q_\infty \Gamma_{\max} \left[1 - \left(\frac{y}{b/2} \right)^2 \right]^{1/2} \qquad (8.38)$$

Figure 8.10 shows an elliptic planform and the spanwise lift distribution $L'(y)$ that is elliptic too. As Eq. (8.24) indicated, the downwash of such a wing is constant, and, combined with

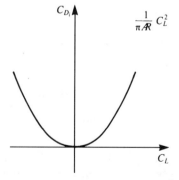

Figure 8.8 Lift polar for an elliptic wing.

8.1 Finite Wing: The Lifting Line Model

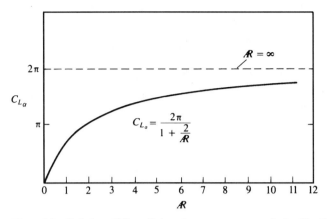

Figure 8.9 Variation of lift coefficient slope versus aspect ratio for thin elliptic wings.

the velocity induced by the bound vortex w_b, must be equal to the upwash of the free-stream $Q_\infty \alpha$ so that the combined normal velocity component is zero, according to Eq. (8.7). (Note that it is possible to have elliptic loading with other than an elliptic planform, but in that case, local twist or camber needs to be adjusted so that w_i will remain constant.)

The section lift coefficient C_l is defined by using the local chord from Eq. (8.31) and is

$$C_l \equiv \frac{L'(y)}{(1/2)\rho Q_\infty^2 c(y)} = \frac{2\Gamma_{max}}{c_0 Q_\infty} = C_L$$

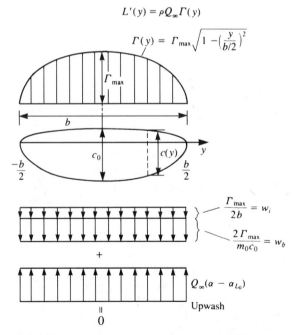

Figure 8.10 Chord and load distribution for a thin elliptic wing. Note that the induced downwash is constant and combined with the downwash of the bound vortex is equal to the free-stream upwash, resulting in zero velocity normal to the wing surface (Eq. (8.7)).

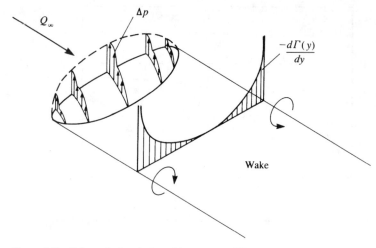

Figure 8.11 Schematic description of the pressure difference and wake vortex distribution of a thin elliptic wing.

Thus, for the elliptic wing, both section lift coefficient and wing lift coefficient are the same, that is,

$$C_l = \frac{2\pi}{1 + 2/\!R}(\alpha - \alpha_{L0}) \equiv C_{l_\alpha}(\alpha - \alpha_{L0}) \tag{8.39}$$

Similarly, the section induced drag coefficient is

$$C_{d_i} \equiv \frac{L'(y)\alpha_i}{(1/2)\rho Q_\infty^2 c(y)} = \frac{1}{\pi}\frac{S}{b^2}C_l^2 = C_{D_i} \tag{8.40}$$

The strength of the circulation in the wake is simply the spanwise derivative of $\Gamma(y)$ (see Eq. (8.21)),

$$\frac{d\Gamma(y)}{dy} = -\frac{4\Gamma_{max}}{b^2}\frac{y}{\sqrt{[1-(y/(b/2))^2]}} \tag{8.41}$$

This spanwise wake vortex strength is shown schematically in Fig. 8.11. It is clear from this figure that near the wingtips, where $|\Gamma(y)/dy|$ is the largest, the wake vortex will be the strongest. Therefore, owing to the induced velocity at the wake it will roll up, mostly near the wingtips, to form two concentrated trailing vortices as shown by the flow visualization in Fig. 8.12. The induced effect of this wake rollup on $\Gamma(y)$ is assumed to be negligible in this model; but this effect can be investigated by the numerical methods of later chapters.

8.1.5 General Spanwise Circulation Distribution

A more general solution for the spanwise circulation $\Gamma(y)$ in Eq. (8.16) can be obtained by describing the unknown distribution in terms of a trigonometric expansion. Using the spanwise coordinate θ, as defined in Eq. (8.23), we select the following Fourier expansion:

$$\Gamma(\theta) = 2bQ_\infty \sum_{n=1}^{\infty} A_n \sin n\theta \tag{8.42}$$

The shapes of the first three symmetric terms in this expansion are shown schematically in

8.1 Finite Wing: The Lifting Line Model

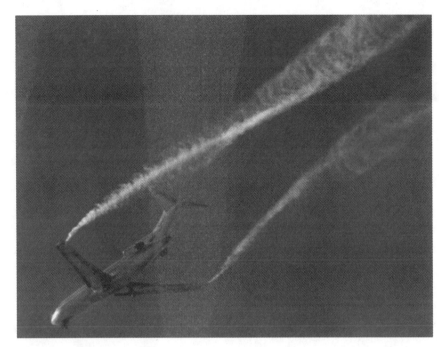

Figure 8.12 Flow visualization of the rollup of the trailing vortices behind an airplane (wing tip vortices made visible by ejecting smoke at the wing tips of a Boeing 727 airplane). (Courtesy of NASA.)

Fig. 8.13, and all terms fulfill Eq. (8.17) at the wingtips:

$$\Gamma(0) = \Gamma(\pi) = 0 \qquad (8.43)$$

Substituting $\Gamma(\theta)$ and $d\Gamma(\theta)/dy$ into Eq. (8.16) yields

$$\frac{-4b}{m_0(\theta)c(\theta)} \sum_{n=1}^{\infty} A_n \sin n\theta$$

$$+ \frac{-b}{2\pi} \int_\pi^0 \frac{\sum_{n=1}^{\infty} A_n n \cos n\theta_0 [1/(-b/2)\sin\theta_0][(-b/2)\sin\theta_0 d\theta_0]}{(b/2)(\cos\theta - \cos\theta_0)}$$

$$+\alpha(\theta) - \alpha_{L0}(\theta) = \frac{-4b}{m_0(\theta)c(\theta)} \sum_{n=1}^{\infty} A_n \sin n\theta - \frac{1}{\pi} \int_0^\pi \frac{\sum_{n=1}^{\infty} nA_n \cos n\theta_0\, d\theta_0}{\cos\theta_0 - \cos\theta}$$

$$+\alpha(\theta) - \alpha_{L0}(\theta) = 0 \qquad (8.44)$$

Using Glauert's integral (Eq. (5.22)) for the second term gives us

$$\frac{-4b}{m_0(\theta)c(\theta)} \sum_{n=1}^{\infty} A_n \sin n\theta - \sum_{n=1}^{\infty} nA_n \frac{\sin n\theta}{\sin\theta} + \alpha(\theta) - \alpha_{L0}(\theta) = 0 \qquad (8.44a)$$

Comparing this result with Eq. (8.16) indicates that the first term is $-\alpha_e$ and the second term is $-\alpha_i$:

$$\alpha_i(\theta) = \sum_{n=1}^{\infty} nA_n \frac{\sin n\theta}{\sin\theta} \qquad (8.45)$$

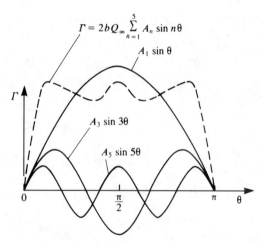

Figure 8.13 Sine series representation of symmetric spanwise circulation distribution $\Gamma(\theta)$, $n = 1, 3, 5$.

Therefore, the section lift and drag coefficients can be readily obtained:

$$C_l = \frac{\rho Q_\infty \Gamma(\theta)}{(1/2)\rho Q_\infty^2 c(\theta)} = \frac{4b}{c(\theta)} \sum_{n=1}^\infty A_n \sin n\theta \tag{8.46}$$

$$C_{d_i} = C_l \alpha_i = \frac{4b}{c(\theta)} \sum_{n=1}^\infty A_n \sin n\theta \left(\sum_{k=1}^\infty k A_k \frac{\sin k\theta}{\sin \theta} \right) \tag{8.47}$$

The wing aerodynamic coefficients are obtained by the spanwise integration of these section coefficients:

$$C_L = \int_{-b/2}^{b/2} \frac{C_l(y)c(y)\,dy}{S} = \frac{4b}{S} \int_0^\pi \sum_{n=1}^\infty A_n \sin n\theta \frac{b}{2} \sin \theta \, d\theta \tag{8.48}$$

$$C_{D_i} = \int_{-b/2}^{b/2} \frac{C_{d_i}(y)c(y)\,dy}{S} = \frac{2b^2}{S} \int_0^\pi \sum_{n=1}^\infty \sum_{k=1}^\infty k A_k A_n \sin k\theta \sin n\theta \, d\theta \tag{8.49}$$

Recall that

$$\int_0^\pi \sin n\theta \sin k\theta \, d\theta = \begin{pmatrix} 0 & \text{for} & n \neq k \\ \pi/2 & \text{for} & n = k \end{pmatrix} \tag{8.50}$$

and for the lift integral only the first term will appear. The lift coefficient becomes

$$C_L = \frac{\pi b^2 A_1}{S} = \pi \mathcal{R} A_1 \tag{8.51}$$

For the drag, only the terms where $n = k$ will be left:

$$C_{D_i} = \frac{\pi b^2}{S} \sum_{n=1}^\infty n A_n^2 = \pi \mathcal{R} \sum_{n=1}^\infty n A_n^2 \tag{8.52}$$

Using the results for the lift, we can rewrite this as

$$C_{D_i} = \frac{\pi^2 \mathcal{R}^2 A_1^2}{\pi \mathcal{R}} \left[1 + \sum_{n=2}^\infty \frac{n A_n^2}{A_1^2} \right] = \frac{C_L^2}{\pi \mathcal{R}} \left[1 + \sum_{n=2}^\infty \frac{n A_n^2}{A_1^2} \right] = (C_{D_i})_{\text{elliptic}}(1 + \delta_1) \tag{8.53}$$

8.1 Finite Wing: The Lifting Line Model

where δ_1 includes the higher order terms for $n = 2, 3, \ldots$ (only the odd terms are considered for a symmetric load distribution). This clearly indicates that for a given wing aspect ratio, the elliptic wing will have the lowest drag coefficient since for a generic wing planform $\delta_1 \geq 0$ and for the elliptic wing $\delta_1 = 0$.

Similarly, the lift coefficient for the general spanwise loading can be formulated as

$$C_L = \pi \mathcal{R} A_1 = m(\alpha - \alpha_{L_0}) \tag{8.54}$$

Assume that the wing is untwisted and therefore $\alpha - \alpha_{L_0} = \text{const}$. Following Glauert (Ref. 5.2, p. 142) we define an equivalent two-dimensional wing that has the same lift coefficient C_L. This wing is now set at an angle of attack $\alpha^* - \alpha_{L_0}$ such that

$$C_L = 2\pi(\alpha^* - \alpha_{L_0}) \tag{8.55}$$

The difference between these two cases is due to the wake-induced angle of attack, which is obtained from these two equations:

$$(\alpha - \alpha_{L_0}) - (\alpha^* - \alpha_{L_0}) = C_L \left[\frac{1}{m} - \frac{1}{2\pi} \right] \equiv \frac{C_L}{\pi \mathcal{R}}(1 + \delta_2) \tag{8.56}$$

and

$$1 + \delta_2 = \pi \mathcal{R} \left[\frac{\alpha - \alpha_{L_0}}{\pi \mathcal{R} A_1} - \frac{1}{2\pi} \right]$$

where $\delta_2 > 0$. Taking A_1 from this relation and substituting into Eq. (8.51) results in

$$C_L = \frac{2\pi(\alpha - \alpha_{L0})}{1 + (2/\mathcal{R})(1 + \delta_2)} \tag{8.57}$$

Thus, for the elliptic wing $\delta_2 = 0$ and also its lift coefficient is higher than for wings with other spanwise load distributions.

8.1.6 Twisted Elliptic Wing

The spanwise loading of wings can be varied by introducing twist to the wing planform. To illustrate the effects of twist, consider a wing with an elliptic chord distribution. For this purpose let us rearrange Eq. (8.44a) such that

$$\sum_{n=1}^{\infty} A_n \sin n\theta \left[\frac{4b}{m_0(\theta)c(\theta)} + \frac{n}{\sin \theta} \right] = \alpha(\theta) - \alpha_{L0}(\theta) \tag{8.58}$$

This is the governing equation for the coefficients for the circulation distribution for the general case that is described using lifting-line theory. Section 8.1.4 presents an exact solution for an untwisted elliptic planform wing (elliptic loading), but solutions for other cases must be obtained numerically using techniques that will be described in later sections. It is of interest to study the effect of wing twist on the solution for a particular geometry (geometric twist occurs for a spanwise variation of angle of attack and aerodynamic twist occurs for a spanwise variation of the zero-lift angle).

Filotas[8.1] has found a closed-form solution for a wing with an elliptic planform and arbitrary twist and that solution will be presented in what follows. Consider an elliptic chord distribution as given in Eq. (8.31):

$$c = c_0 \sin \theta \tag{8.59}$$

and for simplicity let $m_0 = 2\pi$. Then Eq. (8.58) becomes

$$\sum_{n=1}^{\infty} A_n \sin n\theta \left[\frac{R}{2} + n\right] = [\alpha(\theta) - \alpha_{L0}(\theta)] \sin \theta$$

where the aspect ratio of the elliptic wing is $R = 4b/\pi c_0$. Note that the above equation is a Fourier series representation for the right-hand side whose coefficients are given by

$$A_n = \frac{2}{\pi} \frac{1}{R/2 + n} \int_0^{\pi} [\alpha(\theta) - \alpha_{L0}(\theta)] \sin \theta \sin n\theta \, d\theta \tag{8.60}$$

To find the wing lift coefficient, the coefficient A_1 is obtained as

$$A_1 = \frac{2}{\pi} \frac{1}{R/2 + 1} \int_0^{\pi} [\alpha(\theta) - \alpha_{L0}(\theta)] \sin^2 \theta \, d\theta$$

$$= \frac{4}{\pi} \frac{1}{R + 2} \int_0^{\pi} [\alpha(\theta) - \alpha_{L0}(\theta)] \sin^2 \theta \, d\theta$$

and the lift is obtained by using Eq. (8.51):

$$C_L = \frac{4R}{R + 2} \int_0^{\pi} [\alpha(\theta) - \alpha_{L0}(\theta)] \sin^2 \theta \, d\theta \tag{8.61}$$

Example:

As an example consider a wing with a linear twist where

$$\alpha(y) = \alpha \pm \alpha_0 \left|\frac{y}{b/2}\right| = \alpha \pm \alpha_0 |\cos \theta|$$

The effect of the twist can be analyzed by taking the variable part of $\alpha(y)$ only, and adding the contribution of the constant angle of attack later. Therefore, let

$$\alpha(y) = \alpha_0 |\cos \theta|$$

and use Eq. (8.60) to compute the coefficients A_n as

$$\frac{A_n}{\alpha_0} = \frac{4}{\pi} \frac{1}{R/2 + n} \int_0^{\pi/2} \cos \theta \sin \theta \sin n\theta \, d\theta$$

$$= \frac{2}{\pi} \frac{1}{R/2 + n} \int_0^{\pi/2} \sin 2\theta \sin n\theta \, d\theta$$

$$= \frac{2}{\pi} \frac{1}{R/2 + n} \left[\frac{\sin(n - 2)\theta}{2(n - 2)} - \frac{\sin(n + 2)\theta}{2(n + 2)}\right]\Big|_0^{\pi/2}$$

$$= \frac{1}{\pi} \frac{1}{R/2 + n} \left[\frac{\sin(n - 2)\pi/2}{(n - 2)} - \frac{\sin(n + 2)\pi/2}{(n + 2)}\right]$$

Evaluating the individual coefficients for a wing with $R = 6$ and for a twist of $\alpha = \alpha_0 |\cos \theta|$ and substituting into Eq. (8.42) yields

$$\Gamma(\theta) = \frac{2bQ_\infty \alpha_0}{\pi} \left[\frac{1}{3} \sin \theta + \frac{1}{5} \sin 3\theta - \frac{1}{42} \sin 5\theta + \frac{2}{225} \sin 7\theta - \cdots\right]$$

8.1 Finite Wing: The Lifting Line Model

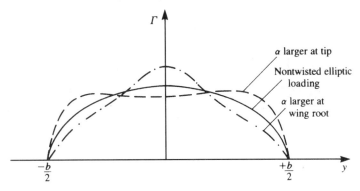

Figure 8.14 Effect of wing twist on the spanwise loading of an elliptic wing.

For a twist of $\alpha = \alpha_0(-|\cos\theta|)$, the circulation is

$$\Gamma(\theta) = \frac{2bQ_\infty\alpha_0}{\pi}\left[-\frac{1}{3}\sin\theta - \frac{1}{5}\sin 3\theta + \frac{1}{42}\sin 5\theta - \frac{2}{225}\sin 7\theta - \cdots\right]$$

These results, combined with an additional constant angle of attack α are plotted schematically in Fig. 8.14, which shows that having a larger angle of attack at the tip will increase the load there. Similarly, larger angles of attack near the wing root will increase the loading there.

8.1.7 Conclusions from Lifting-Line Theory

The most important result of the lifting-line theory is its ability to establish the effect of wing aspect ratio on the lift slope and induced drag. Some of the more important conclusions are:

1. The wing lift slope $dC_L/d\alpha$ decreases as wing aspect ratio becomes smaller (as shown by Eq. (8.36) for an elliptic wing and by Eq. (8.57) for a wing with general spanwise circulation).
2. The induced drag of a wing increases as wing aspect ratio decreases (as shown by Eq. (8.37) for an elliptic wing and by Eq. (8.53) for a wing with general spanwise circulation).
3. A wing with elliptic loading will have the lowest induced drag and the highest lift, as indicated by Eqs. (8.53) and (8.57).
4. This theory also provides valuable information about the wing's spanwise loading and about the existence of the trailing vortex wake.
5. The theory is limited to small disturbances and large aspect ratio and, for example, Eq. (8.6), which requires that the wake be aligned with the local velocity, was not addressed at all (because of the small angle of attack assumption).
6. There are possible modifications to this theory, such as the addition of wing sweep (e.g., Weissinger[8.2]). However, the study of wings with more complex geometry is difficult with this model whereas some of the more refined methods (introduced in the following chapters) are clearly more capable in dealing with this problem.
7. Using the results of this theory we must remember that the drag of a wing includes the induced drag portion (predicted by this model) plus the viscous drag, which must be taken into account.

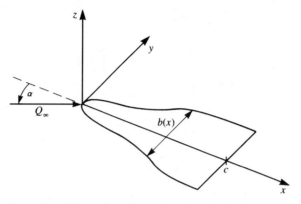

Figure 8.15 Nomenclature for slender, thin, pointed wing.

8.2 Slender Wing Theory

In this chapter three-dimensional solutions that rely on the small disturbance approximation are presented. By assuming that the wing is long and narrow ($\mathcal{R} \ll 1$), and that its angle of attack is small, the special case of slender-wing theory can be developed.

8.2.1 Definition of the Problem

Consider the slender wing of Fig. 8.15 with a span $b(x)$ and root chord c, where both the wing camberline η and its angle of attack are small, that is

$$\tan\alpha \ll 1 \quad \text{and} \quad \frac{\eta}{c} \ll 1$$

and we consider wings with no spanwise camber ($\partial\eta/\partial y = 0$). The flow is assumed to be potential and therefore the equation for the perturbation velocity potential is

$$\nabla^2 \Phi = 0 \tag{8.62}$$

which must be solved together with the boundary condition requiring no flow across the wing solid surface. This will be approximated at $z = 0$ for this case of small angle of attack and the z component of the total velocity $w^*(x, y, 0\pm)$ must be zero:

$$w^*(x, y, 0\pm) = \frac{\partial \Phi}{\partial z}(x, y, 0\pm) - Q_\infty \left(\frac{\partial \eta}{\partial x} - \alpha \right) = 0 \tag{8.63}$$

To solve this problem, we seek singularity elements that create antisymmetry (pressure jump) in the z direction. The doublet solution based on the $\partial/\partial z$ derivative (see Eq. (3.36)) is the most suitable and it is developed for the general lifting surface in Section 4.5. By distributing these doublet elements over the surface of the wing, we obtain the following integral equation (Eq. (4.45)) for the boundary condition of zero normal flow:

$$\frac{1}{4\pi} \int_{\text{wing + wake}} \frac{\mu(x_0, y_0)}{(y - y_0)^2} \left[1 + \frac{(x - x_0)}{\sqrt{(x - x_0)^2 + (y - y_0)^2}} \right] dx_0 \, dy_0$$

$$- Q_\infty \left(\frac{\partial \eta}{\partial x} - \alpha \right) = 0 \tag{8.64}$$

This integral is singular and its principal value must be evaluated. However, before proceeding further, the slender wing assumption allows us to make some simplifications. Since

8.2 Slender Wing Theory

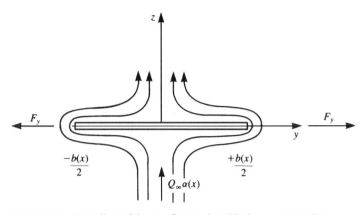

Figure 8.16 Streamlines of the crossflow as viewed in the $x = $ const. plane.

for the slender wing $x \gg y, z$ we can assume that the derivatives are inversely affected:

$$\frac{\partial}{\partial x} \ll \frac{\partial}{\partial y}, \frac{\partial}{\partial z} \tag{8.65}$$

Substituting this into the continuity equation (Eq. 8.62) allows us to consider the first term as negligible, compared to the other derivatives:

$$\nabla^2 \Phi \approx \frac{\partial^2 \Phi}{\partial y^2} + \frac{\partial^2 \Phi}{\partial z^2} = 0 \tag{8.66}$$

This can be interpreted such that the cross-flow effect is dominant, and for any $x = $ const. plane a local two-dimensional solution is sufficient. This is described schematically in Fig. 8.16. Also, for small-disturbance compressible flow (see Section 4.8), this implies that the Mach number dependency is lost and these solutions are applicable to supersonic potential flows as well.

Since the flowfield is now sought in the two-dimensional plane ($x = $ const.), the angle of attack and camber effects can be included in a local angle of attack $\alpha(x)$ such that

$$\alpha(x) \approx \alpha - \frac{\partial \eta}{\partial x}$$

If we recall the slenderness assumption that

$$|x - x_0| \gg |y - y_0|$$

the kernel in the integral of Eq. (8.64) becomes

$$\left[1 + \frac{(x - x_0)}{\sqrt{(x - x_0)^2 + (y - y_0)^2}} \right] \approx \begin{pmatrix} 2 & \text{for} & x > x_0 \\ 0 & \text{for} & x < x_0 \end{pmatrix} \tag{8.67}$$

The physical interpretation of this result is that portions of the wing ahead of a given x section ($x > x_0$) will have influence on the wing, whereas the influence of wing sections and the flowfield behind this x section ($x < x_0$) is negligible – thus the effect of the trailing wake for slender wings is *small*! Substituting this result into the boundary condition (Eq. (8.64)), and recalling that on the wing $z = 0$, we can reduce Eq. (8.64) to

$$\frac{1}{2\pi} \int_{-b(x)/2}^{b(x)/2} \frac{\mu(x, y_0)}{(y - y_0)^2} \, dy_0 = -Q_\infty \alpha(x) \tag{8.68}$$

which must be solved for any $x = $ const. wing station with local span $b(x)$. Note that by selecting the doublet distribution in the two-dimensional cross section, this boundary condition can be independently derived by integrating the two-dimensional doublet induced velocity (Section 3.14).

8.2.2 Solution of the Flow over Slender Pointed Wings

The integral equation (Eq. (8.68)) for the unknown doublet strength contains a strong singularity at $y = y_0$ (see Appendix C for a discussion of the principal value of this integral). Recalling the results of Section 3.14 that a doublet distribution can be replaced by an equivalent vortex distribution [e.g., $d\mu(y)/dy = -\gamma(y)$] allows us to use some of the results of thin-airfoil theory for the crossflow plane solution when the vortex distribution is used instead. The proposed vortex distribution consists of horseshoe-type vortices distributed continuously over the wing. This vortex model is described schematically in the right-hand side of Fig. 8.17, where for the purpose of illustration, discrete horseshoe elements are used instead of the continuous distribution. At any $x = $ const. section the trailing vortices form a two-dimensional vortex distribution of circulation per length $\gamma(y)$ along the strip $-b(x)/2 < y < b(x)/2, z = 0$ as shown in Fig. 8.18. Note that in the cross-flow plane, owing to left/right symmetry, the total circulation is zero, and the lift is generated by the spanwise segments of the horseshoe vortices (as shown in the left-hand side of Fig. 8.17). The perturbation velocity potential for this two-dimensional cross-flow (modeled by the vortex distribution shown in Fig. 8.18, and formulated in Section 3.14) at any x station is

$$\Phi = \frac{1}{2\pi} \int_{-b(x)/2}^{b(x)/2} \gamma(y_0) \tan^{-1} \frac{z}{(y - y_0)} \, dy_0 \qquad (8.69)$$

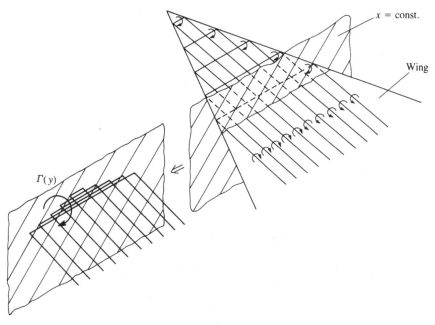

Figure 8.17 Horseshoe model for the slender, thin, pointed wing.

8.2 Slender Wing Theory

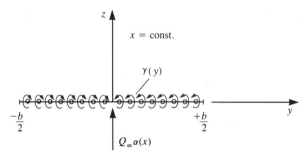

Figure 8.18 Vortex distribution in the crossflow ($x = $ const.) plane.

Observe that the positive vorticity vector in the y–z plane points in the positive x direction, as shown in Fig. 8.18. The velocity components in the $x = $ const. plane, due to this velocity potential, are

$$v(x, y, 0\pm) = \frac{\partial \Phi}{\partial y} = \mp \frac{\gamma(y)}{2} \tag{8.70}$$

$$w(x, y, 0\pm) = \frac{\partial \Phi}{\partial z} = \frac{1}{2\pi} \int_{-b(x)/2}^{b(x)/2} \gamma(y_0) \frac{dy_0}{(y - y_0)} \tag{8.71}$$

Because of the slender-wing assumption, only the local trailing vortex distribution (parallel to the x axis) will affect the near field downwash. By substituting this vortex distribution induced downwash into the wing boundary condition, Eq. (8.63) becomes

$$\frac{1}{2\pi} \int_{-b(x)/2}^{b(x)/2} \gamma(y_0) \frac{dy_0}{(y - y_0)} = -Q_\infty \alpha(x) \tag{8.72}$$

Comparing this form of the boundary condition with the formulation for high aspect ratio wings (Eq. (8.11)) clearly indicates that because the slender-wing assumption was used the effect of the spanwise vortices was neglected.

The solution for the vortex distribution, *at each x station*, is reduced now to the solution of this equation for $\gamma(y)$ with the additional condition that

$$\int_{-b(x)/2}^{b(x)/2} \gamma(y) \, dy = 0 \tag{8.73}$$

Because of the similarity between this integral equation (Eq. (8.72)) and the lifting-line equation (see Eqs. (8.10) and (8.11)), a solution of similar form is proposed. Let the spanwise circulation $\Gamma(x, y)$, at each x section, be an elliptic distribution as in Eq. (8.21):

$$\Gamma(x, y) \equiv \Gamma(y) = \Gamma_{\max} \left[1 - \left(\frac{y}{b(x)/2} \right)^2 \right]^{1/2} \tag{8.74}$$

The physical meaning of this circulation is best described by observing the horseshoe vortex structure shown in Fig. 8.17, where the downwash induced by the spanwise segments of the horseshoe vortices ahead of this x station is neglected when evaluating the boundary conditions. Then if the total circulation ahead of an $x = $ const. chordwise station is replaced by a single spanwise vortex line, as shown in the left side of Fig. 8.17, then its strength will be $\Gamma(y)$.

The spanwise distribution of the trailing vortices (shown in Fig. 8.18) is obtained by differentiating with respect to y (as in Eq. (8.41)):

$$\gamma(y) = -\frac{d\Gamma(y)}{dy} = \frac{4\Gamma_{max}}{b(x)^2} \frac{y}{\sqrt{[1 - (y/b(x)/2)^2]}} \quad (8.75)$$

Substitution into the integral equation, Eq. (8.72), results in

$$\frac{1}{2\pi} \int_{-b(x)/2}^{b(x)/2} \frac{4\Gamma_{max}}{b(x)^2} \frac{y_0}{\sqrt{[1 - (y_0/b(x)/2)^2]}} \frac{dy_0}{(y - y_0)} = -Q_\infty \alpha(x) \quad (8.76)$$

But this integral has already been evaluated in this chapter (see Eq. (8.22)) and resulted in a constant spanwise downwash. With the use of the results of Eqs. (8.22) and (8.24) the spanwise integration yields

$$\int_{-b(x)/2}^{b(x)/2} \frac{y_0}{\sqrt{[1 - (y_0/b(x)/2)^2]}} \frac{dy_0}{(y - y_0)} = -\frac{\pi b(x)}{2} \quad (8.77)$$

and Eq. (8.76) becomes

$$\frac{\Gamma_{max}}{b(x)} = Q_\infty \alpha(x) \quad (8.78)$$

which shows that the spanwise induced downwash due to an elliptic circulation distribution is constant and independent of y. The value of Γ_{max} is easily evaluated now and is

$$\Gamma_{max} = b(x) Q_\infty \alpha(x) \quad (8.79)$$

To establish the relation between the velocity potential and Γ consider a path of integration along the local y axis (for a $x = $ const. section),

$$\Phi(x, y, 0\pm) = \int_{-b(x)/2}^{y} \frac{\mp \gamma(y)}{2} dy = \pm \frac{\Gamma(y)}{2}$$

where the integration starts at the left leading edge of the $x = $ const. station and the integration path is above (0+) or under (0−) the wing. Therefore, the potential jump ($\Delta \Phi$) across the wing and the lift of the wing portion ahead of this x station ($\rho Q_\infty \Gamma(y)$) are elliptic too and we have

$$\Delta \Phi(x = \text{const.}, y) = \Phi(x, y, 0+) - \Phi(x, y, 0-) = 2\Phi(x, y, 0+)$$
$$= \Gamma(x = \text{const.}, y) \equiv \Gamma(y) \quad (8.80)$$

as shown in Fig. 8.19. Note that the local $\Gamma(y)$ is equivalent to the sum of all the spanwise bound vortex segments of the horseshoe elements ahead of it (see left side of Fig. 8.17) and therefore is equivalent to the lift of the wing portion ahead of this x station.

Substituting $\gamma(y)$ and Γ_{max} into Eqs. (8.69)–(8.71), we can obtain the crossflow potential and its derivatives:

$$\Phi(x, y, 0\pm) = \pm Q_\infty \alpha(x) \frac{b(x)}{2} \sqrt{1 - \left(\frac{y}{b(x)/2}\right)^2} = \pm Q_\infty \alpha(x) \sqrt{\left[\frac{b(x)}{2}\right]^2 - y^2} \quad (8.81)$$

$$u(x, y, 0\pm) = \frac{\partial \Phi}{\partial x}(x, y, 0\pm) = \pm Q_\infty \frac{\partial}{\partial x}\left\{\alpha(x) \sqrt{\left[\frac{b(x)}{2}\right]^2 - y^2}\right\} \quad (8.82)$$

8.2 Slender Wing Theory

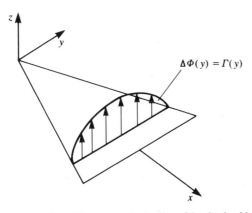

Figure 8.19 Elliptic spanwise loading of the slender thin wing.

This differentiation can be executed only if wing planform shape $b(x)$ and angle of attack $\alpha(x)$ are known. The spanwise velocity component is

$$v(x, y, 0\pm) = \frac{\partial \Phi}{\partial y}(x, y, 0\pm) = \mp \frac{\gamma(y)}{2} = \mp \frac{Q_\infty y \alpha(x)}{\sqrt{[b(x)/2]^2 - y^2}} \tag{8.83}$$

(Note that this result for $\gamma(y)$ can also be obtained by a direct inversion of the integral equation (Eq. (8.72)), which is identical to the integral equation of thin-airfoil theory (Eq. (5.39)) with the camber term deleted. This inversion is given in Eq. (7.18) where the circulation is set equal to zero and $\gamma = 2u_{1L}$.) Based on the boundary conditions stated in Eqs. (8.71)–(8.72) the downwash on the wing is

$$w(x, y, 0\pm) = \frac{\partial \Phi}{\partial z}(x, y, 0\pm) = -Q_\infty \alpha(x) \tag{8.84}$$

The aerodynamic loads will be computed with the use of the linearized Bernoulli equation (Eq. (4.52)). The pressure jump across the wing is given by

$$\Delta p = p(x, y, 0-) - p(x, y, 0+) = \rho Q_\infty \frac{\partial}{\partial x} \Delta \Phi \tag{8.85}$$

and this pressure difference across the wing is then

$$\Delta p = \rho Q_\infty \frac{\partial}{\partial x} \Delta \Phi = 2\rho Q_\infty^2 \frac{\partial}{\partial x} \left\{ \alpha(x) \sqrt{\left[\frac{b(x)}{2}\right]^2 - y^2} \right\}$$

$$= \rho Q_\infty^2 \frac{\partial}{\partial x} \left\{ \alpha(x) b(x) \left[1 - \left(\frac{y}{b(x)/2}\right)^2 \right]^{1/2} \right\} \tag{8.86}$$

For example, let us assume that the wing's angle of attack is constant $\alpha(x) = \alpha$ and for this case the pressure difference becomes

$$\Delta p(x, y) = \frac{\rho}{2} Q_\infty^2 \alpha \frac{b(x) db(x)/dx}{\sqrt{[b(x)/2]^2 - y^2}} \tag{8.87}$$

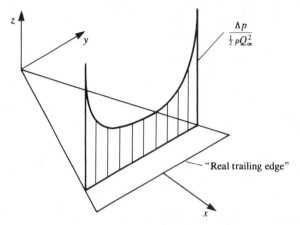

Figure 8.20 Spanwise pressure difference distribution of the slender wing at an $x = $ const. plane.

This spanwise pressure distribution is plotted in Fig. 8.20, and for a delta wing with straight leading edges, the pressure is plotted in Fig. 8.21. It is clear from these figures that this solution has an infinite suction peak along the wing leading edges. It seems as if the trailing edges of a high aspect ratio wing (while being swept backward) were folded into the root-chord and they are not visible, and consequently the lowest Δp at each x station is at the center chord. Also, since the trailing edge is not visible, the Kutta condition is not fulfilled along the "real trailing edge," which resembles the side edges of this imaginary high aspect ratio wing.

The longitudinal wing loading is obtained by an integration of the spanwise pressure difference. This may be simplified by recalling the approach used for Eq. (8.25), indicating

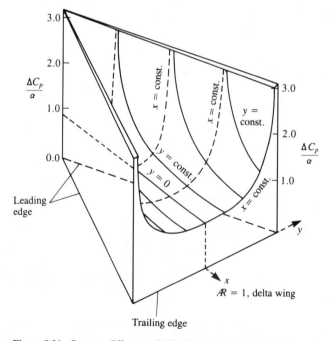

Figure 8.21 Pressure difference distribution on a slender delta wing.

that the following integral is equal to half the area of the corresponding ellipse:

$$\int_{-b(x)/2}^{b(x)/2} \left[1 - \left(\frac{y}{b(x)/2}\right)^2\right]^{1/2} dy = \frac{\pi b(x)}{4} \tag{8.88}$$

Using this result for the spanwise integration we get

$$\frac{dL}{dx} = \int_{-b(x)/2}^{b(x)/2} \Delta p\, dy = \rho Q_\infty^2 \frac{\partial}{\partial x}\left\{\alpha(x)b(x)\int_{-b(x)/2}^{b(x)/2}\left[1-\left(\frac{y}{b(x)/2}\right)^2\right]^{1/2}dy\right\}$$

$$= \frac{\pi \rho Q_\infty^2}{4}\frac{\partial}{\partial x}[\alpha(x)b(x)^2] \tag{8.89}$$

The interesting conclusion from this equation is that if there is no change either in $\alpha(x)$ or in $b(x)$, there will be no lift due to this section. Also, for a wing with linear $b(x)$ (delta wing) and constant α the longitudinal loading is linear too.

The lift of the wing from the tip to a section x is obtained by integrating dL/dx along x:

$$L(x) = \int_0^x \frac{dL}{dx} dx = \frac{\pi}{4}\rho Q_\infty^2[\alpha(x)b(x)^2] \tag{8.90}$$

This means that the lift of the wing up to a given x station depends on the local $\alpha(x)$, $b(x)$, and $db(x)/dx$ only. For the complete wing, therefore, it is a function of its maximum span b and α (at this chordwise station):

$$L = \frac{\pi}{4}\rho Q_\infty^2 \alpha b^2 \tag{8.91}$$

When the wing extends behind its maximum span (and the slope $db(x)/dx$ is negative) the contribution to the lift due to this portion is excluded by this model. Therefore, by using the maximum span in Eq. (8.91) the difficulties for wings having negative $db(x)/dx$ near the trailing edge are avoided.

The spanwise loading, at any x station, is obtained in a similar manner:

$$\frac{dL}{dy} = \rho Q_\infty \Gamma(y) = \rho Q_\infty^2 b(x)\alpha(x)\left\{1 - \left[\frac{y}{b(x)/2}\right]^2\right\}^{1/2} \tag{8.92}$$

which is an elliptic spanwise load distribution, as shown in Fig. 8.19. The lift up to any section x can be obtained by the integration of the spanwise loading, as well:

$$L(x) = \int_{-b/2}^{b/2} \frac{dL}{dy} dy = \frac{\pi}{4}\rho Q_\infty^2[\alpha(x)b(x)^2] \tag{8.93}$$

The lift coefficient is obtained by using Eq. (8.91),

$$C_L = \frac{\pi}{2}\frac{b^2}{S}\alpha = \frac{\pi}{2}\!R\alpha \tag{8.94}$$

and the induced drag coefficient (using Eq. (8.29)) for this elliptic distribution is

$$C_{D_i} = \frac{1}{\pi}\frac{S}{b^2}C_L^2 = C_L\frac{\alpha}{2} \tag{8.95}$$

If the drag force is a result of the pressure distribution only then its magnitude is expected to be $C_L\alpha$, but this result of Eq. (8.95) indicates that the "leading-edge suction" is reducing the drag by a factor of two. This can be shown by observing the suction force acting along the leading edges, as shown schematically in Figure 8.16, which is a result of the rapid turning of the flow at this point. The magnitude of this force was calculated in Section 6.5.3 (Eq. (6.52)) and is positive along the right leading edge,

$$F_y = \frac{\pi \rho b(x)}{4}(Q_\infty \alpha)^2$$

and negative along the left leading edge (here, for simplicity, $\alpha(x) = \alpha$ was assumed). Since this force acts on both leading edges of the wing no net sideforce is created; however, these forces will have a forward pointing component of magnitude T_s:

$$T_s = -2\int_0^c F_y \frac{d(b(x)/2)}{dx} dx = -\frac{\pi}{4}\rho Q_\infty^2 \alpha^2 \int_0^c b(x)\frac{db(x)}{dx} dx$$

$$= -\frac{\pi}{4}\rho Q_\infty^2 \alpha^2 \frac{b^2}{2} = -L\frac{\alpha}{2}$$

Consequently the drag force is the pressure difference integral $L\alpha$ minus the leading edge thrust $L\alpha/2$ and is equal to only one half of $L\alpha$, as obtained in Eq. (8.95).

The pitching moment about the apex of the wing is

$$M_0 = -\int_0^c \frac{dL}{dx} x \, dx = -\frac{\pi}{4}\rho Q_\infty^2 \int_0^c x\frac{d}{dx}[\alpha(x)b(x)^2]\, dx \quad (8.96)$$

Again, to evaluate this integral, the angle of attack and span variation with x are needed. As an example, consider a flat triangular delta wing with a constant angle of attack α where the trailing edge span is $b_{T.E.}$, that is,

$$b(x) = b_{T.E.}\frac{x}{c}$$

Substituting this into Eq. (8.96) gives

$$M_0 = -\frac{\pi}{4}\rho Q_\infty^2 \int_0^c x\frac{d}{dx}\left[\alpha\frac{x^2}{c^2}b_{T.E.}^2\right]dx = -\frac{\pi}{4}\rho Q_\infty^2 \alpha b_{T.E.}^2 \frac{2c}{3} = -L\frac{2c}{3} \quad (8.97)$$

and the center of pressure is at the center of the area,

$$\frac{x_{cp}}{c} = -\frac{M_0}{Lc} = \frac{2}{3} \quad (8.98)$$

8.2.3 The Method of R. T. Jones[8.3]

The results of slender wing theory were obtained by R. T. Jones in a rather simple and elegant manner in 1945. Here we shall follow some of the basic ideas of his method.

First, let us examine the flowfield due to a slender pointed wing in the cross-flow plane (as shown in Fig. 8.22). This plane of observation is fixed to a nonmoving frame of reference, and as the wing moves across it, its momentary cross section increases. Since the flow is attached to the wing, the flowfield in this two-dimensional observation plane is similar to the case of a flow normal to a flat plate (see Fig. 8.23). The velocity potential difference across the plate for this flow, as was shown earlier (Section 6.5.3), is

$$\Delta\Phi = bw\sqrt{1 - \left(\frac{y}{b/2}\right)^2} \quad (8.99)$$

where b is the span of the plate and w is the normal velocity component (in this case $w(x) = Q_\infty \alpha(x)$). However, this two-dimensional flow will not result in any forces because of the symmetry between the upper and lower streamlines. The only way to generate force,

8.2 Slender Wing Theory

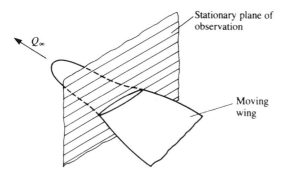

Figure 8.22 A slender wing moving across a stationary plane.

in this situation (with zero net circulation), is to create a change with time (e.g., due to the $\rho(\partial\Phi/\partial t)$ term in the unsteady Bernoulli equation, Eq. (2.35)). Consequently, the R. T. Jones model suggests that the lift will be generated only if the fluid particles will be accelerated, relative to a "ground-fixed" observer.

To demonstrate this principle, consider the two-dimensional plate of Fig. 8.24, as it is being accelerated downward (causing an upwash w). The resulting force per unit length Δx will be

$$\frac{\Delta L}{\Delta x} = \rho \frac{\partial}{\partial t} \int_{-b/2}^{b/2} \Delta \Phi \, dy = \rho \frac{\partial}{\partial t} \int_{-b/2}^{b/2} bw \sqrt{1 - \left(\frac{y}{b/2}\right)^2} \, dy$$

$$= \rho \frac{\partial}{\partial t} \left[wb^2 \frac{\pi}{4} \right] \tag{8.100}$$

This result can be viewed as the "added mass" of the fluid that is being accelerated by the accelerating plate. Following Newton's second law we can calculate the force due to accelerating fluid with added mass m' by a massless plate as

$$\frac{\Delta L}{\Delta x} = \frac{d(m'w)}{dt}$$

Comparing this formulation with Eq. (8.100) we see that the added mass becomes

$$m' = \rho b^2 \frac{\pi}{4} \tag{8.101}$$

which is equivalent to the mass of a fluid cylinder with a diameter of b.

Now, after establishing the added mass approach it is possible to follow the method of R.T. Jones for the slender pointed wing. The lift on the segment of the slender wing that is passing across the plane of observation in Fig. 8.22 will be due to accelerating the added

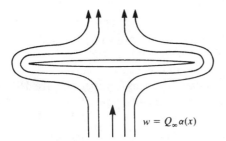

Figure 8.23 Schematic description of the cross-flow streamlines.

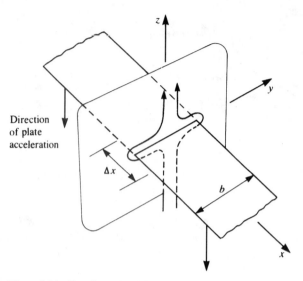

Figure 8.24 Two-dimensional flow resulting from the downward motion of a two-dimensional flat plate.

mass of the fluid:

$$\frac{\Delta L}{\Delta x} = \frac{d(m'w)}{dt} = \frac{d(m'w)}{dx}\frac{dx}{dt} = Q_\infty^2 \alpha(x)\frac{dm'}{dx}$$

where $w(x) = Q_\infty \alpha(x)$, $d\alpha(x)/dx$ is negligible, and $dx/dt = Q_\infty$. Substituting the added mass m' from Eq. (8.101) yields

$$\frac{\Delta L}{\Delta x} = \rho Q_\infty^2 \alpha(x)\frac{\pi}{4}\frac{db(x)^2}{dx} = \rho Q_\infty^2 \alpha(x)\frac{\pi}{2}b(x)\frac{db(x)}{dx} \qquad (8.102)$$

This equation is equivalent to Eq. (8.89) and again states that there will be no lift if $b(x)$ is constant with x.

To obtain the lift, drag, and pitching moment, Eq. (8.102) is integrated, and this yields the same results as presented in the previous section.

8.2.4 Conclusions from Slender Wing Theory

The slender wing solution presented here is based on the small-disturbance assumption, which automatically restricts the range of wing angle of attack. But in this particular case of slender wings, the incidence range is more limited than for high aspect ratio lifting wings because of flow separation along the leading edges. This effect will be discussed in Chapter 15, and in general, these leading-edge separated flow patterns will begin at angles of attack of 5°–10° (depending on leading-edge radius).

The main importance of this slender wing theory is that it provides a *three-dimensional* solution for the limiting case of very small aspect ratio wings. These results can serve as test cases for more complex panel codes, within the limit of small incidence angles.

The slenderness assumption, where one coordinate is larger than the other two, allowed the local treatment of the two-dimensional cross-flow. This logic can be carried over to more advanced methods and also for treating supersonic potential flows. This becomes clear when examining Eq. (4.73), where by omitting the x derivatives, the Mach number dependency is lost too.

8.3 Slender Body Theory

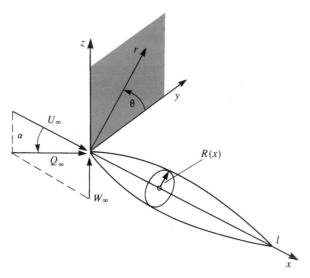

Figure 8.25 Nomenclature used for slender body theory.

The wake influence in this analysis was assumed to be small (negligible), which is, again, a good test case for more advanced panel codes.

8.3 Slender Body Theory

As a final example of classical small-disturbance theories, consider the flow past a slender body of revolution at a small angle of attack α, as shown in Fig. 8.25. It is convenient to use the cylindrical coordinates x, r, θ and then the surface of the slender body of revolution is given as

$$F \equiv r - R(x) = 0 \tag{8.103}$$

If the length of the body is l, slenderness means that the ratio of body radius $R(x)$ to length is small and, for small disturbances, the angle of attack α is small as well, that is,

$$\frac{R(x)}{l} \ll 1, \qquad \alpha \ll 1, \qquad \left|\frac{dR(x)}{dx}\right| \ll 1 \tag{8.104}$$

Laplace's equation for the perturbation potential (in cylindrical coordinates) is given by Eq. (1.33) as

$$\nabla^2 \Phi = \frac{\partial^2 \Phi}{\partial x^2} + \frac{\partial^2 \Phi}{\partial r^2} + \frac{1}{r}\frac{\partial \Phi}{\partial r} + \frac{1}{r^2}\frac{\partial^2 \Phi}{\partial \theta^2} = 0 \tag{8.105}$$

where $r = [y^2 + z^2]^{\frac{1}{2}}$.

In this coordinate system the free-stream velocity is

$$\begin{aligned}\mathbf{Q}_\infty &= U_\infty \mathbf{e}_x + W_\infty \mathbf{e}_z = Q_\infty[\cos\alpha\,\mathbf{e}_x + \sin\alpha(\sin\theta\,\mathbf{e}_r + \cos\theta\,\mathbf{e}_\theta)] \\ &\approx Q_\infty[\mathbf{e}_x + \alpha(\sin\theta\,\mathbf{e}_r + \cos\theta\,\mathbf{e}_\theta)] \end{aligned} \tag{8.106}$$

Following the method of Section 4.2, the zero normal velocity component boundary condition on the body surface is given by $\nabla \Phi^* \cdot \nabla F = 0$, which yields:

$$\frac{\partial \Phi}{\partial r} + Q_\infty \alpha \sin \theta - \left[\frac{\partial \Phi}{\partial x} + Q_\infty\right] \frac{dR(x)}{dx} = 0 \quad \text{for} \quad r = R(x) \tag{8.107}$$

The small-disturbance version of this boundary condition is obtained after neglecting the smaller terms (according to Eq. (8.104)):

$$\frac{\partial \Phi}{\partial r}(x, R, \theta) = Q_\infty R'(x) - Q_\infty \alpha \sin \theta \tag{8.108}$$

where $R'(x) \equiv dR(x)/dx$ and it is noted that the boundary condition has not been transferred to the body axis. The reason for this is that the velocity components of this flow are singular at the axis and the application of the boundary condition must be performed with care.

At this point it can be seen that the small-disturbance flow past a slender body of revolution at angle of attack can be replaced by two component flows, the axisymmetric flow past the body at zero angle of attack with body boundary condition

$$\frac{\partial \Phi}{\partial r}(x, R, \theta) = Q_\infty R'(x) \tag{8.108a}$$

and the flow normal to the body axis with free-stream speed $Q_\infty \alpha$ and body boundary condition

$$\frac{\partial \Phi}{\partial r}(x, R, \theta) = -Q_\infty \alpha \sin \theta \tag{8.108b}$$

In the next two sections these two linear subproblems will be formulated; the complete solution is their sum.

8.3.1   *Axisymmetric Longitudinal Flow Past a Slender Body of Revolution*

The axisymmetric version of Laplace's equation (Eq. (8.105)) is

$$\frac{\partial^2 \Phi}{\partial x^2} + \frac{\partial^2 \Phi}{\partial r^2} + \frac{1}{r}\frac{\partial \Phi}{\partial r} = 0 \tag{8.109}$$

and the body boundary condition is given by Eq. (8.108a). The solution is modeled by a distribution of sources of strength $\sigma(x)$ per length along the body axis on the strip $0 \leq x \leq l$, $z = 0$ (Fig. 8.26), and the problem is essentially the axisymmetric version of the two-dimensional thin-airfoil problem. The perturbation velocity potential and velocity components for this distribution are obtained by integrating the equations of the point source (see

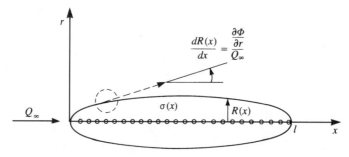

Figure 8.26   Source distribution along the x axis.

8.3 Slender Body Theory

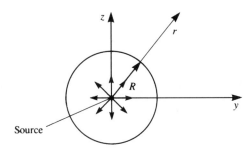

Figure 8.27 Cross-sectional view of the source distribution.

Section 3.4) along the x axis:

$$\Phi(r, x) = \frac{-1}{4\pi} \int_0^l \frac{\sigma(x_0)\, dx_0}{\sqrt{(x - x_0)^2 + r^2}} \tag{8.110}$$

$$q_r(r, x) = \frac{\partial \Phi}{\partial r} = \frac{1}{4\pi} \int_0^l \frac{\sigma(x_0)\, r\, dx_0}{[(x - x_0)^2 + r^2]^{3/2}} \tag{8.111}$$

$$q_x(r, x) = \frac{\partial \Phi}{\partial x} = \frac{1}{4\pi} \int_0^l \frac{\sigma(x_0)(x - x_0)\, dx_0}{[(x - x_0)^2 + r^2]^{3/2}} \tag{8.112}$$

To satisfy the body boundary condition (Eq. (8.108a)), which states that the flow is tangent to the surface,

$$\frac{q_r}{Q_\infty} = R'(x) \quad \text{at} \quad r = R$$

we use the slenderness arguments developed in slender wing theory and consider a mass balance in the cross-flow plane (see Fig. 8.27). Surround the body axis with a circle of radius r. The volume flow (per unit length, Δx) through this circle is equal to the source strength

$$\sigma(x) = 2\pi r q_r \tag{8.113}$$

If Eq. (8.113) is evaluated at $r = R$ and the boundary condition of Eq. (8.108a) is used, the source strength is found to be

$$\sigma(x) = 2\pi R Q_\infty \frac{dR}{dx} = Q_\infty \frac{dS(x)}{dx} \tag{8.114}$$

where $S(x)$ is the body cross-sectional area. The potential and velocity components are then found by substituting $\sigma(x)$ into Eqs. (8.110)–(8.112):

$$\Phi(r, x) = \frac{-Q_\infty}{4\pi} \int_0^l \frac{S'(x_0)\, dx_0}{\sqrt{(x - x_0)^2 + r^2}} \tag{8.115}$$

$$q_r(r, x) = \frac{\partial \Phi}{\partial r} = \frac{Q_\infty}{4\pi} \int_0^l \frac{S'(x_0)\, r\, dx_0}{[(x - x_0)^2 + r^2]^{3/2}} \tag{8.116}$$

$$q_x(r, x) = \frac{\partial \Phi}{\partial x} = \frac{Q_\infty}{4\pi} \int_0^l \frac{S'(x_0)(x - x_0)\, dx_0}{[(x - x_0)^2 + r^2]^{3/2}} \tag{8.117}$$

8.3.2 Transverse Flow Past a Slender Body of Revolution

The governing equation for the perturbation velocity potential is Laplace's equation (Eq. (8.105))

$$\frac{\partial^2 \Phi}{\partial x^2} + \frac{\partial^2 \Phi}{\partial r^2} + \frac{1}{r}\frac{\partial \Phi}{\partial r} + \frac{1}{r^2}\frac{\partial^2 \Phi}{\partial \theta^2} = 0 \tag{8.105}$$

and the body boundary condition is given by Eq. (8.108b),

$$\frac{\partial \Phi}{\partial r}(x, R, \theta) = -Q_\infty \alpha \sin\theta \tag{8.108b}$$

The two-dimensional flow (in the y–z plane) of this problem resembles the flow past a cylinder, which was solved in Section 3.11. Therefore, the solution to this problem is modeled by a distribution of doublets of strength $\mu(x)$ per length on the strip $0 \le x \le l, z = 0$. The doublet axes point in the negative z direction opposing the stream. The velocity potential and velocity components are given by integration of the point elements (see Section 3.5) along the body's length:

$$\Phi(r, \theta, x) = \frac{1}{4\pi}\int_0^l \frac{\mu(x_0) r \sin\theta\, dx_0}{[(x-x_0)^2 + r^2]^{3/2}} \tag{8.118}$$

$$q_r(r, \theta, x) = \frac{\partial \Phi}{\partial r} = \frac{1}{4\pi}\int_0^l \frac{\mu(x_0)\sin\theta\, dx_0}{[(x-x_0)^2 + r^2]^{3/2}} - \frac{3}{4\pi}\int_0^l \frac{\mu(x_0)\sin\theta\, r^2\, dx_0}{[(x-x_0)^2 + r^2]^{5/2}} \tag{8.119}$$

$$q_\theta(r, \theta, x) = \frac{1}{r}\frac{\partial \Phi}{\partial \theta} = \frac{1}{4\pi}\int_0^l \frac{\mu(x_0)\cos\theta\, dx_0}{[(x-x_0)^2 + r^2]^{3/2}} \tag{8.120}$$

$$q_x(r, \theta, x) = \frac{\partial \Phi}{\partial x} = \frac{-3}{4\pi}\int_0^l \frac{\mu(x_0)(x-x_0) r \sin\theta\, dx_0}{[(x-x_0)^2 + r^2]^{5/2}} \tag{8.121}$$

To satisfy the body boundary condition consider the flow in the cross-flow plane (as shown in Fig. 8.28). This is simply the flow past a circular cylinder, and its radial velocity component from Section 3.11 is $(\mu(x)/2\pi)(\sin\theta/R^2(x))$. Thus the boundary condition at $r = R$ becomes

$$\frac{\partial \Phi}{\partial r}(R, \theta) = \frac{\mu(x)}{2\pi}\frac{\sin\theta}{R^2(x)} = -Q_\infty \alpha \sin\theta$$

and the doublet strength is found to be

$$\mu(x) = 2\pi Q_\infty \alpha R^2(x) = 2Q_\infty \alpha S(x) \tag{8.122}$$

For small values of r (including the body surface) the solution is the flow past the circle in the cross-flow plane and the perturbation potential and velocity components are

$$\Phi(r, \theta, x) = Q_\infty \alpha R^2 \frac{\sin\theta}{r} \tag{8.123}$$

$$q_r(r, \theta, x) = -Q_\infty \alpha R^2 \frac{\sin\theta}{r^2} \tag{8.124}$$

$$q_\theta(r, \theta, x) = Q_\infty \alpha R^2 \frac{\cos\theta}{r^2} \tag{8.125}$$

$$q_x(r, \theta, x) = \frac{\partial \Phi}{\partial x} = 2Q_\infty \alpha R R' \frac{\sin\theta}{r} \tag{8.126}$$

8.3 Slender Body Theory

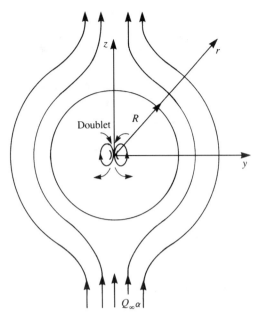

Figure 8.28 Cross-flow model using doublet distribution along the x axis and pointing in the $-z$ direction.

Note that since the doublet strength is a function of x, the streamwise (axial) velocity component is unequal to zero.

8.3.3 Pressure and Force Information

The perturbation velocity field for the flow at angle of attack past a slender body of revolution is obtained by adding the results from Sections 8.3.1 and 8.3.2. The velocity field will be evaluated on the body surface for the determination of the forces and pitching moment.

For the axisymmetric problem of Section 8.3.1, the radial velocity component is given by the boundary condition (Eq. (8.108a)) as

$$q_r = Q_\infty R'(x) \tag{8.127}$$

The axial component of velocity can be determined by taking the limit of Eq. (8.117) as the radial coordinate approaches zero. Let us denote this component as

$$q_x = q_{xA} \tag{8.128}$$

It is given in Karamcheti[1.5] (p. 577) as

$$q_{xA} = \frac{Q_\infty}{2\pi} S''(x) \ln \frac{r}{2} + \frac{Q_\infty}{4\pi} \int_0^l S'''(x_0) \ln|x - x_0|\, dx_0 \tag{8.128a}$$

The radial, tangential, and axial velocity components of the transverse problem of Section 8.3.2 are found by substituting $r = R$ in Eqs. (8.124)–(8.126). The complete velocity distribution on the body surface is obtained by adding the free-stream components from Eq. (8.106) to the perturbation components to get

$$\mathbf{q}(x, r, \theta) = (Q_\infty + q_x,\ Q_\infty \alpha \sin\theta + q_r,\ Q_\infty \alpha \cos\theta + q_\theta) \tag{8.129}$$

Substituting Eq. (8.129) into the pressure coefficient equation yields

$$C_p = 1 - \frac{q^2}{Q_\infty^2} = -\frac{2q_x}{Q_\infty} - \frac{2\alpha}{Q_\infty}(q_r \sin\theta + q_\theta \cos\theta) - \frac{q_r^2 + q_\theta^2 + q_x^2}{Q_\infty^2} \qquad (8.130)$$

By an inspection of the velocity components we see that the magnitude of the squares of the cross-flow plane components is comparable to the magnitude of the axial component itself and therefore the only term in the pressure coefficient equation that can be neglected is $-q_x^2/Q_\infty^2$. The perturbation components from Eqs. (8.124)–(8.126) are substituted into the modified Eq. (8.130) and after some manipulation the pressure coefficient becomes

$$C_p = -\frac{2q_{xA}}{Q_\infty} - (R')^2 - 4\alpha R' \sin\theta + \alpha^2(1 - 4\cos^2\theta) \qquad (8.131)$$

The force acting on the slender body is given by

$$\mathbf{F} = -\int_S p\mathbf{n}\,dS = -\int_0^l \int_0^{2\pi} p\mathbf{n}R\,d\theta\,dx \qquad (8.132)$$

where dx is the slender-body approximation for the length element. The slender-body approximation for the unit normal is

$$\mathbf{n} = \mathbf{e}_r - R'\mathbf{e}_x = -R'\mathbf{e}_x + \cos\theta\,\mathbf{e}_y + \sin\theta\,\mathbf{e}_z \qquad (8.133)$$

and substituting this into Eq. (8.133) yields the force components in the three coordinate directions:

$$F_x = \int_0^l \int_0^{2\pi} R' R p\,d\theta\,dx \qquad (8.134a)$$

$$F_y = -\int_0^l \int_0^{2\pi} R p \cos\theta\,d\theta\,dx \qquad (8.134b)$$

$$F_z = -\int_0^l \int_0^{2\pi} R p \sin\theta\,d\theta\,dx \qquad (8.134c)$$

The rate of change of the force components with respect to x is given by

$$\frac{dF_x}{dx} = -R'R \int_0^{2\pi} p\,d\theta \qquad (8.135a)$$

$$\frac{dF_y}{dx} = -R \int_0^{2\pi} p \cos\theta\,d\theta \qquad (8.135b)$$

$$\frac{dF_z}{dx} = -R \int_0^{2\pi} p \sin\theta\,d\theta \qquad (8.135c)$$

To use the pressure coefficient from Eq. (8.131) in Eqs. (8.135) the pressure is written as

$$p = \frac{1}{2}\rho Q_\infty^2 C_p + p_\infty \qquad (8.136)$$

and after Eqs. (8.131) and (8.136) are substituted into Eqs. (8.135) and the integration is performed, the axial rate of change of the force components becomes

$$\frac{dF_x}{dx} = \left[p_\infty - \frac{1}{2}\rho Q_\infty^2 \left(\alpha^2 + \frac{q_{xA}}{Q_\infty} + (R')^2 \right) \right] S' \qquad (8.137a)$$

$$\frac{dF_y}{dx} = 0 \tag{8.137b}$$

$$\frac{dF_z}{dx} = \rho Q_\infty^2 \alpha S' \tag{8.137c}$$

The side force distribution is zero and therefore the side force is also zero. The normal force distribution is proportional to the angle of attack and rate of change of cross-sectional area, a result obtained by Munk (see Sears[8.4]). An integration in x shows clearly that the normal force on the body is zero if the body's ends are pointed. A similar result can be obtained for the axial force, which is also zero if the body's ends are pointed (see Ward[8.5]).

The moment about the origin is given by

$$M = -\int_0^l \int_0^{2\pi} \mathbf{r} \times \mathbf{n} p R \, d\theta \, dx \tag{8.138}$$

where the position vector $\mathbf{r}$ is seen to be

$$\mathbf{r} = x\mathbf{e}_x + R\mathbf{e}_r = x\mathbf{e}_x + R\cos\theta \mathbf{e}_y + R\sin\theta \mathbf{e}_z \tag{8.139}$$

The components of the moment about the x and z axes are zero from symmetry considerations and the pitching moment about the y axis is

$$M_y = \frac{1}{2}\rho Q_\infty^2 \int_0^l \int_0^{2\pi} (x + RR') \sin\theta \, R C_p \, d\theta \, dx \tag{8.140}$$

With the use of Eq. (8.135c) the pitching moment can be written as

$$M_y = -\int_0^l (x + RR') \frac{dF_z}{dx} \, dx \tag{8.141}$$

The second term in the integrand is neglected as being second order and, after an integration by parts, the pitching moment becomes

$$M_y = -\rho Q_\infty^2 \alpha \int_0^l x S' \, dx = -\rho Q_\infty^2 \alpha \left[xS \big|_0^l - \int_0^l S \, dx \right] = \rho Q_\infty^2 \alpha V \tag{8.142}$$

where V is the body volume

8.3.4 Conclusions from Slender Body Theory

The above results for the aerodynamic forces acting on slender bodies show that for pointed bodies there is no lift and no drag force, but there is an aerodynamic pitching moment. This important result is very useful when checking the accuracy of numerical methods that calculate the lift and drag by integrating the surface pressure over the body (and may result in lift and drag that are different from zero). Lift and drag forces are possible only when the base is not pointed, and a base pressure exists that is different than that predicted by potential flow theory (e.g., due to flow separations). Some methods for the treatment of bodies with blunt bases are presented by Nielsen.[8.6]

8.4 Far Field Calculation of Induced Drag

It is possible to compute the forces acting on a body or wing by applying the integral form of the momentum equation (Eq. (1.19)). For example, the wing shown in

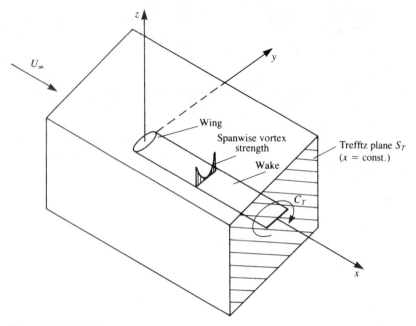

Figure 8.29 Far field control volume used for momentum balance.

Fig. 8.29 is surrounded by a large control volume, and for an inviscid, steady-state flow without body forces, Eq. (1.19) reduces to

$$\int_S \rho \mathbf{q}(\mathbf{q}\cdot\mathbf{n})dS = \mathbf{F} - \int_S p\mathbf{n}\,dS \tag{8.143}$$

where the second term in the right-hand side is the integral of the pressures. A coordinate system is selected such that the x axis is parallel to the free-stream velocity U_∞, and the velocity vector, including the perturbation (u, v, w), becomes

$$\mathbf{q} = (U_\infty + u, v, w)$$

If the x component of the force (drag) is to be computed then Eq. (8.143) becomes

$$D = -\int_S \rho(U_\infty+u)[(U_\infty+u)dy\,dz + v\,dx\,dz + w\,dx\,dy] - \int_S p\,dy\,dz$$

The pressures are found by using Bernoulli's equation:

$$p - p_\infty = \frac{\rho}{2}U_\infty^2 - \frac{\rho}{2}[(U_\infty+u)^2 + v^2 + w^2] = -\rho u U_\infty - \frac{\rho}{2}(u^2+v^2+w^2)$$

Substituting this result into the drag integral yields

$$D = -\rho\int_S U_\infty(U_\infty+u)\,dy\,dz - \rho\int_S (U_\infty+u)(u\,dy\,dz + v\,dx\,dz + w\,dx\,dy)$$
$$+ \rho\int_S u U_\infty\,dy\,dz + \frac{\rho}{2}\int_S (u^2+v^2+w^2)\,dy\,dz \tag{8.144}$$

Note that the second integral will vanish as a result of the continuity equation for the perturbation, and the first and the third will cancel out. Now if the control volume is large then the perturbation velocity components will vanish everywhere but on the wake. If the

8.4 Far Field Calculation of Induced Drag

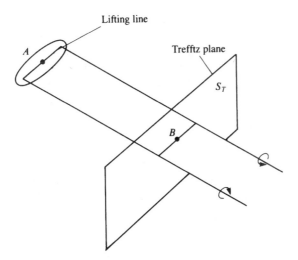

Figure 8.30 Trefftz plane used for the calculation of induced drag.

flow is inviscid, then at this plane S_T shown in Fig. 8.30 (called the *Trefftz plane*) the wake is parallel to the local free stream and will result in velocity components only in the y and z directions (thus $u^2 \ll v^2, w^2$). Therefore, the drag can be obtained by integrating the v and w component on this plane only:

$$D = \frac{\rho}{2} \int_{S_T} (v^2 + w^2) dy\, dz = \frac{\rho}{2} \int_{S_T} \left[\left(\frac{\partial \Phi}{\partial y}\right)^2 + \left(\frac{\partial \Phi}{\partial z}\right)^2 \right] dy\, dz \tag{8.145}$$

where Φ is the perturbation velocity potential. Use of the divergence theorem to transfer the surface integral into a line integral (similar to Eq. (1.20)) results in

$$\int_{S_T} \left[\left(\frac{\partial \Phi}{\partial y}\right)^2 + \left(\frac{\partial \Phi}{\partial z}\right)^2 + \Phi \left(\frac{\partial^2 \Phi}{\partial y^2} + \frac{\partial^2 \Phi}{\partial z^2}\right) \right] dy\, dz = \int_{C_T} \Phi \frac{\partial \Phi}{\partial n} dl$$

The third term in the first integral is canceled since in the two-dimensional Trefftz plane $\nabla^2 \Phi = 0$ and the integration is now limited to a path surrounding the wake (where a potential jump exists). If the wake is modeled by a vortex (or doublet) distribution parallel to the x axis, as in Fig. 8.30, the formulation of Section 3.14 for continuous singularity distributions can be used. Because of the symmetry of the induced velocity above and under the vortex sheet this integral can be reduced to a single spanwise line integral:

$$D = -\frac{\rho}{2} \int_{-b_w/2}^{b_w/2} \Delta \Phi w\, dy = -\frac{\rho}{2} \int_{-b_w/2}^{b_w/2} \Gamma(y) w\, dy \tag{8.146}$$

and the minus sign is a result of the $\partial \Phi / \partial n$ direction pointing inside the circle of integration and b_w is the local wake span. In Eq. (8.146) a horseshoe vortex structure is assumed for the lifting wing, but the wake span is allowed to be different than the wing's span (e.g., due to self-induced wake displacement).

Following the same methodology, the lift force can be derived as

$$L = \rho U_\infty \int_{-b_w/2}^{b_w/2} \Delta \Phi\, dy = \rho U_\infty \int_{-b_w/2}^{b_w/2} \Gamma(y)\, dy \tag{8.147}$$

The above drag formula may be useful in measuring the accuracy of data obtained by numerical integration of the local pressures. As an example for the use of Eq. (8.146), consider the elliptic lifting-line model of Section 8.1. The downwash at the lifting line (point A in Fig. 8.30) due to the elliptic load distribution is constant (Eq. (8.24)):

$$w_i = -Q_\infty \alpha_i = -\frac{\Gamma_{max}}{2b}$$

This was a result observed on the lifting line due to the semi-infinite trailing vortex lines. However, far downstream at a point B (in Fig. 8.30) the downwash is twice as much since to an observer at this point the vortex sheet seems to be infinite in both directions. Using the elliptic distribution $\Gamma(y)$ of Eq. (8.21) and substituting w_i into Eq. (8.146) we find that the drag force becomes

$$D = -\frac{\rho}{2} 2w_i \int_{-b/2}^{b/2} \Gamma(y)\,dy = -\rho w_i \frac{\pi b}{4} \Gamma_{max} = \frac{\pi}{8} \rho \Gamma_{max}^2$$

which is exactly the same result as in Eq. (8.27). Also, in this case a rigid wake model is used and the wake span b_w was assumed to be equal to the wingspan b.

References

[8.1] Filotas, L. T., "Solution of the Lifting Line Equation for Twisted Elliptic Wings," *J. Aircraft*, Vol. 8, No. 10, 1971, pp. 835–836.

[8.2] Weissinger, J., "The Lift Distribution of Swept-Back Wings," NACA TM 1120, 1947.

[8.3] Jones, R. T., "Properties of Low-Aspect-Ratio Pointed Wings at Speeds below and above the Speed of Sound," NACA Rep. 835, May 1945.

[8.4] Sears, W. R., "Small Perturbation Theory," in *General Theory of High Speed Aerodynamics*, Princeton University Press, 1954.

[8.5] Ward, G. N., *Linearized Theory of Steady High-Speed Flow*, Cambridge University Press, 1955, Section 9.8.

[8.6] Nielsen, J. N., *Missile Aerodynamics*, McGraw-Hill, 1960.

Problems

8.1. Consider the Fourier coefficients for the lifting-line circulation in Eq. (8.42). Show that for wing loading symmetrical about the midspan the even coefficients are zero and for antisymmetrical loading the odd coefficients are zero.

8.2. The governing equation for the Fourier coefficients in Problem 8.1 is Eq. (8.58). One method for the numerical solution of this equation is to set all coefficients equal to zero for n greater than some value, say N, and to evaluate the equation for N values of θ. The N linear equations for the unknown coefficients can then be solved using standard techniques. This is called the collocation method. Use the collocation method to find the Fourier coefficients for a flat rectangular wing of aspect ratio 6 for $N = 3, 5, 7$ (two-, three-, and four-term expansions). Calculate the lift and induced drag coefficients for these three cases.

8.3. Find the vortex distribution for slender wing theory by the direct integration of Eq. (8.72) with the use of the results of Section 7.1.

8.4. Consider the flow past a flat elliptic planform wing at angle of attack α. A flap whose extent covers the center half of the wing span is deflected such that the zero-lift angle distribution along the span is given by

$$\alpha_{L_0} = -\beta, \quad -\frac{b}{4} < y < \frac{b}{4}$$

$$\alpha_{L_0} = 0, \quad -\frac{b}{2} < y < -\frac{b}{4} \quad \text{and} \quad \frac{b}{4} < y < \frac{b}{2}$$

where β is constant.

Find the wing lift coefficient and circulation distribution and plot the circulation distribution to study its behavior at the tip of the flap. Use lifting-line theory.

8.5. Study the effect of ground proximity by examining two- and three-dimensional vortex line models. For the two-dimensional case assume that the flat plate is at an angle α to the free-stream U_∞, but both the vortex and the collocation points of the lumped-vortex element are elevated by h above the $z = 0$ plane.

   a. Derive the "zero normal velocity" boundary condition at the collocation point for this lumped-vortex model.

   b. For the three-dimensional case use the vortex line model of Section 8.1.2 and assume that the wing and its image lie in the $z = \pm h$ planes, respectively. Derive an expression for the influence of the image wake system on the wing in terms of h.

   c. Modify the lifting-line equation (Eq. (8.11)) to include ground effect, based on your finding in a and b.

CHAPTER 9

Numerical (Panel) Methods

In the previous chapters the solution to the potential flow problem was obtained by analytical techniques. These techniques (except in Chapter 6) were applicable only after some major geometrical simplifications in the boundary conditions were made. In most of these cases the geometry was approximated by flat, zero-thickness surfaces and for additional simplicity the boundary conditions were transferred, too, to these simplified surfaces (e.g., at $z = 0$).

The application of numerical techniques allows the treatment of more realistic geometries and the fulfillment of the boundary conditions on the actual surface. In this chapter the methodology of some numerical solutions will be examined and applied to various problems. The methods presented here are based on the surface distribution of singularity elements, which is a logical extension of the analytical methods presented in the earlier chapters. Since the solution is now reduced to finding the strength of the singularity elements distributed on the body's surface this approach seems to be more economical, from the computational point of view, than methods that solve for the flowfield in the whole fluid volume (e.g., finite difference methods). Of course this comparison holds for inviscid incompressible flows only, whereas numerical methods such as finite difference methods were basically developed to solve the more complex flowfields where compressibility and viscous effects are not negligible.

9.1 Basic Formulation

Consider a body with known boundaries S_B, submerged in a potential flow, as shown in Fig. 9.1. The flow of interest is in the outer region V where the incompressible, irrotational continuity equation, in the body's frame of reference, in terms of the total potential Φ^* is

$$\nabla^2 \Phi^* = 0 \tag{9.1}$$

Following Green's identity, as presented in Section 3.2, we can construct the general solution to Eq. (9.1) by a sum of source σ and doublet μ distributions placed on the boundary S_B (Eq. (3.13)):

$$\Phi^*(x, y, z) = \frac{-1}{4\pi} \int_{S_B} \left[\sigma \left(\frac{1}{r} \right) - \mu \mathbf{n} \cdot \nabla \left(\frac{1}{r} \right) \right] dS + \Phi_\infty \tag{9.2}$$

Here the vector $\mathbf{n}$ points in the direction of the potential jump μ, which is normal to S_B and positive outside of V (Fig. 9.1), and Φ_∞ is the free-stream potential written as

$$\Phi_\infty = U_\infty x + V_\infty y + W_\infty z \tag{9.3}$$

This formulation does not uniquely describe a solution since a large number of source and doublet distributions will satisfy a given set of boundary conditions (as discussed in Chapter 3). Therefore, a choice has to be made in order to select the desirable combination

9.2 The Boundary Conditions

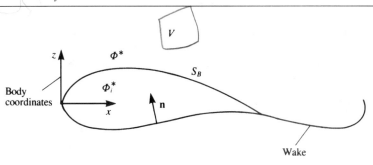

Figure 9.1 Potential flow over a closed body.

of such singularity elements. It is clear from the previous examples (in Chapters 4–8) that for simulating the effect of thickness, source elements can be used, whereas for lifting problems, antisymmetric terms such as the doublet (or vortex) can be used. To uniquely define the solution of this problem, first the boundary conditions of zero flow normal to the surface must be applied. In the general case of three-dimensional flows, specifying the boundary conditions will not immediately yield a unique solution because of two problems. First, a decision has to be made in regard to the "right" combination of source and doublet distributions. Second, some *physical* considerations need to be introduced to fix the amount of circulation around the surface S_B. These considerations deal mainly with the modeling of the wakes and fixing the wake shedding lines and their initial orientation and geometry. (This is the three-dimensional equivalent of a two-dimensional Kutta condition.) However, based on the previous examples, it is likely that the wake will be modeled by thin doublet or vortex sheets (Fig. 9.1) and therefore Eq. (9.2) can be rewritten as

$$\Phi^*(x, y, z) = \frac{1}{4\pi} \int_{\text{body+wake}} \mu \mathbf{n} \cdot \nabla\left(\frac{1}{r}\right) dS - \frac{1}{4\pi} \int_{\text{body}} \sigma \left(\frac{1}{r}\right) dS + \Phi_\infty \quad (9.2a)$$

9.2 The Boundary Conditions

The boundary condition for Eq. (9.1) can directly specify a zero normal velocity component $\partial \Phi^*/\partial n = 0$ on the surface S_B, in which case this "direct" formulation is called the *Neumann problem*. It is possible to specify Φ^* on the boundary, so that indirectly the zero normal flow condition will be met, and this "indirect" formulation is called the *Dirichlet problem*. Of course, a combination of the above boundary conditions is possible, too, and this is called a *mixed boundary condition problem*.

An additional approach would be to search for a singularity distribution that creates enclosed streamlines, equivalent to the geometry of the surface S_B. This method is useful in two dimensions, where the stream function Ψ is well defined (and hence the streamlines $\Psi =$ const. can be easily derived as in Sections 3.10 and 3.11), but for complex, three-dimensional geometries the implementation of this method is difficult and will not be dealt with here.

a. Neumann Boundary Condition

In this case it is required that $\partial \Phi^*/\partial n$ will be specified on the solid boundary S_B, that is,

$$\nabla(\Phi + \Phi_\infty) \cdot \mathbf{n} = 0 \quad (9.4)$$

where Φ is the perturbation potential consisting of the two integral terms in Eq. (9.2a). From this point and on, for convenience, the velocity potential will be split such that Φ_∞ is the free-stream velocity potential (Eq. (9.3)) relative to the origin of the coordinates attached to the surface S_B. The second boundary condition (at the distant, outer boundaries of the flow) requires that the flow disturbance, due to the body's motion through the fluid, should diminish far from the body,

$$\lim_{r \to \infty} \nabla \Phi = 0 \qquad (9.5)$$

where $\mathbf{r} = (x, y, z)$. This condition is automatically met by all the singular solutions considered here. To satisfy the boundary condition of Eq. (9.4) directly, we use the velocity field due to the singularity distribution of Eq. (9.2):

$$\nabla \Phi^*(x, y, z) = \frac{1}{4\pi} \int_{\text{body+wake}} \mu \nabla \left[\frac{\partial}{\partial n} \left(\frac{1}{r} \right) \right] dS - \frac{1}{4\pi} \int_{\text{body}} \sigma \nabla \left(\frac{1}{r} \right) dS + \nabla \Phi_\infty \qquad (9.6)$$

If the singularity distribution strengths σ and μ are known, then Eq. (9.6) describes the velocity field everywhere (of course special treatment is needed when the velocity is evaluated on the surface S_B). Substitution of Eq. (9.6) into the boundary condition in Eq. (9.4) results in

$$\left\{ \frac{1}{4\pi} \int_{\text{body+wake}} \mu \nabla \left[\frac{\partial}{\partial n} \left(\frac{1}{r} \right) \right] dS - \frac{1}{4\pi} \int_{\text{body}} \sigma \nabla \left(\frac{1}{r} \right) dS + \nabla \Phi_\infty \right\} \cdot \mathbf{n} = 0 \qquad (9.7)$$

This equation is the basis for many numerical solutions and should hold for every point on the surface S_B. For example, a certain number of points (called *collocation points*) can be selected on the surface S_B. The boundary condition of Eq. (9.7) is then specified at each of these points in terms of the unknown singularities at all the collocation points. This approach reduces the integral equation (Eq. (9.7)) to a set of algebraic equations. As noted in Chapter 3, the solution at this point is not unique, and the combination of sources and doublets must be specified.

Note that if for an enclosed boundary (e.g., S_B) $\partial \Phi^*/\partial n = 0$, as required by the boundary condition in Eq. (9.4), then the potential inside the body (without internal singularities) will not change (Lamb,[9.1] p. 41), that is,

$$\Phi_i^* = \text{const.} \qquad (9.8)$$

a constant that could be selected as zero. This observation is important since it allows us to specify the boundary condition (Eq. (9.4)) in terms of the potential inside S_B, which is the Dirichlet problem (or Dirichlet boundary condition).

b. Dirichlet Boundary Condition

In this case, the perturbation potential Φ has to be specified everywhere on S_B. Equation (9.2a) does this exactly, and by distributing the singularity elements on the surface, and placing the point (x, y, z) inside the surface S_B, the inner potential Φ_i^* in terms of the surface singularity distributions is obtained:

$$\Phi_i^*(x, y, z) = \frac{1}{4\pi} \int_{\text{body+wake}} \mu \frac{\partial}{\partial n} \left(\frac{1}{r} \right) dS - \frac{1}{4\pi} \int_{\text{body}} \sigma \left(\frac{1}{r} \right) dS + \Phi_\infty \qquad (9.9)$$

Again, these integrals are singular when $r \to 0$, and near this point, their principal value must be evaluated. The zero flow normal to the surface boundary condition (Eq. (9.4)) is defined now using Eq. (9.8). Therefore, the condition $\nabla(\Phi + \Phi_\infty) \cdot \mathbf{n} = 0$, in terms of the velocity potential, becomes

$$\Phi_i^* = (\Phi + \Phi_\infty)_i = \text{const.}$$

or

$$\Phi_i^*(x, y, z) = \frac{1}{4\pi} \int_{\text{body+wake}} \mu \frac{\partial}{\partial n}\left(\frac{1}{r}\right) dS - \frac{1}{4\pi} \int_{\text{body}} \sigma\left(\frac{1}{r}\right) dS + \Phi_\infty = \text{const.} \quad (9.10)$$

Equation (9.10) is the basis for methods utilizing the indirect boundary conditions. However, even at this stage, there are many differences between the various methods of solution, related to setting the value of the inner potential Φ_i^* (in addition to the differences in the source/doublet combinations). For example, by setting $\Phi_i^* = (\Phi + \Phi_\infty)_i = 0$, Eq. (9.10) can be solved on the surface S_B, but the resulting singularity distribution will include Φ_∞ and the strength will be large.

Other values (not necessarily constant) for the inner potential can be specified too and when the inner potential is set to $\Phi_i^* = (\Phi + \Phi_\infty)_i = \Phi_\infty$ (which is equivalent to specifying Eq. (9.10) for the perturbation only in a "ground-fixed frame" where $\Phi_\infty = 0$) then Eq. (9.10) reduces to a simpler form:

$$\frac{1}{4\pi} \int_{\text{body+wake}} \mu \frac{\partial}{\partial n}\left(\frac{1}{r}\right) dS - \frac{1}{4\pi} \int_{\text{body}} \sigma\left(\frac{1}{r}\right) dS = 0 \quad (9.11)$$

To justify the above, consider the Neumann boundary condition (Eq. (9.4)) $\partial \Phi^*/\partial n = 0$, which is equivalent to

$$\frac{\partial \Phi}{\partial n} = -\mathbf{n} \cdot \mathbf{Q}_\infty. \quad (9.4a)$$

Recall that the value for the discontinuity in the normal derivative of the velocity potential as given by Eq. (3.12) is

$$-\sigma = \frac{\partial \Phi^*}{\partial n} - \frac{\partial \Phi_i^*}{\partial n} = \frac{\partial \Phi}{\partial n} - \frac{\partial \Phi_i}{\partial n}$$

and since $\Phi_i = 0$ then also $\partial \Phi_i/\partial n = 0$ on S_B. Consequently, for Eq. (9.11) to be valid, and with the aid of Eq. (9.4a) the source strength is required to be

$$\sigma = \mathbf{n} \cdot \mathbf{Q}_\infty \quad (9.12)$$

where **n** points into the body as in Fig. 9.1.

To define this problem uniquely, the wake doublet distribution should be known or related to the unknown doublets on S_B (Kutta condition). To proceed with the solution, S_B is divided into discrete elements and at each of these elements Eq. (9.10) (or Eq. (9.11)) is evaluated. This results in a set of algebraic equations for the unknown μ distribution. Note that when evaluating the integrals at a point P on the element ($r \to 0$) then $\Phi(P) = \mp \mu/2$ (see Section 3.14).

In this formulation, when Eq. (9.12) is used, the zero normal flow boundary condition information is contained in the source terms and for very thin surfaces the integral may be ill-conditioned and will cause numerical instabilities.

9.3 Physical Considerations

The above mathematical formulation, even after selecting a desirable combination of sources and doublets, and after fulfilling the boundary conditions on the surface S_B, is not

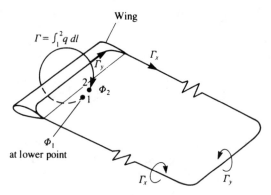

Figure 9.2 Vorticity system created by a finite wing in steady forward flight.

unique. Previous examples showed that for describing the flow over thick bodies without lift the source distribution was sufficient, but for the lifting cases the amount of the circulation was not uniquely defined. Before proceeding further (and using the information developed in Chapter 8), let us examine the case of a lifting wing, as viewed from a large distance (Fig. 9.2). For simplicity, the bound vortex is represented by a concentrated vortex line with the strength Γ ($=\Gamma_x = \Gamma_y$). According to the Helmholtz theorems (Section 2.9) a vortex line cannot start in a fluid and similarly to Eq. (4.64) we can write

$$\frac{\partial \Gamma_x}{\partial x} = \frac{-\partial \Gamma_y}{\partial y} \qquad (9.13)$$

which for the simple case of Fig. 9.2 implies that the problem is modeled by one constant-strength, closed vortex line. Also, the amount of the bound circulation is

$$\Gamma = \int_1^2 \mathbf{q} \cdot d\mathbf{l}$$

where point 1 lies under and point 2 is above the (very) thin wake. These two arguments clearly demonstrate that for the three dimensional lifting problem there is a need to model a wake, since the bound vorticity needs to be continued beyond the wing. Also, as shown in Fig. 9.2, for the wing to have circulation Γ at a spanwise location (see Section 3.14), a discontinuity in the velocity potential near the trailing edge must exist:

$$\Phi_2 - \Phi_1 = \Gamma$$

where Φ_1 is under and Φ_2 is above the wake. Now we are in a position where the additional *physical* conditions, required for a unique solution, can be established in relation to a *wake model*. This model has to specify two additional conditions:

1. To set the wake strength at the trailing edge.
2. To set its shape and location.

a. *Wake Strength*

The simplest solution to this problem is to apply the two-dimensional Kutta condition along the three-dimensional trailing edge (as shown in Fig. 9.3) such that

$$\gamma_{T.E.} = 0 \qquad (9.14)$$

Since, for example, in the two-dimensional case $\partial \mu(x)/\partial x = -\gamma(x)$ (as in Section 3.14) the above condition can be rewritten for the trailing-edge line, such that μ is constant in the

9.3 Physical Considerations

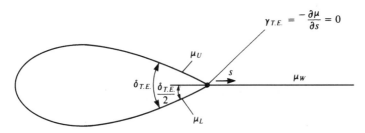

Figure 9.3 Implementation of the Kutta condition when using surface doublet distribution.

wake (μ_w) and equals the value at the trailing-edge ($\mu_{T.E.}$), that is,

$$\mu_{T.E.} = \text{const.} \equiv \mu_W$$

or

$$\mu_U - \mu_L - \mu_W = 0 \tag{9.15a}$$

where μ_U and μ_L are the corresponding upper and lower surface doublet strengths at the trailing edge, as shown in Fig. 9.3. An alternate formulation equates the upper and lower velocities at the trailing edge:

$$\frac{\partial \mu_U}{\partial s} = \frac{\partial \mu_L}{\partial s} \tag{9.15b}$$

where s is a coordinate along the trailing-edge upper and lower surfaces. This formulation is more useful for airfoils with very thin or even cusped trailing edges.[9.2] As an example, the specification of the Kutta condition in terms of constant-strength doublet elements (or vortex rings) is shown in Fig. 9.4 (here for convenience a positive doublet points into the wing). At the wing's trailing edge, the trailing segment of the upper doublet will have a strength of $-\Gamma_U$, the leading vortex segment of the lower surface (which is now inverted) will be $+\Gamma_L$, and the leading segment of the wake vortex is $+\Gamma_W$. Thus, the strength of the wake panel in terms of the local circulation Γ is again

$$-\Gamma_U + \Gamma_L + \Gamma_W = 0$$

or, exactly as in Eq. (9.15a),

$$\Gamma_W = \Gamma_U - \Gamma_L \tag{9.16}$$

In certain situations the shape of the trailing edge is also important. For example, Fig. 9.5a shows a situation where the flow leaves the trailing edge smoothly and parallel to the cusped

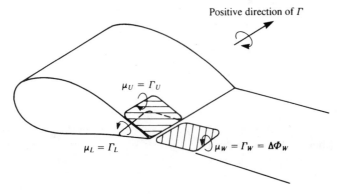

Figure 9.4 Implementation of the Kutta condition when using vortex ring elements.

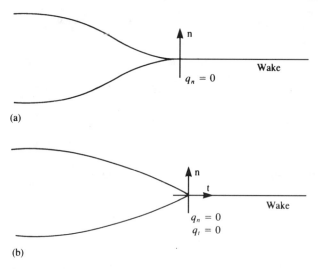

Figure 9.5 Possible conditions that can be applied at (a) cusp and (b) finite angle trailing edges.

trailing edge. In such situations this point is not necessarily a stagnation point and if the velocity formulation is used then only the $q_n = 0$ condition can be used. In the case that the trailing edge has a finite angle (Fig. 9.5b), then in order to have a continuous velocity at this point the condition $q_t = 0$ can also be used.

b. *Wake Shape*

In two dimensions, the trailing vortex segment of the wake is ignored since it has zero vorticity (in steady flow) and it is sufficient to specify the location of the trailing edge where the Kutta condition is met. In three dimensions, the wake influence is more dominant and its geometry clearly affects the solution. To distinguish between the models for bound circulation (which generate the lift) and the circulation shed into the wake, it is logical to assume that the wake should not produce lift – since it is not a solid surface. As an example, let us recall the formulation for the force $\Delta \mathbf{F}$ generated by a vortex sheet γ. The Kutta–Joukowski theorem for lift (Section 3.11) states that

$$\Delta \mathbf{F} = \rho \mathbf{q} \times \gamma \tag{9.17}$$

For a three-dimensional case $\Delta \mathbf{F} = 0$ only if the local flow is parallel to γ (we assume $\gamma \neq 0$). So the condition for the wake geometry is

$$\mathbf{q} \times \gamma_W = 0 \tag{9.18}$$

or

$$\gamma_W \parallel \mathbf{q} \tag{9.18a}$$

that is, the vorticity vector is parallel to the local velocity vector.

An equivalent representation of the wake by a thin doublet sheet is obtained by noting that $\gamma_W = -\nabla \mu_W$ (this will be demonstrated in Chapter 10). If no force is produced by this lifting surface then Eq. (9.18) becomes

$$\mathbf{q} \times \nabla \mu_W = 0 \tag{9.19}$$

9.4 Reduction of the Problem to a Set of Linear Algebraic Equations

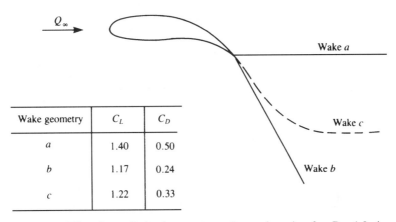

Figure 9.6 Effect of prescribed wake geometry on the aerodynamics of an $\mathcal{R} = 1.5$ wing.

So the condition for the wake panels, in terms of doublets, is

$$\mu_W = \text{const.} \tag{9.19a}$$

and the boundaries of these elements (which are really the vortex lines) should be parallel to the local streamlines, as in Eq. (9.18a). This condition (Eq. (9.18a)) is difficult to satisfy exactly since the wake location is not known in advance. In most cases it is sufficient to assume that the wake leaves the trailing edge at a median angle $\delta_{T.E.}/2$, as shown in Figs. 9.3 and 9.4, whereas for portions of the wake far from the trailing edge, additional effort is required to satisfy the condition of Eq. (9.18).

As an example of the dependence of the solution on the wake initial geometry, the results for a cambered rectangular wing of aspect ratio 1.5 are shown in Fig. 9.6. The solution was obtained by a first-order panel method (VSAERO[9.3]) with 600 panels per semispan and the corresponding lift and drag coefficients are tabulated in the inset to the figure (incidentally, case c is the closest to experimental results).

9.4 Reduction of the Problem to a Set of Linear Algebraic Equations

At this point it is assumed that the problem is unique and that a combination of source/doublet distributions has been selected along with a wake model and the Kutta condition. For the following example $\Phi_i^* = \Phi_\infty$ along with Eq. (9.12) for the source strength will be used and a constant-strength rectilinear panel is assumed (this approach is widely used in many panel codes such as in Ref. 9.3). The body's surface (see Fig. 9.7) is now divided into N surface panels and into N_W additional wake panels. The boundary condition (either Neumann or Dirichlet) will be specified at each of these elements at a collocation point (which for the Dirichlet boundary condition must be specified inside the body where $\Phi_i^* = \Phi_\infty$, e.g., at a point under the center of the panel). In most cases, though, the point may be left on the surface without moving it inside the body. Rewriting, for example, the Dirichlet boundary condition for each of the N collocation points, we can transform Eq. (9.11) into the following form:

$$\sum_{k=1}^{N} \frac{1}{4\pi} \int_{\text{body panel}} \mu \mathbf{n} \cdot \nabla \left(\frac{1}{r}\right) dS + \sum_{\ell=1}^{N_W} \frac{1}{4\pi} \int_{\text{wake panel}} \mu \mathbf{n} \cdot \nabla \left(\frac{1}{r}\right) dS$$

$$- \sum_{k=1}^{N} \frac{1}{4\pi} \int_{\text{body panel}} \sigma \left(\frac{1}{r}\right) dS = 0 \tag{9.20}$$

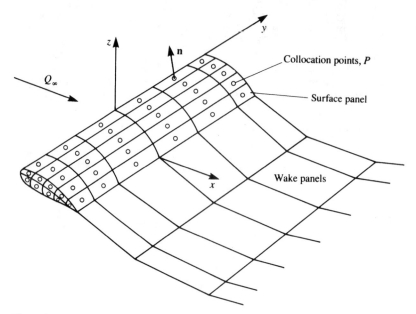

Figure 9.7 Approximation of the body surface by panel elements.

That is, for each collocation point P (shown in Fig. 9.7) the summation of the influences of all k body panels and ℓ wake panels is needed. The integration in Eq. (9.20) is limited now to each individual panel element representing the influence of this panel on point P. For a unit singularity element (σ or μ), this influence depends on the panel's geometry only. The integration can be performed analytically or numerically, prior to this calculation, and for example for a constant-strength μ element shown in Fig. 9.8 the influence of panel k (defined by the four corners 1, 2, 3, and 4) at point P is

$$\frac{1}{4\pi} \int_{1,2,3,4} \frac{\partial}{\partial n}\left(\frac{1}{r}\right) dS \bigg|_k \equiv C_k \tag{9.21}$$

and for a constant-strength σ element

$$\frac{-1}{4\pi} \int_{1,2,3,4} \left(\frac{1}{r}\right) dS \bigg|_k \equiv B_k \tag{9.21a}$$

These integrals are a function of the points 1, 2, 3, 4, and P and an "influence computing routine" can be schematically defined as

$$\begin{pmatrix} x_P, y_P, z_P \\ x_1, y_1, z_1 \\ x_2, y_2, z_2 \\ x_3, y_3, z_3 \\ x_4, y_4, z_4 \\ \mu \end{pmatrix} \Rightarrow \begin{pmatrix} \text{influence} \\ \text{coefficient} \\ \text{calculation} \end{pmatrix} \Rightarrow \begin{pmatrix} \Delta u, \Delta v, \Delta w \\ \Delta \Phi \end{pmatrix}_P \tag{9.22}$$

Of course, in this case $\Delta\Phi_P = C_k$. After computing the influence of each panel on each other panel, Eq. (9.20) for each point P inside the body becomes

$$\sum_{k=1}^{N} C_k \mu_k + \sum_{\ell=1}^{N_W} C_\ell \mu_\ell + \sum_{k=1}^{N} B_k \sigma_k = 0 \quad \text{for each internal point } P \tag{9.23}$$

9.4 Reduction of the Problem to a Set of Linear Algebraic Equations

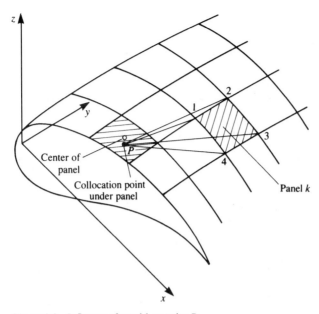

Figure 9.8 Influence of panel k on point P.

This equation is the numerical equivalent of the boundary condition. If the strengths of the sources are selected according to Eq. (9.12) and since the coefficients B_k, which are computed in a manner similar to Eq. (9.22), are known, the source term can be moved to the right-hand side of the equation. Also, by using the Kutta condition, the wake doublets can be expressed in terms of the unknown surface doublets μ_k. For example, in Fig. 9.9 two of the trailing edge (T.E.) doublets μ_r and μ_s (here r, s, and t are some arbitrary counters) are related to the corresponding wake doublet μ_t by Eq. (9.15):

$$\mu_t = \mu_r - \mu_s$$

and hence the influence of the wake element becomes

$$C_t \mu_t = C_t(\mu_r - \mu_s)$$

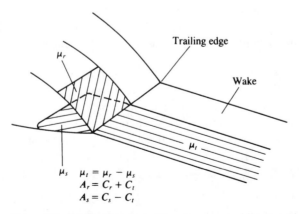

Figure 9.9 Relation between trailing edge upper and lower panel doublet strength and the corresponding wake doublet strength.

This algebraic relation can be substituted into the C_k coefficients of the unknown surface doublet such that

$$A_k = C_k \quad \text{if panel is not at T.E.}$$
$$A_k = C_k \pm C_t \quad \text{if panel is at T.E.}$$

where the $\pm$ sign depends on whether the panel is at the upper or the lower side of the trailing edge (Fig. 9.9). Consequently, for each collocation point P, a linear algebraic equation containing N unknown singularity variables μ_k can be derived:

$$\sum_{k=1}^{N} A_k \mu_k = -\sum_{k=1}^{N} B_k \sigma_k \tag{9.24}$$

Evaluating Eq. (9.24) at each of the N collocation points ($j = 1 \to N$) results in N equations with the N unknown μ_k, in the following form:

$$\begin{pmatrix} a_{11}, a_{12}, \ldots, a_{1N} \\ a_{21}, a_{22}, \ldots, a_{2N} \\ \vdots \quad \vdots \quad \quad \vdots \\ a_{N1}, a_{N2}, \ldots, a_{NN} \end{pmatrix} \begin{pmatrix} \mu_1 \\ \mu_2 \\ \vdots \\ \mu_N \end{pmatrix} = - \begin{pmatrix} b_{11}, b_{12}, \ldots, b_{1N} \\ b_{21}, b_{22}, \ldots, b_{2N} \\ \vdots \quad \vdots \quad \quad \vdots \\ b_{N1}, b_{N2}, \ldots, b_{NN} \end{pmatrix} \begin{pmatrix} \sigma_1 \\ \sigma_2 \\ \vdots \\ \sigma_N \end{pmatrix} \tag{9.25}$$

Note that for evaluating the influence of the panel on itself (a_{kk}, b_{kk}) the integral of the influence coefficients may be singular and its principal value must be evaluated. In this formulation the unknown μ distribution is small (perturbation only) and the numerical solution is believed to be more stable.[9.3] The right-hand side of Eq. (9.25) can be computed since the value of σ_k is known and Eq. (9.25) can be rewritten as

$$\begin{pmatrix} a_{11}, a_{12}, \ldots, a_{1N} \\ a_{21}, a_{22}, \ldots, a_{2N} \\ \vdots \quad \vdots \quad \quad \vdots \\ a_{N1}, a_{N2}, \ldots, a_{NN} \end{pmatrix} \begin{pmatrix} \mu_1 \\ \mu_2 \\ \vdots \\ \mu_N \end{pmatrix} = \begin{pmatrix} \text{RHS}_1 \\ \text{RHS}_2 \\ \vdots \\ \text{RHS}_N \end{pmatrix} \tag{9.25a}$$

where the values of μ_k can be computed by solving this *full-matrix* equation.

Note that the relation $\sigma = \mathbf{Q}_\infty \cdot \mathbf{n}$ of Eq. (9.12) contains the information on the zero normal flow condition for the thickness problem and this formulation will be singular for surfaces approaching zero thickness.

The derivation of the influence coefficient integrals depends on the shape of the panel element (e.g., planar, curved, etc.) and on the singularity distribution (constant or linearly varying strength, etc.). Some examples will be presented in the following chapters.

9.5 Aerodynamic Loads

Once Eq. (9.25) is solved the unknown singularity values are obtained (μ_k in this example). The velocity components are evaluated now in terms of the panel local coordinates (l, m, n) shown in Fig. 9.10. The two tangential perturbation velocity components are

$$q_l = -\frac{\partial \mu}{\partial l}, \quad q_m = -\frac{\partial \mu}{\partial m} \tag{9.26}$$

where the differentiation is done numerically using the values on the neighbor panels. The normal component of the velocity (in this example) is obtained from the local source

9.6 Preliminary Considerations, Prior to Establishing Numerical Solutions

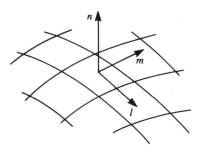

Figure 9.10 Panel local coordinate system for evaluating the tangential velocity components.

strength:

$$q_n = -\sigma \tag{9.27}$$

The total velocity in the local (l, m, n) direction of panel k is

$$\mathbf{Q}_k = (Q_{\infty_l}, Q_{\infty_m}, Q_{\infty_n})_k + (q_l, q_m, q_n)_k \tag{9.28}$$

and of course the normal velocity component on a solid boundary is zero. The pressure coefficient can now be computed for each panel using Eq. (4.53):

$$C_{p_k} = 1 - \frac{Q_k^2}{Q_\infty^2} \tag{9.29}$$

The contribution of this element to the nondimensional fluid dynamic loads is normal to the panel surface and is

$$\Delta C_{F_k} = \frac{\Delta F_k}{(1/2)\rho Q_\infty^2 S}$$

where S is a reference area. In terms of the pressure coefficient the vector form for the panel contribution to the fluid dynamic load becomes

$$\Delta \mathbf{C}_{F_k} = -\frac{C_{p_k} \Delta S_k}{S} \cdot \mathbf{n}_k \tag{9.30}$$

The individual contributions of the panel elements now can be summed to compute the desired aerodynamic forces and moments.

9.6 Preliminary Considerations, Prior to Establishing Numerical Solutions

Prior to establishing a numerical solution, some of the options need to be considered:

  a. *Type of singularity that will be used*: The options usually include sources, doublets, and vortices or any combination of the above.
  b. *Type of boundary conditions*: Velocity or velocity-potential formulation may be used and the corresponding Neumann, Dirichlet, or a combination of such boundary conditions must be selected.
  c. *Wake models*: How and where the Kutta condition will be specified. Also, the shape of the wake is controlled by Eq. (9.18a) and can be set by
     1. Programmer-specified shape based on intuition or on flow visualizations.

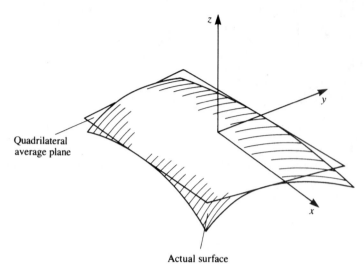

Figure 9.11 Nonplanar surface element and its quadrilateral approximation.

2. Wake relaxation (where the wake points are moved with the local induced velocity, e.g., in Ref. 9.3).
3. Time stepping (where the wake shape is developed by moving the wing from an initial stand-still position, as will be presented in Chapter 13).

d. *Method of discretizing surface and singularity distributions*:
   1. Discretization of geometry: The placing of a simple panel element on an arbitrary three-dimensional configuration is rather difficult. Figure 9.11 describes such a curved surface element with a local coordinate system x, y, z. The shape of the surface can be described as $z = f(x, y)$, but for simplicity it is usually approximated by a piecewise polynomial approximation. For example, if a first-order polynomial is used then the average surface can be described by

   $$z = a_0 + b_1 x + b_2 y$$

   and for a second-order polynomal aproximation

   $$z = a_0 + b_1 x + b_2 y + c_1 x^2 + c_2 xy + c_3 y^2$$

   and so on (where the coefficients a, b, c are constants). Figure 9.11 shows the result of approximating a curved surface element by a first-order plane, while Fig. 9.12 shows the possible consequence of representing a three-dimensional curved surface by such quadrilateral elements. This representation of the geometry may result in difficulties in specifying the boundary conditions, since the "leakage" between the panels can weaken the satisfaction of the zero flow through the boundaries requirement. One possible solution is shown in Fig. 9.13 where the surface is described by five flat subelements (as in the PANAIR code[9.4]).

   2. Discretization of singularity distribution: The strength of the surface distribution of the singularity elements can be represented, too, in terms of a piecewise polynomial approximation. For example, if the doublet distribution on the element of Fig. 9.11 is constant such that

   $$\mu = a_0 = \text{const.}$$

9.6 Preliminary Considerations, Prior to Establishing Numerical Solutions

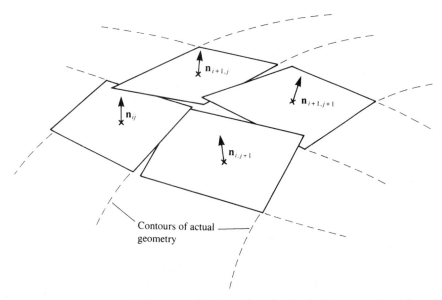

Figure 9.12 Possible difficulty in representing a three-dimensional surface by an array of quadrilateral surface elements.

then this is a zero-order approximation of μ. Similarly, a first-order (or linear) approximation is

$$\mu = a_0 + b_1 x + b_2 y$$

and a second-order (or parabolic) polynomial approximation is

$$\mu = a_0 + b_1 x + b_2 y + c_1 x^2 + c_2 xy + c_3 y^2$$

(Here the coefficients a, b, c are constants, too, and of course are different from the coefficients of the surface approximation).

e. *Considerations of numerical efficiency:* It is clear from the brief discussion on discretization that the computation of the influence coefficients (e.g., Eq. (9.21)) is elaborate. Many methods divide such calculations into near and far field where the far field calculation treats the element as a point singularity (and not as a

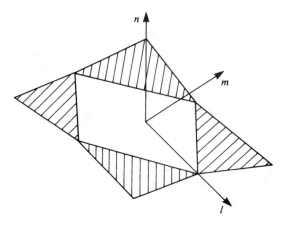

Figure 9.13 Description of a nonplanar panel element by a set of flat subelements.

surface distribution). Typically, the near field is assumed if the distance to a point P is less than 2.5–5 times the larger diagonal of the panel. On the other hand because of the $1/r$ characteristics of the singularity elements, when $r \to 0$ the value of $1/r \to \infty$; therefore, when the point P is too close to the panel (or to a vortex line) cutoff distances are usually applied. (Only the aerodynamic aspects of the numerics are discussed here; other important aspects, e.g., the matrix solver efficiency, are not.)

9.7 Steps toward Constructing a Numerical Solution

When establishing a numerical solution for potential flow a sequence similar to the following is recommended.

a. *Selection of Singularity Element*

The first and one of the most important decisions is the type of singularity element or elements that will be used. This includes the selection of source, doublet, or vortex representation and the method of discretizing these distributions (zero-, first-, second-order, etc.). Also, all of the questions raised in the previous section need to be answered before the actual formulation of the solution can be constructed. Once these decisions have been made an influence routine, similar to the model of Eq. (9.22), needs to be established. This influence computation is a direct function of the element geometry and such a routine outputs the velocity components and the potential ($\Delta u, \Delta v, \Delta w, \Delta \Phi$) induced by the element. In general, the implementation of Eq. (9.22) represents the core of most numerical solutions. Therefore, in the next chapter some of the more frequently used singularity elements will be formulated.

b. *Discretization of Geometry (and Grid Generation)*

Once the basic solution element is selected, the geometry of the problem has to be subdivided (or discretized), such that it will consist of those basic solution elements. In this grid generating process, the elements' corner points and collocation points are defined. The collocation points are points where the boundary conditions, such as the zero normal flow on a solid surface, will be enforced. Figure 9.14a shows how the cambered thin airfoil at an angle of attack can be discretized by using the lumped-vortex element. In this case the camberline is divided into five panels and the locations of the collocation points and of the vortex points are shown in the figure. Similarly, the subdivision of a three-dimensional body into planar surface elements is shown in Fig. 9.14b. (The collocation points are not shown but they are at the center of the panel and may be slightly under the surface.)

It is very important to realize that the grid *does* have an effect on the solution. Typically, a good grid selection will enable convergence to a certain solution when the density is increased (within reason). Moreover, a good grid selection usually will require some preliminary understanding of the problem's fluid dynamics, as will be shown in some of the forthcoming examples.

c. *Influence Coefficients*

In this phase, for each of the elements, an algebraic equation (based on the boundary condition) is derived at the collocation point. To generate the coefficients in an automatic manner, a unit singularity strength is assumed and the element influence routine is called at each of the collocation points (by a DO loop).

9.7 Steps toward Constructing a Numerical Solution

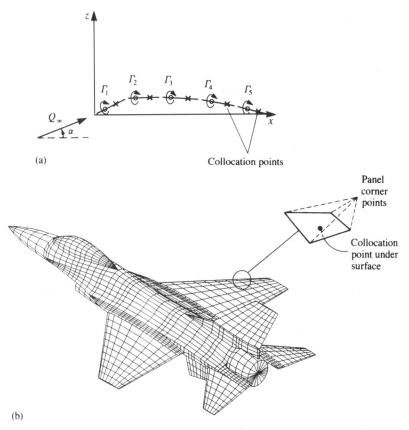

Figure 9.14 Discretization of (a) the geometry of a thin airfoil by using the lumped vortex element and of (b) a three-dimensional body using constant-strength surface doublets and sources.

d. *Establish RHS*

The right-hand side of the matrix equation is the known portion of the free-stream velocity or the potential and requires mainly the computation of geometric quantities (e.g., $-Q_\infty \alpha$).

e. *Solve Linear Set of Equations*

The coefficients and the RHS of the algebraic equations were obtained in the previous steps and now the equations are solved by standard matrix techniques. Here it is assumed that the reader is familiar with such numerical solvers, which can be found in textbooks (e.g., Ref. 9.5 or as the solvers appearing in the student computer programs of Appendix D).

f. *Secondary Computations: Pressures, Loads, Off-Body Velocity, Etc.*

The solution of the matrix equation results in the singularity strengths and the velocity field and any secondary information can be computed now. The pressures will be computed by Bernoulli's equation, and the loads and aerodynamic coefficients by adding up the contributions of the elements. A typical flowchart for such a computer program is shown in Fig. 9.15 where the sequence of computations is close to the above described methodology.

In the following example, the essence of the above steps will be clarified.

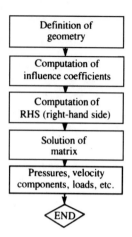

Figure 9.15 Typical flowchart for the numerical solution of the surface singularity distribution problem.

9.8 Example: Solution of Thin Airfoil with the Lumped-Vortex Element

As a first example for demonstrating the principle of numerical solutions, let us consider the solution for the symmetric, thin airfoil. Because the airfoil is thin, no sources will be used, while the doublet distribution will be approximated by two constant-strength doublet elements (μ_1, μ_2 pointing in the $-z$ direction). This element is equivalent to two concentrated vortices at the panel edges (see Fig. 9.16). However, the geometry of the "lumped-vortex" model was developed in Chapter 5, and by placing the vortex at the quarter chord and the collocation point at the three-quarter chord point of the panel the Kutta condition is automatically satisfied. Using this knowledge the equivalent discrete-vortex model (with only two elements) for the airfoil is shown in Fig. 9.17. Also, for the thin lifting surface only the Neumann (velocity) boundary condition can be used, because of the zero thickness of the airfoil. (Note that the doublet representation in Fig. 9.16 clearly indicates the existence of a starting vortex, also shown in Fig. 9.17, at a large distance behind the airfoil.)

a. *Selection of Singularity Element*

For this very simple example the lumped-vortex element is selected and its influence is derived in terms of the geometry involved. Such an element is depicted in Fig. 9.18a; it consists of a concentrated vortex at the panel quarter chord and a collocation point and

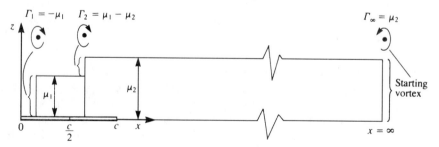

Figure 9.16 Constant-strength doublet element representation of the flat plate at an angle of attack (using two doublet panels pointing in the $-z$ direction).

9.8 Example: Solution of Thin Airfoil with the Lumped-Vortex Element

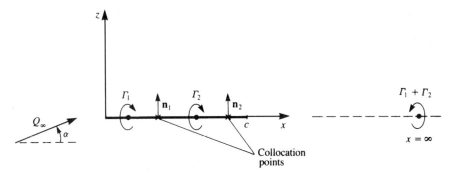

Figure 9.17 Equivalent discrete-vortex model for the flow over a flat plate at an angle of attack (using two elements).

normal vector **n** at the three-quarter chord. It is important to remember that this element is a simplification of the two-dimensional continuous solution and therefore accounts for the Kutta condition at the trailing edge of the airfoil.

If the vortex element of circulation Γ is located at (x_0, z_0), then the velocity induced by this element at an arbitrary point $P(x, z)$, according to the analysis in Section 3.8, will be

$$u = \frac{\Gamma}{2\pi} \frac{(z - z_0)}{(x - x_0)^2 + (z - z_0)^2}$$

$$w = \frac{-\Gamma}{2\pi} \frac{(x - x_0)}{(x - x_0)^2 + (z - z_0)^2}$$

and a matrix version of this equation, which is more useful for computations, is

$$\begin{pmatrix} u \\ w \end{pmatrix} = \frac{\Gamma}{2\pi r^2} \begin{pmatrix} 0 & 1 \\ -1 & 0 \end{pmatrix} \begin{pmatrix} x - x_0 \\ z - z_0 \end{pmatrix} \tag{9.31}$$

where

$$r^2 = (x - x_0)^2 + (z - z_0)^2$$

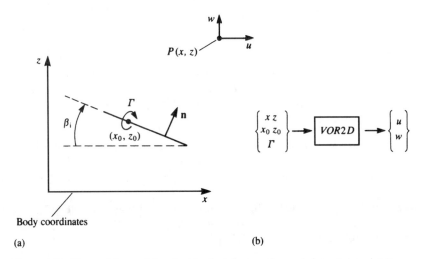

Figure 9.18 Nomenclature and flowchart for the influence of a panel element at a point P.

This can be programmed as an influence coefficient subroutine in the manner shown in Fig. 9.18b. Let us call this routine VOR2D. An algorithm based on Eq. (9.31) will then have the form

$$(u, w) = VOR2D(\Gamma, x, z, x_0, z_0) \tag{9.32}$$

b. *Discretization of Geometry and Grid Generation*

For this example, the thin airfoil case is being solved (Fig. 9.17). For simplicity, only two elements will be used so that no computations are necessary. At this phase the geometrical information on the grid has to be derived. This can be automated by computer routines for more complex situations, but for this case the vortex point locations are

$$(x_{01}, z_{01}) = (c/8, 0)$$
$$(x_{02}, z_{02}) = (5c/8, 0)$$

and the collocation points are

$$(x_{c1}, z_{c1}) = (3c/8, 0)$$
$$(x_{c2}, z_{c2}) = (7c/8, 0)$$

The normal vectors **n** must be evaluated at the collocation points, and for an arbitrary element i we write

$$\mathbf{n}_i = (\sin \beta_i, \cos \beta_i) \tag{9.33}$$

where the angle β_i is defined in Fig. 9.18a. In this particular case, when the airfoil has no camber and is placed on the $z = 0$ plane, both normals are identical:

$$\mathbf{n}_1 = \mathbf{n}_2 = (0, 1)$$

c. *Influence Coefficients*

Here the condition requiring zero velocity normal to the airfoil will be enforced. This boundary condition, according to Eq. (9.4), is

$$(\mathbf{q} + \mathbf{Q}_\infty) \cdot \mathbf{n} = 0 \tag{9.34}$$

The velocity **q** is induced by the unknown vortices, whereas the free-stream normal component can be calculated directly and hence is moved to the right-hand side of the equation:

$$\mathbf{q} \cdot \mathbf{n} = -\mathbf{Q}_\infty \cdot \mathbf{n} \tag{9.34a}$$

Because, in this case, the airfoil was divided into two elements with two unknown vortices of circulation Γ_1, Γ_2, two equations based on the zero flow normal to the airfoil boundary condition will be derived at the collocation points. We define as positive Γ a clockwise rotation, and calculate the velocity induced by a unit strength vortex at point 1 on collocation point 1 with Eq. (9.32):

$$(u_{11}, w_{11}) = VOR2D(1.0, x_{c1}, z_{c1}, x_{01}, z_{01}) = \left(0, -\frac{1}{2\pi \cdot c/4}\right)$$

and the velocity induced at collocation point 1, by a unit vortex at point 2, is

$$(u_{12}, w_{12}) = VOR2D(1.0, x_{c1}, z_{c1}, x_{02}, z_{02}) = \left(0, \frac{1}{2\pi \cdot c/4}\right)$$

The velocity induced at collocation point 2, by a unit vortex at point 1, is

$$(u_{21}, w_{21}) = VOR2D(1.0, x_{c2}, z_{c2}, x_{01}, z_{01}) = \left(0, -\frac{1}{2\pi \cdot 3c/4}\right)$$

9.8 Example: Solution of Thin Airfoil with the Lumped-Vortex Element

and the velocity induced at collocation point 2, by a unit vortex at point 2, is

$$(u_{22}, w_{22}) = VOR2D(1.0, x_{c2}, z_{c2}, x_{02}, z_{02}) = \left(0, -\frac{1}{2\pi \cdot c/4}\right)$$

The influence coefficients a_{ij} are really the normal component of the flow velocity induced by a unit strength vortex element Γ_j at collocation point i:

$$a_{ij} = \mathbf{q}_{ij}(\Gamma_j = 1) \cdot \mathbf{n}_i \tag{9.35}$$

For the current problem, Eq. (9.35) is applied to collocation point 1 and to vortex point 1. Thus

$$a_{11} = (u_{11}, w_{11}) \cdot \mathbf{n}_1 = \left(0, -\frac{1}{2\pi \cdot c/4}\right) \cdot (0, 1) = \frac{-2}{\pi c}$$

Similarly, for the second vortex, we have

$$a_{12} = (u_{12}, w_{12}) \cdot (0, 1) = \frac{2}{\pi c}$$

and for the second collocation point, we get

$$a_{21} = (u_{21}, w_{21}) \cdot (0, 1) = \frac{-2}{3\pi c}$$

$$a_{22} = (u_{22}, w_{22}) \cdot (0, 1) = \frac{-2}{\pi c}$$

Note that the left-hand side of Eq. (9.34a) can be described now as

$$\mathbf{q} \cdot \mathbf{n} = \sum_{j=1}^{2} a_{ij} \Gamma_j \quad \text{for} \quad i = 1, 2 \tag{9.36}$$

d. *Establish RHS*

The solution is based on enforcing the boundary condition of Eq. (9.34a) at the collocation points. Since the product $\mathbf{Q}_\infty \cdot \mathbf{n}$ is known it is transferred to the right-hand side of the equation:

$$\mathbf{q} \cdot \mathbf{n} = -\mathbf{Q}_\infty \cdot \mathbf{n} \equiv \text{RHS} \tag{9.37}$$

It is useful to express the component of the free stream in vector form to allow easy vector operations; for this particular case the right-hand side (RHS) is

$$\text{RHS}_i = -(U_\infty, W_\infty) \cdot \mathbf{n}_i \tag{9.38}$$

where $(U_\infty, W_\infty) = Q_\infty(\cos \alpha, \sin \alpha)$. Computing the RHS vector for the two collocation points results in

$$\text{RHS}_1 = -Q_\infty \sin \alpha$$
$$\text{RHS}_2 = -Q_\infty \sin \alpha$$

e. *Solve Linear Set of Equations*

The results of the previous computations can be summarized as

$$\sum_{j=1}^{2} a_{ij} \Gamma_j = \text{RHS}_i \quad i = 1, 2 \tag{9.39}$$

and explicitly, for this particular case,

$$\begin{pmatrix} -\frac{2}{\pi c} & \frac{2}{\pi c} \\ -\frac{2}{3\pi c} & -\frac{2}{\pi c} \end{pmatrix} \begin{pmatrix} \Gamma_1 \\ \Gamma_2 \end{pmatrix} = -Q_\infty \sin\alpha \begin{pmatrix} 1 \\ 1 \end{pmatrix}$$

which can be solved by standard matrix methods

f. *Secondary Computations: Pressures, Loads, Etc.*
The solution of the above set of algebraic equations is

$$\begin{pmatrix} \Gamma_1 \\ \Gamma_2 \end{pmatrix} = \begin{pmatrix} \frac{3}{4} \\ \frac{1}{4} \end{pmatrix} \pi c Q_\infty \sin\alpha$$

The resulting pressures and loads can be computed by using the Kutta–Joukowski theorem (Section 3.11):

$$\Delta L_i = \rho Q_\infty \Gamma_i$$

and by assuming a constant pressure distribution along the element, the pressure difference becomes

$$\Delta p_i = \rho Q_\infty \Gamma_i / a$$

where a is the panel length. The lift and moment about the airfoil leading edge are then

$$L = \sum_{i=1}^{2} \Delta L_i = \rho Q_\infty^2 \pi c \sin\alpha \tag{9.40}$$

$$M_0 = -\sum_{i=1}^{2} \Delta L_i x_{0_i} = -\rho Q_\infty^2 \pi \frac{c^2}{4} \sin\alpha \tag{9.41}$$

and the nondimensional aerodynamic coefficients are

$$C_l = \frac{L}{\frac{1}{2}\rho Q_\infty^2 c} = 2\pi \sin\alpha \tag{9.42}$$

$$C_{m_0} = \frac{M_0}{\frac{1}{2}\rho Q_\infty^2 c^2} = -\frac{\pi}{2} \sin\alpha \tag{9.43}$$

These results are similar to those for a zero-thickness symmetrical airfoil (Section 5.4) and equal to the exact flat plate solution (Section 6.5.1). The method can easily be extended to various camberline shapes and even multielement lifting airfoils.

Description of more complex numerical methods for solving the potential-flow problem will be presented in the following chapters.

9.9 Accounting for Effects of Compressibility and Viscosity

The potential flow model presented in this chapter results in a very simple mathematical model that can be transformed into a very efficient and economical numerical solution. This led to the development of three-dimensional "panel codes" for arbitrary geometries, and naturally, modifications were sought to improve these methods beyond the limits of incompressible inviscid flows. Some of these modifications are listed here.

a. Effects of Compressibility

The first and most straightforward modification to an incompressible potential-flow based method is to incorporate the effects of "low-speed compressibility" (e.g., for $M_\infty <$ 0.6). This modification can be obtained by using the Prandtl–Glauert rule, as developed in Section 4.8. Thus, small-disturbance flow is assumed, and a compressibility factor β can be defined as

$$\beta = \sqrt{1 - M_\infty^2} \tag{9.44}$$

If the free stream is parallel to the x coordinate then the x coordinate is being stretched with increased Mach number while the y and z coordinates remain unchanged. Consequently, an equivalent incompressible potential $\Phi_{M_\infty=0}$ can be defined such that

$$\Phi_{M_\infty=0} = \Phi_{M_\infty}\left(\frac{x}{\beta}, y, z\right) \tag{9.45}$$

Once the x coordinate is transformed, the equivalent incompressible potential problem is solved as described previously. This procedure results in an increase in the aerodynamic forces (as noted in Section 4.8) which may be approximated by

$$C_L(M_\infty > 0) = \frac{C_L(M_\infty = 0)}{\beta} \tag{9.46}$$

$$C_M(M_\infty > 0) = \frac{C_M(M_\infty = 0)}{\beta} \tag{9.47}$$

b. Effects of Thin Boundary Layers

When analyzing high Reynolds number flows in Section 1.8, it was assumed that the boundary layer is thin and that the boundary conditions are specified on the actual surface of the body. However, by neglecting the viscosity terms in the momentum equation, the information for calculating the viscous surface friction drag is lost too.

It is possible to account for the viscosity effects such as displacement thickness and friction drag by using a boundary layer solution that can be matched with the potential-flow solution (see Chapter 14). Two of the most common methods for combining these two solutions are as follows.

1. The first approach is to use a boundary layer solution, usually a two-dimensional model along a streamline, which is quite effective for simple wings and bodies. The solution begins by solving the inviscid potential flow, which results in the velocity field and the pressure distribution. These data are fed into two-dimensional boundary layer solutions that provide the local wall friction coefficient and the boundary layer thickness (see definitions in Chapter 14). The friction coefficient can then be integrated over the body surface for computing the friction drag. If the displacement thickness effect is sought, then a second iteration of the potential flow computation is needed, but now with modified surface geometry. This modification can be obtained by displacing the body panels according to the local boundary layer displacement, and the procedure can be reiterated until a satisfactory solution is obtained. Some of the principles of a computer program (e.g., the MCAIR panel code) that uses this method are presented in Ref. 9.6.
2. The second approach to incorporate boundary layer solutions into panel codes is to follow the procedure described above, but to account for the displacement effects

by a modification of the boundary conditions instead of a change of the surface geometry. In this case, at each panel the normal flow is given a certain blowing value that accounts for the local boundary layer displacement thickness δ^* (see definition in Chapter 14). The formulation can be derived using the properties of the source distribution of Section 4.4, and the incremental "transpiration velocity"

$$\Delta\sigma_i = \frac{\partial}{\partial s}(q\delta^*) \tag{9.48}$$

is simply added to the source strength obtained from the inviscid model. Here q is the local streamwise velocity component of the potential flow (outside the boundary layer) and the differentiation takes place along a streamline s. Note that as a result of the added transpiration velocity $q_{n_i} = \Delta\sigma_i$, the normal velocity component on the actual surface of the body is nonzero. For more details on this approach see Chapter 14.

c. Models for Wake Rollup, Jets, and Flow Separations

The vortices in the thin wake behind lifting wings tend to follow the local velocity induced by the lifting surface and its wakes. Consequently, the condition stated by Eq. (9.18) results in the wake rollup. This condition causes the shape of the wake to be unknown when the boundary conditions for the potential flow are established. Traditionally, the shape of the wake is assumed to be known (e.g., planar vortex sheet) and after the solution is obtained, the validity of the initial wake shape can be rechecked. In Chapter 15 two methods used by two panel codes will be presented to calculate the wake shape (VSAERO-wake relaxation[9.3] and PMARC-time stepping[9.7,9.8]).

Since the wake is modeled by a doublet/vortex distribution, it is possible to extend this method for modeling jets and even shear layers of separated flows. Details about models to represent the effect of jets can be found in Refs. 9.3 and 9.7; models to treat some effects of flow separation will be presented in Chapter 15.

References

[9.1] Lamb, H., *Hydrodynamics*, Dover, 6th edition, 1945.
[9.2] Yon, S., Katz, J., and Plotkin, A., "Effect of Airfoil (Trailing-Edge) Thickness on the Numerical Solution of Panel Methods Based on the Dirichlet Boundary Condition," *AIAA J.*, Vol. 30, No. 3, March 1992, pp. 697–702.
[9.3] Maskew, B., "Program VSAERO Theory Document," NASA CR 4023, Sept. 1987.
[9.4] Johnson, F. T., "A General Panel Method for the Analysis and Design of Arbitrary Configurations in Incompressible Flows," NASA CR 3079, May, 1980.
[9.5] Gerald, F. C., *Applied Numerical Analysis*, 2nd edition, Addison-Wesley, 1980.
[9.6] Bristow, D. R., and Hawk, J. D., "Subsonic Panel Method for Designing Wing Surfaces from Pressure Distributions," NASA CR-3713, July 1983.
[9.7] Ashby, D., Dudley, M., and Iguchi, S., "Development and Validation of an Advanced Low-Order Panel Method," NASA TM 101024, 1988.
[9.8] Ashby, L. D., Dudley, M. D., Iguchi, S. K., Browne, L., and Katz, J., "Potential Flow Theory and Operation Guide for the Panel Code PMARC," NASA TM 102851, March 1990.

Problems

9.1. Solve the problem of a flat plate at an angle of attack α using the lumped-vortex element. Divide the chord into five equal panels of length a, as shown in Fig. 9.19.

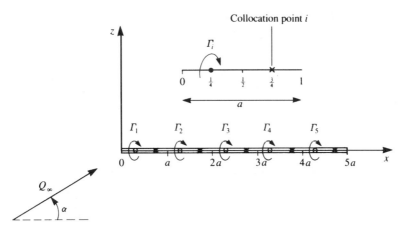

Figure 9.19 Discrete-vortex model for the flat plate at angle of attack.

a. Calculate the influence coefficient matrix a_{ij}. Is this a diagonally dominant matrix?
b. Calculate the lift and moment coefficients. How do these compare with the analytical results of Chapter 5?

9.2. Calculate the lift and moment coefficient (about the origin, $x = 0$) of the two flat plates shown in Fig. 9.20. Use a single-element lumped-vortex model to represent each plate and investigate the effect of the distance between the two plates on their lift (repeat your calculation with gap values of $c/2$, c, $2c$, $4c$).

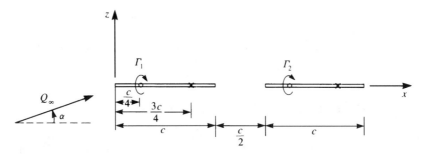

Figure 9.20 Lumped-vortex model for the tandem wing problem.

CHAPTER 10

Singularity Elements and Influence Coefficients

It was demonstrated in the previous chapters that the solution of potential flow problems over bodies and wings can be obtained by the distribution of elementary solutions. The strengths of these elementary solutions of Laplace's equation are obtained by enforcing the zero normal flow condition on the solid boundaries. The steps toward a numerical solution of this boundary value problem are described schematically in Section 9.7. In general, as the complexity of the method is increased, the "element's influence" calculation becomes more elaborate. Therefore, in this chapter, emphasis is placed on presenting some of the typical numerical elements upon which some numerical solutions are based (the list is not complete and an infinite number of elements can be developed). A generic element is shown schematically in Fig. 10.1. To calculate the induced potential and velocity increments at an arbitrary point $P(x_P, y_P, z_P)$ requires information on the element geometry and strength of singularity.

For simplicity, the symbol Δ is dropped in the following description of the singularity elements. However, it must be clear that the values of the velocity potential and velocity components are incremental values and can be added up according to the principle of superposition.

In the following sections some two-dimensional elements will be presented, whose derivation is rather simple. Three-dimensional elements will be presented later and their complexity increases with the order of the polynomial approximation of the singularity strength. Also, the formulation is derived in the panel frame of reference and when these formulas are used in any other "global coordinate system," the corresponding coordinate transformations must be used (for rotations and translations).

10.1 Two-Dimensional Point Singularity Elements

These elements are probably the simplest and easiest to use and also the most efficient in terms of computational effort. Consequently, even when higher order elements are used, if the point of interest is considered to be far from the element, then point elements can be used to describe the "far field" effect. The three point elements that will be discussed are source, doublet and vortex, and their formulation is given in the following sections.

10.1.1 Two-Dimensional Point Source

Consider a point source singularity at (x_0, z_0), with a strength σ, as shown in Fig. 10.2. The increment to the velocity potential $\Delta \Phi$ at a point P (following Section 3.7) is then

$$\Phi(x, z) = \frac{\sigma}{2\pi} \ln \sqrt{(x - x_0)^2 + (z - z_0)^2} \qquad (10.1)$$

10.1 Two-Dimensional Point Singularity Elements

$$\begin{pmatrix} x_P, y_P, z_P \\ \text{Panel geometry} \\ \text{Singularity strength} \end{pmatrix} \Rightarrow \begin{pmatrix} \text{influence} \\ \text{coefficient} \\ \text{calculation} \end{pmatrix} \Rightarrow \begin{pmatrix} \Delta u, \Delta v, \Delta w \\ \Delta \Phi \end{pmatrix}_P$$

Figure 10.1 Schematic description of a generic panel influence coefficient calculation.

and after differentiation of the potential, the velocity components are

$$u = \frac{\partial \Phi}{\partial x} = \frac{\sigma}{2\pi} \frac{x - x_0}{(x - x_0)^2 + (z - z_0)^2} \tag{10.2}$$

$$w = \frac{\partial \Phi}{\partial z} = \frac{\sigma}{2\pi} \frac{z - z_0}{(x - x_0)^2 + (z - z_0)^2} \tag{10.3}$$

10.1.2 Two-Dimensional Point Doublet

Consider a doublet that is oriented in the z direction $[\boldsymbol{\mu} = (0, \mu)]$ as in Section 3.7. If the doublet is located at the point (x_0, z_0), then its incremental influence at point P (Fig. 10.2) is

$$\Phi(x, z) = \frac{-\mu}{2\pi} \frac{z - z_0}{(x - x_0)^2 + (z - z_0)^2} \tag{10.4}$$

and the velocity component increments are

$$u = \frac{\partial \Phi}{\partial x} = \frac{\mu}{\pi} \frac{(x - x_0)(z - z_0)}{[(x - x_0)^2 + (z - z_0)^2]^2} \tag{10.5}$$

$$w = \frac{\partial \Phi}{\partial z} = \frac{-\mu}{2\pi} \frac{(x - x_0)^2 - (z - z_0)^2}{[(x - x_0)^2 + (z - z_0)^2]^2} \tag{10.6}$$

In the case when the basic singularity element is given in a system (x, z) rotated by the angle β relative to the (x^*, z^*) system, as shown in Fig. 10.3, then the velocity components can be found by the transformation

$$\begin{pmatrix} u^* \\ w^* \end{pmatrix} = \begin{pmatrix} \cos \beta & -\sin \beta \\ \sin \beta & \cos \beta \end{pmatrix} \begin{pmatrix} u \\ w \end{pmatrix} \tag{10.7}$$

10.1.3 Two-Dimensional Point Vortex

Consider a point vortex with the strength Γ located at (x_0, z_0). Again using the definitions of the points, as in Fig. 10.2, and the results of Section 3.8, we find that the

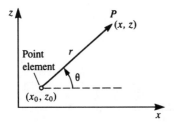

Figure 10.2 The influence of a point singularity element at point P.

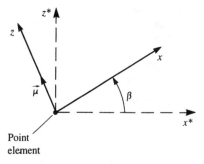

Figure 10.3 Transformation from panel to global coordinate system.

increment to the velocity potential at a point P is

$$\Phi = -\frac{\Gamma}{2\pi}\tan^{-1}\frac{z-z_0}{x-x_0} \tag{10.8}$$

and the increments in the velocity components are

$$u = \frac{\Gamma}{2\pi}\frac{z-z_0}{(x-x_0)^2+(z-z_0)^2} \tag{10.9}$$

$$w = \frac{-\Gamma}{2\pi}\frac{x-x_0}{(x-x_0)^2+(z-z_0)^2} \tag{10.10}$$

Note that all these point elements fulfill the requirements presented in Fig. 10.1. That is, the increments of the velocity components and potential at P depend on the geometry (x, z, x_0, z_0) and the strength of the element.

10.2 Two-Dimensional Constant-Strength Singularity Elements

The discretization of the source, doublet, or vortex distributions in the previous section led to discrete singularity elements that are clearly not a continuous surface representation. A more refined representation of these singularity element distributions can be obtained by dividing the solid surface boundary into elements (panels). One such element is shown schematically in Fig. 10.4, and both the surface shape and the shape of the singularity strength distribution is approximated by a polynomial. In this section, for the surface representation, a straight line will be used. For the singularity strength, only the constant, linearly varying, and quadratically varying strength cases are presented, but the methodology of this section can be applied to higher order elements.

In this section, too, three examples will be presented (source, doublet, and vortex) for evaluating the influence of the generic panel of Fig. 10.4 at an arbitrary point P. For

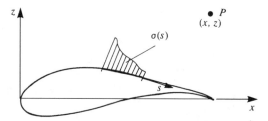

Figure 10.4 A generic surface distribution element.

10.2 Two-Dimensional Constant-Strength Singularity Elements

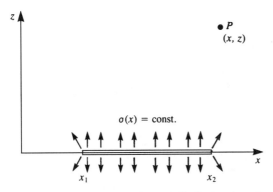

Figure 10.5 Constant-strength source distribution along the x axis.

simplicity, the formulation is derived in a panel-attached coordinate system, and the results need to be transformed back into the global coordinate system of the problem.

10.2.1 Constant-Strength Source Distribution

Consider a source distribution along the x axis as shown in Fig. 10.5. It is assumed that the source strength per length is constant such that $\sigma(x) = \sigma = $ const. The influence of this distribution at a point P is an integral of the influences of the point elements (described in the previous section) along the segment $x_1 \to x_2$:

$$\Phi = \frac{\sigma}{2\pi} \int_{x_1}^{x_2} \ln \sqrt{(x-x_0)^2 + z^2} \, dx_0 \tag{10.11}$$

$$u = \frac{\sigma}{2\pi} \int_{x_1}^{x_2} \frac{x-x_0}{(x-x_0)^2 + z^2} \, dx_0 \tag{10.12}$$

$$w = \frac{\sigma}{2\pi} \int_{x_1}^{x_2} \frac{z}{(x-x_0)^2 + z^2} \, dx_0 \tag{10.13}$$

The integral for the velocity potential (Eq. (10.11)) appears in Appendix B (Eq. (B.11)) (note that $\ln r^2 = 2 \ln r$ is used in the derivation) and in terms of the corner points $(x_1, 0)$, $(x_2, 0)$ of a generic panel element (Fig. 10.6), the distances r_1, r_2, and the angles θ_1, θ_2 it

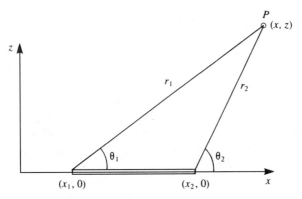

Figure 10.6 Nomenclature for the panel influence coefficient derivation.

becomes

$$\Phi = \frac{\sigma}{4\pi}[(x-x_1)\ln r_1^2 - (x-x_2)\ln r_2^2 + 2z(\theta_2 - \theta_1)] \qquad (10.14)$$

where

$$\theta_k = \tan^{-1}\frac{z}{x-x_k}, \qquad k=1,2 \qquad (10.15)$$

$$r_k = \sqrt{(x-x_k)^2 + z^2}, \qquad k=1,2 \qquad (10.16)$$

The velocity components are obtained by differentiating the potential, and following Appendix B (Eqs. (B.5) and (B.9)), they are

$$u = \frac{\sigma}{2\pi}\ln\frac{r_1}{r_2} = \frac{\sigma}{4\pi}\ln\frac{r_1^2}{r_2^2} \qquad (10.17)$$

$$w = \frac{\sigma}{2\pi}(\theta_2 - \theta_1) \qquad (10.18)$$

Returning to x, z variables we obtain

$$\Phi = \frac{\sigma}{4\pi}\left\{(x-x_1)\ln[(x-x_1)^2 + z^2] - (x-x_2)\ln[(x-x_2)^2 + z^2] \right.$$
$$\left. + 2z\left(\tan^{-1}\frac{z}{x-x_2} - \tan^{-1}\frac{z}{x-x_1}\right)\right\} \qquad (10.19)$$

$$u = \frac{\sigma}{4\pi}\ln\frac{(x-x_1)^2 + z^2}{(x-x_2)^2 + z^2} \qquad (10.20)$$

$$w = \frac{\sigma}{2\pi}\left[\tan^{-1}\frac{z}{x-x_2} - \tan^{-1}\frac{z}{x-x_1}\right] \qquad (10.21)$$

Of particular interest is the case when the point P is on the element (usually at the center). In this case $z = 0\pm$ and the potential becomes

$$\Phi(x, 0\pm) = \frac{\sigma}{4\pi}[(x-x_1)\ln(x-x_1)^2 - (x-x_2)\ln(x-x_2)^2] \qquad (10.22)$$

and at the center of the element it becomes

$$\Phi\left(\frac{x_1+x_2}{2}, 0\pm\right) = \frac{\sigma}{4\pi}(x_2 - x_1)\ln\left(\frac{x_2-x_1}{2}\right)^2 \qquad (10.22a)$$

The x component of the velocity at $z = 0$ becomes

$$u(x, 0\pm) = \frac{\sigma}{2\pi}\ln\frac{(x-x_1)}{|(x-x_2)|} \qquad (10.23)$$

which is zero at the panel center and infinite at the panel edges.

For evaluating the w component of the velocity, it is important to distinguish between the conditions when the panel is approached from its upper or from its lower side. For the case of P being above the panel, $\theta_1 = 0$, while $\theta_2 = \pi$. These conditions are reversed on the lower side and therefore

$$w(x, 0\pm) = \pm\frac{\sigma}{2} \qquad (10.24)$$

This is the same result obtained in Section 3.14 for the source distribution.

10.2 Two-Dimensional Constant-Strength Singularity Elements

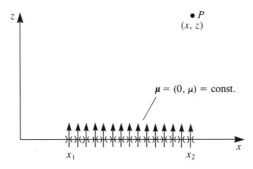

Figure 10.7 Constant-strength doublet distribution along the x axis.

10.2.2 Constant-Strength Doublet Distribution

Consider a doublet distribution along the x axis consisting of elements pointing in the z direction [$\boldsymbol{\mu} = (0, \mu)$], as shown in Fig. 10.7. The influence at a point $P(x, z)$ is an integral of the influences of the point elements between x_1 and x_2:

$$\Phi(x, z) = \frac{-\mu}{2\pi} \int_{x_1}^{x_2} \frac{z}{(x - x_0)^2 + z^2} dx_0 \tag{10.25}$$

and the velocity components are

$$u(x, z) = \frac{\mu}{\pi} \int_{x_1}^{x_2} \frac{(x - x_0)z}{[(x - x_0)^2 + z^2]^2} dx_0 \tag{10.26}$$

$$w(x, z) = \frac{-\mu}{2\pi} \int_{x_1}^{x_2} \frac{(x - x_0)^2 - z^2}{[(x - x_0)^2 + z^2]^2} dx_0 \tag{10.27}$$

Note that the integral for the w component of the source distribution is similar to the potential integral of the doublet. Therefore, the potential at P (by using Eq. (10.21)) is

$$\Phi = \frac{-\mu}{2\pi} \left[\tan^{-1} \frac{z}{x - x_2} - \tan^{-1} \frac{z}{x - x_1} \right] \tag{10.28}$$

Comparison of this expression to the potential of a point vortex (Eq. (10.8)) indicates that this constant doublet distribution is equivalent to two point vortices with opposite sign at the panel edges such that $\Gamma = -\mu$ (see Fig. 10.8). Consequently, the velocity components

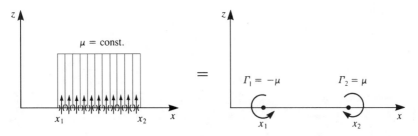

Figure 10.8 Equivalence between a constant-strength doublet panel and two point vortices at the edge of the panel.

are readily available by using Eqs. (10.9) and (10.10):

$$u = \frac{-\mu}{2\pi}\left[\frac{z}{(x-x_1)^2+z^2} - \frac{z}{(x-x_2)^2+z^2}\right] \quad (10.29)$$

$$w = \frac{\mu}{2\pi}\left[\frac{x-x_1}{(x-x_1)^2+z^2} - \frac{x-x_2}{(x-x_2)^2+z^2}\right] \quad (10.30)$$

When the point P is on the element ($z = 0$, $x_1 < x < x_2$) then, following Section 3.14, we have

$$\Phi(x, 0\pm) = \mp\frac{\mu}{2} \quad (10.31)$$

and the velocity components become

$$u(x, 0\pm) = \mp\frac{d\mu(x)}{dx} = 0 \quad (10.32)$$

$$w(x, 0\pm) = \frac{-\mu}{2\pi}\left[\frac{1}{(x-x_1)} - \frac{1}{(x-x_2)}\right] \quad (10.33)$$

and hence the w velocity component is singular at the panel edges.

10.2.3 Constant-Strength Vortex Distribution

Once the influence terms of the constant-strength source element are obtained, owing to the similarity between the source and the vortex distributions, the formulation for this element becomes simple. The constant-strength vortex distribution $\gamma(x) = \gamma = $ const. is placed along the x axis as shown in Fig. 10.9. The influence of this distribution at a point P is an integral of the influences of the point elements between x_1 and x_2. So we have

$$\Phi = -\frac{\gamma}{2\pi}\int_{x_1}^{x_2} \tan^{-1}\frac{z}{x-x_0} dx_0 \quad (10.34)$$

$$u = \frac{\gamma}{2\pi}\int_{x_1}^{x_2} \frac{z}{(x-x_0)^2+z^2} dx_0 \quad (10.35)$$

$$w = -\frac{\gamma}{2\pi}\int_{x_1}^{x_2} \frac{x-x_0}{(x-x_0)^2+z^2} dx_0 \quad (10.36)$$

Details of the integral for the velocity potential appear in Appendix B (Eq. (B.14)), and in terms of the distances and angles of Eqs. (10.15) and (10.16) (as shown in Fig. 10.6) the

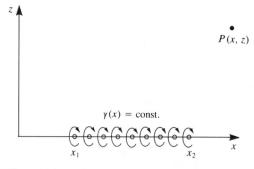

Figure 10.9 Constant-strength vortex distribution along the x axis.

potential becomes

$$\Phi = -\frac{\gamma}{2\pi}\left[(x-x_1)\theta_1 - (x-x_2)\theta_2 + \frac{z}{2}\ln\frac{r_1^2}{r_2^2}\right] \tag{10.37}$$

which in terms of the x, z coordinates is

$$\Phi = -\frac{\gamma}{2\pi}\left[(x-x_1)\tan^{-1}\frac{z}{x-x_1} - (x-x_2)\tan^{-1}\frac{z}{x-x_2} + \frac{z}{2}\ln\frac{(x-x_1)^2+z^2}{(x-x_2)^2+z^2}\right] \tag{10.38}$$

Following the formulation used for the constant-source element, and observing that the u and w velocity components for the vortex distribution are the same as the corresponding w and u components of the source distribution, we obtain

$$u = \frac{\gamma}{2\pi}\left[\tan^{-1}\frac{z}{x-x_2} - \tan^{-1}\frac{z}{x-x_1}\right] \tag{10.39}$$

$$w = \frac{\gamma}{4\pi}\ln\frac{(x-x_2)^2+z^2}{(x-x_1)^2+z^2} \tag{10.40}$$

The influence of the element on itself at $z = 0\pm$ and $(x_1 < x < x_2)$ can be found by approaching from above the x axis. In this case $\theta_1 = 0$, $\theta_2 = \pi$, and

$$\Phi(x, 0+) = -\frac{\gamma}{2\pi}[(x-x_1)0 - (x-x_2)\pi] = \frac{\gamma}{2}(x-x_2) \tag{10.41a}$$

Similarly, when the element is approached from below then

$$\Phi(x, 0-) = -\frac{\gamma}{2}(x-x_2) \tag{10.41b}$$

The x component of the velocity can be found by observing Eq. (10.24) for the source or by recalling Section 3.14:

$$u(x, 0\pm) = \pm\frac{\gamma}{2} \tag{10.42}$$

and the w velocity component is similar to the u component of the source (Eq. (10.23)):

$$w(x, 0\pm) = \frac{\gamma}{4\pi}\ln\frac{(x-x_2)^2}{(x-x_1)^2} \tag{10.43}$$

In most situations the influence is sought at the center of the element where $|r_1| = |r_2|$ and consequently $w(\text{panel} - \text{center}, 0\pm) = 0$.

10.3 Two-Dimensional Linear-Strength Singularity Elements

The representation of a continuous singularity distribution by a series of constant-strength elements results in a discontinuity of the singularity strength at the panel edges. To overcome this problem, a linearly varying strength singularity element can be used. The requirement that the strength of the singularity remains the same at the edge of two neighbor elements results in an additional equation. Therefore with this type of element, for N collocation points $2N$ equations will be formed (see examples in Chapter 11).

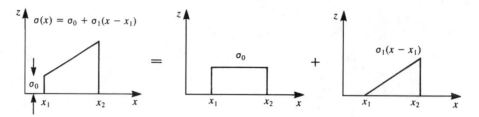

Figure 10.10 Decomposition of a generic linear strength element to constant-strength and linearly varying strength elements.

10.3.1 Linear Source Distribution

Consider a linear source distribution along the x axis ($x_1 < x < x_2$) with a source strength of $\sigma(x) = \sigma_0 + \sigma_1(x - x_1)$, as shown in Fig. 10.10. Based on the principle of superposition, this can be divided into a constant-strength element and a linearly varying strength element with the strength $\sigma(x) = \sigma_1 x$. Therefore, for the general case (as shown in the left-hand side of Fig. 10.10) the results of this section must be added to the results of the constant-strength source element.

The influence of the simplified linear distribution source element, where $\sigma(x) = \sigma_1 x$, at a point P is obtained by integrating the influences of the point elements between x_1 and x_2 (see Fig. 10.11):

$$\Phi = \frac{\sigma_1}{2\pi} \int_{x_1}^{x_2} x_0 \ln \sqrt{(x - x_0)^2 + z^2} \, dx_0 \tag{10.44}$$

$$u = \frac{\sigma_1}{2\pi} \int_{x_1}^{x_2} \frac{x_0(x - x_0)}{(x - x_0)^2 + z^2} \, dx_0 \tag{10.45}$$

$$w = \frac{\sigma_1}{2\pi} \int_{x_1}^{x_2} \frac{x_0 z}{(x - x_0)^2 + z^2} \, dx_0 \tag{10.46}$$

Details of the integration are presented in Appendix B (Eq. (B.17)), and the results are

$$\Phi = \frac{\sigma_1}{4\pi} \left[\frac{x^2 - x_1^2 - z^2}{2} \ln r_1^2 - \frac{x^2 - x_2^2 - z^2}{2} \ln r_2^2 + 2xz(\theta_2 - \theta_1) - x(x_2 - x_1) \right] \tag{10.47}$$

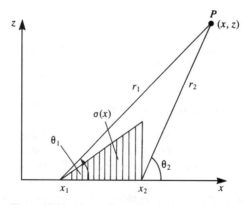

Figure 10.11 Nomenclature for calculating the influence of linearly varying strength source.

where r_1, r_2, θ_1, and θ_2 are defined by Eqs. (10.15) and (10.16). The velocity components are obtained by differentiating the velocity potential (Appendix B, Eqs. (B.18) and (B.19)), which gives

$$u = \frac{\sigma_1}{2\pi}\left[\frac{x}{2}\ln\frac{r_1^2}{r_2^2} + (x_1 - x_2) + z(\theta_2 - \theta_1)\right] \tag{10.48}$$

$$w = \frac{\sigma_1}{4\pi}\left[z\ln\frac{r_2^2}{r_1^2} + 2x(\theta_2 - \theta_1)\right] \tag{10.49}$$

Substitution of r_k and θ_k from Eqs. (10.16) and (10.17) results in

$$\Phi = \frac{\sigma_1}{4\pi}\left[\frac{x^2 - x_1^2 - z^2}{2}\ln[(x-x_1)^2 + z^2] - \frac{x^2 - x_2^2 - z^2}{2}\ln[(x-x_2)^2 + z^2]\right.$$
$$\left. + 2xz\left(\tan^{-1}\frac{z}{x-x_2} - \tan^{-1}\frac{z}{x-x_1}\right) - x(x_2 - x_1)\right] \tag{10.50}$$

$$u = \frac{\sigma_1}{2\pi}\left[\frac{x}{2}\ln\frac{(x-x_1)^2 + z^2}{(x-x_2)^2 + z^2} + (x_1 - x_2) + z\left(\tan^{-1}\frac{z}{x-x_2} - \tan^{-1}\frac{z}{x-x_1}\right)\right] \tag{10.51}$$

$$w = \frac{\sigma_1}{4\pi}\left[z\ln\frac{(x-x_1)^2 + z^2}{(x-x_2)^2 + z^2} + 2x\left(\tan^{-1}\frac{z}{x-x_2} - \tan^{-1}\frac{z}{x-x_1}\right)\right] \tag{10.52}$$

When the point P lies on the element ($z = 0\pm$, $x_1 < x < x_2$), then Eq. (10.50) reduces to

$$\Phi = \frac{\sigma_1}{4\pi}\left[(x^2 - x_1^2)\ln(x - x_1) - (x^2 - x_2^2)\ln|(x - x_2)| - x(x_2 - x_1)\right] \tag{10.53}$$

At the center of the element this reduces to

$$\Phi = \frac{\sigma_1}{4\pi}(x_2^2 - x_1^2)\left(\ln\frac{x_2 - x_1}{2} - \frac{1}{2}\right) \tag{10.53a}$$

Also, on the element

$$u = \frac{\sigma_1}{2\pi}\left[x\ln\frac{x - x_1}{|x - x_2|} + (x_1 - x_2)\right] \tag{10.54}$$

$$w = \pm\frac{\sigma_1}{2}x \tag{10.55}$$

and at the center of the element

$$u = \frac{\sigma_1}{2\pi}(x_1 - x_2) \tag{10.54a}$$

and

$$w = \pm\frac{\sigma_1}{4}(x_2 - x_1) \tag{10.55a}$$

10.3.2 Linear Doublet Distribution

Consider a doublet distribution along the x axis with a strength $\mu(x) = \mu_0 + \mu_1(x - x_1)$, consisting of elements pointing in the z direction [$\mu = (0, \mu)$], as shown in Fig. 10.12. In this case, too, only the linear term ($\mu(x) = \mu_1 x$) is considered and the influence at a point $P(x, z)$ is an integral of the influences of the point elements between

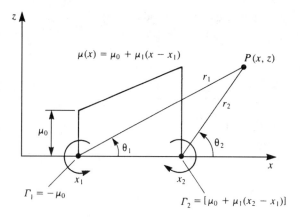

Figure 10.12 Linearly varying strength doublet element.

x_1 and x_2:

$$\Phi(x, z) = \frac{-\mu_1}{2\pi} \int_{x_1}^{x_2} \frac{x_0 z}{(x - x_0)^2 + z^2} \, dx_0 \tag{10.56}$$

$$u(x, z) = \frac{\mu_1}{\pi} \int_{x_1}^{x_2} \frac{x_0(x - x_0)z}{[(x - x_0)^2 + z^2]^2} \, dx_0 \tag{10.57}$$

$$w(x, z) = \frac{-\mu_1}{2\pi} \int_{x_1}^{x_2} \frac{[(x - x_0)^2 - z^2]x_0}{[(x - x_0)^2 + z^2]^2} \, dx_0 \tag{10.58}$$

The integral for the velocity potential is similar to the w velocity component of the linear source (Eq. (10.46)). Therefore, following Eq. (10.49), we obtain

$$\Phi = \frac{-\mu_1}{4\pi} \left[2x(\theta_2 - \theta_1) + z \ln \frac{r_2^2}{r_1^2} \right] \tag{10.59}$$

and in Cartesian coordinates

$$\Phi = \frac{-\mu_1}{4\pi} \left[2x \left(\tan^{-1} \frac{z}{x - x_2} - \tan^{-1} \frac{z}{x - x_1} \right) + z \ln \frac{(x - x_2)^2 + z^2}{(x - x_1)^2 + z^2} \right] \tag{10.60}$$

To obtain the velocity components we observe the similarity between Eq. (10.59) and the potential of a constant-strength vortex distribution (Eq. (10.37)). Replacing μ_1 with $-\gamma$ in Eq. (10.38) yields

$$\Phi^{**} = \frac{\mu_1}{4\pi} \left[2(x - x_1) \tan^{-1} \frac{z}{x - x_1} - 2(x - x_2) \tan^{-1} \frac{z}{x - x_2} + z \ln \frac{(x - x_1)^2 + z^2}{(x - x_2)^2 + z^2} \right] \tag{10.38a}$$

and therefore the potential of the linear doublet distribution of Eq. (10.60) is

$$\Phi = \Phi^{**} + \frac{\mu_1}{2\pi}(x_1 \theta_1 - x_2 \theta_2) \tag{10.61}$$

and the two last terms are potentials of point vortices with strengths $\mu_1 x_1$ and $\mu_1 x_2$ (see Eq. (10.8)). The velocity components therefore are readily available, either by differentiation

10.3 Two-Dimensional Linear-Strength Singularity Elements

of this velocity potential or by using Eqs. (10.39) and (10.9):

$$u = -\frac{\mu_1}{2\pi}\left[\tan^{-1}\frac{z}{x-x_2} - \tan^{-1}\frac{z}{x-x_1}\right] + \frac{\mu_1 x_2}{2\pi}\frac{z}{(x-x_2)^2 + z^2}$$
$$- \frac{\mu_1 x_1}{2\pi}\frac{z}{(x-x_1)^2 + z^2} \tag{10.62}$$

and for the w component using Eqs. (10.40) and (10.10) we get

$$w = -\frac{\mu_1}{4\pi}\ln\frac{(x-x_2)^2 + z^2}{(x-x_1)^2 + z^2} + \frac{\mu_1 x_1}{2\pi}\frac{x-x_1}{(x-x_1)^2 + z^2} - \frac{\mu_1 x_2}{2\pi}\frac{x-x_2}{(x-x_2)^2 + z^2} \tag{10.63}$$

The values of the potential and the velocity components on the element ($z = 0$, $x_1 < x < x_2$) are

$$\Phi = \mp\frac{\mu_1}{2}x \tag{10.64}$$

$$u = \mp\frac{\mu_1}{2} \tag{10.65}$$

$$w = -\frac{\mu_1}{4\pi}\left[\ln\frac{(x-x_2)^2}{(x-x_1)^2} + \frac{2x_1}{x-x_1} - \frac{2x_2}{x-x_2}\right] \tag{10.66}$$

and the w velocity component at the center of the element becomes

$$w = -\frac{\mu_1}{\pi}\left[\frac{x_2 + x_1}{x_2 - x_1}\right] \tag{10.66a}$$

and hence the velocity is singular at the panel edges because of the point vortices there.

Note that for the general element, where $\mu(x) = \mu_0 + \mu_1(x - x_1)$, the potential becomes

$$\Phi = \Phi^{**} - \frac{\mu_0}{2\pi}(\theta_2 - \theta_1) + \frac{\mu_1}{2\pi}(x_1 - x_2)\theta_2 \tag{10.67}$$

and because of the potential jump at the edges of this doublet distribution two concentrated vortices exist. The vortex at x_1 will have a strength of $-\mu_0$ while the one at x_2 will have a strength of $[\mu_1(x_2 - x_1) + \mu_0]$, as shown schematically in Fig. 10.12.

10.3.3 Linear Vortex Distribution

In this case the strength of the vortex distribution varies linearly along the element,

$$\gamma(x) = \gamma_0 + \gamma_1(x - x_1)$$

Again, for simplicity consider only the linear portion where $\gamma(x) = \gamma_1 x$ and γ_1 is a constant. The influence of this vortex distribution at a point P in the x–z plane is obtained by integrating the influences of the point elements between x_1 and x_2:

$$\Phi = -\frac{\gamma_1}{2\pi}\int_{x_1}^{x_2} x_0 \tan^{-1}\frac{z}{x-x_0}\,dx_0 \tag{10.68}$$

$$u = \frac{\gamma_1}{2\pi}\int_{x_1}^{x_2}\frac{x_0 z}{(x-x_0)^2 + z^2}\,dx_0 \tag{10.69}$$

$$w = -\frac{\gamma_1}{2\pi}\int_{x_1}^{x_2}\frac{x_0(x-x_0)}{(x-x_0)^2 + z^2}\,dx_0 \tag{10.70}$$

Using the integral in Appendix B (Eq. (B.22)) we get

$$\Phi = -\frac{\gamma_1}{2\pi}\left[\frac{xz}{2}\ln\frac{r_1^2}{r_2^2} + \frac{z}{2}(x_1 - x_2) + \frac{x^2 - x_1^2 - z^2}{2}\theta_1 - \frac{x^2 - x_2^2 - z^2}{2}\theta_2\right]$$

(10.71)

The velocity components are similar to the integrals of the linear source (Eqs. (10.51) and (10.52)) and are

$$u = -\frac{\gamma_1}{4\pi}\left[z\ln\frac{(x-x_1)^2 + z^2}{(x-x_2)^2 + z^2} - 2x\left(\tan^{-1}\frac{z}{x-x_2} - \tan^{-1}\frac{z}{x-x_1}\right)\right] \quad (10.72)$$

$$w = -\frac{\gamma_1}{2\pi}\left[\frac{x}{2}\ln\frac{(x-x_1)^2 + z^2}{(x-x_2)^2 + z^2} + (x_1 - x_2) + z\left(\tan^{-1}\frac{z}{x-x_2} - \tan^{-1}\frac{z}{x-x_1}\right)\right]$$

(10.73)

When the point P lies on the element ($z = 0\pm$, $x_1 < x < x_2$), then Eq. (10.71) reduces to

$$\Phi = \pm\frac{\gamma_1}{4}(x^2 - x_2^2) \quad (10.74)$$

At the center of the element this reduces to

$$\Phi = \pm\frac{\gamma_1}{16}(x_1^2 + 2x_1x_2 - 3x_2^2) \quad (10.74a)$$

Also, on the element

$$u = \pm\frac{\gamma_1}{2}x \quad (10.75)$$

$$w = \frac{\gamma_1}{2\pi}\left[x\ln\frac{x-x_1}{|x-x_2|} + (x_1 - x_2)\right] \quad (10.76)$$

and at the center of the element (above +, under −)

$$u = \pm\frac{\gamma_1}{4}(x_1 + x_2) \quad (10.75a)$$

$$w = -\frac{\gamma_1}{2\pi}(x_1 - x_2) \quad (10.76a)$$

10.3.4 Quadratic Doublet Distribution

As indicated by Eq. (3.150) in Section 3.14, a quadratic doublet distribution can be replaced by an equivalent linear vortex distribution presented in the previous section. However, in situations when the Dirichlet type boundary condition is applied, it is more convenient to use the corresponding doublet distribution (instead of the linear vortex distribution). Thus, a quadratic doublet distribution along the x axis ($x_1 < x < x_2$) will have a strength distribution of

$$\mu(x) = \mu_0 + \mu_1(x - x_1) + \mu_2(x - x_1)^2$$

where the doublet elements pointing in the z direction [$\mu = (0, \mu)$] are selected as shown in Fig. 10.13. Since the contribution of the constant and linear strength terms were evaluated in the previous sections only the third term ($\mu(x) = \mu_2 x^2$) is considered and the

10.3 Two-Dimensional Linear-Strength Singularity Elements

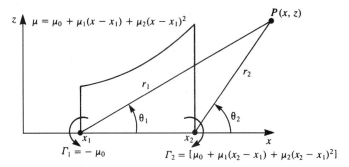

Figure 10.13 Quadratic-strength doublet element.

influence at a point $P(x, z)$ is an integral of the influences of the point elements between x_1 and x_2:

$$\Phi(x, z) = \frac{-\mu_2}{2\pi} \int_{x_1}^{x_2} \frac{x_0^2 z}{(x - x_0)^2 + z^2} \, dx_0 \tag{10.77}$$

$$u(x, z) = \frac{\mu_2}{\pi} \int_{x_1}^{x_2} \frac{(x - x_0) z x_0^2}{[(x - x_0)^2 + z^2]^2} \, dx_0 \tag{10.78}$$

$$w(x, z) = \frac{-\mu_2}{2\pi} \int_{x_1}^{x_2} \frac{[(x - x_0)^2 - z^2] x_0^2}{[(x - x_0)^2 + z^2]^2} \, dx_0 \tag{10.79}$$

The integral for the velocity potential is obtained by introducing the variable $X = x - x_0$ (thus $dX = -dx_0$), which transforms Eq. (10.77) to the form

$$\Phi = \frac{\mu_2}{2\pi} \int_{x - x_1}^{x - x_2} \frac{(x^2 - 2xX + X^2) z}{X^2 + z^2} \, dX$$

The three integrals formed by the terms appearing in the numerator are evaluated in Gradshteyn and Ryzhik[5.7] (pp. 68–69) and yield

$$\Phi(x, z) = \frac{\mu_2}{2\pi} \left[(x^2 - z^2)(\theta_1 - \theta_2) - xz \ln \frac{r_2^2}{r_1^2} + z(x_1 - x_2) \right] \tag{10.80}$$

where the variables r_1, r_2, θ_1, and θ_2 are shown in Fig. 10.13.

Note that Eq. (10.80) can be rewritten as

$$\Phi(x, z) = \Phi^{**} + \frac{\mu_2}{2\pi} \left(x_1^2 \theta_1 - x_2^2 \theta_2 \right) \tag{10.81}$$

such that Φ^{**} is the potential of the equivalent linear vortex distribution of Eq. (10.71) (with $\mu_2 = -\gamma_1/2$):

$$\Phi^{**} = \frac{\mu_2}{2\pi} \left[-xz \ln \frac{r_2^2}{r_1^2} + z(x_2 - x_1) + (x^2 - z^2)(\theta_1 - \theta_2) \right] + \frac{\mu_2}{2\pi} \left[x_2^2 \theta_2 - x_1^2 \theta_1 \right] \tag{10.81a}$$

Therefore, Eq. (10.81) states that the potential of a quadratic doublet distribution is equivalent to the velocity potential of a linearly varying strength vortex distribution plus two concentrated vortices at the panel edges, as shown schematically in Fig. 10.13.

To obtain the velocity components we can use the similarity between Eq. (10.80) and the potential of a linearly varying strength vortex distribution (Eqs. (10.72) and (10.73)).

Replacing μ_2 with $-\gamma_1/2$ in Eqs. (10.72) and (10.73) and adding the velocity components of the two point vortices yields

$$u = \frac{\mu_2}{2\pi}\left[z\ln\frac{r_1^2}{r_2^2} - 2x(\theta_2 - \theta_1) + \frac{zx_2^2}{(x-x_2)^2 + z^2} - \frac{zx_1^2}{(x-x_1)^2 + z^2}\right] \quad (10.82)$$

and the w component of the velocity is

$$w = \frac{\mu_2}{\pi}\left[\frac{x}{2}\ln\frac{r_1^2}{r_2^2} + (x_1 - x_2) + z(\theta_2 - \theta_1) + \frac{x_1^2}{2}\frac{x - x_1}{(x-x_1)^2 + z^2} - \frac{x_2^2}{2}\frac{x - x_2}{(x-x_2)^2 + z^2}\right] \quad (10.83)$$

The value of the potential on the element ($z = 0\pm$, $x_1 < x < x_2$) becomes

$$\Phi(x, 0\pm) = \frac{\mu_2 x^2}{2\pi}(\theta_1 - \theta_2)$$

and above the element $\theta_1 - \theta_2 = -\pi$ whereas under the element $\theta_1 - \theta_2 = \pi$. Consequently,

$$\Phi(x, 0\pm) = \mp\frac{\mu_2 x^2}{2} \quad (10.84)$$

Similarly the velocity components become

$$u(x, 0\pm) = \frac{\mu_2}{2\pi}(-2x)(\theta_2 - \theta_1) = \mp\mu_2 x \quad (10.85)$$

$$w(x, 0\pm) = \frac{\mu_2}{\pi}\left[\frac{x}{2}\ln\frac{(x-x_1)^2}{(x-x_2)^2} + x_1 - x_2 + \frac{x_1^2}{2(x-x_1)} - \frac{x_2^2}{2(x-x_2)}\right] \quad (10.86)$$

and hence the velocity is singular at the panel edges because of the point vortices there.

Note that for the general element, where $\mu(x) = \mu_0 + \mu_1(x - x_1) + \mu_2(x - x_1)^2$, the potential jump at the edges of this doublet distribution results in two concentrated vortices. The vortex at x_1 will have a strength of $-\mu_0$ while the one at x_2 will have a strength of $[\mu_0 + \mu_1(x_2 - x_1) + \mu_2(x_2 - x_1)^2]$, as shown schematically in Fig. 10.13.

10.4 Three-Dimensional Constant-Strength Singularity Elements

In the three-dimensional case, as in the two-dimensional case, the discretization process includes two parts: discretization of the geometry and of the singularity element distribution. If these elements are approximated by polynomials (both geometry and singularity strength) then a first-order approximation to the surface can be defined as a quadrilateral[1] panel, a second-order approximation will be based on parabolic curve-fitting, while a third-order approximation may use a third-order polynomial curve-fitting. Similarly, the strength of the singularity distribution can be approximated (discretized) by constant-strength (zero-order), linearly varying (first-order), or by parabolic (second-order) functions.

The simplest and most basic three-dimensional element will have a quadrilateral geometry and a constant-strength singularity. When the strength of this element (a constant) is

---

[1] A quadrilateral is a flat surface with four straight sides. A rectilinear panel has straight but not necessarily flat sides and can be twisted!

10.4 Three-Dimensional Constant-Strength Singularity Elements

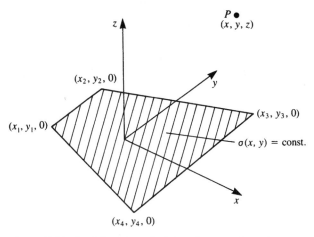

Figure 10.14 Quadrilateral constant-strength source element.

unknown a panel code using N panels can be constructed to solve for these N constants. In the following section, such constant-strength elements will be described.

The derivation is again performed in a local frame of reference, and for a global coordinate system a coordinate transformation is required.

10.4.1 Quadrilateral Source

Consider a surface element with a constant-strength source distribution σ per area bounded by four straight lines as described in Fig. 10.14. The element corner points are designated as $(x_1, y_1, 0), \ldots, (x_4, y_4, 0)$ and the potential at an arbitrary point $P(x, y, z)$, due to this element, is

$$\Phi(x, y, z) = \frac{-\sigma}{4\pi} \int_S \frac{dS}{\sqrt{(x-x_0)^2 + (y-y_0)^2 + z^2}} \tag{10.87}$$

The velocity components can be obtained by differentiating the velocity potential, that is,

$$(u, v, w) = \left(\frac{\partial \Phi}{\partial x}, \frac{\partial \Phi}{\partial y}, \frac{\partial \Phi}{\partial z} \right) \tag{10.88}$$

Execution of the integration within the area bounded by the four straight lines requires a lengthy process and is derived by Hess and Smith.[10.1] Using their results, we can obtain the potential for a planar element as

$$\Phi = \frac{-\sigma}{4\pi} \left\{ \left[\frac{(x-x_1)(y_2-y_1) - (y-y_1)(x_2-x_1)}{d_{12}} \ln \frac{r_1 + r_2 + d_{12}}{r_1 + r_2 - d_{12}} \right. \right.$$
$$+ \frac{(x-x_2)(y_3-y_2) - (y-y_2)(x_3-x_2)}{d_{23}} \ln \frac{r_2 + r_3 + d_{23}}{r_2 + r_3 - d_{23}}$$
$$+ \frac{(x-x_3)(y_4-y_3) - (y-y_3)(x_4-x_3)}{d_{34}} \ln \frac{r_3 + r_4 + d_{34}}{r_3 + r_4 - d_{34}}$$
$$\left. \left. + \frac{(x-x_4)(y_1-y_4) - (y-y_4)(x_1-x_4)}{d_{41}} \ln \frac{r_4 + r_1 + d_{41}}{r_4 + r_1 - d_{41}} \right] \right.$$

$$- |z| \left[\tan^{-1}\left(\frac{m_{12}e_1 - h_1}{zr_1}\right) - \tan^{-1}\left(\frac{m_{12}e_2 - h_2}{zr_2}\right) \right.$$
$$+ \tan^{-1}\left(\frac{m_{23}e_2 - h_2}{zr_2}\right) - \tan^{-1}\left(\frac{m_{23}e_3 - h_3}{zr_3}\right)$$
$$+ \tan^{-1}\left(\frac{m_{34}e_3 - h_3}{zr_3}\right) - \tan^{-1}\left(\frac{m_{34}e_4 - h_4}{zr_4}\right)$$
$$\left. + \tan^{-1}\left(\frac{m_{41}e_4 - h_4}{zr_4}\right) - \tan^{-1}\left(\frac{m_{41}e_1 - h_1}{zr_1}\right) \right] \right\} \quad (10.89)$$

where

$$d_{12} = \sqrt{(x_2 - x_1)^2 + (y_2 - y_1)^2} \quad (10.90a)$$
$$d_{23} = \sqrt{(x_3 - x_2)^2 + (y_3 - y_2)^2} \quad (10.90b)$$
$$d_{34} = \sqrt{(x_4 - x_3)^2 + (y_4 - y_3)^2} \quad (10.90c)$$
$$d_{41} = \sqrt{(x_1 - x_4)^2 + (y_1 - y_4)^2} \quad (10.90d)$$

and

$$m_{12} = \frac{y_2 - y_1}{x_2 - x_1} \quad (10.91a)$$
$$m_{23} = \frac{y_3 - y_2}{x_3 - x_2} \quad (10.91b)$$
$$m_{34} = \frac{y_4 - y_3}{x_4 - x_3} \quad (10.91c)$$
$$m_{41} = \frac{y_1 - y_4}{x_1 - x_4} \quad (10.91d)$$

and

$$r_k = \sqrt{(x - x_k)^2 + (y - y_k)^2 + z^2}, \quad k = 1, 2, 3, 4 \quad (10.92)$$
$$e_k = (x - x_k)^2 + z^2, \quad k = 1, 2, 3, 4 \quad (10.93)$$
$$h_k = (x - x_k)(y - y_k), \quad k = 1, 2, 3, 4 \quad (10.94)$$

The velocity components, based on the results of Ref. 10.1, are

$$u = \frac{\sigma}{4\pi}\left[\frac{y_2 - y_1}{d_{12}}\ln\frac{r_1 + r_2 - d_{12}}{r_1 + r_2 + d_{12}} + \frac{y_3 - y_2}{d_{23}}\ln\frac{r_2 + r_3 - d_{23}}{r_2 + r_3 + d_{23}}\right.$$
$$\left. + \frac{y_4 - y_3}{d_{34}}\ln\frac{r_3 + r_4 - d_{34}}{r_3 + r_4 + d_{34}} + \frac{y_1 - y_4}{d_{41}}\ln\frac{r_4 + r_1 - d_{41}}{r_4 + r_1 + d_{41}}\right] \quad (10.95)$$

$$v = \frac{\sigma}{4\pi}\left[\frac{x_1 - x_2}{d_{12}}\ln\frac{r_1 + r_2 - d_{12}}{r_1 + r_2 + d_{12}} + \frac{x_2 - x_3}{d_{23}}\ln\frac{r_2 + r_3 - d_{23}}{r_2 + r_3 + d_{23}}\right.$$
$$\left. + \frac{x_3 - x_4}{d_{34}}\ln\frac{r_3 + r_4 - d_{34}}{r_3 + r_4 + d_{34}} + \frac{x_4 - x_1}{d_{41}}\ln\frac{r_4 + r_1 - d_{41}}{r_4 + r_1 + d_{41}}\right] \quad (10.96)$$

$$w = \frac{\sigma}{4\pi}\left[\tan^{-1}\left(\frac{m_{12}e_1 - h_1}{zr_1}\right) - \tan^{-1}\left(\frac{m_{12}e_2 - h_2}{zr_2}\right)\right.$$
$$+ \tan^{-1}\left(\frac{m_{23}e_2 - h_2}{zr_2}\right) - \tan^{-1}\left(\frac{m_{23}e_3 - h_3}{zr_3}\right)$$
$$+ \tan^{-1}\left(\frac{m_{34}e_3 - h_3}{zr_3}\right) - \tan^{-1}\left(\frac{m_{34}e_4 - h_4}{zr_4}\right)$$
$$\left. + \tan^{-1}\left(\frac{m_{41}e_4 - h_4}{zr_4}\right) - \tan^{-1}\left(\frac{m_{41}e_1 - h_1}{zr_1}\right)\right] \quad (10.97)$$

The u and v components of the velocity are defined everywhere, but at the edges of the quadrilateral they become infinite. In practice, usually the influence of the element on itself is sought, and near the centroid these velocity components approach zero. The jump in the normal velocity component as $z \to 0$ inside the quadrilateral is similar to the results of Section 4.4:

$$w(z = 0\pm) = \frac{\pm\sigma}{2} \quad (10.98)$$

When the point of interest P lies outside of the quadrilateral then

$$w(z = 0\pm) = 0 \quad (10.99)$$

Far Field: For improved computational efficiency, when the point of interest P is far from the center of the element $(x_0, y_0, 0)$ then the influence of the quadrilateral element with an area of A can be approximated by a point source. The term "far" is controlled by the programmer but usually if the distance is more than 3–5 times the average panel diameter then the simplified approximation is used. Following the formulation of Section 3.4 (in the panel frame of reference) we can calculate the point source influence for the velocity potential as

$$\Phi(x, y, z) = \frac{-\sigma A}{4\pi\sqrt{(x - x_0)^2 + (y - y_0)^2 + z^2}} \quad (10.100)$$

The velocity components of this source element are

$$u(x, y, z) = \frac{\sigma A(x - x_0)}{4\pi[(x - x_0)^2 + (y - y_0)^2 + z^2]^{3/2}} \quad (10.101)$$

$$v(x, y, z) = \frac{\sigma A(y - y_0)}{4\pi[(x - x_0)^2 + (y - y_0)^2 + z^2]^{3/2}} \quad (10.102)$$

$$w(x, y, z) = \frac{\sigma A(z - z_0)}{4\pi[(x - x_0)^2 + (y - y_0)^2 + z^2]^{3/2}} \quad (10.103)$$

A student algorithm for calculating the influence of a quadrilateral constant-strength source element is given in Appendix D, Program No. 12.

10.4.2 Quadrilateral Doublet

Consider the quadrilateral element with a constant doublet distribution shown in Fig. 10.15. Using the doublet element that points in the z direction we can obtain the velocity potential by integrating the point elements:

$$\Phi(x, y, z) = \frac{-\mu}{4\pi}\int_S \frac{z\, dS}{[(x - x_0)^2 + (y - y_0)^2 + z^2]^{3/2}} \quad (10.104)$$

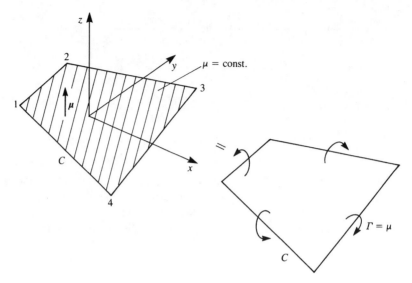

Figure 10.15 Quadrilateral doublet element and its vortex ring equivalent.

This integral for the potential is the same integral as the w velocity component of the quadrilateral source and consequently

$$\Phi = \frac{\mu}{4\pi}\Bigg[\tan^{-1}\left(\frac{m_{12}e_1 - h_1}{zr_1}\right) - \tan^{-1}\left(\frac{m_{12}e_2 - h_2}{zr_2}\right)$$
$$+ \tan^{-1}\left(\frac{m_{23}e_2 - h_2}{zr_2}\right) - \tan^{-1}\left(\frac{m_{23}e_3 - h_3}{zr_3}\right)$$
$$+ \tan^{-1}\left(\frac{m_{34}e_3 - h_3}{zr_3}\right) - \tan^{-1}\left(\frac{m_{34}e_4 - h_4}{zr_4}\right)$$
$$+ \tan^{-1}\left(\frac{m_{41}e_4 - h_4}{zr_4}\right) - \tan^{-1}\left(\frac{m_{41}e_1 - h_1}{zr_1}\right)\Bigg] \quad (10.105)$$

As $z \to 0$

$$\Phi = \mp\frac{\mu}{2} \quad (10.106)$$

The velocity components can be obtained by differentiating the velocity potential,

$$(u, v, w) = \left(\frac{\partial \Phi}{\partial x}, \frac{\partial \Phi}{\partial y}, \frac{\partial \Phi}{\partial z}\right)$$

and following Hess and Smith[10.1] we get

$$u = \frac{\mu}{4\pi}\Bigg[\frac{z(y_1 - y_2)(r_1 + r_2)}{r_1 r_2 \{r_1 r_2 - [(x - x_1)(x - x_2) + (y - y_1)(y - y_2) + z^2]\}}$$
$$+ \frac{z(y_2 - y_3)(r_2 + r_3)}{r_2 r_3 \{r_2 r_3 - [(x - x_2)(x - x_3) + (y - y_2)(y - y_3) + z^2]\}}$$
$$+ \frac{z(y_3 - y_4)(r_3 + r_4)}{r_3 r_4 \{r_3 r_4 - [(x - x_3)(x - x_4) + (y - y_3)(y - y_4) + z^2]\}}$$
$$+ \frac{z(y_4 - y_1)(r_4 + r_1)}{r_4 r_1 \{r_4 r_1 - [(x - x_4)(x - x_1) + (y - y_4)(y - y_1) + z^2]\}}\Bigg] \quad (10.107)$$

10.4 Three-Dimensional Constant-Strength Singularity Elements

$$v = \frac{\mu}{4\pi} \left[\frac{z(x_2 - x_1)(r_1 + r_2)}{r_1 r_2 \{r_1 r_2 - [(x - x_1)(x - x_2) + (y - y_1)(y - y_2) + z^2]\}} \right.$$
$$+ \frac{z(x_3 - x_2)(r_2 + r_3)}{r_2 r_3 \{r_2 r_3 - [(x - x_2)(x - x_3) + (y - y_2)(y - y_3) + z^2]\}}$$
$$+ \frac{z(x_4 - x_3)(r_3 + r_4)}{r_3 r_4 \{r_3 r_4 - [(x - x_3)(x - x_4) + (y - y_3)(y - y_4) + z^2]\}}$$
$$\left. + \frac{z(x_1 - x_4)(r_4 + r_1)}{r_4 r_1 \{r_4 r_1 - [(x - x_4)(x - x_1) + (y - y_4)(y - y_1) + z^2]\}} \right] \quad (10.108)$$

$$w = \frac{\mu}{4\pi} \left[\frac{[(x - x_2)(y - y_1) - (x - x_1)(y - y_2)](r_1 + r_2)}{r_1 r_2 \{r_1 r_2 - [(x - x_1)(x - x_2) + (y - y_1)(y - y_2) + z^2]\}} \right.$$
$$+ \frac{[(x - x_3)(y - y_2) - (x - x_2)(y - y_3)](r_2 + r_3)}{r_2 r_3 \{r_2 r_3 - [(x - x_2)(x - x_3) + (y - y_2)(y - y_3) + z^2]\}}$$
$$+ \frac{[(x - x_4)(y - y_3) - (x - x_3)(y - y_4)](r_3 + r_4)}{r_3 r_4 \{r_3 r_4 - [(x - x_3)(x - x_4) + (y - y_3)(y - y_4) + z^2]\}}$$
$$\left. + \frac{[(x - x_1)(y - y_4) - (x - x_4)(y - y_1)](r_4 + r_1)}{r_4 r_1 \{r_4 r_1 - [(x - x_4)(x - x_1) + (y - y_4)(y - y_1) + z^2]\}} \right] \quad (10.109)$$

On the element, as $z \to 0$

$$u = 0$$
$$v = 0$$

and the z component of the velocity can be computed by the near field formula, which reduces to

$$w = \frac{\mu}{4\pi} \left[\frac{[(x - x_2)(y - y_1) - (x - x_1)(y - y_2)](r_1 + r_2)}{r_1 r_2 \{r_1 r_2 - [(x - x_1)(x - x_2) + (y - y_1)(y - y_2)]\}} \right.$$
$$+ \frac{[(x - x_3)(y - y_2) - (x - x_2)(y - y_3)](r_2 + r_3)}{r_2 r_3 \{r_2 r_3 - [(x - x_2)(x - x_3) + (y - y_2)(y - y_3)]\}}$$
$$+ \frac{[(x - x_4)(y - y_3) - (x - x_3)(y - y_4)](r_3 + r_4)}{r_3 r_4 \{r_3 r_4 - [(x - x_3)(x - x_4) + (y - y_3)(y - y_4)]\}}$$
$$\left. + \frac{[(x - x_1)(y - y_4) - (x - x_4)(y - y_1)](r_4 + r_1)}{r_4 r_1 \{r_4 r_1 - [(x - x_4)(x - x_1) + (y - y_4)(y - y_1)]\}} \right] \quad (10.109a)$$

(Note that here, too, $z_k = 0$ must be used in the r_k terms of Eq. (10.92).)

In Section 10.2.2 it was shown that a two-dimensional constant-strength doublet is equivalent to two equal (and opposite direction) point vortices at the edge of the element. Similarly, in the next section we will show that the constant-strength doublet element is equivalent to a constant-strength vortex ring placed at the panel edges. Therefore, the above formulas for the velocity potential and its derivatives are valid for twisted panels as well (but in this case when the point P lies on the element the u, v velocity components may not be zero).

Far Field: The far field formulas for a quadrilateral doublet with area A can be obtained by using the results of Section 3.5 and are

$$\Phi(x, y, z) = \frac{-\mu A}{4\pi} z[(x - x_0)^2 + (y - y_0)^2 + z^2]^{-3/2} \quad (10.110)$$

$$u = \frac{3\mu A}{4\pi} \frac{(x - x_0)z}{[(x - x_0)^2 + (y - y_0)^2 + z^2]^{5/2}} \quad (10.111)$$

$$v = \frac{3\mu A}{4\pi} \frac{(y-y_0)z}{[(x-x_0)^2 + (y-y_0)^2 + z^2]^{5/2}} \quad (10.112)$$

$$w = -\frac{\mu A}{4\pi} \frac{(x-x_0)^2 + (y-y_0)^2 - 2z^2}{[(x-x_0)^2 + (y-y_0)^2 + z^2]^{5/2}} \quad (10.113)$$

An algorithm for calculating the influence of this quadrilateral constant-strength doublet panel is given in Appendix D, Program No. 12.

10.4.3 Constant Doublet Panel Equivalence to Vortex Ring

Consider the doublet panel of Section 10.4.2 with constant strength μ. Its potential (Eq. (10.104)) can be written as

$$\Phi = -\frac{\mu}{4\pi} \int_S \frac{z \, dS}{r^3}$$

where $r = [(x-x_0)^2 + (y-y_0)^2 + z^2]^{1/2}$. The velocity is

$$\mathbf{q} = \nabla \Phi = -\frac{\mu}{4\pi} \int_S \nabla \frac{z}{r^3} \, dS = \frac{\mu}{4\pi} \int_S \left[\mathbf{i} \frac{\partial}{\partial x_0} \frac{z}{r^3} + \mathbf{j} \frac{\partial}{\partial y_0} \frac{z}{r^3} - \mathbf{k} \left(\frac{1}{r^3} - \frac{3z^2}{r^5} \right) \right] dS$$

where we have used $(\partial/\partial x)(1/r^3) = -(\partial/\partial x_0)(1/r^3)$ and $(\partial/\partial y)(1/r^3) = -(\partial/\partial y_0)(1/r^3)$.

Now, let C represent the curve bounding the panel in Fig. 10.15 and consider a vortex filament of circulation Γ along C. The velocity due to the filament is obtained from the Biot–Savart law (Eq. (2.68)) as

$$\mathbf{q} = \frac{\Gamma}{4\pi} \int_C \frac{d\mathbf{l} \times \mathbf{r}}{r^3}$$

and for $d\mathbf{l} = (dx_0, dy_0)$ and $\mathbf{r} = (x-x_0, y-y_0, z)$ we get

$$\mathbf{q} = \frac{\Gamma}{4\pi} \int_C \left\{ \mathbf{i} \frac{z}{r^3} dy_0 - \mathbf{j} \frac{z}{r^3} dx_0 + \mathbf{k} \left[\frac{(y-y_0)}{r^3} dx_0 - \frac{(x-x_0)}{r^3} dy_0 \right] \right\}$$

Stokes's theorem for the vector $\mathbf{A}$ is

$$\oint_C \mathbf{A} \cdot d\mathbf{l} = \int_S \mathbf{n} \cdot \nabla \times \mathbf{A} \, dS$$

and with $\mathbf{n} = \mathbf{k}$ this becomes

$$\oint_C \mathbf{A} \cdot d\mathbf{l} = \int_S \left(\frac{\partial A_y}{\partial x_0} - \frac{\partial A_x}{\partial y_0} \right) dS$$

Using Stokes's theorem on the above velocity integral we get

$$\mathbf{q} = \frac{\Gamma}{4\pi} \int_S \left[\mathbf{i} \frac{\partial}{\partial x_0} \frac{z}{r^3} + \mathbf{j} \frac{\partial}{\partial y_0} \frac{z}{r^3} - \mathbf{k} \left(\frac{\partial}{\partial x_0} \frac{x-x_0}{r^3} + \frac{\partial}{\partial y_0} \frac{y-y_0}{r^3} \right) \right] dS$$

Upon performing the differentiation, we see that the velocity of the filament is identical to the velocity of the doublet panel if $\Gamma = \mu$.

The above derivation is a simplified version of that by Hess (in Appendix A, Ref. 12.4), who relates a general surface doublet distribution to a corresponding surface vortex distribution

$$\mathbf{q} = -\frac{1}{4\pi} \int_S (\mathbf{n} \times \nabla \mu) \times \frac{\mathbf{r}}{r^3} dS + \frac{1}{4\pi} \int_C \mu \frac{d\mathbf{l} \times \mathbf{r}}{r^3}$$

10.4 Three-Dimensional Constant-Strength Singularity Elements

whose order is one less than the order of the doublet distribution plus a vortex ring whose strength is equal to the edge value of the doublet distribution.

10.4.4 Comparison of Near and Far Field Formulas

To demonstrate the possible range of applicability of the far field approximation, the induced velocity for a unit strength rectangular source or doublet element, shown in Fig. 10.16, is calculated and presented in Figs. 10.17–10.22 (figures based on Browne and Ashby[10.2]). The computed results for the radial velocity component versus distance r/a (where a is the panel length as shown in Fig. 10.16) clearly indicate that the far field and exact formulas converge at about $r/a > 2$ (e.g., Figs. 10.17 or 10.18).

Similar computations for the total velocity induced by a doublet panel are presented in Fig. 10.18, and at $r/a > 2$ the two results seem to be identical.

A velocity survey above the panel (as shown in Fig. 10.16) is presented in Figs. 10.19–10.22. Here the total velocity survey is done in a horizontal plane at an altitude of $z/a = 0.75$ and 3.0, along lines parallel to the panel median and diagonal.

These diagrams clearly indicate that at a height of $z/a = 0.75$ the far field formula (point element) is insufficient for both the doublet and source elements. However, at a distance greater than $z/a = 3$ the difference is small and numerical efficiency justifies the use of the far field formulas.

10.4.5 Constant-Strength Vortex Line Segment

Early numerical solutions for lifting flows were based on vortex distribution solutions of the lifting surface equations (Section 4.5). The three-dimensional solution of such a problem is possible by using constant-strength vortex-line segments, which can be used to model the wing or the wake. The velocity induced by such a vortex segment of circulation Γ was developed in Sections 2.11 and 2.12 and Eq. (2.68b) states

$$\Delta \mathbf{q} = \frac{\Gamma}{4\pi} \frac{d\mathbf{l} \times \mathbf{r}}{r^3} \qquad (10.114)$$

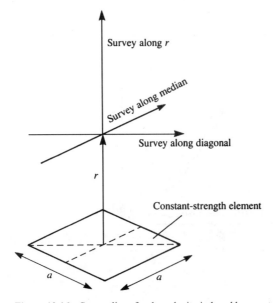

Figure 10.16 Survey lines for the velocity induced by a rectangular, flat element.

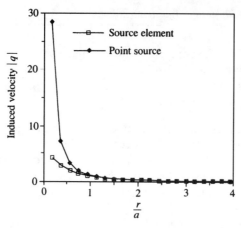

Figure 10.17 Comparison between the velocity induced by a rectangular source element and an equivalent point source versus height r/a.

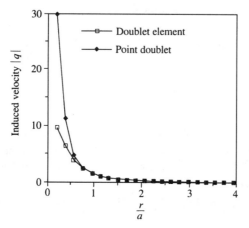

Figure 10.18 Comparison between the velocity induced by a rectangular doublet element and an equivalent point doublet versus height r/a.

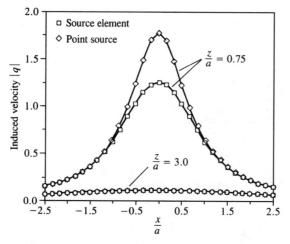

Figure 10.19 Comparison between the velocity induced by a rectangular source element and an equivalent point source along a horizontal survey line (median).

10.4 Three-Dimensional Constant-Strength Singularity Elements

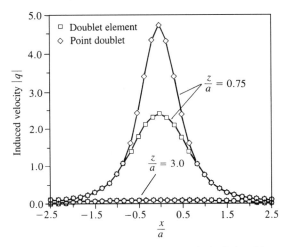

Figure 10.20 Comparison between the velocity induced by a rectangular doublet element and an equivalent point doublet along a horizontal survey line (median).

If the vortex segment points from point 1 to point 2, as shown in Fig. 10.23, then the velocity at an arbitrary point P can be obtained by Eq. (2.72):

$$\mathbf{q}_{1,2} = \frac{\Gamma}{4\pi} \frac{\mathbf{r}_1 \times \mathbf{r}_2}{|\mathbf{r}_1 \times \mathbf{r}_2|^2} \mathbf{r}_0 \cdot \left(\frac{\mathbf{r}_1}{r_1} - \frac{\mathbf{r}_2}{r_2} \right) \tag{10.115}$$

For numerical computation in a Cartesian system where the (x, y, z) values of the points 1, 2, and P are given, the velocity can be calculated by the following steps:

1. Calculate $\mathbf{r}_1 \times \mathbf{r}_2$:

$$(\mathbf{r}_1 \times \mathbf{r}_2)_x = (y_p - y_1) \cdot (z_p - z_2) - (z_p - z_1) \cdot (y_p - y_2)$$
$$(\mathbf{r}_1 \times \mathbf{r}_2)_y = -(x_p - x_1) \cdot (z_p - z_2) + (z_p - z_1) \cdot (x_p - x_2)$$
$$(\mathbf{r}_1 \times \mathbf{r}_2)_z = (x_p - x_1) \cdot (y_p - y_2) - (y_p - y_1) \cdot (x_p - x_2)$$

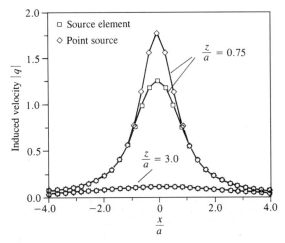

Figure 10.21 Comparison between the velocity induced by a rectangular source element and an equivalent point source along a horizontal survey line (diagonal).

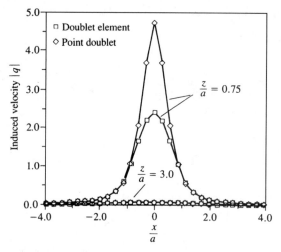

Figure 10.22 Comparison between the velocity induced by a rectangular doublet element and an equivalent point doublet along a horizontal survey line (diagonal).

Also, the absolute value of this vector product is

$$|\mathbf{r}_1 \times \mathbf{r}_2|^2 = (\mathbf{r}_1 \times \mathbf{r}_2)_x^2 + (\mathbf{r}_1 \times \mathbf{r}_2)_y^2 + (\mathbf{r}_1 \times \mathbf{r}_2)_z^2$$

2. Calculate the distances r_1, r_2:

$$r_1 = \sqrt{(x_p - x_1)^2 + (y_p - y_1)^2 + (z_p - z_1)^2}$$

$$r_2 = \sqrt{(x_p - x_2)^2 + (y_p - y_2)^2 + (z_p - z_2)^2}$$

3. Check for singular conditions.

(Since the vortex solution is singular when the point P lies on the vortex a special treatment is needed in the vicinity of the vortex segment–which for numerical purposes is assumed to have a very small radius ϵ.)

IF (r_1, or r_2, or $|\mathbf{r}_1 \times \mathbf{r}_2|^2 < \epsilon$)
THEN $u = v = w = 0$

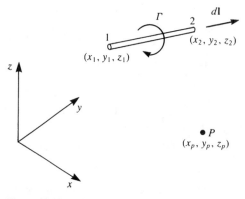

Figure 10.23 Influence of a straight vortex line segment at point P.

where ϵ is the vortex core size (which can be as small as the truncation error) or else u, v, w can be estimated by assuming solid body rotation or any other (more elaborate) vortex core model (see Section 2.5.1 of Ref. 10.3).

4. Calculate the dot product:

$$\mathbf{r}_0 \cdot \mathbf{r}_1 = (x_2 - x_1)(x_p - x_1) + (y_2 - y_1)(y_p - y_1) + (z_2 - z_1)(z_p - z_1)$$
$$\mathbf{r}_0 \cdot \mathbf{r}_2 = (x_2 - x_1)(x_p - x_2) + (y_2 - y_1)(y_p - y_2) + (z_2 - z_1)(z_p - z_2)$$

5. The resulting velocity components are

$$u = K \cdot (\mathbf{r}_1 \times \mathbf{r}_2)_x$$
$$v = K \cdot (\mathbf{r}_1 \times \mathbf{r}_2)_y$$
$$w = K \cdot (\mathbf{r}_1 \times \mathbf{r}_2)_z$$

where

$$K = \frac{\Gamma}{4\pi |\mathbf{r}_1 \times \mathbf{r}_2|^2} \left(\frac{\mathbf{r}_0 \cdot \mathbf{r}_1}{r_1} - \frac{\mathbf{r}_0 \cdot \mathbf{r}_2}{r_2} \right)$$

For computational purposes these steps can be included in a subroutine (e.g., VORTXL – vortex line) that will calculate the induced velocity (u, v, w) at a point $P(x, y, z)$ as a function of the vortex line strength and its edge coordinates, such that

$$(u, v, w) = \text{VORTXL}(x, y, z, x_1, y_1, z_1, x_2, y_2, z_2, \Gamma) \tag{10.116}$$

As an example for programming this algorithm see subroutine VORTEX (VORTEX $\equiv$ VORTXL) in Program No. 13 in Appendix D.

10.4.6 Vortex Ring

Based on the subroutine of Eq. (10.116), a variety of elements can be defined. For example, the velocity induced by a rectilinear vortex ring (shown in Fig. 10.24) can be computed by calling this routine four times for the four segments. Note that this velocity calculation is equivalent to the result for a constant-strength doublet.

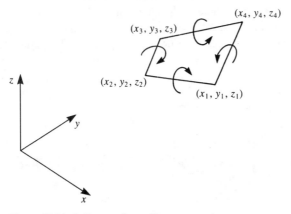

Figure 10.24 Influence of a rectilinear vortex ring.

To obtain the velocity induced by the four segments of a rectilinear vortex ring with circulation Γ calculate

$$(u_1, v_1, w_1) = \text{VORTXL}\,(x, y, z, x_1, y_1, z_1, x_2, y_2, z_2, \Gamma)$$
$$(u_2, v_2, w_2) = \text{VORTXL}\,(x, y, z, x_2, y_2, z_2, x_3, y_3, z_3, \Gamma)$$
$$(u_3, v_3, w_3) = \text{VORTXL}\,(x, y, z, x_3, y_3, z_3, x_4, y_4, z_4, \Gamma)$$
$$(u_4, v_4, w_4) = \text{VORTXL}\,(x, y, z, x_4, y_4, z_4, x_1, y_1, z_1, \Gamma)$$

and the induced velocity at $P(x, y, z)$ is

$$(u, v, w) = (u_1, v_1, w_1) + (u_2, v_2, w_2) + (u_3, v_3, w_3) + (u_4, v_4, w_4)$$

This can be programmed into a subroutine such that

$$\begin{pmatrix} u \\ v \\ w \end{pmatrix} = \text{VORING} \begin{pmatrix} x & y & z \\ x_1 & y_1 & z_1 \\ x_2 & y_2 & z_2 \\ x_3 & y_3 & z_3 \\ x_4 & y_4 & z_4 \\ & \Gamma & \end{pmatrix} \tag{10.117}$$

In most situations the vortex rings are placed on a patch with i, j indices, as shown in Fig. 10.25. In this situation the input to this subroutine can be abbreviated by identifying each panel by its i, j-th corner point:

$$(u, v, w) = \text{VORING}\,(x, y, z, i, j, \Gamma_{ij}) \tag{10.117a}$$

From the programming point of view this routine simplifies the scanning of the vortex rings on the patch. However, the inner vortex segments are scanned twice, which makes the computation less efficient. This can be improved for larger codes when computer run time is more important that programming simplicity.

Note that this formulation is valid everywhere (including the center of the element) but is singular on the vortex ring. Such a routine is used in Program No. 13 in Appendix D.

10.4.7 Horseshoe Vortex

A simplified case of the vortex ring is the horseshoe vortex. In this case the vortex line is assumed to be placed in the x–y plane as shown in Fig. 10.26. The two trailing vortex segments are placed parallel to the x axis at $y = y_a$ and at $y = y_b$, and the leading segment is placed parallel to the y axis between the points (x_a, y_a) and (x_a, y_b). The induced velocity

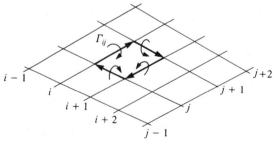

Figure 10.25 The method of calculating the influence of a vortex ring by adding the influence of the straight vortex segment elements.

10.4 Three-Dimensional Constant-Strength Singularity Elements

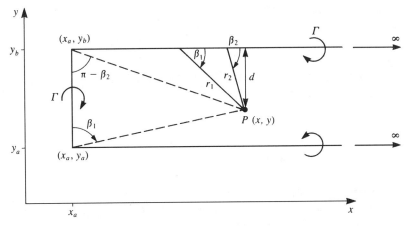

Figure 10.26 Nomenclature used for deriving the influence of a horseshoe vortex element.

in the x–y plane will have only a component in the negative z direction and can be computed by using Eq. (2.69) for a straight vortex segment:

$$w(x, y, 0) = \frac{-\Gamma}{4\pi d}(\cos \beta_1 - \cos \beta_2) \qquad (10.118)$$

where the angles and their cosines are shown in Fig. 10.26. The negative sign is a result of the θ velocity component pointing in the $-z$ direction. For the vortex segment parallel to the x axis, and beginning at $y = y_b$, the corresponding angles are given by

$$\cos \beta_1 = \frac{x - x_a}{\sqrt{(x - x_a)^2 + (y - y_b)^2}}$$

$$\cos \beta_2 = \cos \pi = -1$$

For the finite-length segment, parallel to the y axis,

$$\cos \beta_1 = \frac{y - y_a}{\sqrt{(x - x_a)^2 + (y - y_a)^2}}$$

$$\cos \beta_2 = -\cos(\pi - \beta_2) = \frac{y - y_b}{\sqrt{(x - x_a)^2 + (y - y_b)^2}}$$

For the lower segment beginning at $y = y_a$ the angles are

$$\cos \beta_1 = \cos 0 = 1$$

$$\cos \beta_2 = \frac{x - x_a}{\sqrt{(x - x_a)^2 + (y - y_a)^2}}$$

The downwash due to the horseshoe vortex is now

$$w(x, y, 0) = \frac{-\Gamma}{4\pi} \left\{ \frac{1}{x - x_a} \left[\frac{y_b - y}{\sqrt{(x - x_a)^2 + (y - y_b)^2}} + \frac{y - y_a}{\sqrt{(x - x_a)^2 + (y - y_a)^2}} \right] \right.$$
$$+ \frac{1}{y_b - y} \left[1 + \frac{x - x_a}{\sqrt{(x - x_a)^2 + (y - y_b)^2}} \right]$$
$$\left. + \frac{1}{y - y_a} \left[1 + \frac{x - x_a}{\sqrt{(x - x_a)^2 + (y - y_a)^2}} \right] \right\} \qquad (10.119)$$

After some manipulations we get

$$w(x, y, 0) = \frac{-\Gamma}{4\pi(y - y_a)} \left[1 + \frac{\sqrt{(x - x_a)^2 + (y - y_a)^2}}{x - x_a} \right]$$
$$+ \frac{\Gamma}{4\pi(y - y_b)} \left[1 + \frac{\sqrt{(x - x_a)^2 + (y - y_b)^2}}{x - x_a} \right] \qquad (10.119a)$$

When $x = x_a$, the limit of Eq. (10.119) becomes

$$w(x_a, y, 0) = \frac{-\Gamma}{4\pi} \left[\frac{1}{y - y_a} + \frac{1}{y_b - y} \right] \qquad (10.119b)$$

where the finite-length segment does not induce downwash on itself.

The velocity potential of the horseshoe vortex may be obtained by reducing the results of a constant-strength doublet panel (Section 10.4.2) or by integrating the potential of a point doublet element. The potential of such a point doublet placed at $(x_0, y_0, 0)$ and pointing in the z direction, as derived in Section 3.5 (or in Eq. (10.110)), is

$$\Phi = \frac{-\Gamma}{4\pi} \frac{z}{r^3}$$

where $r = [(x - x_0)^2 + (y - y_0)^2 + z^2]^{1/2}$. To obtain the potential due to the horseshoe element at an arbitrary point P, this point doublet must be integrated over the area enclosed by the horseshoe element:

$$\Phi = \frac{-\Gamma}{4\pi} \int_{y_a}^{y_b} dy_0 \int_{x_a}^{\infty} \frac{z\, dx_0}{[(x - x_0)^2 + (y - y_0)^2 + z^2]^{3/2}}$$

The result is given by Moran[5.1] (p. 445) as

$$\Phi = \frac{-\Gamma}{4\pi} \int_{y_a}^{y_b} \frac{z(x_0 - x)\, dy_0}{[(y - y_0)^2 + z^2][(x - x_0)^2 + (y - y_0)^2 + z^2]^{1/2}} \bigg|_{x_a}^{\infty}$$

$$= \frac{-\Gamma}{4\pi} \int_{y_a}^{y_b} \frac{z\, dy_0}{[(y - y_0)^2 + z^2]} \left[1 + \frac{x - x_a}{[(x - x_a)^2 + (y - y_0)^2 + z^2]^{1/2}} \right]$$

$$= \frac{-\Gamma}{4\pi} \left\{ \tan^{-1} \frac{y_0 - y}{z} + \tan^{-1} \frac{(y_0 - y)(x - x_a)}{z[(x - x_a)^2 + (y - y_0)^2 + z^2]^{1/2}} \right\} \bigg|_{y_a}^{y_b}$$

$$= \frac{-\Gamma}{4\pi} \left\{ \tan^{-1} \frac{z}{y - y_b} - \tan^{-1} \frac{z}{y - y_a} \right.$$

$$\left. + \tan^{-1} \frac{(y_0 - y)(x - x_a)}{z[(x - x_a)^2 + (y - y_0)^2 + z^2]^{1/2}} \bigg|_{y_a}^{y_b} \right\} \qquad (10.120)$$

Note that we have used Eq. (B.10) from Appendix B to evaluate the limits of the first term.

10.5 Three-Dimensional Higher Order Elements

The surface shape and singularity strength distribution over an arbitrarily shaped panel can be approximated by a polynomial of a certain degree. The surface of such an

10.5 Three-Dimensional Higher Order Elements

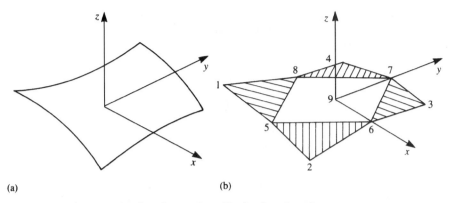

Figure 10.27 Approximation of a curved panel by five flat subpanels.

arbitrary panel as shown in Fig. 10.27a can be approximated by a "zero-order" flat plane

$$z = a_0$$

by a first-order surface

$$z = a_0 + b_1 x + b_2 y$$

by a second-order surface

$$z = a_0 + b_1 x + b_2 y + c_1 x^2 + c_2 xy + c_3 y^2$$

or by any higher order approximations. Evaluation of the influence coefficients in a closed form is possible,[10.1] though, only for flat surfaces, and an approximation of a curved panel by five flat subpanels is shown in Fig. 10.27b. This approach is used in the code PANAIR,[9.4] and for demonstrating a higher order element let us describe this element.

For the singularity distribution a first-order source and a second-order doublet is used, and in the following paragraph the methodology is briefly described.

a. Influence of Source Distribution

The source distribution on this element is approximated by a first-order polynomial:

$$\sigma(x_0, y_0) = \sigma_0 + \sigma_x x_0 + \sigma_y y_0 \tag{10.121}$$

where (x_0, y_0) are the panel local coordinates, σ_0 is the source strength at the origin, and $\sigma_0, \sigma_x,$ and σ_y are three constants. The contribution of this source distribution to the potential $\Delta\Phi$ and to the induced velocity $\Delta(u, v, w)$ (in the panel frame of reference) can be evaluated by performing the integral

$$\Delta\Phi(x, y, z) = \frac{1}{4\pi} \int_{\text{panel}} \frac{-\sigma(x_0, y_0)\, dS}{\sqrt{(x-x_0)^2 + (y-y_0)^2 + z^2}} \tag{10.122}$$

and then differentiating to get the velocity components

$$\Delta(u, v, w) = \left(\frac{\partial \Delta\Phi}{\partial x}, \frac{\partial \Delta\Phi}{\partial y}, \frac{\partial \Delta\Phi}{\partial z} \right) \tag{10.123}$$

The result of this integration depends solely on the geometry of the problem and can be evaluated for an arbitrary field point. Some details of this calculation are provided by Johnson[9.4] and can be reduced to a form that depends on the panel corner point values (the corner point numbering sequence is shown in Fig. 10.27b). Thus, in terms of these corner point values the influence of the panel becomes

$$\Delta\Phi = F_S(\sigma_1, \sigma_2, \sigma_3, \sigma_4, \sigma_9) = f_S(\sigma_0, \sigma_x, \sigma_y) \tag{10.124}$$

$$\Delta(u, v, w) = G_S(\sigma_1, \sigma_2, \sigma_3, \sigma_4, \sigma_9) = g_S(\sigma_0, \sigma_x, \sigma_y) \tag{10.125}$$

where the functions F_S, G_S, f_S, and g_S are linear matrix manipulations and $\sigma_5, \sigma_6, \sigma_7$, and σ_8 are not used. Also, note that σ_0, σ_x, and σ_y are the three basic unknowns for each panel and $\sigma_1, \ldots, \sigma_9$ can be evaluated based on these values (so that for each panel only three unknown values are left).

b. *Influence of Doublet Distribution*

To model the two components of vorticity on the panel surface a second-order doublet is used:

$$\mu(x_0, y_0) = \mu_0 + \mu_x x_0 + \mu_y y_0 + \mu_{xx} x_0^2 + \mu_{xy} x_0 y_0 + \mu_{yy} y_0^2 \tag{10.126}$$

The potential due to a doublet distribution whose axis points in the z direction (see Section 3.5) is

$$\Delta\Phi(x, y, z) = \frac{-1}{4\pi} \int_S \frac{\mu(x_0, y_0) \cdot z \, dS}{[(x - x_0)^2 + (y - y_0)^2 + z^2]^{3/2}} \tag{10.127}$$

and the induced velocity is

$$(u, v, w) = \left(\frac{\partial \Delta\Phi}{\partial x}, \frac{\partial \Delta\Phi}{\partial y}, \frac{\partial \Delta\Phi}{\partial z}\right) \tag{10.128}$$

These integrals can be evaluated (see Johnson[9.4]) in terms of the panel corner points (points 1–9 in Fig. 10.27b) and the result can be presented as

$$\Delta\Phi = F_D(\mu_1, \mu_2, \mu_3, \mu_4, \mu_5, \mu_6, \mu_7, \mu_8, \mu_9)$$
$$= f_D(\mu_0, \mu_x, \mu_y, \mu_{xx}, \mu_{xy}, \mu_{yy}) \tag{10.129}$$

$$\Delta(u, v, w) = G_D(\mu_1, \mu_2, \mu_3, \mu_4, \mu_5, \mu_6, \mu_7, \mu_8, \mu_9)$$
$$= g_D(\mu_0, \mu_x, \mu_y, \mu_{xx}, \mu_{xy}, \mu_{yy}) \tag{10.130}$$

where the functions F_D, G_D, f_D, and g_D are linear matrix manipulations, which depend on the geometry only. Also, note that $\mu_0, \mu_x, \mu_y, \mu_{xx}, \mu_{xy}$, and μ_{yy} are the five basic unknowns for each panel and $\mu_1, \ldots, \mu_9$ can be evaluated based on these values (so that for each panel only five unknown doublet parameters are left).

For more details on higher order elements, see Ref. 9.4.

References

[10.1] Hess, J. L., and Smith, A. M. O., "Calculation of Potential Flow About Arbitrary Bodies," *Progress in Aeronautical Sciences*, Vol. 8, 1967, pp. 1–138.

[10.2] Browne, L. E., and Ashby, D. L., "Study of the Integration of Wind-Tunnel and Computational Methods for Aerodynamic Configurations," NASA TM 102196, July 1989.

[10.3] Sarpkaya, T., "Computational Methods with Vortices–The 1988 Freeman Scholar Lecture," *Journal of Fluids Engineering*, Vol. 111, March 1989, pp. 5–52.

Problems

10.1. Find the x component of velocity u for the constant-strength source distribution by a direct integration of Eq. (10.12).

10.2. Find the velocity potential for the constant doublet distribution by a direct integration of Eq. (10.25).

10.3. Consider the horseshoe vortex of Section 10.4.7 lying in the x–y plane. For the case where the leading segment lies on the x axis ($x_a = 0$) find the velocity induced at a point, whose coordinates are x, y, and z, that lies above the plane of the horseshoe.

CHAPTER 11

Two-Dimensional Numerical Solutions

The principles of singular element based numerical solutions were introduced in Chapter 9 and the first examples are provided in this chapter. The following two-dimensional examples will have all the elements of more refined three-dimensional methods, but because of the simple two-dimensional geometry, the programming effort is substantially less. Consequently, such methods can be developed in a short time for investigating improvements in larger codes and are also suitable for homework assignments and class demonstrations.

Based on the level of approximation of the singularity distribution, surface geometry, and type of boundary conditions, numerous computational methods can be constructed, some of which are presented in Table 11.1. We will not attempt to demonstrate all the possible combinations but will try to cover some of the most frequently used methods (denoted by the word "example" in Table 11.1), including discrete singular elements and constant-strength, linear, and quadratic elements (as an example for higher order singularity distributions). The different approaches in specifying the zero normal velocity boundary condition will be exercised and mainly the outer Neumann normal velocity and the internal Dirichlet boundary conditions will be used (and there are additional options, e.g., an internal Neumann condition). In terms of the surface geometry, for simplicity, only the flat panel element will be used here and in areas of high surface curvature the solution can be improved by using more panels.

In this chapter and in the following ones the primary concern is the simplicity of the explanation and the ease of constructing the numerical technique, while numerical efficiency considerations are secondary. Consequently, the numerical economy of the methods presented can be improved (with some compromise in regard to the ease of code readability). Also, the methods are presented in their simplest form and each can be further developed to match the requirements of a particular problem. Such improvement can be obtained by changing grid spacing and density, location of collocation points, or wake model, or altering the method of enforcing the boundary conditions and of enforcing the Kutta condition.

Also, it is recommended that one read this chapter sequentially since the first methods will be described with more details. As the chapter evolves, some redundant details are omitted and the description may appear inadequate without reading the previous sections.

11.1 Point Singularity Solutions

The basic idea behind point singularity solutions is presented schematically in Fig. 11.1. If an exact solution in a form of a continuous singularity distribution (e.g., a vortex distribution $\gamma(x)$) exists, then it can be divided into several finite segments (e.g., the segment between x_1 and x_2). The local average strength of the element is then $\Gamma_0 = \int_{x_1}^{x_2} \gamma(x)dx$ and it can be placed at a point x_0 within the interval x_1–x_2. A discrete element numerical solution can be obtained by specifying N such unknown element strengths and then establishing N equations for their solution. This can be done by specifying the boundary conditions at N

11.1 Point Singularity Solutions

Table 11.1. *List of possible two-dimensional panel methods and of those tested in this chapter*

Singularity distribution		Boundary conditions		Surface paneling flat/high-order
		Neumann (external)	Dirichlet (internal)	
Point	source	example		flat
	doublet			
	vortex	example		flat
Constant strength	source	example	example	flat
	doublet	example	example	flat
	vortex	example		flat
Linear strength	source	example	example	flat
	doublet	example	example	flat
	vortex			
Quadratic strength	source			
	doublet		example	flat
	vortex			

collocation points along the boundary. Furthermore, when constructing the solution, some of the considerations mentioned in Section 9.3 (e.g., in regard to the Kutta condition and the wake) must be addressed.

As a first example for this very simple approach the lifting and thickness problems of thin airfoils are solved based on models (such as the lumped-vortex element) generated during examination of the analytical solutions in Chapter 5.

11.1.1 Discrete Vortex Method

The discrete vortex method presented here for solving the thin lifting airfoil problem is based on the lumped-vortex element and serves for solving numerically the integral equation (Eq. (5.39)) presented in Chapter 5. The advantage of the numerical approach is that the boundary conditions can be specified on the airfoil's camber surface without a need for the small-disturbance approximation. Also, two-dimensional interactions, such as those due to ground effect or multielement airfoils, can be studied with great ease.

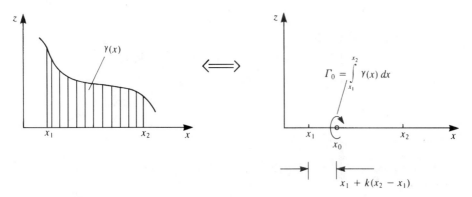

Figure 11.1 Discretization of a continuous singularity distribution.

This method was introduced as an example in Section 9.8 and therefore its principles will be discussed here only briefly. To establish the procedure for the numerical solution, the six steps presented in Section 9.7 are followed.

a. *Choice of Singularity Element*

For this discrete vortex method the lumped-vortex element is selected and its influence is given by Eq. (9.31) (or Eqs. (10.9) and (10.10)):

$$\begin{pmatrix} u \\ w \end{pmatrix} = \frac{\Gamma_j}{2\pi r_j^2} \begin{pmatrix} 0 & 1 \\ -1 & 0 \end{pmatrix} \begin{pmatrix} x - x_j \\ z - z_j \end{pmatrix} \tag{11.1}$$

where

$$r_j^2 = (x - x_j)^2 + (z - z_j)^2$$

Thus, the velocity at an arbitrary point (x, z) due to a vortex element of circulation Γ_j located at (x_j, z_j) is given by Eq. (11.1). This can be included in a subroutine, which will be called VOR2D:

$$(u, w) = \text{VOR2D}(\Gamma_j, x, z, x_j, z_j) \tag{11.2}$$

Such a subroutine is included in Program No. 2 in Appendix D.

b. *Discretization and Grid Generation*

At this phase the thin-airfoil camberline (Fig. 11.2) is divided into N subpanels, which may be equal in length. The N vortex points (x_j, z_j) will be placed at the quarter-chord point of each planar panel (Fig. 11.2). The zero normal flow boundary condition can be fulfilled on the camberline at the three-quarter point of each panel. These N collocation points (x_i, z_i) and the corresponding N normal vectors $\mathbf{n}_i$ along with the vortex points can be computed numerically or supplied as an input file. Note that by discretizing the camberline as shown in Fig. 11.2, we end up with only the panel edges remaining on the original camberline. For convenience, the normal vector is evaluated at the actual camberline and the effect of this choice will be investigated at the end of this section. Consequently, the normal vectors $\mathbf{n}_i$, pointing outward at each of these points, are approximated by using the

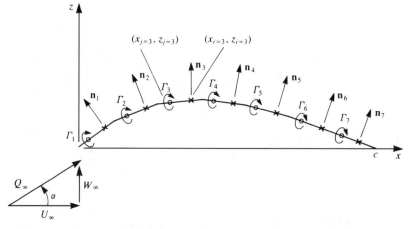

Figure 11.2 Discrete vortex representation of the thin, lifting airfoil model.

11.1 Point Singularity Solutions

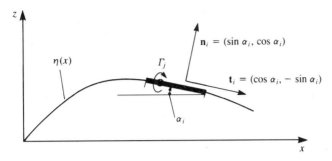

Figure 11.3 Nomenclature used in defining the geometry of a point singularity based surface panel.

surface shape $\eta(x)$, as shown in Fig. 11.3:

$$\mathbf{n}_i = \frac{(-d\eta/dx, 1)}{\sqrt{(d\eta/dx)^2 + 1}} = (\sin \alpha_i, \cos \alpha_i) \tag{11.3}$$

where the angle α_i is defined as shown in Fig. 11.3. Similarly the tangential vector $\mathbf{t}_i$ is

$$\mathbf{t}_i = (\cos \alpha_i, -\sin \alpha_i) \tag{11.3a}$$

Since the lumped-vortex element is based on the Kutta condition, the last panel will inherently fulfill this requirement, and no additional specification of this condition is needed.

c. Influence Coefficients

The normal velocity component at each point on the camberline is a combination of the self-induced velocity and the free-stream velocity. Therefore, the zero normal flow boundary condition can be presented as

$$\mathbf{q} \cdot \mathbf{n} = 0 \quad \text{on solid surface}$$

Division of the velocity vector into the self-induced and free-stream components yields

$$(u, w) \cdot \mathbf{n} + (U_\infty, W_\infty) \cdot \mathbf{n} = 0 \quad \text{on solid surface} \tag{11.4}$$

where the first term is the velocity induced by the singularity distribution on itself (hence "self-induced part") and the second term is the free-stream component $\mathbf{Q}_\infty = (U_\infty, W_\infty)$, as shown in Fig. 11.2.

The self-induced part can be represented by a combination of *influence coefficients*, while the free-stream contribution is known and will be transferred to the right-hand side of the boundary condition. To establish the self-induced portion of the normal velocity, at each collocation point, consider the velocity induced by the jth vortex element at the first collocation point (in order to get the influence due to a unit strength Γ_j assume $\Gamma_j = 1$ in Eq. (11.2)):

$$(u, w)_{1j} = \text{VOR2D}(\Gamma_j = 1, x_1, z_1, x_j, z_j) \tag{11.2a}$$

The influence coefficient a_{ij} is defined as the velocity component normal to the surface, due to a unit strength singularity element. Consequently, the contribution of a *unit strength* singularity element j, at collocation point 1, is

$$a_{1j} = (u, w)_{1j} \cdot \mathbf{n}_1 \tag{11.5}$$

The induced normal velocity component q_{n1}, at point 1, due to all the elements is therefore

$$q_{n1} = a_{11}\Gamma_1 + a_{12}\Gamma_2 + a_{13}\Gamma_3 + \cdots + a_{1N}\Gamma_N$$

Note that the strength of Γ_j is unknown at this point.

Fulfillment of the boundary condition on the surface requires that at each collocation point the normal velocity component will vanish. Specification of this condition (as in Eq. (11.4)) for the first collocation point yields

$$a_{11}\Gamma_1 + a_{12}\Gamma_2 + a_{13}\Gamma_3 + \cdots + a_{1N}\Gamma_N + (U_\infty, W_\infty) \cdot \mathbf{n}_1 = 0$$

But, as mentioned earlier, the last term (free-stream component) is known and can be transferred to the right-hand side of the equation. Consequently, the right hand side (RHS) is defined as

$$\text{RHS}_i = -(U_\infty, W_\infty) \cdot \mathbf{n}_i \tag{11.6}$$

Specifying the boundary condition for each of the N collocation points results in the following set of algebraic equations:

$$\begin{pmatrix} a_{11} & a_{12} & \cdots & a_{1N} \\ a_{21} & a_{22} & \cdots & a_{2N} \\ a_{31} & a_{32} & \cdots & a_{3N} \\ \vdots & \vdots & \ddots & \vdots \\ a_{N1} & a_{N2} & \cdots & a_{NN} \end{pmatrix} \begin{pmatrix} \Gamma_1 \\ \Gamma_2 \\ \Gamma_3 \\ \vdots \\ \Gamma_N \end{pmatrix} = \begin{pmatrix} \text{RHS}_1 \\ \text{RHS}_2 \\ \text{RHS}_3 \\ \vdots \\ \text{RHS}_N \end{pmatrix}$$

This influence coefficient calculation procedure can be accomplished by using two "DO loops" where the outer loop scans the collocation points and the inner scans the vortices.

```
        DO 1 i = 1, N    (collocation point loop)
        DO 1 j = 1, N    (vortex point loop)
                (u, w)_ij = VOR2D(Γ = 1.0, x_i, z_i, x_j, z_j)
                a_ij = (u, w)_ij · n_i
1       CONTINUE
C       END DO LOOP
```

d. *Establish RHS Vector*

The right-hand side vector, which is the normal component of the free stream, can be computed within the outer loop of the previously described DO loops by using Eq. (11.6),

$$\text{RHS}_i = -(U_\infty, W_\infty) \cdot \mathbf{n}_i$$

where $(U_\infty, W_\infty) = Q_\infty(\cos\alpha, \sin\alpha)$. If we use the formulation of Eq. (11.3) for the normal vector, the RHS becomes

$$\text{RHS}_i = -Q_\infty(\cos\alpha \sin\alpha_i + \sin\alpha \cos\alpha_i) = -Q_\infty[\sin(\alpha + \alpha_i)] \tag{11.6a}$$

Note that α is the free-stream angle of attack (Fig. 11.2) and α_i is the ith panel inclination.

e. *Solve Linear Set of Equations*

The results of the previous computations can be summarized (for each collocation point i) as

$$\sum_{j=1}^{N} a_{ij}\Gamma_j = \text{RHS}_i \tag{11.7}$$

11.1 Point Singularity Solutions

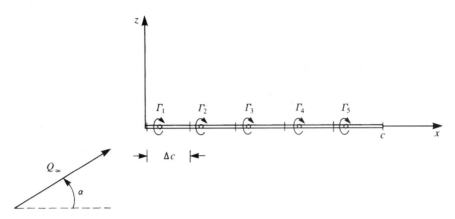

Figure 11.4 Representation of a lifting flat plate by five discrete vortices.

For example consider the case of a flat plate (shown in Fig. 11.4) where only five equal length elements ($\Delta c = c/5$) were used. Equation (11.7) for the five panels becomes

$$\frac{1}{\pi \Delta c}\begin{pmatrix} -1 & 1 & \frac{1}{3} & \frac{1}{5} & \frac{1}{7} \\ -\frac{1}{3} & -1 & 1 & \frac{1}{3} & \frac{1}{5} \\ -\frac{1}{5} & -\frac{1}{3} & -1 & 1 & \frac{1}{3} \\ -\frac{1}{7} & -\frac{1}{5} & -\frac{1}{3} & -1 & 1 \\ -\frac{1}{9} & -\frac{1}{7} & -\frac{1}{5} & -\frac{1}{3} & -1 \end{pmatrix}\begin{pmatrix} \Gamma_1 \\ \Gamma_2 \\ \Gamma_3 \\ \Gamma_4 \\ \Gamma_5 \end{pmatrix} = -Q_\infty \sin\alpha \begin{pmatrix} 1 \\ 1 \\ 1 \\ 1 \\ 1 \end{pmatrix}$$

This linear set of algebraic equations is diagonally dominant and can be solved by standard matrix methods. Its solution is

$$\begin{pmatrix} \Gamma_1 \\ \Gamma_2 \\ \Gamma_3 \\ \Gamma_4 \\ \Gamma_5 \end{pmatrix} = \pi \Delta c Q_\infty \sin\alpha \begin{pmatrix} 2.46092 \\ 1.09374 \\ 0.70314 \\ 0.46876 \\ 0.27344 \end{pmatrix}$$

and is shown schematically in Fig. 11.5. Note that the total circulation is $\pi c Q_\infty \sin\alpha$, which is the exact result.

f. *Secondary Computations: Pressures, Loads, Velocities, Etc.*

The resulting pressures and loads for this case can be computed by using the Kutta–Joukowski theorem for each panel j. Thus the lift and pressure difference are

$$\Delta L_j = \rho Q_\infty \Gamma_j \tag{11.8}$$

$$\Delta p_j = \rho Q_\infty \frac{\Gamma_j}{\Delta c} \tag{11.9}$$

where Δc is the panel length. The total lift and moment (about the leading edge) per unit

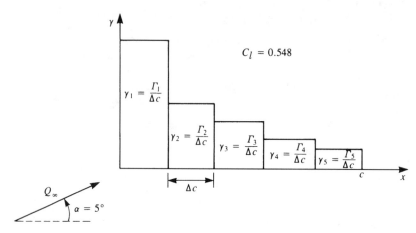

Figure 11.5 Graphic representation of the computed vorticity distribution with a five-element discrete-vortex method.

span are obtained by summing the contribution of each element:

$$L = \sum_{j=1}^{N} \Delta L_j \tag{11.10}$$

$$M_0 = -\sum_{j=1}^{N} \Delta L_j (x_j \cos \alpha) \tag{11.11}$$

while the nondimensional coefficients become

$$C_l = \frac{L}{(1/2)\rho Q_\infty^2 c} \tag{11.12}$$

$$C_{m_0} = \frac{M_0}{(1/2)\rho Q_\infty^2 c^2} \tag{11.13}$$

The following examples are presented to demonstrate possible applications of this method.

Example 1: Thin Airfoil with Parabolic Camber

Consider the thin airfoil with parabolic camber of Section 5.4, where the camberline shape is

$$\eta(x) = 4\epsilon \frac{x}{c}\left[1 - \frac{x}{c}\right]$$

For small values of $\epsilon < 0.1c$ the numerical results are close to the analytic results as shown in Fig. 11.6 (here actually $\epsilon = 0.1$ was used). This example can also be used to investigate the effect of the small-disturbance approximation (for the boundary conditions) on the pressure distribution, as shown by Figs. 11.7 and 11.8. For the numerical solution the vortices were placed on the camberline where the boundary condition was satisfied. For the analytical solution (and for the second numerical solution, aimed at simulating the analytical solution) the vortex distribution and the boundary condition were specified on the x axis. The analytical pressure distribution can be obtained by substituting the coefficients A_0 and A_1 from Section 5.4 into Eqs. (5.44a) and (5.48), which gives

$$\Delta C_p = \frac{2\gamma}{Q_\infty} = 4\left[\frac{1 + \cos\theta}{\sin\theta}\alpha + \frac{4\epsilon}{c}\sin\theta\right]$$

11.1 Point Singularity Solutions

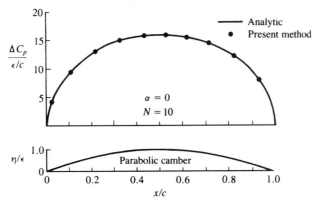

Figure 11.6 Chordwise pressure difference for a thin airfoil with parabolic camber at zero angle of attack ($\epsilon = 0.1$).

This can be rewritten in terms of the x coordinate by using Eq. (5.45) (e.g., $\sin\theta = 2[(x/c)(1 - x/c)]^{1/2}$ and $\cos\theta = 1 - 2x/c$):

$$\Delta C_p = 4\sqrt{\frac{c-x}{x}}\alpha + 32\frac{\epsilon}{c}\sqrt{\frac{x}{c}\left(1 - \frac{x}{c}\right)}$$

The effect of angle of attack is shown in Fig. 11.8 where a fairly large angle ($\alpha = 10°$) is used. Note the large suction peak at the leading edge, which is exaggerated by the thin airfoil solution. In general, Figs. 11.7 and 11.8 demonstrate that both thin-airfoil theory and the lumped-vortex panel method yield similar results. A simple computer program using the principles of this section is presented in Appendix D, Program No. 2.

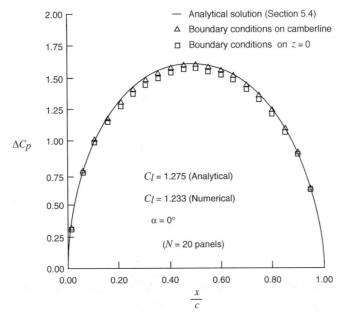

Figure 11.7 Effect of small-disturbance boundary condition on the computed pressure difference on a thin parabolic camber airfoil ($\alpha = 0, \epsilon = 0.1$).

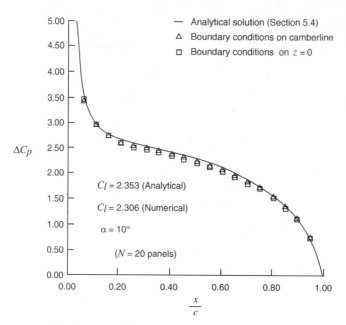

Figure 11.8 Effect of small-disturbance boundary condition on the computed pressure difference on a thin parabolic camber airfoil ($\alpha = 10°$, $\epsilon = 0.1$).

Example 2: Two-Element Airfoil

The advantage of this numerical solution technique is that it is not limited to the restrictions of small-disturbance boundary conditions. For example, a two-element airfoil with large deflection can be analyzed (and the results will have physical meaning when the actual flow is attached).

Figure 11.9 shows the geometry of the two-element airfoil made up of circular arcs and the pressure difference distributions. The interaction is shown by the plots of the close and separated elements (far from each other). When the elements are apart, the lift of the first element decreases while that of the second increases.

Example 3: Sensitivity to Grid

After this first set of numerical examples, some possible pitfalls of the numerical approach can be observed (and hopefully avoided later).

First note the method of paneling the gap region in the previous example of the two-element airfoil (Fig. 11.10). If very few elements are used, then it is always advised to align the vortex points with vortex points and collocation points with collocation points. We must remember that a numerical solution depends on the model and the grid (and hence is not *unique*). The *convergence* of a method can be tested by increasing the number of panels, which should result in a converging solution. Therefore, it is always advisable to use smaller panels than the typical length of the geometry that we are modeling. In the case of the two-element airfoil, the typical distance is the gap clearance, and (if possible with the more refined methods) paneling this area by elements of at least one-tenth the size of the gap is recommended.

11.1 Point Singularity Solutions

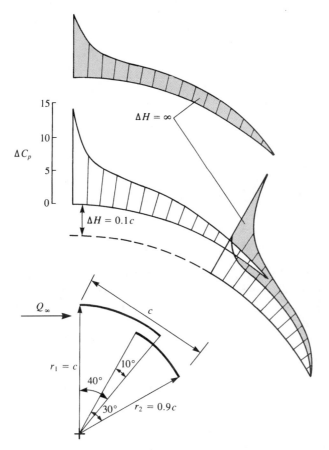

Figure 11.9 Effect of airfoil/flap proximity on their chordwise pressure difference. ΔH is the vertical spacing between the two elements.

Another important observation can be made by trying to calculate the velocity induced by the five-element vortex representation of the flat plate of Fig. 11.4. If the velocity survey is performed at $z = 0.05c$, then the wavy lines shown in Fig. 11.11 are obtained. This waviness will disappear at larger distances, and in any computation careful investigation is needed for the near and far field effects of a particular singular element distribution.

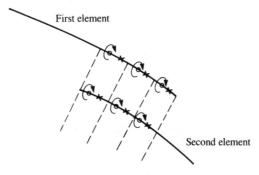

Figure 11.10 Method of paneling the gap region of a two-element airfoil (discrete-vortex model).

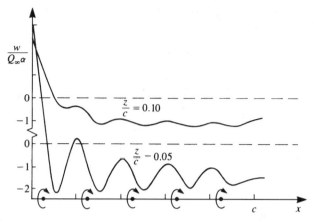

Figure 11.11 Survey of induced normal velocity above a thin airfoil (as shown in Fig. 11.4) modeled by discrete vortices.

Of course, in the previous examples, fairly accurate solutions were obtained with very few panels. This is because the lumped-vortex element induces the same downwash at the collocation point $(3a/4, 0)$ of a panel of length a as the exact flat plate solution, as depicted by Fig. 11.12.

11.1.2 Discrete Source Method

Based on the principles of the previous section, let us develop a discrete source method for solving the symmetric, nonzero-thickness airfoil problem of Section 5.1 (at $\alpha = 0$). For developing this method, too, let us follow the six steps suggested in Section 9.7 and apply them to the solution of the thin symmetric airfoil.

a. *Selection of Singularity Element*

The results of Chapter 5 indicate that the solution of the thin symmetric airfoil problem can be based on (discrete) source elements. The velocity induced by such an element placed at (x_j, z_j) and with a strength of σ_j is given by Eqs. (10.2) and (10.3) and

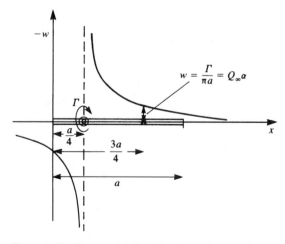

Figure 11.12 Downwash induced by a lumped-vortex element.

11.1 Point Singularity Solutions

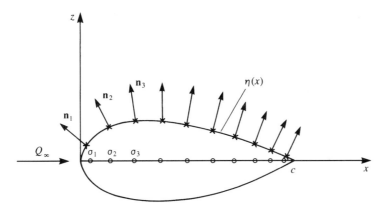

Figure 11.13 Discrete source model of symmetric airfoil at zero angle of attack.

can be expressed in matrix form as

$$\begin{pmatrix} u \\ w \end{pmatrix} = \frac{\sigma_j}{2\pi r_j^2} \begin{pmatrix} 1 & 0 \\ 0 & 1 \end{pmatrix} \begin{pmatrix} x - x_j \\ z - z_j \end{pmatrix} \qquad (11.14)$$

where

$$r_j^2 = (x - x_j)^2 + (z - z_j)^2$$

The above calculation can be included in a subroutine such that

$$(u, w) = \text{SORC2D}(\sigma_j, x, z, x_j, z_j) \qquad (11.15)$$

and (x, z) is the field point of interest.

b. Discretization of Geometry

First and most important is the definition of the coordinate system, which is shown in Fig. 11.13. Since the problem is symmetric, the unknown σ_j elements are placed along the x axis, at the center of N equal segments at $x_{j=1}, x_{j=2}, x_{j=3}, \ldots, x_{j=N}$.

Next, the collocation points need to be specified. In this case it is possible to leave these points on the airfoil surface as shown in Fig. 11.13, and the values of these points $(x_{i=1}, z_{i=1}), (x_{i=2}, z_{i=2}), \ldots, (x_{i=N}, z_{i=N})$ need to be established. The normal $\mathbf{n}_i$ pointing outward, at each of these points, is found from the surface shape $\eta(x)$, as defined by Eq. (11.3). As is demonstrated by the example at the end of this section, the solution can be improved considerably by moving the first and the last collocation points toward the leading and trailing edges, respectively (see Fig. 11.14).

c. Influence Coefficients

In this phase the zero normal flow boundary condition is implemented in a manner depicted by Eq. (11.4). For example, the velocity induced by the jth source element at the first collocation point can be obtained by using Eq. (11.15) and is

$$(u, w)_{1j} = \text{SORC2D}(\sigma_j, x_1, z_1, x_j, z_j) \qquad (11.16)$$

The influence coefficient a_{ij} is defined as the self-induced velocity component, of a unit strength source, normal to the surface. Consequently, the contribution of a unit strength singularity element $\sigma_j = 1$, at collocation point 1, is

$$a_{1j} = (u, w)_{1j} \cdot \mathbf{n}_1$$

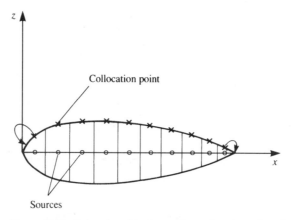

Figure 11.14 Relocation of the first and last collocation point to improve numerical solution with the discrete source method.

The induced normal velocity component q_{n1}, at point 1, due to all the elements is then

$$q_{n1} = a_{11}\sigma_1 + a_{12}\sigma_2 + a_{13}\sigma_3 + \cdots + a_{1N}\sigma_N$$

and the strength of σ_j is unknown at this point.

d. *Establish Boundary Condition (RHS)*

Fulfilling the boundary condition on the surface requires that at each collocation point the normal velocity component will vanish. Specifying this condition for the first collocation point yields

$$a_{11}\sigma_1 + a_{12}\sigma_2 + a_{13}\sigma_3 + \cdots + a_{1N}\sigma_N + (U_\infty, W_\infty) \cdot \mathbf{n}_1 = 0$$

where of course $W_\infty = 0$. But the last term (free-stream component) is known and can be transferred to the right-hand side of the equation. Using the definition of Eq. (11.6) for the right hand side we get

$$\text{RHS}_i = -(U_\infty, W_\infty) \cdot \mathbf{n}_i = -U_\infty \sin\alpha_i \qquad (11.6b)$$

If we specify the boundary condition for each of the collocation points we obtain a set of algebraic equations similar to those of the previous discrete vortex example:

$$\begin{pmatrix} a_{11} & a_{12} & \cdots & a_{1N} \\ a_{21} & a_{22} & \cdots & a_{2N} \\ a_{31} & a_{32} & \cdots & a_{3N} \\ \vdots & \vdots & \ddots & \vdots \\ a_{N1} & a_{N2} & \cdots & a_{NN} \end{pmatrix} \begin{pmatrix} \sigma_1 \\ \sigma_2 \\ \sigma_3 \\ \vdots \\ \sigma_N \end{pmatrix} = \begin{pmatrix} \text{RHS}_1 \\ \text{RHS}_2 \\ \text{RHS}_3 \\ \vdots \\ \text{RHS}_N \end{pmatrix}$$

This procedure is automated by a double DO loop where the collocation points are scanned first and then at each collocation point the influences of the singularity elements are scanned.

e. *Solve Equations*

The above set of algebraic equations can be solved for σ_i by using standard methods of linear algebra. It is assumed here that the reader is familiar with such methods, and as an example a direct solver can be found in the computer programs of Appendix D.

11.1 Point Singularity Solutions

f. *Calculation of Pressures and Loads*

Once the strength of the sources σ_j is known, the total tangential velocity Q_t at each collocation point can be calculated using Eq. (11.15) and Eq. (11.3a):

$$Q_{t_i} = \left[\sum_{j=1}^{N} (u, w)_{ij} + (U_\infty, W_\infty) \right] \cdot \mathbf{t}_i \tag{11.17}$$

The pressure coefficient then becomes

$$C_p = 1 - \frac{Q_t^2}{Q_\infty^2} \tag{11.18}$$

Since this flow is symmetric, neither lift nor drag will be produced (based on the conclusions of Section 5.1). Therefore, no further load calculations are included for this case. Also, note that for a closed body the net flow generated inside the body must be zero ($\sum_{i=1}^{N} \sigma_i = 0$), and this condition may be useful for evaluating numerical results.

Example 1: Fifteen-Percent-Thick Symmetric Airfoil

The above method is applied to the 15%-thick van de Vooren airfoil of Section 6.6. If the collocation points are left above the source points, as in Fig. 11.13, then the results shown by the triangles in Fig. 11.15 are obtained. This solution, clearly, is highly inaccurate near the leading edge. However, by moving the front collocation point more forward (to the 0.1 panel length location) and the rear collocation point closer to the trailing edge (to the 0.9 panel length location), as shown in Fig. 11.14 (and not moving the source points), we obtain a much improved solution. This solution, when compared with the exact solution of Section 6.6, is satisfactory over most of the region, excluding some minor problems near the trailing edge (Fig. 11.15).

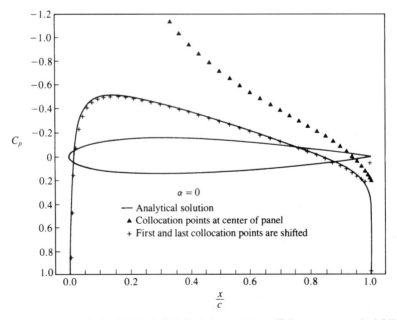

Figure 11.15 Calculated and analytical chordwise pressure coefficient on a symmetric airfoil ($\alpha = 0$): ▲ with collocation points above source points (triangles), and + with front collocation point moved forward and rear collocation point moved backward by 0.9 panel length, respectively.

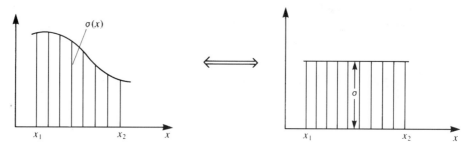

Figure 11.16 Constant-strength singularity approximation for a continuous strength distribution.

11.2 Constant-Strength Singularity Solutions (Using the Neumann B.C.)

A more refined discretization of a continuous singularity distribution is the element with a constant strength. This type of element is shown schematically in Fig. 11.16, and it is assumed that $\sigma \approx 1/(x_2 - x_1) \int_{x_1}^{x_2} \sigma(x) dx$, and as $(x_2 - x_1) \to 0$ the approximation seems to improve. In this case, too, only one constant (the strength of the element) is unknown, and by dividing the surface into N panels and specifying the boundary conditions on each of the collocation points, N linear algebraic equations can be constructed.

In principle, the point singularity methods are satisfactory in estimating the zero-thickness camberline lift, but they are inadequate near the stagnation points of a thick airfoil. The constant-strength methods are capable of more accuracy near the stagnation points and can be used to model closed surfaces with thickness resulting in a more detailed pressure distribution, which is essential for airfoil shape design.

11.2.1 Constant Strength Source Method

The constant-strength source methods that will be presented here are capable of calculating the pressures on a nonlifting thick airfoil and were among the first successful codes used.[10.1] For explaining the method, we shall follow the basic six step procedure.

a. Selection of Singularity Element

Consider the constant-strength source element of Section 10.2.1, where the panel is based on a flat surface element. To establish a normal-velocity boundary condition based method, only the induced velocity formulas are used (Eqs. (10.17) and (10.18)). The parameters θ and r are shown in Fig. 11.17, and the velocity components $(u, w)_p$ in the directions of the panel coordinate system are

$$u_p = \frac{\sigma}{4\pi} \ln \frac{r_1^2}{r_2^2} \tag{11.19}$$

$$w_p = \frac{\sigma}{2\pi}(\theta_2 - \theta_1) \tag{11.20}$$

In terms of the panel x, z variables these equations become

$$u_p = \frac{\sigma}{4\pi} \ln \frac{(x - x_1)^2 + z^2}{(x - x_2)^2 + z^2} \quad \text{(panel coordinates)} \tag{11.21}$$

$$w_p = \frac{\sigma}{2\pi}\left[\tan^{-1}\frac{z}{x - x_2} - \tan^{-1}\frac{z}{x - x_1}\right] \quad \text{(panel coordinates)} \tag{11.22}$$

11.2 Constant-Strength Singularity Solutions (Using the Neumann B.C.)

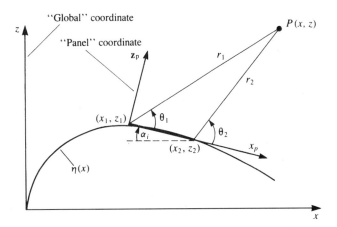

Figure 11.17 Nomenclature for a planar surface panel.

Note that for simplicity, the subscript p was omitted in these equations since in general it is obvious that the panel coordinates must be used (however, when the equations depend on the r, θ variables only, as in this case, the global x, z coordinates can be used as well). This approach will be taken in all following sections when presenting the influence terms of the panels. To transform these velocity components into the directions of the x, z global coordinates, a rotation by the panel orientation angle α_i is performed such that

$$\begin{pmatrix} u \\ w \end{pmatrix} = \begin{pmatrix} \cos \alpha_i & \sin \alpha_i \\ -\sin \alpha_i & \cos \alpha_i \end{pmatrix} \begin{pmatrix} u_p \\ w_p \end{pmatrix} \tag{11.23}$$

For later applications when the coordinates of the point P must be transformed into the panel coordinate system the following transformation can be used:

$$\begin{pmatrix} x \\ z \end{pmatrix}_p = \begin{pmatrix} \cos \alpha_i & -\sin \alpha_i \\ \sin \alpha_i & \cos \alpha_i \end{pmatrix} \begin{pmatrix} x - x_0 \\ z - z_0 \end{pmatrix} \tag{11.23a}$$

In this case (x_0, z_0) are the coordinates of the panel origin in the global coordinate system x, z and the subscript p stands for panel coordinates.

This procedure (e.g., Eqs. (11.21) and (11.22) and the transformation of Eq. (11.23)) can be included in an induced-velocity subroutine SOR2DC (where C stands for constant), which will compute the velocity (u, w) at an arbitrary point (x, z) in the global coordinate system due to the jth element whose endpoints are identified by the j and the $j + 1$ counters:

$$(u, w) = \text{SOR2DC}(\sigma_j, x, z, x_j, z_j, x_{j+1}, z_{j+1}) \tag{11.24}$$

b. Discretization of Geometry

As an example for this method, the 15%-thick symmetric airfoil of Section 6.6 is considered. In most cases involving thick airfoils, a more dense paneling is used near the leading and trailing edges. A frequently used method for dividing the chord into panels with larger density near the edges is shown in Fig. 11.18. If ten chordwise panels are needed, then the semicircle is divided by this number; thus $\Delta \beta = \pi/10$. The corresponding x stations are found by using the following cosine spacing formula:

$$x = \frac{c}{2}(1 - \cos \beta) \tag{11.25}$$

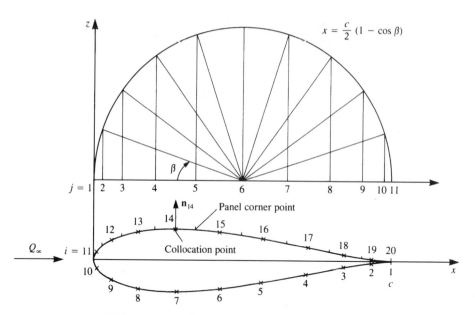

Figure 11.18 "Full-cosine" method of spacing the panels on the airfoil's surface.

Once the x axis is divided into N panels with strength σ_j, the $N+1$ panel corner points $(x_{j=1}, z_{j=1}), (x_{j=2}, z_{j=2}), \ldots, (x_{j=N+1}, z_{j=N+1})$ are computed. The collocation points can be placed at the center of each panel (shown by the x mark on the airfoil surface in Fig. 11.18) and the values of these points $(x_{i=1}, z_{i=1}), (x_{i=2}, z_{i=2}), \ldots, (x_{i=N}, z_{i=N})$ are computed too. The normal $\mathbf{n}_i$, which points outward at each of these points, is found from the surface shape $\eta(x)$, as defined by Eq. (11.3).

c. *Influence Coefficients*

In this phase the zero normal flow boundary condition is implemented. For example, the velocity induced by the jth source element at the first collocation point can be obtained by using Eq. (11.24) and is

$$(u, w)_{1j} = \text{SOR2DC}(\sigma_j, x_1, z_1, x_j, z_j, x_{j+1}, z_{j+1}) \tag{11.26}$$

The influence coefficient a_{ij} is defined as the velocity component normal to the surface. Consequently, the contribution of a unit strength singularity element j, at collocation point 1, is

$$a_{1j} = (u, w)_{1j} \cdot \mathbf{n}_1 \tag{11.27}$$

Note that a closer observation of Eqs. (11.3) and (11.23) shows that the normal velocity component at the ith panel is found by rotating the velocity induced by a unit strength j element by $(\alpha_j - \alpha_i)$; therefore

$$a_{1j} = [-\sin(\alpha_j - \alpha_1), \cos(\alpha_j - \alpha_1)] \begin{pmatrix} u_{1j} \\ w_{1j} \end{pmatrix}_p \tag{11.27a}$$

Here the velocity components $(u, w)_p$ are obtained from Eqs. (11.21) and (11.22). To

11.2 Constant-Strength Singularity Solutions (Using the Neumann B.C.)

evaluate the influence of the element on itself, recall Eq. (10.24):

$$w_p(x, 0\pm) = \pm\frac{\sigma}{2} \tag{11.28}$$

Based on this equation, the boundary condition (e.g., in Eq. (11.4)) will be specified at a point slightly above the surface ($z = 0+$ in the panel frame of reference). Consequently, when $i = j$ the influence coefficient becomes

$$a_{ii} = \frac{1}{2} \tag{11.29}$$

d. *Establish Boundary Condition (RHS)*

Specifying the boundary condition, as stated in Eq. (11.4), at collocation point 1, results in the following algebraic equation:

$$\frac{1}{2}\sigma_1 + a_{12}\sigma_2 + a_{13}\sigma_3 + \cdots + a_{1N}\sigma_N + (U_\infty, W_\infty) \cdot \mathbf{n}_1 = 0$$

where of course $W_\infty = 0$ for the symmetric airfoil case. The free-stream normal velocity component is transferred to the right-hand side and the vector RHS_i is found, as in the previous example (by using Eq. (11.6b)):

$$\text{RHS}_i = -U_\infty \sin \alpha_i$$

Both the influence coefficients and the RHS vector can be computed by a double DO loop where the collocation points are scanned first (and the RHS_i vector is calculated) and then at each collocation point the influences of the singularity elements are scanned.

e. *Solve Equations*

Specifying the boundary condition for each ($i = 1 \to N$) of the collocation points results in a set of algebraic equations with the unknown σ_j ($j = 1 \to N$). These equations will have the form

$$\begin{pmatrix} \frac{1}{2} & a_{12} & \cdots & \cdots & a_{1N} \\ a_{21} & \frac{1}{2} & \cdots & \cdots & a_{2N} \\ a_{31} & a_{32} & \frac{1}{2} & \cdots & a_{3N} \\ \vdots & \vdots & \cdots & \ddots & \vdots \\ a_{N1} & a_{N2} & \cdots & \cdots & \frac{1}{2} \end{pmatrix} \begin{pmatrix} \sigma_1 \\ \sigma_2 \\ \sigma_3 \\ \vdots \\ \sigma_N \end{pmatrix} = \begin{pmatrix} \text{RHS}_1 \\ \text{RHS}_2 \\ \text{RHS}_3 \\ \vdots \\ \text{RHS}_N \end{pmatrix}$$

The above set of algebraic equations has a well-defined diagonal and can be solved for σ_j by using standard methods of linear algebra.

f. *Calculation of Pressures and Loads*

Once the strength of the sources σ_j is known, the velocity at each collocation point can be calculated using Eq. (11.24) and the pressure coefficient can be calculated by using Eq. (11.18).

Note that this method is derived here for nonlifting shapes and the Kutta condition is not used. Consequently, the circulation of the airfoil will be zero and hence no lift and drag will be produced. However, the pressure distribution is well predicted as shown in the following Example 1.

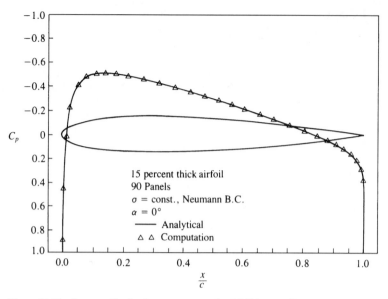

Figure 11.19 Pressure distribution on a symmetric airfoil (at $\alpha = 0$).

The numeric formulation presented here does not assume a symmetric solution. But, as it appears, the solution is symmetric (about the x axis) and the number of unknowns can be reduced to $N/2$ by a minor modification in the process of the influence coefficient calculation (in Eq. (11.27)). In this case the velocity induced by the panel $(u, w)_{ij}$ and by its mirror image $(u, w)^i_{ij}$ will be calculated by using Eq. (11.24):

$$(u, w)_{ij} = \text{SOR2DC}(\sigma_j = 1, x_i, z_i, x_j, z_j, x_{j+1}, z_{j+1})$$

$$(u, w)^i_{ij} = \text{SOR2DC}(\sigma_j = 1, x_i, z_i, x_j, -z_j, x_{j+1}, -z_{j+1})$$

and the influence coefficient a_{ij} is then

$$a_{ij} = \left[(u, w)_{ij} + (u, w)^i_{ij}\right] \cdot \mathbf{n}_i$$

The rest of the procedure is unchanged, but with this modification we end up with only $N/2$ unknowns σ_j (e.g., for the upper surface only).

Example 1: Pressure Distribution on a Symmetric Airfoil

The above method is applied to the 15%-thick symmetric van de Vooren airfoil of Section 6.6. The computed pressure distribution is shown by the triangles in Fig. 11.19 and they agree well with the exact analytical results, including those at the leading and trailing edge regions.

Note that in this case, too, for a closed body the sum of the sources must be zero ($\sum_{i=1}^{N} \sigma_i = 0$), and this condition may be useful for evaluating numerical results.

A sample student computer program used for this calculation is provided in Appendix D (Program No. 3).

11.2.2 *Constant-Strength Doublet Method*

The simplest two-dimensional panel code that can calculate the flow over thick lifting airfoils is based on the constant-strength doublet. The surface pressure distribution computed by this method is satisfactory on the surface, but since this element is equivalent

11.2 Constant-Strength Singularity Solutions (Using the Neumann B.C.)

to two concentrated vortices at the edges of the element, near-field off-surface velocity computations will have the same fluctuations as shown in Fig. 11.11 (but the velocity calculated at the collocation point and the resulting pressure distribution are correct).

a. *Selection of Singularity Element*

Consider the constant-strength doublet element of Section 10.2.2 pointing in the positive z direction, where the panel is based on a flat element. To establish a normal-velocity boundary condition based method, the induced velocity formulas of Eqs. (10.29) and (10.30) are used (which are equivalent to two point vortices with a strength μ at the panel edges):

$$u_p = \frac{\mu}{2\pi} \left[\frac{z}{(x-x_1)^2 + z^2} - \frac{z}{(x-x_2)^2 + z^2} \right] \quad \text{(panel coordinates)} \quad (11.30)$$

$$w_p = \frac{-\mu}{2\pi} \left[\frac{x-x_1}{(x-x_1)^2 + z^2} - \frac{x-x_2}{(x-x_2)^2 + z^2} \right] \quad \text{(panel coordinates)} \quad (11.31)$$

Here, again, the velocity components $(u, w)_p$ are in the direction of the panel local coordinates, which need to be transformed back to the x, z system by Eq. (11.23).

This procedure can be included in an induced-velocity subroutine DUB2DC (where C stands for constant), which will compute the velocity (u, w) at an arbitrary point (x, z) due to the jth element:

$$(u, w) = \text{DUB2DC}(\mu_j, x, z, x_j, z_j, x_{j+1}, z_{j+1}) \quad (11.32)$$

b. *Discretization of Geometry*

The panel corner points and collocation points are generated exactly as in the previous section (see Fig. 11.18). However, in this lifting case, a wake panel (shown in Fig. 11.20) has to be specified. This doublet element will have a strength μ_W and extends to $x = \infty$. In practice, the far portion (starting vortex) of the wake will have no influence and can be "placed" far downstream (e.g., at $(\infty, 0)$).

c. *Influence Coefficients*

To obtain the normal component of the velocity at a collocation point (e.g., the first point) due to the jth doublet element, Eq. (11.32) is used:

$$(u, w)_{1j} = \text{DUB2DC}(\mu_j, x_1, z_1, x_j, z_j, x_{j+1}, z_{j+1}) \quad (11.33)$$

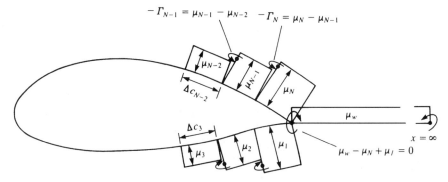

Figure 11.20 Schematic description of constant-strength doublet panel elements near an airfoil's trailing edge.

The influence coefficients a_{ij} are defined as the velocity components normal to the surface. Consequently, the contribution of a unit strength singularity element j, at collocation point 1, is

$$a_{1j} = (u, w)_{1j} \cdot \mathbf{n}_1$$

Similarly to the case of the constant-source method, the influence coefficients can be found by using Eq. (11.27):

$$a_{1j} = [-\sin(\alpha_j - \alpha_1), \cos(\alpha_j - \alpha_1)] \begin{pmatrix} u_{1j} \\ w_{1j} \end{pmatrix}_p \quad (11.27)$$

where α_1 and α_j are the first and the jth panel angles, as defined in Fig. 11.17, and $(u_{1j}, w_{1j})_p$ are the velocity components of Eqs. (11.30) and (11.31) due to a unit strength element, as measured in the panel frame of reference.

To evaluate the influence of the element on itself, at the center of the panel, we recall Eqs. (10.32) and (10.33):

$$u_p(x, 0\pm) = 0 \quad (11.34)$$

$$w_p(x, 0\pm) = \frac{-\mu}{\pi} \frac{2}{(x_2 - x_1)} \quad (11.35)$$

Consequently, when $i = j$ the influence coefficient becomes

$$a_{ii} = \frac{-2}{\pi \Delta c_i} \quad (11.35a)$$

where Δc_i is the ith panel chord.

d. *Establish Boundary Condition (RHS)*

The free-stream normal velocity component RHS_i is found as in the previous examples (e.g., by using Eq. (11.6)).

e. *Solve Equations*

Specification of the boundary condition of Eq. (11.4) for each ($i = 1 \to N$) of the collocation points results in a set of algebraic equations with the unknown μ_j ($j = 1 \to N$). However, the equivalent vortex representation in Fig. 11.20 reveals that the strength of the vortex at the trailing edge is $-\Gamma = \mu_1 - \mu_N$. Since the Kutta condition requires the circulation at the trailing edge to be zero, we must add a wake panel to cancel this vortex:

$$(\mu_1 - \mu_N) + \mu_W = 0 \quad (11.36)$$

A combination of this equation with the N boundary conditions results in $N + 1$ linear equations:

$$\begin{pmatrix} a_{11} & a_{12} & \cdots & a_{1N} & a_{1W} \\ a_{21} & a_{22} & \cdots & a_{2N} & a_{2W} \\ \vdots & \vdots & \ddots & \vdots & \vdots \\ a_{N1} & a_{N2} & \cdots & a_{NN} & a_{NW} \\ 1 & 0 & 0 \cdots & -1 & 1 \end{pmatrix} \begin{pmatrix} \mu_1 \\ \mu_2 \\ \vdots \\ \mu_N \\ \mu_W \end{pmatrix} = \begin{pmatrix} RHS_1 \\ RHS_2 \\ \vdots \\ RHS_N \\ 0 \end{pmatrix}$$

This system of equations is the numerical equivalent of the boundary condition (Eq. (11.4)) and is well defined and will have a stable solution.

f. Calculation of Pressures and Loads

Once the strength of the doublets μ_j is known, the perturbation tangential velocity component at each collocation point can be calculated by summing the induced velocities of all the panels, using Eq. (11.33). The tangential velocity at collocation point i is then

$$q_{t_i} = \sum_{j=1}^{N+1}(u,w)_{ij} \cdot \mathbf{t}_i \qquad (11.37)$$

where $(u,w)_{ij}$ is the result of Eq. (11.33), $\mathbf{t}_i$ is the local tangent vector defined by Eq. (11.3a), and the $(N+1)$-th component is due to the wake. Note that to evaluate the tangential velocity component induced by a panel on itself Eq. (3.141) can be used:

$$q_t = -\frac{1}{2}\frac{\partial \mu(l)}{\partial l} \quad \text{on panel} \qquad (11.38)$$

where l represents distance along a surface line. So when evaluating the tangential component of the perturbation velocity the result of Eq. (11.38) must be included (when $i = j$ in Eq. (11.37)). The pressure coefficient can be computed by using Eq. (11.18),

$$C_{p_j} = 1 - \frac{(Q_{t_\infty} + q_t)_j^2}{Q_\infty^2} \qquad (11.39)$$

where

$$(Q_{t_\infty})_j = \mathbf{t}_j \cdot \mathbf{Q}_\infty \qquad (11.39a)$$

Note that the local lift can be calculated, too, by using the Kutta–Joukowski formula for a point vortex:

$$\Delta L_j = \rho Q_\infty \Gamma_j = -\rho Q_\infty (\mu_{j+1} - \mu_j) \qquad (11.40)$$

where the minus sign is used for doublet panels pointing outside the airfoil. This formulation should be equivalent to the result that we get by assuming constant pressure on the panel, namely

$$\Delta C_{l_j} = -C_{p_j} \Delta l_j \cos\alpha_j / c \qquad (11.41)$$

where Δl_j and α_j are shown in Fig. 11.21. The total lift and moment are obtained by summing the contribution of each element:

$$L = \sum_{j=1}^{N} \Delta L_j \qquad (11.42)$$

$$M_0 = -\sum_{j=1}^{N} \Delta L_j (x_j \cos\alpha) \qquad (11.43)$$

and the nondimensional coefficients can be calculated by using Eqs. (11.12) and (11.13). Note that by observing the wake vortex at $x = \infty$ in Fig. 11.20 and recalling Kelvin's theorem (Eq. (2.16)), we can compute the total lift simply from the wake doublet strength as

$$L = -\rho Q_\infty \mu_W \qquad (11.42a)$$

Example: Lifting Thick Airfoil

The above method was used for computing the pressure distribution over the 15%-thick van de Vooren airfoil of Section 6.6, as shown in Fig. 11.22. The data agree

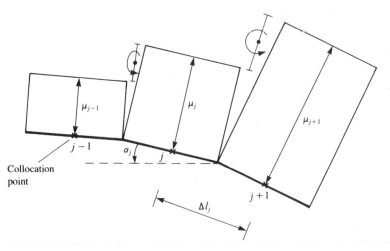

Figure 11.21 Typical segment of constant-strength doublet panels on the airfoil's surface.

satisfactorily with the analytic solution for both the 0° (Fig. 11.22a) and 5° angle-of-attack conditions (Fig. 11.22b). A slight disagreement is visible near the maximum suction area and near the rear stagnation point. These results can be improved by moving the grid and the collocation points near these areas, but such an optimization procedure is not carried out here. The solution near the trailing edge can also be improved by using the velocity formulation (Eq. (9.15b)) for the Kutta condition.

The computer program used for this example is included in Appendix D (Program No. 4).

11.2.3 Constant-Strength Vortex Method

The constant-strength vortex distribution was shown to be equivalent to a linear-strength doublet distribution (Section 10.3.2) and therefore is expected to improve the solution of the flow over thick bodies. However, this method is more difficult to use successfully compared to the other methods presented here. One of the problems arises from the fact that the self-induced effect (Eq. (10.43)) of this panel is zero at the center of the element (and the influence coefficient matrix, without a pivoting scheme, will have a zero diagonal). Also, when using the Kutta condition at an airfoil's trailing edge (Fig. 11.23) the requirement that $\gamma_1 + \gamma_N = 0$ eliminates the lift of the two trailing-edge panels. Consequently, if N panels are used, then only $N - 2$ independent equations can be used and the scheme can not work without certain modifications to the method. One such modification is presented in Ref. 5.1 (pp. 281–282) where additional conditions are found by minimizing a certain error function. In this section, we try to use an approach similar to the previous source and doublet methods, and only the specifications of the boundary conditions will be modified. We will follow the basic six-step procedure of the previous sections.

a. *Selection of Singularity Element*

Consider the constant-strength vortex element of Section 10.2.3, where the panel is based on a flat surface element. To establish a normal-velocity boundary condition based

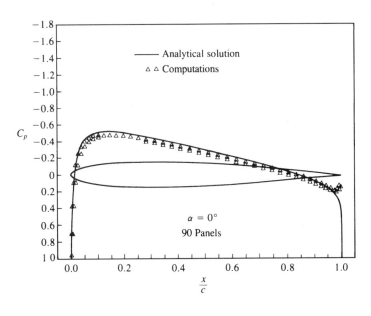

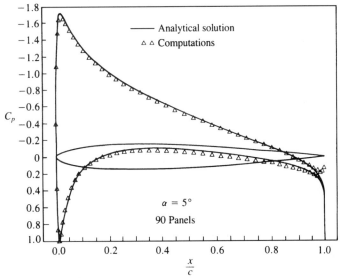

Figure 11.22 Chordwise pressure distribution on a symmetric airfoil at angles of attack of 5° and 0°.

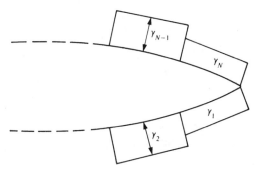

Figure 11.23 Constant-strength vortex panels near the trailing edge of an airfoil.

method, only the induced-velocity formulas are used (Eqs. (10.39) and (10.40)):

$$u_p = \frac{\gamma}{2\pi}\left[\tan^{-1}\frac{z-z_2}{x-x_2} - \tan^{-1}\frac{z-z_1}{x-x_1}\right] \quad \text{(panel coordinates)} \qquad (11.44)$$

$$w_p = -\frac{\gamma}{4\pi}\ln\frac{(x-x_1)^2+(z-z_1)^2}{(x-x_2)^2+(z-z_2)^2} \quad \text{(panel coordinates)} \qquad (11.45)$$

Here, again, the velocity components $(u, w)_p$ are in the direction of the panel local coordinates, which need to be transformed back to the x, z system by Eq. (11.23).

This procedure can be included in an induced-velocity subroutine VOR2DC (where C stands for constant), which will compute the velocity (u, w) at an arbitrary point (x, z) due to the jth element:

$$(u, w) = \text{VOR2DC}(\gamma_j, x, z, x_j, z_j, x_{j+1}, z_{j+1}) \qquad (11.46)$$

b. *Discretization of Geometry*

To generate the panel corner points $(x_{j=1}, z_{j=1}), (x_{j=2}, z_{j=2}), \ldots, (x_{j=N+1}, z_{j=N+1})$, collocation points $(x_{i=1}, z_{i=1}), (x_{i=2}, z_{i=2}), \ldots, (x_{i=N}, z_{i=N})$ placed at the center of each panel, and the normal vectors $\mathbf{n}_i$, the procedure of the previous section can be used (see Fig. 11.18).

c. *Influence Coefficients*

A possible modification of the boundary condition, which will eliminate the zero self-induced effect, is to use an internal zero tangential velocity boundary condition. This is based on Eq. (9.8), which states that inside an enclosed body $\Phi_i^* = $ const. Consequently, the normal and tangential derivatives of the total potential inside the body are zero:

$$\frac{\partial \Phi^*}{\partial n} = \frac{\partial \Phi^*}{\partial l} = 0 \qquad (11.47)$$

In this particular case the inner tangential velocity condition will be used and at each panel

$$(U_\infty + u, W_\infty + w)_i \cdot (\cos\alpha_i, -\sin\alpha_i) = 0 \qquad (11.47a)$$

To specify this condition at each of the collocation points (which are now at the center of the panel and slightly inside), the tangential velocity component is obtained by using Eq. (11.46). For example, the velocity at a collocation point 1 due to the jth vortex element is

$$(u, w)_{1j} = \text{VOR2DC}(\gamma_j, x_1, z_1, x_j, z_j, x_{j+1}, z_{j+1}) \qquad (11.48)$$

The influence coefficient a_{ij} is now defined as the velocity component tangent to the surface. Consequently, the contribution of a unit strength singularity element j, at collocation point 1, is

$$a_{1j} = (u, w)_{1j} \cdot (\cos\alpha_1, -\sin\alpha_1)$$

where α_1 is the orientation of the panel (of the collocation point) as shown in Fig. 11.17. The general influence coefficient is then

$$a_{ij} = (u, w)_{ij} \cdot (\cos\alpha_i, -\sin\alpha_i) \qquad (11.49)$$

11.2 Constant-Strength Singularity Solutions (Using the Neumann B.C.)

Use of this boundary condition ensures a nonzero value for the self-induced influence of the panel. At the center of the panel, Eqs. (10.42) and (10.43) are recalled,

$$u_p(x, 0\pm) = \pm\frac{\gamma}{2}$$
$$w_p(x, 0\pm) = 0$$

Consequently, when $i = j$ the influence coefficient becomes

$$a_{ii} = -\frac{1}{2} \tag{11.50}$$

d. Establish Boundary Condition (RHS)

The free-stream tangential velocity component RHS_i is found by

$$RHS_i = -(U_\infty, W_\infty) \cdot (\cos\alpha_i, -\sin\alpha_i) \tag{11.51}$$

Note that in this case the free stream may have an angle of attack. The numerical procedure (using the double DO loop routine) for calculating the influence coefficients and the RHS_i vector is the same as for the previous methods.

e. Solve Equations

Specifying the boundary condition for each ($i = 1 \to N$) of the collocation points results in a set of algebraic equations with the unknowns γ_j ($j = 1 \to N$). In addition the Kutta condition needs to be specified at the trailing edge:

$$\gamma_1 + \gamma_N = 0 \tag{11.52}$$

But now we have $N + 1$ equations with only N unknowns. Therefore, one of the equations must be deleted (e.g., the kth equation) and by adding the Kutta condition the following matrix equation is obtained:

$$\begin{pmatrix} a_{11} & \cdots & & \cdots & a_{1N} \\ a_{21} & a_{22} & & \cdots & a_{2N} \\ \vdots & \vdots & \ddots & & \vdots \\ a_{N-1,1} & a_{N-1,2} & \cdots & & a_{N-1,N} \\ 1 & 0 & 0 & \cdots & 1 \end{pmatrix} \begin{pmatrix} \gamma_1 \\ \gamma_2 \\ \gamma_3 \\ \vdots \\ \gamma_N \end{pmatrix} = \begin{pmatrix} RHS_1 \\ RHS_2 \\ \vdots \\ RHS_{N-1} \\ 0 \end{pmatrix}$$

f. Calculation of Pressures and Loads

Once the strength of the vortices γ_j is known, the velocity at each collocation point can be calculated using Eq. (11.48) and the pressure coefficient can be calculated by using Eq. (11.18) (note that the tangential perturbation velocity at each panel is $\gamma_j/2$):

$$C_p = 1 - \left[\frac{Q_\infty \cos(\alpha + \alpha_j) + \gamma_j/2}{Q_\infty}\right]^2 \tag{11.53}$$

where $\mathbf{Q}_\infty \cdot \mathbf{t}_j = Q_\infty \cos(\alpha + \alpha_j)$. The aerodynamic loads can be calculated by adding the pressure coefficient or by using the Kutta–Joukowski theorem. Thus the lift of the jth panel is

$$\Delta L_j = \rho Q_\infty \gamma_j \Delta c_j$$

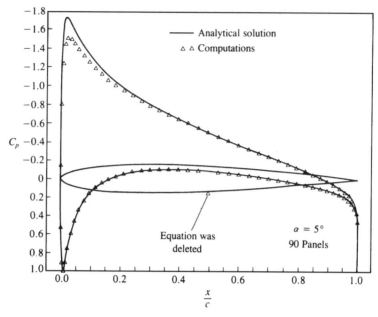

Figure 11.24 Chordwise pressure distribution on a symmetric airfoil at angle of attack of 5° using constant-strength vortex panels.

where Δc_j is the panel length. The total lift and moment are obtained by summing the contribution from each element,

$$L = \sum_{j=1}^{N} \Delta L_j \tag{11.54}$$

$$M_0 = -\sum_{j=1}^{N} \Delta L_j (x_j \cos \alpha) \tag{11.55}$$

and the nondimensional coefficients can be calculated by using Eqs. (11.12) and (11.13).

Example: Symmetric Thick Airfoil at Angle of Attack

The above method is applied to the 15%-thick, symmetric, van de Vooren airfoil of Section 6.6. The computed pressure distribution is shown by the triangles in Fig. 11.24 and they agree fairly well with the exact analytical results. The point where the computations disagree is where one equation was deleted. This can easily be corrected by a local smoothing procedure, but the purpose of this example is to highlight this problem. From the practical point of view it is better to use panels with a higher order (e.g., linear) vortex distribution or any of the following methods.

The sample student computer program used for this calculation is provided in Appendix D (Program No. 5).

11.3 Constant-Potential (Dirichlet Boundary Condition) Methods

In the previous examples the direct, zero normal velocity (Neumann) boundary condition was used. In this section similar methods will be formulated based on the constant-potential method (Dirichlet boundary condition). This condition was described in detail in

11.3 Constant-Potential (Dirichlet Boundary Condition) Methods

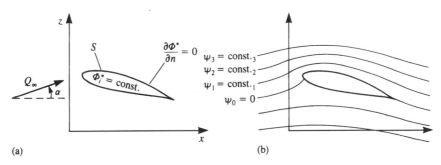

Figure 11.25 Methods of fulfilling the zero normal velocity boundary condition on a solid surface.

Section 9.2 and in principle it states that if $\partial \Phi^*/\partial n = 0$ on the surface of a closed body then the internal potential Φ_i^* must stay constant (Fig. 11.25a):

$$\Phi_i^* = \text{const.} \tag{11.56}$$

It is possible to specify this boundary condition in terms of the stream function Ψ (Fig. 11.25b) and in this case the body shape is enclosed by the stagnation streamline where $\Psi = \text{const.}$ (which may be selected as zero). Many successful numerical methods are based on the stream function and they are very similar to the methods described in this chapter. Also, the stream function can describe flows that are rotational, but an equivalent three-dimensional formulation of such methods is nonexistent. Because of the lack of three-dimensional capability, only the velocity potential based solutions will be discussed here.

Following Chapter 9, the velocity potential can be divided into a free-stream potential Φ_∞ and perturbation potential Φ, and the zero normal velocity boundary condition on a solid surface (internal Dirichlet condition) is

$$\Phi_i^* = (\Phi + \Phi_\infty)_i = \text{const.}$$

If we place the singularity distribution on the boundary S (and following the two-dimensional equivalent of Eq. (9.10) – see Eq. (3.17)) this internal boundary condition becomes

$$\Phi_i^*(x, z) = \frac{1}{2\pi} \int_S \left[\sigma \ln r - \mu \frac{\partial}{\partial n} (\ln r) \right] dS + \Phi_\infty = \text{const.} \tag{11.57}$$

and when the point (x, z) is on the surface then the coefficient $1/2\pi$ becomes $1/\pi$.

This formulation is not unique and the combination of source and doublet distributions must be fixed. For example, source-only or doublet-only solutions can be used with this internal boundary condition, but when using both types of singularity, the strength of one must be prescribed. Also, any vortex distribution can be replaced by an equivalent doublet distribution, and therefore solutions based on vortices can be used too.

To construct a numerical solution the surface S is divided into N panels and the integration is performed for each panel such that

$$\sum_{j=1}^{N} \frac{1}{2\pi} \int_{\text{panel}} \sigma \ln r \, dS - \sum_{j=1}^{N} \frac{1}{2\pi} \int_{\text{panel}} \mu \frac{\partial}{\partial n} (\ln r) \, dS + \Phi_\infty = \text{const.}$$

The integration is limited now to each individual panel element, and for constant, linear, and quadratic strength elements this was done in Chapter 10. For example, in the case of constant-strength singularity elements on each panel the influence of panel j at a point P is

$$-\frac{1}{2\pi}\int_{\text{panel}}\frac{\partial}{\partial n}(\ln r)\,dS\Big|_j \equiv C_j \tag{11.58}$$

In the case of a doublet distribution and for a source distribution

$$\frac{1}{2\pi}\int_{\text{panel}}(\ln r)\,dS\Big|_j \equiv B_j \tag{11.59}$$

Once these influence integrals have been evaluated (as in Chapter 10) the boundary condition inside the surface (at any point) becomes

$$\sum_{j=1}^{N} B_j \sigma_j + \sum_{j=1}^{N} C_j \mu_j + \Phi_\infty = \text{const.} \quad \text{for each collocation point} \tag{11.60}$$

Specifying this boundary condition on N collocation points allows N linear equations to be created.

11.3.1 Combined Source and Doublet Method

As the first example for this approach let us use the combination of source and doublet elements on the surface. This means that each panel will have a local source and doublet strength of its own. Since Eq. (11.60) is not unique, either the source or the doublet values must be specified. Here the inner potential is selected to be equal to Φ_∞ and for this case the source strength is given by Eq. (9.12) as

$$\sigma_j = \mathbf{n}_j \cdot \mathbf{Q}_\infty \tag{11.61}$$

Since the value of the inner perturbation potential was set to zero (or $\Phi_i^* = \Phi_\infty$) Eq. (11.60) reduces to

$$\sum_{j=1}^{N} B_j \sigma_j + \sum_{j=1}^{N} C_j \mu_j = 0 \tag{11.62}$$

and μ_j represents the jump in the perturbation potential. This equation (boundary condition) is specified at each collocation point *inside* the body, providing a linear algebraic equation for this point. The steps toward establishing such a numerical solution are as follows:

a. *Selection of Singularity Element*

The velocity potential at an arbitrary point P (not on the surface) due to a constant-strength source was derived in the panel's frame of reference in Eq. (10.19):

$$\Phi = \frac{\sigma}{4\pi}\left\{(x-x_1)\ln[(x-x_1)^2+z^2] - (x-x_2)\ln[(x-x_2)^2+z^2] \right. \\ \left. + 2z\left(\tan^{-1}\frac{z}{x-x_2} - \tan^{-1}\frac{z}{x-x_1}\right)\right\} \quad \text{(panel coordinates)} \tag{11.63}$$

and that due to a constant-strength doublet in Eq. (10.28):

$$\Phi = \frac{-\mu}{2\pi}\left[\tan^{-1}\frac{z}{x-x_2} - \tan^{-1}\frac{z}{x-x_1}\right] \quad \text{(panel coordinates)} \tag{11.64}$$

11.3 Constant-Potential (Dirichlet Boundary Condition) Methods

These equations can be included in two subroutines that calculate the potential at point (x, z) due to the source and doublet element j:

$$\Delta\Phi_s = \text{PHICS}(\sigma_j, x, z, x_j, z_j, x_{j+1}, z_{j+1}) \tag{11.65}$$

$$\Delta\Phi_d = \text{PHICD}(\mu_j, x, z, x_j, z_j, x_{j+1}, z_{j+1}) \tag{11.66}$$

These subroutines will include the transformation of the point (x, z) into the panel coordinates (e.g., in Eq. (11.23a)) and it is assumed that these potential increments are expressed in terms of the global x, z coordinates. However, since the influence coefficients depend on view angles and distances between points (see Fig. 10.6), the transformation of $\Delta\Phi$ back to the global coordinate system can be skipped.

b. *Discretization of Geometry*

The $N + 1$ panel corner points and N collocation points are generated in a manner similar to the previous example of the constant-strength source (Fig. 11.18). However, now the internal Dirichlet boundary condition will be applied and therefore the collocation points must be placed inside the body. (Usually an inward displacement of 0.05 panel lengths is sufficient, but attention is needed near the trailing edge so that the collocation point is not placed outside the body. In the case where the self-induced influence is specified by a separate formula, then for simplicity, the collocation point can be left at the center of the panel surface without the inward displacement.)

c. *Influence Coefficients*

The increment in the velocity potential at collocation point i due to a unit strength constant-source element of panel j is obtained by using Eq. (11.65):

$$b_{ij} = \text{PHICS}(\sigma_j = 1, x_i, z_i, x_j, z_j, x_{j+1}, z_{j+1}) \tag{11.67}$$

and that due to the same panel but with a unit strength doublet is

$$c_{ij} = \text{PHICD}(\mu_j = 1, x_i, z_i, x_j, z_j, x_{j+1}, z_{j+1}) \tag{11.68}$$

Note that this calculation is simpler (requiring less algebraic operations) than comparable calculations using the velocity boundary condition, which require the computation of two velocity components and a multiplication by the local normal vector.

Also, the influence of the doublet panel on itself (using Eq. (10.31)) is

$$c_{ii} = \frac{1}{2} \tag{11.69}$$

and for the source the self-induced effect can be calculated by using Eq. (11.67).

Determination of the influence of the doublets at each of the collocation points will result in an $N \times N$ influence matrix, with $N + 1$ unknowns (where the wake doublet μ_W is the $(N + 1)$-th unknown). The additional equation is provided by using the Kutta condition (see Fig. 11.20):

$$(\mu_1 - \mu_N) + \mu_W = 0 \tag{11.36}$$

Combining this equation with the influence matrix will result in $N + 1$ linear equations for the influence of the doublets:

$$\sum_{j=1}^{N+1} C_{ij}\mu_j = \begin{pmatrix} c_{11} & c_{12} & \cdots & c_{1N} & c_{1W} \\ c_{21} & c_{22} & \cdots & c_{2N} & c_{2W} \\ \vdots & \vdots & \ddots & \vdots & \vdots \\ c_{N1} & c_{N2} & \cdots & c_{NN} & c_{NW} \\ 1 & 0 & 0 & \cdots & -1 & 1 \end{pmatrix} \begin{pmatrix} \mu_1 \\ \mu_2 \\ \vdots \\ \mu_N \\ \mu_W \end{pmatrix}$$

If we replace μ_W with $\mu_N - \mu_1$ from Eq. (11.36), we can reduce the order of the above matrix to N. The first row, for example, will have the form

$$(c_{11} - c_{1W})\mu_1 + c_{12}\mu_2 + \cdots + (c_{1N} + c_{1W})\mu_N$$

and only the first and the Nth columns will change because of the term $\pm c_{iW}$. We can rewrite the doublet influence such that

$$\begin{aligned} a_{ij} &= c_{ij}, & j &\neq 1, N \\ a_{i1} &= c_{i1} - c_{iW}, & j &= 1 \\ a_{iN} &= c_{iN} + c_{iW}, & j &= N \end{aligned} \quad (11.70)$$

With this definition of the doublet coefficients and with the b_{ij} coefficients of the source influence, Eq. (11.62), specified for each collocation point $1 \to N$, the matrix equation will have the form

$$\begin{pmatrix} a_{11}, & a_{12}, & \ldots, & a_{1N} \\ a_{21}, & a_{22}, & \ldots, & a_{2N} \\ \vdots & \vdots & & \vdots \\ a_{N1}, & a_{N2}, & \ldots, & a_{NN} \end{pmatrix} \begin{pmatrix} \mu_1 \\ \mu_2 \\ \vdots \\ \mu_N \end{pmatrix} + \begin{pmatrix} b_{11}, & b_{12}, & \ldots, & b_{1N} \\ b_{21}, & b_{22}, & \ldots, & b_{2N} \\ \vdots & \vdots & & \vdots \\ b_{N1}, & b_{N2}, & \ldots, & b_{NN} \end{pmatrix} \begin{pmatrix} \sigma_1 \\ \sigma_2 \\ \vdots \\ \sigma_N \end{pmatrix} = 0 \quad (11.71)$$

d. *Establish RHS Vector*

If we specify the source strength at the collocation point, according to Eq. (11.61), the second matrix multiplication can be executed. Then this known part is moved to the right-hand side of the equation. Thus

$$\begin{pmatrix} \text{RHS}_1 \\ \text{RHS}_2 \\ \vdots \\ \text{RHS}_N \end{pmatrix} = - \begin{pmatrix} b_{11}, & b_{12}, & \ldots, & b_{1N} \\ b_{21}, & b_{22}, & \ldots, & b_{2N} \\ \vdots & \vdots & & \vdots \\ b_{N1}, & b_{N2}, & \ldots, & b_{NN} \end{pmatrix} \begin{pmatrix} \sigma_1 \\ \sigma_2 \\ \vdots \\ \sigma_N \end{pmatrix} \quad (11.72)$$

e. *Solve Equations*

At this phase the N equations will have the form

$$\begin{pmatrix} a_{11}, & a_{12}, & \ldots, & a_{1N} \\ a_{21}, & a_{22}, & \ldots, & a_{2N} \\ \vdots & \vdots & & \vdots \\ a_{N1}, & a_{N2}, & \ldots, & a_{NN} \end{pmatrix} \begin{pmatrix} \mu_1 \\ \mu_2 \\ \vdots \\ \mu_N \end{pmatrix} = \begin{pmatrix} \text{RHS}_1 \\ \text{RHS}_2 \\ \vdots \\ \text{RHS}_N \end{pmatrix} \quad (11.73)$$

with N unknown values μ_j, which can be computed by solving this *full-matrix* equation.

11.3 Constant-Potential (Dirichlet Boundary Condition) Methods

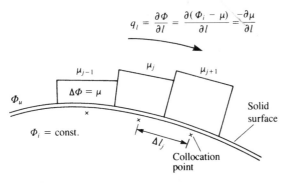

Figure 11.26 Doublet panels on the surface of a solid boundary.

f. *Calculation of Pressures and Loads*

Once the strength of the doublets μ_j is known, the potential outside the surface can be calculated. This is shown schematically in Fig. 11.26, which indicates that the internal perturbation potential Φ_i is constant (and equal to zero) and the external potential Φ_u is equal to the internal potential plus the local potential jump $-\mu$ across the solid surface,

$$\Phi_u = \Phi_i - \mu \tag{11.74}$$

The local external tangential velocity component above each collocation point can be calculated by differentiating the velocity potential along the tangential direction:

$$Q_t = \frac{\partial \Phi_u^*}{\partial l} \tag{11.75}$$

where l is a line along the surface. For example, the simplest numerical interpretation of this formula is

$$Q_{t_j} = \frac{\mu_j - \mu_{j+1}}{\Delta l_j} + Q_{t_\infty} \tag{11.76}$$

where Δl_j is the distance between the two adjacent collocation points, as shown in the figure. This formulation is more accurate at the jth panel second corner point and can be used to calculate the velocity at this point. The pressure coefficient can be computed by using Eq. (11.18):

$$C_{p_j} = 1 - \frac{Q_{t_j}^2}{Q_\infty^2} \tag{11.77}$$

The contribution to the lift coefficient is then

$$\Delta C_{l_j} = -C_{p_j} \Delta l_j \cos \alpha_j / c \tag{11.78}$$

where Δl_j and α_j are shown in Fig. 11.21. The total lift and moment are obtained by summing the contribution of each element:

$$L = \sum_{j=1}^{N} \Delta L_j \tag{11.79}$$

$$M_0 = -\sum_{j=1}^{N} \Delta L_j (x_j \cos \alpha) \tag{11.80}$$

and the nondimensional coefficients can be calculated by using Eqs. (11.12) and (11.13).

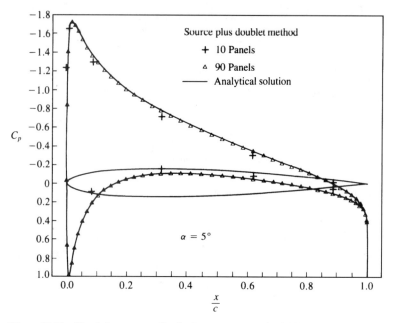

Figure 11.27 Chordwise pressure distribution on a symmetric airfoil, using 10 and 90 panels (combined source/doublet method with Dirichlet boundary condition).

Example: Lifting Thick Airfoil

A short computer program (Program No. 9 in Appendix D) was prepared to demonstrate the above method and the same airfoil geometry was used as for the previous examples. Table 11.2 shows the numeric equivalent of Eq. (11.71) for $N = 10$ panels, along with the RHS vector and the solution vector (of the doublet strengths). These results are plotted in Fig. 11.27, which shows that even with such a low number of panels a fairly reasonable solution is obtained. When using a larger number of panels ($N = 90$) the solution is very close to the analytic solution, both at the leading and trailing edges. As mentioned earlier, the potential based influence computations (Eqs. (11.67) and (11.68)) and the pressure calculations (of Eq. (11.76) or Eq. (11.38)) seem to be computationally more efficient than those of the previous methods.

11.3.2 Constant-Strength Doublet Method

An even simpler method for lifting airfoils can be derived by setting the source strengths to zero in Eq. (11.60). The value of the constant for the internal potential is selected to be zero (since a choice similar to that of the previous section of $\Phi_i = \Phi_\infty$ will result in the trivial solution). Consequently, the boundary condition describing the internal potential (Eq. (11.60)) reduces to

$$\sum_{j=1}^{N} C_j \mu_j + \Phi_\infty = 0 \tag{11.81}$$

where

$$\Phi_\infty = U_\infty x + W_\infty z \tag{11.82}$$

11.3 Constant-Potential (Dirichlet Boundary Condition) Methods

Table 11.2. *Influence matrix for the airfoil shown in Fig. 11.27 using ten panels ($\alpha = 5°$, constant-strength source and doublets with the Dirichlet boundary conditions)*

$$\begin{bmatrix} 0.48 & 0.00 & 0.01 & 0.01 & 0.01 & 0.01 & 0.01 & 0.01 & 0.04 & 0.43 \\ -0.02 & 0.50 & 0.01 & 0.01 & 0.01 & 0.01 & 0.02 & 0.06 & 0.34 & 0.06 \\ -0.02 & 0.01 & 0.50 & 0.02 & 0.02 & 0.03 & 0.07 & 0.27 & 0.08 & 0.03 \\ -0.01 & 0.01 & 0.02 & 0.50 & 0.04 & 0.08 & 0.23 & 0.10 & 0.03 & 0.02 \\ 0.00 & 0.01 & 0.02 & 0.06 & 0.50 & 0.24 & 0.12 & 0.03 & 0.01 & 0.01 \\ 0.01 & 0.01 & 0.03 & 0.12 & 0.24 & 0.50 & 0.06 & 0.02 & 0.01 & 0.00 \\ 0.02 & 0.03 & 0.10 & 0.23 & 0.08 & 0.04 & 0.50 & 0.02 & 0.01 & -0.01 \\ 0.03 & 0.08 & 0.27 & 0.07 & 0.03 & 0.02 & 0.02 & 0.50 & 0.01 & -0.02 \\ 0.06 & 0.34 & 0.06 & 0.02 & 0.01 & 0.01 & 0.01 & 0.01 & 0.50 & -0.02 \\ 0.43 & 0.04 & 0.01 & 0.01 & 0.01 & 0.01 & 0.01 & 0.01 & 0.00 & 0.48 \end{bmatrix} \begin{bmatrix} \mu_1 \\ \mu_2 \\ \mu_3 \\ \mu_4 \\ \mu_5 \\ \mu_6 \\ \mu_7 \\ \mu_8 \\ \mu_9 \\ \mu_{10} \end{bmatrix}$$

$$+ \begin{bmatrix} -0.08 & 0.00 & 0.09 & 0.10 & 0.05 & 0.05 & 0.10 & 0.09 & 0.00 & -0.06 \\ 0.00 & -0.11 & 0.04 & 0.08 & 0.04 & 0.04 & 0.08 & 0.04 & -0.05 & 0.00 \\ 0.04 & 0.03 & -0.11 & 0.02 & 0.03 & 0.03 & 0.03 & -0.02 & 0.04 & 0.04 \\ 0.05 & 0.09 & 0.03 & -0.11 & 0.00 & 0.00 & -0.02 & 0.04 & 0.10 & 0.05 \\ 0.06 & 0.12 & 0.08 & -0.02 & -0.07 & -0.04 & 0.00 & 0.08 & 0.12 & 0.06 \\ 0.06 & 0.12 & 0.08 & 0.00 & -0.04 & -0.07 & -0.02 & 0.08 & 0.12 & 0.06 \\ 0.05 & 0.10 & 0.04 & -0.02 & 0.00 & 0.00 & -0.11 & 0.03 & 0.09 & 0.05 \\ 0.04 & 0.04 & -0.02 & 0.03 & 0.03 & 0.03 & 0.02 & -0.11 & 0.03 & 0.04 \\ 0.00 & -0.05 & 0.04 & 0.08 & 0.04 & 0.04 & 0.08 & 0.04 & -0.11 & 0.00 \\ -0.06 & 0.00 & 0.09 & 0.10 & 0.05 & 0.05 & 0.10 & 0.09 & 0.00 & -0.08 \end{bmatrix} \begin{bmatrix} -0.07 \\ -0.05 \\ 0.02 \\ 0.18 \\ 0.61 \\ 0.46 \\ 0.00 \\ -0.15 \\ -0.22 \\ -0.24 \end{bmatrix}_{\sigma_i} = 0$$

The solution is:

$$\begin{bmatrix} \mu_1 \\ \mu_2 \\ \mu_3 \\ \mu_4 \\ \mu_5 \\ \mu_6 \\ \mu_7 \\ \mu_8 \\ \mu_9 \\ \mu_{10} \end{bmatrix} = \begin{bmatrix} -0.1795420 \\ -0.1691454 \\ -0.1914112 \\ -0.2294960 \\ -0.2296733 \\ -7.9113595 \times 10^{-2} \\ 9.1565706 \times 10^{-2} \\ 0.2577208 \\ 0.3524704 \\ 0.3651389 \end{bmatrix} \quad \text{RHS}_i = \begin{bmatrix} 0.8 \\ 0.7 \\ 0.01 \\ -0.07 \\ -0.13 \\ -0.12 \\ -0.05 \\ 0.03 \\ 0.08 \\ 0.09 \end{bmatrix}$$

Note that now μ will represent the potential jump from zero to Φ_u on the boundary (see Fig. 11.26) and therefore Φ_u is the local *total* potential (whereas in the previous example μ was the jump in the perturbation potential only).

Equation (11.81) can be specified at each collocation point *inside* the body, providing a linear algebraic equation for this point. The steps toward establishing such a numeric solution are very similar to the previous method.

a. Selection of Singularity Element

For this case a constant-strength doublet element is used and the potential at an arbitrary point P (not on the surface) due to a constant-strength doublet is given by Eq. (11.64) and by the routine of Eq. (11.66):

$$\Delta\Phi_d = \text{PHICD}(\mu_j, x, z, x_j, z_j, x_{j+1}, z_{j+1}) \tag{11.66}$$

b. Discretization of Geometry

The $N+1$ panel corner points and N collocation points are generated in a manner similar to the previous example and a typical grid is shown in Fig. 11.18. Since in this case the internal Dirichlet boundary condition is used the collocation points must be placed inside the body with a small inward displacement under the panel center (although this inward displacement can be skipped if the self-induced influence is specified separately).

c. Influence Coefficients

The increment in the velocity potential at collocation point i due to a unit strength constant doublet element of panel j is given by Eq. (11.68):

$$c_{ij} = \text{PHICD}(\mu_j = 1, x_i, z_i, x_j, z_j, x_{j+1}, z_{j+1}) \tag{11.68}$$

The construction of the doublet influence matrix and the inclusion of the Kutta condition (and the wake doublet μ_W) is exactly the same as in the previous example. Thus, after substituting the Kutta condition ($\mu_W = \mu_N - \mu_1$), the c_{ij} influence coefficients become the a_{ij} coefficients (see Eq. (11.70)). If we use these results, Eq. (11.81), when specified at each collocation point, will have the form

$$\begin{pmatrix} a_{11}, & a_{12}, & \ldots, & a_{1N} \\ a_{21}, & a_{22}, & \ldots, & a_{2N} \\ \vdots & \vdots & & \vdots \\ a_{N1}, & a_{N2}, & \ldots, & a_{NN} \end{pmatrix} \begin{pmatrix} \mu_1 \\ \mu_2 \\ \vdots \\ \mu_N \end{pmatrix} + \begin{pmatrix} \Phi_{\infty_1} \\ \Phi_{\infty_2} \\ \vdots \\ \Phi_{\infty_N} \end{pmatrix} = 0 \tag{11.83}$$

d. Establish RHS Vector

The second term in this equation is known and can be transferred to the right-hand side of the equation. The RHS vector then becomes

$$\begin{pmatrix} \text{RHS}_1 \\ \text{RHS}_2 \\ \vdots \\ \text{RHS}_N \end{pmatrix} = - \begin{pmatrix} \Phi_{\infty_1} \\ \Phi_{\infty_2} \\ \vdots \\ \Phi_{\infty_N} \end{pmatrix} \tag{11.84}$$

and the Φ_{∞_j} term is calculated by using Eq. (11.82).

e. Solve Equations

At this phase the N equations will have the form similar to Eq. (11.73) and can be solved for the N unknown values μ_j.

f. Calculation of Pressures and Loads

Once the strength of the doublets μ_j is known, the potential outside the surface can be calculated based on the principle shown schematically in Fig. 11.26 (but now $\Phi_i^* = 0$). Equation (11.75) is still the basis for calculating the local velocity but now the external potential Φ_u is equal to the local *total* potential jump $-\mu$ across the solid surface. Thus, the local external tangential velocity above each collocation point can be calculated by differentiating the velocity potential along the tangential direction, and Eq. (11.76) will have the form

$$Q_{tj} = \frac{\mu_j - \mu_{j+1}}{\Delta l_j} \tag{11.85}$$

11.3 Constant-Potential (Dirichlet Boundary Condition) Methods

Table 11.3. *Influence matrix for the airfoil shown in Fig. 11.27 using ten panels ($\alpha = 5°$, constant-strength doublets only, with the Dirichlet boundary conditions)*

$$\begin{bmatrix} 0.50 & 0.00 & 0.01 & 0.01 & 0.01 & 0.01 & 0.01 & 0.01 & 0.04 & 0.40 & 0.02 \\ 0.00 & 0.50 & 0.01 & 0.01 & 0.01 & 0.01 & 0.02 & 0.06 & 0.34 & 0.04 & 0.02 \\ 0.00 & 0.01 & 0.50 & 0.02 & 0.02 & 0.03 & 0.07 & 0.27 & 0.08 & 0.01 & 0.02 \\ 0.00 & 0.01 & 0.02 & 0.50 & 0.04 & 0.08 & 0.23 & 0.10 & 0.03 & 0.01 & 0.01 \\ 0.00 & 0.01 & 0.02 & 0.06 & 0.50 & 0.24 & 0.12 & 0.03 & 0.01 & 0.00 & 0.00 \\ 0.00 & 0.01 & 0.03 & 0.12 & 0.24 & 0.50 & 0.06 & 0.02 & 0.01 & 0.00 & 0.00 \\ 0.01 & 0.03 & 0.10 & 0.23 & 0.08 & 0.04 & 0.50 & 0.02 & 0.01 & 0.00 & -0.01 \\ 0.01 & 0.08 & 0.27 & 0.07 & 0.03 & 0.02 & 0.02 & 0.50 & 0.01 & 0.00 & -0.02 \\ 0.04 & 0.34 & 0.06 & 0.02 & 0.01 & 0.01 & 0.01 & 0.01 & 0.50 & 0.00 & -0.02 \\ 0.40 & 0.04 & 0.01 & 0.01 & 0.01 & 0.01 & 0.01 & 0.01 & 0.00 & 0.50 & -0.02 \\ -1.00 & 0.00 & 0.00 & 0.00 & 0.00 & 0.00 & 0.00 & 0.00 & 0.00 & 1.00 & -1.00 \end{bmatrix} \begin{bmatrix} \mu_1 \\ \mu_2 \\ \mu_3 \\ \mu_4 \\ \mu_5 \\ \mu_6 \\ \mu_7 \\ \mu_8 \\ \mu_9 \\ \mu_{10} \\ \mu_w \end{bmatrix} = \begin{bmatrix} 0.88 \\ 0.49 \\ -0.08 \\ -0.61 \\ -0.92 \\ -0.91 \\ -0.59 \\ -0.06 \\ 0.50 \\ 0.88 \\ 0.00 \end{bmatrix}$$

By substituting $\mu_w = \mu_{10} - \mu_1$ we get

$$\begin{bmatrix} 0.48 & 0.00 & 0.01 & 0.01 & 0.01 & 0.01 & 0.01 & 0.01 & 0.04 & 0.43 \\ -0.02 & 0.50 & 0.01 & 0.01 & 0.01 & 0.01 & 0.02 & 0.06 & 0.34 & 0.06 \\ -0.02 & 0.01 & 0.50 & 0.02 & 0.02 & 0.03 & 0.07 & 0.27 & 0.08 & 0.03 \\ -0.01 & 0.01 & 0.02 & 0.50 & 0.04 & 0.08 & 0.23 & 0.10 & 0.03 & 0.02 \\ 0.00 & 0.01 & 0.02 & 0.06 & 0.50 & 0.24 & 0.12 & 0.03 & 0.01 & 0.01 \\ 0.01 & 0.01 & 0.03 & 0.12 & 0.24 & 0.50 & 0.06 & 0.02 & 0.01 & 0.00 \\ 0.02 & 0.03 & 0.10 & 0.23 & 0.08 & 0.04 & 0.50 & 0.02 & 0.01 & -0.01 \\ 0.03 & 0.08 & 0.27 & 0.07 & 0.03 & 0.02 & 0.02 & 0.50 & 0.01 & -0.02 \\ 0.06 & 0.34 & 0.06 & 0.02 & 0.01 & 0.01 & 0.01 & 0.01 & 0.50 & -0.02 \\ 0.43 & 0.04 & 0.01 & 0.01 & 0.01 & 0.01 & 0.01 & 0.01 & 0.00 & 0.48 \end{bmatrix} \begin{bmatrix} \mu_1 \\ \mu_2 \\ \mu_3 \\ \mu_4 \\ \mu_5 \\ \mu_6 \\ \mu_7 \\ \mu_8 \\ \mu_9 \\ \mu_{10} \end{bmatrix} = \begin{bmatrix} 0.88 \\ 0.49 \\ -0.08 \\ -0.61 \\ -0.92 \\ -0.91 \\ -0.59 \\ -0.06 \\ 0.50 \\ 0.88 \end{bmatrix}$$

The solution is:

$$\begin{bmatrix} \mu_1 \\ \mu_2 \\ \mu_3 \\ \mu_4 \\ \mu_5 \\ \mu_6 \\ \mu_7 \\ \mu_8 \\ \mu_9 \\ \mu_{10} \end{bmatrix} = \begin{bmatrix} -0.6970030 \\ -0.3259878 \\ 0.2767572 \\ 0.8576335 \\ 1.182451 \\ 1.016438 \\ 0.5045773 \\ -0.2092538 \\ -0.8805848 \\ -1.270261 \end{bmatrix}$$

$$\mu_w = \mu_{10} - \mu_1 = -0.5732538$$

The pressure coefficient and the fluid dynamic loads can be calculated now using the formulation of the previous section (Eqs. (11.77)–(11.80)).

Example 1: Lifting Thick Airfoil

This constant-strength doublet method is applied to the same problem of the previous section and the resulting pressure distribution with 10 and 90 panels is very close to the results presented in Fig. 11.27. It seems that this method is as effective as the combined source and doublet method and it does not have the matrix multiplication of the source matrix (less numerics). The influence coefficients for the $N = 10$ panel case are presented in Table 11.3 along with the solution vector. This information is presented since it was found that such data

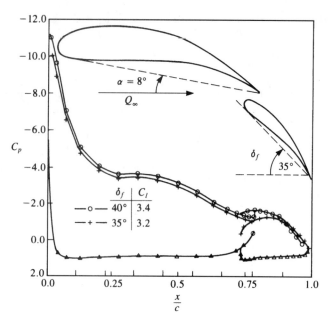

Figure 11.28 Effect of flap deflection on the chordwise pressure distribution of a two-element airfoil. (Triangles represent lower surface results for both flap angles.)

(as in Tables 11.2 and 11.3) are extremely useful in the early stages of code development and validation. Table 11.3 presents the doublet influence coefficients before and after the inclusion of the Kutta condition and also the magnitude of the solution vector μ_j, which is larger in this case than in the case shown in Table 11.2. This is a result of the unknowns μ_j representing here the total velocity potential whereas in the combined doublet/source case the doublet represents the perturbation potential only. Maskew[9.3] claims that since the unknown μ_j are larger in the doublet-only solution, there is a numerical advantage (in terms of convergence for a large number of panels) for using the combined source/doublet method.

Example 2: Two-Element Airfoil

To model multielement airfoils the Kutta condition must be specified, separately, for each element. This method is then applied to the two-element airfoil shown in Fig. 11.28. The effect of a 5° flap deflection on the chordwise pressure distribution is shown in the lower part of the figure. In general the effect of flap deflection is to increase the lift of the main planform more than the lift of the flap itself.

A sample student computer program for this method is presented in Appendix D (Program No. 8).

11.4 Linearly Varying Singularity Strength Methods (Using the Neumann B.C.)

As an example of higher order paneling methods using the Neumann boundary condition, the linear source and vortex formulations will be presented. Since the linear doublet distribution is equal to the constant-strength vortex distribution, only the above two methods will be studied. Here the panel surface is assumed to be planar and the singularity

11.4 Linearly Varying Singularity Strength Methods (Using the Neumann B.C.)

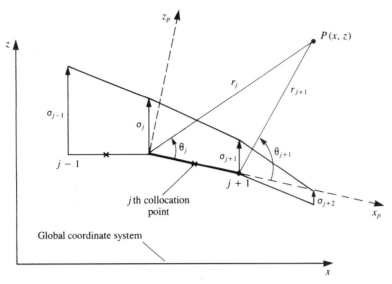

Figure 11.29 Nomenclature for a linear-strength surface singularity element.

will change linearly along the panel. Consequently, the singularity strength on each panel includes two unknowns and additional equations need to be formulated.

11.4.1 Linear-Strength Source Method

The source-only based method will be applicable only to nonlifting configurations and is considered to be a more refined model than the one based on constant-strength source elements. The basic six step procedure follows.

a. *Selection of Singularity Element*

A segment of the discretized singularity distribution on a solid surface is shown in Fig. 11.29. To establish a normal-velocity boundary condition based method (see Eq. (11.4)), the induced-velocity formulas of a constant- and a linear-strength source distribution are combined (Eqs. (10.17) and (10.48), and Eqs. (10.18) and (10.49)). The parameters θ and r are shown in Fig. 11.29, and the velocity $(u, w)_p$, measured in the panel local coordinate system $(x, z)_p$, has components

$$u_p = \frac{\sigma_0}{4\pi} \ln \frac{r_1^2}{r_2^2} + \frac{\sigma_1}{2\pi} \left[\frac{x - x_1}{2} \ln \frac{r_1^2}{r_2^2} + (x_1 - x_2) + z(\theta_2 - \theta_1) \right]$$

(in panel coordinates) (11.86)

$$w_p = \frac{\sigma_0}{2\pi} (\theta_2 - \theta_1) + \frac{\sigma_1}{4\pi} \left[z \ln \frac{r_2^2}{r_1^2} + 2(x - x_1)(\theta_2 - \theta_1) \right]$$

(in panel coordinates) (11.87)

where the subscripts 1 and 2 refer to the panel edges j and $j + 1$, respectively. In these equations σ_0 and σ_1 are the source strength values, as shown in Fig. 11.30. If the strength of σ at the beginning of each panel is set equal to the strength of the source at the end point of the previous panel (as shown in Fig. 11.29), a continuous source distribution is obtained.

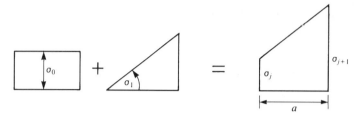

Figure 11.30 Decomposition of a generic linear-strength singularity element.

Now, if the unknowns are the panel edge values of the source distribution ($\sigma_j, \sigma_{j+1}, \ldots$ as in Fig. 11.29) then for N surface panels on a closed body the number of unknowns is $N + 1$. The relation between the source strengths of the elements shown in Fig. 11.30 and the panel edge values is

$$\sigma_j = \sigma_0 \quad (11.88a)$$

$$\sigma_{j+1} = \sigma_0 + \sigma_1 a \quad (11.88b)$$

where a is the panel length, and for convenience the induced-velocity equations are rearranged in terms of the panel-edge source strengths σ_j and σ_{j+1} (and the subscripts 1 and 2 are replaced with the j and $j + 1$ subscripts, respectively):

$$u_p = \frac{\sigma_j(x_{j+1} - x_j) + (\sigma_{j+1} - \sigma_j)(x - x_j)}{2\pi(x_{j+1} - x_j)} \ln \frac{r_j}{r_{j+1}}$$
$$- \frac{z}{2\pi}\left(\frac{\sigma_{j+1} - \sigma_j}{x_{j+1} - x_j}\right)\left[\frac{(x_{j+1} - x_j)}{z} + (\theta_{j+1} - \theta_j)\right] \quad \text{(panel coordinates)}$$

$$(11.89)$$

$$w_p = -\frac{z}{2\pi}\left(\frac{\sigma_{j+1} - \sigma_j}{x_{j+1} - x_j}\right) \ln \frac{r_{j+1}}{r_j}$$
$$+ \frac{\sigma_j(x_{j+1} - x_j) + (\sigma_{j+1} - \sigma_j)(x - x_j)}{2\pi(x_{j+1} - x_j)}(\theta_{j+1} - \theta_j) \quad \text{(panel coordinates)}$$

$$(11.90)$$

Note that Eqs. (11.89) and (11.90) can be divided into velocity induced by σ_j and by σ_{j+1} such that

$$(u, w)_p = (u^a, w^a)_p + (u^b, w^b)_p \quad (11.91)$$

where the superscript $()^a$ and $()^b$ represent the contribution due to the leading and trailing singularity strengths, respectively. If we rearrange Eqs. (11.89) and (11.90) we can separate the $()^a$ part of the velocity components,

$$u_p^a = \frac{\sigma_j(x_{j+1} - x)}{2\pi(x_{j+1} - x_j)} \ln \frac{r_j}{r_{j+1}} + \frac{z}{2\pi}\left(\frac{\sigma_j}{x_{j+1} - x_j}\right)\left[\frac{(x_{j+1} - x_j)}{z} + (\theta_{j+1} - \theta_j)\right]$$

$$(11.92a)$$

$$w_p^a = \frac{z}{2\pi}\left(\frac{\sigma_j}{x_{j+1} - x_j}\right) \ln \frac{r_{j+1}}{r_j} + \frac{\sigma_j(x_{j+1} - x)}{2\pi(x_{j+1} - x_j)}(\theta_{j+1} - \theta_j) \quad (11.92b)$$

11.4 Linearly Varying Singularity Strength Methods (Using the Neumann B.C.)

from the $(\ )^b$ part of the velocity components,

$$u_p^b = \frac{\sigma_{j+1}(x-x_j)}{2\pi(x_{j+1}-x_j)} \ln \frac{r_j}{r_{j+1}} - \frac{z}{2\pi}\left(\frac{\sigma_{j+1}}{x_{j+1}-x_j}\right)\left[\frac{(x_{j+1}-x_j)}{z} + (\theta_{j+1}-\theta_j)\right]$$

(11.92c)

$$w_p^b = -\frac{z}{2\pi}\left(\frac{\sigma_{j+1}}{x_{j+1}-x_j}\right)\ln \frac{r_{j+1}}{r_j} + \frac{\sigma_{j+1}(x-x_j)}{2\pi(x_{j+1}-x_j)}(\theta_{j+1}-\theta_j)$$

(11.92d)

To transform these velocity components back to the x, z coordinates, a rotation by the panel orientation angle α_i is performed as given by Eq. (11.23):

$$\begin{pmatrix} u \\ w \end{pmatrix} = \begin{pmatrix} \cos\alpha_i & \sin\alpha_i \\ -\sin\alpha_i & \cos\alpha_i \end{pmatrix}\begin{pmatrix} u \\ w \end{pmatrix}_p$$

(11.23)

This procedure can be included in an induced-velocity subroutine SOR2DL (where L stands for linear), which will compute the velocity (u, w) at an arbitrary point (x, z) in the global coordinate system due to the jth element:

$$\begin{pmatrix} u, w \\ u^a, w^a \\ u^b, w^b \end{pmatrix} = \text{SOR2DL}(\sigma_j, \sigma_{j+1}, x, z, x_j, z_j, x_{j+1}, z_{j+1})$$

(11.93)

The four additional velocity components (u^a, w^a, u^b, w^b) will be a byproduct of subroutine SOR2DL.

b. Discretization of Geometry

The panel corner points, collocation points, and normal vectors are computed as in the previous methods.

c. Influence Coefficients

In this phase the zero normal flow boundary condition is implemented. For example, the velocity induced by the jth element with a unit strength σ_j and σ_{j+1}, at the first collocation point, can be obtained by using Eq. (11.93):

$$\begin{pmatrix} u, w \\ u^a, w^a \\ u^b, w^b \end{pmatrix}_{1j} = \text{SOR2DL}(\sigma_j = 1, \sigma_{j+1} = 1, x_1, z_1, x_j, z_j, x_{j+1}, z_{j+1})$$

(11.93a)

where the superscripts $(\ )^a$ and $(\ )^b$ represent the contributions due to the leading and trailing singularity strengths, respectively. This example indicates that the velocity at each collocation point is influenced by the two edges of the jth panel. Thus, adding the influence of the $(j+1)$-th panel and each subsequent panel gives the local induced velocity at the first collocation point

$$(u, w)_1 = (u^a, w^a)_{11}\sigma_1 + [(u^b, w^b)_{11} + (u^a, w^a)_{12}]\sigma_2$$
$$+ [(u^b, w^b)_{12} + (u^a, w^a)_{13}]\sigma_3 + \cdots$$
$$+ [(u^b, w^b)_{1,N-1} + (u^a, w^a)_{1N}]\sigma_N + (u^b, w^b)_{1N}\sigma_{N+1}$$

This equation can be reduced to a form

$$(u, w)_1 = (u, w)_{11}\sigma_1 + (u, w)_{12}\sigma_2 + \cdots + (u, w)_{1,N+1}\sigma_{N+1}$$

such that for the first and last terms

$$(u, w)_{11} = (u^a, w^a)_{11} \sigma_1 \tag{11.94a}$$

$$(u, w)_{1,N+1} = (u^b, w^b)_{1N} \sigma_{N+1} \tag{11.94b}$$

and for all other terms

$$(u, w)_{1,j} = [(u^b, w^b)_{1,j-1} + (u^a, w^a)_{1,j}] \sigma_j \tag{11.94c}$$

From this point on the procedure is similar to the constant-strength source method. The influence coefficient is calculated when $\sigma_j = 1$ and

$$a_{ij} = (u, w)_{i,j} \cdot \mathbf{n}_i \tag{11.95}$$

For each collocation point there will be $N + 1$ such coefficients and unknowns σ_j.

d. Establish Boundary Condition (RHS)

The free-stream normal velocity component RHS_i is found, as in the case of the discrete source (by using Eq. (11.6b)), at the collocation point:

$$\text{RHS}_i = -U_\infty \sin \alpha_i$$

where α_i is the panel inclination angle depicted by Fig. 11.3.

e. Solve Equations

Specification of the boundary condition for each ($i = 1 \to N$) of the collocation points results in N linear algebraic equations with the unknowns σ_j ($j = 1 \to N + 1$). The additional equation can be found by requiring that the flow leaves parallel to the trailing edge; thus

$$\sigma_1 + \sigma_{N+1} = 0 \tag{11.96}$$

Another option that will yield similar results is to establish an additional collocation point slightly behind the trailing edge and require that the velocity will be zero there (stagnation point for finite-angle trailing edges). Consequently, the set of equations to be solved becomes

$$\begin{pmatrix} a_{11} & a_{12} & \cdots & a_{1,N+1} \\ a_{21} & a_{22} & \cdots & a_{2,N+1} \\ a_{31} & a_{32} & \cdots & a_{3,N+1} \\ \vdots & \vdots & \ddots & \vdots \\ a_{N1} & a_{N2} & \cdots & a_{N,N+1} \\ 1 & 0 & 0 \cdots 0 & 1 \end{pmatrix} \begin{pmatrix} \sigma_1 \\ \sigma_2 \\ \sigma_3 \\ \vdots \\ \sigma_N \\ \sigma_{N+1} \end{pmatrix} = \begin{pmatrix} \text{RHS}_1 \\ \text{RHS}_2 \\ \text{RHS}_3 \\ \vdots \\ \text{RHS}_N \\ 0 \end{pmatrix}$$

The above set of algebraic equations has a well-defined diagonal and can be solved for σ_j by using standard methods of linear algebra.

f. Calculation of Pressures and Loads

Once the strength of the sources σ_j is known, the velocity at each collocation point can be calculated using Eq. (11.93) and the pressure coefficient can be calculated by using Eq. (11.18).

The formulation presented here is for a nonlifting symmetric wing or body. For this case, the number of unknowns can be reduced to $N/2$ (e.g., to the number of upper surface elements only) by the minor modification presented at the end of Section 11.2.1.

11.4 Linearly Varying Singularity Strength Methods (Using the Neumann B.C.)

Note that for a closed body the sum of the sources must be zero. However, this condition is not an independent one and cannot be used instead of the additional equation (Eq. (11.96)).

A computer program based on this method is presented in Appendix D (Program No. 6) and the computed pressure coefficients for the airfoil of Section 6.6 result in a pressure distribution similar to that of Fig. 11.19.

11.4.2 Linear-Strength Vortex Method

The constant-strength vortex method of Section 11.2.3 posed some difficulties that can be corrected by using the linear-strength vortex element. To describe the method let us follow the basic six-step procedure:

a. *Selection of Singularity Element*

The linearly varying source distribution shown in Fig. 11.29 includes the same nomenclature that is used for the linearly varying strength vortex panel. The velocity components $(u, w)_p$ in the direction of the panel coordinates were obtained in Sections 10.2 and 10.3 (Eqs. (10.39), (10.40), (10.72), and (10.73)):

$$u_p = \frac{\gamma_0}{2\pi}\left[\tan^{-1}\frac{z}{x-x_2} - \tan^{-1}\frac{z}{x-x_1}\right] \quad \text{(panel coordinates)}$$

$$+ \frac{\gamma_1}{4\pi}\left[z\ln\frac{(x-x_1)^2+z^2}{(x-x_2)^2+z^2} + 2x\left(\tan^{-1}\frac{z}{x-x_2} - \tan^{-1}\frac{z}{x-x_1}\right)\right] \quad (11.97)$$

$$w_p = -\frac{\gamma_0}{4\pi}\ln\frac{(x-x_1)^2+z^2}{(x-x_2)^2+z^2} \quad \text{(panel coordinates)}$$

$$-\frac{\gamma_1}{2\pi}\left[\frac{x}{2}\ln\frac{(x-x_1)^2+z^2}{(x-x_2)^2+z^2} + (x_1-x_2)\right.$$

$$\left. + z\left(\tan^{-1}\frac{z}{x-x_2} - \tan^{-1}\frac{z}{x-x_1}\right)\right] \quad (11.98)$$

where the subscripts 1 and 2 refer to the panel edges j and $j+1$, respectively. As in the case of the linearly varying strength source, it is useful to rearrange these equations in terms of their edge vortex strengths γ_j and γ_{j+1} (see Eqs. (11.88a, b)) and to use the subscripts j and $j+1$ instead of 1 and 2:

$$u_p = \frac{z}{2\pi}\left(\frac{\gamma_{j+1}-\gamma_j}{x_{j+1}-x_j}\right)\ln\frac{r_{j+1}}{r_j}$$

$$+ \frac{\gamma_j(x_{j+1}-x_j)+(\gamma_{j+1}-\gamma_j)(x-x_j)}{2\pi(x_{j+1}-x_j)}(\theta_{j+1}-\theta_j) \quad \text{(panel coordinates)}$$

$$(11.99)$$

$$w_p = -\frac{\gamma_j(x_{j+1}-x_j)+(\gamma_{j+1}-\gamma_j)(x-x_j)}{2\pi(x_{j+1}-x_j)}\ln\frac{r_j}{r_{j+1}}$$

$$+ \frac{z}{2\pi}\left(\frac{\gamma_{j+1}-\gamma_j}{x_{j+1}-x_j}\right)\left[\frac{(x_{j+1}-x_j)}{z} + (\theta_{j+1}-\theta_j)\right] \quad \text{(panel coordinates)}$$

$$(11.100)$$

These two equations combined with the transformation of Eq. (11.23) can be included in a subroutine VOR2DL such that

$$\begin{pmatrix} u, w \\ u^a, w^a \\ u^b, w^b \end{pmatrix}_{ij} = \text{VOR2DL}(\gamma_j, \gamma_{j+1}, x_i, z_i, x_j, z_j, x_{j+1}, z_{j+1}) \tag{11.101}$$

where the superscripts $(\ )^a$ and $(\ )^b$ represent the contributions due to the leading and trailing singularity strengths, respectively. For simplicity, this procedure is not repeated here but can be obtained simply by taking all terms multiplied by γ_j in Eqs. (11.99) and (11.100) to produce the $(\ )^a$ component and all terms multiplied by γ_{j+1} to produce the $(\ )^b$ component (as was done in the case of the linearly varying strength source – see Eqs. (11.91) and (11.92)). This decomposition of the velocity components is automatically calculated by the subroutine described by Eq. (11.101), and

$$(u, w) = (u^a, w^a) + (u^b, w^b)$$

b. *Discretization of Geometry*

The panel corner points, collocation points, and normal vectors are computed as in the previous methods.

c. *Influence Coefficients*

In this phase the zero normal flow boundary condition (Eq. (11.4)) is implemented. For example, the self-induced velocity due to the jth element with a unit strength γ_j and γ_{j+1}, at the first collocation point, can be obtained by using Eq. (11.101):

$$\begin{pmatrix} u, w \\ u^a, w^a \\ u^b, w^b \end{pmatrix}_{1j} = \text{VOR2DL}(\gamma_j = 1, \gamma_{j+1} = 1, x_1, z_1, x_j, z_j, x_{j+1}, z_{j+1}) \tag{11.102}$$

Similarly to the case of the linearly varying strength source the velocity at each collocation point is influenced by the two edges of the jth panel. Thus adding the influence of the $(j+1)$-th panel and each subsequent panel gives the local self-induced velocity at the first collocation point

$$(u, w)_1 = (u^a, w^a)_{11}\gamma_1 + [(u^b, w^b)_{11} + (u^a, w^a)_{12}]\gamma_2$$
$$+ [(u^b, w^b)_{12} + (u^a, w^a)_{13}]\gamma_3 + \cdots$$
$$+ [(u^b, w^b)_{1,N-1} + (u^a, w^a)_{1N}]\gamma_N + (u^b, w^b)_{1N}\gamma_{N+1}$$

This equation can be reduced to a form

$$(u, w)_1 = (u, w)_{11}\gamma_1 + (u, w)_{12}\gamma_2 + \cdots + (u, w)_{1,N+1}\gamma_{N+1}$$

such that for the first and last terms

$$(u, w)_{11} = (u^a, w^a)_{11}\gamma_1 \tag{11.103a}$$
$$(u, w)_{1,N+1} = (u^b, w^b)_{1N}\gamma_{N+1} \tag{11.103b}$$

and for all other terms

$$(u, w)_{1,j} = [(u^b, w^b)_{1,j-1} + (u^a, w^a)_{1,j}]\gamma_j \tag{11.103c}$$

11.4 Linearly Varying Singularity Strength Methods (Using the Neumann B.C.)

From this point on the procedure is similar to the linearly varying strength source method. The influence coefficient is calculated when $\gamma_j = 1$ and

$$a_{ij} = (u, w)_{i,j} \cdot \mathbf{n}_i \tag{11.104}$$

For each collocation point there will be $N + 1$ such coefficients and unknowns γ_j.

d. *Establish Boundary Condition (RHS)*

The free-stream normal velocity component RHS_i is found as in the case of discrete vortex (by using Eq. (11.6a)):

$$\text{RHS}_i = -(U_\infty, W_\infty) \cdot (\cos\alpha_i, -\sin\alpha_i)$$

e. *Solve Equations*

Specification of the boundary condition for each ($i = 1 \to N$) of the collocation points results in N linear algebraic equations with the unknowns γ_j ($j = 1 \to N + 1$). The additional equation can be found by specifying the Kutta condition at the trailing edge:

$$\gamma_1 + \gamma_{N+1} = 0 \tag{11.105}$$

Consequently, the set of equations to be solved becomes

$$\begin{pmatrix} a_{11} & a_{12} & \cdots & a_{1,N+1} \\ a_{21} & a_{22} & \cdots & a_{2,N+1} \\ a_{31} & a_{32} & \cdots & a_{3,N+1} \\ \vdots & \vdots & \ddots & \vdots \\ a_{N1} & a_{N2} & \cdots & a_{N,N+1} \\ 1 & 0 & 0 \cdots 0 & 1 \end{pmatrix} \begin{pmatrix} \gamma_1 \\ \gamma_2 \\ \gamma_3 \\ \vdots \\ \gamma_N \\ \gamma_{N+1} \end{pmatrix} = \begin{pmatrix} \text{RHS}_1 \\ \text{RHS}_2 \\ \text{RHS}_3 \\ \vdots \\ \text{RHS}_N \\ 0 \end{pmatrix}$$

The above set of algebraic equations has a well-defined diagonal and can be solved for γ_j by using standard methods of linear algebra.

f. *Calculation of Pressures and Loads*

Once the strength of the vortices γ_j is known, the perturbation velocity at each collocation point can be calculated using the results for a vortex distribution (e.g., Eq. (3.147)):

$$Q_{t_j} = (Q_{t_\infty})_j + \frac{\gamma_j + \gamma_{j+1}}{4} \tag{11.106}$$

and the pressure coefficient can be calculated by using Eq. (11.18):

$$C_p = 1 - \frac{Q_t^2}{Q_\infty^2}$$

The lift of the panel can be computed from this pressure distribution or by using the Kutta–Joukowski theorem:

$$\Delta L_j = \rho Q_\infty \frac{\gamma_j + \gamma_{j+1}}{2} \Delta c_j \tag{11.107}$$

where Δc_j is the panel length. The total lift and moment are obtained by summing the contribution of the individual elements (as in Eqs. (11.54) and (11.55)).

A computer program based on this method is presented in Appendix D (Program No. 7), and the computed pressure coefficients for the same airfoil of Section 6.6, used for the previous examples, at an angle of attack of 5° are presented in Fig. 11.31.

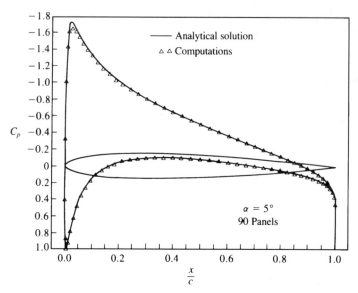

Figure 11.31 Chordwise pressure distribution along a symmetric airfoil (using the linearly varying strength vortex method).

11.5 Linearly Varying Singularity Strength Methods (Using the Dirichlet B.C.)

In this section linearly varying strength doublet and source elements will be used to formulate methods based on the Dirichlet boundary condition. Since a linear vortex distribution is equivalent to a quadratic doublet distribution (which will be described in Section 11.6), such vortex methods are not presented here.

11.5.1 *Linear Source/Doublet Method*

For this example, the approach of Section 11.3 is used and a combination of linearly varying strength sources and doublets will be distributed on the solid boundaries S (Fig. 11.32). Following Section 11.3.1 we select the velocity potential within the volume enclosed by S as $\Phi_i^* = \Phi_\infty$ and the boundary condition at each collocation point is reduced to the form given by Eq. (11.62):

$$\sum_{j=1}^{N_1} B_j \sigma_j + \sum_{j=1}^{N_1} C_j \mu_j = 0 \tag{11.62}$$

where N_1 is the number of singularity strength parameters. To establish a solution based on this equation using linearly varying singularity distributions, let us follow the basic six-step procedure.

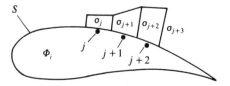

Figure 11.32 Linear-strength singularity element model for a closed body.

11.5 Linearly Varying Singularity Strength Methods (Using the Dirichlet B.C.)

a. *Selection of Singularity Element*

The potential at an arbitrary point P (not on the surface) due to a linearly varying strength source $\sigma(x) = \sigma_0 + \sigma_1 x$ was derived in Eqs. (10.14) and (10.47) (in the panel coordinate system) and is

$$\Phi = \frac{\sigma_0}{4\pi}\left[(x - x_1)\ln r_1^2 - (x - x_2)\ln r_2^2 + 2z(\theta_2 - \theta_1)\right]$$

$$+ \frac{\sigma_1}{4\pi}\left[\frac{x^2 - x_1^2 - z^2}{2}\ln r_1^2 - \frac{x^2 - x_2^2 - z^2}{2}\ln r_2^2\right.$$

$$\left. + 2xz(\theta_2 - \theta_1) - x(x_2 - x_1)\right] \quad \text{(panel coordinates)} \quad (11.108)$$

Of particular interest is the case when the point P lies on the element (usually at the center). In this case $z = 0\pm$ and the potential becomes

$$\Phi\left(\frac{x_1 + x_2}{2}, 0\pm\right) = \frac{\sigma_0}{2\pi}(x_2 - x_1)\ln\left(\frac{x_2 - x_1}{2}\right)^2$$

$$+ \frac{\sigma_1}{4\pi}(x_2^2 - x_1^2)\left(\ln\frac{x_2 - x_1}{2} - \frac{1}{2}\right) \quad \text{(panel coordinates)}$$

$$(11.108a)$$

Using the transformation of Eq. (11.23a) (from panel to global coordinates and back) and the above equations for the velocity potential we can formulate a subroutine (e.g., PHILS). From the computational point of view it is more useful to evaluate the panel influence based on its edge values. Since the source values are assumed to be known, based on Fig. 11.30 we can write for the jth panel

$$\sigma_{0_j} = \sigma_j \quad (11.109)$$

$$\sigma_{1_j} = \frac{\sigma_{j+1} - \sigma_j}{x_{j+1} - x_j} \quad (11.109a)$$

where σ_j and σ_{j+1} are the source values at the panel's two edges. So based on the formulas for the velocity potential (Eqs. (11.108) and (11.108a)), including the transformation between the panel and global coordinate systems as in Eq. (11.23a)), and on the substitution of Eqs. (11.109) and (11.109a) the influence subroutine for the linearly varying source is defined as

$$\Phi_S = \text{PHILS}(\sigma_j, \sigma_{j+1}, x, z, x_j, z_j, x_{j+1}, z_{j+1}) \quad (11.110)$$

Here the potential due to the jth element is a function of the coordinates and the panel edge singularity strengths.

Next, the potential at an arbitrary field point P due to the linearly varying strength doublet $\mu(x) = \mu_0 + \mu_1 x$ is obtained by combining Eqs. (10.28) and (10.59):

$$\Phi = \frac{-\mu_0}{2\pi}(\theta_2 - \theta_1) - \frac{\mu_1}{4\pi}\left[2x(\theta_2 - \theta_1) + z\ln\frac{r_2^2}{r_1^2}\right] \quad \text{(panel coordinates)}$$

$$(11.111)$$

When the point P is on the element ($z = 0$, $x_1 < x < x_2$), this reduces to the results of

Eqs. (10.31) and (10.64):

$$\Phi(x, 0\pm) = \mp\left(\frac{\mu_0}{2} + \frac{\mu_1}{2}x\right) \quad \text{(panel coordinates)} \tag{11.111a}$$

Similarly to the source case, a substitution for the doublet strength parameters μ_0 and μ_1, in terms of the panel edge values, gives

$$\mu_{0_j} = \mu_j \tag{11.112}$$

$$\mu_{1_j} = \frac{\mu_{j+1} - \mu_j}{x_{j+1} - x_j} \tag{11.112a}$$

These equations combined with Eq. (11.111) result in

$$\Phi = \frac{-\mu_j}{2\pi}(\theta_{j+1} - \theta_j) - \frac{\mu_{j+1} - \mu_j}{4\pi(x_{j+1} - x_j)}\left[2x(\theta_{j+1} - \theta_j) + z\ln\frac{r_{j+1}^2}{r_j^2}\right]$$

(panel coordinates) $\tag{11.113}$

and now the panel influence depends on the leading and trailing singularity strengths μ_j, μ_{j+1}. It is useful to rearrange this equation so that the first term Φ^a is a function of μ_j and the second part Φ^b is a function of μ_{j+1}. Thus Eq. (11.113) is rewritten as

$$\Phi = \Phi^a + \Phi^b \tag{11.113a}$$

and

$$\Phi^a = \frac{-\mu_j}{2\pi}\left\{\theta_{j+1} - \theta_j - \frac{1}{(x_{j+1} - x_j)}\left[x(\theta_{j+1} - \theta_j) + \frac{z}{2}\ln\frac{r_{j+1}^2}{r_j^2}\right]\right\}$$

(panel coordinates) $\tag{11.114}$

$$\Phi^b = \frac{-\mu_{j+1}}{2\pi(x_{j+1} - x_j)}\left[x(\theta_{j+1} - \theta_j) + \frac{z}{2}\ln\frac{r_{j+1}^2}{r_j^2}\right] \quad \text{(panel coordinates)} \tag{11.115}$$

Again, using the transformation of Eq. (11.23a) (from global to panel coordinates) and the above equations for the velocity potential we can formulate an influence subroutine such that

$$\begin{pmatrix}\Phi \\ \Phi^a \\ \Phi^b\end{pmatrix}_D = \text{PHILD}(\mu_j, \mu_{j+1}, x, z, x_j, z_j, x_{j+1}, z_{j+1}) \tag{11.116}$$

These subroutines (for the source and doublet elements) compute the potential at a point P due to the jth panel and the superscripts $(\;)^a$ and $(\;)^b$ in the case of the doublet element represent the panel influence contributions due to the leading and trailing doublet strengths, respectively.

b. Discretization of Geometry

The $N+1$ panel corner points and N collocation points are generated in a manner similar to the previous examples (see Fig. 11.18). In this case of the internal Dirichlet boundary condition the collocation points must be placed inside the body. This small inward displacement of the collocation point can be skipped if the panel self-induced influence is specified in a separate formula.

11.5 Linearly Varying Singularity Strength Methods (Using the Dirichlet B.C.)

c. Influence Coefficients

The increment in the velocity potential at collocation point i due to a linearly varying strength doublet (of panel j) is computed by using Eq. (11.116):

$$\begin{pmatrix} \Phi \\ \Phi^a \\ \Phi^b \end{pmatrix}_D = \text{PHILD}[(\mu_j = 1, \mu_{j+1} = 1), x_i, z_i, x_j, z_j, x_{j+1}, z_{j+1}] \quad (11.116a)$$

Note that the contribution due to the panel edge singularity strengths is automatically computed (as in Eqs. (11.114) and (11.115)). Thus, for the first collocation point the doublet influence due to the first panel is

$$\Phi_{11}^a \mu_1 + \Phi_{11}^b \mu_2$$

and the influence due to all the doublet panels is

$$\Phi_{D1} = \left(\Phi_{11}^a \mu_1 + \Phi_{11}^b \mu_2\right) + \left(\Phi_{12}^a \mu_2 + \Phi_{12}^b \mu_3\right) + \cdots \\ + \left(\Phi_{1N}^a \mu_N + \Phi_{1N}^b \mu_{N+1}\right) + \Phi_{1W} \mu_W \quad (11.117)$$

where μ_W is the constant-strength wake doublet element (as in Section 11.3.1). The strength of this wake doublet element is set by applying the Kutta condition at the trailing edge such that

$$\mu_W = \mu_{N+1} - \mu_1 \quad (11.118)$$

It is possible to add additional conditions at the trailing edge such as the requirement that the tangential velocity components on the upper and lower surfaces will be equal (or the upper and lower doublet gradients will be equal and opposite in sign). In terms of the four nearest (to the trailing edge) corner-point doublet values this condition becomes

$$\mu_1 - \mu_2 = \mu_N - \mu_{N+1} \quad (11.118a)$$

Equation (11.117) for the potential can be formulated for each collocation point resulting in N equations (for N panels). But a closer examination of the problem reveals that at this phase we have N equations with $N+2$ unknowns $\mu_1, \ldots, \mu_{N+1}, \mu_W$. An additional equation is obtained by specifying the Kutta condition at the trailing edge (Eq. (11.118)). The last "missing" equation can be obtained by specifying the $\Phi_i^* = \Phi_\infty$ Dirichlet condition on an additional point inside the body. This equation will have the form of the regular N boundary conditions (e.g., as in Eq. (11.62)) and for best results it should be specified near the trailing edge (e.g., on the camberline, inside a thick airfoil, or between the upper and lower trailing edge collocation points).

To construct the influence coefficient matrix the potential at any point inside the body due to the doublet distribution can be expressed in terms of the influence coefficients c_{ij}, as in Eq. (11.62), such that the first term in each row (as in Eq. (11.117)) is

$$c_{i1} = \Phi_{i1}^a$$

the $(N+1)$-th term in each row is

$$c_{i,N+1} = \Phi_{iN}^b$$

and the last, $(N+2)$-th, term is the wake contribution:

$$c_{i,N+2} \equiv c_{iW} = \Phi_{iW}$$

(Since the wake is modeled by a constant-strength doublet panel its influence coefficient is calculated by using Eq. (11.66), which is presented in Section 11.3.1.) All the other elements will include the influence of the two neighbor panels:

$$c_{i,j} = \Phi^b_{i,j-1} + \Phi^a_{i,j}, \qquad i \neq 1, N+1, N+2$$

These $N+1$ influence equations and the $(N+2)$-th Kutta condition can be summarized in a matrix as

$$\sum_{j=1}^{N+2} c_{ij}\mu_j = \begin{pmatrix} c_{11} & c_{12} & \cdots & c_{1,N+1} & c_{1W} \\ c_{21} & c_{22} & \cdots & c_{2,N+1} & c_{2W} \\ \vdots & \vdots & \ddots & \vdots & \vdots \\ c_{N+1,1} & c_{N+1,2} & \cdots & c_{N+1,N+1} & c_{N+1,W} \\ 1 & 0 & 0 & \cdots & -1 & 1 \end{pmatrix} \begin{pmatrix} \mu_1 \\ \mu_2 \\ \vdots \\ \mu_{N+1} \\ \mu_W \end{pmatrix}$$

(11.119)

It is possible to substitute the last equation (the Kutta condition) into the previous equations as was done in Section 11.3.1 and reduce the order of the matrix by one to $N+1$ (but for simplicity it is not done for this case).

In this example for the $(N+1)$-th equation, the boundary condition was specified at an additional collocation point inside the body. It is possible to use Eq. (11.118a) instead of this alternative and then Eq. (11.119) will have the form

$$\sum_{j=1}^{N+2} c_{ij}\mu_j = \begin{pmatrix} c_{11} & c_{12} & \cdots & c_{1,N+1} & c_{1W} \\ c_{21} & c_{22} & \cdots & c_{2,N+1} & c_{2W} \\ \vdots & \vdots & \ddots & \vdots & \vdots \\ c_{N,1} & c_{N,2} & \cdots & c_{N,N+1} & c_{N,W} \\ 1 & 0 & 0 & \cdots & -1 & 1 \\ -1 & 1 & 0 & \cdots & 1 & -1 & 0 \end{pmatrix} \begin{pmatrix} \mu_1 \\ \mu_2 \\ \vdots \\ \vdots \\ \mu_{N+1} \\ \mu_W \end{pmatrix}$$

(11.119a)

Here the first N equations are a statement of the doublet contribution to the Dirichlet boundary condition on the N collocation points and the last two equations are forms of the Kutta condition (Eqs. (11.118) and (11.118a)).

d. *Establish RHS Vector*

The combination of source and doublet distributions of Eq. (11.62) is not unique and as in Section 11.3.1 the strength of the sources will be specified by using Eq. (11.61) (note that in this case the panel edge source values σ_j are evaluated at the panel edges),

$$\sigma_j = \mathbf{n}_j \cdot \mathbf{Q}_\infty \tag{11.61}$$

Since the source contribution to the velocity potential is now known, the potential due to all the N panels can be computed by using Eq. (11.110) such that

$$\text{RHS}_i = -\sum_{j=1}^{N} \text{PHILS}(\sigma_j, \sigma_{j+1}, x_i, z_i, x_j, z_j, x_{j+1}, z_{j+1}) \tag{11.120}$$

11.5 Linearly Varying Singularity Strength Methods (Using the Dirichlet B.C.)

The right-hand side vector RHS_i is now evaluated at the N collocation points plus at the $(N+1)$-th point (which is needed when using Eq. (11.119) and was selected inside the body and near the trailing edge).

e. Solve Equations

Substituting the influence coefficients of the doublets and the RHS vector into the boundary condition (Eq. (11.62)) and when using an additional collocation point based equation we get

$$\begin{pmatrix} c_{11} & c_{12} & \cdots & c_{1,N+1} & c_{1W} \\ c_{21} & c_{22} & \cdots & c_{2,N+1} & c_{2W} \\ \vdots & \vdots & \ddots & \vdots & \vdots \\ c_{N+1,1} & c_{N+1,2} & \cdots & c_{N+1,N+1} & c_{N+1,W} \\ 1 & 0 & 0 & \cdots & -1 & 1 \end{pmatrix} \begin{pmatrix} \mu_1 \\ \mu_2 \\ \vdots \\ \mu_{N+1} \\ \mu_W \end{pmatrix} = \begin{pmatrix} RHS_1 \\ RHS_2 \\ \vdots \\ RHS_{N+1} \\ 0 \end{pmatrix}$$
(11.121)

In the case that both of Eqs. (11.118) and (11.118a) are applied at the trailing edge then with the use of Eq. (11.119a) the equations to be solved become

$$\begin{pmatrix} c_{11} & c_{12} & \cdots & c_{1,N+1} & c_{1W} \\ c_{21} & c_{22} & \cdots & c_{2,N+1} & c_{2W} \\ \vdots & \vdots & \ddots & \vdots & \vdots \\ c_{N,1} & c_{N,2} & \cdots & c_{N,N+1} & c_{N,W} \\ 1 & 0 & 0 & \cdots & -1 & 1 \\ -1 & 1 & 0 & \cdots & 1 & -1 & 0 \end{pmatrix} \begin{pmatrix} \mu_1 \\ \mu_2 \\ \vdots \\ \mu_{N+1} \\ \mu_W \end{pmatrix} = \begin{pmatrix} RHS_1 \\ RHS_2 \\ \vdots \\ RHS_N \\ 0 \\ 0 \end{pmatrix}$$
(11.121a)

Either of these full-matrix equations with $N+2$ unknown values μ_j can be solved by standard matrix solvers.

f. Calculation of Pressures and Loads

Once the values of the doublet parameters $(\mu_1, \ldots, \mu_{N+1})$ are known, the tangential velocity component at each collocation point can be calculated by differentiating the local potential:

$$Q_{t_j} = (Q_{t_\infty})_j + \frac{\mu_j - \mu_{j+1}}{x_{j+1} - x_j} \tag{11.122}$$

and the pressure coefficient can be calculated by using Eq. (11.18):

$$C_p = 1 - \frac{Q_t^2}{Q_\infty^2} \tag{11.18}$$

The lift and pitching moment of the panel can be obtained by using the method described by Eqs. (11.78)–(11.80).

A computer program based on this method was formulated and the computed pressure coefficients for the airfoil of Section 6.6 at an angle of attack of 5° are very similar to the data presented in Fig. 11.31. The linearly varying strength doublet portion of this method can be found in Program No. 10 in Appendix D.

As an additional example, the pressure distribution on a four-element airfoil was calculated by using this method and is presented in Fig. 11.33. Experimental data (which are

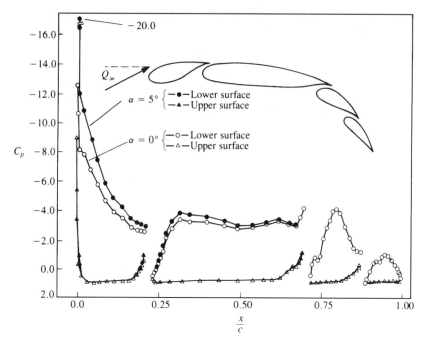

Figure 11.33 Effect of angle of attack on the chordwise pressure distribution of a four-element airfoil.

not presented here) agrees well with these results excluding the large suction peak at the leading edge (which was near $C_p = -13$ in the experimental data). The effect of an angle of attack change is depicted in this figure and it seems that mostly the forward elements are affected by such a change.

11.5.2 Linear Doublet Method

The method of the previous section can be further simplified by equating the source strengths to zero in Eq. (11.60). The value of the constant for the internal potential is selected to be zero and consequently the boundary condition describing the internal potential (Eq. (11.60)) reduces to

$$\sum_{j=1}^{N_1} C_j \mu_j + \Phi_\infty = 0 \qquad (11.81)$$

where N_1 is the number of singularity strength parameters and

$$\Phi_\infty = U_\infty x + W_\infty z \qquad (11.82)$$

Again, note that now $-\mu$ will represent the potential jump from zero to Φ_u on the boundary (see Fig. 11.26) and therefore Φ_u is the local *total* potential (whereas in the previous example $-\mu$ was the jump in the perturbation potential only).

Equation (11.81) can be specified at each collocation point *inside* the body, providing a linear algebraic equation for this point. The steps toward establishing such a numeric solution are very similar to the previous method:

11.5 Linearly Varying Singularity Strength Methods (Using the Dirichlet B.C.)

a. Selection of Singularity Element

The potential at an arbitrary point P due to the jth linearly varying strength doublet element with the two edge values of μ_j and μ_{j+1} is given by Eq. (11.116):

$$\begin{pmatrix} \Phi \\ \Phi^a \\ \Phi^b \end{pmatrix}_D = \text{PHILD}(\mu_j, \mu_{j+1}, x, z, x_j, z_j, x_{j+1}, z_{j+1}) \tag{11.116}$$

Recall that the superscripts $(\ )^a$ and $(\ )^b$ represent the panel influence contributions due to the leading and trailing doublet strengths, respectively.

b. Discretization of Geometry

The $N + 1$ panel corner points and N collocation points are generated in a manner similar to the previous examples (see Fig. 11.18). In this case of the internal Dirichlet boundary condition the collocation points must be placed inside the body. This small inward displacement of the collocation point can be skipped if the panel self-induced influence is specified in a separate formula.

c. Influence Coefficients

The influence of this doublet panel is calculated exactly as in the previous section. The velocity potential at each point is the sum of all the individual panel influences. For example, for the first panel it is given by Eq. (11.117):

$$\Phi_{D1} = (\Phi^a_{11}\mu_1 + \Phi^b_{11}\mu_2) + (\Phi^a_{12}\mu_2 + \Phi^b_{12}\mu_3) + \cdots$$
$$+ (\Phi^a_{1N}\mu_N + \Phi^b_{1N}\mu_{N+1}) + \Phi_{1W}\mu_W \tag{11.117}$$

where μ_W is the constant-strength wake doublet element (as in Section 11.3.1). The strength of this wake doublet element is set by applying the Kutta condition at the trailing edge and is given by Eq. (11.118).

Defining the influence coefficients c_{ij} as in the previous section we can summarize the following $N + 1$ influence relations (where the $(N + 1)$-equation is based on an additional boundary condition inside the body) and the $(N + 2)$-th Kutta condition in a matrix as in Eq. (11.119):

$$\sum_{j=1}^{N+2} C_{ij}\mu_j = \begin{pmatrix} c_{11} & c_{12} & \cdots & c_{1,N+1} & c_{1W} \\ c_{21} & c_{22} & \cdots & c_{2,N+1} & c_{2W} \\ \vdots & \vdots & \ddots & \vdots & \vdots \\ c_{N+1,1} & c_{N+1,2} & \cdots & c_{N+1,N+1} & c_{N+1,W} \\ 1 & 0 & 0 & \cdots & -1 & 1 \end{pmatrix} \begin{pmatrix} \mu_1 \\ \mu_2 \\ \vdots \\ \mu_{N+1} \\ \mu_W \end{pmatrix}$$

(11.119)

Substituting this into the boundary condition (Eq. (11.81)) results in

$$\begin{pmatrix} c_{11} & c_{12} & \cdots & c_{1,N+1} & c_{1W} \\ c_{21} & c_{22} & \cdots & c_{2,N+1} & c_{2W} \\ \vdots & \vdots & \ddots & \vdots & \vdots \\ c_{N+1,1} & c_{N+1,2} & \cdots & c_{N+1,N+1} & c_{N+1,W} \\ 1 & 0 & 0 & \cdots & -1 & 1 \end{pmatrix} \begin{pmatrix} \mu_1 \\ \mu_2 \\ \vdots \\ \mu_{N+1} \\ \mu_W \end{pmatrix} + \begin{pmatrix} \Phi_{\infty_1} \\ \Phi_{\infty_2} \\ \vdots \\ \Phi_{\infty_{N+1}} \\ 0 \end{pmatrix} = 0$$

(11.123)

d. *Establish RHS Vector*

The second term in this equation is known and can be transferred to the right-hand side of the equation. The RHS vector then becomes

$$\begin{pmatrix} \text{RHS}_1 \\ \text{RHS}_2 \\ \vdots \\ \text{RHS}_{N+1} \\ 0 \end{pmatrix} = - \begin{pmatrix} \Phi_{\infty_1} \\ \Phi_{\infty_2} \\ \vdots \\ \Phi_{\infty_{N+1}} \\ 0 \end{pmatrix} \qquad (11.124)$$

and the Φ_{∞_j} term is calculated by using Eq. (11.82).

e. *Solve Equations*

Substituting the influence coefficients of the doublets and the RHS vector into boundary condition of Eq. (11.81) we get

$$\begin{pmatrix} c_{11} & c_{12} & \cdots & c_{1,N+1} & c_{1W} \\ c_{21} & c_{22} & \cdots & c_{2,N+1} & c_{2W} \\ \vdots & \vdots & \ddots & \vdots & \vdots \\ c_{N+1,1} & c_{N+1,2} & \cdots & c_{N+1,N+1} & c_{N+1,W} \\ 1 & 0 & 0 & \cdots & -1 & 1 \end{pmatrix} \begin{pmatrix} \mu_1 \\ \mu_2 \\ \vdots \\ \mu_{N+1} \\ \mu_W \end{pmatrix} = \begin{pmatrix} \text{RHS}_1 \\ \text{RHS}_2 \\ \vdots \\ \text{RHS}_{N+1} \\ 0 \end{pmatrix}$$

$$(11.125)$$

In the case that both of Eqs. (11.118) and (11.118a) are applied at the trailing edge then the boundary condition is specified only at the N collocation points (see previous section). Consequently, with the use of Eq. (11.119a) the equations to be solved become

$$\begin{pmatrix} c_{11} & c_{12} & \cdots & c_{1,N+1} & c_{1W} \\ c_{21} & c_{22} & \cdots & c_{2,N+1} & c_{2W} \\ \vdots & \vdots & \ddots & \vdots & \vdots \\ c_{N,1} & c_{N,2} & \cdots & c_{N,N+1} & c_{N,W} \\ 1 & 0 & 0 & \cdots & -1 & 1 \\ -1 & 1 & 0 & \cdots & 1 & -1 & 0 \end{pmatrix} \begin{pmatrix} \mu_1 \\ \mu_2 \\ \vdots \\ \mu_{N+1} \\ \mu_W \end{pmatrix} = \begin{pmatrix} \text{RHS}_1 \\ \text{RHS}_2 \\ \vdots \\ \text{RHS}_N \\ 0 \\ 0 \end{pmatrix} \qquad (11.125a)$$

Either of these full-matrix equations with $N + 2$ unknown values μ_j can be solved by standard matrix solvers. Note that in this case, too (compared to Eq. (11.121)), the doublet represents the jump in the total potential (and not the perturbation only).

f. *Calculation of Pressures and Loads*

Once the values of the doublet parameters $(\mu_1, \ldots, \mu_{N+1})$ are known, the tangential velocity component at each collocation point can be calculated by differentiating the local potential. For example, such a two-point formula is

$$Q_{t_j} = \frac{\mu_j - \mu_{j+1}}{x_{j+1} - x_j} \qquad (11.126)$$

and the pressure coefficient can be calculated by using Eq. (11.18). The lift and pitching moment of the panel can be obtained by using the method described by Eqs. (11.78)–(11.80).

This method seems to involve less numerical calculations than the equivalent linear doublet/source method and therefore will require somewhat less computational time. (A computer program based on this method is presented in Appendix D, Program No. 10).

11.6 Methods Based on Quadratic Doublet Distribution (Using the Dirichlet B.C.)

To demonstrate some of the techniques needed for higher order methods, the linearly varying elements were used in the previous section. The last method to be tested then should be a linearly varying strength vortex panel based method using the Dirichlet boundary condition. However, such a linearly varying vortex element is equivalent to a quadratic doublet distribution, which will be used to formulate the next two methods.

Also, for the constant-strength singularity distribution based elements, for N panel elements, N boundary condition based equations were constructed. Combined with the Kutta condition a set of $N + 1$ algebraic equations were obtained, including the last unknown, which was the wake doublet strength μ_W. For higher order methods, the number of unknowns increase with the order of approximation and, therefore, additional equations must be specified. In the case of the linear methods, $N - 1$ additional equations (-1 since the trailing-edge point is excluded) were obtained by equating the neighbor panel strengths, which is equivalent to requiring a continuous singularity distribution strength. In this section a generic approach is provided to obtain the additional equations that are needed as the order of approximation for the singularity strength distribution increases. Usually these equations are based on requirements such as smooth first, second, and higher derivatives at the panel corner points between the two neighbor panels, but there are different methods of obtaining these additional equations that can be optimized for certain problems. Therefore, the approach presented here is mainly to demonstrate the method, but improvement of these methods to tailor them for specific problems is encouraged.

11.6.1 Linear Source/Quadratic Doublet Method

For this example, too, the approach of Section 11.3 is used and a combination of linearly varying strength sources and vortices will be distributed on the solid boundaries S (Fig. 11.34) and the vortex distribution is replaced by a quadratic doublet distribution. Following Section 11.3.1 we select the velocity potential within the volume enclosed by S as $\Phi_i^* = \Phi_\infty$ and the boundary condition at each collocation point is reduced to the form given by Eq. (11.62):

$$\sum_{j=1}^{N_1} B_j \sigma_j + \sum_{j=1}^{N_1} C_j \mu_j = 0 \qquad (11.62)$$

where N_1 is the number of singularity strength parameters. In this section a solution based on this equation using linearly varying source and quadratic doublet distributions will be established. The wake, however, will be modeled by a constant-strength doublet as shown

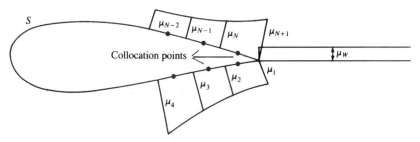

Figure 11.34 Doublet distribution near the trailing edge of an airfoil.

in Fig. 11.34. Now, to develop the method of solution, let us follow the basic six-step procedure:

a. Selection of Singularity Element

The potential at an arbitrary point P due to a linearly varying strength source element with the edge values of σ_j and σ_{j+1} was derived in the previous section and is given by Eq. (11.110):

$$\Phi_S = \text{PHILS}(\sigma_j, \sigma_{j+1}, x_i, z_i, x_j, z_j, x_{j+1}, z_{j+1}) \tag{11.110}$$

Similarly, the potential at point P due to a quadratic doublet distribution element where the doublet strength (in panel local coordinates) varies as

$$\mu(x) = \mu_0 + \mu_1 x + \mu_2 x^2$$

is obtained by combining Eqs. (10.28), (10.59), and (10.80):

$$\Phi = \frac{-\mu_0}{2\pi}(\theta_2 - \theta_1) - \frac{\mu_1}{4\pi}\left[2x(\theta_2 - \theta_1) + z \ln \frac{r_2^2}{r_1^2}\right]$$
$$+ \frac{\mu_2}{2\pi}\left[(x^2 - z^2)(\theta_1 - \theta_2) - xz \ln \frac{r_2^2}{r_1^2} + z(x_1 - x_2)\right] \quad \text{(panel coordinates)} \tag{11.127}$$

where the variables r_1, r_2, θ_1, and θ_2 are shown in Fig. 11.35.

When the point P is on the element ($z = 0\pm$, $x_1 < x < x_2$) then this reduces to a combination of Eqs. (10.31), (10.64), and (10.84):

$$\Phi(x, 0\pm) = \mp\left(\frac{\mu_0}{2} + \frac{\mu_1}{2}x + \frac{\mu_2 x^2}{2}\right) \quad \text{(panel coordinates)} \tag{11.128}$$

Using the transformation of Eq. (11.23a) (from global to panel coordinates and back) and the above equations for the velocity potential we can formulate an influence subroutine for the quadratic doublet element such that

$$\begin{pmatrix} \Phi \\ \Phi^a \\ \Phi^b \\ \Phi^c \end{pmatrix}_D = \text{PHIQD}[(\mu_0, \mu_1, \mu_2)_j, x, z, x_j, z_j, x_{j+1}, z_{j+1}] \tag{11.129}$$

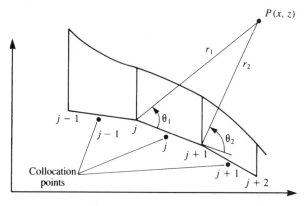

Figure 11.35 Nomenclature for a quadratic doublet panel element.

This subroutine computes the potential at a point P due to the jth panel and the superscripts $(\ )^a$, $(\ )^b$, and $(\ )^c$ represent the panel influence contributions due to the terms in Eq. (11.127) multiplied by μ_0, μ_1, and μ_2, respectively. The increment in the potential at point P is therefore the sum of these three terms,

$$\Phi = \Phi^a + \Phi^b + \Phi^c \tag{11.130}$$

It is assumed that these three components of the velocity potential are automatically computed by this subroutine.

b. Discretization of Geometry

The $N+1$ panel corner points and N collocation points are generated in a manner similar to the previous examples (see Fig. 11.18). In this case of the internal Dirichlet boundary condition the collocation points must be placed inside the body. This small inward displacement of the collocation point can be skipped if the panel self-induced influence is specified in a separate formula.

c. Influence Coefficients

The increment in the velocity potential at collocation point i due to a quadratic doublet element is computed by using Eq. (11.129) (note that a unit strength was assigned to all three doublet distribution coefficients μ_0, μ_1, and μ_2):

$$\begin{pmatrix} \Phi \\ \Phi^a \\ \Phi^b \\ \Phi^c \end{pmatrix}_D = \text{PHIQD}[(\mu_0 = 1, \mu_1 = 1, \mu_2 = 1)_j, x, z, x_j, z_j, x_{j+1}, z_{j+1})] \tag{11.129a}$$

For example, calculation of the velocity potential increment by this formula for the first collocation point due to the first panel yields

$$(\Phi_1^a \mu_0 + \Phi_1^b \mu_1 + \Phi_1^c \mu_2)_1 \equiv a_{11}\mu_{01} + b_{11}\mu_{11} + c_{11}\mu_{12}$$

Here the first subscript of the doublets is used as the doublet parameter counter (0, 1, and 2) while the second subscript is the influencing panel counter. Similarly, for the coefficients a, b, and c, the first counter is the collocation point counter and the second is the influencing panel counter. The potential at this first collocation point due to all the doublet panels is

$$\Phi_{D1} = (a_{11}\mu_{01} + b_{11}\mu_{11} + c_{11}\mu_{21}) + (a_{12}\mu_{02} + b_{12}\mu_{12} + c_{12}\mu_{22}) + \cdots$$
$$+ (a_{1N}\mu_{0N} + b_{1N}\mu_{1N} + c_{1N}\mu_{2N}) + A_{1W}\mu_W \tag{11.131}$$

where μ_W is the constant-strength wake doublet element (as in Section 11.3.1) and A_{1W} is the wake influence coefficient, which can be calculated by using Eq. (11.66). The strength of this wake doublet element is set by applying the Kutta condition at the trailing edge and is given by Eq. (11.118):

$$\mu_W = \mu_U - \mu_L \tag{11.132}$$

where μ_U and μ_L are the upper and lower panel doublet strengths at the trailing edge, respectively.

Equation (11.131) for the potential can be formulated for each collocation point, resulting in N equations (for N panels) with $3N$ unknowns $(\mu_0, \mu_1, \mu_2)_j$ and one more unknown, which is the wake doublet strength μ_W. Additional equations can be obtained by requiring that the doublet distribution and its gradient will be continuous between neighboring elements. Applying this to the point between the jth and the $(j+1)$-th panel for the doublet strength we get

$$\mu_{0j} + \mu_{1j}\Delta c_j + \mu_{2j}(\Delta c_j)^2 = \mu_{0,j+1} \tag{11.133a}$$

and for the doublet gradient

$$\mu_{1j} + 2\mu_{2j}\Delta c_j = \mu_{1,j+1} \tag{11.133b}$$

where Δc_j is the panel length. Equations (11.133a,b) can be applied to all panel corner points, excluding the trailing edge, and thereby result in $2N - 2$ equations. The last three equations are found as follows:

1. Equation (11.133b) can be applied at the trailing edge, so that the upper and lower doublet gradients will be equivalent:

$$\mu_{1N} + 2\mu_{2N}\Delta c_N = \mu_{11} \tag{11.134}$$

2. The Kutta condition is specified at the trailing edge by using Eq. (11.132):

$$\mu_W = \mu_{0N} + \mu_{1N}\Delta c_N + \mu_{2N}(\Delta c_N)^2 - \mu_{01} \tag{11.135}$$

3. The last "missing" equation can be obtained by specifying the $\Phi_i^* = \Phi_\infty$ Dirichlet boundary condition (Eq. 11.62) on an additional point inside the body. This equation will have the form of the regular N boundary conditions (e.g., as in Eq. (11.131)) and for best results it needs to be specified near the trailing edge (e.g., on the camberline toward the trailing edge, inside a thick airfoil, or between the upper and lower trailing edge collocation points).

d. *Establish RHS Vector*

The boundary condition given by Eq. (11.62) is specified at the $N + 1$ collocation points and the strength of the sources will be specified by using Eq. (11.61) (note that in this case the panel edge source values σ_j are evaluated at the panel edges):

$$\sigma_j = \mathbf{n}_j \cdot \mathbf{Q}_\infty \tag{11.61}$$

Since the source contribution to the velocity potential is now known, the potential due to all the N panels (with $N + 1$ panel edge source values) can be computed by using Eq. (11.120) such that

$$\text{RHS}_i = -\sum_{j=1}^{N} \text{PHILS}(\sigma_j, \sigma_{j+1}, x_i, z_i, x_j, z_j, x_{j+1}, z_{j+1}) \tag{11.120}$$

The right-hand side vector RHS_i is now evaluated at the N collocation points plus at the $(N + 1)$-th point (which was selected inside the body and near the trailing edge).

e. *Solve Equations*

At this point it is possible to establish $N + 1$ equations based on the boundary condition (Eq. (11.62)) plus two equations based on the trailing-edge conditions (Eqs. (11.134)

11.6 Methods Based on Quadratic Doublet Distribution (Using the Dirichlet B.C.) 319

and (11.135)) and they will have the following form:

$$
\begin{pmatrix}
a_{11}\,b_{11}\,c_{11} & a_{21}\,b_{21}\,c_{21} & \cdots & a_{1,N}\,b_{1,N}\,c_{1,N} & A_{1W} \\
a_{21}\,b_{21}\,c_{21} & a_{22}\,b_{22}\,c_{22} & \cdots & a_{2,N}\,b_{2,N}\,c_{2,N} & A_{2W} \\
\vdots & \vdots & \ddots & \vdots & \vdots \\
a_{N+1,1}b_{N+1,1}c_{N+1,1} & abc_{N+1,1} & \cdots & a_{N+1,N}b_{N+1,N}c_{N+1,N} & A_{N+1,W} \\
0,\,-1,\,0 & 0 & \cdots & 0,\,1,\,2\Delta c_N & 0 \\
-1,\,0,\,0 & 0 & \cdots & 1,\,\Delta c_N,\,(\Delta c_N)^2 & -1
\end{pmatrix}
\begin{pmatrix}
\mu_{01} \\ \mu_{11} \\ \mu_{21} \\ \mu_{02} \\ \mu_{12} \\ \mu_{22} \\ \vdots \\ \mu_{2N} \\ \mu_W
\end{pmatrix}
$$

$$
=
\begin{pmatrix}
\mathrm{RHS}_1 \\ \mathrm{RHS}_2 \\ \vdots \\ \mathrm{RHS}_{N+1} \\ 0 \\ 0
\end{pmatrix}
\tag{11.136}
$$

In this equation the matrix will have $3N+1$ columns (same as the μ vector) and $N+3$ rows (same as the RHS vector). The additional $2(N-1)$ equations can be obtained by using Eqs. (11.133)–(11.135). However, from the computational point of view it is desirable to reduce the influence matrix size, which can be done by substituting the above equations backward while calculating the influence matrix.

It is possible to rearrange Eqs. (11.133a,b) to create a regression formula such that

$$
\mu_{1j} = \frac{2}{\Delta c_j}(\mu_{0,j+1} - \mu_{0j}) - \mu_{1,j+1} \tag{11.137a}
$$

$$
\mu_{2j} = \frac{1}{2\Delta c_j}(\mu_{1,j+1} - \mu_{1j}) \tag{11.137b}
$$

Using these regression equations, all μ_{1j} and μ_{2j} unknowns can be eliminated (by performing column operations on the matrix), excluding the last two (μ_{1N} and μ_{2N}). These algebraic operations can be performed automatically when assembling the matrix coefficient and will reduce Eq. (11.136) to the form

$$
\begin{pmatrix}
A_{11} & A_{12} & A_{13} & \cdots & A_{1,N+2} & A_{1W} \\
A_{21} & A_{22} & A_{23} & \cdots & A_{2,N+2} & A_{2W} \\
\vdots & \vdots & \vdots & \ddots & \vdots & \vdots \\
A_{N+1,1} & A_{N+1,2} & A_{N+1,3} & \cdots & A_{N+1,N+2} & A_{N+1,W} \\
0 & A_{N+2,2} & 0 & \cdots & 2\Delta c_N & 0 \\
-1 & A_{N+3,2} & 0 & \cdots & (\Delta c_N)^2 & -1
\end{pmatrix}
\begin{pmatrix}
\mu_{01} \\ \mu_{02} \\ \mu_{03} \\ \vdots \\ \mu_{0N} \\ \mu_{1N} \\ \mu_{2N} \\ \mu_W
\end{pmatrix}
=
\begin{pmatrix}
\mathrm{RHS}_1 \\ \mathrm{RHS}_2 \\ \vdots \\ \mathrm{RHS}_{N+1} \\ 0 \\ 0
\end{pmatrix}
\tag{11.138}
$$

where now we have $N+3$ equations with $N+3$ unknowns $\mu_{01}, \mu_{02}, \ldots, \mu_{0N}, \mu_{1N}, \mu_{2N}, \mu_W$ and A_{ij} are the new coefficients obtained after the resubstitution process.

Solution of the full-matrix equation (Eq. (11.138)) with $N+3$ unknown values μ_j can be obtained by standard matrix solvers. Then the local panel doublet distribution can be obtained by using the same regression formulation that was used to reduce the number of columns in the influence coefficient matrix.

f. Calculation of Pressures and Loads

Once the values of the doublet parameters ($\mu_{01}, \mu_{11}, \mu_{02}, \ldots, \mu_{0N}, \mu_{1N}, \mu_{2N}$) are known, the panel doublet distribution can be calculated (with Eqs. (11.132)). Then the perturbation velocity at each collocation point can be calculated by using the tangential derivative of the doublet at the panel center and the total tangential velocity becomes

$$Q_{t_j} = Q_{t_\infty} - \left(\mu_{1j} + 2\mu_{2j}\frac{\Delta c_j}{2}\right) \qquad (11.139)$$

and the pressure coefficient can be calculated by using Eq. (11.18). The lift and pitching moment of the panel can be obtained by using the method described by Eqs. (11.78)–(11.80).

A computer program based on the quadratic doublet method is presented in Appendix D (Program No. 11) and the formulation of the doublet-only method is very similar to this one. The pressure coefficients computed with this method for the airfoil of Section 6.6 at an angle of attack of 5° are very similar to the data presented in Fig. 11.31.

11.6.2 Quadratic Doublet Method

The method of the previous section can be further simplified by equating the source strengths to zero in Eq. (11.60). The value of the constant for the internal potential is selected to be zero and consequently the boundary condition describing the internal potential (Eq. (11.60)) reduces to

$$\sum_{j=1}^{N_1} C_j \mu_j + \Phi_\infty = 0 \qquad (11.81)$$

where N_1 is the number of singularity strength parameters and

$$\Phi_\infty = U_\infty x + W_\infty z \qquad (11.82)$$

Again, note that now μ will represent the potential jump from zero to Φ_u on the boundary (see Fig. 11.26) and therefore Φ_u is the local *total* potential (whereas in the previous example μ was the jump in the perturbation potential only).

Equation (11.81) can be specified at each collocation point *inside* the body, providing a linear algebraic equation for this point. The steps toward establishing such a numeric solution are very similar to the previous method.

a. Selection of Singularity Element

The potential at an arbitrary point P due to the jth doublet element with the three doublet parameters μ_0, μ_1, and μ_2 is given by Eq. (11.129):

$$\begin{pmatrix} \Phi \\ \Phi^a \\ \Phi^b \\ \Phi^c \end{pmatrix}_D = \text{PHIQD}[(\mu_0, \mu_1, \mu_2)_j, x, z, x_j, z_j, x_{j+1}, z_{j+1}] \qquad (11.129)$$

Recall that the superscripts $(\,)^a$, $(\,)^b$, and $(\,)^c$ represent the panel influence contributions due to the three coefficients describing the panel doublet distribution.

11.6 Methods Based on Quadratic Doublet Distribution (Using the Dirichlet B.C.)

b. *Discretization of Geometry*

The $N+1$ panel corner points and N collocation points are generated in a manner similar to the previous examples (see Fig. 11.18). In this case of the internal Dirichlet boundary condition the collocation points must be placed inside the body. This small inward displacement of the collocation point can be skipped if the panel self-induced influence is specified in a separate formula.

c. *Influence Coefficients*

The influence of this doublet panel is calculated exactly as in the previous section. The velocity potential at each point is the sum of all the individual panel influences. For example, for the first collocation point due to the first panel it is

$$\left(\Phi_1^a \mu_0 + \Phi_1^b \mu_1 + \Phi_1^c \mu_2\right)_1 \equiv a_{11}\mu_{01} + b_{11}\mu_{11} + c_{11}\mu_{12}$$

and the potential at this collocation point due to all the doublet panels is given by Eq. (11.131):

$$\Phi_{D1} = (a_{11}\mu_{01} + b_{11}\mu_{11} + c_{11}\mu_{21}) + (a_{12}\mu_{02} + b_{12}\mu_{12} + c_{12}\mu_{22}) + \cdots$$
$$+ (a_{1N}\mu_{0N} + b_{1N}\mu_{1N} + c_{1N}\mu_{2N}) + A_{1W}\mu_W \quad (11.131)$$

where μ_W is the constant-strength wake doublet element (as in Section 11.3.1) and A_{1W} is the wake influence coefficient, which can be calculated by using Eq. (11.66). The strength of this wake doublet element is set by applying the Kutta condition at the trailing edge and is given by Eq. (11.135).

Using the backward substitution process described in the previous section the potential at the $N+1$ collocation points (the additional point is inside the body and near the trailing edge) will have the form

$$\begin{pmatrix} \Phi_1 \\ \Phi_2 \\ \vdots \\ \Phi_{N+1} \\ 0 \\ 0 \end{pmatrix} = \begin{pmatrix} A_{11} & A_{12} & A_{13} & \cdots & A_{1,N+2} & A_{1W} \\ A_{21} & A_{22} & A_{23} & \cdots & A_{2,N+2} & A_{2W} \\ \vdots & \vdots & \vdots & \ddots & \vdots & \vdots \\ A_{N+1,1} & A_{N+1,2} & A_{N+1,3} & \cdots & A_{N+1,N+2} & A_{N+1,W} \\ 0 & A_{N+2,2} & 0 & \cdots & 2\Delta c_N & 0 \\ -1 & A_{N+3,2} & 0 & \cdots & (\Delta c_N)^2 & -1 \end{pmatrix} \begin{pmatrix} \mu_{01} \\ \mu_{02} \\ \mu_{03} \\ \vdots \\ \mu_{0N} \\ \mu_{1N} \\ \mu_{2N} \\ \mu_W \end{pmatrix}$$
(11.140)

Here the last two equations are the trailing-edge conditions, based on Eqs. (11.134) and (11.135), and the coefficients A_{ij} are the result of the backward substitution as described in the previous section. Substitution of this into the boundary condition (Eq. (11.81)) results in

$$\begin{pmatrix} A_{11} & A_{12} & A_{13} & \cdots & A_{1,N+2} & A_{1W} \\ A_{21} & A_{22} & A_{23} & \cdots & A_{2,N+2} & A_{2W} \\ \vdots & \vdots & \vdots & \ddots & \vdots & \vdots \\ A_{N+1,1} & A_{N+1,2} & A_{N+1,3} & \cdots & A_{N+1,N+2} & A_{N+1,W} \\ 0 & A_{N+2,2} & 0 & \cdots & 2\Delta c_N & 0 \\ -1 & A_{N+3,2} & 0 & \cdots & (\Delta c_N)^2 & -1 \end{pmatrix} \begin{pmatrix} \mu_{01} \\ \mu_{02} \\ \mu_{03} \\ \vdots \\ \mu_{0N} \\ \mu_{1N} \\ \mu_{2N} \\ \mu_W \end{pmatrix} + \begin{pmatrix} \Phi_{\infty_1} \\ \Phi_{\infty_2} \\ \vdots \\ \Phi_{\infty_{N+1}} \\ 0 \\ 0 \end{pmatrix} = 0$$
(11.141)

where again the last two equations are the trailing-edge conditions and Φ_{∞_j} are known (e.g., from Eq. (11.82)).

d. Establish RHS Vector

The second term in Eq. (11.141) is known and can be transferred to the right-hand side of the equation. The RHS vector then becomes

$$\begin{pmatrix} \text{RHS}_1 \\ \text{RHS}_2 \\ \vdots \\ \text{RHS}_{N+1} \\ \text{RHS}_{N+2} \\ \text{RHS}_{N+3} \end{pmatrix} = - \begin{pmatrix} \Phi_{\infty_1} \\ \Phi_{\infty_2} \\ \vdots \\ \Phi_{\infty_{N+1}} \\ 0 \\ 0 \end{pmatrix} \quad (11.142)$$

and the Φ_{∞_j} terms are calculated by using Eq. (11.82).

e. Solve Equations

Substituting the influence coefficients of the doublets and the RHS vector into the boundary condition of Eq. (11.141) we get

$$\begin{pmatrix} A_{11} & A_{12} & A_{13} & \cdots & A_{1,N+2} & A_{1W} \\ A_{21} & A_{22} & A_{23} & \cdots & A_{2,N+2} & A_{2W} \\ \vdots & \vdots & \vdots & \ddots & \vdots & \vdots \\ A_{N+1,1} & A_{N+1,2} & A_{N+1,3} & \cdots & A_{N+1,N+2} & A_{N+1,W} \\ 0 & A_{N+2,2} & 0 & \cdots & 2\Delta c_N & 0 \\ -1 & A_{N+3,2} & 0 & \cdots & (\Delta c_N)^2 & -1 \end{pmatrix} \begin{pmatrix} \mu_{01} \\ \mu_{02} \\ \mu_{03} \\ \vdots \\ \mu_{0N} \\ \mu_{1N} \\ \mu_{2N} \\ \mu_W \end{pmatrix} = \begin{pmatrix} \text{RHS}_1 \\ \text{RHS}_2 \\ \vdots \\ \text{RHS}_{N+1} \\ 0 \\ 0 \end{pmatrix}$$

(11.143)

This full-matrix equation with $N+3$ unknown values μ_j can be solved by standard matrix solvers. Note that in this case (compared to Eq. (11.136)) the doublet represents the jump in the total potential (and not the perturbation only).

f. Calculation of Pressures and Loads

Once the values of the doublet parameters ($\mu_{01}, \ldots, \mu_{0N}, \mu_{1N}, \mu_{2N}$) are known, each panel doublet distribution can be obtained by using the backward substitution equations (e.g., Eqs. (11.134), (11.135), and (11.137)). Then the velocity at each collocation point can be calculated by differentiating the local potential:

$$Q_{t_j} = -\left(\mu_{1j} + 2\mu_{2j}\frac{\Delta c_j}{2}\right) \quad (11.144)$$

and the pressure coefficient can be calculated by using Eq. (11.18). The lift and pitching moment of the panel can be obtained by using the method described by Eqs. (11.78)–(11.80).

This method seems to involve less numerical calculations than the equivalent quadratic doublet/linear source method and therefore will require somewhat less computational time. (A computer program based on this method is presented in Appendix D, Program No. 11.)

11.7 Some Conclusions about Panel Methods

The examples presented in this chapter indicate that most methods can yield reasonable results. The methods were presented in their simplest form and their computational efficiency usually can be improved. For example, when calculating the influence of the panels, terms that depend on panel cornerpoint geometry are calculated twice (for each of the neighbor panels) and this redundancy can easily be corrected in the computer programming phase.

It seems that in terms of ease of construction and the least computational effort the constant-strength doublet method with the internal velocity potential boundary condition is the most successful. Also, in general, the use of the velocity potential boundary condition will result in fewer numerical manipulations and hence less computational time.

The use of higher order methods requires more computational effort and is justified when the velocity near the body must be continuous (as inside the gaps of multielement airfoils). However, constant-strength singularity element based methods can give good results, too, when a sufficient number of panels are used (see Fig. 11.28).

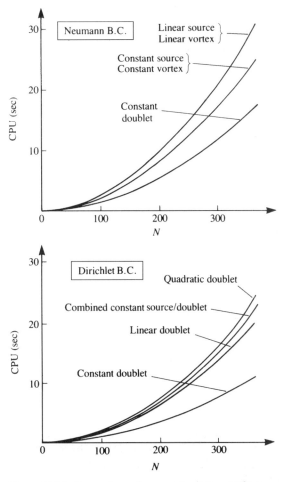

Figure 11.36 Comparison of computational time (CPU, in seconds of VAX-6000-320 computer) among the various panel methods versus number of panels.

All of the methods presented for the solution of lifting flows can be extended to include several bodies (or airfoils) and then for each element a separate Kutta condition is used. As an example, the chordwise pressure distribution along a four-element airfoil is presented in Fig. 11.33. The computation was done with a linear-strength source and doublet combination using the internal Dirichlet boundary condition.

Most of the methods presented here were investigated in Ref. 11.1 and the computation times (CPU, measured in seconds of VAX-6000-320 computer) versus number of panels N are presented in Fig. 11.36. In these data the matrix inversion time (which has the same order of magnitude) was subtracted to increase the resolution in the figure. These data indicate that the constant doublet method with the Dirichlet boundary condition is the fastest, and computational effort increases with increasing the order of the method. (However, it seems that low and higher order methods can converge to solutions of similar quality, in terms of circulation and lift, with a similar number of panels.)

It is noted, too, that each computational method depends on the grid and on various other parameters. Therefore each technique must be validated first before it can be applied to unknown cases. As an example, the sensitivity of the linear doublet (with the internal Dirichlet boundary condition) to panel density is presented in Fig. 11.37. The very low density of five upper and five lower panels resulted in a crude solution, which improved considerably by doubling the number of panels. When panel density was increased to 70, results similar to those presented in Fig. 11.31 were obtained.

Another example for the sensitivity of the methods to geometrical details is presented in Fig. 11.38. Here the $(N + 1)$-th collocation point for the quadratic doublet (with Dirichlet B.C.) is moved inside the airfoil. In Fig. 11.38a this collocation point is placed near the trailing edge and the results for both $\alpha = 0°$ and $5°$ are good. (Bringing this point too close to the trailing-edge panel collocation points, though, may cause the matrix to be ill-conditioned

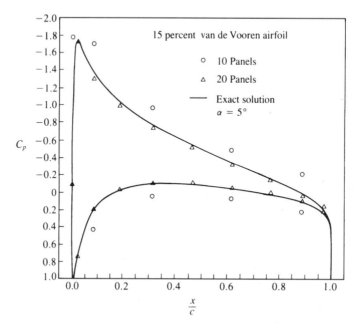

Figure 11.37 Effect of panel density on the computed pressure distribution, using very few panels (linear doublet method with the Dirichlet B.C.).

11.7 Some Conclusions about Panel Methods

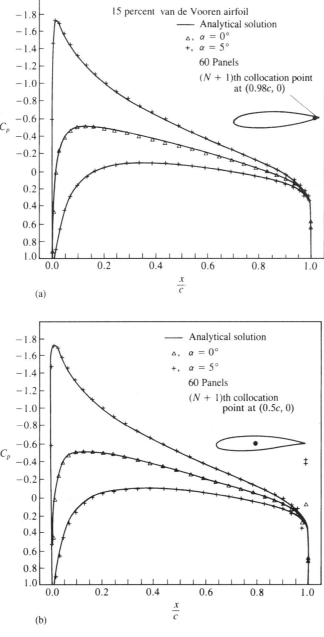

Figure 11.38 Effect of placing the $(N+1)$th collocation point inside the 15% thick van de Vooren airfoil (using a quadratic doublet method with the Dirichlet B.C.).

for large panel numbers.) But if this point is placed at the center of the airfoil (as shown in Fig. 11.38b) the results near the trailing edge become erratic.

Another interesting problem arises when attempting to model airfoils with cusped (very thin) trailing edges. The geometry of such an airfoil is presented in the inset to Fig. 11.39a, and more information on this 15%-thick airfoil is provided in Sections 6.6 and 6.7.

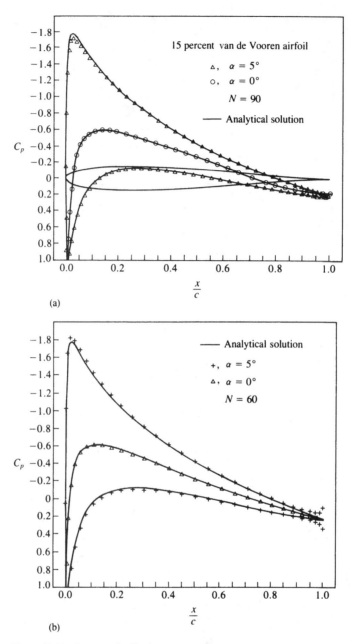

Figure 11.39 Pressure distribution on a cusped trailing edge 15%-thick van de Vooren airfoil using: (a) linear vortex method with Neumann B.C., and (b) constant-strength source/doublet method with the Dirichlet B.C.

Most methods will have problems near the trailing edge because of the very tight placing of the collocation points. This is illustrated in Fig. 11.39b, where the data were calculated with a constant-strength source/doublet method using the Dirichlet B.C. Such problems can be cured by modeling a finite angle there (instead of zero angle) and this can be achieved by simply having larger trailing-edge panels. Also, numerical

11.7 Some Conclusions about Panel Methods

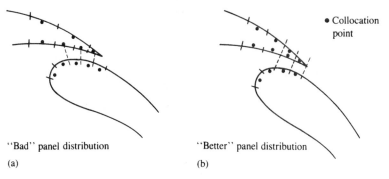

"Bad" panel distribution
(a)

"Better" panel distribution
(b)

Figure 11.40 Recommended and not recommended options for panel distributions inside the gap of a two-element airfoil.

experimentation reported in Ref. 9.2 indicated that when using the velocity formulation for the Kutta condition (see Eq. 9.15b), the magnitude of these problems near the trailing edge was considerably reduced. The linear-strength vortex method with the Neumann B.C. seemed to be the only method that was not sensitive to this cusped trailing-edge problem (see Fig. 11.39a).

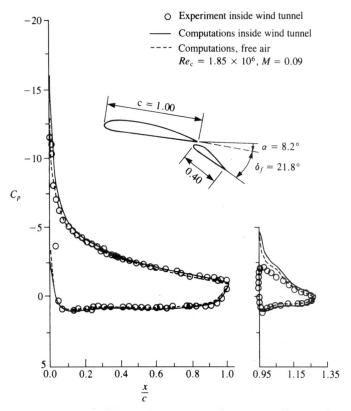

Figure 11.41 Two-dimensional experimental and computed (constant-source/doublet, with Dirichlet B.C.) chordwise pressure distribution on a NACA 4412 wing and a NACA 4415 flap (flap chord is 40% of wing chord). Experiments from Adair, D., and Horne, W. C., "Turbulent Separated Flow in the Vicinity of a Single-Slotted Airfoil Flap," *AIAA Paper 88-0613*, Jan. 1988.

In conclusion, most methods will work and can be tailored to particular needs, and in many cases, problems can be avoided by simple means such as selecting a better spacing of the panels. As a final example to enforce this statement consider the panel distribution inside the gap of a two-element airfoil, as shown in Fig. 11.40a. Since the lower surface collocation points are close to the panel corner points the influence of this panel can be overestimated (e.g., in the case of a vortex being at the panel corner point). A simple rearrangement of the panels, as shown in Fig. 11.40b, can improve the solution in this area.

Finally, before concluding this chapter we must note that the present analysis is based on potential flow theory and, for example, the calculated drag coefficient is zero. However, the viscous boundary layer does result in certain values of drag coefficient (even at zero lift) and a large selection of such information is provided by Abbot and Doenhoff.[11.2] The effect of viscosity on the pressure distribution (for the smaller angles of attack) is usually small but at larger angles of attack flow separation may cause the pressure distribution to change considerably (for more details see Chapters 14 and 15). As an example, the calculated (by constant-strength doublet/source method with Dirichlet B.C.) and experimental pressure distribution on a NACA 4412 airfoil with a NACA 4415 flap is presented in Fig. 11.41. In this condition the airfoil is near stall, that is, the flow on the front airfoil is attached and on the flap a limited trailing-edge separation is present. Near the leading edge the calculated suction peak is larger than the experimental data and on the flap it is considerably less because of the trailing-edge separation. Also, in general, even for the attached flow case the experimental circulation is slightly less, as indicated by the comparison between the experimental and calculated data on the first airfoil element in Fig. 11.41.

References

[11.1] Yon, S., "Formulation of and Comparison among Various Two-Dimensional Panel Methods," M.S. Thesis, Dept. of Aerospace Engineering and Engineering Mechanics, San Diego State University, Summer 1990.

[11.2] Abbott, I. H., and Von Doenhoff, A. E., *Theory of Wing Sections*, Dover, 1959.

Problems

This chapter allows the formulation and study of a large number of interesting problems. If the geometry is limited to airfoils, then a generic problem will have the following form:

a. Construct a numerical technique, based on one or more of the methods in this chapter.

b. Calculate the pressure distribution, lift and moment coefficient, and center of pressure and:

  I. Compare with results presented in Ref. 11.2.

  II. Study one or more of the aerodynamic problems presented in Fig. 11.42.

  III. Study effect of grid density, compare between various methods, etc.

Ample useful information on airfoil shapes, zero-angle pressure distributions, lift, and drag is presented in Ref. 11.2, and the use of this information for homework problems and student projects is recommended.

As an example consider the following possible problems:

11.1. Investigate the problems of ground effect, tandem wings, and biplane as shown in Figs. 11.42a–c. Use five-point vortex elements to model a flat plate (say at $\alpha = 5°$)

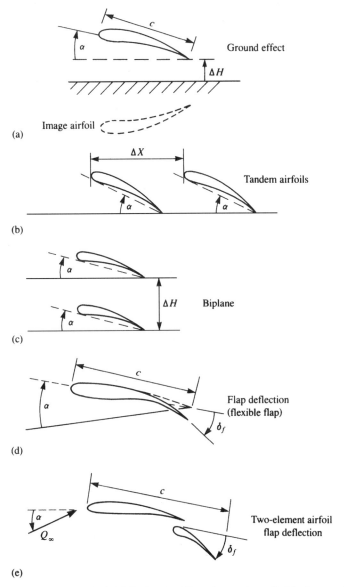

Figure 11.42 Typical airfoil-related problems that can be studied (as homework problems) by using the methods of this chapter.

and investigate the change in the pressure distribution and lift as the distance (ΔH or Δx in Fig. 11.42a–c) is being increased.

11.2. Construct a constant-strength source based computer program to calculate the pressure distribution over a symmetric NACA 0012 airfoil and compare to the results of Ref. 11.2 (p. 321; also, the airfoil coordinates appear on this page).

11.3. Develop a constant-strength source/doublet method (using the Dirichlet B.C.) to study the flow over a NACA 63_2-415 airfoil (see Ref. 11.2, pp. 418 and 528).

Calculate the pressure distribution over the airfoil, center of pressure, and lift, as a function of angle of attack. (For a more advanced assignment study the effect of a 20% trailing-edge flap with $\delta_f = 10°$; see Fig. 11.42d.)

11.4. Extend the computer program of Problem 11.3 to the case of a two-element airfoil (Fig. 11.42e, where the flap chord is one half of the main airfoil's chord). Use the NACA 63_2-415 airfoil section for both elements and check the effect of flap deflection on lift, pitching moment, and center of pressure.

CHAPTER 12

Three-Dimensional Numerical Solutions

Three dimensional numerical solutions based on surface singularity distributions are similar, in principle, to methods presented for the two-dimensional case. From the theoretical aspect, only the wake and the trailing-edge conditions (three-dimensional Kutta condition) will require some additional attention. The most difficult aspect in three dimensions, though, is the modeling of the geometry, especially when arbitrary geometry capability is sought.

In the first part of this chapter the geometry (of the wing) is kept relatively simple and the aerodynamics of a thin lifting surface is modeled. In principle, this simple method has all the elements of the more complex panel methods and is capable of modeling the effect of wing planform shape on the fluid dynamic loads. In addition, the examples that are being presented require only limited programming effort and, therefore, are suitable for classroom instruction. Furthermore, the introduction in class of the numerical lifting-line model (Section 12.1), next to Prandtl's lifting-line model of Section 8.1, provides additional insight and a clear explanation of the spanwise integral equation.

In the second part of this chapter the principles of panel codes capable of solving the flow over bodies with arbitrary three-dimensional geometry will be presented. Over the years many such methods were developed and improved, but recent trends indicate an increased use of the approach that is based on the combination of surface source and doublet distributions with the inner potential boundary condition (for closed bodies). Consequently, only this approach will be presented through a brief description of one low-order and one high-order panel method.

In terms of classroom instruction it is recommended at this phase that use be made of one of the commercially available panel codes and that students be trained first to use the pre- and post-processor. This graphic pre-processor generates the surface grid (panel model) that is used to define the input to the computer program. The post-processor is usually a graphic utility that allows for a rapid analysis of the three-dimesional results by using extensive graphic representations. It is believed that after studying and preparing examples with the lifting surface code in this chapter (Section 12.3) students can safely proceed to use a larger panel code since at this phase they must have a deep understanding of the formulation and the construction of these codes.

12.1   Lifting-Line Solution by Horseshoe Elements

As a first example, consider the numerical solution of the lifting-line problem of Section 8.1. This can help us to understand the assumptions and limitations of the single vortex line method, which in this numerical form can be extended easily to include effects of wing sweep, dihedral, or even side slip. For simplicity (and in the spirit of Prandtl's model) only one chordwise vortex is used here but the method can easily be extended to include more chordwise vortices. The small-disturbance assumption of Chapter 8 still holds for this case, and a thin lifting wing with large aspect ratio ($\mathcal{R} > 4$) is assumed. This problem is

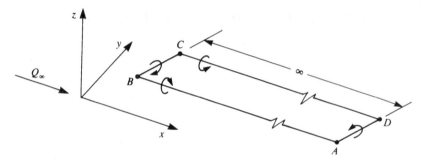

Figure 12.1 "Horseshoe" model of a lifting wing.

stated in terms of a vortex distribution in Section 4.5 and the following horseshoe model can be considered as the simplest approach to its solution.

In regard to solving Laplace's equation, the vortex line is a solution of this equation and the only boundary condition that needs to be satisfied is the zero normal flow across the thin wing's solid surface:

$$\nabla(\Phi + \Phi_\infty) \cdot \mathbf{n} = 0 \qquad (12.1)$$

In the classical case of Prandtl's lifting-line model, the wing is placed on the x–y plane and then this boundary condition requires that the sum of the normal velocity component induced by the wing's bound vortices w_b, by the wake w_i and by the free-stream velocity $\mathbf{Q}_\infty$, will be zero (see also Eq. (8.2a)):

$$w_b + w_i + Q_\infty \alpha = 0 \qquad (12.2)$$

Based on the proposed horseshoe element and on the above boundary condition, let us construct a numerical solution, following the six-step procedure of Chapter 9.

a. Choice of Singularity Element

To solve this problem the horseshoe element shown in Fig. 12.1 is selected. This element consists of a straight bound vortex segment (*BC* in Fig. 12.1) that models the lifting properties and of two semi-infinite trailing vortex lines that model the wake. The segment *BC* does not necessarily have to be parallel to the *y* axis, but at the element tips the vortex is shed into the flow where it must be parallel to the streamlines so that no force will act on the trailing vortices. In order not to violate the Helmholtz condition, these vortex elements are viewed as the near portions of vortex rings whose starting vortices extend far back, so that the effect of this segment (*AD* in Fig. 12.1) is negligible. The requirement that the far wake must be parallel to the free stream poses some modeling difficulties (which were not raised at all when constructing the classical lifting-line model). This is illustrated in Fig. 12.2a, which shows that the trailing wake has to be bent near the trailing edge to meet this free wake condition. Another possibility is shown in Fig. 12.2b, where the simple horseshoe vortex is kept, but the trailing segments are not shed at the trailing edge. Of course the very small angle of attack assumption (as in the case of the lifting-line model) allows the placing of the wake on the x–y plane of the body coordinate system as shown in Fig. 12.3. Since in this section the numerical solution of the lifting-line model is attempted, we shall adopt the model shown in Fig. 12.3, which assumes small angles of attack. However, the method can easily be modified to use the wake model as presented in Fig. 12.2a, and an even more detailed model will be presented in Section 12.3.

12.1 Lifting-Line Solution by Horseshoe Elements

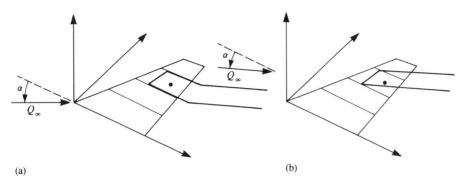

Figure 12.2 Difficulties of meeting the "wake parallel to local velocity" condition by a single horseshoe vortex representation.

The method by which the thin wing planform is divided into elements is shown in Fig. 12.3 and a typical spanwise element is shown in Fig. 12.4. Here, based on the results of the lumped-vortex model, the bound vortex is placed at the panel quarter chord line and the collocation point is at the center of the panel's three-quarter chord line. The strength of the vortex Γ is assumed to be constant for the horseshoe element and a positive circulation is defined as shown in the figure. Since this element is based on the lumped-vortex model, which includes the two-dimensional Kutta condition, it is assumed that this three-dimensional model accounts (in an approximate way) for the Kutta condition:

$$\gamma_{T.E.} = 0 \tag{12.3}$$

where the subscript *T.E.* stands for trailing edge. The velocity induced by such an element at an arbitrary point $P(x, y, z)$, shown in Fig. 12.4, can be computed by three applications

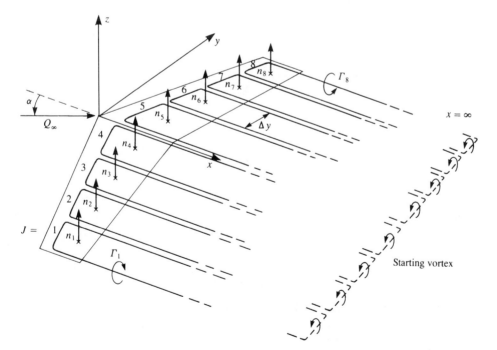

Figure 12.3 Horseshoe vortex lattice model for solving the lifting-line problem.

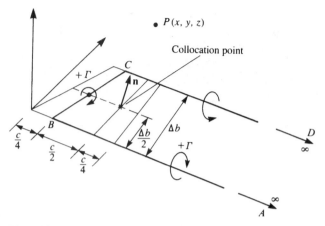

Figure 12.4 A spanwise horseshoe vortex element.

of the vortex line routine VORTXL (Eq. (10.116)) of Section 10.4.5:

$$(u_1, v_1, w_1) = \text{VORTXL}(x, y, z, x_A, y_A, z_A, x_B, y_B, z_B, \Gamma)$$
$$(u_2, v_2, w_2) = \text{VORTXL}(x, y, z, x_B, y_B, z_B, x_C, y_C, z_C, \Gamma) \quad (12.4)$$
$$(u_3, v_3, w_3) = \text{VORTXL}(x, y, z, x_C, y_C, z_C, x_D, y_D, z_D, \Gamma)$$

At this point, let us follow the small-disturbance lifting-line approach and assume

$$y_A = y_B, \qquad y_C = y_D, \quad \text{and} \quad x_A = x_D \to \infty$$

Of course ∞ means that the influence of the vortex line beyond x_A or x_D is negligible, which from the practical point of view means at least twenty wing spans behind the wing. It is possible, at this point, to align the wake with the free stream by adjusting the points at $x = \infty$ (e.g., $z_A = x_A \sin\alpha$ and $z_D = x_D \sin\alpha$). It is also possible to use the model of Fig. 12.2a, which requires breaking the two trailing vortex segments into two segments each and computing their induced velocity in a similar manner.

The velocity induced by the three vortex segments is then

$$(u, v, w) = (u_1, v_1, w_1) + (u_2, v_2, w_2) + (u_3, v_3, w_3) \quad (12.4a)$$

It is convenient to include these computations (Eqs. (12.4) and (12.4a)) in a subroutine such that

$$(u, v, w) = \text{HSHOE}(x, y, z, x_A, y_A, z_A, x_B, y_B, z_B, x_C, y_C, z_C, x_D, y_D, z_D, \Gamma)$$

It is recommended at this point that the trailing vortex wake-induced downwash $(u, v, w)^*$ be separated from the velocity induced by the bound vortex segments and saved. This information is needed for the induced-drag computations and if done at this phase will only slightly increase the computational effort. The influence of the trailing segment is obtained by simply omitting the (u_2, v_2, w_2) part from Eq. (12.4a):

$$(u, v, w)^* = (u_1, v_1, w_1) + (u_3, v_3, w_3) \quad (12.5)$$

So, at this point it is assumed that $(u, v, w)^*$ is automatically obtained as a byproduct of subroutine HSHOE.

12.1 Lifting-Line Solution by Horseshoe Elements

b. *Discretization and Grid Generation*

At this phase the wing is divided into N spanwise elements as shown by Fig. 12.3 (with the panel side edge assumed to be parallel to the x axis). For this example the span is divided equally into $N = 8$ panels, and the spanwise counter j will have values between 1 and N. Also, geometrical information such as the panel area S_j, normal vector $\mathbf{n}_j$, and the coordinates of the collocation points (x_i, y_i, z_i) are calculated at this phase. For example, if the panel is approximated by a flat plate then the normal $\mathbf{n}_j$ is a function of the local angle α_j as defined in Fig. 11.3 or Fig. 11.17:

$$\mathbf{n}_j = (\sin \alpha_j, \cos \alpha_j) \tag{12.6}$$

c. *Influence Coefficients*

To fulfill the boundary conditions, Eq. (12.2) is specified at each of the collocation points (see Fig. 12.3). The velocity induced by the horseshoe vortex element no. 1 at collocation point no. 1 (hence the use of the index 1,1) can be computed by using the HSHOE routine developed before:

$$(u, v, w)_{11} = \text{HSHOE}(x_1, y_1, z_1, x_{A1}, y_{A1}, z_{A1}, x_{B1}, y_{B1}, z_{B1},$$
$$x_{C1}, y_{C1}, z_{C1}, x_{D1}, y_{D1}, z_{D1}, \Gamma = 1.0)$$

Note that $\Gamma = 1$ is used to evaluate the influence coefficient due to a unit strength vortex. Similarly, the velocity induced by the second vortex at the first collocation point will be

$$(u, v, w)_{12} = \text{HSHOE}(x_1, y_1, z_1, x_{A2}, y_{A2}, z_{A2}, x_{B2}, y_{B2}, z_{B2},$$
$$x_{C2}, y_{C2}, z_{C2}, x_{D2}, y_{D2}, z_{D2}, \Gamma = 1.0)$$

The no normal flow across the wing boundary condition (Eq. (12.2)), at this point, can be rewritten for the first collocation point as

$$[(u, v, w)_{11}\Gamma_1 + (u, v, w)_{12}\Gamma_2 + (u, v, w)_{13}\Gamma_3 + \cdots$$
$$+ (u, v, w)_{1N}\Gamma_N + (U_\infty, V_\infty, W_\infty)] \cdot \mathbf{n}_1 = 0$$

and the strengths of the vortices Γ_j are not known at this phase. Establishing the same procedure for each of the collocation points results in the discretized form of the boundary condition:

$$\begin{aligned}
a_{11}\Gamma_1 + a_{12}\Gamma_2 + a_{13}\Gamma_3 + \cdots + a_{1N}\Gamma_N &= -\mathbf{Q}_\infty \cdot \mathbf{n}_1 \\
a_{21}\Gamma_1 + a_{22}\Gamma_2 + a_{23}\Gamma_3 + \cdots + a_{2N}\Gamma_N &= -\mathbf{Q}_\infty \cdot \mathbf{n}_2 \\
a_{31}\Gamma_1 + a_{32}\Gamma_2 + a_{33}\Gamma_3 + \cdots + a_{3N}\Gamma_N &= -\mathbf{Q}_\infty \cdot \mathbf{n}_3 \\
&\vdots \\
a_{N1}\Gamma_1 + a_{N2}\Gamma_2 + a_{N3}\Gamma_3 + \cdots + a_{NN}\Gamma_N &= -\mathbf{Q}_\infty \cdot \mathbf{n}_N
\end{aligned}$$

where the influence coefficients are defined as

$$a_{ij} \equiv (u, v, w)_{ij} \cdot \mathbf{n}_i \tag{12.7}$$

The normal velocity components of the free-stream flow $\mathbf{Q}_\infty \cdot \mathbf{n}_i$ are known and moved to the right-hand side of the equation:

$$\text{RHS}_i \equiv -(U_\infty, V_\infty, W_\infty) \cdot \mathbf{n}_i \tag{12.8}$$

We now have a set of N linear algebraic equations with N unknown Γ_j that can be solved by standard matrix solution techniques.

For example, for the case of a planar wing with constant angle of attack α, this results in the following set of equations:

$$\begin{pmatrix} a_{11} & a_{12} & \cdots & a_{1N} \\ a_{21} & a_{22} & \cdots & a_{2N} \\ a_{31} & a_{32} & \cdots & a_{3N} \\ \vdots & & \ddots & \vdots \\ a_{N1} & a_{N2} & \cdots & a_{NN} \end{pmatrix} \begin{pmatrix} \Gamma_1 \\ \Gamma_2 \\ \Gamma_3 \\ \vdots \\ \Gamma_N \end{pmatrix} = -Q_\infty \sin\alpha \begin{pmatrix} 1 \\ 1 \\ 1 \\ \vdots \\ 1 \end{pmatrix}$$

In practice it is recommended that two DO loops be used to automate the computation of the a_{ij} coefficients. The first will scan the collocation points, and the inner loop will scan the vortex elements for each collocation point:

$$\text{DO } 1 \ i = 1, N \quad \text{(collocation point loop)}$$
$$\text{RHS}_i = -\mathbf{Q}_\infty \cdot \mathbf{n}_i$$
$$\text{DO } 1 \ j = 1, N \quad \text{(vortex element loop)}$$
$$(u, v, w)_{ij} = \text{HSHOE}(x_i, y_i, z_i, x_{Aj}, y_{Aj}, z_{Aj}, x_{Bj}, y_{Bj}, z_{Bj},$$
$$x_{Cj}, y_{Cj}, z_{Cj}, x_{Dj}, y_{Dj}, z_{Dj}, \Gamma = 1.0)$$
$$a_{ij} = (u, v, w)_{ij} \cdot \mathbf{n}_i$$
$$b_{ij} = (u, v, w)^*_{ij} \cdot \mathbf{n}_i$$

1   END

Here b_{ij} is the normal component of the wake-induced downwash that will be used for the induced-drag computations and $(u, v, w)^*_{ij}$ is given by Eq. (12.5).

d. *Establish RHS Vector*

The right-hand side vector, Eq. (12.8), is actually the normal component of the free stream, which can be computed within the outer DO loop of the influence coefficient computations (as shown above). However, if one plans to upgrade the code by including unsteady effects or the simulation of normal "transpiration" flows, then this calculation should be done separately.

e. *Solve Linear Set of Equations*

The solution of the above described problem can be obtained by standard matrix methods. Furthermore, since the influence of such an element on itself is the largest, the matrix will have a dominant diagonal, and the solution is stable.

f. *Secondary Computations: Pressures, Loads, Velocities, Etc.*

The solution of the above set of equations results in the vector $(\Gamma_1, \Gamma_2, \ldots, \Gamma_N)$. The lift of each bound vortex segment is obtained by using the Kutta–Joukowski theorem:

$$\Delta L_j = \rho Q_\infty \Gamma_j \Delta y_j \tag{12.9}$$

where Δy_j is the panel bound vortex projection normal to the free stream (see Fig. 12.4 where the panel width $\Delta b = \Delta y$). The induced-drag computation is somwhat more complex. Following the lifting-line results of Eq. (8.20a), we have

$$\Delta D_j = -\rho w_{\text{ind}_j} \Gamma_j \Delta y_j \tag{12.10}$$

12.1 Lifting-Line Solution by Horseshoe Elements

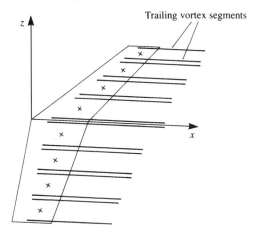

Figure 12.5 Array of the trailing vortex segments responsible for the induced downwash on the three-dimensional wing.

where the induced downwash w_{ind_j} at each collocation point j is computed by summing the velocity induced by all the trailing vortex segments (see Fig. 12.5). This can be done during the phase of the influence coefficient computations or even later, by using the HSHOE routine with the influence of the bound vortex segment turned off. This procedure can be summarized by the following matrix formulation where all the b_{ij} and the Γ_j are known:

$$\begin{pmatrix} w_{\text{ind}_1} \\ w_{\text{ind}_2} \\ w_{\text{ind}_3} \\ \vdots \\ w_{\text{ind}_N} \end{pmatrix} = \begin{pmatrix} b_{11} & b_{12} & \cdots & b_{1N} \\ b_{21} & b_{22} & \cdots & b_{2N} \\ b_{31} & b_{32} & \cdots & b_{3N} \\ \vdots & \vdots & \ddots & \vdots \\ b_{N1} & b_{N2} & \cdots & b_{NN} \end{pmatrix} \begin{pmatrix} \Gamma_1 \\ \Gamma_2 \\ \Gamma_3 \\ \vdots \\ \Gamma_N \end{pmatrix}$$

The total lift and drag are then calculated by summing the individual panel contributions:

$$D = \sum_{j=1}^{N} \Delta D_j$$

$$L = \sum_{j=1}^{N} \Delta L_j$$

The induced drag can be calculated, too, by using Eq. (8.146) in the Trefftz plane, which is selected to be far behind the trailing edge and normal to the free stream. Since the wake is force free, the trailing vortex lines will be normal to this plane and their induced velocity can be calculated by using the two-dimensional formula (e.g., Eqs. (3.81) and (3.82)). Consequently, the wake-induced downwash at each of the trailing vortex lines is

$$w_{\text{ind}_j} = \frac{-1}{2\pi} \sum_{i=1}^{N_W} \frac{x_j - x_i}{(z_j - z_i)^2 + (x_j - x_i)^2}$$

where N_W is the number of trailing vortex lines and the influence of a vortex line on itself is set to zero. Once the induced downwash at each of the vortex lines is obtained, the induced drag is evaluated by applying Eq. (8.146):

$$D = -\frac{\rho}{2} \int_{-b_w/2}^{b_w/2} \Gamma(y) w \, dy = -\frac{\rho}{2} \sum_{i=1}^{N_W} \Gamma_j w_{\text{ind}_j} \Delta y_j \qquad (12.10a)$$

If wake rollup routines are used then the results of Eq. (12.10a) may not be unique owing to the nonunique placing of the wake trailing vortices. Therefore, it is recommended that one calculate first the wing circulation with the rolled up wake, and for this induced velocity and drag calculation one should then use the spacing Δy_j of the vortex lines, as released at the trailing edge. (This is the simplest approximation for a force-free wake since many wake rollup routines may not converge to this condition.) Moreover, note that Eq. (12.10a) is similar to Eq. (12.10) but it has a coefficient of $\frac{1}{2}$, which is a result of the first being evaluated at the Trefftz plane (where the trailing vortices seem to be two dimensional) whereas Eq. (12.10) is evaluated at the spanwise bound-vortex line (and there the trailing vortices are observed to be semi-infinite).

This first simple example presented a numerical solution for the lifting-line model, and inclusion of wing sweep and dihedral effects can be done as a homework assignment. Some of the limitations with regard to the wake model and the trailing-edge conditions will be studied in the vortex-ring model that will be presented next. Also, the method presented here does not take advantage of the wing symmetry to reduce computational effort. This important modification is discussed in the following section.

12.2 Modeling of Symmetry and Reflections from Solid Boundaries

In situations when symmetry exists between the left and right halves of the body's surface, or when ground proximity is modeled, a rather simple method can be used to include these features in the numerical scheme. In terms of programming simplicity these modifications will affect only the influence coefficient calculation section of the code.

For example, consider the symmetric wing (left to right), shown in Fig. 12.6, where only the right-hand half of the wing must be modeled. The influence of a panel j at point P can be obtained by any of the influence routines of Chapter 10. For this example, let us use the HSHOE routine of the previous section. Thus the velocity induced at point P by the jth element (with the four corner points A, B, C, and D) is

$$(u_i, v_i, w_i) = \text{HSHOE}(x, y, z, x_{Aj}, y_{Aj}, z_{Aj}, x_{Bj}, y_{Bj}, z_{Bj}, x_{Cj},$$
$$y_{Cj}, z_{Cj}, x_{Dj}, y_{Dj}, z_{Dj}, \Gamma_j)$$

But because of the left/right symmetry, the image panel in the left half wing in Fig. 12.6 will have the same strength, and its effect can be evaluated by calling the influence of the

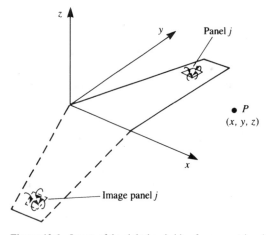

Figure 12.6 Image of the right-hand side of a symmetric wing model.

actual vortex at point $(x, -y, z)$. Note that the sign was changed for the y coordinate. Thus

$$(u_{ii}, v_{ii}, w_{ii}) = \text{HSHOE}(x, -y, z, x_{Aj}, y_{Aj}, z_{Aj}, x_{Bj}, y_{Bj}, z_{Bj}, x_{Cj},$$
$$y_{Cj}, z_{Cj}, x_{Dj}, y_{Dj}, z_{Dj}, \Gamma_j)$$

and the velocity induced by the two equal strength elements at point P is

$$(u, v, w) = (u_i + u_{ii}, v_i - v_{ii}, w_i + w_{ii}) \qquad (12.11)$$

Note that for simplicity, instead of moving the four corner points into the image plane (a total of sixteen numbers), only the y value of point P (one number) was moved; this change is corrected by the minus sign added to the v component of the resulting image velocity.

This procedure can reduce the number of unknowns by half, and only the vortices of the right semiwing need to be modeled. Therefore, when scanning the elements of the semispan in the influence coefficient step the coefficients a_{ij} are modified (see Eq. (12.7)) such that

$$a_{ij} = (u, v, w)_{ij} \cdot \mathbf{n}_i = (u_i + u_{ii}, v_i - v_{ii}, w_i + w_{ii})_{ij} \cdot \mathbf{n}_i \qquad (12.12)$$

The inclusion of ground effect can be achieved by using the same method. In this situation (described in Fig. 12.7) the ground plane is simulated by modeling a mirror image wing under the x–y plane. Again, the velocity at a point P induced by the elements of the real wing (u_g, v_g, w_g) and of the imaginary wing (u_{gg}, v_{gg}, w_{gg}) are added up. If we use the HSHOE routine to demonstrate this principle, the upper element induced velocity is

$$(u_g, v_g, w_g) = \text{HSHOE}(x, y, z, x_{Aj}, y_{Aj}, z_{Aj}, x_{Bj}, y_{Bj}, z_{Bj}, x_{Cj},$$
$$y_{Cj}, z_{Cj}, x_{Dj}, y_{Dj}, z_{Dj}, \Gamma_j)$$

and the velocity induced by the same element but at a point $(x, y, -z)$ is

$$(u_{gg}, v_{gg}, w_{gg}) = \text{HSHOE}(x, y, -z, x_{Aj}, y_{Aj}, z_{Aj}, x_{Bj}, y_{Bj}, z_{Bj},$$
$$x_{Cj}, y_{Cj}, z_{Cj}, x_{Dj}, y_{Dj}, z_{Dj}, \Gamma_j)$$

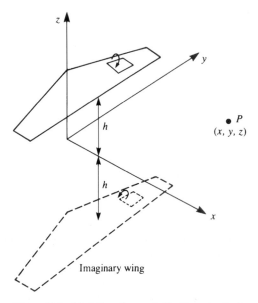

Figure 12.7 Modeling of ground effect by using the image technique.

and the combined influence is

$$(u, v, w) = (u_g + u_{gg}, v_g + v_{gg}, w_g - w_{gg}) \tag{12.13}$$

The coefficient a_{ij} that includes the ground effect is

$$a_{ij} = (u, v, w)_{ij} \cdot \mathbf{n}_i = (u_g + u_{gg}, v_g + v_{gg}, w_g - w_{gg})_{ij} \cdot \mathbf{n}_i \tag{12.14}$$

Note that the wing in Fig. 12.7 is raised in the x, y, z system and the ground plane is assumed to be at the $z = 0$ plane.

Using this method for computing the flow over a symmetric wing in ground proximity reduces the number of unknown elements by a factor of 4. Because much of the computational effort is spent on the matrix inversion, which increases at a rate of N^2, the use of this reflection technique can reduce computation time by approximately 1/16! Examples for incorporating this technique into a computer program are presented in the next section and in Appendix D, Programs No. 13, 14, and 16.

12.3 Lifting-Surface Solution by Vortex Ring Elements

In this section the three-dimensional thin lifting surface problem will be solved using vortex ring elements. The main advantage of using these elements is that they require very little programming effort (although computational efficiency can be further improved). Additionally, the exact boundary conditions will be satisfied on the actual wing surface, which can have camber and various planform shapes.

As with the previous example, this singularity element is based on the vortex line solution of the incompressible continuity equation. The boundary condition that must be satisfied by the solution is the zero normal flow across the thin wing's solid surface:

$$\nabla(\Phi + \Phi_\infty) \cdot \mathbf{n} = 0 \tag{12.1}$$

In the small-disturbance lifting surface formulation of Section 4.5, this boundary condition was expressed in terms of a surface-vortex distribution (Eq. (4.50)) as

$$\frac{-1}{4\pi} \int_{\text{wing+wake}} \frac{\gamma_y(x - x_0) - \gamma_x(y - y_0)}{[(x - x_0)^2 + (y - y_0)^2]^{3/2}} \, dx_0 \, dy_0 = Q_\infty \left(\frac{\partial \eta}{\partial x} - \alpha \right) \tag{12.15}$$

Note that in Eq. (12.15) the small-disturbance approximation to the boundary condition was satisfied on the wing surface projected onto the x–y plane whereas in the following example the actual boundary condition (Eq. (12.1)) will be implemented.

In order to solve this lifting surface problem numerically, the wing is divided into elements containing vortex ring singularities as shown in Fig. 12.8. The solution procedure is as follows.

a. Choice of Singularity Element

The method by which the thin wing planform is divided into panels is shown in Fig. 12.8 and some typical panel elements are shown in Fig. 12.9. The leading segment of the vortex ring is placed on the panel's quarter chord line and the collocation point is at the center of the three-quarter chord line. The normal vector $\mathbf{n}$ is defined at this point, too. A positive Γ is defined here according to the right-hand rotation rule (for the leading segment), as shown in the figure.

From the numerical point of view these vortex ring elements are stored in rectangular patches (arrays) with i, j indexing as shown by Fig. 12.10. The velocity induced at an

12.3 Lifting-Surface Solution by Vortex Ring Elements

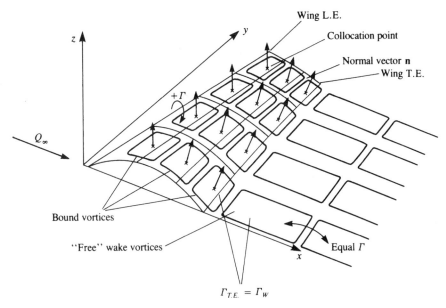

Figure 12.8 Vortex ring model for a thin lifting surface.

arbitrary point $P(x, y, z)$, by a typical vortex ring at location i, j, can be computed by applying the vortex line routine VORTXL (Eq. (10.116)) to the ring's four segments:

$(u_1, v_1, w_1) = \text{VORTXL}(x, y, z, x_{i,j}, y_{i,j}, z_{i,j}, x_{i,j+1}, y_{i,j+1}, z_{i,j+1}, \Gamma_{i,j})$

$(u_2, v_2, w_2) = \text{VORTXL}(x, y, z, x_{i,j+1}, y_{i,j+1}, z_{i,j+1}, x_{i+1,j+1},$
$\qquad y_{i+1,j+1}, z_{i+1,j+1}, \Gamma_{i,j})$

$(u_3, v_3, w_3) = \text{VORTXL}(x, y, z, x_{i+1,j+1}, y_{i+1,j+1}, z_{i+1,j+1}, x_{i+1,j},$
$\qquad y_{i+1,j}, z_{i+1,j}, \Gamma_{i,j})$

$(u_4, v_4, w_4) = \text{VORTXL}(x, y, z, x_{i+1,j}, y_{i+1,j}, z_{i+1,j}, x_{i,j}, y_{i,j}, z_{i,j}, \Gamma_{i,j})$

The velocity induced by the four vortex segments is then

$$(u, v, w) = (u_1, v_1, w_1) + (u_2, v_2, w_2) + (u_3, v_3, w_3) + (u_4, v_4, w_4) \quad (12.16)$$

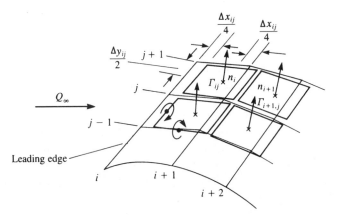

Figure 12.9 Nomenclature for the vortex ring elements.

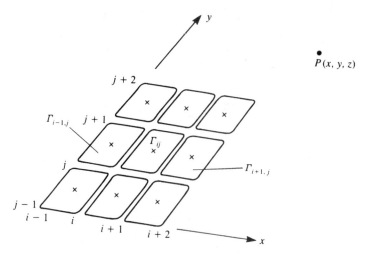

Figure 12.10 Arrangement of vortex rings in a rectangular array.

It is convenient to include these computations in a subroutine (see Eq. (10.117)) such that

$$(u, v, w) = \text{VORING}(x, y, z, i, j, \Gamma) \tag{12.17}$$

Note that in this formulation it is assumed that by specifying the i, j counters, the (x, y, z) coordinates at the four corners of this panel are automatically identified (see Fig. 12.10).

The use of this subroutine can considerably shorten the programming effort, however, for the segment between two such rings the induced velocity is computed twice. For the sake of simplicity this routine will be used for this problem, but more advanced programming can easily correct this loss of computational efficiency.

It is recommended at this point, too, that one calculate the velocity induced by the trailing vortex segments only (the vortex lines parallel to the free stream, as in Fig. 12.5). This information is needed for the induced-drag computations and if done at this phase will only slightly increase the computational effort. The influence of the trailing segments is obtained by simply omitting the $(u_1, v_1, w_1) + (u_3, v_3, w_3)$ part from Eq. (12.16):

$$(u, v, w)^* = (u_2, v_2, w_2) + (u_4, v_4, w_4) \tag{12.18}$$

So, at this point it is assumed that $(u, v, w)^*$ is automatically obtained as a byproduct of subroutine VORING.

b. Discretization and Grid Generation

The method by which the thin wing planform is divided into elements is shown in Fig. 12.8 and some typical panel elements are shown in Fig. 12.9. Also, only the wing semispan is modeled and the mirror image method will be used to account for the other semispan. The leading segment of the vortex ring is placed on the panel's quarter chord line and the collocation point is at the center of the three-quarter chord line. The normal vector **n** is defined at this point, as shown in Fig. 12.9. A positive Γ is defined here as the right-hand rotation, as shown in the figure. For the pressure distribution calculations we need the local circulation, which for the leading edge panel is equal to Γ_i but for all the elements behind it is equal to the difference $\Gamma_i - \Gamma_{i-1}$. In the case of increased surface curvature the above described vortex rings will not be placed exactly on the lifting surface, and a finer grid needs to be used, or the wing surface can be redefined accordingly. By placing the leading

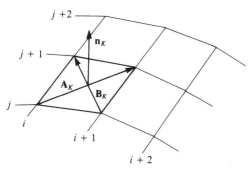

Figure 12.11 Definition of wing outward normal.

segment of the vortex ring at the quarter chord line of the panel the two-dimensional Kutta condition is satisfied along the chord (recall the lumped-vortex element). Also, along the wing trailing edges, the trailing vortex of the last panel row (which actually simulates the starting vortex) must be canceled to satisfy the three-dimensional trailing-edge condition:

$$\gamma_{T.E.} = 0 \tag{12.19}$$

For steady-state flow this is done by attempting to align the wake vortex panels parallel to the local streamlines, and their strength is equal to the strength of the shedding panel at the trailing edge (see Fig. 12.8 where $\Gamma_{T.E.} = \Gamma_W$, for each row).

For this example (in Fig. 12.8) the chord is divided equally into $M = 3$ panels and the semispan is divided equally into $N = 4$ panels. Therefore, the chordwise counter i will have values from 1 to M and the spanwise counter j will have values between 1 and N. Also, geometrical information such as the vortex ring corner points, panel area S_k, normal vector $\mathbf{n}_k$, and the coordinates of the collocation points are calculated at this phase (note that the panel sequential counter k will have values between 1 and $M \times N$). A simple and fairly general method for evaluating the normal vector is shown in Fig. 12.11. The panel opposite corner points define two vectors $\mathbf{A}_k$ and $\mathbf{B}_k$, and their vector product will point in the direction of $\mathbf{n}_k$:

$$\mathbf{n}_k = \frac{\mathbf{A}_k \times \mathbf{B}_k}{|\mathbf{A}_k \times \mathbf{B}_k|} \tag{12.20}$$

This method is used in Program No. 13 in Appendix D; however, it is possible to evaluate the normal vector on the actual wing surface (if an analytic description is available).

The results of the grid generating phase are shown schematically in Fig. 12.12. For more information about generating panel corner points, collocation points, area, and normal

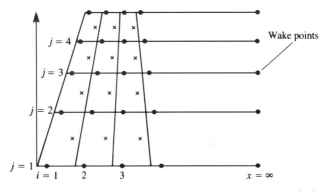

Figure 12.12 Array of wing and wake panel corner points (dots) and of collocation points (× symbols).

```
                K=0
                DO 11 I=1,M              chordwise loop (scan collocation points)
                DO 11 J=1,N              spanwise  loop (scan collocation points)
                K=K+1
                L=0
                DO 10 I1=1,M             chordwise loop (scan vortex rings)
                DO 10 J1=1,N             spanwise  loop (scan vortex rings)
                L=L+1
                CALL VORING(QC(I,J,1),QC(I,J,2),QC(I,J,3),I1,J1,GAMA=1,U,V,W)
         C      ADD INFLUENCE OF WING'S OTHER HALF
                CALL VORING(QC(I,J,1),-QC(I,J,2),QC(I,J,3),I1,J1,GAMA=1,U1,V1,W1)
                U2=U+U1
                V2=V-V1
                W2=W+W1
         C      ADD INFLUENCE OF WAKE
                IF(I1.LT.M) GOTO 10
                CALL VORING(QC(I,J,1), QC(I,J,2),QC(I,J,3),I1+1,J1,GAMA=1,U3,V3,W3)
                CALL VORING(QC(I,J,1),-QC(I,J,2),QC(I,J,3),I1+1,J1,GAMA=1,U4,V4,W4)
                U2=U2+U3+U4
                V2=V2+V3-V4
                W2=W2+W3+W4
         10     A(K,L)=U2*AL(I,J)+V2*AM(I,J)+W2*AN(I,J)
         C      A(K,L) is influence coefficient and
         C      (AL(I,J),AM(I,J),AN(I,J)) is the normal vector of panel (I,J)
         11     CONTINUE
```

Figure 12.13 Example of a double "DO loop" to calculate the influence coefficients of a vortex ring model (FORTRAN).

vector, see the student computer Program No. 13 in Appendix D (and subroutine PANEL for the use of Eq. (12.20)).

c. Influence Coefficients

The influence coefficient calculation proceeds in a manner similar to the methods presented so far, but in this three-dimensional case more attention is needed to the scanning sequence of the surface panels.

Let us establish a collocation point scanning procedure that takes the first chordwise row where $i = 1$ and scans spanwise with $j = 1 \to N$ and so on (see Fig. 12.10). This procedure can be described by the two DO loops shown in Fig. 12.13. As the panel scanning begins, a sequential counter assigns a value K to each panel (the sequence of K is shown in Fig. 12.14), which will have values from 1 to $M \times N$.

Let us assume that the collocation point scanning has started and $K = 1$ (which is point $(i = 1, j = 1)$ on Fig. 12.12). The velocity induced by the first vortex ring is then

$$(u_i, v_i, w_i)_{11} = \text{VORING}(x, y, z, i = 1, j = 1, \Gamma = 1.0)$$

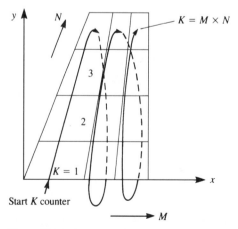

Figure 12.14 Sequence of scanning the wing panels (with the counter K).

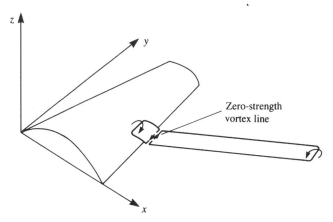

Figure 12.15 Method of attaching a vortex wake panel to fulfill the Kutta condition.

and that from its image on the left semispan is

$$(u_{ii}, v_{ii}, w_{ii})_{11} = \text{VORING}(x, -y, z, i=1, j=1, \Gamma=1.0)$$

and the velocity induced by the unit strength Γ_1 vortex and its image at collocation point 1 is

$$(u, v, w)_{11} = (u_i + u_{ii}, v_i - v_{ii}, w_i + w_{ii})_{11} \tag{12.21}$$

Note that the subscript $(\ )_{11}$ represents the influence of the first vortex at the first collocation point, and both counters can have values from 1 to $M \times N$. Also, a unit strength vortex is used in the process of evaluating the influence coefficient a_{11}, which is

$$a_{11} = (u, v, w)_{11} \cdot \mathbf{n}_1$$

To scan all the vortex rings influencing this point, an inner scanning loop is needed with the counter $L = 1 \to N \times M$ (see Fig. 12.13). Thus, at this point, the K counter is at point 1, and the L counter will scan all the vortex rings on the wing surface, and all the influence coefficients a_{1L} are computed (also, in Eq. (12.21) the $(\ )_{11}$ index means $K = 1, L = 1$):

$$a_{1L} = (u, v, w)_{1L} \cdot \mathbf{n}_1 \tag{12.22}$$

When a particular vortex ring is at the trailing edge, a "free wake" vortex ring with the same strength is added to cancel the spanwise starting vortex line (as shown in Fig. 12.15). Therefore, when the influence of such a trailing-edge panel vortex is calculated ($i = M$, in the inner vortex-ring loop in Fig. 12.13) the contribution of this segment is added. For example, in Fig. 12.8 the first wake panel is encountered when $i = 3$ (or the L counter is equal to 9). If the wake grid is added into the $M + 1$ corner point array (as shown in Fig. 12.12 where this point is added at $x = \infty$) then the velocity due to the $i = 3, j = 1$ (or $L = 9$) panel is

$$(u, v, w)_{19} = \text{VORING}(x_1, y_1, z_1, i=3, j=1, \Gamma=1.0)$$

and that due to the attached wake is

$$(u, v, w)_{19W} = \text{VORING}(x_1, y_1, z_1, i=3+1, j=1, \Gamma=1.0)$$

When the wing is symmetric as in this case and only the right half is paneled, then the (u, v, w) velocity components of the trailing-edge and wake panels include the influence of the left-hand side image (as in Eq. (12.21)). The corresponding influence coefficient is

$$a_{19} = [(u, v, w)_{19} + (u, v, w)_{19W}] \cdot \mathbf{n}_1 \tag{12.22a}$$

As mentioned before, parallel to the computation of the a_{KL} coefficients, the normal velocity component induced by the streamwise segments can also be computed by using the $(u, v, w)^*$ portion as in Eq. (12.5). For the first element then

$$b_{1L} = (u, v, w)_{1L}^* \cdot \mathbf{n}_1 \tag{12.23}$$

This procedure continues until all the collocation points have been scanned; a FORTRAN example is presented in Fig. 12.13.

d. *Establish RHS*

The RHS vector is computed as before by scanning each of the collocation points on the wing:

$$\text{RHS}_K = -\mathbf{Q}_\infty \cdot \mathbf{n}_K \tag{12.24}$$

e. *Solve Linear Set of Equations*

Once the computations of the influence coefficients and the right-hand side vector are completed, the zero normal flow boundary condition on each of the collocation points will result in the following set of algebraic equations:

$$\begin{pmatrix} a_{11} & a_{12} & \cdots & a_{1m} \\ a_{21} & a_{22} & \cdots & a_{2m} \\ a_{31} & a_{32} & \cdots & a_{3m} \\ \vdots & & \ddots & \vdots \\ a_{m1} & a_{m2} & \cdots & a_{mm} \end{pmatrix} \begin{pmatrix} \Gamma_1 \\ \Gamma_2 \\ \Gamma_3 \\ \vdots \\ \Gamma_m \end{pmatrix} = \begin{pmatrix} \text{RHS}_1 \\ \text{RHS}_2 \\ \text{RHS}_3 \\ \vdots \\ \text{RHS}_m \end{pmatrix}$$

Here K is the vertical collocation point counter and L is the horizontal vortex ring counter and the order of this matrix is $m = M \times N$.

f. *Secondary Computations: Pressures, Loads, Velocities, Etc.*

The solution of the above set of equations results in the vector $(\Gamma_1, \ldots, \Gamma_K, \ldots, \Gamma_m)$. If the counter K is resolved back to the original i, j counters then the lift of each bound vortex segment is obtained by using the Kutta–Joukowski theorem:

$$\Delta L_{ij} = \rho Q_\infty (\Gamma_{i,j} - \Gamma_{i-1,j}) \Delta y_{ij}, \quad i > 1 \tag{12.25}$$

and when the panel is at the leading edge ($i = 1$) then

$$\Delta L_{ij} = \rho Q_\infty \Gamma_{i,j} \Delta y_{ij}, \quad i = 1 \tag{12.25a}$$

The pressure difference across this panel is

$$\Delta p_{ij} = \frac{\Delta L_{ij}}{\Delta S_{ij}} \tag{12.26}$$

where ΔS_{ij} is the panel area and Δy_{ij} is the panel width (similar to Δb in Fig. 12.4).

The induced-drag computation is somewhat more complex. In this case

$$\Delta D_{ij} = -\rho w_{\text{ind}_{i,j}} (\Gamma_{i,j} - \Gamma_{i-1,j}) \Delta y_{ij}, \quad i > 1 \tag{12.27}$$

$$\Delta D_{ij} = -\rho w_{\text{ind}_{i,j}} \Gamma_{i,j} \Delta y_{i,j}, \quad i = 1 \tag{12.27a}$$

where the induced downwash at each collocation point i, j is computed by summing up the velocity induced by all the trailing vortex segments (see Fig. 12.5 for the horseshoe vortex element case). This can be done during the phase of the influence coefficient computation (Eq. (12.23)) by using the VORING routine with the influence of the bound vortex segments turned off. This procedure can be summarized by the following matrix formulation where all the b_{KL} and the Γ_K are known:

$$\begin{pmatrix} w_{\text{ind}_1} \\ w_{\text{ind}_2} \\ w_{\text{ind}_3} \\ \vdots \\ w_{\text{ind}_m} \end{pmatrix} = \begin{pmatrix} b_{11} & b_{12} & \cdots & b_{1m} \\ b_{21} & b_{22} & \cdots & b_{2m} \\ b_{31} & b_{32} & \cdots & b_{3m} \\ \vdots & & \ddots & \vdots \\ b_{m1} & b_{m2} & \cdots & b_{mm} \end{pmatrix} \begin{pmatrix} \Gamma_1 \\ \Gamma_2 \\ \Gamma_3 \\ \vdots \\ \Gamma_m \end{pmatrix}$$

and again $m = N \times M$. The total lift and drag are then calculated by summing the individual panel contributions:

$$D = \sum_{i=1}^{M} \sum_{j=1}^{N} \Delta D_{ij}$$

$$L = \sum_{i=1}^{M} \sum_{j=1}^{N} \Delta L_{ij}$$

The induced drag can also be calculated by using Eq. (8.146) in the Trefftz plane, through the discretization of Eq. (12.10a):

$$D = -\frac{\rho}{2} \sum_{k=1}^{N_W} \Gamma_k w_{\text{ind}_k} \Delta y_k$$

Here the counter k scans the trailing-edge vortices and N_W is the number of trailing-edge vortices. Since the wake is force free, the trailing vortex lines will be normal to this plane and their induced velocity w_{ind_k} can be calculated by using the two-dimensional formula (e.g., Eqs. (3.81) and (3.82)). Similarly to the lifting-line case, if wake rollup routines are used it is recommended for this induced velocity and drag calculation, that one first calculate the wing circulation with the rolled up wake and to use the spacing Δy_{ij} of the vortex lines, as released at the trailing edge. (This is the simplest approximation for a force-free wake since many wake rollup routines may not converge to this condition.)

Example: Planar Wing

Consider a planar wing planform, where the leading, trailing and side edges are made of straight lines and the wing has no camber. By using the method of this section the lift slope C_{L_α} can be calculated. The general effect of wing aspect ratio R and sweep Λ is summarized in Fig. 12.16 (computed results are the same as those drawn by Jones and Cohen). The two-dimensional values of the lift slope are shown at the right-hand side of the figure where $R = \infty$. For the two-dimensional unswept wing $C_{L_\alpha} = 2\pi$, as obtained in Chapter 5. The effect of leading-edge sweep is to reduce this lift slope. Similarly, because of the increased downwash of the trailing vortices, smaller aspect ratio wings will have smaller lift slope.

The effect of leading-edge sweep on the spanwise loading is shown in Fig. 12.17 for an $R = 4$, planar wing. Aft swept wings will have more lift toward their tips while forward swept wings will have larger loading near the root. This effect can

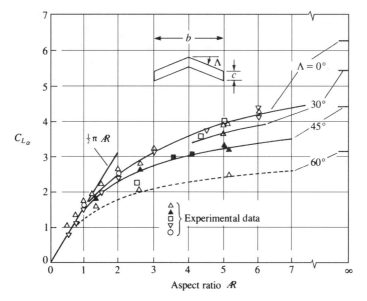

Figure 12.16 Effect of aspect ratio on the lift coefficient slope of untapered planar wings. From Jones, R. T., and Cohen, D., "High Speed Wing Theory," Princeton Aeronautical Paperback, No. 6, 1960, Princeton University Press, Princeton, N. J.

be explained by observing the downwash induced by the right wing vortex on the left half wing (Fig. 12.18). This downwash is larger near the wing centerline and decreases toward the wingtip. In the case of the forward swept wing, an upwash at the wing centerline will increase the lift there.

From the wing structural point of view, for the same lift, the root bending moments will be smaller for a forward swept wing than for a wing with the same aft sweep. Moreover, for such untwisted wings the stall will be initialized at the root section of the forward swept wing, which will create smaller rolling moments

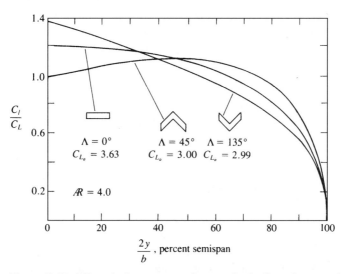

Figure 12.17 Effect of wing sweep on the spanwise loading of untapered planar wings. From Ref. 13.13, reprinted with permission of ASME.

12.3 Lifting-Surface Solution by Vortex Ring Elements

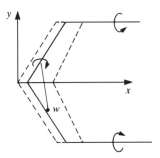

Figure 12.18 Schematic description of the effect of wing's leading-edge sweep.

(due to possible asymmetry of the stall) than in the case of a comparable aft swept wing. The main reason that most high-speed wings use aft sweep is the aeroelastic divergence of the classical wing structures. (This problem can be avoided by tailoring the torsional properties of composite structures.)

Wing root bending can be reduced, too, by tapering the wing. The taper ratio λ is defined as the ratio of tip to root chords:

$$\lambda = \frac{c(y = b/2)}{c(y = 0)} \quad (12.28)$$

The spanwise loading of an untwisted wing with various taper ratios is shown in Fig. 12.19. As was noted the load is decreasing toward the tip but the local lift

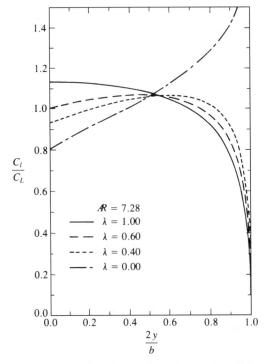

Figure 12.19 Effect of taper ratio on the spanwise variation of the lift coefficient for untwisted wings. From Bertin, J. J., and Smith, L. M., *Aerodynamics for Engineers*, 2nd edition, 1989, Prentice-Hall, p. 258. Reprinted by permission of Prentice-Hall, Inc. Englewood Cliffs, N.J.

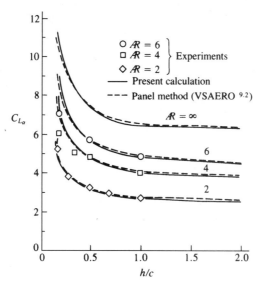

Figure 12.20 Effect of ground proximity on the lift coefficient slope of rectangular wings. From Ref. 13.13, reprinted with permission of ASME.

coefficient (divided by the local chord) is increasing with a reduction in taper ratio. This means that the tip of such wings will stall first, an unfavorable behavior that can be corrected by twist (which reduces the angle of attack toward the tip).

The method presented here can model ground proximity. Figure 12.20 presents the effect of distance from the ground for unswept rectangular wings. The increase in the lift slope in the proximity of the ground is present also for the smaller aspect ratio wings. In the case of the finite wing the image trailing wake induces an upwash on the wing that results in an additional gain in the lift due to ground proximity.

The effect of wing dihedral (see inset in Fig. 12.21) in ground proximity is shown in Fig. 12.21. Far from the ground the dihedral (as the sweep) reduces the

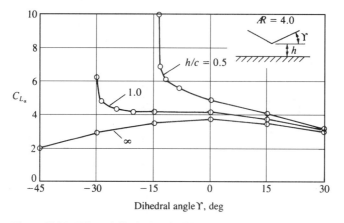

Figure 12.21 Effect of dihedral on the lift coefficient slope of rectangular wings in ground effect. From Kalman, T. P., Rodden, W. P., and Giesing, J. P., "Application of the Doublet-Lattice Method to Nonplanar Configurations in Subsonic Flow," *J. Aircraft*, Vol. 8, No. 6, 1971. Reprinted with permission. Copyright AIAA.

lift slope. However, near the ground, especially for negative values of dihedral (anhedral), the increase in lift of the wing portion near the ground is large, as shown in the figure.

12.4 Introduction to Panel Codes: A Brief History

Observing the brief history of potential flow solutions, along with the methodology presented in Chapters 3–5, implies that the trend is toward using surface distributions of elementary solutions with gradually increasing complexity. So in principle, if a problem can be solved by distributing the unknown quantity on the boundary surface rather than in the entire volume surrounding the body (as in finite difference methods), then a faster numerical solution is obtainable. This observation is true for most practical inviscid flow problems (e.g., lift of wings in attached flows etc.).

This reduction of the three-dimensional computational domain to a two-dimensional one (on a three-dimensional boundary) led to the rapid development of computer codes for the implementation of panel methods, some of which are listed in Table 12.1. Probably the first successful three-dimensional panel code is known as the Hess code[12.1] (or Douglas–Neumann), which was developed by the Douglas Aircraft Company and used a Neumann velocity boundary condition. This method was based on flat source panels and had a true three-dimensional capability for nonlifting potential flows.

The Woodward I code,[12.2] which originated in the Seattle area, was capable of solving lifting flows for thick airplane-like configurations. This code also had a supersonic potential

Table 12.1. *Chronological list of some panel methods and their main features*

Method	Geometry of panel	Singularity distribution	Boundary conditions	Remarks
1962, Douglas-Neumann[12.1]	Flat	Constant source	Neumann	
1966, Woodward I[12.2]	Flat	Linear sources Constant vortex	Neumann	$M > 1$
1973, USSAERO[12.3]	Flat	Linear sources Linear vortex	Neumann	$M > 1$
1972, Hess I[12.4]	Flat	Constant source Constant doublet	Neumann	
1980, MCAIR[12.5]	Flat	Constant source Quadratic doublet	Dirichlet	Coupling with B. L. design mode
1980, SOUSSA[12.6]	Parabolic	Constant source Constant doublet	Dirichlet	Linearized unsteady
1981, Hess II[12.7]	Parabolic	Linear source Quadratic doublet	Neumann	
1981, PAN AIR[12.8,12.9]	Flat subpanels	Linear source Quadratic doublet	both	$M > 1$
1982, VSAERO[12.10,12.11]	Flat	Constant source and doublet	both	Coupling with B. L., wake rollup
1983, QUADPAN[12.12]	Flat	Constant source and doublet	Dirichlet	
1987, PMARC[12.13,12.14]	Flat	Constant source Constant doublet	both	Unsteady wake rollup

flow solution option, which increased its applicability. The method was later improved and was released as the USSAERO code[12.3] (or the Woodward II code). At about the same time Hess added doublet elements to his nonlifting method so that he could solve for flows with lift; this code[12.4] was widely used by the industry and was called the Hess I code.

All of the computer codes listed in Table 12.1 had the capability to correct for low-speed compressibility effects by using the Prandtl–Glauert transformation (as in Section 4.8).

The above computer codes were considered to be the first-generation panel programs, but as computer technology evolved, more complex algorithms could be developed based on higher order approximations to the panel surface and singularity distribution. For example, the MCAIR code,[12.5] which evolved into a high-order singularity method, had two new interesting features. One was an inverse two-dimensional solution for multielement airfoils with prescribed pressure distribution. The second option was an iterative coupling with a boundary layer procedure. Pressure and velocity data from the potential flow solution were fed into a boundary layer analysis that estimated the displacement thickness and surface friction. During the next iteration of the potential solver the three-dimensional panel geometry was modified to include the added displacement thickness of the boundary layer.

At about the same time the SOUSSA code[12.6] was developed and it used the Dirichlet boundary condition (as did MCAIR) and had the additional feature of an unsteady oscillatory mode. Meanwhile, John Hess of the McDonnell Douglas Aircraft Co. had updated the Hess I code to the Hess II code,[12.7] which now had parabolic panel shape and higher order singularity distributions.

During this second-generation panel code development period, the largest effort was invested in the PAN AIR code,[12.8,12.9] which was developed for NASA by the Boeing Co. The basic panel element in this code had five, flat, subelements with higher order singularity distribution; boundary conditions were usually Dirichlet, but on selected areas the Neumann condition could be used as well. This code also had the capability for solving the supersonic potential flow equations.

Until the early 1980s most panel codes were limited (along with the availability of mainframe computers) to the larger aerospace companies. However, computer technology rapidly evolved and cost decreased in these years, so that it was economically logical for smaller companies (e.g., general aviation contractors, boat builders, race-car teams, etc.) to use this technology. The first panel code commercially available to the smaller industries was VSAERO[12.10,12.11] (which was developed under a grant from NASA Ames Research Center). This code can be viewed as the beginning of a third period in the development of panel codes, since it returned to a simpler, first-order panel and singularity elements. This code used the Dirichlet boundary condition for thick bodies and the Neumann condition for thin vortex-lattice panels. Interaction with several methods of boundary layer solutions along streamlines was used, but the displacement thickness effect was corrected by adding sources (blowing or transpiration), rather than adjusting the panel geometry (as in MCAIR). In addition, a wake rollup routine was added that computed the induced velocity on the wake and moved the wake vortices to a new "force-free" position. Following the success of this code (due to computational economy) the Lockheed company developed a similar method, called QUADPAN.[12.12]

At this point it seems that the theory of panel methods has matured and most of the effort is invested in pre- and post-processing (automatic generation of surface grids and graphical representation of results). Also being developed are interactive airfoil and wing design routines, where the designer can modify interactively the body's geometry in order to obtain a desirable pressure distribution.

Table 12.2. *Claimed advantages of low- and high-order panel codes*

	Low-order methods	High-order methods
Derivation of influence coefficients	simple derivation	more complex derivation
Computer programming	relatively simple coding	requires more coding effort
Program size	short (fits minicomputers)	longer (will run on mainframes only)
Run cost	low	considerably higher
Accuracy	less – for same number of panels (but more accurate for same run time)	higher accuracy for a given number of panels
Sensitivity to gaps in paneling	not very sensitive*	not allowed
Extension to $M > 1$	possible	simple (for arbitrary geometry)

* This is a major advantage for the comparatively untrained user. Also this feature allows for an easy treatment of very narrow gaps where viscous effects control the otherwise high speed inviscid flow (see example in Section 12.7).

Some of the other improvements of these methods, during the second half of the 1980s, included the addition of an unsteady motion option[12.13] and an overall numeric optimization of the method (in terms of computer memory requirement and efficiency of matrix solver). Such a code is PMARC[12.14] (Panel Method Ames Research Center), which was developed at NASA Ames and is now suitable for home computers.

This recent trend of some code developers toward the use of low-order methods, and the fact that many different methods are now being used, led to several comparison studies. For instance, the study of Margason et al.[12.15] indicates that low-order methods are clearly faster and cheaper to operate. Some of the claimed advantages of each of the methods are listed in Table 12.2 and the decision of which one to choose for a particular application is not obvious. It is important to point out that, "any method will provide good results after validating it through a large number of test cases." (Dr. John Hess).

12.5 First-Order Potential-Based Panel Methods

As an example of three-dimensional first-order panel methods, some of the features are discussed, following the six steps used for the previous computational methods. It is recommended at this phase that students use one of the available panel codes along with its graphical pre- and post-processor. It is useful to become familiar first with the pre-processor and the grid generation process, through homework assignments, and only later devote more time to the aerodynamic results. In the following discussion, some of the features of a first-order method (e.g., VSAERO[12.10,12.11] and PMARC[12.14]) are described.

a. Choice of Singularity Elements

The basic panel element used in this method has a constant-strength source or doublet, and the surface is also planar (but doublet panels that are equivalent to a vortex ring and can be twisted). Following the formulation of Section 9.4, the Dirichlet boundary condition on a thick body can be reduced to the form of Eq. (9.23), which states that the perturbation potential inside the body is zero:

$$\sum_{k=1}^{N} C_k \mu_k + \sum_{\ell=1}^{N_W} C_\ell \mu_\ell + \sum_{k=1}^{N} B_k \sigma_k = 0 \tag{12.29}$$

This equation will be evaluated for each collocation point inside the body and the influence coefficients C_k, C_ℓ of the body and wake doublets and B_k of the sources are calculated by the formulas of Section 10.4.

Both the VSAERO[12.10,12.11] and PMARC[12.14] computer programs allow additional modeling of zero-thickness surfaces by vortex lattice grids, which are treated in a manner described in Section 12.2. On these surfaces the zero normal velocity boundary condition is used, which results in a similar set of equations on the collocation point of panel i:

$$\sum_{k=1}^{N} C_k^* \mu_k + \sum_{\ell=1}^{N_W} C_\ell^* \mu_\ell + \sum_{k=1}^{N} B_k^* \sigma_k = -\mathbf{Q}_\infty \cdot \mathbf{n}_i \qquad (12.30)$$

The B_k^*, C_k^* induced velocity coefficients are given, too, in Section 10.4.

b. *Discretization and Grid Generation*

In this phase the shape of the body is divided into surface panel elements as shown in Fig. 12.22. It is useful to have a graphic representation of the grid so that possible input errors such as gaps between the panels and misplaced corner points can be corrected. The grid is usually constructed of rectangular subgrids (patches) and some of the patches forming the model of Fig. 12.22 are shown as well. Note that triangular panels, as in the nose cone area, are actually rectangular panels with two coinciding corners. At this phase the panel corner points, collocation points (which may be on the surface or slightly inside the body), and the outward normal vectors $\mathbf{n}_k$ are identified and the counter k for each panel is assigned. A typical example of generating a wing grid and its unfolded patch are shown in Fig. 12.23.

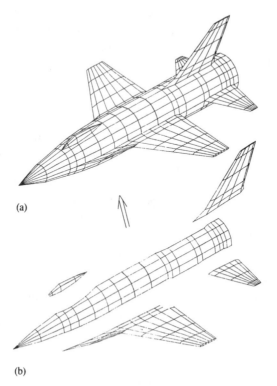

(a)

(b)

Figure 12.22 Representation of the surface geometry of a generic airplane by subarrays (patches): (a) complete model; (b) separate patches.

12.5 First-Order Potential-Based Panel Methods

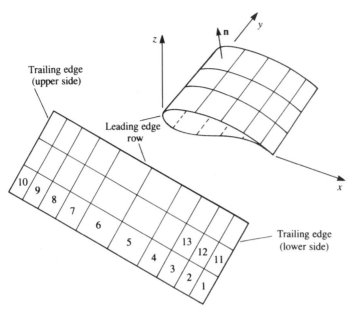

Figure 12.23 Method of storing the grid information on a wing patch (and identifying the wing's outer surface).

c. *Influence Coefficients*

At this phase the boundary conditions of Eq. (12.29) (or (12.30)) are evaluated and for this example we shall use only the Dirichlet boundary condition. As was noted earlier, Eq. (12.29) is not unique and the combination of sources and doublets must be selected. For example, fixing the source strengths as

$$\sigma_k = \mathbf{n}_k \cdot \mathbf{Q}_\infty \tag{12.31}$$

will result in a set of equations with the doublet strengths as the unknowns. The above selection of the source strength is based on the results of Section 4.4 and includes most of the normal velocity component required for the zero normal flow boundary condition (in the nonlifting case). Consequently, the unknown μ_k strengths will be smaller.

So, at this point, the potential at the collocation point of each panel (inside the body) is influenced by all the N other panels and the coefficients appearing in Eq. (12.29) can be calculated. Now, let us consider a wake panel that is shed by an upper panel with a counter l and a lower panel with a counter m, as shown in Fig. 12.24. Equation (12.29) for the first collocation point can be derived as

$$C_{11}\mu_1 + \cdots + C_{1l}\mu_l + \cdots + C_{1m}\mu_m + \cdots + C_{1N}\mu_N$$
$$+ \sum_{p=1}^{N_W} C_{1p}\mu_p + \sum_{k=1}^{N} B_{1k}\sigma_k = 0 \tag{12.32}$$

The influence of this particular wake panel at point 1, when singled out from the $\sum_{p=1}^{N_W} C_{1p}\mu_p$ term, is then

$$C_{1p}(\mu_l - \mu_m) \tag{12.33}$$

where the counter p scans the wake panels. But this second summation of the wake influences in Eq. (12.29) does not contain additional unknown values of μ. Therefore, the results of this second summation can be resubstituted into the equation, using the results of the

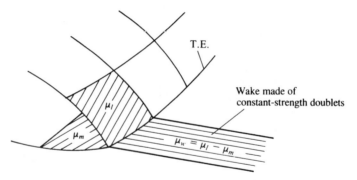

Figure 12.24 A typical wake panel shed by the trailing edge upper and lower panels.

Kutta condition (Eq. (12.33)). In the particular case of Fig. 12.24, the equation for the first point becomes

$$C_{11}\mu_1 + \cdots + (C_{1l} + C_{1p})\mu_l + \cdots + (C_{1m} - C_{1p})\mu_m + \cdots$$
$$+ C_{1N}\mu_N + \sum_{k=1}^{N} B_{1k}\sigma_k = 0$$

Consequently, this equation can be simplified to a form

$$\sum_{k=1}^{N} A_{1k}\mu_k = -\sum_{k=1}^{N} B_{1k}\sigma_k \qquad (12.34)$$

where $A_{1k} = C_{1k}$ if no wake is shed from this panel and $A_{1k} = C_{1k} \pm C_{1p}$ if it is shedding a wake panel. This equation now has the form

$$\begin{pmatrix} a_{11}, a_{12}, \ldots, a_{1N} \\ a_{21}, a_{22}, \ldots, a_{2N} \\ \vdots \qquad \vdots \\ a_{N1}, a_{N2}, \ldots, a_{NN} \end{pmatrix} \begin{pmatrix} \mu_1 \\ \mu_2 \\ \vdots \\ \mu_N \end{pmatrix} = - \begin{pmatrix} b_{11}, b_{12}, \ldots, b_{1N} \\ b_{21}, b_{22}, \ldots, b_{2N} \\ \vdots \qquad \vdots \\ b_{N1}, b_{N2}, \ldots, b_{NN} \end{pmatrix} \begin{pmatrix} \sigma_1 \\ \sigma_2 \\ \vdots \\ \sigma_N \end{pmatrix} \qquad (12.35)$$

which is a set of N linear equations for the N unknown μ_k (σ_k is known from Eq. (12.31)). Notice that on the diagonal, $a_{kk} = 1/2$, except when the panel is at the trailing edge.

d. *Establish RHS*

The right-hand-side matrix multiplication can be carried out since the strengths of the sources are known. This procedure establishes the RHS vector and Eq. (12.35) reduces to the form

$$\begin{pmatrix} a_{11}, a_{12}, \ldots, a_{1N} \\ a_{21}, a_{22}, \ldots, a_{2N} \\ \vdots \qquad \vdots \\ a_{N1}, a_{N2}, \ldots, a_{NN} \end{pmatrix} \begin{pmatrix} \mu_1 \\ \mu_2 \\ \vdots \\ \mu_N \end{pmatrix} = \begin{pmatrix} \text{RHS}_1 \\ \text{RHS}_2 \\ \vdots \\ \text{RHS}_N \end{pmatrix} \qquad (12.36)$$

e. *Solution of Linear Equations*

The above matrix is full and has a nonzero diagonal and so a stable numerical solution is possible. Usually when the number of panels is low (e.g., less than 500) a direct solver can be used. However, as the number of panels increases (up to about 10,000), iterative

12.5 First-Order Potential-Based Panel Methods

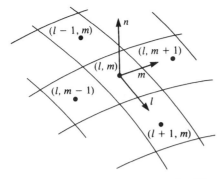

Figure 12.25 Nomenclature used for the differentiation of the velocity potential for local tangential velocity calculations.

solvers are used so that only one row of the matrix occupies the computer memory during the solution.

f. *Computation of Velocities, Pressures, and Loads*

One of the advantages of the velocity potential formulation is that the computation of the surface velocity components and pressures is determinable by the local properties of the solution (velocity potential in this case). The perturbation velocity components on the surface of a panel can be obtained by Eqs. (9.26), in the tangential directions:

$$q_l = -\frac{\partial \mu}{\partial l}, \qquad q_m = -\frac{\partial \mu}{\partial m} \tag{12.37}$$

and in the normal direction:

$$q_n = \sigma \tag{12.37a}$$

where l, m are the local tangential coordinates (see Fig. 12.25). For example, the perturbation velocity component in the l direction can be formulated (e.g., by using central differences) as

$$q_l = \frac{1}{2\Delta l}(\mu_{l-1} - \mu_{l+1}) \tag{12.38}$$

where Δl is the panel length in the l direction. In most cases the panels do not have equal sizes and instead of this simple formula, a more elaborate one can be used (sometimes only the term Δl is modified). The total velocity at collocation point k is the sum of the free stream plus the perturbation velocity:

$$\mathbf{Q}_k = (Q_{\infty_l}, Q_{\infty_m}, Q_{\infty_n})_k + (q_l, q_m, q_n)_k \tag{12.39}$$

where l_k, m_k, n_k are the local panel coordinate directions (shown in Fig. 12.25) and of course the total normal velocity component on the surface is zero. The pressure coefficient can now be computed for each panel using Eq. (4.53):

$$C_{p_k} = 1 - \frac{Q_k^2}{Q_\infty^2} \tag{12.40}$$

The contribution of this element to the aerodynamic loads $\Delta \mathbf{F}_k$ is

$$\Delta \mathbf{F}_k = -C_{p_k}\left(\frac{1}{2}\rho Q_\infty^2\right)\Delta S_k \mathbf{n}_k \tag{12.41}$$

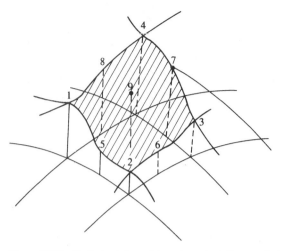

Figure 12.26 Typical points used to evaluate the influence of a higher order singularity distribution.

The total aerodynamic loads are then obtained by adding the contributions of the individual panels (multiplied first by the lift, drag, and side-force direction vectors). A sample computer program based on this method is provided in Appendix D, Program No. 14.

In many situations off-body velocity field information is also required. This type of calculation can be done by using the velocity influence formulas of Chapter 10 (since the strengths σ and μ are known at this point).

12.6 Higher Order Panel Methods

The mathematical principle behind these methods is similar to that of the low-order methods, but the complexity of the element in terms of its geometry and singularity distribution is increased. The boundary conditions to be solved are still Eq. (12.29) (Dirichlet) and Eq. (12.30) (Neumann) or a combination of both. (That is, on some panels the Neumann and on the other panels the Dirichlet condition will be used – but not both conditions on the same panel.) The influence coefficients are more complex and they depend on more than one singularity parameter (whereas only one such a parameter was required for a constant-strength source or doublet element). In the following section a brief description of such a method is presented and more details on one of these methods (PAN AIR) is provided in Refs. 12.8 and 12.9.

a. *Choice of Singularity Elements*

Using a first-order source and second-order doublet distribution as described in Section 10.5 allows us to determine the influence of each panel in terms of its values at its nine points (as shown in Fig. 12.26). The surface is divided into five flat subelements and the relative location of these points on these surfaces is shown in Fig. 10.27. The influence of the panel's subelements can be summarized as

$$\Delta \Phi = F_S(\sigma_1, \sigma_2, \sigma_3, \sigma_4, \sigma_9) = f_S(\sigma_0, \sigma_x, \sigma_y) \tag{12.42}$$

$$\Delta(u, v, w) = G_S(\sigma_1, \sigma_2, \sigma_3, \sigma_4, \sigma_9) = g_S(\sigma_0, \sigma_x, \sigma_y) \tag{12.43}$$

for the first-order source element and

$$\Delta \Phi = F_D(\mu_1, \mu_2, \mu_3, \mu_4, \mu_5, \mu_6, \mu_7, \mu_8, \mu_9) = f_D(\mu_0, \mu_x, \mu_y, \mu_{xx}, \mu_{xy}, \mu_{yy}) \tag{12.44}$$

12.6 Higher Order Panel Methods

$$\Delta(u, v, w) = G_D(\mu_1, \mu_2, \mu_3, \mu_4, \mu_5, \mu_6, \mu_7, \mu_8, \mu_9)$$
$$= g_D(\mu_0, \mu_x, \mu_y, \mu_{xx}, \mu_{xy}, \mu_{yy}) \quad (12.45)$$

for the second-order doublet. The subscripts 1 through 9 denote the strength of the singularity distribution at this point according to the sequence in Fig. 12.26. Note that for a source five unknowns are used, but by assuming a linear strength distribution this can be reduced by algebraic manipulations to three (e.g., $f_S = \sigma_0 + \sigma_x x + \sigma_y y$). Similarly, by assuming a parabolic distribution for the doublet strength the number of unknowns is reduced to six per panel (e.g., $g_D = \mu_0 + \mu_x x + \mu_y y + \mu_{xy} xy + \mu_{xx} x^2 + \mu_{yy} y^2$).

b. *Discretization and Grid Generation*

The grid generation procedure is similar to the procedure described for the zero-order method, but now all nine nodal points are stored in the memory. Also, gaps in the geometry are not allowed since a continuous geometry is assumed.

c. *Influence Coefficients*

Again we shall follow the case where the strength of the source (for thick bodies) is set by Eq. (12.31)

$$\sigma_k = \mathbf{n}_k \cdot \mathbf{Q}_\infty$$

The Dirichlet boundary condition can be reduced then to the form

$$\sum_{k=1}^{6N} C_k \mu_k + \sum_{\ell=1}^{N_W} C_\ell \mu_\ell + \sum_{k=1}^{3N} B_k \sigma_k = 0 \quad (12.46)$$

or if the Neumann condition is used then on the ith collocation point

$$\sum_{k=1}^{6N} C_k^* \mu_k + \sum_{\ell=1}^{N_W} C_\ell^* \mu_\ell + \sum_{k=1}^{3N} B_k^* \sigma_k = -\mathbf{Q}_\infty \cdot \mathbf{n}_i \quad (12.47)$$

In principle, for N panels we have $6N$ unknown doublet strengths, but by matching the magnitudes (or slopes) of the neighbor panels, $5N$ very simple additional equations can be obtained (see for example the two-dimensional case in Section 11.6.1). These neighbor panel relations are resubstituted into Eq. (12.46) or Eq. (12.47) such that for N panels N linear algebraic equations must be solved. Also, as before, the wake doublets μ_ℓ do not contain any new unknowns and based on the corresponding doublet values of the wake shedding panels the wake influence can be substituted into the C_k, C_k^* coefficients. Thus for each panel i (when using the internal Dirichlet boundary condition)

$$\sum_{k=1}^{N} A_{ik} \mu_k = -\sum_{k=1}^{N} B_{ik} \sigma_k \quad (12.48)$$

where the collocation point counter $i = 1 \to N$.

d. *Establish RHS*

The right-hand side of this equation includes the known source strengths (for the Dirichlet boundary condition) and the free-stream component normal to the surface (for the Neumann boundary condition case) and can be computed. The additional $2N$ equations for the source corner point values are obtained by matching the source strength at the panel edges.

e. *Solution of Linear Equations*
The solution is the same as for low-order methods.

f. *Computation of Velocities, Pressures, and Loads*
The local tangential velocity is calculated by using Eq. (12.37), but since at each panel there are nine values of μ a finer arithmetic scheme is used for calculating the local gradients of the velocity potential. Once the velocity components are found the local pressure coefficient and the aerodynamic loads are found by using Eqs. (12.40) and (12.41).

More details on such high-order panel codes can be found in Refs. 9.4 and 12.9.

12.7   Sample Solutions with Panel Codes

Panel methods have the advantage of modeling the flow over complex three-dimensional configurations. However, the first thing to remember is that the method is based on potential flow solutions and therefore its forte is in solving *attached* flowfields. In the case of such attached flowfields the calculated pressure distribution and the lift will be close to the experimental results, but for the drag force only the lift-induced drag portion is provided by the potential flow solution and an estimation of the viscous drag is required. For flows with considerable areas of flow separations the method usually can point toward areas of large pressure gradients that cause the flow separations, but the computed pressure distributions will be wrong. The following examples will show some of the cases where such methods can provide useful engineering information, along with some cases where the effects of viscosity become more important.

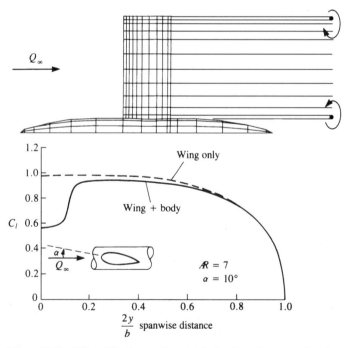

Figure 12.27   Effect of fuselage on the spanwise loading of a rectangular wing.

12.7 Sample Solutions with Panel Codes

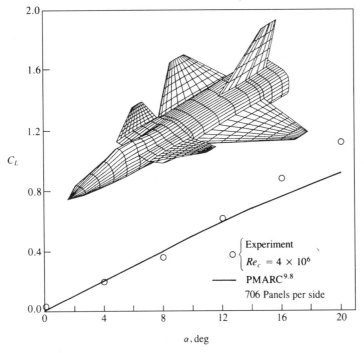

Figure 12.28 Comparison between calculated and experimental lift coefficients for a generic fighter aircraft configuration. (Experimental results taken from Stoll, F., and Koenig, D. G., "Large Scale Wind-Tunnel Investigation of a Close-Coupled Canard-Delta-Wing Fighter Model through High Angles of Attack," *AIAA Paper 83-2554*.)

Example 1: Wing Body Combination

All classical methods (e.g., lifting surface) were capable of modeling simple lifting surfaces only with some estimation of wing/fuselage juncture effects. Panel methods, in contrast, can solve the flow over fairly complex wing/fuselage combinations. For example, Fig. 12.27 shows a typical case where the lift near the centerline is reduced due to the presence of the fuselage. The wake vortex originating near the wing/fuselage juncture, whose circulation is opposite in direction to the tip vortex, must be modeled carefully (so that it will not intersect the fuselage). The location of this vortex is important, too, since it may affect the flow on the rest of the aircraft and may cause flow separations on the aft section of the fuselage and on the tail. Some methods allow the "stitching" of the wing-root vortex to the fuselage for better modeling of the lift "carry-over" from the wing to the fuselage.

Example 2: Lift of High-Speed Airplane Configurations

Airplanes that operate at higher speeds where compressibility effects are not negligible usually encounter low-speed flight conditions during takeoff and landing. For these conditions panel methods can provide useful aerodynamic information. As an example the calculated and experimental lift coefficients for two such aircraft configurations are provided in Figs. 12.28 and 12.29. Both figures indicate that at the lower angles of attack (less than 15° in this case) the calculations agree fairly well with the experiments. However, at larger angles of attack, leading-edge vortex

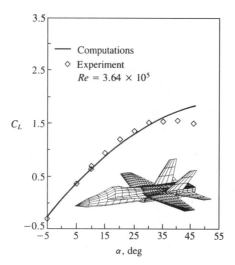

Figure 12.29 Comparison between calculated and measured lift coefficients for the McDonnell Douglas F/A-18 airplane (using 668 panels per side). From Ref. 12.16. Reprinted with permission. Copyright AIAA.

lift (e.g., for $\alpha > 15°$ in Fig. 12.28 and for $15° < \alpha < 30°$ in Fig. 12.29) can cause additional lift, and such vortex lift models were not introduced in this chapter. Moreover, the flow over these complete configurations is usually very complex and many regional flow separations and vortex flows exist. Therefore, even if the results presented in these two figures agree reasonably with the computations, the computations can serve mainly as a first-order prediction tool; final validation usually would require extensive testing.

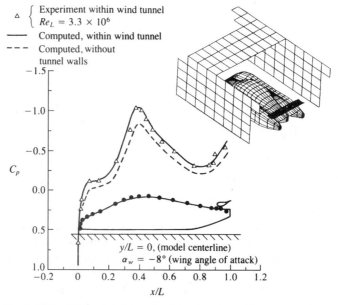

Figure 12.30 Effect of wind-tunnel walls on the pressure distribution on the upper surface centerline (shown by the black dots) of a streamlined automobile (model frontal area to wind-tunnel cross-sectional area ratio was about 13%). From Ref. 12.17.

12.7 Sample Solutions with Panel Codes

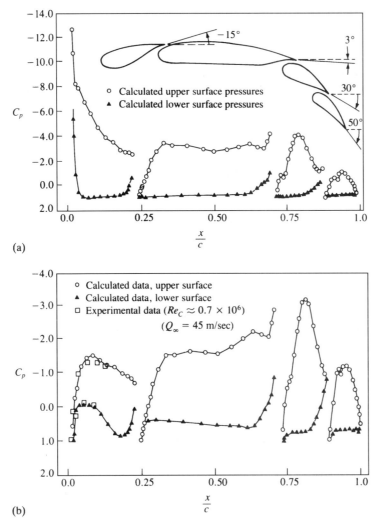

Figure 12.31 Pressure distribution on a two-dimensional four-element airfoil (a) and at the centerline of an $\mathcal{R} = 1.5$, rectangular wing, having the same airfoil section and angle of attack (b). From Ref. 12.18. Reprinted with permission. Copyright AIAA.

Example 3: Wind-Tunnel Wall Interference Corrections

Wind tunnels provide a well-controlled environment where a variety of tests, such as measurement of aerodynamic pressures and loads, can be carried out. However, model designers usually prefer larger models and therefore, in many cases, the effect of the test section walls is not negligible. The most obvious interference between the model and the wind-tunnel test section walls is called "solid blockage," in which the presence of the model inside the wind tunnel reduces the flow cross-sectional area and, according to Bernoulli's principle, the flow speed will increase there. Since the local velocity at the test section is higher than it would be in a free flow outside the wind tunnel the aerodynamic coefficients are overestimated. In addition to this "blockage effect" there is a "reflection effect" that changes the lift of lifting surfaces near solid boundaries (as in the case of ground effect). Figure 12.30

shows the increase in the suction peaks (and velocity) over the upper surface of a streamlined automobile model; and this increase was well predicted by the panel method. In this case, for theoretical purposes the aft section of the vehicle was highly streamlined and flow separations there were minimal, but a generalization of this approach to other bluff body shapes must be approached with extra care.

In general, the wind-tunnel wall corrections are obtained by two sequential computations where in the first the flowfield over the model within the wind-tunnel test section is computed and in the second computation the wind-tunnel walls are removed. The differences between these two cases provides the potential flow effect of wind-tunnel boundaries on model lift and blockage (Refs. 12.16 and 12.17). Note that this wind-tunnel wall correction method inherently includes effects of lift and blockage, and it provides more details than previous semi-empirical methods (for the complete wind-tunnel wall effect, though, the viscous effects should be included, too.)

Example 4: High-Lift, Low Aspect Ratio Multielement Wing

As the wing aspect ratio becomes small, two-dimensional airfoil analysis may not be applicable and a considerable difference exists between the two- and three-dimensional chordwise pressure distributions. Consequently, a two-dimensional development of such an airfoil section, without considering the complete three-dimensional analysis, is not recommended.[12.18] As an example, typical computational results for the two-dimensional pressure distribution on a four-element airfoil are shown in Fig. 12.31a. The computed pressure distribution, at the centerline of a rectangular wing ($AR = 1.5$), having the same airfoil section and the same attitude, are presented in Fig. 12.31b. The most obvious differences between the two cases of Fig. 12.31 are the threefold reduction in the C_p range of the three-dimensional data and the change in the shape of the pressure distribution when compared with the two-dimensional case. Also, in the three-dimensional case, pressure gradients are the strongest near the second flap (from the trailing edge) and with increased angle of attack, flow separation can be initiated here (and not near the leading edge, as it seems from the two-dimensional data).

Example 5: Wake Length

One of the objectives of this section is to highlight some possible errors in modeling the potential flow problems. One such frequent problem arises when the wake of a wing is too short (in steady-state flow). This problem occurs when, owing to the need to present the wing and the wake in the same scale, the wake model becomes short. Since in most situations the wake is modeled by constant-strength doublet elements a starting vortex is present at the edge of the far field panel as shown in Fig. 12.32. This vortex induces a downwash on the wing thereby reducing its lift. The problem can easily be cured by a longer wake, which should be at least 20 chord lengths behind the wing trailing edge (of course this distance depends on wing aspect ratio; see also the effect of a starting vortex in Example 1, Section 13.12).

Example 6: Modeling of Gap in Wing

Example 4 indicated that the modeling of a gap between two airfoils is obtained in a satisfactory manner. However, if the gap is parallel to the streamlines, as in the case of the chordwise gap between a main wing and a flap, establishing

12.7 Sample Solutions with Panel Codes

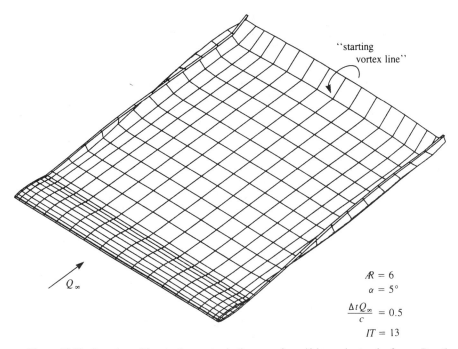

Figure 12.32 Location of the starting vortex in the case of specifying a short wake for an $R = 6$ rectangular wing in a steady-state flow. (13 time steps).

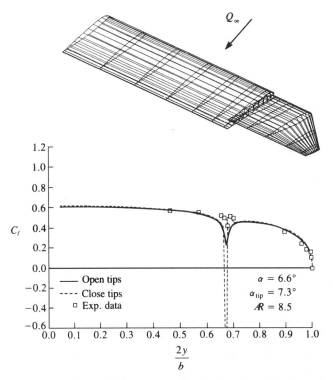

Figure 12.33 Effect of a gap between a floating wing tip and the wing on the spanwise loading. From Ref. 12.19.

a good panel model may be difficult. Since the potential flow does not account for viscosity, the velocity inside a narrow gap will increase to unrealistic values, which in reality are reduced by the viscous friction. This problem is demonstrated by Fig. 12.33 (taken from Ref. 12.19) where such a side gap between a large aspect ratio wing and a moving wing tip (shown in Fig. 12.33) is calculated by different panel models. When the two wing parts were modeled as two separate, closed bodies, the high speed within the gap resulted in large suction peaks (shown by the broken lines) that are different from the experimental data. The solid line shows the case where the wing and the wingtip closures were removed from the gap (thereby leaving the two bodies open in the gap region) and the results now agree more with the experimental data. Since this example was executed with a first-order panel method (Ref. 12.11) which is not sensitive to gaps in the panel model this problem was partially resolved, but this approach is not applicable to higher order panel methods. (Note that the accurate potential solution is practically inaccurate in this case and the removal of the wing side edge closures inside the gap should be viewed as a viscous flow effect modeling exercise!)

References

[12.1] Hess, J. L., and Smith, A. M. O., "Calculation of Non-Lifting Potential Flow About Arbitrary Three-dimensional Bodies," Douglas-McDonnell Rep. No. E.S. 40622, March 15, 1962; also *Journal of Ship Research*, No. 8, pp. 22–44, 1964.

[12.2] Woodward, F. A., "Analysis and Design of Wing-Body Combinations at Subsonic and Supersonic Speeds," *Journal of Aircraft*, Vol. 5, No. 6, 1968, pp. 528–534.

[12.3] Woodward, F. A., "An Improved Method for the Aerodynamic Analysis of Wing-Body-Tail Configurations in Subsonic and Supersonic Flow," NASA CR-2228, May 1973.

[12.4] Hess, J. L., "Calculation of Potential Flow About Arbitrary Three-Dimensional Lifting Bodies," Final Technical Report MDC J5679-01, McDonnell Douglas, Long Beach, California, Oct. 1972.

[12.5] Bristow, D. R., "Development of Panel Methods for Subsonic Analysis and Design," NASA CR 3234, 1980.

[12.6] Morino, L., "Steady, Oscillatory, and Unsteady Subsonic and Supersonic Aerodynamics – Production Version (SOUSSA-P1,1) Vol. 1, Theoretical Manual," NASA CR 159130, 1980.

[12.7] Hess, J. L., and Friedman, D. M., "An Improved High-Order Panel Method for Three-Dimensional Lifting Flow," Douglas Aircraft Co. Rep. No. NADC-79277-60, 1981.

[12.8] Carmichael, R. L., and Ericson, L. L., "PAN AIR – A Higher Order Panel Method for Predicting Subsonic or Supersonic Linear Potential Flows About Arbitrary Configurations," *AIAA Paper 81-1255*, June 1981.

[12.9] Magnus, A. E., and Epton, M. A., "PAN AIR – A Computer Program for Predicting Subsonic or Supersonic Linear Potential Flows About Arbitrary Configurations Using a Higher Order Panel Method, Volume 1 – Theory Document," NASA CR 3251, Nov. 1983.

[12.10] Maskew, B., "Prediction of Subsonic Aerodynamic Characteristics: A Case for Low-Order Panel Methods," *AIAA Journal*, Vol. 19, No. 2, 1982, pp. 157–163.

[12.11] Maskew, B., "Program VSAERO, A Computer Program for Calculating the Nonlinear Aerodynamic Characteristics of Arbitrary Configurations," NASA CR-166476, Nov. 1982.

[12.12] Youngren, H. H., Bouchard, E. E., Coopersmith, R. M., and Miranda, L. R., "Comparison of Panel Method Formulations and Its Influence on the Development of QUADPAN, an Advanced Low-Order Method," *AIAA Paper 83-1827*, 1983.

[12.13] Katz, J., and Maskew B., "Unsteady Low-Speed Aerodynamic Model for Complete Aircraft Configurations," *AIAA Paper 86-2180*, Aug. 1986; also, *Journal of Aircraft*, Vol. 25, No. 4, 1988, pp. 302–310.

[12.14] Ashby, D. L., Dudley, M. R., and Iguchi, S. K., "Development and Validation of an Advanced Low-Order Panel Method," NASA TM 101024, Oct. 1988.

[12.15] Margason, R. J., Kjelgaard, S. O., Sellers, W. L., III., Morris, C. E. K., Jr., Walkey, K. B., and Shields, E. W., "Subsonic Panel Methods – A Comparison of Several Production Codes," *AIAA Paper 85-0280*, Jan. 1985.

[12.16] Browne, L., and Katz, J., "Application of Panel Methods to Wind-Tunnel Wall Interference Corrections," *AIAA Paper 90-0007*, Jan. 1990.

[12.17] Katz, J., "Integration of Computational Methods into Automotive Wind Tunnel Testing," *SAE Paper 89-0601*, presented at the SAE Int. Conference, Feb. 1989.

[12.18] Katz, J., "Aerodynamics of High-Lift, Low Aspect-Ratio Unswept Wings," *AIAA Journal*, Vol. 27, No. 8, 1989, pp. 1123–1124.

[12.19] Martin, D. M., and Fortin, P. E., "VSAERO Analysis of Tip Planforms for the Free-Tip Rotor," NASA CR-177487, June 1988.

Problems

Three-dimensional solutions that were presented in this chapter usually require more effort than an average homework problem and the following suggested examples are more suitable for midterm or final projects.

12.1. Construct a computational method based on the lifting-line model of Section 12.1 and study the effect of wing sweep on the spanwise loading. Use at least ten spanwise elements (per semispan), and assume a constant chord c and wing aspect ratio of 8.

12.2. Use the method of Problem 12.1 to study the spanwise load distribution on the wing planforms presented in Fig. 12.34.

12.3. Use the method of Problem 12.1 to study the effect of taper ratio on the spanwise load distribution of an unswept leading-edge wing. Assume wing span $b = 6$, root chord $c_0 = 1$, and $\Lambda = 1.0, 0.6,$ and 0.2.

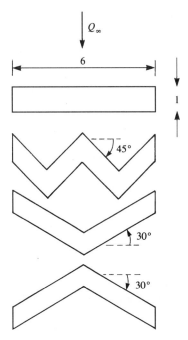

Figure 12.34 Various wing planform shapes to be used for homework problems.

12.4. Construct a computational method based on the lifting-surface model of Section 12.3. Use at least ten spanwise (per semispan) and five chordwise panels, and assume a constant chord c and wing aspect ratio of 8.
  a. Study the effect of wing sweep on the spanwise loading.
  b. Study the spanwise load distribution on the wing planforms presented in Fig. 12.34.
  c. Study the effect of taper ratio on the spanwise load distribution of an unswept leading-edge wing. Assume wing span $b = 6$, root chord $c_0 = 1$, and $\Lambda = 1.0, 0.6$, and 0.2.

12.5. Use the pre-processor to the panel code that is available in your institute to generate the grid for the wing configurations of Fig. 12.34 (use at least 200 panels per semispan). Assume all wings have a NACA 63_2-415 airfoil section.

12.6. Using the three-dimensional panel code available in your institute solve the flow over the wing planforms of Problem 12.5 and present chordwise pressure distributions at five equally spaced spanwise stations. Also plot the spanwise loading for each wing and compare their root bending moment M_x.

12.7. Use the three-dimensional panel code available in your institute for the following exercises:
  a. Study the effect of wing sweep on the spanwise loading. Assume a rectangular wing with aspect ratio of $\!R = 8$, constant chord, and leading-edge sweep of $\Lambda = 0°, 20°, 40°$, and $60°$.
  b. Study the effect of taper ratio on the spanwise load distribution of an unswept leading-edge wing. Assume wing span $b = 6$, root chord $c_0 = 1$ and $\lambda = 1.0, 0.6$, and 0.2.

For all these cases present the chordwise pressure distribution at five equally spaced spanwise stations. Also plot the spanwise loading for each wing and compare their root bending moment M_x.

CHAPTER 13

Unsteady Incompressible Potential Flow

We have seen in the previous chapters that in an incompressible, irrotational fluid the velocity field can be obtained by solving the continuity equation. However, the incompressible continuity equation does not directly include time-dependent terms, and the time dependency is introduced through the boundary conditions. Therefore, the first objective is to demonstrate that the methods of solution that were developed for steady flows can be used with only small modifications. These modifications will include the treatment of the "zero normal flow on a solid surface" boundary conditions and the use of the unsteady Bernoulli equation. Furthermore, as a result of the nonuniform motion, the wake becomes more complex than in the corresponding steady flow case and it should be properly accounted for. Consequently, this chapter is divided into three parts, as follows:

a. Formulation of the problem and of the proposed modifications for converting steady-state flow methods to treat unsteady flows (Sections 13.1–13.6).
b. Examples of converting analytical models to treat time-dependent flows (e.g., thin lifting airfoil and slender wing in Sections 13.8–13.9).
c. Examples of converting numerical models to treat time-dependent flows (Sections 13.10–13.13).

For the numerical examples only the simplest models are presented; however, application of the approach to any of the other methods of Chapter 11 is strongly recommended (e.g., can be given as a student project).

In the general case of the arbitrary motion of a solid body submerged in a fluid (e.g., a maneuvering wing or aircraft) the motion path is determined by the combined dynamic and fluid dynamic equations. However, this chapter will deal with the loads generated by the fluid only and therefore the path along which the body (or the wing or aircraft) moves is assumed to be prescribed.

13.1 Formulation of the Problem and Choice of Coordinates

When treating time-dependent motions of bodies, the selection of the coordinate systems becomes very important. It is useful to describe the unsteady motion of the surface on which the "zero normal flow" boundary condition is applied in a body-fixed coordinate system (x, y, z); see for example the maneuvering wing depicted in Fig. 13.1. The motion of the origin O of this coordinate system (x, y, z) is then prescribed in an inertial frame of reference (X, Y, Z) and is assumed to be known (as shown in Fig. 13.1). For simplicity, assume that at $t = 0$ the inertial frame (X, Y, Z) coincides with the frame (x, y, z). Then, at $t > 0$, the relative motion of the origin $(\ )_o$ of the body-fixed frame of reference is prescribed by its location $\mathbf{R}_0(t)$, and the instantaneous orientation $\Theta(t)$, where (ϕ, θ, ψ) are the Euler[13.1] rotation angles:

$$(X_0, Y_0, Z_0) = \mathbf{R}_0(t) \qquad (13.1)$$

$$(\phi, \theta, \psi) = \Theta(t) \qquad (13.2)$$

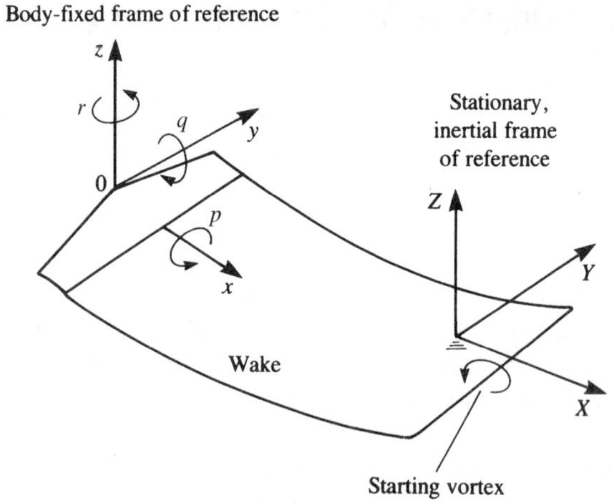

Figure 13.1 Inertial and body coordinates used to describe the motion of the body.

For example, in the case of a constant-velocity flow of speed U_∞ in the positive x direction (in the wing's frame of reference in Fig. 13.1) the function $\mathbf{R}_0(t)$ will be

$$(X_0, Y_0, Z_0) = (-U_\infty t, 0, 0)$$

which means that the wing is being translated in the negative X direction.

The fluid surrounding the body is assumed to be inviscid, irrotational, and incompressible over the entire flowfield, excluding the body's solid boundaries and its wakes. Therefore, a velocity potential $\Phi(X, Y, Z)$ can be defined in the inertial frame and the continuity equation, in this frame of reference, becomes

$$\nabla^2 \Phi = 0 \quad \text{(in } X, Y, Z \text{ coordinates)} \tag{13.3}$$

and the first boundary condition requiring zero normal velocity across the body's solid boundaries is

$$(\nabla \Phi + \mathbf{v}) \cdot \mathbf{n} = 0 \quad \text{(in } X, Y, Z \text{ coordinates)} \tag{13.4}$$

Here $-\mathbf{v}$ is the surface's velocity and $\mathbf{n} = \mathbf{n}(X, Y, Z, t)$ is the vector normal to this moving surface, as viewed from the inertial frame of reference. (Note that $\mathbf{v}$ is defined with the minus sign so that the undisturbed flow velocity will be positive in the body's frame of reference.) Since Eq. (13.3) does not depend directly on time, the time dependency is introduced through this boundary condition (e.g., the location and orientation of $\mathbf{n}$ can vary with time). It is interesting to point out that Φ is the total velocity potential, but as a result of its definition in a frame that is attached to the undisturbed fluid its magnitude is small (in fact it is similar to the perturbation potential of Section 4.2).

The second boundary condition requires that the flow disturbance, due to the body's motion through the fluid, should diminish far from the body (or wing in Fig. 13.1),

$$\lim_{|\mathbf{R}-\mathbf{R}_0| \to \infty} \nabla \Phi = 0 \tag{13.5}$$

where $\mathbf{R} = (X, Y, Z)$.

For the unsteady flow case the use of the Kelvin condition will supply an additional equation that can be used to determine the streamwise strength of the vorticity shed into the wake. In general, the Kelvin condition states that in the potential flow region the angular

13.1 Formulation of the Problem and Choice of Coordinates

momentum cannot change, and thus the circulation Γ around a fluid curve enclosing the wing and its wake is conserved:

$$\frac{d\Gamma}{dt} = 0 \quad \text{(for any } t\text{)} \tag{13.6}$$

The solution of this problem, which becomes time dependent because of the boundary condition (Eq. (13.4)), is easier in the body-fixed coordinate system. Consequently, a transformation f between the two coordinate systems has to be established, based on the flight path information of Eqs. (13.1) and (13.2), that is,

$$\begin{pmatrix} x \\ y \\ z \end{pmatrix} = f(X_0, Y_0, Z_0, \phi, \theta, \psi) \begin{pmatrix} X \\ Y \\ Z \end{pmatrix} \tag{13.7}$$

Such a transformation should include the translation and the rotation of the (x, y, z) system, and, for example, may have the form (Ref. 13.1, pp. 312–313)

$$\begin{pmatrix} x \\ y \\ z \end{pmatrix} = \begin{pmatrix} 1 & 0 & 0 \\ 0 & \cos\phi(t) & \sin\phi(t) \\ 0 & -\sin\phi(t) & \cos\phi(t) \end{pmatrix} \begin{pmatrix} \cos\theta(t) & 0 & -\sin\theta(t) \\ 0 & 1 & 0 \\ \sin\theta(t) & 0 & \cos\theta(t) \end{pmatrix}$$

$$\times \begin{pmatrix} \cos\psi(t) & \sin\psi(t) & 0 \\ -\sin\psi(t) & \cos\psi(t) & 0 \\ 0 & 0 & 1 \end{pmatrix} \begin{pmatrix} X - X_0 \\ Y - Y_0 \\ Z - Z_0 \end{pmatrix} \tag{13.7a}$$

Similarly, the kinematic velocity $\mathbf{v}$ of the undisturbed fluid due to the motion of the wing in Fig. 13.1, as viewed in the body frame of reference, is given by

$$\mathbf{v} = -[\mathbf{V}_0 + \mathbf{\Omega} \times \mathbf{r}] \tag{13.8}$$

where $\mathbf{V}_0$ is the velocity of the (x, y, z) system's origin, and must be resolved into the instantaneous (x, y, z) directions,

$$\mathbf{V}_0 = (\dot{X}_0, \dot{Y}_0, \dot{Z}_0) \tag{13.9a}$$

Here $\mathbf{r} = (x, y, z)$ is the position vector and $\mathbf{\Omega}$ is the rate of rotation of the body's frame of reference, as shown in Fig. 13.1,

$$\mathbf{\Omega} = (p, q, r) \tag{13.9b}$$

where (p, q, r) are the angular velocity components, as shown in Fig. 13.1. In situations when an additional relative motion $\mathbf{v}_{\text{rel}}$ within the (x, y, z) system is desired (e.g., small-amplitude oscillation of the wing or its flap, in addition to the average motion of the body system) then Eq. (13.8) becomes

$$\mathbf{v} = -[\mathbf{V}_0 + \mathbf{v}_{\text{rel}} + \mathbf{\Omega} \times \mathbf{r}] \tag{13.8a}$$

and

$$\mathbf{v}_{\text{rel}} = (\dot{x}, \dot{y}, \dot{z}) \tag{13.9c}$$

To an observer in the (x, y, z) frame, the velocity direction is opposite to the flight direction (as derived in the X, Y, Z frame) and therefore the minus sign appears in Eq. (13.8).

The proper transformation of Eqs. (13.3)–(13.5) into the body's frame of reference requires the evaluation of the various derivatives in the (x, y, z) system. This can be found using the standard chain rule differentiation. For example, the $\partial/\partial X$ term becomes

$$\frac{\partial}{\partial X} = \frac{\partial x}{\partial X}\frac{\partial}{\partial x} + \frac{\partial y}{\partial X}\frac{\partial}{\partial y} + \frac{\partial z}{\partial X}\frac{\partial}{\partial z}$$

Here, $\partial x/\partial X$, $\partial y/\partial X$, and $\partial z/\partial X$ include the information about the instantaneous orientation of the body-fixed frame of reference (it is also assumed that the time t is the same in both frames). For example, consider the case when the body frame of reference of Fig. 13.1 translates to the left and only one degree of rotation with θ is allowed (so that $\phi = \psi = 0$ and the y and Y axes remain parallel). In this situation, Eq. (13.7) will provide the transformation between the (x, y, z) and (X, Y, Z) coordinates, which will depend on one rotation only, and the above chain differentiation results in

$$\frac{\partial}{\partial X} = \cos\theta \frac{\partial}{\partial x} + \sin\theta \frac{\partial}{\partial z}$$

$$\frac{\partial}{\partial Y} = \frac{\partial}{\partial y} \tag{13.10}$$

$$\frac{\partial}{\partial Z} = -\sin\theta \frac{\partial}{\partial x} + \cos\theta \frac{\partial}{\partial z}$$

The time derivative in the (x, y, z) system can be obtained by the chain rule (and is similar to Eq. (13.8) less the relative velocity term $\mathbf{v}_{\text{rel}}$):

$$\frac{\partial}{\partial t}\bigg|_{\text{inertial}} = -[\mathbf{V}_0 + \mathbf{\Omega} \times \mathbf{r}] \cdot \left(\frac{\partial}{\partial x}, \frac{\partial}{\partial y}, \frac{\partial}{\partial z}\right) + \frac{\partial}{\partial t}\bigg|_{\text{body}} \tag{13.11}$$

But it is possible to transform Eqs. (13.3), and (13.4) into the body's frame of reference without explicitly knowing Eq. (13.7) and still arrive at the same conclusions. For example, at any moment the continuity equation is independent of the coordinate system orientation and the mass should be conserved. Therefore, the quantity $\nabla^2 \Phi$ is independent of the instantaneous coordinate system and the continuity equation in terms of (x, y, z) remains unchanged (the reader is encouraged to prove this by using the chain rule):

$$\nabla^2 \Phi = 0 \quad (\text{in } x, y, z \text{ coordinates}) \tag{13.12}$$

Also, the two boundary conditions (Eq. (13.4) and (13.5)) should state the same physical conditions. The gradient $\nabla\Phi$ will have the same magnitude and the kinematic velocity $\mathbf{v}$ is given by Eq. (13.8). Therefore, the zero-velocity normal to a solid surface boundary condition, in the body frame, becomes

$$(\nabla\Phi + \mathbf{v}) \cdot \mathbf{n} = 0 \quad (\text{in } x, y, z \text{ coordinates}) \tag{13.13}$$

Here $\mathbf{n}$ is the normal to the body's surface, in terms of the body coordinates (x, y, z).

If we use Eq. (13.8a) with $\mathbf{v}_{\text{rel}}$ representing the motion of the body in the (x, y, z) coordinates, Eq. (13.13) becomes

$$(\nabla\Phi - \mathbf{V}_0 - \mathbf{v}_{\text{rel}} - \mathbf{\Omega} \times \mathbf{r}) \cdot \mathbf{n} = 0 \quad (\text{in } x, y, z \text{ coordinates}) \tag{13.13a}$$

Note that this boundary condition can be derived directly by using Eq. (2.27) of Chapter 2. According to that terminology, the surface is defined in the body frame of reference by F as

$$F = z - \eta(x, y, t) = 0$$

Application of Eq. (2.27) $((DF/Dt) = 0)$ in the (X, Y, Z) coordinate system where the particles move with a velocity $\nabla \Phi$ yields

$$\frac{DF}{Dt}\bigg|_{X,Y,Z} = \frac{\partial F}{\partial t} + \nabla \Phi \cdot \nabla F = 0$$

Transferring this expression into the (x, y, z) system requires the transformation of the $\partial F/\partial t$ term by using Eq. (13.11) and with the second term remaining unchanged this becomes

$$\frac{\partial F}{\partial t} - (\mathbf{V}_0 + \mathbf{\Omega} \times \mathbf{r}) \cdot \nabla F + \nabla \Phi \cdot \nabla F = 0$$

Recalling that in the body's frame of reference $\partial F/\partial t = -\partial \eta/\partial t$ and according to Eq. (2.26) $\nabla F/|\nabla F| = \mathbf{n}$ we get

$$[\nabla \Phi - (\mathbf{V}_0 + \mathbf{\Omega} \times \mathbf{r})] \cdot \mathbf{n} - \frac{\partial \eta/\partial t}{|\nabla F|} = 0$$

The last term represents the relative motion within the body's frame of reference and by exchanging $(\partial \eta/\partial t)/|\nabla F|$ with a possible three-dimensional relative motion $\mathbf{v}_{\text{rel}} \cdot \mathbf{n}$ we get

$$[\nabla \Phi - (\mathbf{V}_0 + \mathbf{\Omega} \times \mathbf{r})] \cdot \mathbf{n} - \mathbf{v}_{\text{rel}} \cdot \mathbf{n} = 0$$

which is identical to the previous result of Eq. (13.13a).

In the case of more complex flowfields, when the modeling of nonzero velocity components across the boundaries is desired (e.g., engine inlet/exit flows or simulation of boundary layer displacement by blowing) a transpiration velocity V_N can be added:

$$(\nabla \Phi - \mathbf{V}_0 - \mathbf{v}_{\text{rel}} - \mathbf{\Omega} \times \mathbf{r}) \cdot \mathbf{n} = V_N \quad (13.13b)$$

The most important conclusion from these results (Eqs. (13.12) and (13.13)) is that for incompressible flows the instantaneous solution is independent of time derivatives. That is, since the speed of sound is assumed to be infinite, the influence of the momentary boundary condition is immediately radiated across the whole fluid region. Therefore, steady-state solution techniques can be used to treat the time-dependent problem by substituting the instantaneous boundary condition (Eq. (13.13)) at each moment. The wake shape, however, does depend on the time history of the motion and consequently an appropriate vortex wake model has to be developed.

For many situations involving lifting problems the wake separation line has to be prescribed. As in the case of the steady-state flows the Kutta condition is assumed to be valid for the time-dependent case as well (for attached flows with reduced frequencies of less than 1 where the reduced frequency $k = \omega L/2U_\infty$ is defined in a manner similar to Eq. (1.52)). Therefore, along trailing edges of lifting surfaces, the velocity has to be finite (to fix the rear stagnation line) and

$$\nabla \Phi < \infty \quad \text{(at trailing edges)} \quad (13.14)$$

13.2 Method of Solution

The continuity equation (Eq. (13.12)) is exactly the same as the corresponding steady-state equation and consequently solution methods similar to those presented in the previous (steady-state flow) chapters can be used. Recalling the formulation, based on Green's identity (Section 3.3), we can construct the general solution to Eq. (13.12) by integrating the contribution of the basic solutions of source σ and doublet μ distributions over the body's surface and its wakes:

$$\Phi(x, y, z) = \frac{1}{4\pi} \int_{body+wake} \mu \mathbf{n} \cdot \nabla\left(\frac{1}{r}\right) dS - \frac{1}{4\pi} \int_{body} \sigma \left(\frac{1}{r}\right) dS \qquad (13.15)$$

This formulation does not include directly a vortex distribution; however, it was demonstrated earlier (e.g., in Section 10.4.3) that doublet distributions can be exchanged with equivalent vortex distributions. Also, from this point and on, the velocity potential Φ is considered to be specified in terms of the body's coordinate system.

This singular element solution automatically fulfills the boundary condition of Eq. (13.5). To satisfy the boundary condition of Eq. (13.13), Eq. (13.15) is differentiated with respect to the body coordinates. The resulting velocity induced by the combination of the doublet and source distributions is then

$$\nabla \Phi = \frac{1}{4\pi} \int_{body+wake} \mu \nabla\left[\frac{\partial}{\partial n}\left(\frac{1}{r}\right)\right] dS - \frac{1}{4\pi} \int_{body} \sigma \nabla\left(\frac{1}{r}\right) dS \qquad (13.16)$$

In order to establish the Neumann form of the boundary value problem, the local velocity at each point on the body has to satisfy the zero normal flow condition across the body's surface (Eq. (13.13) or, in the case of transpiration, Eq. (13.13b)). Substitution of Eq. (13.16) into Eq. (13.13a) allows us to form the final integral equation with the unknown μ and σ distributions:

$$\left\{ \frac{1}{4\pi} \int_{body+wake} \mu \nabla\left[\frac{\partial}{\partial n}\left(\frac{1}{r}\right)\right] dS - \frac{1}{4\pi} \int_{body} \sigma \nabla\left(\frac{1}{r}\right) dS \right. $$
$$\left. - \mathbf{V}_0 - \mathbf{v}_{rel} - \mathbf{\Omega} \times \mathbf{r} \right\} \cdot \mathbf{n} = 0 \qquad (13.17)$$

For thick bodies, this condition of zero normal flow across solid boundaries can be defined by using the Dirichlet formulation of Section 9.2. In this case the inner perturbation potential is assumed to be constant such that

$$\Phi_i = \text{const.}$$

By selecting $\Phi_i = 0$ for the velocity potential (observe that the problem is formulated in the inertial (X, Y, Z) frame of reference where $\Phi_\infty = 0$, and the magnitude of Φ corresponds to the perturbation potential[1] in the steady-state flow case) a formulation similar to Eq. (9.11) is obtained:

$$\frac{1}{4\pi} \int_{body+wake} \mu \frac{\partial}{\partial n}\left(\frac{1}{r}\right) dS - \frac{1}{4\pi} \int_{body} \sigma \left(\frac{1}{r}\right) dS = 0 \qquad (13.18)$$

Equations (13.17) and (13.18) still do not uniquely describe a solution since a large number of source and doublet distributions will satisfy a set of such boundary conditions. It is possible to set the doublet strength or the source strength to zero, in a manner similar to the thick and thin wing cases (as in Chapter 11). A frequently followed choice for panel methods (e.g., PMARC[9.7,9.8]) is to set the value of the source distribution equal to the local kinematic velocity (the time-dependent equivalent of the free-stream velocity). To justify this, observe the Neumann boundary condition of Eq. (13.13a), which states that on the solid boundary,

$$\frac{\partial \Phi}{\partial n} = (\mathbf{V}_0 + \mathbf{v}_{rel} + \mathbf{\Omega} \times \mathbf{r}) \cdot \mathbf{n}$$

---

[1] For convenience, therefore, in this chapter Φ is often referred to as a perturbation potential.

13.3 Additional Physical Considerations

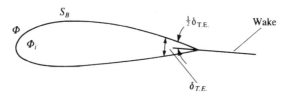

Figure 13.2 Possible assumption for wake shape near the trailing edge.

But Eq. (3.12) states that the jump in the local normal velocity component is

$$-\sigma = \frac{\partial \Phi}{\partial n} - \frac{\partial \Phi_i}{\partial n}$$

and since $\Phi_i = 0$ then also $\partial \Phi_i / \partial n = 0$ on the solid boundary S_B. By substituting $\partial \Phi / \partial n$, the source strength becomes

$$\sigma = -\mathbf{n} \cdot (\mathbf{V}_0 + \mathbf{v}_{\text{rel}} + \mathbf{\Omega} \times \mathbf{r}) \qquad (13.19)$$

13.3 Additional Physical Considerations

The above mathematical formulation, even after selecting a desirable combination of sources and doublets and after fulfilling the boundary conditions on the surface S_B (in Fig. 13.2), is not unique. As we have seen in the previous chapters, for lifting flow conditions the magnitude of circulation depends on the wake shape and on the location of the wake shedding line and, therefore, an appropriate wake model needs to be established. Following the practice of the previous chapters, we will base the wake model on some additional *physical* conditions (e.g., the Kutta condition) as described next.

a. *Wake Strength*

The simplest solution to this problem is to apply the two-dimensional Kutta condition along the trailing edges (*T.E.*) of lifting wings (see Fig. 13.2):

$$\gamma_{T.E.} = 0 \qquad (13.20)$$

The validity of this assumption depends on the component of the kinematic velocity normal to the trailing edge, which must be much smaller than the characteristic velocity (e.g., Q_∞) for Eq. (13.20) to be valid (see additional discussion of this condition in Section 13.11).

Also, the Kelvin condition can be used to calculate the change in the wake circulation:

$$\frac{d\Gamma}{dt} = 0 \qquad (13.6)$$

b. *Wake Shape*

Following the requirement of Section 9.3, that the wake is force free, the Kutta–Joukowski theorem (Section 3.11) states that

$$\mathbf{Q} \times \boldsymbol{\gamma}_W = 0 \qquad (13.21)$$

When the wake is modeled by a vortex distribution of strength γ_W, Eq. (13.21) can be interpreted as a requirement that the velocity should be parallel to the circulation vector,

$$\boldsymbol{\gamma}_W \| \mathbf{Q} \qquad (13.21a)$$

Furthermore, in most cases the trailing edge has a finite angle and an additional assumption has to be made about the angle at which the wake leaves the trailing edge. In these cases it is usually sufficient to assume that the wake leaves the trailing edge at a median angle $\delta_{T.E.}/2$, as shown in Fig. 13.2.

13.4 Computation of Pressures

Solution of Eq. (13.12) will provide the velocity potential and the velocity components. Once the flowfield is determined the resulting pressures can be computed by the Bernoulli equation. In the inertial frame of reference this equation is [note that $(\nabla\Phi)^2 = \nabla\Phi \cdot \nabla\Phi = (u^2 + v^2 + w^2)$]

$$\frac{p_\infty - p}{\rho} = \frac{(\nabla\Phi)^2}{2} + \frac{\partial \Phi}{\partial t} = \frac{1}{2}\left[\left(\frac{\partial \Phi}{\partial X}\right)^2 + \left(\frac{\partial \Phi}{\partial Y}\right)^2 + \left(\frac{\partial \Phi}{\partial Z}\right)^2\right] + \frac{\partial \Phi}{\partial t}$$
(in X, Y, Z coordinates)   (13.22)

The magnitude of the velocity $\nabla\Phi$ is independent of the frame of reference (only the resolution of the velocity vector into its components is affected) and therefore the form of the first term in this equation remains unchanged. The time derivative of the velocity potential, however, is affected by the frame of reference and must be evaluated by using Eq. (13.11); therefore, the pressure difference $p_\infty - p$ will have the form[13.2]

$$\frac{p_\infty - p}{\rho} = \frac{1}{2}\left[\left(\frac{\partial \Phi}{\partial x}\right)^2 + \left(\frac{\partial \Phi}{\partial y}\right)^2 + \left(\frac{\partial \Phi}{\partial z}\right)^2\right] - (\mathbf{V}_0 + \mathbf{\Omega} \times \mathbf{r}) \cdot \nabla\Phi + \frac{\partial \Phi}{\partial t}$$
(in x, y, z coordinates)   (13.23)

In the case of three-dimensional panel methods it is often simpler to use the instantaneous Bernoulli equation, in its original form (Eq. (2.35)):

$$\frac{p_{\text{ref}} - p}{\rho} = \frac{Q^2}{2} - \frac{v_{\text{ref}}^2}{2} + \frac{\partial \Phi}{\partial t} \tag{13.24}$$

Here Q and p are the local fluid velocity and pressure values, p_{ref} is the far field reference pressure, and v_{ref} is the magnitude of the kinematic velocity as given in Eq. (13.8):

$$\mathbf{v}_{\text{ref}} = -[\mathbf{V}_0 + \mathbf{\Omega} \times \mathbf{r}] \tag{13.25}$$

It is often convenient to express this kinematic velocity in the direction of the moving (x, y, z) frame as $[U(t), V(t), W(t)]$, which can be obtained by a simple transformation f_1 (which is a function of the momentary rotation angles ϕ, θ, ψ, resembling Eq. (13.7a)):

$$\begin{pmatrix} U \\ V \\ W \end{pmatrix} = f_1(\phi, \theta, \psi) \begin{pmatrix} v_{\text{ref}_X} \\ v_{\text{ref}_Y} \\ v_{\text{ref}_Z} \end{pmatrix} \tag{13.26}$$

The total velocity at an arbitrary point (or collocation point k in the case of a numerical solution) on the body is the sum of the local kinematic velocity (e.g., the reference velocity in Eq. (13.25)) plus the perturbation velocity,

$$\mathbf{Q}_k = \left(v_{\text{ref}_l}, v_{\text{ref}_m}, v_{\text{ref}_n}\right)_k + (q_l, q_m, q_n)_k \tag{13.27}$$

where $(l, m, n)_k$ are the local tangential and normal directions (see Fig. 9.10) and the components of $\mathbf{v}_{\text{ref}}$ in these directions are obtained by a transformation similar to Eq. (13.26).

The local perturbation velocity is $(q_l, q_m, q_n) = (\partial\Phi/\partial l, \partial\Phi/\partial m, \partial\Phi/\partial n)$ and of course the relative normal velocity component on the solid body is zero. The pressure coefficient can now be computed for each panel as

$$C_p = \frac{p - p_{\text{ref}}}{(1/2)\rho v_{\text{ref}}^2} = 1 - \frac{Q^2}{v_{\text{ref}}^2} - \frac{2}{v_{\text{ref}}^2}\frac{\partial\Phi}{\partial t} \qquad (13.28)$$

or if we use Eq. (13.23) then the pressure coefficient becomes

$$C_p = \frac{p - p_{\text{ref}}}{(1/2)\rho v_{\text{ref}}^2} = -\frac{(\nabla\Phi)^2}{v_{\text{ref}}^2} + \frac{2}{v_{\text{ref}}^2}[\mathbf{V}_0 + \mathbf{\Omega} \times \mathbf{r}] \cdot \nabla\Phi - \frac{2}{v_{\text{ref}}^2}\frac{\partial\Phi}{\partial t} \qquad (13.28a)$$

Note that in situations such as the forward flight of a helicopter rotor v_{ref} can be selected as the forward flight speed or the local blade speed at each section on the rotor blade. Consequently, different values of the pressure coefficient will be obtained – and this matter is usually left to be determined by the particular application.

The contribution of an element with an area of ΔS_k to the aerodynamic loads $\Delta \mathbf{F}_k$ is

$$\Delta \mathbf{F}_k = -C_{p_k}\left(\frac{1}{2}\rho v_{\text{ref}}^2\right)_k \Delta S_k \mathbf{n}_k \qquad (13.29)$$

(Note that v_{ref} here has a subscript k, which means that it depends on the body coordinates. This assignment is usually not recommended but may be used in cases such as the forward flight of a helicopter rotor.)

Once the potential field and the velocity field are obtained, the corresponding pressure field is calculated using Eq. (13.28a) and additional information such as forces, moments, surface velocity surveys, etc. can be obtained.

13.5 Examples for the Unsteady Boundary Condition

As a first example, let us investigate several simple motions and the corresponding derivation of the boundary conditions. Consider a flat plate at an angle of attack α moving at a constant velocity U_∞ in the negative X direction, as shown in Fig. 13.3. The translation of the origin is

$$\mathbf{V}_0 = (\dot{X}_0, \dot{Y}_0, \dot{Z}_0) = (-U_\infty, 0, 0)$$

and the rotation is

$$\mathbf{\Omega} = 0$$

The vector $\mathbf{n}$ on the flat plate and in the body frame is

$$\mathbf{n} = (\sin\alpha, 0, \cos\alpha)$$

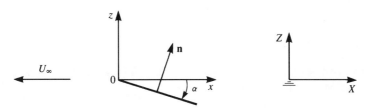

Figure 13.3 Translation of a flat plate placed in the (x, y, z) system with a speed of $(-U_\infty, 0, 0)$.

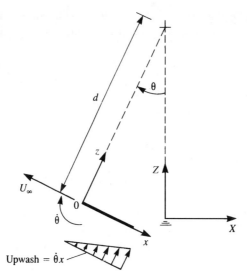

Figure 13.4 Motion of a flat plate along a circular arc with radius d.

Substitution of these values into the boundary condition (Eq. (13.13a)) results in

$$(\nabla \Phi - \mathbf{V}_0 - \Omega \times \mathbf{r}) \cdot \mathbf{n} = \left(\frac{\partial \Phi}{\partial x} + U_\infty, 0, \frac{\partial \Phi}{\partial z}\right) \cdot (\sin\alpha, 0, \cos\alpha) = 0$$

and by assuming that $\alpha \ll 1$ and $\partial\Phi/\partial x \ll U_\infty$, we reduce this to the classical result

$$\frac{\partial \Phi}{\partial z} = -\left(U_\infty + \frac{\partial \Phi}{\partial x}\right) \tan\alpha \approx -U_\infty \alpha \qquad (13.30)$$

For the second example consider a one degree of freedom motion along a circular arc (Fig. 13.4) with a radius of d. If the origin $(\ )_0$ is moving with a speed of U_∞, then

$$(\dot{X}_0, \dot{Y}_0, \dot{Z}_0) = (-U_\infty \cos\theta, 0, U_\infty \sin\theta)$$

But it is easy to observe that $\mathbf{V}_0$, at any moment, resolved in the direction of the (x, y, z) system, will be

$$\mathbf{V}_0 = (-U_\infty, 0, 0)$$

The rotation for this case is $(q = \dot{\theta})$

$$\Omega = (0, \dot{\theta}, 0)$$

where

$$\dot{\theta} = \frac{U_\infty}{d}$$

and

$$\Omega \times \mathbf{r} = (\dot{\theta}z, 0, -\dot{\theta}x)$$

Also, the normal is $\mathbf{n} = (0, 0, 1)$. The boundary condition for a flat plate (at zero angle of attack in its coordinate system) is then

$$(\nabla\Phi - \mathbf{V}_0 - \Omega \times \mathbf{r}) \cdot \mathbf{n} = \left(\frac{\partial \Phi}{\partial x} - \dot{\theta}z + U_\infty, 0, \frac{\partial \Phi}{\partial z} + \dot{\theta}x\right) \cdot (0, 0, 1) = 0$$

13.5 Examples for the Unsteady Boundary Condition

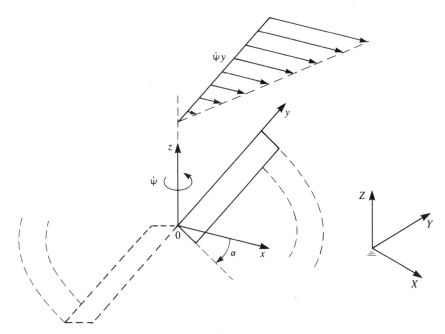

Figure 13.5 Description of the motion of a helicopter rotor in hover.

which results in

$$\frac{\partial \Phi}{\partial z} = -\dot{\theta}x = -U_\infty \frac{x}{d} \tag{13.31}$$

Note that there is an upwash of $w = \dot{\theta}x$ due to this motion (see Fig. 13.4), resulting in lift (even though $\alpha = 0$ in the flat plate's coordinate system). Furthermore, it is clear from this example that if the wing's leading edge is placed at $x = -c$ then instead of an upwash the wing will be subject to a downwash (or negative lift). Therefore, the location of the rotation axis is very important in motions with body rotations.

Next consider the rotation of a helicopter rotor in hover (Fig. 13.5) at a rate $r = \dot{\psi}$. For this case

$$\mathbf{V}_0 = (0, 0, 0)$$
$$\mathbf{\Omega} = (0, 0, \dot{\psi})$$
$$\mathbf{\Omega} \times \mathbf{r} = (-\dot{\psi}y, \dot{\psi}x, 0)$$
$$\mathbf{n} = (\sin \alpha, 0, \cos \alpha)$$

The boundary condition for a flat plate rotor, at an angle of attack α, is

$$(\nabla \Phi - \mathbf{V}_0 - \mathbf{\Omega} \times \mathbf{r}) \cdot \mathbf{n} = \left(\frac{\partial \Phi}{\partial x} + \dot{\psi}y, \frac{\partial \Phi}{\partial y} - \dot{\psi}x, \frac{\partial \Phi}{\partial z} \right) \cdot (\sin \alpha, 0, \cos \alpha) = 0$$

which results in

$$\frac{\partial \Phi}{\partial z} = \left(\dot{\psi}y + \frac{\partial \Phi}{\partial x} \right) \tan \alpha \tag{13.32}$$

Note that to the rotor blade sections the oncoming velocity seems to increase with the radius $|y|$, and consequently most of the loads will be generated close to the tips.

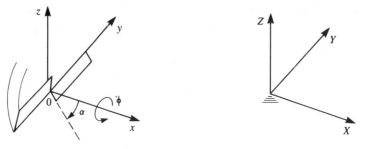

Figure 13.6 Description of the motion of a rotating propeller.

Similarly, the boundary condition for a statically spinning propeller blade can be established by rotating the blades at a rate $p = \dot{\phi}$ about the x axis as shown in Fig. 13.6 (observe the definition of α in Fig. 13.6 for this example).

For this case

$$\mathbf{V}_0 = (0, 0, 0)$$
$$\mathbf{\Omega} = (\dot{\phi}, 0, 0)$$
$$\mathbf{\Omega} \times \mathbf{r} = (0, -\dot{\phi}z, \dot{\phi}y)$$
$$\mathbf{n} = (\sin\alpha, 0, \cos\alpha)$$

and α is defined here in a similar manner to the wing at angle of attack. The boundary condition for a flat propeller blade is

$$(\nabla\Phi - \mathbf{V}_0 - \mathbf{\Omega} \times \mathbf{r}) \cdot \mathbf{n} = \left(\frac{\partial\Phi}{\partial x}, \frac{\partial\Phi}{\partial y} + \dot{\phi}z, \frac{\partial\Phi}{\partial z} - \dot{\phi}y\right) \cdot (\sin\alpha, 0, \cos\alpha) = 0$$

which results in

$$\frac{\partial\Phi}{\partial z} = -\frac{\partial\Phi}{\partial x}\tan\alpha + \dot{\phi}y \tag{13.33}$$

Again, most of the load will be generated at the tips where $\dot{\phi}|y|$ is the largest. Also, if the propeller advances at a speed of U_∞ parallel to its x axis, then the boundary condition becomes

$$\frac{\partial\Phi}{\partial z} = -\left(\frac{\partial\Phi}{\partial x} + U_\infty\right)\tan\alpha + \dot{\phi}y \tag{13.34}$$

13.6 Summary of Solution Methodology

The results of Sections 13.1–13.4 indicate that the unsteady flow method of solution is very similar to the steady-state methods presented in the previous chapters. Therefore, it is possible to use those steady-state methods with only a few small modifications. These modifications can be limited to three sections of the analytical or the numerical model and in general are:

1. Update of the "zero normal flow on a solid surface" boundary condition to include the kinematic velocity components (as in Eq. (13.13a)).
2. Similarly, use of the modified unsteady Bernoulli equation (as in Eq. (13.23)).
3. Construction of a wake model, based on the requirements of Section 13.3.

13.7 Sudden Acceleration of a Flat Plate

The first two modifications are very minor and local (in terms of computing); however, the third modification is more elaborate and in the following sections a simple vortex wake model will be used.

Based on these conclusions the rest of this chapter is concerned with presenting examples for the transformation of some analytical and numerical methods into the time-dependent mode.

13.7 Sudden Acceleration of a Flat Plate

One of the simplest and yet the most basic examples of unsteady aerodynamics is the sudden acceleration motion of an airfoil.[13.3] This example will be studied first by the simple lumped-vortex method so that the most basic differences between this case and the steady flow case will be highlighted.

Consider the thin, uncambered airfoil shown in the upper part of Fig. 13.7 to be at rest at $t < 0$. Then at $t = 0$ the airfoil is suddenly accelerated to a constant velocity U_∞. The boundary condition for $t > 0$ and for small α, according to Eq. (13.30), is

$$\frac{\partial \Phi}{\partial z} = -\left(U_\infty + \frac{\partial \Phi}{\partial x}\right)\tan\alpha \approx -U_\infty \alpha, \quad t > 0$$

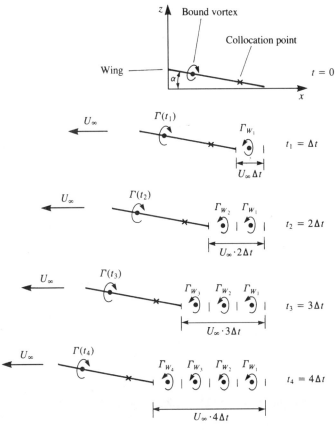

Figure 13.7 Development of the wake after the plunging motion of a flat plate as modeled by a single lumped-vortex element.

Now let us represent the airfoil by a lumped-vortex element. By doing so, a single vortex solution is selected instead of the more general form of Eq. (13.15) that includes doublets and sources. Also, by placing the vortex at the quarter chord, the Kutta condition is assumed to be satisfied for this case.

At this point also a wake model has to be established and here a discrete vortex wake is selected (Fig. 13.7). Now at a time $t_1 = \Delta t$, the airfoil already has traveled a distance $U_\infty \Delta t$ and its circulation is $\Gamma(t_1)$. Recalling Kelvin's theorem (Eq. (13.6)), we see that the airfoil bound circulation has to be canceled with a starting vortex (Fig. 13.7 at $t_1 = \Delta t$). The concentrated wake vortex has to be placed along the path traveled by the trailing edge, during this interval, so that the discretization effect will be minimal. At this point the middle of the interval is selected and the effect of this choice can be demonstrated later.

The zero normal flow boundary condition is satisfied at the collocation point at the plate's three-quarter chord point (as shown by the x in Fig. 13.7) where the downwash induced by the bound vortex is $-\Gamma(t_1)/2\pi(c/2)$ and the downwash induced by the first wake vortex is approximated by $\Gamma_{W_1}/\{2\pi[(c/4) + (U_\infty \Delta t/2)]\}$, assuming $\alpha \ll 1$. The boundary condition $\partial \Phi/\partial z = -U_\infty \alpha$ for the first time step then becomes

$$\frac{-\Gamma(t_1)}{2\pi(c/2)} + \frac{\Gamma_{W_1}}{2\pi[(c/4) + (U_\infty \Delta t/2)]} = -U_\infty \alpha$$

This equation can be rewritten in the form

$$w_b + w_W + U_\infty \alpha = 0$$

which indicates that the sum of the normal velocity induced by the airfoil w_b, the wake w_W, and the free stream must be zero. An additional equation is obtained by using the Kelvin condition:

$$\frac{d\Gamma}{dt} = \Gamma(t_1) + \Gamma_{W_1} = 0$$

Note that Γ is considered positive for right-hand rotation, and in Fig. 13.7 for illustration purposes and with the knowledge of the solution, the wake vortex is drawn in the negative direction.

The above set of two equations with two unknowns is solved for $\Gamma(t_1)$ and Γ_{W_1}. Now, after the second time step the airfoil has moved to a new location, as shown in the figure (for $t_2 = 2\Delta t$). It was assumed in Section 2.9 that for high Reynolds number flows, vortex decay is negligible (and zero for irrotational flow) and therefore the strength of Γ_{W_1} will not change with time. It is possible to calculate the induced velocity at the wake vortex and then move it along the local streamline, but for simplicity it is assumed here that its location remains unchanged (in the inertial frame).

At $t = 2\Delta t$, the two equations describing the no normal flow across the airfoil boundary condition and the Kelvin condition are

$$\frac{-\Gamma(t_2)}{\pi c} + \frac{\Gamma_{W_2}}{2\pi[(c/4) + (U_\infty \Delta t/2)]} + \frac{\Gamma_{W_1}}{2\pi[(c/4) + (U_\infty 3\Delta t/2)]} = -U_\infty \alpha$$

$$\Gamma(t_2) + \Gamma_{W_2} + \Gamma_{W_1} = 0$$

This set is solved for $\Gamma(t_2)$ and Γ_{W_2}, while Γ_{W_1} is known from the previous calculation at $t = t_1$.

13.7 Sudden Acceleration of a Flat Plate

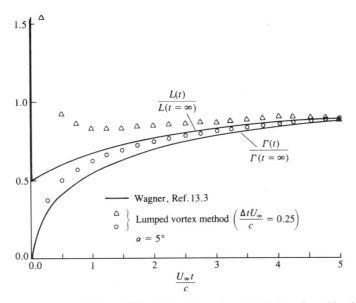

Figure 13.8 Variation of lift and circulation after the initiation of a sudden forward motion of a two-dimensional flat plate.

At $t = 3\Delta t$, the two equations can be written in a similar manner:

$$\frac{-\Gamma(t_3)}{\pi c} + \frac{\Gamma_{W_3}}{2\pi[(c/4) + (U_\infty \Delta t/2)]} + \frac{\Gamma_{W_2}}{2\pi[(c/4) + (U_\infty 3\Delta t/2)]}$$
$$+ \frac{\Gamma_{W_1}}{2\pi[(c/4) + (U_\infty 5\Delta t/2)]} = -U_\infty \alpha$$
$$\Gamma(t_3) + \Gamma_{W_3} + \Gamma_{W_2} + \Gamma_{W_1} = 0$$

Again this set is solved for $\Gamma(t_3)$ and Γ_{W_3}, while Γ_{W_2} and Γ_{W_1} are known from the previous calculations.

In this simplified analysis it was assumed that α is very small and the distance to the collocation points can be approximated by $[(c/4) + (U_\infty \Delta t/2)]$, $[(c/4) + (U_\infty 3\Delta t/2)]$, ... and that the induced velocity is normal to the surface. Of course, numerically this error can easily be corrected.

The results of this computation for $\Gamma(t)$ (shown by the circles), along with a more accurate solution, are presented in Fig. 13.8. The circulation at $t = 0$ is zero since the airfoil is still at rest. At $t > 0$ the circulation increases but is far less than the steady-state value due to the downwash of the starting vortex. In this two-dimensional case the increase of the circulation is slow and this transient growth extends to infinity.

To compute the lift, the small disturbance approximation ($U_\infty \gg \nabla\Phi$) is applied to the unsteady Bernoulli equation (Eq. (13.23)),

$$p_\infty - p = \rho\left[(U_\infty, 0, 0) \cdot \nabla\Phi + \frac{\partial \Phi}{\partial t}\right]$$

and the pressure difference between the airfoil's upper and lower surface is

$$\Delta p = 2\rho\left(U_\infty \cdot \frac{\gamma}{2} + \frac{\partial \Phi}{\partial t}\right) = \rho U_\infty \gamma(x) + \rho\frac{\partial}{\partial t}\int_0^x \gamma(x)\,dx \quad (13.35)$$

Here we used the results of Eq. (3.147) for a planar vortex distribution where the upper surface induced velocity is $\nabla\Phi = (\gamma/2, 0, 0)$.

For the lumped-vortex method there is only one airfoil vortex and therefore the lift L' per unit span is

$$L' = \int_0^c \Delta p \, dx = \rho \left[U_\infty \Gamma(t) + \frac{\partial}{\partial t} \Gamma(t) \cdot c \right] \quad (13.36)$$

Results of this simple model (triangular symbols) along with the exact solution of Wagner[13.3] are presented in Fig. 13.8. The lift at $t = 0+$ is exactly half of the steady-state lift, but this lift is not due to the airfoil circulation (circulatory force) but due to the acceleration portion of the lift that results from the change in the upwash ($\partial \Phi / \partial t$). The magnitude of this force due to fluid acceleration becomes smaller with the reduced influence of the starting vortex. At $t = 0$ when the airfoil was suddenly accelerated from rest the lift was *infinite*, owing to this acceleration term as shown in Fig. 13.8.

It is interesting to point out that for a two-dimensional airfoil there is a drag force during the transient. This drag force will have two components, owing to the two terms of the lift in Eq. (13.36). The first is due to the wake-induced downwash, which rotates the circulatory lift term by an induced angle w_W/U_∞. The second is due to the fluid acceleration $\partial \Phi / \partial t$, which acts normal to the flat plate and its contribution to the drag is α times the second lift term in Eq. (13.36). Consequently, the drag force per unit span is

$$D = \rho \left[w_W(x,t) \Gamma(t) + \frac{\partial}{\partial t} \Gamma(t) c \alpha \right] \quad (13.37)$$

Here the first term is due to the wake-induced downwash $w_W(x,t)$, which in the lumped-vortex case is evaluated at the three-quarter chord point. The second term is due to the fluid acceleration and its center of pressure is at the center of the $\int_0^c \dot{w}(x,t) dx$ term, which is closer to the aft section of the airfoil and varies with time. Calculated drag coefficients with this method and with a more accurate method[13.2] are presented in Fig. 13.9.

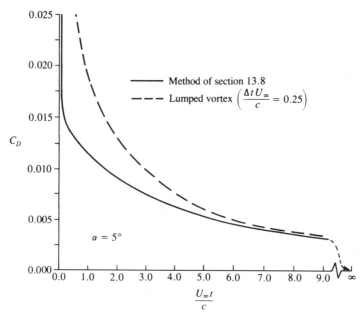

Figure 13.9 Variation of drag after the initiation of a sudden forward motion of a two-dimensional flat plate.

13.7 Sudden Acceleration of a Flat Plate

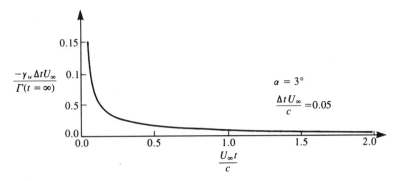

Figure 13.10 Wake circulation distribution behind a flat plate that was suddenly set into motion.

While examining the wake vorticity as presented in Fig. 13.10, it can be observed that the first vortices are the strongest and that all vortices have counterclockwise (negative) values. Consequently, when the wake is allowed to roll up, as a result of the velocity field induced by the wake and the airfoil, the shape shown in Fig. 13.11 will be obtained.

A simple computer program based on the formulation of this section is presented in Appendix D, Program No. 15.

13.7.1 The Added Mass

In situations when a body is accelerated in a fluid, it is possible to use Newton's second law, in its simplest form, to compute the force F acting on the body. For example, if the body's motion is parallel to the z axis then

$$F = m_{\text{tot}} \frac{dw}{dt}$$

Of course the mass m_{tot} consists of the mass of the body m and of the fluid m' that is being accelerated as well. Unfortunately, the evaluation of the added mass (m') is not always easy since in most situations the local fluid acceleration may be caused by effects other than the body's motion (e.g., in the case of a time-dependent wake-induced downwash). However, in the case of a constant acceleration of a flat plate normal to itself, in a fluid at rest, the added mass can be more easily evaluated (see also Section 8.2.3).

Consider the flat plate of Fig. 13.12 accelerating in a fluid, such that the acceleration $\dot{w}$ is constant (see also Fig. 8.24). Since the continuity equation is independent of time, at each frozen moment the potential will be similar to the steady-state flow potential of the

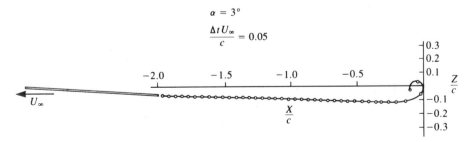

Figure 13.11 Wake rollup behind a two-dimensional flat plate that was suddenly set into motion.

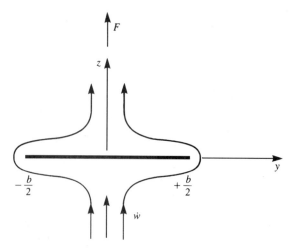

Figure 13.12 A flat plate in a sinking motion.

flow normal to the plate. From Section 6.5.3 the potential for such a flow is

$$\Phi^{\pm} = \pm w\sqrt{\left(\frac{b}{2}\right)^2 - y^2} = \pm w\frac{b}{2}\sqrt{1 - \left(\frac{y}{b/2}\right)^2} \tag{13.38}$$

where $+$ is used for the potential above the plate and $-$ is used for the potential below it. This is an elliptic distribution of the potential, similar to the one obtained in Chapter 8 for the lifting-line and slender wing theories. Because of the antisymmetry between the upper and lower surfaces

$$\Delta\Phi(y) = 2\Phi^{+}(y) = wb\sqrt{1 - \left(\frac{y}{b/2}\right)^2} \tag{13.39}$$

It is interesting to point out that $\Delta\Phi$ in this equation can be replaced by Γ, and its derivative is the spanwise circulation $\gamma(y)$ as shown in Fig. 13.13. Examination of the terms in the pressure equation (Eq. (13.23)) reveals that for this motion $(\mathbf{V}_0 + \mathbf{\Omega} \times \mathbf{r}) \cdot \nabla\Phi = (0, 0, w) \cdot \nabla\Phi = 0$ since $\nabla\Phi$ will have a y component only, on the plate's surface. (This can be deduced from Fig. 13.13, too, by observing that the "traditional" lift $\boldsymbol{\gamma} \times (0, 0, w)$ has no vertical component.) Therefore, the pressure difference is due to the velocity potential's

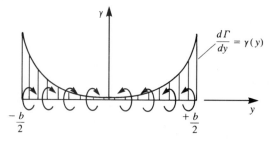

Figure 13.13 Spanwise circulation on a flat plate in a sinking motion.

time derivative $\partial \Phi/\partial t$, only, in the unsteady Bernoulli equation, and is

$$\Delta p = \rho \frac{\partial}{\partial t} \Delta \Phi = \rho b \sqrt{1 - \left(\frac{y}{b/2}\right)^2} \frac{\partial w(t)}{\partial t} \quad (13.40)$$

The integral is easily evaluated by recalling that the area of the ellipse is $\pi b/4$ and the lift L' of a massless plate becomes

$$L' = \int_{-b/2}^{b/2} \Delta p \, dy = \frac{\pi}{4} \rho b^2 \frac{\partial w(t)}{\partial t} = \frac{\pi}{4} \rho b^2 \dot{w} \quad (13.41)$$

Note that because of the left/right symmetry, the center of pressure is at $y = 0$. Also, the lift is created only if the plate is under acceleration and the amount of fluid being accelerated (added mass) is equal to the mass of a fluid cylinder with a diameter of b ($m' = \rho b^2 \pi/4$), and in summary we can write

$$L' = m' \dot{w} \quad (13.41a)$$

13.8 Unsteady Motion of a Two-Dimensional Thin Airfoil

As is indicated in Section 13.6, steady-state flow methods can be extended to treat the time-dependent problem with only a few modifications. Following this methodology, in this section we treat the time-dependent equivalent of the small-disturbance, thin, lifting airfoil in steady flow (Sections 5.2–5.4). One of the more difficult aspects of this unsteady problem is the modeling of the vortex wake's shape and strength, which depend on the time history of the motion. By selecting a discretized vortex wake model at the early stages of this discussion we limit ourselves mostly to numerical solutions. Nonetheless, this approach allows for a simple formulation (which is clear and easy to explain to students) that avoids an elaborate mathematical treatment of the wake influence integral.

The two-dimensional, thin lifting surface with a chord length of c is shown schematically in Fig. 13.14. At $t = 0$ the airfoil is at rest in the inertial system X, Z, and at $t > 0$ it moves along a time-dependent curved path, S. (Note that the fluid in Fig. 13.14 is basically at

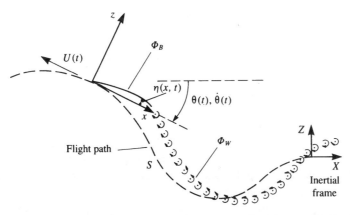

Figure 13.14 Nomenclature for the unsteady motion of a two-dimensional thin airfoil. (Note that the motion is observed from the X, Z coordinate system and the airfoil moves toward the left side of the page.)

rest and the airfoil moves toward the left of the page.) For convenience, the coordinates x, z are selected such that the origin is placed on the path S, and the x coordinate axis is always tangent to the path. The airfoil shape (camberline) is given in this coordinate system by $\eta(x, t)$, which is considered to be small ($\eta/c \ll 1$). Since small-disturbance motion is assumed, the path radius of curvature ϱ is also much larger than the chord c (or $c/\varrho = \dot\theta c/U(t) \ll 1$).

13.8.1 Kinematics

In Section 13.1 (Eq. (13.12)) it was shown that the continuity equation in the moving frame of reference x, z remains as

$$\nabla^2 \Phi = 0 \tag{13.42}$$

where Φ is the equivalent of the steady-state perturbation potential.

The time-dependent version of the boundary condition requiring no normal flow across the surface is given by Eq. (13.13a):

$$(\nabla\Phi - \mathbf{V}_0 - \mathbf{v}_{\text{rel}} - \mathbf{\Omega} \times \mathbf{r}) \cdot \mathbf{n} = 0 \tag{13.43}$$

In this section the instantaneous shape of the airfoil is given by $\eta(x, t)$ and therefore the vector $\mathbf{n}$ normal to the surface is

$$\mathbf{n} = \frac{(-\partial\eta/\partial x, 0, 1)}{\sqrt{(\partial\eta/\partial x)^2 + 1}} \tag{13.44}$$

To establish the kinematic relations for the airfoil's motion (according to Eq. (13.8)) the instantaneous velocity and orientation of the x, z system can be described by a flight velocity $U(t)$ and a rotation $\dot\theta$ about the y coordinate. Note that the x coordinate was selected such that the instantaneous velocity of the origin $(\ )_0$ (of Eq. (13.9a)), resolved into the directions of the x, z coordinate system, is

$$\mathbf{V}_0 = [-U(t), 0, 0] \tag{13.45}$$

The instantaneous rotation is then

$$\mathbf{\Omega} = [0, \dot\theta(t), 0] \tag{13.46}$$

Also, by allowing a relative motion of the chordline within the coordinate system x, z we get the relative velocity from Eq. (13.9c) of

$$\mathbf{v}_{\text{rel}} = \left(0, 0, \frac{\partial\eta}{\partial t}\right) \tag{13.47}$$

At this point, it is convenient to divide the velocity potential Φ into an airfoil potential Φ_B and to a wake potential Φ_W (for example, if a time-stepping numerical approach is used, then the strengths of the wake singularities are assumed to be known and only the airfoil's singularity distribution Φ_B must be obtained):

$$\Phi = \Phi_B + \Phi_W \tag{13.48}$$

Evaluating the product

$$\mathbf{\Omega} \times \mathbf{r} = (\dot\theta z, 0, -\dot\theta x)$$

13.8 Unsteady Motion of a Two-Dimensional Thin Airfoil

and substituting Eqs. (13.45)–(13.47) into the boundary condition (Eq. (13.43)) we obtain

$$\left(\frac{\partial \Phi_B}{\partial x} + \frac{\partial \Phi_W}{\partial x} + U - \dot{\theta}z, 0, \frac{\partial \Phi_B}{\partial z} + \frac{\partial \Phi_W}{\partial z} + \dot{\theta}x - \frac{\partial \eta}{\partial t}\right) \cdot \left(-\frac{\partial \eta}{\partial x}, 0, 1\right) = 0 \tag{13.49}$$

This can be rearranged in terms of the boundary condition for the unknown potential Φ_B:

$$\frac{\partial \Phi_B}{\partial z} = \left(\frac{\partial \Phi_B}{\partial x} + \frac{\partial \Phi_W}{\partial x} + U - \dot{\theta}z\right)\frac{\partial \eta}{\partial x} - \frac{\partial \Phi_W}{\partial z} - \dot{\theta}x + \frac{\partial \eta}{\partial t} \equiv W(x, t) \tag{13.50}$$

The main advantage of this formulation lies in the previous assumption that if the wake potential is known (and it is usually known from the previous time steps, when a time-stepping solution is applied) then the solution can be reduced to solving an equivalent steady-state flow problem, at each time step. For example, if this model is compared to the thin, lifting airfoil of Section 5.2 then the boundary condition of Eq. (13.50) is equivalent to the steady-state boundary condition of Eq. (5.29). Therefore, by exchanging the local downwash $W(x, t)$ with the right-hand side of Eq. (5.29), the methods of solution developed in Chapter 5 can be applied at each moment. Also, note that the boundary condition (Eq. (13.50)) is not reduced to its small-disturbance approximation yet and can be specified (numerically) on the airfoil's surface and not on the $z = 0$ plane.

13.8.2 Wake Model

As was discussed in Sections 4.7 and 13.3, the wake shed from the trailing edges of lifting surfaces can be modeled by doublet or vortex distributions. In the two-dimensional case the unsteady airfoil's wake will be shed only if the airfoil's circulation varies with time (Kelvin's condition). Therefore, if the airfoil circulation is varying continuously, then a continuous vortex sheet will be shed at the trailing edge, as shown in Fig. 13.15a. For simplicity, a discrete-vortex model of this continuous vortex sheet is approximated here, as shown in Fig. 13.15b. The strength of each vortex Γ_{W_i} is equal to the vorticity shed during the corresponding time step Δt such that $\Gamma_{W_i} = \int_{t-\Delta t}^{t} \gamma_W(t) U(t) dt$. Consequently,

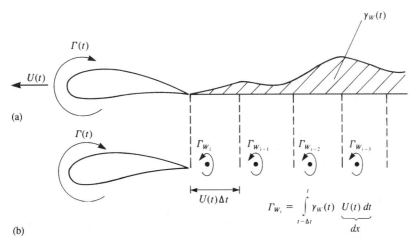

Figure 13.15 Discretization of the wake's continuous vortex distribution by the use of discrete vortices.

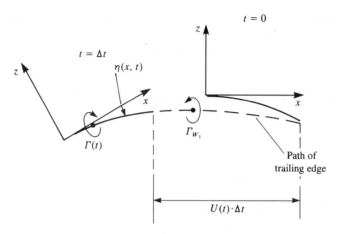

Figure 13.16 Method of placing the latest discrete-vortex wake on the path covered by the trailing edge during the current time step. (Note that in this figure, too, the fluid is at rest and the airfoil moves toward the left side of the page.)

for each vortex element, its strength and location must be specified. In regard to specifying wake vortex location, consider the first wake element after the beginning of the motion, as shown in Fig. 13.16. The wake was probably shed at the airfoil's trailing edge, which moved during the latest time step along the dashed line. (Note, again, that in this figure the fluid is stationary and the airfoil moves to the left.) So at first estimate it will be placed on this line. The distance and relative angle to the trailing edge are important numerical parameters, and usually the wake vortex location should be aligned with the trailing edge and be placed closer to the latest position of the trailing edge. (This is so, since the discrete vortex when placed at the middle of this interval is an approximation that underestimates the induced velocity when compared with the continuous wake vortex sheet result. This is mainly due to the small distance (zero distance) of the continuous wake from the trailing edge during the time interval, compared to the relatively larger distance of an equivalent discrete vortex with similar vorticity placed amid the interval of the latest time step. A typical numerical approach to correct for this wake-discretization error is to place the latest vortex closer to the trailing edge (e.g., within the range of 0.2–0.3 $U(t)\Delta t$, as shown in Fig. 13.16).)

The strength of the latest vortex element is calculated by using Kelvin's condition (Eq. (13.6)), which states that

$$\frac{d\Gamma}{dt} = \frac{d\Gamma(t)}{dt} + \frac{d\Gamma_W}{dt} = 0 \tag{13.51}$$

where $\Gamma(t)$ is the airfoil's circulation and Γ_W is the wake's circulation, respectively. For the first time step

$$\Gamma(t) + \Gamma_{W_1} = 0$$

and for the ith time step

$$\Gamma_{W_i} = -[\Gamma(t_i) - \Gamma(t_{i-1})] \tag{13.52a}$$

or by assuming that the Kelvin condition was met at the previous time step then

$$\Gamma_{W_i} = -\left[\Gamma(t_i) + \sum_{k=1}^{i-1} \Gamma_{W_k}\right] \tag{13.52b}$$

13.8 Unsteady Motion of a Two-Dimensional Thin Airfoil

It must be noted, too, that the Helmholtz theorems of Section 2.9 imply that there is no vortex decay. That is, if a wake vortex element is shed from the trailing edge its strength will be conserved (which is a good practical approximation for most high Reynolds number flows).

Since the vortex wake is force free, each vortex must move with the local stream velocity (Eq. (13.21a)). The local velocity is a result of the velocity components induced by the wake and airfoil (wing) and is usually measured in the inertial frame of reference X, Z. To achieve the vortex wake rollup, at each time step the combined airfoil and wake-induced velocity $(u, w)_i$ is calculated and then the vortex elements are moved by

$$(\Delta x, \Delta z)_i = (u, w)_i \Delta t \tag{13.53}$$

In this simple scheme the velocity components and vortex positions of the current time step are used. But more refined techniques can be applied here to improve the wake shape near the trailing edge (by using information from the current and previous time steps).

13.8.3 Solution by the Time-Stepping Method

The above presentation of the thin airfoil fluid dynamics is formulated as an initial value problem. Typical initial conditions can be a steady-state motion or a start from rest. In the latter case there is no wake at $t = 0$ and the first wake element is formed at $t = \Delta t$. Consider the airfoil after the first time step Δt, as shown in Fig. 13.17a. The problem at this moment is to obtain the flowfield details and to calculate the pressure difference across the airfoil, in the presence of one wake vortex – which represents the vorticity shed at the trailing edge since the initiation of the motion. To apply the results of the small-disturbance solutions of Sections 5.2 and 5.3, the airfoil's camber and angle of attack are assumed to be small $\eta/c \ll 1$, and the path curvature must be large ($\dot{\theta} c/U \ll 1$), so that the boundary conditions can be transferred to the $z = 0$ plane. If the lifting thin airfoil is modeled by a chordwise vortex distribution $\gamma(x, t)$, then at the first time step this problem resembles exactly the model shown in Fig. 5.7, but with an additional wake vortex Γ_{W_1} (Fig. 13.17a). The downwash induced by the airfoil bound circulation $\gamma(x, t)$ is given by Eq. (5.38) (note

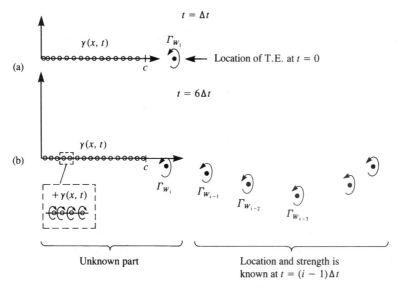

Figure 13.17 Representation of the lifting thin wing by a continuous vortex distribution and the wake by discrete vortex elements.

that the boundary conditions are transferred to the $z = 0$ plane):

$$\frac{\partial \Phi_B}{\partial z}(x,t)_{z=0} = \frac{-1}{2\pi}\int_0^c \gamma(x_0, t)\frac{dx_0}{x - x_0} \qquad (13.54a)$$

At any later time the downwash of the N_W discrete vortices of the wake (see Fig. 13.17b) on the airfoil can be summed up numerically by using, for example, Eq. (11.1):

$$\frac{\partial \Phi_W}{\partial z}(x,t)_{z=0} = \sum_{k=1}^{N_W} \frac{-\Gamma_k}{2\pi} \frac{x - x_k}{(x - x_k)^2 + (z - z_k)^2} \qquad (13.54b)$$

where k is the counter of the wake vortices. The downwash of the bound vortex distribution of Eq. (13.54a) must be equal to the right-hand side of the boundary condition in Eq. (13.50):

$$W(x,t) = \left(\frac{\partial \Phi_B}{\partial x} + \frac{\partial \Phi_W}{\partial x} + U - \dot{\theta}\eta\right)\frac{\partial \eta}{\partial x} - \frac{\partial \Phi_W}{\partial z} - \dot{\theta}x + \frac{\partial \eta}{\partial t}$$

$$\approx U\frac{\partial \eta}{\partial x} - \frac{\partial \Phi_W}{\partial z} - \dot{\theta}x + \frac{\partial \eta}{\partial t} \qquad (13.55)$$

where the smaller terms were neglected. Substitution of this approximate value of $W(x,t)$ and $\partial \Phi_B/\partial z$ from Eq. (13.54a) into the boundary condition (Eq. (13.50)) results in

$$\frac{-1}{2\pi}\int_0^c \gamma(x_0, t)\frac{dx_0}{x - x_0} = U(t)\frac{\partial \eta(x,t)}{\partial x} - \frac{\partial \Phi_W}{\partial z}(x,t) - \dot{\theta}(t)x + \frac{\partial \eta(x,t)}{\partial t},$$
$$0 < x < c \qquad (13.56)$$

which is the time-dependent equivalent of the steady-state boundary condition (Eq. (5.39)) and must hold for each point x on the airfoil's chord. In addition, the Kutta condition is assumed to be valid for this flow (recall that $\dot{\theta}c/U \ll 1$):

$$\gamma(c, t) = 0 \qquad (13.57)$$

If the right-hand side of Eq. (13.56) is known then the solution for the vortex distribution is given in Section 5.3. In fact, all terms appearing in the right-hand side are known (will have numeric value) at any time t, excluding the latest wake vortex influence, which depends on the solution $\gamma(x, t)$. This difficulty can be overcome by assuming that the strength of the latest wake vortex is also known, and adjusting for this assumption later.

The classical approach of Glauert, presented in Section 5.3, approximates $\gamma(x, t)$ by a chordwise trigonometric expansion at any time t. This requires the transformation of the equations into trigonometric variables as appear in Eq. (5.45):

$$x = \frac{c}{2}(1 - \cos\vartheta) \qquad (13.58)$$

Based on this transformation a solution similar to the vortex distribution of Eq. (5.48) is proposed for the time-dependent problem (at each frozen moment, t):

$$\gamma(\vartheta, t) = 2U(t)\left[A_0(t)\frac{1 + \cos\vartheta}{\sin\vartheta} + \sum_{n=1}^{\infty} A_n(t)\sin(n\vartheta)\right] \qquad (13.59)$$

Substitution of this proposed solution into the boundary condition (see details in Section 5.3, Eqs. (5.49) to (5.53)) results in

$$\frac{W(x,t)}{U(t)} = -A_0(t) + \sum_{n=1}^{\infty} A_n(t)\cos(n\vartheta) \qquad (13.60)$$

13.8 Unsteady Motion of a Two-Dimensional Thin Airfoil

which is a trigonometric expansion of the momentary chordwise downwash. The coefficients A_n were found in Section 5.3 (Eqs. (5.51) and (5.52)) and are

$$A_0(t) = -\frac{1}{\pi} \int_0^\pi \frac{W(x,t)}{U(t)} d\vartheta, \quad n = 0 \tag{13.61}$$

$$A_n = \frac{2}{\pi} \int_0^\pi \frac{W(x,t)}{U(t)} \cos n\vartheta \, d\vartheta, \quad n = 1, 2, 3, \ldots \tag{13.62}$$

So at this point, if the momentary chordwise downwash $W(x,t)$ is known, then the momentary circulation distribution on the airfoil is known, too, from Eqs. (13.59), (13.61), and (13.62). However, as mentioned before, the strength of the latest vortex Γ_{W_i} (Fig. 13.17b) in the downwash term is unknown, but it can be calculated by using Kelvin's condition (Eq. (13.51)). A simple iterative scheme to calculate the strength of this vortex is as follows: At a given time step assume that the strength of the latest vortex Γ_{W_i} is known (0 or $\Gamma(t)/2$ are reasonable initial assumptions). Then the total circulation (which must be zero for the converged solution) can be expressed as

$$f(\Gamma) = \Gamma(t) + \Gamma_{W_i} + \sum_{k=1}^{i-1} \Gamma_{W_k} \quad \{= 0 \text{ for the converged solution}\} \tag{13.63}$$

where the first term is the airfoil's circulation, the second term is the latest wake vortex, and the last term is the circulation of all the other wake vortices (which are known from the previous time steps). The circulation of the airfoil is obtained by using Eq. (5.58):

$$\Gamma(t) = \int_0^c \gamma(x,t)\,dx = \int_0^\pi \gamma(\vartheta,t)\frac{c}{2}\sin\vartheta\,d\vartheta$$

$$= 2U(t)\int_0^\pi \left[A_0(t)\frac{1+\cos\vartheta}{\sin\vartheta} + \sum_{n=1}^\infty A_n(t)\sin(n\vartheta)\right]\frac{c}{2}\sin\vartheta\,d\vartheta$$

$$= U(t)c\pi\left[A_0(t) + \frac{A_1(t)}{2}\right] \tag{13.64}$$

The iterations for determining the strength of the latest vortex element (using a simple Newton–Raphson iteration) will have the form

$$(\Gamma_{W_i})_{j+1} = (\Gamma_{W_i})_j - \frac{f(\Gamma_{W_i})_j}{f'(\Gamma_{W_i})_j} \tag{13.65}$$

where j is the iteration counter and the derivative $f'(\Gamma)$ is approximated by

$$f'(\Gamma)_j = \frac{[f(\Gamma)_j - f(\Gamma)_{j-1}]}{(\Gamma_{W_i})_j - (\Gamma_{W_i})_{j-1}}$$

The solution of the momentary airfoil's vortex distribution $\Gamma(t)$ can be summarized such that, first, at a given time step t_i the downwash $W(x,t)$ is calculated by Eq. (13.55). By assuming an initial vortex strength Γ_{W_i} for the most recently shed trailing-edge vortex we can calculate the wake influence via Eq. (13.54b). So now the chordwise downwash $W(x,t)$ can be calculated at any point along the chord and, for example, can be evaluated at say fifty nodal points on the chordline. This allows the numerical computation of the coefficients $A_n(t)$ and $f(\Gamma)$ (in Eqs. (13.61)–(13.63)). Then using Eq. (13.65) the next value of the latest wake vortex is obtained, and this short iterative process (beginning with the downwash calculation and ending with the value of the chordwise vortex distribution) is repeated. This simple iteration scheme will usually converge within 3–5 iterations and this will conclude the solution of the vortex distribution for this time step.

13.8.4 Fluid Dynamic Loads

The fluid dynamic pressures and loads generated by the airfoil can be calculated by using the unsteady Bernoulli equation:

$$\frac{p_\infty - p}{\rho} = \frac{1}{2}\left[\left(\frac{\partial \Phi}{\partial x}\right)^2 + \left(\frac{\partial \Phi}{\partial y}\right)^2 + \left(\frac{\partial \Phi}{\partial z}\right)^2\right]$$
$$- (\mathbf{V}_0 + \mathbf{\Omega} \times \mathbf{r}) \cdot \nabla \Phi + \frac{\partial \Phi}{\partial t} \quad (13.23)$$

By recalling Eqs. (13.45) and (13.46), we obtain for the terms inside the second parentheses

$$-(\mathbf{V}_0 + \mathbf{\Omega} \times \mathbf{r}) = [U(t) - \dot{\theta}(t)\eta, 0, \dot{\theta}(t)x]$$

If the reduced frequency ($\dot{\theta}c/U$) is small and the point of interest is on the airfoil surface then the pressure equation becomes

$$\frac{p_\infty - p}{\rho} \approx U(t)\frac{\partial \Phi}{\partial x} + \dot{\theta}(t)x\frac{\partial \Phi}{\partial z} + \frac{\partial \Phi}{\partial t} \quad (13.66)$$

In cases when $\partial \Phi/\partial z$ has the same value both above and below the chordline (e.g., for thin airfoils) then the second term is the same above and below the thin surface and this term does not contribute to the pressure difference. For these cases the pressure equation can be approximated as

$$\frac{p_\infty - p}{\rho} \approx U(t)\frac{\partial \Phi}{\partial x} + \frac{\partial \Phi}{\partial t} \quad (13.66a)$$

where the first term is similar to the steady-state term, but for the time-dependent case also the change in the potential contributes to the pressures (due to the acceleration of the fluid).

The pressure difference across the airfoil Δp (positive Δp is in the $+z$ direction) is then two times the pressure increment at any one side of the thin surface:

$$\Delta p = p_l - p_u = 2\rho\left[U(t)\frac{\partial \Phi}{\partial x} + \frac{\partial \Phi}{\partial t}\right]_l = \rho\left[U(t)\frac{\partial}{\partial x}\Delta\Phi + \frac{\partial}{\partial t}\Delta\Phi\right] \quad (13.67)$$

where $\Delta\Phi(x, t) = \Phi(x, 0+, t) - \Phi(x, 0-, t) = \int_0^x \gamma(x_0, t)dx_0 = \Gamma(x, t)$ and therefore the pressure difference in terms of the airfoil chordwise circulation γ becomes

$$\Delta p = \rho\left[U(t)\gamma(x, t) + \frac{\partial}{\partial t}\int_0^x \gamma(x_0, t)\,dx_0\right] \quad (13.67a)$$

The normal force on the thin airfoil is then

$$L' \equiv F_z = \int_0^c \Delta p\,dx = \int_0^c \rho\left[U(t)\gamma(x, t) + \rho\frac{\partial}{\partial t}\Gamma(x, t)\right]dx$$
$$= \rho U(t)\Gamma(t) + \rho\int_0^c \frac{\partial}{\partial t}\Gamma(x, t)\,dx \quad (13.68)$$

where the first term is due to the instantaneous circulation (and similar to the steady-state circulatory term) and the second term includes the contribution of the time dependency.

13.8 Unsteady Motion of a Two-Dimensional Thin Airfoil

To evaluate the time derivative of the velocity potential in terms of the coefficients A_n (appearing in Eq. (13.59)) recall that $\Phi = \int \mathbf{q} \cdot d\mathbf{l}$ (or $\Delta\Phi = \int \gamma \, dl$). Then

$$\frac{\partial}{\partial t}\Delta\Phi(x,t) = \frac{\partial}{\partial t}\int_0^x \gamma(x_0,t)\,dx_0 = \frac{\partial}{\partial t}\int_0^\vartheta \gamma(\vartheta_0,t)\frac{c}{2}\sin\vartheta_0\,d\vartheta_0$$

$$= \frac{\partial}{\partial t}\left\{2U(t)\int_0^\vartheta \left[A_0(t)\frac{1+\cos\vartheta_0}{\sin\vartheta_0} + \sum_{n=1}^\infty A_n(t)\sin(n\vartheta_0)\right]\frac{c}{2}\sin\vartheta_0\,d\vartheta_0\right\}$$

where $x = \frac{c}{2}(1-\cos\vartheta)$. With the use of the integrals (see Ref. 5.7, p. 139)

$$\int_0^\vartheta \sin^2\vartheta_0\,d\vartheta_0 = \frac{\vartheta}{2} - \frac{1}{4}\sin 2\vartheta$$

$$\int_0^\vartheta \sin n\vartheta_0 \sin\vartheta_0\,d\vartheta_0 = \frac{\sin(n-1)\vartheta}{2(n-1)} - \frac{\sin(n+1)\vartheta}{2(n+1)}$$

and after some algebra we get

$$\frac{\partial}{\partial t}\Delta\Phi(\vartheta,t) = 2\left\{B_0(\vartheta + \sin\vartheta) + B_1\left(\frac{\vartheta}{2} - \frac{1}{4}\sin 2\vartheta\right)\right.$$
$$\left. + \sum_{n=2}^\infty B_n\left[\frac{\sin(n-1)\vartheta}{2(n-1)} - \frac{\sin(n+1)\vartheta}{2(n+1)}\right]\right\} \quad (13.69)$$

where

$$B_n = \frac{c}{2}\frac{\partial}{\partial t}[A_n(t)U(t)], \quad n = 0, 1, 2, 3, \ldots \quad (13.70)$$

For a given airfoil geometry, the mean camberline $\eta(x,t)$ is a known value and the coefficients $A_0(t), A_1(t), A_2(t), \ldots$ can be computed by Eqs. (13.61) and (13.62) (assuming that the wake influence is known). The pressure difference of Eq. (13.67) can be evaluated since all terms in this equation depend on the coefficients $A_n(t)$. The force in the z direction is then

$$L' \equiv F_z = \int_0^c \Delta p\,dx = 2\rho \int_0^\pi \left\{U^2(t)\left[A_0(t)\frac{1+\cos\vartheta}{\sin\vartheta} + \sum_{n=1}^\infty A_n(t)\sin(n\vartheta)\right]\right.$$
$$+ B_0(\vartheta + \sin\vartheta) + B_1\left(\frac{\vartheta}{2} - \frac{1}{4}\sin 2\vartheta\right)$$
$$\left. + \sum_{n=2}^\infty B_n\left[\frac{\sin(n-1)\vartheta}{2(n-1)} - \frac{\sin(n+1)\vartheta}{2(n+1)}\right]\right\}\frac{c}{2}\sin\vartheta\,d\vartheta$$

These integrals are similar to those treated in Section 5.3 and after their evaluation we get

$$L'(t) = \rho c\left\{\frac{3\pi}{2}B_0 + \frac{\pi}{2}B_1 + \frac{\pi}{4}B_2 + \pi U^2 A_0 + \frac{\pi}{2}U^2 A_1\right\} \quad (13.71)$$

In terms of the A_n's (using Eq. (13.70)), we get

$$L'(t) = \pi\rho c\left\{\left[U^2 A_0 + \frac{3c}{4}\frac{\partial}{\partial t}(UA_0)\right]\right.$$
$$\left. + \left[U^2\frac{A_1}{2} + \frac{c}{4}\frac{\partial}{\partial t}(UA_1) + \frac{c}{8}\frac{\partial}{\partial t}(UA_2)\right]\right\} \quad (13.71a)$$

and it is clear that the velocity and the coefficients are a function of time (i.e., $U \equiv U(t)$, $A_n = A_n(t)$).

The pitching moment about the airfoil's leading edge is

$$M_0(t) = -\int_0^c \Delta p x \, dx = -\int_0^c \rho \left[U(t) \frac{\partial}{\partial x} \Delta \Phi + \frac{\partial}{\partial t} \Delta \Phi \right] x \, dx$$

$$= -2\rho \int_0^\pi \left\{ U^2(t) \left[A_0(t) \frac{1 + \cos \vartheta}{\sin \vartheta} + \sum_{n=1}^\infty A_n(t) \sin(n\vartheta) \right] \right.$$

$$+ B_0(\vartheta + \sin \vartheta) + B_1 \left(\frac{\vartheta}{2} - \frac{1}{4} \sin 2\vartheta \right)$$

$$\left. + \sum_{n=2}^\infty B_n \left[\frac{\sin(n-1)\vartheta}{2(n-1)} - \frac{\sin(n+1)\vartheta}{2(n+1)} \right] \right\} \frac{c}{2} (1 - \cos \vartheta) \frac{c}{2} \sin \vartheta \, d\vartheta$$

and after an evaluation of these integrals we get

$$M_0 = -\frac{cL'}{2} + \frac{\rho c^2}{4} \left[U^2 \left(A_0 \pi + A_2 \frac{\pi}{2} \right) - \frac{\pi}{2} B_0 - \frac{\pi}{2} B_1 + \frac{\pi}{8} B_3 \right]$$

With the use of Eq. (13.71) for L', the moment about $x = 0$ becomes

$$M_0(t) = -\rho c^2 \frac{\pi}{2} \left[\frac{U^2}{2} \left(A_0 + A_1 - \frac{A_2}{2} \right) + \frac{7}{4} B_0 + \frac{3}{4} B_1 + \frac{1}{4} B_2 - \frac{1}{16} B_3 \right] \tag{13.72}$$

and in terms of the A_n's

$$M_0(t) = -\rho c^2 \frac{\pi}{2} \left[\frac{U^2}{2} A_0 + \frac{7c}{8} \frac{\partial}{\partial t} (U A_0) + \frac{U^2}{2} A_1 + \frac{3c}{8} \frac{\partial}{\partial t} (U A_1) \right.$$

$$\left. - \frac{U^2}{4} A_2 + \frac{c}{8} \frac{\partial}{\partial t} (U A_2) - \frac{c}{32} \frac{\partial}{\partial t} (U A_3) \right] \tag{13.72a}$$

Example 1: Small-Amplitude Oscillations of a Thin Airfoil

One of the simplest and yet important examples is the small-amplitude unsteady motion of a flat plate airfoil, which was analyzed by Theodorsen[13.4] and by von Karman and Sears.[13.5] For this case let us assume that the (x, z) frame of reference in Fig. 13.14 moves to the left of the page at a constant speed $U(t) = U = $ const. in an otherwise stationary fluid. Also, the (x, z) frame does not rotate for this example ($\theta = \dot{\theta} = 0$) and the small-amplitude unsteady motion will be introduced through the v_{rel} term (or the $\partial \eta / \partial t$ term in Eq. (13.55)) in the boundary conditions.

The time-dependent chordline position can be represented by a vertical displacement $h(t)$ (positive in the z direction) and by an instantaneous angle of attack $\alpha(t)$ (Fig. 13.18). The chordline shape is then

$$\eta = h - \alpha(x - a)$$

where a is the pitching axis location. For simplicity, first, we shall assume that the pitching axis is at the origin ($a = 0$), and h is the vertical displacement of the leading edge. The position η then becomes

$$\eta = h - \alpha x$$

13.8 Unsteady Motion of a Two-Dimensional Thin Airfoil

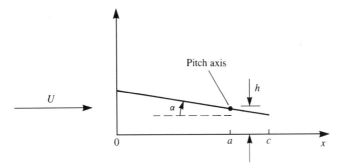

Figure 13.18 Nomenclature for the oscillatory pitching and heaving motion of a flat plate.

Further assume that the vertical displacement is small (e.g., $\eta \ll c$). The derivatives of η are

$$\frac{\partial \eta}{\partial t} = \dot{h} - \dot{\alpha} x$$

$$\frac{\partial \eta}{\partial x} = -\alpha$$

where the dot denotes a time derivative. Substituting this into the downwash $W(x, t)$ term of Eq. (13.55) we get

$$W(x, t) = -U\alpha + \dot{h} - \dot{\alpha} x - \frac{\partial \Phi_W}{\partial z}$$

Since the wake effect is a function of the motion history let us concentrate first on the loads due to the motion only. This portion of the downwash, $W^*(x, t)$, is then

$$W^*(x, t) = -U\alpha + \dot{h} - \dot{\alpha} x = -U\alpha + \dot{h} - \frac{c}{2}\dot{\alpha} + \frac{c}{2}\dot{\alpha} \cos \vartheta$$

and here x was replaced by the trigonometric variable ϑ, using Eq. (13.58). Substitution of this term of the downwash into Eqs. (13.61) and (13.62) provides the values of the A_n coefficients:

$$A_0 = \frac{1}{U}\left(U\alpha - \dot{h} + \frac{c}{2}\dot{\alpha}\right)$$

$$A_1 = \frac{\dot{\alpha} c}{2U}$$

$$A_2 = A_3 = \cdots = A_N = 0$$

The circulation due to the downwash W^* can be obtained by recalling the results of Eq. (5.58):

$$\Gamma^*(t) = \int_0^c \gamma(x, t)\, dx = \pi c U\left(A_0 + \frac{A_1}{2}\right)$$

and after substitution of the A_n coefficients the circulation becomes

$$\Gamma^*(t) = \pi c\left(U\alpha - \dot{h} + \frac{3}{4} c\dot{\alpha}\right)$$

The lift per unit span is then obtained from Eq. (13.71a):

$$L^* = \rho U \Gamma + \pi \rho c^2 U\left(\frac{3}{4}\frac{\partial A_0}{\partial t} + \frac{1}{4}\frac{\partial A_1}{\partial t}\right)$$

and in terms of the displacement h and the angle of attack, α, we get

$$L^* = \pi\rho U c\left(U\alpha - \dot{h} + \frac{3}{4}c\dot{\alpha}\right) + \pi\rho c^2\left[\frac{3}{4}(U\dot{\alpha} - \ddot{h}) + \frac{c}{2}\ddot{\alpha}\right]$$

In the derivation of this expression the Kutta condition was satisfied, but the downwash of the unsteady wake was not included. Theodorsen[13.4] and von Karman and Sears[13.5] showed that for a small-amplitude oscillatory motion the final result will include similar terms and the effect of the wake is to reduce the lift due to the first term in L^* by a factor of $C(k)$, which is called the *lift deficiency factor*. Now, if we consider harmonic heave and pitch oscillations such that

$$h = h_0 \sin \omega t$$

$$\alpha = \alpha_0 \sin \omega t$$

then the lift per unit span becomes

$$L' = \pi\rho U c C(k)\left[U\alpha - \dot{h} + \frac{3}{4}c\dot{\alpha}\right] + \pi\rho\frac{c^2}{4}\left[U\dot{\alpha} - \ddot{h} + \frac{c}{2}\ddot{\alpha}\right] \tag{13.73a}$$

This equation includes the effects of the periodic wake and some of the constants in the second (*added mass*) term are different from those in L^*. Moreover, the added mass part of this solution does not satisfy the Kutta condition and therefore this term differs in its definition from the second term in L^*. In the case when the pitch axis is moved to a location a (and also h is measured at this point), as shown in Fig. 13.18, then the lift per unit span will have a form similar to the results of Ref. 13.4:

$$L' = \pi\rho U c C(k)\left[U\alpha - \dot{h} + \left(\frac{3}{4} - \frac{a}{c}\right)c\dot{\alpha}\right] + \pi\rho\frac{c^2}{4}\left[(U\dot{\alpha} - \ddot{h}) + c\left(\frac{1}{2} - \frac{a}{c}\right)\ddot{\alpha}\right] \tag{13.73b}$$

This can be rewritten as

$$L' = L'_1 + L'_2$$

where L'_1 is similar to the circulatory lift term in a steady motion and L'_2 is the lift due to the acceleration (added mass).

Figure 13.19 shows a plot of the lift deficiency factor $C(k)$ versus the reduced frequency k, which is defined similarly to the nondimensional number of Eq. (1.52):

$$k = \frac{\omega c}{2U}$$

As Fig. 13.19 indicates, the wake has a delaying effect on the circulatory part of the lift such that

$$L'_1(t) = L'_1 \sin(\omega t - \varpi)$$

and ϖ represents the time shift effect of the wake (note that ϖ changes with the reduced frequency as shown in Fig. 13.19).

After a similar treatment of the moment about the leading edge we get

$$M_0 = -\frac{\pi\rho c^2}{4}\left\{-\frac{c}{2}\ddot{h} + \frac{3Uc}{4}\dot{\alpha} + \frac{9}{32}c^2\ddot{\alpha} + UC(k)\left[-\dot{h} + U\alpha + \frac{3c}{4}\dot{\alpha}\right]\right\} \tag{13.74a}$$

13.8 Unsteady Motion of a Two-Dimensional Thin Airfoil

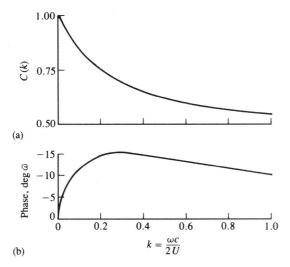

Figure 13.19 Graphic description of Theodorsen's[13.4] lift deficiency and phase lag functions (following p. 482 in W. Johnson, *Helicopter Theory*, Princeton University Press, 1980).

Again, when the pitch axis is moved to a point a (Fig. 13.18) then the pitching moment about this point is

$$M = -\frac{\pi \rho c^2}{4}\left\{c\left(\frac{a}{c}-\frac{1}{2}\right)\ddot{h} + Uc\left(\frac{3}{4}-\frac{a}{c}\right)\dot{\alpha} + \frac{c^2}{4}\left(\frac{9}{8}+\frac{4a^2}{c^2}-\frac{4a}{c}\right)\ddot{\alpha} \right.$$
$$\left. -\left(\frac{4a}{c}-1\right)UC(k)\left[-\dot{h}+U\alpha+c\left(\frac{3}{4}-\frac{a}{c}\right)\dot{\alpha}\right]\right\} \quad (13.74b)$$

The most important portion of Theodorsen's analysis is that the basic nature of the unsteady effect can be briefly summarized by the $C(k)$ diagrams in Fig. 13.19. As the reduced frequency k increases the magnitude of the $\rho U \Gamma$ term in the lift is reduced. Additionally, the lift lag initially increases with the reduced frequency, but for $k > 0.4$ a gradual decrease in the phase shift is shown.

The above model for the small-amplitude oscillation of a thin airfoil is useful in estimating the unsteady loads in cases such as wing flutter or propulsion. The propulsion effect due to the heaving oscillations of a flat plate is shown schematically in Fig. 13.20. Recall that the leading-edge suction causes the circulatory part of the lift ($\rho U \Gamma$) to become normal to the instantaneous motion path, which clearly results in a propulsive (forward pointing) component. If the heaving motion is relatively slow then the second term in Eq. (13.73) is relatively small, too, and high propulsive efficiencies[13.2] can be obtained.

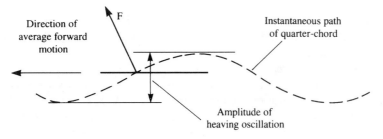

Figure 13.20 Schematic description of the propulsion effect due to heaving oscillations of a flat plate.

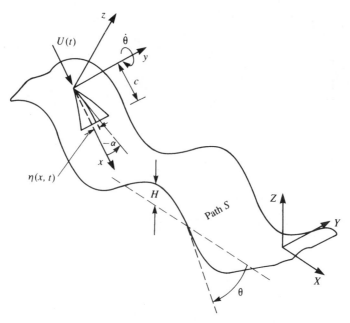

Figure 13.21 Nomenclature for the unsteady motion of a slender thin wing along the path S. (Note that the motion is observed from the X, Y, Z coordinate system where the fluid is at rest and the wing moves toward the upper left side of the page.)

13.9 Unsteady Motion of a Slender Wing

As the simplest example for the conversion of a three-dimensional wing theory to the time-dependent mode, consider the planar motion of a slender wing in the x, z plane.[13.6] (This is a three degrees of freedom motion with $\dot{X}, \dot{Z}$, and $\dot{\theta}$ as shown in Fig. 13.21.) Since for a slender wing the longitudinal dimension is much larger than the other two dimensions ($x \gg y, z$) we can assume that the derivatives are inversely affected such that:

$$\frac{\partial}{\partial x} \ll \frac{\partial}{\partial y}, \frac{\partial}{\partial z} \tag{13.75}$$

As in the case of the steady-state flow over slender wings and bodies, substitution of this condition into the continuity equation (Eq. (13.12)) allows us to neglect the first term, compared to the other derivatives:

$$\nabla^2 \Phi \approx \frac{\partial^2 \Phi}{\partial y^2} + \frac{\partial^2 \Phi}{\partial z^2} = 0 \tag{13.76}$$

This suggests that the cross-flow effect is dominant, and for any $x = $ const. station, a local two-dimensional solution is sufficient. An interesting aspect of this simplification is that the wake influence is negligible, too, as long as the longitudinal time variations (e.g., wing's forward acceleration) are small.

The slender, thin lifting surface with a chord length of c is shown schematically in Fig. 13.21. At $t = 0$ the wing is at rest in the inertial system (X, Y, Z), and at $t > 0$ it moves along a time-dependent curved path, S (for this particular case S is assumed to be two dimensional). For convenience, the coordinates x, z are selected such that the origin O is placed on the path S, and the x coordinate axis is always tangent to the path. The wing shape (camberline) is given in this coordinate system by $\eta(x, t)$, which is considered to be

13.9 Unsteady Motion of a Slender Wing

small ($\eta/c \ll 1$, since small-disturbance motion is assumed). Also, the normal component of the kinematic velocity is small (i.e., $\dot{\theta}c/U(t) \ll 1$).

The time-dependent version of the boundary condition requiring no normal flow across the surface (at any $x =$ const. station) for this case is given by Eq. (13.50):

$$\frac{\partial \Phi}{\partial z} = W(x, t) \tag{13.77}$$

where Φ is the wing's perturbation potential and the subscript B is not used in the case of the slender wing (since the wake effect was neglected).

13.9.1 Kinematics

In the body coordinate system shown in Fig. 13.21, the chordwise downwash $W(x, t)$ (assuming small-disturbance flow) is given by Eq. (13.55):

$$W(x, t) = U\frac{\partial \eta}{\partial x} - \frac{\partial \Phi_W}{\partial z} - \dot{\theta}x + \frac{\partial \eta}{\partial t} \tag{13.55}$$

where the smaller terms were neglected. As a result of the slenderness assumption, the wake influence can also be neglected and the chordwise downwash becomes

$$W(x, t) = U\frac{\partial \eta}{\partial x} - \dot{\theta}x + \frac{\partial \eta}{\partial t} \tag{13.78}$$

Let us now follow Section 8.2.2 and model the cross-flow (shown in Fig. 8.18) at any x station by a vortex distribution $\gamma(y, t)$. The perturbation velocity potential is given then by Eq. (8.69):

$$\Phi(x, y, z, t) = \frac{1}{2\pi} \int_{-b(x)/2}^{b(x)/2} \gamma(y_0, t) \tan^{-1} \frac{z}{(y - y_0)} dy_0 \tag{13.79}$$

The velocity components in the $x =$ const. plane, due to this velocity potential, are

$$v(x, y, 0\pm, t) = \frac{\partial \Phi}{\partial y} = \mp \frac{\gamma(y, t)}{2} \tag{13.80}$$

$$w(x, y, 0\pm, t) = \frac{\partial \Phi}{\partial z} = \frac{1}{2\pi} \int_{-b(x)/2}^{b(x)/2} \gamma(y_0, t) \frac{dy_0}{(y - y_0)} \tag{13.81}$$

It is evident in these formulas that because of the slender wing assumption, only the local spanwise vortex distribution will affect the near field downwash. By substituting this vortex distribution–induced downwash $w(x, y, 0\pm, t)$ into the wing boundary condition (Eq. (13.77)) we get for each $x =$ const. section on the wing

$$\frac{1}{2\pi} \int_{-b(x)/2}^{b(x)/2} \gamma(y_0) \frac{dy_0}{(y - y_0)} = U\frac{\partial \eta}{\partial x} - \dot{\theta}x + \frac{\partial \eta}{\partial t} \tag{13.82}$$

and it is clear that γ, U, η, and $\dot{\theta}$ are functions of time. A comparison of this form of the boundary condition with the formulation for high aspect ratio wings (Eq. (8.11)) clearly indicates that as a result of the slender wing assumption the effect of the vortex lines parallel to the y axis (including the time-dependent portion of the wake) were neglected.

13.9.2 Solution of the Flow over the Unsteady Slender Wing

Solution of the vortex distribution for any given time t, at each $x =$ const. station, is reduced now to solving Eq. (13.82) for $\gamma(y) \equiv \gamma(y, t)$. Because of the similarity between this case and the steady-state slender wing case we know that the spanwise circulation (load) distribution is elliptic, as in Eq. (8.74):

$$\Gamma(y) = \Gamma_{\max}\left[1 - \left(\frac{y}{b(x)/2}\right)^2\right]^{1/2} \tag{13.83}$$

and again $\Gamma(y) \equiv \Gamma(y, t)$. The spanwise vorticity distribution (shown in Fig. 8.18) is obtained by differentiating with respect to y (as in Eq. (8.41) or (8.75)):

$$\gamma(y) = -\frac{d\Gamma(y)}{dy} = \frac{4\Gamma_{\max}}{b(x)^2}\frac{y}{\sqrt{[1 - (y/b(x)/2)^2]}} \tag{13.84}$$

Substitution of this into the integral equation, Eq. (13.82), results in

$$\frac{1}{2\pi}\int_{-b(x)/2}^{b(x)/2}\frac{4\Gamma_{\max}}{b(x)^2}\frac{y_0}{\sqrt{[1-(y_0/b(x)/2)^2]}}\frac{dy_0}{(y-y_0)} = U\frac{\partial\eta}{\partial x} - \dot{\theta}x + \frac{\partial\eta}{\partial t} \tag{13.85}$$

But the left-hand side integral has already been evaluated in Chapter 8 (see Eq. (8.77)), resulting in $(4\Gamma_{\max}/b(x)^2)(-\pi b(x)/2)$. Substitution of this result into Eq. (13.85) yields

$$\frac{1}{2\pi}\frac{4\Gamma_{\max}}{b(x)^2}\left[\frac{-\pi b(x)}{2}\right] = U\frac{\partial\eta}{\partial x} - \dot{\theta}x + \frac{\partial\eta}{\partial t}$$

and after rearranging the terms we get

$$\frac{\Gamma_{\max}}{b(x)} = -U\frac{\partial\eta}{\partial x} + \dot{\theta}x - \frac{\partial\eta}{\partial t} = -W(x, t) \tag{13.86}$$

which shows that the spanwise induced downwash due to an elliptic lift distribution is constant and independent of y. The value of $\Gamma_{\max}$ (at each x station and time t) is easily evaluated now and is

$$\Gamma_{\max} = -b(x)W(x, t) \tag{13.87}$$

Recalling that the velocity potential can be defined by a path of integration along the local y axis (for an $x = $ const. section), we have

$$\Phi(x, y, 0\pm, t) = \int_{-b(x)/2}^{y}\frac{\mp\gamma(y)}{2}dy = \pm\frac{\Gamma(y)}{2}$$

where the integration starts at the left leading edge of the $x = $ const. station and the integration path is above $(0+)$ or below $(0-)$ the wing.

By substituting $\gamma(y)$ and $\Gamma_{\max}$ into Eqs. (13.79)–(13.81), we can obtain the cross-flow potential and its derivatives:

$$\Phi(x, y, 0\pm, t) = \mp W(x,t)\frac{b(x)}{2}\sqrt{1 - \left[\frac{y}{b(x)/2}\right]^2} = \mp W(x,t)\sqrt{\left[\frac{b(x)}{2}\right]^2 - y^2} \tag{13.88}$$

$$u(x, y, 0\pm, t) = \frac{\partial\Phi}{\partial x}(x, y, 0\pm, t) = \mp\frac{\partial}{\partial x}\left\{W(x,t)\sqrt{\left[\frac{b(x)}{2}\right]^2 - y^2}\right\} \tag{13.89}$$

This differentiation can be executed only if the wing planform shape $b(x)$ and chordwise

13.9 Unsteady Motion of a Slender Wing

downwash $W(x, t)$ are known. The other derivatives of the velocity potential are

$$v(x, y, 0\pm, t) = \frac{\partial \Phi}{\partial y}(x, y, 0\pm, t) = \mp\frac{\gamma(y)}{2} = \pm\frac{W(x,t)y}{\sqrt{[b(x)/2]^2 - y^2}} \quad (13.90)$$

and based on Eq. (13.77), the downwash on the wing is

$$w(x, y, 0\pm, t) = \frac{\partial \Phi}{\partial z}(x, y, 0\pm, t) = W(x, t) \quad (13.91)$$

Once the velocity field is obtained the pressure distribution on the wing can be calculated by using the Bernoulli equation. If the reduced frequencies and the accelerations in the z direction are small, then the pressure is given by Eq. (13.66a):

$$\frac{p_\infty - p}{\rho} \approx U(t)\frac{\partial \Phi}{\partial x} + \frac{\partial \Phi}{\partial t} \quad (13.66a)$$

where the first term is similar to the steady-state (circulatory) term, and the second term is a result of the change of the flow with time. The pressure difference across the thin wing is then

$$\Delta p = p(x, y, 0-, t) - p(x, y, 0+, t) = 2\rho\left[U\frac{\partial \Phi}{\partial x} + \frac{\partial \Phi}{\partial t}\right]_{z=0+}$$

$$= \rho U\frac{\partial}{\partial x}\Delta\Phi + \rho\frac{\partial}{\partial t}\Delta\Phi \quad (13.92)$$

since $\Delta\Phi = 2\Phi(x, y, 0+, t)$. Substitution of the results for the velocity potential and its derivatives yields

$$\Delta p = -2\rho U \frac{\partial}{\partial x}\left\{W(x, t)\frac{b(x)}{2}\sqrt{1 - \left[\frac{y}{b(x)/2}\right]^2}\right\}$$

$$- 2\rho\frac{\partial}{\partial t}\left\{W(x, t)\frac{b(x)}{2}\sqrt{1 - \left[\frac{y}{b(x)/2}\right]^2}\right\} \quad (13.93)$$

The longitudinal wing loading is obtained by integrating the spanwise pressure difference and by recalling the result of Eq. (8.88) that

$$\int_{-b(x)/2}^{b(x)/2}\left[1 - \left(\frac{y}{b(x)/2}\right)^2\right]^{1/2} dy = \frac{\pi b(x)}{4} \quad (13.94)$$

With this in mind,

$$\frac{dL}{dx} = \int_{-b(x)/2}^{b(x)/2} \Delta p\, dy = -\rho U(t)\frac{\partial}{\partial x}\left\{W(x, t)b(x)\int_{-b(x)/2}^{b(x)/2}\left[1 - \left(\frac{y}{b(x)/2}\right)^2\right]^{1/2} dy\right\}$$

$$- \rho\frac{\partial}{\partial t}\left\{W(x, t)b(x)\int_{-b(x)/2}^{b(x)/2}\left[1 - \left(\frac{y}{b(x)/2}\right)^2\right]^{1/2} dy\right\}$$

$$= -\frac{\pi\rho U(t)}{4}\frac{\partial}{\partial x}[W(x, t)b(x)^2] - \frac{\pi\rho}{4}\frac{\partial}{\partial t}[W(x, t)b(x)^2] \quad (13.95)$$

This is the unsteady version of the slender wing lift (see Section 8.2.3). A similar formulation was derived by Lighthill[13.7] when studying the swimming of slender fish in small-amplitude motion.

In regard to the momentary drag force, recall that the x axis of the coordinate system used for this problem remains parallel to the flight path (Fig. 13.21) and the normal component of the force was designated as lift. Similarly, it is possible to define the axial component of the force as drag. Since at each moment for a given set of boundary conditions the potential flow problem is independent of time (excluding the wake influence – which is neglected in this example) the drag component due to the circulatory part of the force can be approximated by the steady-state results of Eq. (8.95). Consequently, owing to the leading-edge suction, the drag due to the circulatory lift (the first term of Eq. (13.95)) is half of the projected pressure difference component:

$$\frac{1}{2} \frac{\pi \rho U}{4} \frac{\partial}{\partial x}[W(x,t)b(x)^2]$$

while the drag due to the fluid acceleration (second term in Eq. (13.95)) is not reduced by the leading-edge suction. Thus the instantaneous drag force becomes

$$\frac{dD}{dx} = \int_{-\frac{b(x)}{2}}^{\frac{b(x)}{2}} \Delta p \frac{\partial \eta}{\partial x} dy \approx -\frac{\pi \rho}{4} \frac{\partial \eta}{\partial x}\left[\frac{U}{2}\frac{\partial}{\partial x} + \frac{\partial}{\partial t}\right][W(x,t)b(x)^2] \qquad (13.96)$$

Example 1: Heaving Oscillations of a Slender Delta Wing

As one of the simplest examples let us consider the small-amplitude heaving oscillations of a slender delta wing. The x, z coordinate system is selected such that it moves to the left of the page at a constant velocity $U(t) = U_\infty$. The wing remains parallel to the x axis, but it oscillates up and down at a frequency ω and amplitude h_0 (see Fig. 13.22). The small displacement of the wing relative to the x axis is then

$$\eta(x, t) = h_0 \sin \omega t$$

Then with $\mathbf{v}_{\text{rel}} = (0, 0, \partial \eta / \partial t)$ the time-dependent downwash $W(x, t)$ of Eq. (13.55) becomes

$$W(x, t) = \frac{\partial \eta}{\partial t} = h_0 \omega \cos \omega t$$

The longitudinal loading dL/dx is obtained from Eq. (13.95):

$$\frac{dL}{dx} = -\frac{\pi \rho U_\infty}{4} \frac{\partial}{\partial x}[h_0 \omega b(x)^2 \cos \omega t] - \frac{\pi \rho}{4} \frac{\partial}{\partial t}[h_0 \omega b(x)^2 \cos \omega t]$$

$$= -\frac{\pi \rho U_\infty}{4} \frac{\partial}{\partial x}[b(x)^2] h_0 \omega \cos \omega t + \frac{\pi \rho}{4} b(x)^2 h_0 \omega^2 \sin \omega t$$

For a flat triangular delta wing with a chord c and trailing edge span of $b_{T.E.}$ the

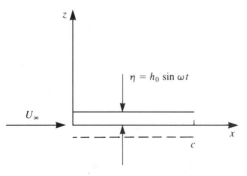

Figure 13.22 Heaving oscillations of a wing with an amplitude of h_0.

13.9 Unsteady Motion of a Slender Wing

span $b(x)$ becomes

$$b(x) = b_{T.E.} \frac{x}{c}$$

and the longitudinal lift distribution is

$$\frac{dL}{dx} = -\frac{\pi \rho U_\infty}{2} \frac{b_{T.E.}^2}{c^2} x h_0 \omega \cos \omega t + \frac{\pi \rho}{4} \frac{(b_{T.E.} x)^2}{c^2} h_0 \omega^2 \sin \omega t$$

The lift of the wing is obtained by integrating $\frac{dL}{dx}$ along the chord:

$$L = \int_0^c \frac{dL}{dx} dx = -\frac{\pi \rho U_\infty}{4} b_{T.E.}^2 h_0 \omega \cos \omega t + \frac{\pi \rho}{12} b_{T.E.}^2 c h_0 \omega^2 \sin \omega t \quad (13.97)$$

The pitching moment about the apex ($x = 0$) is

$$M_0 = -\int_0^c \frac{dL}{dx} x \, dx = -\frac{\pi \rho U_\infty}{6} b_{T.E.}^2 c h_0 \omega \cos \omega t + \frac{\pi \rho}{16} b_{T.E.}^2 c^2 h_0 \omega^2 \sin \omega t \quad (13.98)$$

As we can see the loads on the wing are created by two terms that have a phase shift between them. The lift of the wing, for example, can be divided such that

$$L_1 = -\frac{\pi \rho U_\infty}{4} b_{T.E.}^2 h_0 \omega \cos \omega t = -\frac{\pi \rho U_\infty}{4} b_{T.E.}^2 \dot{\eta}$$

$$L_2 = \frac{\pi \rho}{12} b_{T.E.}^2 c h_0 \omega^2 \sin \omega t = -\frac{\pi \rho}{12} b_{T.E.}^2 c \ddot{\eta}$$

The time-dependent vertical displacement of the wing $\eta(t)$ and these two terms of the lift (L_1, L_2) are shown schematically in Fig. 13.23. The first term L_1 resembles the steady-state (circulatory) term and the lift is a result of the instantaneous effective angle of attack. This lags in phase with the motion such that when the wing moves downward it creates lift L_1 and vice versa. The second term L_2 is a result of the wing vertical acceleration (added mass) and is in phase with the motion.

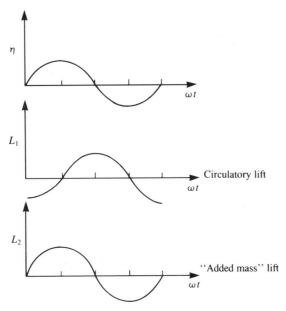

Figure 13.23 Schematic description of the heaving motion and the two parts of the lift during one cycle.

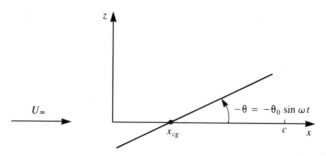

Figure 13.24 Nomenclature used to define the pitch oscillations of the slender wing.

Example 2: Pitch Oscillations of a Slender Delta Wing

Another simple example is the small-amplitude pitching oscillations of a slender delta wing. The origin of the x, z coordinate system is now moving to the left of the page at a constant velocity $U(t) = U_\infty$. The wing is pitching about the point x_{cg} with a frequency ω and a small amplitude θ_0 (shown in Fig. 13.24):

$$\theta = \theta_0 \sin \omega t$$

The wing's chordline $\eta(x, t)$ is given then by

$$\eta(x, t) = -(x - x_{cg}) \tan \theta \approx -(x - x_{cg})\theta = -(x - x_{cg})\theta_0 \sin \omega t$$

By substituting the derivatives of η into Eq. (13.50) we obtain the time-dependent downwash

$$W(x, t) = -U_\infty \theta_0 \sin \omega t - \theta_0 \omega (x - x_{cg}) \cos \omega t$$

Note that the same result can be obtained from Eq. (13.55) by placing the chordline on the x axis and pitching the x–z plane with $\dot{\theta}$. The longitudinal loading dL/dx is obtained from Eq. (13.95):

$$\frac{dL}{dx} = \frac{-\pi \rho U_\infty}{4} \frac{\partial}{\partial x}\{[-U_\infty \theta_0 \sin \omega t - \theta_0 \omega (x - x_{cg}) \cos \omega t]b(x)^2\}$$

$$- \frac{\pi \rho}{4} \frac{\partial}{\partial t}\{[-U_\infty \theta_0 \sin \omega t - \theta_0 \omega (x - x_{cg}) \cos \omega t]b(x)^2\}$$

For a flat triangular delta wing with a chord c and trailing-edge span of $b_{T.E.}$ the local span $b(x)$ becomes

$$b(x) = b_{T.E.} \frac{x}{c}$$

and the longitudinal lift distribution is

$$\frac{dL}{dx} = \frac{\pi \rho U_\infty}{4} \frac{b_{T.E.}^2}{c^2} [2x U_\infty \theta_0 \sin \omega t + \theta_0 \omega (3x^2 - 2x x_{cg}) \cos \omega t]$$

$$+ \frac{\pi \rho}{4} \frac{b_{T.E.}^2}{c^2} [U_\infty \theta_0 \omega x^2 \cos \omega t - \theta_0 \omega^2 (x^3 - x^2 x_{cg}) \sin \omega t]$$

The lift of the wing is obtained by integrating dL/dx along the chord:

$$L = \int_0^c \frac{dL}{dx} dx = \frac{\pi \rho U_\infty}{4} \frac{b_{T.E.}^2}{c^2} [U_\infty \theta_0 c^2 \sin \omega t + \theta_0 \omega (c^3 - c^2 x_{cg}) \cos \omega t]$$

$$+ \frac{\pi \rho}{4} \frac{b_{T.E.}^2}{c^2} \left[U_\infty \theta_0 \omega \frac{c^3}{3} \cos \omega t - \theta_0 \omega^2 \left(\frac{c^4}{4} - \frac{c^3}{3} x_{cg} \right) \sin \omega t \right] \quad (13.99)$$

The pitching moment about the apex ($x = 0$) is

$$M_0 = -\int_0^c \frac{dL}{dx} x \, dx$$

$$= -\frac{\pi \rho U_\infty}{4} \frac{b_{T.E.}^2}{c^2} \left[U_\infty \theta_0 \frac{2c^3}{3} \sin \omega t + \theta_0 \omega \left(\frac{3c^4}{4} - \frac{2c^3}{3} x_{cg} \right) \cos \omega t \right]$$

$$- \frac{\pi \rho}{4} \frac{b_{T.E.}^2}{c^2} \left[U_\infty \theta_0 \omega \frac{c^4}{4} \cos \omega t - \theta_0 \omega^2 \left(\frac{c^5}{5} - \frac{c^4}{4} x_{cg} \right) \sin \omega t \right] \quad (13.100)$$

As we can see in this case, too, the loads on the wing can be divided into three terms that have a phase shift between them. The three components can be rewritten as

$$L_1 = \frac{\pi \rho b_{T.E.}^2}{4} U_\infty^2 \theta$$

$$L_2 = \frac{\pi \rho b_{T.E.}^2}{4} U_\infty \left[(c - x_{cg}) + \frac{c}{3} \right] \dot{\theta}$$

$$L_3 = \frac{\pi \rho b_{T.E.}^2}{4} \left(\frac{c^2}{4} - \frac{c}{3} x_{cg} \right) \ddot{\theta}$$

and

$$L = L_1 + L_2 + L_3$$

It is clear that the first term is a result of the instantaneous angle of attack while the second term is a result of the downwash caused by the pitch rotation. This part is a function of the pitch axis. The last term is due to the acceleration (added mass) and depends, too, on the location of x_{cg}.

The damping of the wing due to a constant pitch motion can be found by integrating L_2 only in Eq. (13.100):

$$\frac{\partial M_0}{\partial \dot{\theta}} = -\frac{\pi \rho U_\infty}{4} \frac{b_{T.E.}^2}{c^2} \left(\frac{3c^4}{4} - \frac{2c^3}{3} x_{cg} \right) - \frac{\pi \rho}{4} \frac{b_{T.E.}^2}{c^2} U_\infty \frac{c^4}{4}$$

$$= -\frac{\pi \rho U_\infty}{4} b_{T.E.}^2 \left(c^2 - \frac{2c}{3} x_{cg} \right) \quad (13.101)$$

13.10 Algorithm for Unsteady Airfoil Using the Lumped-Vortex Element

As was mentioned earlier, with only a few minor modifications, steady-state solution techniques can be updated to treat unsteady flows. As a first numerical example, the discrete vortex model of thin lifting airfoils (Section 11.1.1) will be modified. There are three areas of the program that will be affected:

1. The normal velocity component on the solid boundary should include the components of the unsteady motion as well (as in Eq. (13.13a)) – this is a minor and local modification to the steady-state program.
2. Similar corrections due to the unsteady motion should be included in the pressure calculations (using the modified Bernoulli equation, Eq. (13.23)) – again this is a limited local modification.

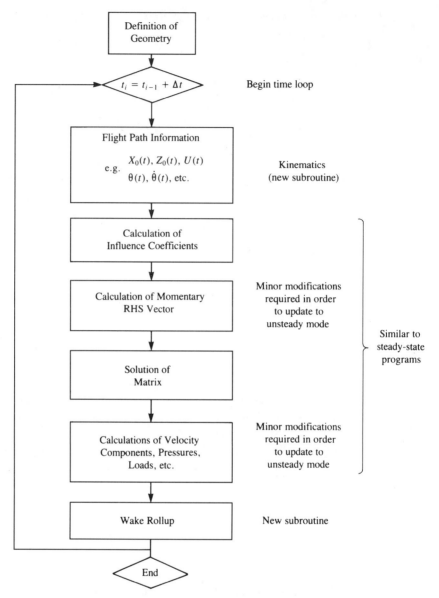

Figure 13.25 Schematic flowchart for the numerical solution of the unsteady wing problem.

3. An unsteady wake model has to be established (e.g., the time-stepping discrete-vortex model as presented in Section 13.8.2). Note that such a wake model can be added on, in a simple manner, to the steady-state solvers.

(In this example the discrete vortex model for the thin lifting airfoil of Section 11.1.1 will be transformed to the unsteady mode. But any of the methods presented in Chapter 11 can be modified easily and can be given as a student project.)

The mechanics of such an upgrade are demonstrated by the generic flowchart of Fig. 13.25. A comparison of this diagram with the steady-state diagram of Fig. 9.15 reveals that, first, a time-stepping loop has to be established (only one programming statement). Then a new element appears (the flight path information block), which has all the kinematic

13.10 Algorithm for Unsteady Airfoil Using the Lumped-Vortex Element

information on the body's or wing's motion. The rest of the program will follow the methodology of Chapter 11, and the only additional block is the "wake rollup" block, which will perform the wake rollup at each time step. Consequently, we shall follow the same sequence used in the previous numerical chapters:

a. *Choice of Singularity Element*

For this discrete-vortex method the lumped-vortex element is selected and its influence is given in Section 11.1.1:

$$\begin{pmatrix} u \\ w \end{pmatrix} = \frac{\Gamma_j}{2\pi r_j^2} \begin{pmatrix} 0 & 1 \\ -1 & 0 \end{pmatrix} \begin{pmatrix} x - x_j \\ z - z_j \end{pmatrix} \quad (13.102)$$

where

$$r_j^2 = (x - x_j)^2 + (z - z_j)^2$$

Thus, the velocity at an arbitrary point (x, z) due to a vortex element Γ_j located at (x_j, z_j) is given by this equation. This can be included in a subroutine, which was defined by Eq. (11.2):

$$(u, w) = \text{VOR2D}(\Gamma_j, x, z, x_j, z_j) \quad (11.2)$$

Using this lumped-vortex element, the airfoil will be represented by a set of discrete vortices placed on the camberline, as shown in Fig. 13.26. If the airfoil's circulation changes with time, then vortex wake elements are shed at the trailing edge and the wake will be modeled by using the same discrete-vortex model (Fig. 13.26).

b. *Kinematics*

Let us establish an inertial frame of reference X, Z, shown in Fig. 13.27, such that this frame of reference is stationary while the airfoil is moving to the left of the page. Next, the airfoil's camberline is placed in a moving frame of reference x, z with the leading edge at the origin. The flight path of the origin and the orientation of the x, z system are prescribed as

$$\begin{aligned} X_0 &= X_0(t) \\ Z_0 &= Z_0(t) \\ \theta &= \theta(t) \end{aligned} \quad (13.103)$$

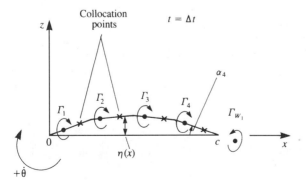

Figure 13.26 Discrete vortex model for the unsteady thin airfoil problem (shown during the first time step, $t = \Delta t$).

and the instantaneous velocity of the origin and its rotation $\dot{\theta}$ about the y axis are

$$\dot{X}_0 = \dot{X}_0(t)$$
$$\dot{Z}_0 = \dot{Z}_0(t) \tag{13.104}$$
$$\dot{\theta} = \dot{\theta}(t)$$

For example, if the airfoil moves to the left at a constant speed of U_∞ and sinks at a speed of W_∞ then

$$X_0 = -U_\infty t, \qquad \dot{X}_0(t) = -U_\infty$$
$$Z_0 = -W_\infty t, \qquad \dot{Z}_0(t) = -W_\infty \tag{13.105}$$
$$\theta = 0, \qquad \dot{\theta} = 0$$

Or if the airfoil flies at a constant forward speed U_∞, and performs pitch oscillations at a frequency ω about the y axis, then

$$X_0 = -U_\infty t, \qquad \dot{X}_0(t) = -U_\infty$$
$$Z_0 = 0, \qquad \dot{Z}_0(t) = 0 \tag{13.106}$$
$$\theta = \sin \omega t, \qquad \dot{\theta} = \omega \cos \omega t$$

It is useful to establish a transformation between the two coordinate systems shown in the figure such that

$$\begin{pmatrix} X \\ Z \end{pmatrix} = \begin{pmatrix} \cos \theta(t) & \sin \theta(t) \\ -\sin \theta(t) & \cos \theta(t) \end{pmatrix} \begin{pmatrix} x \\ z \end{pmatrix} + \begin{pmatrix} X_0 \\ Z_0 \end{pmatrix} \tag{13.107}$$

and similarly the transformed velocity components are

$$\begin{pmatrix} \dot{X} \\ \dot{Z} \end{pmatrix} = \begin{pmatrix} \cos \theta(t) & \sin \theta(t) \\ -\sin \theta(t) & \cos \theta(t) \end{pmatrix} \begin{pmatrix} \dot{x} \\ \dot{z} \end{pmatrix} \tag{13.108}$$

The inverse transformation is also useful, and the velocity components U_t, W_t observed in the x, z frame due to the translation of the origin are

$$\begin{pmatrix} U_t \\ W_t \end{pmatrix} = \begin{pmatrix} \cos \theta(t) & -\sin \theta(t) \\ \sin \theta(t) & \cos \theta(t) \end{pmatrix} \begin{pmatrix} -\dot{X}_0 \\ -\dot{Z}_0 \end{pmatrix} \tag{13.109}$$

c. Discretization and Grid Generation

At this phase the thin-airfoil camberline (Fig. 13.26) is divided into N subpanels, which may be equal in length. The N vortex points (x_j, z_j) will be placed at the quarter chord of each planar panel and the zero normal flow boundary condition can be fulfilled on the camberline at the three-quarter point of each panel. These N collocation points (x_i, z_i) and the corresponding N normal vectors $\mathbf{n}_i$ along with the vortex points can be computed numerically or supplied as an input file. The normal $\mathbf{n}_i$ pointing outward at each of these points is found in the x, z frame from the surface shape $\eta(x)$, as shown in Fig. 13.27:

$$\mathbf{n}_i = \frac{(-d\eta/dx, 1)}{\sqrt{(d\eta/dx)^2 + 1}} = (\sin \alpha_i, \cos \alpha_i) \tag{13.110}$$

where the angle α_i is shown in Fig. 13.26 for panel number 4. Similarly, the tangential vector τ_i is

$$\tau_i = (\cos \alpha_i, -\sin \alpha_i) \tag{13.111}$$

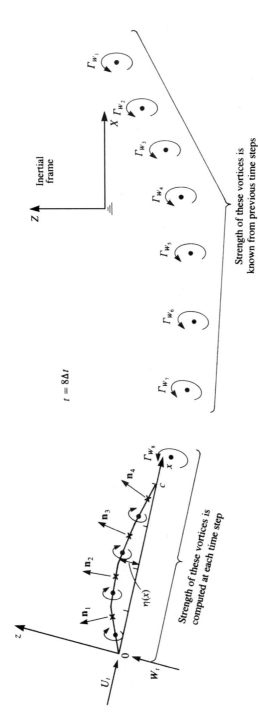

Figure 13.27 Discrete-vortex model for the unsteady thin airfoil problem after eight time steps.

Since the lumped-vortex element inherently fulfills the Kutta condition for each panel, no additional specification of this condition is required.

If the airfoil's geometry does not change with time (e.g., in the case of a flexible airfoil), then the calculation of the vortex points and collocation points can be done before the beginning of the time loop, as shown in Fig. 13.25 (where the box "definition of geometry" is placed outside of the time-stepping loop).

d. *Influence Coefficients*

At this point, the zero normal flow across the solid surface boundary condition is implemented. To specify this condition, the kinematic conditions need to be known and the time-stepping loop (shown in Fig. 13.25) is initiated. Let us select I_t as the time step counter, so that the momentary time is

$$t = I_t \cdot \Delta t$$

For simplicity, we assume that at $t = 0$ the two coordinate systems x, z and X, Z coincided and the airfoil was at rest (hence there was no wake). Consequently, the calculation begins at $t = \Delta t$ (Fig. 13.26) and the wake at this moment consists of a single vortex Γ_{W_1}, which is placed along the path of the trailing edge (see also Fig. 13.16). The location of the trailing edge at $t = 0$ and at $t = \Delta t$ is obtained by using the transformations of Eq. (13.107). The wake vortex is then placed usually at 0.2–0.3 of the distance covered by the trailing edge during the latest time step (see the discussion on this subject in the beginning of Section 13.8.2)

In general, the normal velocity component at each point on the camberline is a combination of the self-induced velocity, the kinematic velocity, and the wake-induced velocity. The self-induced part can be represented by a combination of influence coefficients, as in the steady-state flow case. If the shape of the airfoil $\eta(x)$ remains constant with time then these coefficients will be evaluated only once. The normal velocity component due to the motion of the wing is known from the kinematic equations (Eq. (13.13a)) and will be transferred to the right-hand side (RHS) of the equation. The velocity induced by the most recent wake vortex is unknown and will be resolved by adding an additional equation (the Kelvin condition). The strength of the other wake vortices is known from the previous time steps (for the general case when $I_t > 1$, but since at $t = \Delta t$ only one wake vortex is present, the rest of the wake contribution for the first time step is zero) and their effect on the normal velocity will be transferred to the right-hand side, as well.

To formulate the momentary boundary condition, let us use Eq. (13.13a), and for simplicity we allow only one component of the relative velocity $\mathbf{v}_{\text{rel}} = (0, \partial \eta / \partial t)$ limited to small amplitudes, within the coordinate system x, z. Also, it is convenient to divide the perturbation potential into an airfoil potential Φ_B and a wake potential Φ_W and both parts of the velocity potential will be modeled by discrete-vortex elements of circulation Γ. Consequently, the boundary condition of Eq. (13.13a) becomes

$$(\nabla \Phi_B + \nabla \Phi_W - \mathbf{V}_0 - \mathbf{v}_{\text{rel}} - \mathbf{\Omega} \times \mathbf{r}) \cdot \mathbf{n} = 0 \qquad (13.112)$$

To establish the self-induced portion of the normal velocity ($\nabla \Phi_B \cdot \mathbf{n}$ in Eq. (13.112)), at each collocation point, consider the velocity induced by the airfoil's Γ_jth element at the first collocation point (in order to get the influence due to a unit strength Γ_j assume $\Gamma_j = 1$):

$$(u, w)_{1j} = \text{VOR2D}(\Gamma_j = 1, x_1, z_1, x_j, z_j) \qquad (13.113)$$

The influence coefficient a_{ij} is defined as the velocity component induced by the airfoil's unit strength Γ_j element, normal to the surface (at collocation point i). Consequently, the contribution of a unit strength singularity element j, at collocation point 1, is

$$a_{1j} = (u, w)_{1j} \cdot \mathbf{n}_1 \tag{13.114}$$

The induced normal velocity component q_{n1}, at collocation point 1, due to all N vortex elements and the latest wake vortex is therefore

$$q_{n1} = a_{11}\Gamma_1 + a_{12}\Gamma_2 + a_{13}\Gamma_3 + \cdots + a_{1N}\Gamma_N + a_{1W}\Gamma_{W_t}$$

Note that the strengths of the airfoil vortices Γ_j and of the latest wake vortex Γ_{W_t} are unknown at this point (see also Fig. 13.27 where $\Gamma_{W_t} = \Gamma_{W_8}$).

For the boundary condition on the surface to be fulfilled requires that at each collocation point the normal velocity component will vanish (Eq. (13.112)). Specifying this condition for the ith collocation point we obtain

$$a_{i1}\Gamma_1 + a_{i2}\Gamma_2 + a_{i3}\Gamma_3 + \cdots + a_{iN}\Gamma_N + a_{iW}\Gamma_{W_t}$$
$$+ [U(t) + u_W, W(t) + w_W]_i \cdot \mathbf{n}_i = 0 \tag{13.115}$$

and here the terms $(-\mathbf{V}_0 - \mathbf{v}_{\text{rel}} - \mathbf{\Omega} \times \mathbf{r})$ were replaced by an equivalent tangential and normal velocity $[U(t), W(t)]_i$ representing the kinematic velocity due to the motion of the airfoil, and $(u_W, w_W)_i$ are the velocity components induced by the the wake vortices (except the latest wake vortex – shown in Fig. 13.27). The wake influence can be calculated using Eq. (13.102) since the location of all wake vortex points is known. The time-dependent kinematic velocity components $U(t), W(t)$ (see also Eq. (13.49)) are calculated with the help of Eq. (13.109):

$$\begin{pmatrix} U(t) \\ W(t) \end{pmatrix} = \begin{pmatrix} \cos\theta(t) & -\sin\theta(t) \\ \sin\theta(t) & \cos\theta(t) \end{pmatrix} \begin{pmatrix} -\dot{X}_0 \\ -\dot{Z}_0 \end{pmatrix} + \begin{pmatrix} -\dot{\theta}\eta \\ \dot{\theta}x - \frac{\partial \eta}{\partial t} \end{pmatrix} \tag{13.116}$$

Since these terms are known at each time step, they can be transferred to the right-hand side of the equation. Consequently, the right-hand side (RHS) is defined as

$$\text{RHS}_i = -[U(t) + u_W, W(t) + w_W]_i \cdot \mathbf{n}_i \tag{13.117}$$

When we specify the boundary condition for each of the collocation points we obtain the following set of algebraic equations:

$$\begin{pmatrix} a_{11} & a_{12} & \cdots & a_{1N} & a_{1W} \\ a_{21} & a_{22} & \cdots & a_{2N} & a_{2W} \\ \vdots & \vdots & \ddots & \vdots & \vdots \\ a_{N1} & a_{N2} & \cdots & a_{NN} & a_{NW} \\ 1 & 1 & \cdots & 1 & 1 \end{pmatrix} \begin{pmatrix} \Gamma_1 \\ \Gamma_2 \\ \vdots \\ \Gamma_N \\ \Gamma_{W_t} \end{pmatrix} = \begin{pmatrix} \text{RHS}_1 \\ \text{RHS}_2 \\ \vdots \\ \text{RHS}_N \\ \Gamma(t - \Delta t) \end{pmatrix} \tag{13.118}$$

Note that for the lumped-vortex element the Kutta condition is not stated explicitly. The last equation represents the Kelvin condition:

$$\Gamma(t) - \Gamma(t - \Delta t) + \Gamma_{W_t} = 0 \tag{13.119}$$

and the instantaneous airfoil circulation is the sum of all the airfoil's vortices:

$$\Gamma(t) = \sum_{j=1}^{N} \Gamma_j \tag{13.120}$$

and $\Gamma(t - \Delta t)$ is the circulation measured at the previous time step. This influence coefficient calculation procedure can be accomplished by using two DO loops, where the outer loop scans the collocation points (x_i, z_i) and the inner scans the vortices Γ_j. If the airfoil chord geometry does not vary with time the influence coefficient calculation needs to be carried out only once at the beginning of the computations.

e. *Establish RHS Vector*

The right-hand-side vector, which is the normal component of the kinematic velocity and the wake-induced velocity, can be computed within the outer loop of the previously described DO loops by using Eq. (13.117).

f. *Solve Linear Set of Equations*

The results of the previous calculations (shown by Eq. (13.118)) can be summarized in indicial form (for each collocation point i) as

$$\sum_{j=1}^{N+1} a_{ij} \Gamma_j = \text{RHS}_i \tag{13.121}$$

Again, if the shape of the airfoil remains unchanged then the matrix inversion occurs only once. For time steps larger then 1 the calculation is reduced to

$$\Gamma_j = \sum_{i=1}^{N+1} a_{ij}^{-1} \text{RHS}_i \tag{13.122}$$

where a_{ij}^{-1} are the coefficients of the inverted matrix.

g. *Computation of Velocity Components, Pressures, and Loads*

The resulting pressures and loads can be computed by using the Bernoulli equation (Eq. (13.24)) near the panel surface:

$$\frac{p_{\text{ref}} - p}{\rho} = \frac{Q^2}{2} - \frac{v_{\text{ref}}^2}{2} + \frac{\partial \Phi}{\partial t}$$

The pressure difference between the camberline upper and lower surfaces is then

$$\Delta p = p_l - p_u = \rho \left[\left(\frac{Q_t^2}{2} \right)_u - \left(\frac{Q_t^2}{2} \right)_l + \left(\frac{\partial \Phi}{\partial t} \right)_u - \left(\frac{\partial \Phi}{\partial t} \right)_l \right] \tag{13.123}$$

and the tangential velocity Q_t is found from

$$Q_{t_j} = [U(t) + u_W, W(t) + w_W]_j \cdot \boldsymbol{\tau}_j \pm \frac{\partial \Phi}{\partial \tau_j} \tag{13.124}$$

where the $\pm$ sign stands for above and below the surface, respectively. The tangential derivative of the thin airfoil potential can be approximated as

$$\pm \frac{\partial \Phi}{\partial \tau_j} = \pm \frac{\gamma}{2} \approx \pm \frac{\Gamma_j}{2 \Delta l_j} \tag{13.125}$$

and here Δl_j is the jth panel length. (Note that here $\partial/\partial \tau_j$ is used for tangential derivative and $\partial/\partial t$ for a derivative with respect to time.)

The velocity-potential time derivative, obtained using the definition $\Phi^\pm = \pm \int_0^x (\gamma/2)\, dl$, is

$$\pm \frac{\partial \Phi_j}{\partial t} = \pm \frac{\partial}{\partial t} \sum_{k=1}^{j} \frac{\Gamma_k}{2} \tag{13.126}$$

13.10 Algorithm for Unsteady Airfoil Using the Lumped-Vortex Element

so that the local potential is the sum of the vortices from the leading edge to the jth vortex point. After substituting these terms into Eq. (13.123), we find that the pressure difference between the airfoil's upper and lower surfaces becomes

$$\Delta p_j = \rho \left[(U(t) + u_W, W(t) + w_W)_j \cdot \boldsymbol{\tau}_j \frac{\Gamma_j}{\Delta l_j} + \frac{\partial}{\partial t} \sum_{k=1}^{j} \Gamma_k \right] \qquad (13.127)$$

For example, in the case of a translatory motion parallel to the x axis, as in Eq. (13.105), and for a flat thin airfoil placed on the x axis the normal and tangential vectors become

$$\mathbf{n}_i = (0, 1), \qquad \boldsymbol{\tau}_j = (1, 0)$$

The upper and lower tangential velocity components are then

$$Q_{t_j} = U_\infty \pm \frac{\Gamma_j}{2\Delta l_j}$$

and the pressure difference becomes

$$\Delta p_j = \rho \left(U_\infty \frac{\Gamma_j}{\Delta l_j} + \frac{\partial}{\partial t} \sum_{k=1}^{j} \Gamma_k \right)$$

The total lift and moment are obtained by integrating the pressure difference along the chordline:

$$L \equiv F_z = \sum_{j=1}^{N} \Delta p_j \Delta l_j \cos \alpha_j \qquad (13.128)$$

$$M_0 = -\sum_{j=1}^{N} \Delta p_j \cos \alpha_j \Delta l_j x_j \qquad (13.129)$$

The drag of the two-dimensional airfoil during the unsteady motion is due to the induced angle caused by the wake and due to the added mass effect (caused by the relative fluid acceleration; see also discussion leading to Eq. (13.37)):

$$D = \sum_{j=1}^{N} \rho \left(w_{W_j} \Gamma_j + \frac{\partial}{\partial t} \sum_{k=1}^{j} \Gamma_k \Delta l_k \sin \alpha_k \right) \qquad (13.130)$$

Here the first term is due to the wake-induced downwash w_{W_j} which in the lumped-vortex element case is evaluated at the panel's three-quarter chord point. The second term is due to the fluid acceleration (second term in Eq. (13.127)) and its center of pressure is assumed to act at the panel center.

h. Vortex Wake Rollup

Since the vortex wake is force free, each vortex must move with the local stream velocity (Eq. (13.21a)). The local velocity is a result of the velocity components induced by the wake and airfoil and is usually measured in the inertial frame of reference X, Z. To achieve the vortex wake rollup, at each time step the induced velocity $(u, w)_i$ at each vortex wake point is calculated and then the vortex elements are moved by

$$(\Delta x, \Delta z)_i = (u, w)_i \Delta t \qquad (13.131)$$

The velocity induced at each wake vortex point is a combination of the airfoil Γ_j and wake Γ_k vortices and can be obtained by using the same influence routine (Eq. (11.2)):

$$(u, w)_i = \sum_{j=1}^{N} \text{VOR2D}(\Gamma_j, x_{W_i}, z_{W_i}, x_j, z_j)$$
$$+ \sum_{k=1}^{N_W} \text{VOR2D}(\Gamma_{W_k}, x_{W_i}, z_{W_i}, x_{W_k}, z_{W_k}) \quad (13.132)$$

In this simple scheme the velocity components of the current or previous time step (or a combination thereof) were used. However, more refined techniques can be applied here to improve the wake shape near the trailing edge.

Summary

The solution procedure is described schematically by the flowchart of Fig. 13.25. In principle at each time step the motion kinematics is calculated (Eqs. (13.103) and (13.104)), the location of the latest wake vortex is established, and the RHS_i vector is calculated. Then, during the first time step the influence coefficients appearing in Eq. (13.118) are calculated and the matrix equation is solved. At later time steps, the airfoil vortex distribution can be calculated by the momentary RHS_j vector, using Eq. (13.122). Once the vortex distribution is obtained the pressures and loads are calculated, using Eqs. (13.128)–(13.130). To conclude the time step, the wake vortex locations are updated using the velocity induced by the flowfield from Eq. (13.132).

As an example, consider the sudden acceleration of a flat plate (discussed in Section 13.7) placed along the x axis. The angle of attack α can be obtained by rotating the plate frame of reference by $\theta = \alpha$ relative to the direction of $\mathbf{Q}_\infty$. For this case at $t > 0$ the velocity of the origin is $(\dot{X}_0, \dot{Z}_0) = (-Q_\infty, 0)$ and Eq. (13.116) results in the following free-stream components:

$$\begin{pmatrix} U(t) \\ W(t) \end{pmatrix} = Q_\infty \begin{pmatrix} \cos\alpha \\ \sin\alpha \end{pmatrix}$$

Since the normal vector to the flat plate is $\mathbf{n} = (0, 1)$ the right-hand side (downwash) vector of Eq. (13.117) becomes

$$\text{RHS}_i = -(Q_\infty \cos\alpha + u_W, Q_\infty \sin\alpha + w_W) \cdot (0, 1) = -(Q_\infty \sin\alpha + w_W)$$

The wake-induced downwash is obtained by using Eq. (13.113) and then at each moment Eq. (13.118) is solved for the airfoil vortex distribution. The pressure difference is then obtained by using Eq. (13.127). Results of this computation for the case of the sudden acceleration of a flat plate are presented in Figs. 13.8–13.10 along with the results obtained with the one-element lumped-vortex method.

Program No. 15 in Appendix D includes most of the elements of this method except the matrix inversion phase and may be useful in developing a computer program based on this method. The matrix inversion is included, though, in Program No. 16, which is a more complicated three-dimensional model.

13.11 Some Remarks about the Unsteady Kutta Condition

The potential flow examples, as presented in Chapters 3 and 4, indicate that the solution for lifting flows is not unique for a given set of boundary conditions. This difficulty was resolved by requiring that the flow leave smoothly at the trailing edge of two-dimensional

13.11 Some Remarks about the Unsteady Kutta Condition

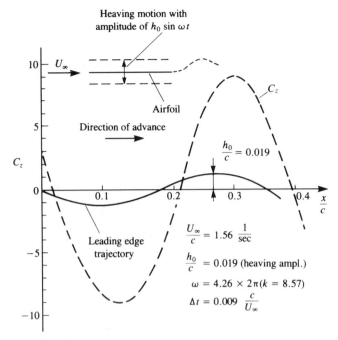

Figure 13.28 Variation of the vertical displacement and normal force coefficient during one heaving cycle. From Katz, J., and Weihs, D., "Behavior of Vortex Wakes from Oscillating Airfoils," *J. Aircraft*, Vol. 15, No. 12, 1978. Reprinted with permission. Copyright AIAA.

airfoils, thereby fixing the amount of circulation generated by the airfoil. The above two-dimensional Kutta condition was almost automatically extended to the three-dimensional steady-state case and in this chapter was used for unsteady flows as well. Although from the mathematical point of view a condition to fix the amount of circulation is required, it is not obvious that this condition is the best candidate. However, prior to arriving at any conclusion in this regard, let us use the method of this section to study the wake rollup behind a thin airfoil undergoing a small-amplitude heaving oscillation.

Consider the small-amplitude heaving oscillation of the flat plate shown in the inset to Fig. 13.28. Assume that the motion of the origin of the x, z coordinates is given by

$$X_0 = -U_\infty t, \qquad \dot{X}_0(t) = -U_\infty$$
$$Z_0 = -h_0 \sin \omega t, \qquad \dot{Z}_0(t) = -h_0 \omega \cos \omega t$$
$$\theta = 0, \qquad \dot{\theta} = 0$$

The time-dependent kinematic velocity components of Eq. (13.116) then become

$$\begin{pmatrix} U(t) \\ W(t) \end{pmatrix} = \begin{pmatrix} U_\infty \\ h_0 \omega \cos \omega t \end{pmatrix}$$

Since the normal vector to the flat plate is $\mathbf{n} = (0, 1)$ the right-hand-side (downwash) vector of Eq. (13.117) becomes

$$\text{RHS}_i = -(h_0 \omega \sin \omega t + w_W)$$

The wake-induced downwash is obtained by using Eq. (13.113) and then at each moment Eq. (13.118) is solved for the airfoil vortex distribution. The pressure difference is then

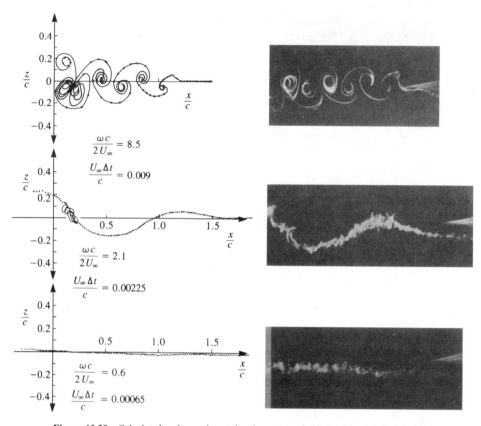

Figure 13.29 Calculated and experimental wake patterns behind a thin airfoil undergoing heaving oscillations at various frequencies. From Katz, J., and Weihs, D., "Behavior of Vortex Wakes from Oscillating Airfoils," *J. Aircraft*, Vol. 15, No. 12, 1978. Reprinted with permission. Copyright AIAA.

obtained by using Eq. (13.127) and the wake rollup is obtained by moving the wake vortices with the local induced velocity (Eq. (13.132)). The rest of the details are as presented in the previous section and results of the wake rollup computation for the flat plate oscillating at various frequencies is shown in Fig. 13.29. The comparison in this figure indicates that up to a high reduced frequency of $\omega c/2Q_\infty = 8.5$ the calculated wake shape is similar to the results of flow visualizations. Since the wake shape is a direct result of the airfoil's circulation history and the calculated wake shape is similar to the experimentally observed shape, it is safe to assume that the calculated airfoil's lift history is similar to the experimental one (which was not measured in this case). As an example, the vertical load C_z on the airfoil during one cycle is presented in Fig. 13.28 next to the motion history (note the phase shift due to the wake influence).

Now that we have generated a good example in favor of the unsteady Kutta condition let us investigate some possible parameters affecting its validity. It is clear that conditions such as very high oscillation frequency, large amplitudes, and large angles of attack will cause some trailing-edge separation. Such local flow separation automatically violates the Kutta condition, but in practice it may not have a noticeable effect on the lift, although it may cause a lag in the aerodynamic loads. Experimental investigations of the unsteady Kutta condition[13.8–13.11] usually indicate that the streamlines do not leave parallel to the trailing edge at reduced frequencies of $\omega c/2Q_\infty > 0.6$, but the lift and pressure distributions are not affected in a visible manner even at higher frequencies. These experiments were based on small-amplitude oscillations of airfoils where the trailing-edge vertical displacement was small.

So, based on the indirect results of Fig. 13.29 and some cited references we can try to establish some guidelines for the boundaries for the validity of the unsteady Kutta condition. First and most important, large angles of attack where trailing-edge separation begins to develop must be avoided. Also, it is clear that as the reduced frequency increases the "allowed" trailing-edge displacement amplitude (e.g., h_0 in the previous example) must be smaller. So, for example, with $h_0 = 0.1c$ and with reduced frequencies of up to $\omega c/2Q_\infty = 1.0$ calculations based on the Kutta condition may provide reasonable load calculations. The vertical kinematic velocity of the trailing edge (e.g., $\dot{h}_0/U_\infty$ in the previous example) is an important parameter, too, and in the case of the highest frequency oscillation in Fig. 13.29 it has a value of about 0.35. Hence, in addition to the previously mentioned limits on the reduced frequency and trailing-edge amplitude, if we limit ourselves to trailing-edge vertical displacement velocity of $\dot{h}_0/U_\infty \ll 1$, then for practical purposes we can assume that the unsteady Kutta condition is valid. Furthermore, we must remember that characteristic airplane maneuvers will fall into a category where the reduced frequency is far less than 1, and therefore the use of the Kutta condition is justified in most cases. However, for a rapidly pitching helicopter rotor in forward flight this may not be the case!

The above discussion was aimed primarily at the lift calculation; however, the lag (due to viscous effects) in the adjustment of the flow at the trailing edge may cause some lags in the aerodynamic loads, and this effect is still not explored sufficiently.

13.12 Unsteady Lifting-Surface Solution by Vortex Ring Elements

The method of transforming steady-state flow based numerical solutions into the time-dependent mode is described schematically in Fig. 13.25 and in the introduction to Section 13.10. In this section the same approach is applied to the three-dimensional thin lifting surface problem. In this example, as in the case of the lifting surface of Section 12.3, the wing's bound circulation and the vortex wake will be modeled by vortex ring elements (see Fig. 13.30). The main advantage of using vortex ring elements is that they require little programming effort (although computational efficiency can be further improved). Also, in this numerical example, the boundary conditions are specified on the actual wing surface, which can have camber and various planform shapes.

The solution is again based on the time-stepping technique, and at the beginning of the motion only the wing-bound vortex rings exist (upper part of Fig. 13.30). Note that the closing segment of the trailing-edge vortex elements in Fig. 13.30 will represent the starting vortex. Consequently, during the first time step there will be no wake panels and if the wing is represented by K unknown vortex rings ($K = 8$ in Fig. 13.30) then by specifying the zero normal flow boundary condition on the K collocation points, a solution at $t = \Delta t$ is possible. With this model, therefore, we do not have to add an additional equation to enforce the Kelvin condition since the vortex ring model inherently fulfills this condition. During the second time step, the wing is moved along its flight path and each trailing-edge vortex panel sheds a wake panel with a vortex strength equal to its circulation in the previous time step (lower part of Fig. 13.30). Also, during this second time step, there will be only one row of wake vortices, but with a known strength. Therefore, the wing K bound vortices can

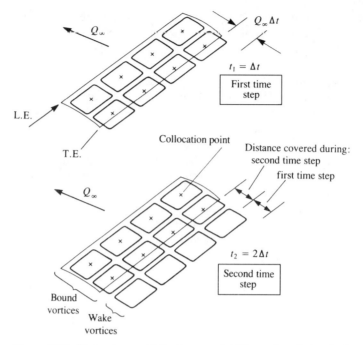

Figure 13.30 Vortex ring model for the unsteady lifting surface during the first time step (upper figure) and during the second time step (lower figure).

be calculated for this time step, too, by specifying the boundary condition on the same K collocation points. This time-stepping methodology can then be continued for any type of flight path and at each time step the vortex wake corner points can be moved by the local velocity, so that wake rollup can be simulated. One of the advantages of this wake model is its simplicity, it being equal in its formulation to the wing's bound vortex rings (or constant-strength doublet panels). Consequently, this wake model can be used for more advanced panel methods (and is actually used by the code PMARC[9.7,9.8,12.13]).

It is recommended that prior to reading this section the reader should be familiar with the steady-state solution presented in Section 12.3. The presentation of the unsteady version of this method then proceeds with the same sequence (steps) of the previous sections. So, in general, the wing is divided into panel elements containing vortex ring singularities as shown in Fig. 13.30 and the solution procedure is as follows:

a. *Choice of Singularity Element*

The method by which the thin-wing planform is divided into panels is similar to what was described in Section 12.3 and shown schematically in Figs. 13.30 and 13.31. The leading segment of the vortex ring is placed on the panel's quarter chord line and the collocation point is at the center of the three-quarter chord line. The normal vector **n** is defined at this point, too, which falls at the center of the vortex ring. A positive Γ is defined here according to the right-hand rotation rule, as shown by the arrows in Fig. 13.31.

From the numerical point of view these vortex ring elements are stored in rectangular patches (arrays) with i, j indexing as shown by Fig. 13.31 (see also Fig. 12.10). The velocity induced at an arbitrary point $P(x, y, z)$, by a typical vortex ring (see Fig. 13.32) at location ij, can be computed by applying the vortex line routine VORTXL (Eq. (10.106)) to the

13.12 Unsteady Lifting-Surface Solution by Vortex Ring Elements

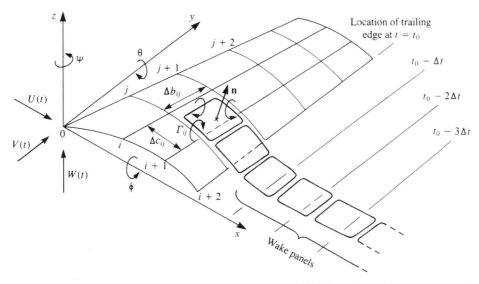

Figure 13.31 Nomenclature for the unsteady motion of a thin lifting surface along a predetermined path. (Note that in this figure, too, the fluid is at rest and the airfoil moves toward the left side of the page.)

ring's four segments:

$$(u_1, v_1, w_1) = \text{VORTXL}(x, y, z, x_{i,j}, y_{i,j}, z_{i,j}, x_{i,j+1}, y_{i,j+1}, z_{i,j+1}, \Gamma_{i,j})$$

$$(u_2, v_2, w_2) = \text{VORTXL}(x, y, z, x_{i,j+1}, y_{i,j+1}, z_{i,j+1},$$
$$x_{i+1,j+1}, y_{i+1,j+1}, z_{i+1,j+1}, \Gamma_{i,j})$$

$$(u_3, v_3, w_3) = \text{VORTXL}(x, y, z, x_{i+1,j+1}, y_{i+1,j+1},$$
$$z_{i+1,j+1}, x_{i+1,j}, y_{i+1,j}, z_{i+1,j}, \Gamma_{i,j})$$

$$(u_4, v_4, w_4) = \text{VORTXL}(x, y, z, x_{i+1,j}, y_{i+1,j}, z_{i+1,j}, x_{i,j}, y_{i,j}, z_{i,j}, \Gamma_{i,j})$$

The velocity induced by the four vortex segments is then

$$(u, v, w) = (u_1, v_1, w_1) + (u_2, v_2, w_2) + (u_3, v_3, w_3) + (u_4, v_4, w_4) \quad (13.133)$$

It is convenient to include these computations in a subroutine (see Eqs. (10.117) and (12.17)).

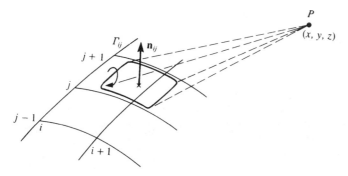

Figure 13.32 Typical vortex ring element and its influence at point P.

such that

$$(u, v, w) = \text{VORING}(x, y, z, i, j, \Gamma) \quad (13.134)$$

Note that in this formulation it is assumed that by specifying the i, j counters, the (x, y, z) coordinates of this panel are automatically identified (see Fig. 12.10).

The use of this subroutine can considerably shorten the programming effort; however, for the vortex segment between two adjacent vortex rings the velocity induced by the vortex segment is computed twice. For the sake of simplicity this routine will be used for this problem, but more advanced programming can easily correct this compromise of computational efficiency.

It is recommended at this point, too, that one calculate the velocity induced by the trailing vortex segments only (the vortex lines parallel to the free stream, as in Fig. 12.5). This information is needed for the induced-drag computations and if done at this point will only slightly increase the computational effort. The influence of the trailing segments is obtained by simply omitting the $(u_1, v_1, w_1) + (u_3, v_3, w_3)$ part from Eq. (12.133):

$$(u, v, w)^* = (u_2, v_2, w_2) + (u_4, v_4, w_4) \quad (13.135)$$

and it is assumed that $(u, v, w)^*$ is automatically obtained as a byproduct of subroutine VORING.

b. Kinematics

Let us establish an inertial frame of reference X, Y, Z, as shown in Fig. 13.1, such that this frame of reference is stationary while the wing is moving to the left of the page. The flight path of the origin and the orientation of the x, y, z system is assumed to be known and is prescribed as

$$X_0 = X_0(t), \quad Y_0 = Y_0(t), \quad Z_0 = Z_0(t) \quad (13.136)$$

$$\phi = \phi(t), \quad \theta = \theta(t), \quad \psi = \psi(t) \quad (13.136a)$$

and the momentary velocity of the origin and its rotations about the axes are

$$\dot{X}_0 = \dot{X}_0(t), \quad \dot{Y}_0 = \dot{Y}_0(t), \quad \dot{Z}_0 = \dot{Z}_0(t) \quad (13.137)$$

$$p = p(t), \quad q = q(t), \quad r = r(t) \quad (13.137a)$$

For example, if the wing moves to the left (parallel to the X axis) at a constant speed of U_∞ and has a constant sideslip (parallel to the Y axis) of V_∞, then

$$X_0 = -U_\infty t, \quad Y_0 = -V_\infty t, \quad Z_0 = 0 \quad (13.138)$$

$$\phi = \theta = \psi = 0 \quad (13.138a)$$

and

$$\dot{X}_0 = -U_\infty, \quad \dot{Y}_0 = -V_\infty, \quad \dot{Z}_0 = 0 \quad (13.139)$$

$$p = q = r = 0 \quad (13.139a)$$

It is useful to establish a transformation between the two coordinate systems (Eq. (13.7a)) that depends on the translation of the origin and the orientation of the body frame of reference:

$$\begin{pmatrix} x \\ y \\ z \end{pmatrix} = f(\phi, \theta, \psi) \begin{pmatrix} X - X_0 \\ Y - Y_0 \\ Z - Z_0 \end{pmatrix} \quad (13.140)$$

This transformation, without the translation part, can be used also for the velocity transformation. To define such a three-dimensional transformation uniquely, the order of rotation must be specified. For example, if we adopt the order of rotation such that first we rotate about the z axis, then about the y axis, and finally about the x axis then the transformation becomes:

a. rotation by ψ (sideslip):

$$\begin{pmatrix} U1 \\ V1 \\ W1 \end{pmatrix} = \begin{pmatrix} \cos\psi(t) & \sin\psi(t) & 0 \\ -\sin\psi(t) & \cos\psi(t) & 0 \\ 0 & 0 & 1 \end{pmatrix} \begin{pmatrix} -\dot{X}_0 \\ -\dot{Y}_0 \\ -\dot{Z}_0 \end{pmatrix}$$

b. rotation by θ (angle of attack):

$$\begin{pmatrix} U2 \\ V2 \\ W2 \end{pmatrix} = \begin{pmatrix} \cos\theta(t) & 0 & -\sin\theta(t) \\ 0 & 1 & 0 \\ \sin\theta(t) & 0 & \cos\theta(t) \end{pmatrix} \begin{pmatrix} U1 \\ V1 \\ W1 \end{pmatrix}$$

c. rotation by ϕ (roll angle):

$$\begin{pmatrix} U3 \\ V3 \\ W3 \end{pmatrix} = \begin{pmatrix} 1 & 0 & 0 \\ 0 & \cos\phi(t) & \sin\phi(t) \\ 0 & -\sin\phi(t) & \cos\phi(t) \end{pmatrix} \begin{pmatrix} U2 \\ V2 \\ W2 \end{pmatrix} \quad (13.140a)$$

where $U3, V3, W3$ are the velocity components observed in the x, y, z frame due to the translation of the origin.

The time-dependent kinematic velocity components $U(t), V(t), W(t)$, in the x, y, z frame, are then a combination of the translation velocity and the rotation of the body frame of reference:

$$\begin{pmatrix} U(t) \\ V(t) \\ W(t) \end{pmatrix} = \begin{pmatrix} U3 \\ V3 \\ W3 \end{pmatrix} + \begin{pmatrix} -qz + ry \\ -rx + pz \\ -py + qx - \frac{\partial \eta}{\partial t} \end{pmatrix} \quad (13.141)$$

Since the instantaneous rotation and translation rates are known, these kinematic terms are known, too, at each time step.

c. Discretization and Grid Generation

The method by which the thin wing planform is divided into elements is the same as presented in Section 12.3 and is shown in Figs. 13.30 and 13.31. Some typical panel elements are also shown in Figs. 12.8–12.10. Also, if only the wing's semispan is modeled then the mirror image method must be used to account for the other semispan. The leading segment of the vortex ring is placed on the panel's quarter chord line and the collocation point is at the center of the three-quarter chord line (see Fig. 13.31). The lifting surface shape is usually given by $z = \eta(x, y)$ and is divided into N spanwise and M chordwise panels. Using a procedure such as shown in Fig. 12.13 allows the scanning and calculation of geometrical information such as the panel area S_{ij}, normal vector $\mathbf{n}_{ij}$, and the coordinates of the collocation points. A simple and fairly general method for evaluating the normal vector is shown in Fig. 12.11. The panel opposite corner points define two vectors $\mathbf{A}$ and

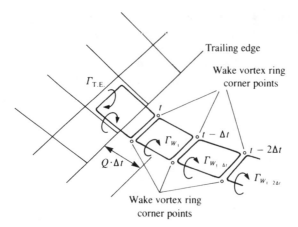

Figure 13.33 Nomenclature for the wake shedding procedure at a typical trailing-edge panel.

B, and their vector product will point in the direction of **n**:

$$\mathbf{n} = \frac{\mathbf{A} \times \mathbf{B}}{|\mathbf{A} \times \mathbf{B}|}$$

A positive Γ for a vortex ring is defined here using the right-hand rotation convention, as shown in Fig. 13.31. For increased surface curvature the above described vortex rings will not be placed exactly on the lifting surface and the normal vectors may be offset somewhat, and a finer grid needs to be used, or the wing surface can be redefined accordingly. By placing the leading segment of the vortex ring at the quarter chord line of the panel the two-dimensional Kutta condition is satisfied along the chord (recall the lumped-vortex element).

At this point the wake shedding procedure must be addressed. Consider a typical trailing-edge vortex ring placed on the last panel row (as shown in Fig. 13.33). The trailing segment (parallel to the trailing edge) is placed in the interval covered by the trailing edge during the latest time step (of length $Q \cdot \Delta t$). Usually it must be placed closer to the trailing edge within 0.2–0.3 of the above distance (see discussion about this topic at the beginning of Section 13.8.2). The wake vortex ring corner points must be created at each time step, such that at the first time step only the two aft points of the vortex ring are created. Therefore, during the first time step there are no free wake elements and the trailing vortex segment of the trailing-edge vortex ring represents the starting vortex. During the second time step the wing trailing edge has advanced and a wake vortex ring can be created using the new aft points of the trailing-edge vortex ring and the two points where these points were during the previous time step (see Fig. 13.30). This shedding procedure is repeated at each time step and a set of new trailing-edge wake vortex rings is created (wake shedding procedure).

The strength of the most recently shed wake vortex ring (Γ_{W_t} in Fig. 13.33) is set equal to the strength of the shedding vortex $\Gamma_{T.E. \cdot t-\Delta t}$ (placed at the trailing edge) in the previous time step (as if it was shed from the trailing edge and left to flow with the local velocity):

$$\Gamma_{W_t} = \Gamma_{T.E. \cdot t-\Delta t} \qquad (13.142)$$

Once the wake vortex is shed, its strength is unchanged (recall the Helmholtz theorems in Section 2.9), and the wake vortex carries no aerodynamic loads (and therefore moves with the local velocity).

This is the unsteady equivalent of the Kutta condition. For the steady-state flow conditions, all wake panels shed from a particular trailing-edge panel will have the same vortex strength, which is equal to the strength of the shedding panel. Thus, the spanwise oriented vortex lines of the adjacent vortex rings will cancel each other and only a horseshoe-like vortex will remain.

d. *Influence Coefficients*

At this point, the zero normal flow across the solid surface boundary condition is implemented. To specify this condition, the kinematic conditions must be known and the time-stepping loop (shown in Fig. 13.25) is initiated. Let us select again I_t as the time-step counter, so that the momentary time is

$$t = I_t \cdot \Delta t$$

Let us assume that at $t = 0$ the two coordinate systems x, y, z and X, Y, Z coincided and the wing was at rest. The calculation is initiated at $t = \Delta t$ and the wake at this moment consists of the vortex line created by the trailing segments of the trailing-edge vortex rings (Fig. 13.30). The location of the trailing edge at $t = 0$ and at $t = \Delta t$, needed for specifying the wake panels' corner points, is obtained by using coordinate transformations such as Eq. (13.140).

The normal velocity component at each point on the camberline is a combination of the self-induced velocity, the kinematic velocity, and the wake-induced velocity. The self-induced part can be represented by a combination of influence coefficients, as in the steady-state flow case. If the shape of the wing $\eta(x, y)$ remains constant with time then these coefficients will be evaluated only once. The normal velocity component due to the motion of the wing is known from the kinematic equations (Eqs. (13.13a) or (13.141)) and will be transferred to the right-hand side (RHS_K) of the equation. The strength of the other wake vortices is known from the previous time steps and the wake-induced normal velocity on each panel will be transferred to the right-hand side, too.

Let us establish a collocation point scanning procedure (similar to that of Section 12.3) that takes the first chordwise row where $i = 1$ and scans spanwise with $j = 1 \to N$ and so on (recall that we have M chordwise panels – see Fig. 12.10). This procedure can be described by two DO loops shown in Fig. 12.13. As the panel scanning begins, a sequential counter assigns a value K to each panel (the sequence of K is shown in Fig. 12.14); K will have values from 1 to $M \times N$.

Once the collocation point scanning has started, $K = 1$ (which is point ($i = 1, j = 1$) on Fig. 12.12. The velocity induced by the first vortex ring is then

$$(u, v, w)_{11} = \text{VORING}(x, y, z, i = 1, j = 1, \Gamma_1 = 1.0)$$

Note that a unit strength vortex is used for evaluating the influence coefficient a_{11}, which is

$$a_{11} = (u, v, w)_{11} \cdot \mathbf{n}_1 \tag{13.143}$$

To scan all the vortex rings influencing this point, an inner scanning loop is needed with the counter $L = 1 \to N \times M$ (see Fig. 12.13). Thus, at this point, the K counter is at point 1, and the L counter will scan all the vortex rings on the wing surface, and all the influence coefficients a_{1L} are computed (also, in Eq. (13.143) the $()_{11}$ index means $K = 1, L = 1$):

$$a_{1L} = (u, v, w)_{1L} \cdot \mathbf{n}_1$$

and for the K, Lth panel

$$a_{KL} = (u, v, w)_{KL} \cdot \mathbf{n}_K \tag{13.143a}$$

As mentioned before, parallel to the computation of the a_{KL} coefficients, the normal velocity component induced by the streamwise segments of the wing vortex rings can also be computed. These b_{KL} coefficients, which will be used for the induced-drag calculation, are calculated by using Eq. (13.135):

$$b_{KL} = (u, v, w)^*_{KL} \cdot \mathbf{n}_K \tag{13.144}$$

and it is assumed that these coefficients are a byproduct of the a_{KL} calculations and do not require additional computational effort.

This procedure continues until all the collocation points have been scanned. A FORTRAN example for this influence coefficient calculation is presented in Fig. 12.13 of Chapter 12.

e. *Establish RHS Vector*

Specifying the zero normal velocity boundary condition on the surface ($Q_{n_K} = 0$) on an arbitrary collocation point K we obtain

$$Q_{n_K} = a_{K1}\Gamma_1 + a_{K2}\Gamma_2 + a_{K3}\Gamma_3 + \cdots + a_{Km}\Gamma_m$$
$$+ [U(t) + u_W, V(t) + v_W, W(t) + w_W]_K \cdot \mathbf{n}_K = 0$$

where $[U(t), V(t), W(t)]_K$ are the time-dependent kinematic velocity components due to the motion of the wing (Eq. (13.141)), $(u_W, v_W, w_W)_K$ are the velocity components induced by the wake vortices, and $m = M \times N$. The wake influence can be calculated using Eq. (13.134) since the location of all vortex points is known (including the wake vortex points). Since these terms are known at each time step, they can be transferred to the right-hand side of the equation. Consequently, the right-hand side is defined as

$$\text{RHS}_K = -[U(t) + u_W, V(t) + v_W, W(t) + w_W]_K \cdot \mathbf{n}_K \tag{13.145}$$

f. *Solve Linear Set of Equations*

Once the computations of the influence coefficients and the right-hand side vector are completed, the zero normal flow boundary condition on all the wing's collocation points will result in the following set of algebraic equations:

$$\begin{pmatrix} a_{11} & a_{12} & \cdots & a_{1m} \\ a_{21} & a_{22} & \cdots & a_{2m} \\ a_{31} & a_{32} & \cdots & a_{3m} \\ \vdots & \vdots & \ddots & \vdots \\ a_{m1} & a_{m2} & \cdots & a_{mm} \end{pmatrix} \begin{pmatrix} \Gamma_1 \\ \Gamma_2 \\ \Gamma_3 \\ \vdots \\ \Gamma_m \end{pmatrix} = \begin{pmatrix} \text{RHS}_1 \\ \text{RHS}_2 \\ \text{RHS}_3 \\ \vdots \\ \text{RHS}_m \end{pmatrix}$$

(Recall that K is the vertical and L is the horizontal matrix counter and the order of this matrix is $m = M \times N$.)

The results of this matrix equation can be summarized in indicial form (for each collocation point K) as

$$\sum_{L=1}^{m} a_{KL}\Gamma_L = \text{RHS}_K \tag{13.146}$$

If the shape of the wing remains unchanged then the matrix inversion occurs only once. For

13.12 Unsteady Lifting-Surface Solution by Vortex Ring Elements

time steps larger then 1 the calculation is reduced to

$$\Gamma_K = \sum_{L=1}^{m} a_{KL}^{-1} \text{RHS}_L \tag{13.147}$$

where a_{KL}^{-1} are the coefficients of the inverted matrix.

g. *Computation of Velocity Components, Pressures, and Loads*

For the pressure distribution calculations the local circulation is needed. For the leading-edge panel this is equal to Γ_{ij} but for all the elements behind it the circulation is equal to the difference $\Gamma_{ij} - \Gamma_{i-1,j}$. The fluid dynamic loads then can be computed by using the Bernoulli equation (Eq. (13.24)) and the pressure difference is given by Eq. (13.123):

$$\Delta p = p_l - p_u = \rho \left[\left(\frac{Q_t^2}{2} \right)_u - \left(\frac{Q_t^2}{2} \right)_l + \left(\frac{\partial \Phi}{\partial t} \right)_u - \left(\frac{\partial \Phi}{\partial t} \right)_l \right]$$

The tangential velocity due to the wing vortices will have two components on the wing, and it can be approximated by the two directions i, j on the surface as

$$\pm \frac{\partial \Phi}{\partial \tau_i} = \pm \frac{\gamma}{2} \approx \pm \frac{\Gamma_{i,j} - \Gamma_{i-1,j}}{2 \Delta c_{ij}} \tag{13.148a}$$

$$\pm \frac{\partial \Phi}{\partial \tau_j} \approx \pm \frac{\Gamma_{i,j} - \Gamma_{i,j-1}}{2 \Delta b_{ij}} \tag{13.148b}$$

where $\pm$ represents the upper and lower surfaces, respectively, and Δc_{ij} and Δb_{ij} are the panel lengths in the ith and jth directions, respectively. Similarly, τ_i and τ_j are the panel tangential vectors in the i and j directions (of course, these vectors are different for each panel and the ij subscript from $\tau_{i_{ij}}$ is dropped for the sake of simplicity).

The velocity-potential time derivative is obtained by using the definition $\Phi^\pm = \pm \int_0^x (\gamma/2) \, dl$ and by integrating from the leading edge. Since for this vortex ring model $\Delta \Phi = \Gamma$ then

$$\pm \frac{\partial \Phi_{ij}}{\partial t} = \pm \frac{\partial}{\partial t} \frac{\Gamma_{ij}}{2} \tag{13.149}$$

Substitution of these terms into the pressure difference equation results in

$$\Delta p_{ij} = \rho \left\{ [U(t) + u_W, V(t) + v_W, W(t) + w_W]_{ij} \cdot \tau_i \frac{\Gamma_{i,j} - \Gamma_{i-1,j}}{\Delta c_{ij}} \right.$$
$$\left. + [U(t) + u_W, V(t) + v_W, W(t) + w_W]_{ij} \cdot \tau_j \frac{\Gamma_{i,j} - \Gamma_{i,j-1}}{\Delta b_{ij}} + \frac{\partial}{\partial t} \Gamma_{ij} \right\} \tag{13.150}$$

The contribution of this panel to the loads, resolved along the three body axes, is then

$$\Delta \mathbf{F} = -(\Delta p \Delta S)_{ij} \mathbf{n}_{ij} \tag{13.151}$$

The total forces and moments are then obtained by adding the contribution of the individual panels.

The total force obtained by this pressure difference integration will have some of the thin lifting surface problems since it does not account for the leading-edge suction force. In general, the lifting properties of the wing will be estimated adequately by this method but the induced drag will be overestimated. Also, in the case of an arbitrary motion, the definition of lift and drag is more difficult and even the definition of a reference velocity (e.g., free-stream velocity) is not always simple. For example, presenting the pressure coefficient data on a helicopter blade in forward flight can be based on the local blade velocity or on the helicopter flight speed. So for the simplicity of this discussion on the induced drag, we shall limit the motion of the lifting surface such that it moves forward along a straight line without side slip (but the forward speed may vary). The induced drag is then the force component parallel to the flight direction, and each panel contribution is

$$\Delta D_{ij} = \rho \left[(w_{\text{ind}} + w_W)_{ij} (\Gamma_{ij} - \Gamma_{i-1,j}) \Delta b_{ij} + \frac{\partial}{\partial t} \Gamma_{ij} \Delta S_{ij} \sin \alpha_{ij} \right] \quad (13.152)$$

and if the panel is at the leading edge then

$$\Delta D_{ij} = \rho \left[(w_{\text{ind}} + w_W)_{ij} \Gamma_{ij} \Delta b_{ij} + \frac{\partial}{\partial t} \Gamma_{ij} \Delta S_{ij} \sin \alpha_{ij} \right] \quad (13.152a)$$

where α_{ij} is the panel's angle of attack relative to the free-stream direction. The first term here is due to the downwash induced by the wing's streamwise vortex lines w_{ind} and due to the wake w_W; the second term is due to the fluid acceleration. The induced downwash w_{ind} at each collocation point i, j is computed by summing the velocity induced by all the trailing segments of the wing-bound vortices. This can be done during the phase of the influence coefficient computation (Eq. (13.144)) by using the VORING routine with the influence of the spanwise vortex segments turned off. This procedure can be summarized by the following matrix formulations where all the b_{KL} and the Γ_K are known:

$$\begin{pmatrix} w_{\text{ind}-1} \\ w_{\text{ind}-2} \\ w_{\text{ind}-3} \\ \vdots \\ w_{\text{ind}-m} \end{pmatrix} = \begin{pmatrix} b_{11} & b_{12} & \cdots & b_{1m} \\ b_{21} & b_{22} & \cdots & b_{2m} \\ b_{31} & b_{32} & \cdots & b_{3m} \\ \vdots & \vdots & \ddots & \vdots \\ b_{m1} & b_{m2} & \cdots & b_{mm} \end{pmatrix} \begin{pmatrix} \Gamma_1 \\ \Gamma_2 \\ \Gamma_3 \\ \vdots \\ \Gamma_m \end{pmatrix}$$

where again $m = N \times M$.

The main difficulty in the induced-drag calculation for a general motion lies in the identification of the force component that will be designated as drag. Once this problem is resolved then the above method can be extended to more complex wing motions (and then angles such as α_{ij} in Eq. (13.152) must be defined).

h. *Vortex Wake Rollup*

Since the vortex wake is force free, each vortex must move with the local stream velocity (Eq. (13.21a)). The local velocity is a result of the velocity components induced by the wake and wing and is usually measured in the inertial frame of reference X, Y, Z, at each vortex ring corner point. To achieve the vortex wake rollup, at each time step the induced velocity $(u, v, w)_\ell$ is calculated and then the vortex elements are moved by

$$(\Delta x, \Delta y, \Delta z)_\ell = (u, v, w)_\ell \Delta t \quad (13.153)$$

13.12 Unsteady Lifting-Surface Solution by Vortex Ring Elements

The velocity induced at each wake vortex point is a combination of the wing and wake influence and can be obtained by using the same influence routine (Eq. (13.134)):

$$(u, v, w)_\ell = \sum_{K=1}^{m} \text{VORING}(x_\ell, y_\ell, z_\ell, i, j, \Gamma_K)$$
$$+ \sum_{k=1}^{N_W} \text{VORING}(x_\ell, y_\ell, z_\ell, i_W, j_W, \Gamma_{Wk}) \qquad (13.154)$$

and there are m wing panels and N_W wake panels.

In the case of a strong wake rollup the size of the wake vortex ring can increase (or be stretched) and if a vortex line segment length increases its strength must be reduced (from the angular momentum point of view). For the methods presented in this section it is assumed that this stretching is small and therefore is not accounted for.

Summary

The solution procedure is described schematically by the flowchart of Fig. 13.25. In principle at each time step the motion kinematics is calculated (Eqs. (13.142)), the location of the latest wake vortex ring is established, and the RHS_i vector is calculated. The influence coefficients appearing in Eq. (13.146) are calculated only during the first time step and the matrix is inverted. At later time steps, the wing vortex distribution can be calculated by the momentary RHS_j vector, using Eq. (13.147). Once the vortex distribution is obtained the pressures and loads are calculated, using Eq. (13.151). At the end of each time step, the wake vortex ring corner point locations are updated using the velocity induced by the flowfield.

A student program based on this algorithm is enclosed in Appendix D (Program No. 16).

Example 1: Sudden Acceleration of an Uncambered Rectangular Wing into a Constant-Speed Forward Flight

In this case the coordinate system is selected such that the x coordinate is parallel to the motion and the kinematic velocity components of Eq. (13.141) become $[U(t), 0, 0]$. The angle of attack effect is taken care of by pitching the wing in the body frame of reference and for the planar wing then all the normal vectors will be $\mathbf{n} = (\sin\alpha, 0, \cos\alpha)$. Consequently, the RHS vector of Eq. (13.145) becomes

$$\text{RHS}_K = -\{[U(t) + u_W]\sin\alpha + w_W \cos\alpha\}_K \qquad (13.155)$$

and here the wake influence will change with time. The rest of the time-stepping solution is as described previously in this section. For the numerical investigation the wing is divided into four chordwise and thirteen spanwise equally spaced panels, and the time step is $U_\infty \Delta t/c = 1/16$.

Following the results of Ref. 13.13 we present the transient lift coefficient variation with time for rectangular wings with various aspect ratios in Fig. 13.34. The duration of the first time step actually represents the time of the acceleration during which the $\partial\Phi/\partial t$ term is large. Immediately after the wing has reached its steady-state speed (U_∞) the lift drops because of the influence of the starting vortex and most of the lift is a result of the $\partial\Phi/\partial t$ term (because of the change in the downwash of the starting vortex–see also Fig. 13.8). Also, the initial lift loss and the length of the transient seems to decrease with a reduction in the wing aspect ratio (because of the presence of the trailing vortex wake).

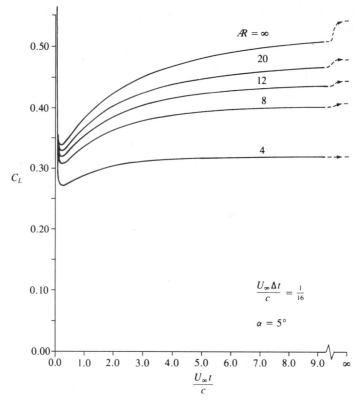

Figure 13.34 Transient lift coefficient variation with time for uncambered, rectangular wings that were suddenly set into a constant-speed forward flight. Calculation is based on four chordwise and thirteen spanwise panels and $U_\infty \Delta t/c = 1/16$.

The transient drag coefficient variation with time for the same rectangular wings is presented in Fig. 13.35. Recall that this is the inviscid (induced) drag and it is zero for the two-dimensional wing ($R = \infty$). Consequently, the larger aspect ratio wings will experience the largest increase in the drag owing to the downwash of the starting vortices. The length of the transient is similar to the results of the previous figure, that is, a smaller aspect ratio wing will reach steady state in a shorter distance (in chord lengths).

The above drag calculation results allow us to investigate the components of Eq. (13.152). For example, Fig. 13.36 depicts the drag C_{D_1} due to induced downwash (first term in Eq. (13.152)) and due to the fluid acceleration term C_{D_2} (which is the second term in Eq. (13.152)) for a rectangular wing with an aspect ratio of 8. At the beginning of the motion, most of the drag is due to the $\partial \Phi/\partial t$ term, but later the steady-state induced-drag portion develops to its full value.

The effect of wing aspect ratio on the nondimensional transient lift of uncambered, rectangular wings that were suddenly set into a constant-speed forward flight is shown in Fig. 13.37. This figure is very useful for validating a new calculation scheme, and the results are sensitive to the spacing of the latest wake vortex from the trailing-edge (actually this is one method to establish the distance of the trailing-edge vortex behind the trailing edge for a given time step). For comparison the results of Wagner[13.3] for the two-dimensional case are presented as well (in his

13.12 Unsteady Lifting-Surface Solution by Vortex Ring Elements

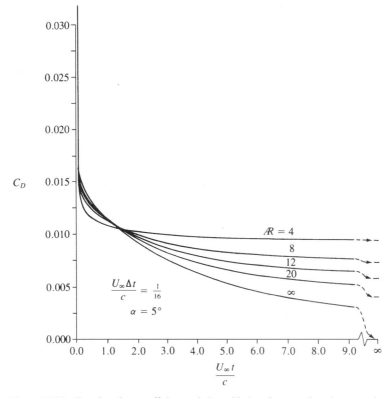

Figure 13.35 Transient drag coefficient variation with time for uncambered, rectangular wings that were suddenly set into a constant-speed forward flight. Calculation is based on four chordwise and thirteen spanwise panels and $U_\infty \Delta t/c = 1/16$.

case the acceleration time is zero and the lift at $t = 0+$ is ∞). It is clear from this figure, too, that both the length of the transient and the loss of initial lift decrease with decreasing wing aspect ratio. The difference between the computed curve and the classical results of Wagner can be attributed to the finite acceleration rate during the first time step. The effect of this finite acceleration is to increase the lift sharply during the acceleration and to increase it moderately later (this effect of finite acceleration is discussed in Ref. 13.14).

Example 2: Heaving Oscillations of a Rectangular Wing

As a final example this method is used to simulate the heaving oscillations of a rectangular wing near the ground. The boundary conditions for this case were established exactly as in the example of Section 13.9.1. The ground effect is obtained by the mirror image method and the results[13.13] for a planar rectangular wing with $R = 4$ are presented in Fig. 13.38. The upper portion shows the effect of frequency on the lift without the presence of the ground, and the loads increase with increased frequency. The lower portion of the figure depicts the loads for the same motion, but with the ground effect. This case was generated to study the loads on the front wing of a race car due to the heaving oscillation of the body, and the data indicate that the ground effect does magnify the amplitude of the aerodynamic loads.

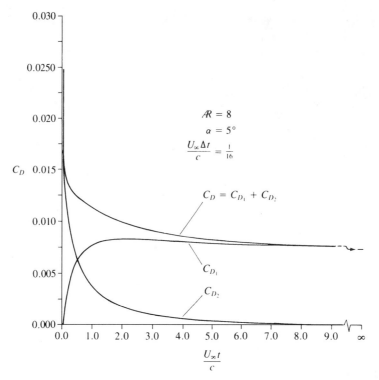

Figure 13.36 Separation of the transient drag coefficient into a part due to induced downwash C_{D_1} and due to fluid acceleration C_{D_2}. Calculation is based on four chordwise and thirteen spanwise panels and $U_\infty \Delta t/c = 1/16$.

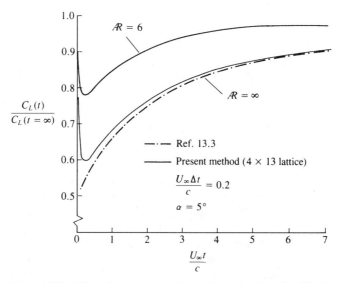

Figure 13.37 Effect of aspect ratio on the nondimensional transient lift of uncambered, rectangular wings that were suddenly set into a constant-speed forward flight. Calculation is based on four chordwise and thirteen spanwise panels.

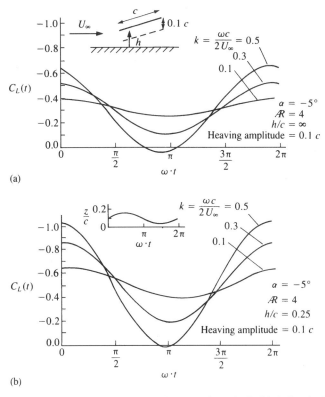

Figure 13.38 Effect of ground proximity on the periodic lift during the heaving oscillation of an aspect ratio = 4 rectangular wing.

13.13 Unsteady Panel Methods

Principles of converting a steady-state, potential flow solution into a time-dependent mode were summarized in Section 13.6 and were demonstrated in the sections that followed. The complexity of these examples in terms of geometry increased gradually and in the previous section the three-dimensional thin lifting surface problem was illustrated. Although this method was capable of estimating the fluid dynamic lift, the calculation of the drag was indirect and inefficient from the computational point of view. Therefore, a similar conversion of a three-dimensional panel method into the unsteady mode can provide, first of all, the capability of treating thick and complex body shapes, and in addition the fluid dynamic loads will be obtained by a direct integration of the pressure coefficients. Since the pressure coefficient is obtained by a local differentiation of the velocity potential (and not by summing the influence of all the panels) this approach yields an improved numerical efficiency. In addition, the drag force is obtained as a component of the pressure coefficient integration and there is no need for a complicated estimation of the leading-edge suction force.

The following example is based on the conversion of a steady-state panel method using constant-strength source and doublet elements[12.12] (described in Section 12.5), which resulted in the time-dependent version[12.14] presented in this section. Familiarity with Sections 12.5 and 13.12 is also advised since some details mentioned in these sections are described here only briefly.

The method of the conversion is described schematically in Fig. 13.25, and the potential flow solution will be included in a time-stepping loop that will start at $t = 0$. During each of the following time steps the strength of the latest wake row is computed by using the Kutta condition, and the previously shed wake vortex strengths will remain unchanged. Thus, at each time step, for N panels, N equations will result with N unknown doublet strengths. If the geometry of the body does not change with time then the matrix is inverted only once. In a case when the body geometry does change (e.g., when a propeller rotates relative to a wing) the influence coefficients and matrix inversion (or portions of it) are recalculated at each time step. The description of the method, based on the eight-step procedure, is then as follows.

a. *Choice of Singularity Element*

The basic panel element used in this method has a constant-strength source and/or doublet, and the surface is also planar (but the doublet panels that are equivalent to a vortex ring can be twisted). Following the formulation of Section 9.4, we can reduce the Dirichlet boundary condition on a thick body (e.g., Eq. (13.18)) to the following form (see Eq. (12.29)):

$$\sum_{k=1}^{N} C_k \mu_k + \sum_{\ell=1}^{N_W} C_\ell \mu_\ell + \sum_{k=1}^{N} B_k \sigma_k = 0 \qquad (13.156)$$

which condition must hold at any moment t. This equation will be evaluated for each collocation point inside the body and the influence coefficients C_k, C_ℓ of the body and wake doublets, respectively, and B_k of the sources are calculated by the formulas of Section 10.4. (In this example only the Dirichlet boundary condition is described but with a similar treatment the Neumann condition can be applied to part or all of the panels.)

b. *Kinematics*

Let us establish an inertial frame of reference X, Y, Z, as shown in Fig. 13.39, such that this frame of reference is stationary while the airplane is moving to the left of the page. The flight path of the origin and the orientation of the x, y, z system is assumed to be known and the boundary condition (Eq. (13.13a)) on the solid surface becomes

$$\frac{\partial \Phi}{\partial n} = (\mathbf{V}_0 + \mathbf{v}_{\text{rel}} + \boldsymbol{\Omega} \times \mathbf{r}) \cdot \mathbf{n} \qquad (13.157)$$

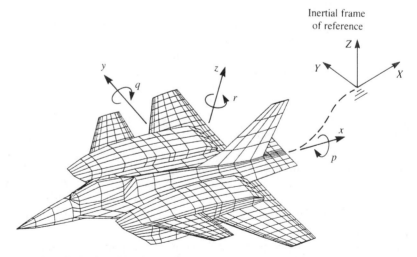

Figure 13.39 Body and inertial coordinate systems used to describe the motion of the body.

13.13 Unsteady Panel Methods

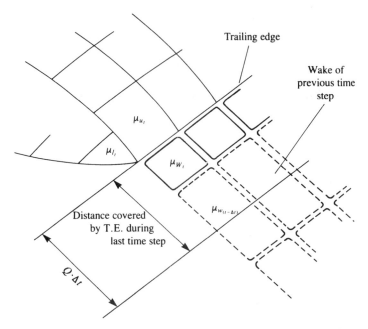

Figure 13.40 Schematic description of a wing's trailing edge and the latest wake row of the unsteady wake.

The kinematic velocity components at each point in the body frame due to the motion $(\mathbf{V}_0 + \mathbf{v}_{rel} + \mathbf{\Omega} \times \mathbf{r})$ are given as $[U(t), V(t), W(t)]$ by Eq. (13.141). If the combined source/doublet method is used (see Section 13.2) then the Dirichlet boundary condition requires that the source strength is given by Eq. (13.19):

$$\sigma = -\mathbf{n} \cdot (\mathbf{V}_0 + \mathbf{v}_{rel} + \mathbf{\Omega} \times \mathbf{r}) \tag{13.19}$$

c. Discretization and Grid Generation

In this phase the geometry of the body is divided into surface panel elements (see for example Fig. 12.22). The panel corner points, collocation points (usually slightly inside the body), and the outward normal vectors $\mathbf{n}_k$ are identified while the counter k for each panel is assigned. A typical example of generating a wing grid and the unfolded patch is shown in Fig. 12.23.

The wake shedding procedure is described schematically by Fig. 13.40. A typical trailing-edge segment is shown with momentary upper μ_u and lower μ_l doublet strengths. The Kutta condition requires that the vorticity at the trailing edge remains zero:

$$\mu_{W_t} = (\mu_u - \mu_l)_t \tag{13.158}$$

Thus, the strength of the latest wake panel μ_{W_t} is directly related to the wing's (or body's) unknown doublets. Note that the spanwise segment (parallel to the trailing edge) of the latest wake panel (which is actually equivalent to a vortex ring) is placed in the interval covered by the trailing edge during the latest time step (of length $Q \cdot \Delta t$). Usually it must be placed closer to the trailing edge, within 0.2–0.3 of the above distance (see discussion about this topic at the beginning of Section 13.8.2). During the second time step the wing trailing edge has advanced and a new wake panel row can be created using the new aft points of the trailing edge. The previous $(t - \Delta t)$-th wake row will remain momentarily in its previous location (as observed in the inertial frame) so that a continuous wake sheet is formed. The wake corner points then will be moved with the local velocity, in the wake rollup calculation phase. Once the wake panel is shed, its strength is unchanged (recall the Helmholtz theorems in Section 2.9), and the wake vortex carries no aerodynamic loads (and therefore moves with the local velocity). Thus the strengths of all the previous wake panels are known from previous time steps. This shedding procedure is repeated at each time step and a row of new trailing-edge wake vortex rings are created (wake shedding procedure).

d. *Influence Coefficients*

To specify the time-dependent boundary condition the kinematic conditions need to be known (from Eq. (13.141)) and a time-stepping loop (shown in Fig. 13.25) is initiated with I_t as the time-step counter:

$$t = I_t \cdot \Delta t$$

Let us assume that at $t = 0$ the two coordinate systems x, y, z and X, Y, Z in Fig. 13.39 coincided and the wing was at rest. The calculation is initiated at $t = \Delta t$ and the wake at this moment consists of one wake panel row (the wake panel row adjacent to the trailing edge in Fig. 13.40). The Dirichlet boundary condition (Eq. (13.156)) when specified, for example, at the ith panel's collocation point (inside the body) is influenced by all the N body and N_W wake panels and will have the form

$$\sum_{k=1}^{N} C_{ik}\mu_k + \sum_{\ell=1}^{N_W} C_{i\ell}\mu_\ell + \sum_{k=1}^{N} B_{ik}\sigma_k = 0 \tag{13.156a}$$

But the strength of all the wake panels is related to the unknown doublet values of the trailing-edge upper and lower panels, via the Kutta condition (Eq. (13.158)). Therefore, by resubstituting the trailing-edge condition (see also a similar explanation in Section 12.5), we can reduce this boundary condition to include only the body's unknown doublets, and for the first time step it becomes

$$\sum_{k=1}^{N} A_{ik}\mu_k + \sum_{k=1}^{N} B_{ik}\sigma_k = 0, \quad t = \Delta t \tag{13.159}$$

where $A_{ik} = C_{ik}$ if no wake is shed from this panel and $A_{1k} = C_{ik} \pm C_{i\ell}$ if it is shedding a wake panel.

During the subsequent time steps wake panels will be shed, but, as noted, their strength is known from the previous computations. Thus, Eq. (13.159) is valid only for the first time step, and for $t > \Delta t$ the influence of these wake doublets μ_W (excluding the latest row) must be included in the boundary condition. So for all the other time steps Eq. (13.156a) will have the form

$$\sum_{k=1}^{N} A_{ik}\mu_k + \sum_{\ell=1}^{M_W} C_{i\ell}\mu_\ell + \sum_{k=1}^{N} B_{ik}\sigma_k = 0, \quad t > \Delta t \tag{13.160}$$

Note that now the wake counter M_W does not include the latest wake row.

e. *Establish RHS Vector*

Since the source value is set by the value of the local kinematic velocity (Eq. (13.19)), the second and third terms in Eq. (13.160) are known at each time step and, therefore, can be transferred to the right-hand side of the equation. The RHS vector is then

13.13 Unsteady Panel Methods

defined as

$$\begin{pmatrix} \text{RHS}_1 \\ \text{RHS}_2 \\ \vdots \\ \text{RHS}_N \end{pmatrix} = - \begin{pmatrix} c_{11}, c_{12}, \ldots, c_{1M_W} \\ c_{21}, c_{22}, \ldots, c_{2M_W} \\ \vdots \quad \vdots \\ c_{N1}, c_{N2}, \ldots, c_{NM_W} \end{pmatrix} \begin{pmatrix} \mu_{1_w} \\ \mu_{2_w} \\ \vdots \\ \mu_{M_w} \end{pmatrix}$$

$$- \begin{pmatrix} b_{11}, b_{12}, \ldots, b_{1N} \\ b_{21}, b_{22}, \ldots, b_{2N} \\ \vdots \quad \vdots \\ b_{N1}, b_{N2}, \ldots, b_{NN} \end{pmatrix} \begin{pmatrix} \sigma_1 \\ \sigma_2 \\ \vdots \\ \sigma_N \end{pmatrix} \quad (13.161)$$

(Again, note that μ_{l_w} and σ_k are known.) In the case when the body geometry is not changing with time the b_{kl} coefficients are calculated only once, but the c_{kl} coefficients of the wake must be recomputed at each moment because of the wake's time-dependent rollup.

f. Solve Set of Linear Equations

Once the the momentary RHS vector is established, the boundary condition, when specified at the body's N collocation points, will have the form

$$\begin{pmatrix} a_{11}, a_{12}, \ldots, a_{1N} \\ a_{21}, a_{22}, \ldots, a_{2N} \\ \vdots \quad \vdots \\ a_{N1}, a_{N2}, \ldots, a_{NN} \end{pmatrix} \begin{pmatrix} \mu_1 \\ \mu_2 \\ \vdots \\ \mu_N \end{pmatrix} = \begin{pmatrix} \text{RHS}_1 \\ \text{RHS}_2 \\ \vdots \\ \text{RHS}_N \end{pmatrix} \quad (13.162)$$

This matrix has a nonzero diagonal ($a_{kk} = \frac{1}{2}$, when the panel is not at the trailing edge) and has a stable numerical solution.

The results of this matrix equation can be summarized in indicial form (for each collocation point k) as

$$\sum_{l=1}^{N} a_{kl} \mu_l = \text{RHS}_k \quad (13.163)$$

If the shape of the body remains unchanged then the matrix inversion occurs only once. For time steps larger then 1 the calculation is reduced to

$$\mu_k = \sum_{l=1}^{N} a_{kl}^{-1} \text{RHS}_l \quad (13.164)$$

where a_{kl}^{-1} are the coefficients of the inverted matrix. In situations when a large number of panels are used (more than 2,000) then from the computational point of view it is often more economical to iterate for a new instantaneous solution of Eq. (13.163), at each time step, than to store the large inverted matrix a_{kl}^{-1} in the memory.

g. Computation of Velocity Components, Pressures, and Loads

One of the advantages of the velocity potential formulation is that the computation of the surface velocities and pressures is determinable by the local properties of the solution (velocity potential in this case). The perturbation velocity components on the surface of a panel can be obtained by Eqs. (9.26) in the tangential direction,

$$q_l = -\frac{\partial \mu}{\partial l}, \quad q_m = -\frac{\partial \mu}{\partial m} \quad (13.165a)$$

and in the normal direction (similar to Eq. (9.27))

$$q_n = \sigma \tag{13.165b}$$

where l, m are the local tangential coordinates (see Fig. 12.25). For example, the perturbation velocity component in the l direction can be formulated (e.g., by using central differences) as

$$q_l = \frac{1}{2\Delta l}(\mu_{l-1} - \mu_{l+1}) \tag{13.166}$$

In most cases the panels do not have equal sizes and instead of this simple formula, a more elaborate differentiation must be used. The total velocity at collocation point k is the sum of the kinematic velocity plus the perturbation velocity:

$$\mathbf{Q}_k = [U(t), V(t), W(t)]_k \cdot (l, m, n)_k + (q_l, q_m, q_n)_k \tag{13.167}$$

where l_k, m_k, n_k are the local panel coordinate directions (shown in Fig. 12.25) and of course the normal velocity component for a solid surface is zero. The pressure coefficient can now be computed for each panel using Eq. (13.28):

$$C_p = \frac{p - p_{\text{ref}}}{(1/2)\rho v_{\text{ref}}^2} = 1 - \frac{Q^2}{v_{\text{ref}}^2} - \frac{2}{v_{\text{ref}}^2}\frac{\partial \Phi}{\partial t} \tag{13.168}$$

Here $\mathbf{Q}$ and p are the local fluid velocity and pressure values, $\partial\Phi/\partial t = \partial\mu/\partial t$ (since $\Phi_i = 0$), p_{ref} is the far field reference pressure, and v_{ref} can be taken as the kinematic velocity as appears in Eq. (13.8):

$$\mathbf{v}_{\text{ref}} = -[\mathbf{V}_0 + \mathbf{\Omega} \times \mathbf{r}] \tag{13.169}$$

or as the translation velocity of the origin $\mathbf{V}_0$. For nonlifting bodies the use of Eq. (13.128a) instead of Eq. (13.168) is recommended when the body's rotation axis is parallel to the direction of motion. (In the case of more complex motion the use of the pressure equation and the selection of $\mathbf{v}_{\text{ref}}$ should be investigated more carefully.)

The contribution of an element with an area of ΔS_k to the aerodynamic loads $\Delta \mathbf{F}_k$ is then

$$\Delta \mathbf{F}_k = -C_{p_k}\left(\frac{1}{2}\rho v_{\text{ref}}^2\right)_k \Delta S_k \mathbf{n}_k \tag{13.170}$$

In many situations off-body velocity field information is required as well. This type of calculation can be done by using the velocity influence formulas of Chapter 10 (and the singularity distribution strengths of σ and of μ are known at this point).

h. Vortex Wake Rollup

Since the wake is force free, each wake panel (or wake vortex ring) must move with the local stream velocity (Eq. (13.21a)). The local velocity is a result of the kinematic motion and the velocity components induced by the wake and body and is usually measured in the inertial frame of reference X, Y, Z, at each panel's corner points. This velocity can be calculated (using the velocity influence formulas of Section 10.4.1 for the doublet and of Section 10.4.2 for the source panels) since the strength of all the singularity elements in the field is known at this point of the calculation.

To achieve the wake rollup, at each time step, the induced velocity $(u, v, w)_\ell$ at each wake panel corner point ℓ is calculated in the stationary inertial frame and then the vortex elements are moved by

$$(\Delta x, \Delta y, \Delta z)_\ell = (u, v, w)_\ell \Delta t \tag{13.171}$$

13.13 Unsteady Panel Methods

Summary

The time-stepping solution is best described by the block diagram in Fig. 13.25. For cases with fixed geometry (e.g., a maneuvering airplane) the geometrical information, such as panel corner points, collocation points, and normal vectors, must be calculated first. Then the time-stepping loop begins and based on the motion kinematics the geometry of the wake panel row adjacent to the trailing edge is established. Once the geometry of the trailing-edge area is known the influence coefficients a_{kl} of Eq. (13.162) can be calculated. The same kinematic velocity information (e.g., Eq. (13.141)) allows the body's source strength (Eq. (13.19)) and the RHS vector of Eq. (13.161) to be obtained. Next, the unknown doublet distribution is obtained and the surface velocity components and pressures are calculated. Prior to advancing to the next time step, the wake rollup procedure is performed and then the time is increased by Δt, the body is moved along the flight path, and the next time step is treated in a similar manner.

Some examples of using the unsteady, constant-strength singularity element based panel method of Ref. 12.13 are presented in the following paragraphs.

Example 1: Large-Amplitude Pitch Oscillation of a NACA 0012 Airfoil

The previous examples on the pitch oscillations of an airfoil were obtained by thin airfoil methods that do not provide the detailed pressure distribution on the surface. In this case the computations are based on a thick airfoil model and the two-dimensional results were obtained by using a large aspect ratio ($\mathcal{R} = 1{,}000$) rectangular wing. The lift and pitching moment histograms, during a fairly large amplitude pitch oscillation cycle, of this NACA 0012 two-dimensional airfoil are presented in Fig. 13.41. Comparison is made with experimental results of Ref. 13.15 for oscillations about the airfoil's quarter chord. The computations are

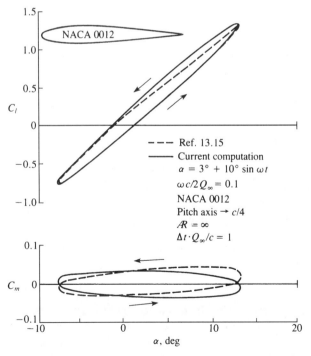

Figure 13.41 Lift and pitching-moment loops for the pitch oscillation of a NACA 0012 airfoil. From Ref. 12.13. Reprinted with permission. Copyright AIAA.

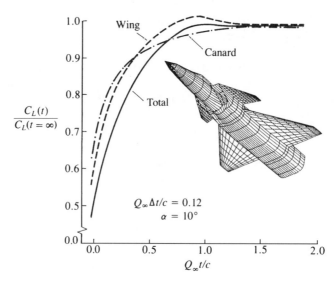

Figure 13.42 Lift coefficient variation after an airplane model was suddenly set into a constant-speed forward motion. From Ref. 12.13. Reprinted with permission. Copyright AIAA.

reasonably close to the experimental values of the lift coefficient through the cycle. During the pitchdown motion, however, a limited flow separation reduces the lift of the airfoil in the experimental data. The shape of the pitching-moment loops is close to the experimental result with a small clockwise rotation. This is a result of the inaccuracy of computing the airfoil's center of pressure, since only nine chordwise panels were used.

This example indicates, too, that if the flow stays attached over the airfoil then the Kutta condition based load calculation is applicable to engineering analysis even for these large trailing edge displacements.

Example 2: Sudden Acceleration of an Airplane Configuration

The transient load on a thin airfoil that was suddenly set into motion was first reported during the 1920s[13.3] and only recently with the use of panel methods could this type of analysis be applied to more realistic airplane configurations. Such computation for a complex aircraft shape is presented in Fig. 13.42, and the panel grid consists of 706 panels per side of the model. The transient lift growth of this wing/canard combination differs somewhat from the monotonic lift increase of a single lifting surface as presented in Fig. 13.37. At the first moment the lift of the wing and canard grow at about the same rate, with the lift of the wing being slightly lower because of the canard-induced downwash. Then the wing's lift increases beyond its steady-state value, since the canard wake has not yet reached the wing. At about $Q_\infty t/c \approx 1.0$ the canard wake reaches the wing and its influence begins to reduce the wing's lift. This behavior results in the lift overshoot, as shown in the figure.

Example 3: Helicopter Rotor

The flexibility of this method can be demonstrated by rotating a pair of high aspect ratio, untwisted wings around the z axis, to simulate rotor aerodynamics. The trailing-edge vortices behind this two-bladed rotor, which was impulsively set into motion, are presented in Fig. 13.43. Similar information on wake trajectory and

13.13 Unsteady Panel Methods

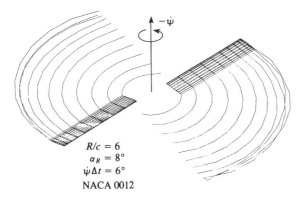

Figure 13.43 Panel model of a two-bladed rotor and its wake in hover, after one-quarter revolution. From Ref. 12.13. Reprinted with permission. Copyright AIAA.

rollup, for more complex rotorcraft geometries and motions (including forward flight), can easily be calculated by this technique. The spanwise lift distribution on one rotor blade of Fig. 13.43, after one-quarter revolution ($\Delta\psi = 90°$), is presented in Fig. 13.44. The rotor for this example is untwisted and has a collective pitch angle of $\alpha_R = 8°$, to duplicate the geometry of the rotor tested by Caradonna and Tung.[13.16] The large difference between this spanwise loading ($\Delta\psi = 90°$) and the experimental loading measured in Ref. 13.16, for a hovering rotor, is due to the undeveloped wake. This solution can be considerably improved by allowing about eight revolutions of the rotor, so that the wake-induced flow will develop. This spiral vortex wake-induced downwash did reduce the spanwise lift distribution on the wake to values that are close to those measured by Caradonna and Tung,[13.16] as shown in Fig. 13.44 (by the "steady hover" line). Figure 13.45 presents the corresponding chordwise pressures for three blade stations. The computed pressures fall close to the measurements of Ref. 13.16 and the small deviations could be a result of the sparse panel grid used or could be caused by experimental errors.

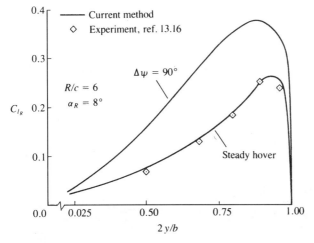

Figure 13.44 Spanwise load distribution on the rotor blades of Fig. 13.43. From Ref. 12.13. Reprinted with permission. Copyright AIAA.

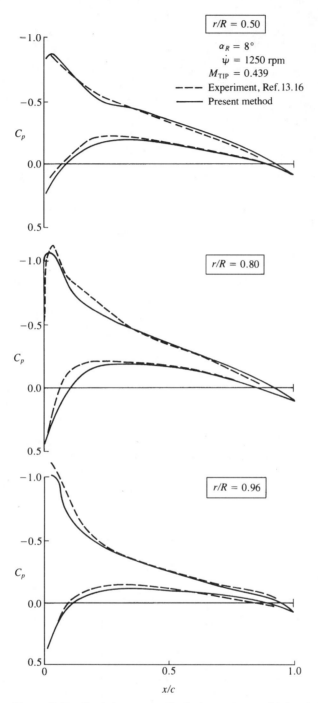

Figure 13.45 Chordwise pressure distribution on the rotor blades of Fig. 13.43. From Ref. 12.13. Reprinted with permission. Copyright AIAA.

13.13 Unsteady Panel Methods

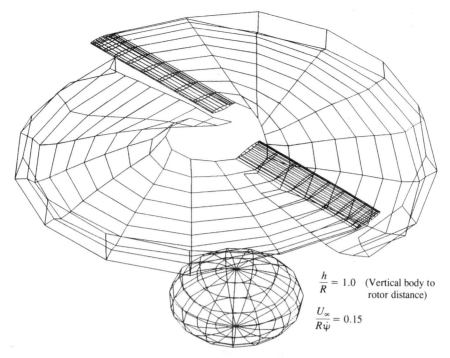

Figure 13.46 Wake shape behind a two-bladed rotor and a body in forward flight, after one-half revolution.

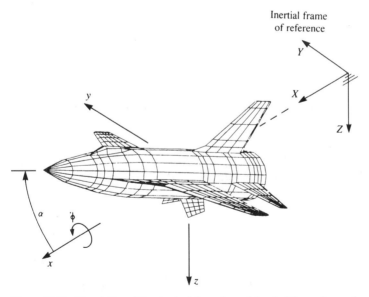

Figure 13.47 Description of the standard dynamic model and of the coning motion. From Katz, J., "Numerical Simulation of Aircraft Rotary Aerodynamics," *AIAA Paper 88-0399*, 1988. Reprinted with permission. Copyright AIAA.

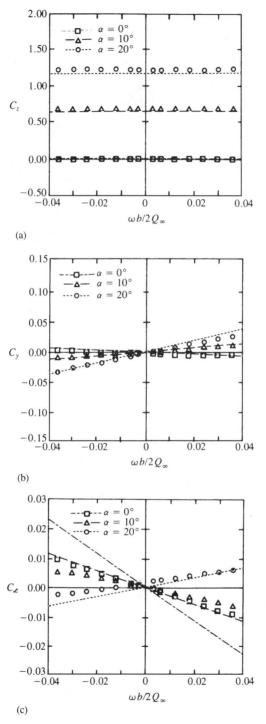

Figure 13.48 Comparison between measured and calculated normal C_z, side forces C_y, and rolling moment C_ℓ in a coning motion (without side slip). Symbols represent experimental data of Ref. 13.17. From Katz, J., "Numerical Simulation of Aircraft Rotary Aerodynamics," *AIAA Paper 88-0399*, 1988. Reprinted with permission. Copyright AIAA.

Increasing the complexity of the motion is fairly simple. The forward flight of this rotor with a generic body is shown in Fig. 13.46

Example 4: Coning Motion of a Generic Airplane

The coning motion is described schematically in Fig. 13.47 for the generic airplane geometry modeled by 718 panels per side. In principle the x coordinate of the body system translates forward at a constant speed Q_∞ and the model angle of attack is set within this frame of reference. The rotation $\dot{\phi}$ is performed about the x axis, as shown in the figure. Computed and experimental normal force C_z, side force C_y, and rolling moment C_ℓ are presented in Figs. 13.48a and 13.48b. The aircraft model was rotated about its center of gravity at a rate of up to $\omega b/2U = 0.04$. This rate is fairly low, but representative of possible aircraft flight conditions, and was selected to match the experiments of Ref. 13.17. The normal force is not affected by this low rotation rate and both experimental and computed lines are close to being horizontal. For higher angles of attack, the computational results are lower than the experimental data owing to the vortex lift of the strakes. The side force, in this type of motion, is influenced by the side slip of the vertical and horizontal tail surfaces. Consequently, the computed values of C_y, for the above angle-of-attack range, are close to the experimental data.

The computed rolling moment of the configuration C_ℓ at $\alpha = 0$ (Fig. 13.48c) is much larger than shown by the experiment. However, the computation does capture the fact that the trend of the curve slope (which is really the roll damping) becomes negative at the larger angles of attack. This slope is also a function of the distance between the wing's center of pressure and the rotation axis, and the error in computing this distance is probably the reason for the larger (computed) rolling moments.

References

[13.1] Etkin, B., and Reid, L. D., *Dynamics of Flight: Stability and Control*, John Wiley & Sons, 3rd edition, 1996, pp. 98–99.

[13.2] Katz, J., and Weihs, D., "Hydrodynamic Propulsion by Large Amplitude Oscillation of an Airfoil with Chordwise Flexibility," *J. Fluid Mech.*, Vol. 88, Pt. 3, 1978, pp. 485–497.

[13.3] Wagner, H., "Uber die Entstehung des Dynamischen Autriebes von Tragflugeln," Z.F.A.M.M., Vol. 5, No. 1, Feb. 1925, pp. 17–35.

[13.4] Theodorsen, T., "General Theory of Aerodynamic Instability and the Mechanism of Flutter," NACA Rep. 496, 1935.

[13.5] Von Karman, T., and Sears, W. R., "Airfoil Theory for Non-Uniform Motion," *Journal of the Aeronautical Sciences*, Vol. 5, No. 10, 1938, pp. 379–390.

[13.6] Katz, J., and Weihs, D., "Large Amplitude Unsteady Motion of a Flexible Slender Propulsor," *J. Fluid Mech.*, Vol. 90, Pt. 4, 1979, pp. 713–723.

[13.7] Lighthill, M. J., "Note on the Swimming of Slender Fish," *J. Fluid Mechanics*, Vol. 9, 1960, pp. 305–317.

[13.8] Archibald, F. S., "Unsteady Kutta Condition at High Values of the Reduced Frequency," *AIAA J.*, Vol. 12, No. 1, 1974, pp. 43–48.

[13.9] Satyanarayana, B., and Davis, S., "Experimental Studies of Unsteady Trailing Edge Conditions," *AIAA J.*, Vol. 16, No. 2, 1978, pp. 125–129.

[13.10] Fleeter, S., "Trailing Edge Condition for Unsteady Flows at High Reduced Frequency," *AIAA Paper 79-0152*, Jan. 1979.

[13.11] Poling, D. R., and Telionis, D. P., "The Response of Airfoils to Periodic Disturbances – The Unsteady Kutta Condition," *AIAA J.*, Vol. 24, No. 2, 1986, pp. 193–199.

[13.12] Katz, J., and Weihs, D., "Wake Rollup and the Kutta Condition for Airfoils Oscillating at High Frequency," *AIAA J.*, Vol. 19, No. 12, 1981, pp. 1604–1606.

[13.13] Katz, J., "Calculation of the Aerodynamic Forces on Automotive Lifting Surfaces," *ASME J. Fluids Eng.*, Vol. 107, 1985, pp. 438–443.

[13.14] Katz, J., and Weihs, D., "The Effect of Chordwise Flexibility on the Lift of a Rapidly Accelerated Airfoil," *Aeronaut. Quarterly*, Feb., 1979, pp. 360–369.

[13.15] McCroskey, W. J., McAlister, K. W., Carr, L. W., Pucci, S. L., Lambert, O., and Indergrand, R. F., "Dynamic Stall on Advanced Airfoil Sections," *J. Amer. Helicopter Soc.*, July 1981, pp. 40–50.

[13.16] Caradonna, F. X., and Tung, C., "Experimental and Analytical Studies of a Model Helicopter Rotor in Hover," *NASA TM-81232*, 1981.

[13.17] Jermey, C., and Schiff, L. B., "Wind Tunnel Investigation of the Aerodynamic Characteristics of the Standard Dynamic Model in Coning Motion at Mach 0.6," *AIAA Paper 85-1828*, Aug. 1985.

Problems

13.1. Consider a two-dimensional version of the relative motion described in Fig. 13.1, between a body fixed frame of reference (x, z) and an inertial frame (X, Z) such that

$$(X_0, Z_0) = (-U_\infty t, -W_\infty t)$$

$$\theta = \sin \omega t$$

and

$$(\dot{X}_0, \dot{Z}_0) = (-U_\infty, -W_\infty)$$

$$\dot{\theta} = \omega \cos \omega t$$

a. Use the chain rule to evaluate the derivatives $\partial/\partial X$, $\partial/\partial Z$, and $\partial/\partial t$ in terms of the body coordinates.

b. Using your results from (a) transform the Bernoulli equation

$$\frac{p_\infty - p}{\rho} = \frac{(\nabla \Phi)^2}{2} + \frac{\partial \Phi}{\partial t} \tag{13.172}$$

into the (x, z) frame of reference.

13.2. The two-dimensional flat plate, shown in Fig. 13.49, is initially at rest and at $t = 0+$ it moves suddenly forward at a constant speed U_∞. Obtain the time-dependent circulation $\Gamma(t)$ and lift $L(t)$ of the flat plate using two chordwise lumped-vortex elements (select the vortex and collocation points as suggested in Section 11.1.1) with a discrete-vortex model for the wake and present your results graphically (as in Fig. 13.8).

a. Study the effect of time step $U_\infty \Delta t/c$ in the range $U_\infty \Delta t/c = 0.02$–$0.2$. Note that a smaller time step simulates a faster acceleration to the terminal speed U_∞ and therefore has a physical effect on the results (in addition to the numerical effect).

b. Study the effect of wake vortex positioning by placing the latest trailing-edge vortex at the beginning, center, and end of the interval covered by the trailing edge during the latest time step (see Fig. 13.49). Compare your results with the more accurate calculations in Fig. 13.34 (for $R = \infty$).

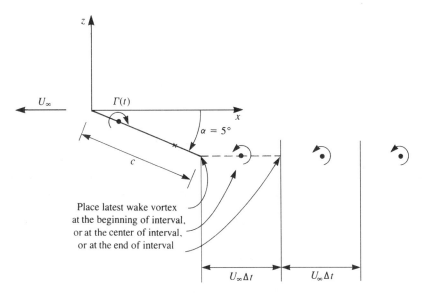

Figure 13.49 Nomenclature for the suddenly accelerated flat plate.

13.3. Use the flat plate model of the previous example to study the constant acceleration of a flat plate. Assume that the forward speed is $U(t) = at$ and the time step is $a\Delta t^2/2c = 0.1$. Calculate the time-dependent circulation $\Gamma(t)$ and lift $L(t)$ of the flat plate for several values of the acceleration a and present your results graphically (as in Fig. 13.8). For simplicity, place the latest vortex shed from the trailing edge at one third of the distance covered by the trailing edge during that time step.

13.4. Convert any of the two-dimensional panel codes of Chapter 11 (e.g., a constant-strength doublet method) to the unsteady mode and validate it by calculating the lift and circulation after a sudden acceleration. (This problem requires a larger effort and can be given as a final project.)

CHAPTER 14

The Laminar Boundary Layer

The discussion so far has focused mainly on the potential flow model whose solution provides a useful but restricted description of the flow. For practical problems such as the flow over an airfoil, however, effects of the viscous flow near the solid surface must be included. The objective of this chapter, therefore, is to explain how a viscous boundary layer model can be combined with the inviscid flow model to provide a more complete representation of the flowfield. These principles can be demonstrated by using the laminar boundary layer model, which provides all the necessary elements for combining the viscous and inviscid flow models. We must remember, though, that the Reynolds number of the flow over actual airplanes or other vehicles is such that large portions of the flow are turbulent, and the solely laminar flow model must be augmented to reflect this. However, the principles of the matching process remain similar. Extensions of this laminar boundary layer based approach to flows with transition, to turbulent boundary layers, or to cases with flow separation, and other aspects of airfoil design, will be discussed briefly in Chapter 15. (Although in these cases the viscous flow model may change substantially from the laminar model, the viscous–inviscid coupling strategy remains unchanged.)

Boundary layer theory is a very wide topic and there are several textbooks that focus solely on this subject (e.g., see Ref. 1.6). Since the main topic in this book is the potential flow solution, no attempt is made to provide a comprehensive description of this field apart from the elements needed to explain the coupling process. Consequently, the discussion in this chapter is limited to describing the basics of the boundary layer model and the elements necessary to explain the concept of combining the inner viscous and the outer inviscid flows. With these two (inner and outer flow) models in mind, the information sought from the viscous boundary layer solution in this chapter is:

1. The scale, or thickness, of the boundary layer and its streamwise growth.
2. Displacement effects (to the inviscid model) resulting from the slower velocity inside the viscous layer.
3. Skin-friction and resulting drag estimates, which cannot be calculated by the inviscid flow.
4. Hints about more advanced topics such as boundary layer transition and flow separation.

14.1   The Concept of the Boundary Layer

The focus of this book, as discussed earlier, is on low-speed aerodynamics. In Section 1.8 it was established that such flows are mostly inviscid, apart from thin regions near a solid surface where the viscous effects are not negligible. This observation has allowed us to calculate the flowfield past an airfoil under the assumption of potential flow (and to obtain a satisfactory representation of the pressure field and resulting forces and moments). One result of this calculation is the nonzero tangential velocity component on the airfoil surface, which we denote as $U_e = U_e(x, t)$, where x is the coordinate along the surface. A

14.1 The Concept of the Boundary Layer

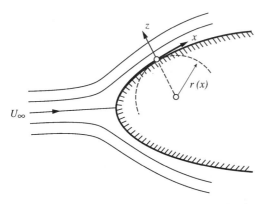

Figure 14.1 Coordinate system used for the boundary layer along a curved surface.

typical situation within the flowfield near a solid boundary is described schematically in Fig. 1.14. In the actual flow, the no-slip condition requires the velocity at the surface to be zero and so it is necessary to insert a thin layer of fluid adjacent to the wall where the flow is assumed to be viscous and where the tangential velocity component grows from zero at the wall to U_e at the edge. This layer of rapid change in the tangential velocity component is called the *boundary layer* and in Section 1.8 we concluded that for high Reynolds number flows its thickness $\delta = \delta(x, t)$ is much smaller than the characteristic length, L, along the solid surface:

$$\delta \ll L$$

To demonstrate the properties of the boundary layer, we limit the discussion to the two-dimensional case and for additional simplicity, we consider the continuity and momentum equations with constant properties ($\mu = $ const. and $\rho = $ const.). Since most practical solid surfaces are not flat, let us begin with a curvilinear coordinate system, to allow for the inclusion of moderately curved surfaces such as the upper surface of an airfoil. The selected coordinate system is shown in Fig. 14.1 with the abscissa, x, selected along the solid surface, and the ordinate, z, normal to it. If we denote the local radius of curvature of the surface as $r(x)$ (see Fig. 14.1), then the local curvature k is given as

$$k(x) = \frac{1}{r(x)}$$

and a new variable h may be used (to simplify the algebraic operations) such that

$$h = h(x) = 1 + zk(x) \tag{14.1}$$

The continuity equation (Eq. (1.23)) and the two components of the momentum equation (Eq. (1.30)), without the body forces, written in the curvilinear system (see Rosenhead,[14.1] p. 201, or Schlichting,[1.6] p. 68) are

$$\frac{\partial u}{\partial x} + \frac{\partial hw}{\partial z} = 0 \tag{14.2}$$

$$\frac{\partial u}{\partial t} + u\frac{\partial u}{\partial x} + w\frac{\partial hu}{\partial z} = \frac{-1}{\rho}\frac{\partial p}{\partial x} + \frac{h\mu}{\rho}\frac{\partial}{\partial z}\left[\frac{1}{h}\left(\frac{\partial hu}{\partial z} - \frac{\partial w}{\partial x}\right)\right] \tag{14.3}$$

$$\frac{\partial w}{\partial t} + u\frac{\partial w}{\partial x} + hw\frac{\partial w}{\partial z} - ku^2 = \frac{-h}{\rho}\frac{\partial p}{\partial z} - \frac{h\mu}{\rho}\frac{\partial}{\partial x}\left[\frac{1}{h}\left(\frac{\partial hu}{\partial z} - \frac{\partial w}{\partial x}\right)\right] \tag{14.4}$$

These equations can be simplified by a combination of a dimensional analysis, similar to the one used in Section 1.7, and a consideration of the order of magnitude of the terms. For this analysis a set of characteristic quantities may be defined again; however, now we make a distinction between the streamwise and the normal directions:

$$x^* = \frac{x}{L}, \quad z^* = \frac{z}{\delta}, \quad k^* = kL$$
$$u^* = \frac{u}{U}, \quad w^* = \frac{w}{W}$$
$$p^* = \frac{p}{p_0} \tag{14.5}$$
$$t^* = \frac{t}{T} = \frac{t}{L/U}$$

where W is the characteristic speed in the z direction. The flow equations are now rewritten in terms of the new variables, and the first, the continuity equation, becomes

$$\frac{U}{L}\frac{\partial u^*}{\partial x^*} + \left(1 + z^* k^* \frac{\delta}{L}\right)\frac{W}{\delta}\frac{\partial w^*}{\partial z^*} = 0 \tag{14.6a}$$

Since the surface curvature is not large, we can rewrite Eq. (14.1) as

$$h = 1 + z^* k^* \frac{\delta}{L} \approx 1$$

and with this simplification, Eq. (14.6a) becomes

$$\frac{U}{L}\frac{\partial u^*}{\partial x^*} + \frac{W}{\delta}\frac{\partial w^*}{\partial z^*} = 0 \tag{14.6b}$$

If we assume that all nondimensional variables of Eq. (14.5) are of $O(1)$ inside the boundary layer, then for both terms in the continuity equation to be of the same order it is necessary that U/L be of the order of W/δ. Therefore, if $\delta \ll L$, then it follows that $W \ll U$, and the order of magnitude of W is determined as

$$\frac{W}{U} = O\left(\frac{\delta}{L}\right)$$

Introducing the nondimensional variables into the momentum equation in the x direction (similarly to the treatment of Eq. (1.60)) we obtain

$$\Omega\frac{\partial u^*}{\partial t^*} + u^*\frac{\partial u^*}{\partial x^*} + \frac{W}{U}\frac{L}{\delta}w^*\frac{\partial u^*}{\partial z^*} = -Eu\frac{\partial p^*}{\partial x^*} + \frac{1}{Re}\left(\frac{L^2}{\delta^2}\frac{\partial^2 u^*}{\partial z^{*2}} - \frac{L}{\delta}\frac{W}{U}\frac{\partial^2 w^*}{\partial x^* \partial z^*}\right) \tag{14.7}$$

where all nondimensional numbers are defined as before (e.g., Eu is the Euler number). All three terms on the left-hand side of this equation appear to have the same order of magnitude while the last (second viscous) term can be clearly neglected in comparison with the first viscous term. If we recall our basic assumption that, inside the boundary layer, the inertia terms (left-hand side of Eq. (14.7)) are of the same order of magnitude as the viscous terms, then the remaining viscous term is of $O(1)$, therefore,

$$\frac{1}{Re}\left(\frac{L^2}{\delta^2}\right) \approx O(1)$$

14.1 The Concept of the Boundary Layer

and it follows that

$$\frac{\delta}{L} = O(Re^{-1/2}) \quad \text{and} \quad \frac{W}{U} = O(Re^{-1/2}) \tag{14.8}$$

Consequently, only one term, the second viscous term, is neglected in this equation! Substitution of the nondimensional quantities into the momentum equation in the z direction results in

$$\frac{W}{U}\Omega\frac{\partial w^*}{\partial t^*} + \frac{W}{U}u^*\frac{\partial w^*}{\partial x^*} + \frac{W^2}{U^2}\frac{L}{\delta}w^*\frac{\partial w^*}{\partial z^*} - k^*u^{*2}$$
$$= -\frac{L}{\delta}Eu\frac{\partial p^*}{\partial z^*} - \frac{1}{Re}\frac{\partial}{\partial x^*}\left(\frac{L}{\delta}\frac{\partial u^*}{\partial z^*} - \frac{W}{U}\frac{\partial w^*}{\partial x^*}\right) \tag{14.9}$$

Again, all inertia terms on the left-hand side are of the same order of magnitude ($O(\delta/L)$) and are considerably smaller than the pressure term, which is multiplied by L/δ. The viscous terms are of the the same order as the inertia terms ($O(\delta/L)$) since according to Eq. (14.8), $1/Re = O(\delta^2/L^2)$. Therefore, all inertia and viscous terms appearing in this equation are much smaller than the pressure term and can be neglected.

Rearranging the remaining terms in the continuity and momentum equations indicates that for the continuity equation all terms appear to be of the same order of magnitude; therefore, it remains in its previous form:

$$\frac{\partial u}{\partial x} + \frac{\partial w}{\partial z} = 0 \tag{14.10}$$

For the momentum equation in the x direction only one viscous term is neglected, and so we have

$$\frac{\partial u}{\partial t} + u\frac{\partial u}{\partial x} + w\frac{\partial u}{\partial z} = \frac{-1}{\rho}\frac{\partial p}{\partial x} + \frac{\mu}{\rho}\frac{\partial^2 u}{\partial z^2} \tag{14.11}$$

while, in the z direction, all terms but the normal pressure gradient become negligible, implying that the normal pressure gradient itself is equal to zero, as well:

$$0 = -\frac{\partial p}{\partial z} \tag{14.12}$$

Equations (14.10)–(14.12) define the classical two-dimensional boundary layer equations proposed by the German scientist Prandtl (1874–1953) in 1904. At the wall, the no-slip condition yields

$$z = 0, \quad u = w = 0 \tag{14.13a}$$

and at the edge of the boundary layer ($z = \delta$) the tangential velocity component must approach the inviscid surface value of $U_e(x, t)$, that is,

$$z = \delta, \quad u = U_e(x, t) \tag{14.13b}$$

Solution of this Prandtl's boundary layer model will be discussed in Section 14.3. In addition, Eq. (14.12) states that the pressure across the boundary layer (normal to the surface) is constant and therefore this pressure is taken to be the inviscid pressure evaluated on the surface. Application of the momentum equation (Eq. (14.11)) outside the boundary layer yields

$$\frac{\partial U_e}{\partial t} + U_e\frac{\partial U_e}{\partial x} = \frac{-1}{\rho}\frac{\partial p}{\partial x} \tag{14.14}$$

This value can be inserted into Eq. (14.11) so that the pressure p is no longer an unknown in the problem. With the above assumptions and for the case of steady-state flow, Eq. (14.11) reduces to

$$u\frac{\partial u}{\partial x} + w\frac{\partial u}{\partial z} = U_e \frac{\partial U_e}{\partial x} + \frac{\mu}{\rho}\frac{\partial^2 u}{\partial z^2} \tag{14.11a}$$

14.2 Boundary Layer on a Curved Surface

Note that the wall curvature does not appear in the boundary layer equations (Eqs. (14.10)–(14.12)) so that the equations seem to be written in Cartesian coordinates. For the analysis in the present section the solution process for the flow past an airfoil proceeds as follows: First, solve the inviscid problem, and, second, solve the boundary layer equations to model the effects of viscosity. This solution process represents a high Reynolds number alternative to the solution of the complete Navier–Stokes equations and, with the extension to include transition and turbulence, has been the approach used almost extensively for airfoil design. There are nonetheless many practical aerodynamic applications where the Prandtl boundary layer equations are not sufficient to adequately describe the flow (as described in Section 14.6), but solution techniques in the spirit of Prandtl's analysis have been developed to successfully extend the applicability of the boundary layer concept (and some of these are discussed in Section 14.8). However, for the description of the high Reynolds number flowfield past an airfoil there is not universal agreement on one set of governing equations (based on first principles) or on one solution technique.

To provide guidance in the understanding and evaluation of the available boundary layer solution techniques, it is useful to extend the boundary layer equations to second order (in the spirit of a perturbation expansion as discussed in Chapter 7) and in this way to study the effects of curvature as well as to continue the iterative process of the interaction between inviscid and viscous solutions. This extension is treated in Van Dyke[14.2] and Schlichting[1.6] (Chapter IX) and we will proceed in a similar way. For the analysis in this section, we will consider steady flow (e.g., $\partial/\partial t = 0$). A suitable small perturbation parameter for this problem (see Eq. (14.8)) is defined as

$$\epsilon = Re^{-1/2} \tag{14.15}$$

The nondimensional variables in Eq. (14.5) are introduced with $\delta = \epsilon$ and $W = \epsilon U$. We now construct an asymptotic (inner) expansion in the boundary layer and keep the first two terms:

$$u = u_1 + \epsilon u_2 \tag{14.16a}$$
$$w = w_1 + \epsilon w_2 \tag{14.16b}$$
$$p = p_1 + \epsilon p_2 \tag{14.16c}$$

Next, substitute the above equations into the nondimensional continuity and Navier–Stokes equations (Eqs. (14.6a), (14.7), and (14.9)). If terms of like order are collected in each equation, the terms corresponding to the largest order are the classical boundary layer equations, which in dimensional form are given in Eqs. (14.10)–(14.12). If terms of the next largest order are now equated (terms of $O(\epsilon)$) the second-order, steady-state boundary layer equations are obtained and in dimensional form are (see Van Dyke[14.2])

$$\frac{\partial u_2}{\partial x} + \frac{\partial w_2}{\partial z} = -k\frac{\partial}{\partial z}(zw_1) \tag{14.17a}$$

14.2 Boundary Layer on a Curved Surface

$$u_1 \frac{\partial u_2}{\partial x} + u_2 \frac{\partial u_1}{\partial x} + w_1 \frac{\partial u_2}{\partial z} + w_2 \frac{\partial u_1}{\partial z} + \frac{1}{\rho}\frac{\partial p_2}{\partial x} - \frac{\mu}{\rho}\frac{\partial^2 u_2}{\partial z^2}$$
$$= k\left[\frac{\partial}{\partial z}\left(z\frac{\partial u_1}{\partial z}\right) - w_1 \frac{\partial}{\partial z}(zu_1)\right] \tag{14.17b}$$

$$\frac{1}{\rho}\frac{\partial p_2}{\partial z} = ku_1^2 \tag{14.17c}$$

Note that the terms appearing on the right-hand side of the equations are proportional to the curvature k.

To determine appropriate boundary conditions for the second-order equations, we need to introduce an asymptotic expansion for the outer inviscid flow and match this with the inner solution (as shown, for example, in Chapter 7). In the inviscid flow, we expect that changes in the x and z directions are comparable in size and therefore we can use the nondimensional Navier–Stokes equations as they appear in Eqs. (14.7) and (14.9) but with $\delta = L$ and $W = U$. Let us construct an outer asymptotic expansion for the inviscid flow:

$$u_0 = U_1 + \epsilon U_2 \tag{14.18a}$$
$$w_0 = W_1 + \epsilon W_2 \tag{14.18b}$$
$$p_0 = P_1 + \epsilon P_2 \tag{14.18c}$$

We expect that the first-order outer solution is the inviscid solution we have already obtained and the second-order solution will be a correction due to the presence of the boundary layer. When the above expansions (Eqs. (14.18a–c)) are substituted into the Navier–Stokes equations (e.g., Eq. (1.60)), it is seen that the viscous terms on the right-hand side (which are multiplied by $1/Re$) vanish to second order so that both the first- and second-order problems are inviscid.

To obtain the appropriate boundary conditions for the second-order problems, we need to match the two solutions (inner expansion of outer solution to the outer expansion of inner solution). In other words, the limit of the boundary layer solution as the edge of the boundary layer is approached should be equivalent to the limit of the inviscid flow solution as the surface is approached.

First we will match the tangential velocity component. Consider the outer solution to second order. If we write its expansion as z approaches zero and keep terms to $O(\epsilon)$ we get

$$U_1(x,0) + \epsilon\left[U_2(x,0) + z\frac{\partial U_1(x,0)}{\partial z}\right] \tag{14.19}$$

Consider the inner solution as z goes to a value z_e just outside the boundary layer. We get

$$u_1(x,z) + \epsilon u_2(x,z), \qquad z \to z_e \tag{14.20}$$

Matching the last two equations term by term yields

$$u_1(x,z) = U_1(x,0), \qquad z \to z_e \tag{14.21a}$$
$$u_2(x,z) = U_2(x,0) + z\frac{\partial U_1(x,0)}{\partial z}, \qquad z \to z_e \tag{14.21b}$$

Next, matching the normal components of velocity in a similar way leads to the following results:

$$W_1(x,0) = 0 \tag{14.22a}$$
$$W_2(x,0) = w_1(x,z) - z\frac{\partial W_1(x,0)}{\partial z}, \qquad z \to z_e \tag{14.22b}$$

And finally, matching the pressure expansions leads to

$$p_1(x, z) = P_1(x, 0) \quad \text{as} \quad z \to z_e \tag{14.23a}$$

$$p_2(x, z) = P_2(x, 0) + z\frac{\partial P_1(x, 0)}{\partial z} \quad \text{as} \quad z \to z_e \tag{14.23b}$$

(Note that the results for the boundary layer variables at first order (Eqs. (14.21a) and (14.23a)) are simply those for the Prandtl boundary layer.) To complete the second-order outer problem definition, we need to evaluate the $W_2(x, 0)$ term. Based on Eq. (14.22b), when $z = z_e$

$$W_2(x, 0) = w_1(x, z_e) - z_e\frac{\partial W_1(x, 0)}{\partial z}$$

Substituting Eq. (14.18a and b) into the outer continuity equation (Eq. 14.2) we have

$$\frac{\partial W_1(x, 0)}{\partial z} = -\frac{dU_e}{dx} \tag{14.24}$$

Now substitute Eq. (14.24) into Eq. (14.22b) to get

$$W_2(x, 0) = z_e\frac{dU_e}{dx} + w_1(x, z_e) \tag{14.25}$$

To evaluate $w_1(x, z_e)$ consider the first-order inner continuity equation

$$\frac{\partial u_1}{\partial x} + \frac{\partial w_1}{\partial z} = 0$$

Integrate this equation across the boundary layer to get

$$w_1(x, z_e) = -\int_0^{z_e} \frac{\partial u_1}{\partial x} dz$$

$$= -\int_0^{z_e} \left(\frac{\partial u_1}{\partial x} + \frac{\partial U_e}{\partial x}\right) dz - z_e\frac{\partial U_e}{\partial x}$$

Using the Leibnitz rule[14.3] we can rewrite this expression as

$$\frac{d}{dx}\int_0^{z_e}(U_e - u_1)\,dz - \frac{dz_e}{dx}(U_e - u_1)_{z=z_e} - z_e\frac{dU_e}{dx}$$

At this point it is useful to introduce the displacement thickness δ^*, the physical meaning of which will be explained later:

$$\delta^* = \int_0^{z_e}\left(1 - \frac{u_1}{U_e}\right) dz \tag{14.26}$$

Using the definition of δ^* to simplify the previous term we get

$$w_1(x, z_e) = \frac{d}{dx}(U_e\delta^*) - z_e\frac{dU_e}{dx}$$

Substitution of this result into Eq. (14.25) results in

$$W_2(x, 0) = \frac{d}{dx}(U_e\delta^*) \tag{14.27}$$

Equations (14.26) and (14.27) provide the first clue about the effect of the boundary layer on the inviscid solution. The displacement thickness is described schematically in Fig. 14.2 and it indicates the extent to which the surface would have to be displaced in order to be

14.2 Boundary Layer on a Curved Surface

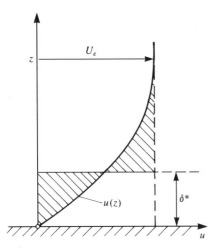

Figure 14.2 Illustration of the displacement thickness δ^* in a boundary layer. (Note that the area enclosed by the two shaded triangular regions should be equal.)

left with the same flow rate of the viscous flow, but with an inviscid velocity profile (of $u(x, z) = U_e(x)$, which is constant inside the boundary layer). Consequently, the boundary of the surface for the (outer) potential flow boundary conditions must be raised by δ^*, as shown in Fig. 14.3.

The second-order outer problem can now be seen as the flow past the airfoil with its surface raised by the displacement thickness. Note that the displacement thickness is nonzero in the wake behind the airfoil so that the new body is semi-infinite in length. The equivalent result from the matching condition (Eq. (14.27)) is simply a statement that the new surface (airfoil plus displacement thickness) is a streamline of the flow. This version of the boundary condition is preferable since it uses the airfoil geometry (with a *transpiration velocity*) so that the geometry is fixed from iteration to iteration. The influence coefficients in a panel method then only need to be calculated once.

Now that the second-order outer problem is defined, the appropriate matching conditions ($u_2(x, z_e)$ and $p_2(x, z_e)$) for the second-order boundary layer problem must be addressed. Consider Eq. (14.21b) for the tangential velocity component at the edge of the boundary layer, $u_2(x, z_e)$. Using the curvilinear form for the vorticity ζ_y (taken from Van Dyke[14.2]), for the first-order outer flow, we get

$$\zeta_y = \frac{\partial U_1}{\partial z} + \frac{kU_1}{h} - \frac{1}{h}\frac{\partial W_1}{\partial x} \tag{14.28}$$

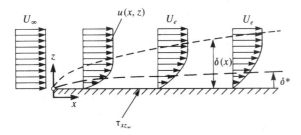

Figure 14.3 Nomenclature for the boundary layer flow over a flat plate at zero incidence.

which must equal zero. If we apply this equation at the wall $z = 0$, where $W_1 = 0$, then $h = 1$ and $\partial W_1/\partial x = 0$, and therefore the result is

$$\frac{\partial U_1(x, 0)}{\partial z} = -kU_1(x, 0)$$

Substituting this result into Eq. (14.21b) at the boundary layer edge we get

$$u_2(x, z_e) = U_2(x, 0) - kzU_1(x, 0) \tag{14.29}$$

The pressure at the boundary layer edge is (Eq. (14.23b))

$$p_2(x, z_e) = P_2(x, 0) + z\frac{\partial P_1(x, 0)}{\partial z} \tag{14.30}$$

To evaluate the second term in this equation, the first-order, outer, inviscid, normal momentum equation (Eq. (14.4), but, without the viscous terms) is evaluated at the wall to give

$$\frac{\partial P_1(x, 0)}{\partial z} = kU_1^2(x, 0)$$

Substituting this result into Eq. (14.30) we obtain

$$p_2(x, z_e) = P_2(x, 0) + kzU_1^2(x, 0) \tag{14.31}$$

The normal momentum equation (Eq. (14.17c)) can now be integrated across the boundary layer to yield $p_2(x, z)$ to use in the streamwise momentum equation (Eq. (14.17b)).

A new sequential process is now available that includes the effects of displacement thickness and curvature. Once the first iteration (inviscid flow plus traditional boundary layer) is completed, the inviscid flow is updated (second-order outer) to include displacement thickness interaction and then the viscous flow is updated (second-order inner) to include the effect of curvature. In theory, the solution process now is correct to second order. This second-order boundary layer theory is not currently being used in the calculation of airfoil flows. The purpose in introducing it here is to demonstrate the appearance of the effect of curvature (the right-hand side terms in Eqs. (14.17a–c)) and to provide a formal derivation of the matching condition of Eq. (14.27). Typically, for practical airfoil flows, the curvature effects are negligible (except perhaps in the neighborhood of the leading and trailing edge) and the main correction term (which is small) is due to the displacement thickness interaction. In some analyses, the pressure variation across the boundary layer due to curvature (Eq. (14.17c)) is included.

Traditionally, the boundary layer iteration process has proceeded as follows:

1. Solve the inviscid flow.
2. Now knowing the external velocity and pressure, solve Prandtl's boundary layer model.
3. Solve the inviscid flow with displacement thickness.
4. Update the boundary layer model with first-order boundary layer equations.

The above process, using the traditional boundary layer equations, proceeds sequentially, starting with the inviscid flow calculation. This interaction is called a "weak interaction." Typically, iteration is not necessary to provide a good approximation to the flow. Before we consider situations where the weak interaction methods described above become invalid, let us study some basic solutions of the traditional boundary layer equations (Eqs. (14.10)–(14.12)).

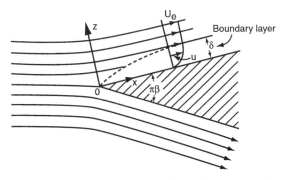

Figure 14.4 Coordinate system used for the flow over a wedge.

14.3 Similar Solutions to the Boundary Layer Equations

Shortly after Prandtl formulated the boundary layer equations (in 1904), two of his students, Blasius and Hiemenz, obtained solutions to two very important examples, the flat-plate boundary layer and stagnation point flow. These solutions are members of a class of similar solutions to the equations that exist for flows lacking an appropriate characteristic length. We will develop the governing equations for this class of "wedge flows" (see the schematic description in Fig. 14.4) and then consider the above special cases.

We start with the boundary layer equations (Eqs. (14.10)–(14.12)) for steady-state flow:

$$\frac{\partial u}{\partial x} + \frac{\partial w}{\partial z} = 0 \tag{14.10}$$

$$\frac{\partial u}{\partial x} + w\frac{\partial u}{\partial z} = U_e\frac{\partial U_e}{\partial x} + \frac{\mu}{\rho}\frac{\partial^2 u}{\partial z^2} \tag{14.11a}$$

$$0 = -\frac{\partial p}{\partial z} \tag{14.12}$$

where the streamwise pressure gradient term in Eq. (14.11) has been replaced by $U_e(\partial U_e/\partial x)$, and in addition, Eq. (14.12) is satisfied as well. The boundary conditions are

$$u = w = 0 \quad \text{for} \quad z = 0$$
$$u = U_e \quad \text{for} \quad z = z_e = \infty$$

A similar solution is one where the velocity profile at any streamwise station, scaled by the velocity in the outer flow, can be represented as a function of one suitably scaled transverse coordinate η,

$$\eta = \frac{z}{g} \tag{14.32}$$

where g is a measure of the boundary layer thickness. The similar velocity profile is now

$$\frac{u}{U_e} = f'(\eta) \tag{14.33}$$

and for convenience in integration, it is selected to be the derivative of the function f. This assumption, in effect, reduces the two-dimensional problem to a one-dimensional one in the transverse direction. It is convenient to introduce the stream function as defined in Section 2.13. In this case the stream function Ψ can be obtained by integrating the u velocity component:

$$\Psi = \int u\, dz = gU_e \int f'\, d\eta = U_e g f \tag{14.34}$$

Note that the introduction of the stream function automatically satisfies the continuity equation, Eq. (14.10).

Based on the properties of the stream function, the velocity components and their derivatives (for insertion into Eq. (14.11)) can be obtained. The u velocity component is

$$u = \frac{\partial \Psi}{\partial z} = \frac{\partial \Psi}{\partial \eta}\frac{\partial \eta}{\partial z} = U_e f' \tag{14.35}$$

$$w = -\frac{\partial \Psi}{\partial x} = -f\frac{d}{dx}(U_e g) - U_e g f'\frac{\partial \eta}{\partial x} = -f\frac{d}{dx}(U_e g) + U_e \eta f' g \tag{14.36}$$

where $\partial \eta/\partial x = -\eta g'/g$. The first and second derivatives of the u velocity components are

$$\frac{\partial u}{\partial x} = U_e' f' + U_e f''\frac{\partial \eta}{\partial x} = U_e' f' - U_e f'' \eta \frac{g'}{g}$$

$$\frac{\partial u}{\partial z} = U_e f''\frac{\partial \eta}{\partial z} = U_e \frac{f''}{g} \tag{14.37}$$

$$\frac{\partial^2 u}{\partial z^2} = U_e \frac{f'''}{g^2} \tag{14.38}$$

When the above results are substituted into the axial momentum equation (Eq. (14.11a)) and the resulting equation is multiplied by $g^2/\nu U_e$, we get

$$f''' + \frac{g^2 U_e'}{\nu}(1 - f'^2) + \frac{g}{\nu}\frac{d}{dx}(U_e g) f f'' = 0 \tag{14.39}$$

This equation can be further simplified by defining the coefficients α and β:

$$f''' + \beta(1 - f'^2) + \alpha f f'' = 0 \tag{14.39a}$$

where the coefficients are

$$\alpha = \frac{g}{\nu}\frac{d}{dx}(U_e g), \qquad \beta = \frac{g^2 U_e'}{\nu} \tag{14.40}$$

For the boundary layer equations to have a similar solution, Eq. (14.39a) must be an ordinary differential equation for f. Therefore both α and β must be constant. Note that

$$2\alpha - \beta = \frac{1}{\nu}\frac{d}{dx}(g^2 U_e)$$

An integration yields

$$(2\alpha - \beta)x = \frac{g^2 U_e}{\nu}$$

and therefore

$$g = \sqrt{\frac{(2\alpha - \beta)\nu x}{U_e}} \tag{14.41}$$

Without loss of generality we take $\alpha = 1$ and

$$g = \sqrt{\frac{(2 - \beta)\nu x}{U_e}} \tag{14.41a}$$

14.3 Similar Solutions to the Boundary Layer Equations

We are now in a position to determine the form that U_e must take to result in similar solutions to the boundary layer equations. Solving for U'_e from the equation for β (Eq. (14.40)) and using Eq. (14.41a) for g we get

$$\frac{U'_e}{U_e} = \frac{\beta}{2-\beta}\frac{1}{x}$$

and integration leads to

$$U_e = U_1 x^m \tag{14.42}$$

where U_1 is a constant and $m = \beta/(2-\beta)$. This inviscid outer flow velocity represents the symmetric flow of a uniform stream approaching a wedge of included angle $\pi\beta$ (see Fig. 14.4) when $0 < \beta < 2$ and the flow around an expansion corner of angle $\pi\beta/2$ when $-2 < \beta < 0$. Note that these solutions may be thought of as local solutions, valid in the neighborhood of $x = 0$, since U_e (in Eq. (14.42)) becomes unbounded as x becomes large.

The mathematical problem (differential equation and boundary conditions) for $f(\eta)$ is now

$$f''' + ff'' + \beta(1 - f'^2) = 0 \tag{14.43}$$
$$f(0) = f'(0) = 0, \qquad f'(\infty) = 1 \tag{14.44}$$

Equation (14.43) is called the Falkner–Skan[14.4] equation and is a third-order, nonlinear, ordinary differential equation. The Blasius semi-infinite flat-plate example is recovered for $\beta = m = 0$ and the Hiemenz stagnation flow for $\beta = m = 1$. In this latter case the wedge angle becomes equal to π and similar conditions are depicted by Fig. 3.6, and the corresponding inviscid velocity field is given by Eq. (3.55). Numerical solutions to the above differential equation are available, and tabulated results can be found in White[14.5] (p. 246). The importance of this solution is that it allows the calculation of the effect of the external pressure gradient on the velocity profile inside the boundary layer. One case, with negative β, is of particular interest since it provides some clues about flow separation. This case is depicted schematically in Fig. 14.5, showing the flow around a convex corner. In this case the flow accelerates around the corner with an adverse pressure gradient. When $\beta = -0.199$ (or $m = -0.0905$), corresponding to a turning angle of $18°$ in the figure, the flow separates; that is, $(\partial u/\partial z)_{z=0} = 0$ ($f''(0) = 0$, or $\tau_w = 0$ according to Eq. (1.12)). This example highlights the role of the adverse pressure gradient in contributing to the boundary layer separation.

Consider the Blasius solution for the flow of a uniform stream of speed $U_1 = U_e = $ const. along a semi-infinite flat plate. The coordinate x is measured from the leading edge (see Fig. 14.3). The differential equation for f (with $\beta = 0$) becomes

$$f''' + ff'' = 0 \tag{14.45}$$

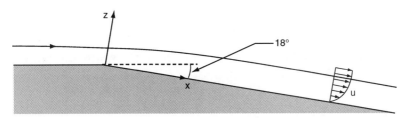

Figure 14.5 Falkner–Skan flow past a convex corner (the angle $18°$ represents the limit at which the flow will separate).

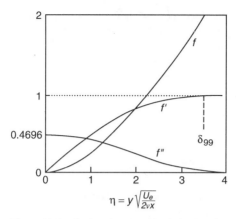

Figure 14.6 The function f and its derivatives for the Blasius solution of the flow over a flat plate.

where now $\eta = z/g$ and substituting g from Eq. (14.41a) we get

$$\eta = \frac{z}{g} = \frac{z}{\sqrt{2\nu x/U_e}} \tag{14.46}$$

Equation (14.45) was solved by Blasius using a power series expansion technique (see details in Yuan,[1,2] pp. 310–311), but here we provide a graphical description of this solution in Fig. 14.6 (based on the numerical results of White,[14.5] pp. 236–237). For example, from the graph we get the value of $f''(0) = 0.4696$. With this information the wall shear stress τ_w can be calculated as

$$\tau_w = \mu \left(\frac{\partial u}{\partial z} \right)_{z=0} = \frac{\mu U_e f''(0)}{\sqrt{2\nu x/U_e}}$$
$$= \frac{0.332}{\sqrt{U_e x/\nu}} \rho U_e^2 = \frac{0.332}{\sqrt{Re_x}} \rho U_e^2 \tag{14.47}$$

and here Re_x is defined as $Re_x = U_e x/\nu$. The friction coefficient becomes

$$C_f = \frac{\tau_w}{(1/2)\rho U_e^2} = \frac{0.664}{\sqrt{Re_x}} \tag{14.48}$$

The total friction force F (or drag D), per unit width, for one side of the plate is obtained by integrating along the plate:

$$F = \int_0^L \tau_w \, dx = 0.664 \rho U_e^2 \sqrt{\frac{\nu L}{U_e}}$$

and in nondimensional form, the total force coefficient, C_F, which may be viewed as the average friction coefficient for the plate, is

$$C_F = \frac{F}{(1/2)\rho U_e^2 L} = \frac{0.664 \rho U_e^2 \sqrt{\nu L/U_e}}{(1/2)\rho U_e^2} = \frac{1.328}{\sqrt{Re_L}} \tag{14.49}$$

14.3 Similar Solutions to the Boundary Layer Equations

The boundary layer thickness (defined as the transverse location where $u/U_e = 0.99$ and $\eta \approx 3.6$) is now

$$\frac{\delta_{99}}{x} = \frac{5.00}{\sqrt{Re_x}} \tag{14.50}$$

The displacement thickness is calculated by using Eq. (14.26):

$$\delta^* = \int_0^\infty \left(1 - \frac{u}{U_e}\right) dz = \sqrt{\frac{2\nu x}{U_e}} \int_0^\infty (1 - f'(\eta))\, d\eta$$

$$= \sqrt{\frac{2\nu x}{U_e}} \lim_{\eta \to \infty} [\eta - f(\eta)] = 1.217 \sqrt{\frac{2\nu x}{U_e}} \tag{14.51}$$

where the value of $[\eta - f(\eta)] = 1.217$ is taken from the tabulated results for f, f', and f'' in White.[14.5] Consequently, the displacement thickness growth is

$$\frac{\delta^*}{x} = \frac{1.721}{\sqrt{Re_x}} \tag{14.52}$$

The loss of momentum due to the presence of the boundary layer can be characterized by a momentum thickness θ in analogy with the definition of the displacement thickness:

$$\theta = \int_0^\infty \frac{u}{U_e} \left(1 - \frac{u}{U_e}\right) dz \tag{14.53}$$

The value of θ for the present case can be evaluated by using the same numerical values of the function f:

$$\theta = \sqrt{\frac{2\nu x}{U_e}} \int_0^\infty f'[1 - f'(\eta)]\, d\eta = 0.470 \sqrt{\frac{2\nu x}{U_e}} \tag{14.54}$$

and the momentum thickness growth is

$$\frac{\theta}{x} = \frac{0.664}{\sqrt{Re_x}} \tag{14.55}$$

Note that in some of the treatments of the Blasius solution in the literature η is defined without the 2 under the square-root sign and the differential equation (Eq. (14.45)) then has a 2 in front of the second term.

The Blasius solution for the boundary layer velocity profile is given by f' in Fig. 14.6. Comparison of this curve with experimental data in Fig. 14.7 shows good correlation, for the case of the flat plate. Even if the Reynolds number is low (e.g., $Re < 2,000$), this type of solution can be considered as local only, and for the case of the flow over a low Reynolds number airfoil, the velocity profile varies along the chord and the similarity assumption cannot be used.

As a final example for this group of solutions, consider the flow toward a stagnation point and let $U_e = ax$ (see Eq. (14.42) for $\beta = m = 1$). This solution is often used to provide the initial conditions (at the forward stagnation point) for the numerical solution for the boundary layer flow past an airfoil with a round leading edge. The differential equation (Eq. (14.39)) becomes

$$f''' + ff'' + 1 - f'^2 = 0 \tag{14.56}$$

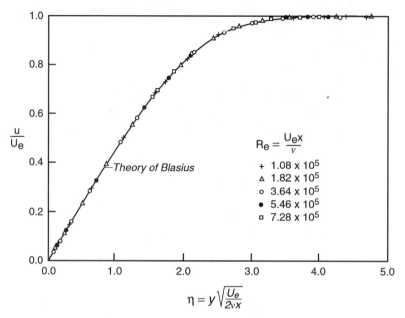

Figure 14.7 Comparison of Blasius theory for the velocity profile in the boundary layer with the experimental results of Nikuradse (from Ref. 1.6, p. 142).

where $\eta = z/[\nu/a]^{1/2}$. Based on the tabulated results for the solution of this equation taken from White[14.5] (Table 4.2, p. 246) the boundary layer thickness is constant and is

$$\delta = 2.4\sqrt{\frac{\nu}{a}}$$

The displacement and momentum thicknesses are also constant and given by

$$\delta^* = 0.6479\sqrt{\frac{\nu}{a}} \qquad (14.57)$$

$$\theta = 0.2924\sqrt{\frac{\nu}{a}} \qquad (14.58)$$

At this point it is useful to revisit the Falkner–Skan family of similarity solutions. In the discussion on Eq. (14.42) and (14.43) it was pointed out that for various values of β, different similarity profiles can be obtained. Three of these velocity profiles are presented in Fig. 14.8, which can be used to demonstrate the effect of the external pressure field on the boundary layer. The case for $\beta = 0$ is of course the similarity solution of Blasius and is the same as the one described in Fig. 14.6 or 14.7. The case for $\beta = 1$ can be used to show the effect of favorable pressure gradient, $(\partial p/\partial x) < 0$, on the boundary layer, which energizes it and reduces the displacement and momentum thicknesses. The opposite is true for boundary layers with adverse pressure gradient, $\partial p/\partial x > 0$, and the limiting case of *flow separation* is decribed by the case of $\beta = -0.199$. Note that at this condition, $(\partial u/\partial z)_{z=0} = 0$ and the skin friction τ_w is zero as well. For larger values of the adverse pressure gradient the flow near the wall will reverse, and the flow is called *separated* while in the case of $\beta = -0.199$ the flow is about to separate (but still flows parallel to the solid surface).

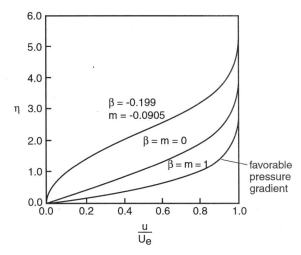

Figure 14.8 Falkner–Skan solutions for the boundary layer velocity profiles with and without external pressure gradients.

14.4 The von Karman Integral Momentum Equation

For many engineering applications, it is not necessary to obtain the details of the flow variables inside the boundary layer. For example, for design analyses, we need the wall shear stress to calculate the drag force on the airfoil and the displacement thickness to allow for coupling with the outer flow. These variables can be found from a solution of an integral version of the boundary layer equation that is attributed to von Karman. In spite of this approach being considered as "approximate," it does not rely on the similarity assumption and the shape of the boundary layer velocity profile along the surface can change significantly. Consequently, this method can be easily extended beyond the laminar flow discussion in this chapter to include effects such as boundary layer transition and separation.

For a derivation of the integral momentum equation, consider the continuity and boundary layer equations. It is possible to arrive at the von Karman integral equation by simply integrating Eq. (14.11) with respect to z from $z = 0$ to $z = \delta$:

$$\frac{\partial}{\partial t}\int_0^\delta u\,dz + \int_0^\delta u\frac{\partial u}{\partial x}\,dz + \int_0^\delta w\frac{\partial u}{\partial z}\,dz = \frac{-1}{\rho}\int_0^\delta \frac{\partial p}{\partial x}\,dz + \frac{\mu}{\rho}\int_0^\delta \frac{\partial^2 u}{\partial z^2}\,dz \tag{14.59}$$

The last term on the left-hand side can be integrated by parts:

$$\int_0^\delta w\frac{\partial u}{\partial z}\,dz = \int_0^\delta \frac{\partial(uw)}{\partial z}\,dz - \int_0^\delta u\frac{\partial w}{\partial z}\,dz$$

$$= uw\big|_0^\delta - \int_0^\delta u\frac{\partial w}{\partial z}\,dz$$

However, $w_{z=0} = 0$, and therefore the first term is $U_e w_{z=\delta}$, and this term becomes

$$\int_0^\delta w\frac{\partial u}{\partial z}\,dz = U_e\int_0^\delta \frac{\partial w}{\partial z}\,dz - \int_0^\delta u\frac{\partial w}{\partial z}\,dz$$

With the use of the continuity equation to replace the derivatives of w with those of u, this becomes

$$\int_0^\delta w \frac{\partial u}{\partial z} dz = -U_e \int_0^\delta \frac{\partial u}{\partial x} dz + \int_0^\delta u \frac{\partial u}{\partial x} dz \qquad (14.60)$$

Substituting this into Eq. (14.59), and keeping in mind that at $z = \delta$, $\partial u/\partial z = 0$ and $\tau = 0$, we get

$$\frac{\partial}{\partial t}\int_0^\delta u\, dz + 2\int_0^\delta u \frac{\partial u}{\partial x} dz - U_e \int_0^\delta \frac{\partial u}{\partial x} dz = \frac{-1}{\rho}\frac{dp}{dx}\delta - \frac{\tau_w}{\rho}$$

and finally

$$\frac{\partial}{\partial t}\int_0^\delta u\, dz + \int_0^\delta \frac{\partial u^2}{\partial x} dz - U_e \int_0^\delta \frac{\partial u}{\partial x} dz = \frac{-1}{\rho}\frac{dp}{dx}\delta - \frac{\tau_w}{\rho} \qquad (14.61)$$

With the use of the Leibnitz rule, the second and third terms in this equation become

$$\int_0^\delta \frac{\partial u^2}{\partial x} dz = \frac{d}{dx}\int_0^\delta u^2\, dz - U_e^2 \frac{d\delta}{dx}$$

$$U_e \int_0^\delta \frac{\partial u}{\partial x} dz = U_e \frac{d}{dx}\int_0^\delta u\, dz - U_e^2 \frac{d\delta}{dx}$$

and Eq. (14.61) reduces to

$$\frac{\partial}{\partial t}\int_0^\delta u\, dz + \frac{d}{dx}\int_0^\delta u^2\, dz - U_e \frac{d}{dx}\int_0^\delta u\, dz = \frac{-1}{\rho}\frac{dp}{dx}\delta - \frac{\tau_w}{\rho} \qquad (14.62)$$

This is the von Karman integral momentum equation for the boundary layer.

This can be further simplified by assuming steady-state conditions, $\partial/\partial t = 0$, and by eliminating the pressure term using the momentum equation (Eq. (14.14)) outside the boundary layer. We then have

$$U_e \frac{\partial U_e}{\partial x} = \frac{-1}{\rho}\frac{\partial p}{\partial x} \qquad (14.63)$$

Substitution of these results into Eq. (14.62) results in

$$\frac{d}{dx}\int_0^\delta u^2\, dz - U_e \frac{d}{dx}\int_0^\delta u\, dz - U_e \frac{dU_e}{dx}\delta = -\frac{\tau_w}{\rho} \qquad (14.64)$$

But

$$-\frac{d}{dx}\left(\int_0^\delta u U_e\, dz\right) dx = U_e \frac{d}{dx}\left(\int_0^\delta u\, dz\right) dx + \frac{dU_e}{dx}\left(\int_0^\delta u\, dz\right) dx$$

and with this in mind Eq. (14.64) can be rearranged as

$$\frac{d}{dx}\int_0^\delta (U_e - u)u\, dz + \frac{dU_e}{dx}\int_0^\delta (U_e - u)\, dz = \frac{\tau_w}{\rho} \qquad (14.65)$$

Now recall the definitions of the displacement thickness δ^*:

$$\delta^* = \int_0^\infty \left(1 - \frac{u}{U_e}\right) dz \simeq \int_0^{z_e}\left(1 - \frac{u}{U_e}\right) dz = \int_0^\delta \left(1 - \frac{u}{U_e}\right) dz \qquad (14.26)$$

Note that $z_e \geq \delta$, as discussed earlier (in Section 14.2). Similarly, for the momentum

14.4 The von Karman Integral Momentum Equation

thickness θ we get

$$\theta = \int_0^\delta \left(1 - \frac{u}{U_e}\right)\frac{u}{U_e} dz \tag{14.53}$$

An important observation is that the momentum thickness at a streamwise station is proportional to the drag from the leading edge to that station. If the drag per unit width D is defined as the momentum deficiency $\rho \int_0^\delta (U_e - u)u\, dz$, then by comparing with Eq. (14.53) we get

$$D = \rho \int_0^\delta (U_e - u)u\, dz = \rho U_e^2 \theta \tag{14.66}$$

With the aid of these two boundary layer thicknesses, Eq. (14.65) can be rewritten as

$$\frac{d}{dx}(U_e^2 \theta) + U_e \delta^* \frac{dU_e}{dx} = \frac{\tau_w}{\rho} \tag{14.67a}$$

or

$$\frac{d\theta}{dx} + (2\theta + \delta^*)\frac{1}{U_e}\frac{dU_e}{dx} = \frac{\tau_w}{\rho U_e^2} \tag{14.67b}$$

This equation is an ordinary differential equation and contains two integral quantities, the displacement and momentum thicknesses. When using the shape factor H and the friction coefficient C_f, a dimensionless version of the integral momentum equation is obtained:

$$\frac{d\theta}{dx} + (H + 2)\frac{\theta}{U_e}\frac{dU_e}{dx} = \frac{C_f}{2} \tag{14.67c}$$

where

$$C_f = \frac{\tau_w}{\frac{1}{2}\rho U_e^2} \tag{14.68}$$

$$H = \frac{\delta^*}{\theta} \tag{14.69}$$

Based on the results of the Blasius solution, the momentum thickness is smaller than the displacement thickness, and therefore the shape factor is larger than 1 (actually $H = 2.59$ for the Blasius solution).

A more intuitive method, based on the control volume approach, for developing the boundary layer integral formulation is worth presenting. Consider the two-dimensional segment of the boundary layer as depicted by Fig. 14.9. The rectangular control volume

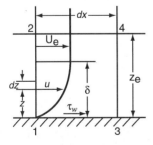

Figure 14.9 Schematic description of the control volume used to formulate the integral boundary layer equations.

(actually control surface) is bounded by the four corners 1–2–3–4. The plane 2–4 is placed above the boundary layer, at $z = z_e$, where there are no transverse changes in the u component of the velocity; and a constant outer speed U_e prevails. The mass flow rate entering the control element through plane 1–2 is therefore

$$\rho \int_0^{z_e} u\, dz$$

The flow rate leaving through plane 3–4 can be approximated by using the first term of a Taylor series:

$$\rho \int_0^{z_e} u\, dz + \rho \frac{d}{dx}\left(\int_0^{z_e} u\, dz\right) dx$$

Since there is no flow across the wall (plane 1–3), the net change in the mass flow rate must have entered through plane 2–4 and is

$$\rho \frac{d}{dx}\left(\int_0^{z_e} u\, dz\right) dx \tag{14.70}$$

In a similar manner, the momentum in the x direction entering across plane 1–2 is

$$\rho \int_0^{z_e} u^2\, dz \tag{14.71}$$

and that leaving through plane 3–4 is

$$\rho \int_0^{z_e} u^2\, dz + \rho \frac{d}{dx}\left(\int_0^{z_e} u^2\, dz\right) dx \tag{14.72}$$

The mass flow rate entering through plane 2–4, expressed by Eq. (14.70), has a constant speed of U_e outside the control surface. Therefore, the momentum entering this plane is

$$\rho U_e \frac{d}{dx}\left(\int_0^{z_e} u\, dz\right) dx \tag{14.73}$$

The time rate of the momentum change within the control surface is

$$\rho \frac{\partial}{\partial t}\left(\int_0^{z_e} u\, dz\right) dx \tag{14.74}$$

Thus, the net rate of change of the momentum in the x direction is due to the change with time (Eq. (14.74)) and due to the difference between the momentum leaving (Eq. (14.72)) and entering (Eqs. (14.71) and (14.73)) the control surface:

$$\rho \frac{\partial}{\partial t}\left(\int_0^{z_e} u\, dz\right) dx + \rho \int_0^{z_e} u^2\, dz + \rho \frac{d}{dx}\left(\int_0^{z_e} u^2\, dz\right) dx$$
$$- \rho \int_0^{z_e} u^2\, dz - \rho U_e \frac{d}{dx}\left(\int_0^{z_e} u\, dz\right) dx$$
$$= \rho \frac{\partial}{\partial t}\left(\int_0^{z_e} u\, dz\right) dx + \rho \frac{d}{dx}\left(\int_0^{z_e} u^2\, dz\right) dx - \rho U_e \frac{d}{dx}\left(\int_0^{z_e} u\, dz\right) dx$$
$$\tag{14.75}$$

According to the momentum principle (Eq. (1.17)), this change in the linear momentum must be equal to the forces acting on the control surface. Since the body forces were neglected the only forces acting are the pressure and the laminar shear stress on the wall. Using Eq. (1.12), we obtain the shear force on the wall, along the segment 1–3:

$$-\tau_w \, dx = -\mu \frac{\partial u}{\partial z}\bigg|_{z=0} dx \tag{14.76}$$

Since according to Eq. (14.12) the pressure is independent of z, then $p(x)$ is a function of x only. Consequently, the pressure force on segment 1–2 is pz_e and on segment 3–4 is

$$-\left(p + \frac{dp}{dx} dx\right) z_e$$

and the net force acting on the control surface is the sum of the shear and pressure forces:

$$-\tau_w \, dx - \frac{dp}{dx} dx z_e \tag{14.77}$$

Equating the forces in Eq. (14.77) with the change in the momentum in Eq. (14.75) results in

$$\rho \frac{\partial}{\partial t}\left(\int_0^{z_e} u \, dz\right) dx + \rho \frac{d}{dx}\left(\int_0^{z_e} u^2 \, dz\right) dx$$
$$- \rho U_e \frac{d}{dx}\left(\int_0^{z_e} u \, dz\right) dx = -\tau_w \, dx - \frac{dp}{dx} dx z_e$$

Now if we let $z_e \to \delta$ and divide by $\rho \, dx$, we obtain the von Karman integral equation for the boundary layer:

$$\frac{\partial}{\partial t} \int_0^\delta u \, dz + \frac{d}{dx} \int_0^\delta u^2 \, dz - U_e \frac{d}{dx} \int_0^\delta u \, dz = -\frac{\tau_w}{\rho} - \frac{1}{\rho}\frac{dp}{dx}\delta \tag{14.62}$$

14.5   Solutions Using the von Karman Integral Equation

The integral momentum equation (e.g., Eq. (14.67)) has three unknowns (displacement thickness, momentum thickness, and wall shear stress) and cannot be solved without additional information. The necessary information is obtained from an assumed profile family $f(\eta)$, which in principle defines the entire boundary layer velocity field $u(x, z)$:

$$\frac{u(x, z)}{U_e} = f(\eta; \mathcal{P})$$

$$\eta = \frac{z}{\delta(x)}$$

The transverse length scale $\delta(x)$ is comparable to the local boundary layer thickness, and $\mathcal{P}$ is a profile parameter that determines the shape of the local velocity profile. Different integral boundary layer calculation methods make different choices for δ and $\mathcal{P}$. Two examples will follow; the first is in the spirit of Pohlhousen (see Ref. 1.2, Section 9.6) and the second is the one proposed by Thwaites.[14.6]

14.5.1 Approximate Polynomial Solution

The following example demonstrates the simplicity of the integral approach; by suggesting an approximate velocity distribution within the boundary layer, parameters such as the boundary layer thicknesses and skin-friction coefficient can be readily calculated. For example, even simple polynomial velocity profiles can be used:

$$\frac{u}{U_e} = f(\eta) = a_0 + a_1\eta + a_2\eta^2 + \cdots, \quad 0 \leq \eta \leq 1$$

and for $\eta > 1$, $f(\eta) = 1$. Pohlhausen (see Ref. 1.2, Section 9.6) used a fourth-order polynomial to develop a set of solutions including the effect of the pressure gradient inside the boundary layer. In the spirit of his solution we demonstrate the case for the steady-state boundary layer along a flat plate, without pressure gradient. The proposed velocity function is then

$$\frac{u}{U_e} = a_1\eta + a_2\eta^2 + a_3\eta^3 + a_4\eta^4 \tag{14.78}$$

The boundary conditions for the original boundary layer problem (Eqs. (14.10)–(14.12)) are

$$\text{at} \quad z = 0, \quad u = w = 0$$

$$\text{at} \quad z = \delta, \quad u = U_e, \quad \frac{\partial u}{\partial z} = 0$$

and the requirement for smooth transition at the outer edge of the boundary layer forces $\partial u/\partial z = 0$ at $z = \delta$. Additional boundary conditions can be generated by observing the change of the streamwise momentum inside the boundary layer (Eq. (14.11a)), but, without the pressure gradient. Thus, assuming $\partial U_e/\partial x = 0$, combined with the previous boundary conditions, results in (see Eq. (14.11a))

$$\text{at} \quad z = 0, \quad \frac{\partial^2 u}{\partial z^2} = 0$$

$$\text{at} \quad z = \delta, \quad \frac{\partial^2 u}{\partial z^2} = 0$$

Applying these additional conditions to the velocity function of Eq. (14.78) we get

$$\frac{u}{U_e} = 2\eta - 2\eta^3 + \eta^4 \tag{14.79}$$

To solve the problem for δ we substitute the velocity profile into von Karman's integral equation, but first the wall shear stress is calculated:

$$\tau_w = \mu \left(\frac{\partial u}{\partial z}\right)\bigg|_{z=0} = \mu \frac{2U_e}{\delta} \tag{14.80}$$

Substituting this and the velocity profile into Eq. (14.62), without the pressure term, we obtain

$$\frac{d}{dx} \int_0^\delta U_e^2 \left[2\left(\frac{z}{\delta}\right) - 2\left(\frac{z}{\delta}\right)^3 + \left(\frac{z}{\delta}\right)^4\right]^2 dz$$

$$- U_e \frac{d}{dx} \int_0^\delta U_e \left[2\left(\frac{z}{\delta}\right) - 2\left(\frac{z}{\delta}\right)^3 + \left(\frac{z}{\delta}\right)^4\right] dz = -\nu \frac{2U_e}{\delta}$$

Evaluating the two integrals, we get

$$U_e^2 \frac{d}{dx}(0.5825\delta) - U_e^2 \frac{d}{dx}(0.7000\delta) = -\nu \frac{2U_e}{\delta}$$

14.5 Solutions Using the von Karman Integral Equation

and after rearranging

$$\delta \frac{d\delta}{dx} = 17.021 \frac{\nu}{U_e}$$

Integrating with $\delta = 0$ at $x = 0$ we get

$$\delta = 5.836 \frac{x}{\sqrt{Re_x}} \tag{14.81}$$

With this solution for δ and with the velocity profile of Eq. (14.78), $u(x, z)$ is known everywhere and the values for the boundary layer thicknesses are calculated:

$$\delta^* = \int_0^\infty \left(1 - \frac{u}{U_e}\right) dz = 1.751 \frac{x}{\sqrt{Re_x}}$$

$$\theta = \int_0^\infty \left(1 - \frac{u}{U_e}\right) \frac{u}{U_e} dz = 0.685 \frac{x}{\sqrt{Re_x}}$$

or

$$\frac{\delta^*}{x} = \frac{1.751}{\sqrt{Re_x}} \tag{14.82}$$

$$\frac{\theta}{x} = \frac{0.685}{\sqrt{Re_x}} \tag{14.83}$$

and the skin-friction coefficient is

$$C_f = \frac{\tau_w}{(1/2)\rho U_e^2} = \frac{0.6854}{\sqrt{Re_x}} \tag{14.84}$$

This method can easily be extended to include the effect of streamwise pressure gradient, a feature required for airfoil analysis (see, for example, Schlichting,[1.6] pp. 206–211).

14.5.2 The Correlation Method of Thwaites

As a second example of an integral boundary layer method, the method due to Thwaites[14.6] will be presented. The presentation follows the approach in White[14.5] (pp. 268–270). Start with the dimensionless form of the integral momentum equation (Eq. (14.67c)) and multiply by $U_e \theta / \nu$ to get

$$\frac{U_e \theta}{\nu} \frac{d\theta}{dx} + (H + 2) \frac{\theta^2}{\nu} \frac{dU_e}{dx} = \frac{\tau_w \theta}{\mu U_e} \tag{14.85}$$

The term on the right-hand side, the shear correlation function, is denoted by S and is a function of the profile shape alone. The profile parameter $\mathcal{P}$ is denoted by λ and defined by

$$\lambda = \frac{\theta^2}{\nu} \frac{dU_e}{dx} \tag{14.86}$$

Consequently, H and S are now functions of λ:

$$S(\lambda) = \frac{\tau_w \theta}{\mu U_e} \tag{14.87a}$$

$$H(\lambda) = \frac{\delta^*}{\theta} \tag{14.87b}$$

With these definitions and recalling that $\theta\, d\theta = d(\theta^2/2)$ we have for the momentum integral equation

$$U_e \frac{d}{dx} \frac{\lambda}{dU_e/dx} = 2\{S(\lambda) - \lambda[H(\lambda) + 2]\} = F(\lambda) \tag{14.88}$$

Thwaites[14.6] used a compilation of the available analytical and experimental results to obtain the following curve fit for the function F:

$$F(\lambda) = 0.45 - 6.0\lambda \tag{14.89}$$

Using this result, Eq. (14.88) is integrated to yield the momentum thickness (for more details see White,[14.5] p. 269):

$$\theta^2 = \frac{0.45\nu}{U_e^6(x)} \int_0^x U_e^5(x)\, dx \tag{14.90}$$

This result provides an approximate solution for steady, two-dimensional, incompressible laminar boundary layers.

To develop a solution to a particular problem, we start with the inviscid solution U_e. Equation (14.90) is then integrated to obtain θ and, along with U_e, we find λ from Eq. (14.86). Curve fits for S and H are given in Cebeci and Bradshaw[14.7] (p. 110) as follows: for $0 < \lambda < 0.1$

$$\begin{aligned} S &= 0.22 + 1.57\lambda - 1.80\lambda^2 \\ H &= 2.61 - 3.75\lambda + 5.24\lambda^2 \end{aligned} \tag{14.91a}$$

and for $-0.1 < \lambda < 0$

$$\begin{aligned} S &= 0.22 + 1.402\lambda + 0.018 \frac{\lambda}{(\lambda + 0.107)} \\ H &= 2.088 + \frac{0.0731}{(\lambda + 0.14)} \end{aligned} \tag{14.91b}$$

For each point x along the surface, the skin-friction coefficient is found from Eq. (14.87a) and the corresponding value of S:

$$C_f = \frac{2\mu}{\rho U_e \theta} S(\lambda) \tag{14.92}$$

As two examples of the Thwaites[14.6] method, consider the flat plate and the stagnation point boundary layer solutions of the previous section. First, for the Blasius flat-plate boundary layer, the momentum thickness and displacement thickness are

$$\frac{\delta^*}{x} = \frac{1.721}{\sqrt{Re_x}} \tag{14.52}$$

$$\frac{\theta}{x} = \frac{0.664}{\sqrt{Re_x}} \tag{14.55}$$

For the Thwaites result, let $U_e = U_1$ and integrate to get

$$\theta^2 = 0.45 \frac{\nu x}{U_1}$$

and therefore

$$\frac{\theta}{x} = \frac{0.6708}{\sqrt{Re_x}} \tag{14.93}$$

Since the outer velocity is constant, $\lambda = 0$, and from the curve fit (Eq. (14.91a)), $H = 2.61$. The displacement thickness is then obtained from Eq. (14.87b) as

$$\frac{\delta^*}{x} = \frac{1.7507}{\sqrt{Re_x}} \tag{14.94}$$

For the stagnation point flow, the momentum thickness and the displacement thickness are

$$\delta^* = 0.6479 \left(\frac{\nu}{a}\right)^{1/2} \tag{14.57}$$

$$\theta = 0.2924 \left(\frac{\nu}{a}\right)^{1/2} \tag{14.58}$$

Using the Thwaites method for this case, let $U_e = ax$ (see Eq. (14.42) for $m = 1$) and integrate Eq. (14.90) to get

$$\theta^2 = 0.075 \frac{\nu}{a}$$

and therefore

$$\theta = 0.2739 \left(\frac{\nu}{a}\right)^{1/2} \tag{14.95}$$

The profile parameter, Eq. (14.86), is then given by $\lambda = .075$ and from the curve fit, Eq. (14.91a), we get $H = 2.358$. The displacement thickness is then given by Eq. (14.87b):

$$\delta^* = 0.6459 \left(\frac{\nu}{a}\right)^{1/2} \tag{14.96}$$

For the two cases, the displacement thicknesses differ by less than 2% and the momentum thicknesses by less than 1% from the corresponding similarity solutions.

14.6 Weak Interactions, the Goldstein Singularity, and Wakes

The iteration process, which starts with a solution of the inviscid flow equations and proceeds sequentially to a solution of the boundary layer equations, has been called a weak interaction. The pressure distribution and external velocity $U_e(x)$ are prescribed for the boundary layer calculation. We have studied similar solutions (Falkner–Skan) and approximate solutions using an integral version of the boundary layer momentum equation. To continue the iteration the inviscid flow equations are solved with the displacement thickness interaction included and therefore the boundary layer solution must include results for the trailing wake.

Goldstein[14.8] studied analytically the boundary layer flow past a finite length flat plate aligned parallel to a uniform stream and showed that a singularity was present at the trailing edge and that the boundary layer equations could not be integrated into the wake. This result would clearly also apply to an airfoil with a cusped trailing edge. The source of the difficulty was attributed to the discontinuity in boundary conditions at the trailing edge (no slip on the plate side and vanishing shear stress on the wake side).

In addition to the above problem with the weak interaction version of the equations at the trailing edge, researchers developing numerical methods[14.9] to integrate the boundary layer equations discovered that the calculations could not be continued past a streamwise location where the wall shear stress approached zero. The boundary layer equations are parabolic partial differential equations where the flow variables are independent of downstream conditions for positive values of the streamwise velocity component. The equations represent a balance of inertia, friction, and pressure. Since the flow momentum is small in the boundary layer, an adverse pressure gradient can lead to the skin friction approaching zero followed by a region of *reversed* flow. Under these conditions, the streamlines near the wall would appear to *separate* from the wall. This singularity at a point of flow separation was also studied analytically by Goldstein.[14.10] Using Goldstein's results, Brown and Stewartson[14.11] showed that the slope of the displacement thickness grows without bound as the separation point is approached.

For an airfoil with a finite trailing-edge angle (e.g., the van de Vooren airfoil in Chapter 6), the trailing edge is a stagnation point in the inviscid flow. The numerical solutions in Chapter 11 show a steep decrease in surface speed as the trailing edge is approached, which corresponds to a sharp increase in pressure. This strong adverse pressure gradient in the neighborhood of the trailing edge in all likelihood will lead to flow separation upstream of the edge (see Stewartson[14.12]).

It therefore appears that even for airfoil flows without separation (attached flows) the boundary layer equations (in their weak interaction version) cannot be integrated beyond the trailing edge and that a second iteration is not possible. In addition, the equations cannot be integrated beyond a point of flow separation. The singularities at a separation point and the trailing edge were originally thought to indicate that the boundary layer equations of Prandtl were invalid at these points. It is now known (see, for example, Stewartson[14.12]) that a regular solution of the boundary layer equations is possible in the vicinity of the point of vanishing skin friction if the pressure and outer flow velocity are not prescribed in advance. It is the weak interaction version of the equations that is invalid and a strong interaction version must be used. (Of course, the original assumptions involved in the derivation of the boundary layer equations must still be met; therefore only flows with mild separation can be treated.) In this version, the inviscid and viscous equations cannot be solved sequentially but must be coupled in some fashion. Catherall and Mangler[14.13] were the first to integrate the boundary layer equations through a separation point. In their method the displacement thickness was prescribed. Goldstein[14.8] derived a solution for the development of the near wake downstream of the trailing edge of a finite flat plate but details in the neighborhood of the trailing edge were not provided. Stewartson[14.14] and Messiter[14.15] independently derived a local solution that provided the bridge between the Blasius solution upstream of the trailing edge and the Goldstein near wake solution downstream of the edge. This solution, valid in a streamwise region of extent $O(Re^{-3/8})$, provides the displacement thickness interaction by an asymptotic matching of flows in three transverse layers starting at the plate (and hence giving it the name "triple-deck" theory).

The extension of the boundary layer into the wake behind a flat plate is shown schematically in Fig. 14.10. At the vicinity of the trailing edge the above discussed difficulties

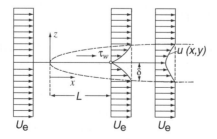

Figure 14.10 Wake model behind a flat plate.

prevail; however, far behind the plate the streamwise velocity component approaches the free-stream value and a solution can be obtained from a linearized version of the boundary layer equations. The streamwise velocity component u in the far wake ($x \gg L$) can be written (see Schlichting,[1.6] p. 177)

$$\frac{u}{U_e} = u_c e^{-z^2 U_e/4xv} \qquad (14.97a)$$

where the wake centerline velocity u_c is

$$u_c = C\sqrt{\frac{L}{x}} \qquad (14.97b)$$

Here C is a constant, and Eq. (14.97b) describes the decay of the centerline velocity while the exponential term in Eq. (14.97a) represents a Gaussian velocity profile behind the plate. The force per unit span due to this velocity defect is

$$F = \rho U_e \int_{-\infty}^{\infty} u \, dz = C 2\rho U_e^2 \sqrt{\frac{\pi v L}{U_e}}$$

The drag coefficient C_d for both sides of the plate is then

$$C_d = \frac{F}{(1/2)\rho U_e^2 L}$$

and by equating these two expressions we get

$$C = \frac{C_d}{4}\sqrt{\frac{U_e L}{\pi v}}$$

and the velocity profile becomes

$$\frac{u}{U_e} = C_d \sqrt{\frac{U_e L}{16\pi v}\frac{L}{x}} e^{-z^2 U_e/4xv} = \frac{C_d}{4}\sqrt{\frac{Re_L}{\pi}}\sqrt{\frac{L}{x}} \cdot e^{-z^2 U_e/4xv} \qquad (14.98)$$

Using the results from the Blasius flat plate (e.g., Eq. (14.55) and (14.66)) for both sides to evaluate C_d we get

$$\frac{u}{U_e} = \frac{0.664}{\sqrt{\pi}}\sqrt{\frac{L}{x}} e^{-z^2 U_e/4xv} \qquad (14.99)$$

and this formulation is considered to be valid for $x > 3L$. Of course, this simple wake model serves only to demonstrate the principle of convecting the skin-friction effects behind the body. For most practical cases, however, effects such as transition to turbulent flow, trailing-edge separation, or shear layer instability will complicate the flow within the wake.

14.7 Two-Equation Integral Boundary Layer Method

Traditional boundary layer methods are tied to the approximation of a thin viscous layer and the fact that the pressure and external velocity are known (from a previous inviscid solution). For limited regions of separation or in the neighborhood of the trailing edge the basic assumptions of boundary layer theory remain valid, but we have shown that we can not specify the pressure and external velocity. The boundary layer adjusts through the displacement effect and interactive approaches are needed. The boundary layer must be solved in an inverse mode (external velocity not specified) rather than in the traditional (direct) mode. One-equation integral methods (such as the method due to Thwaites) tie the local profile shape to the local pressure gradient and are therefore not suited for flows with strong interaction.

Two-equation integral boundary layer methods eliminate the direct link between the profile shape and pressure gradient and are therefore satisfactory for treating flows with strong interaction. A profile family $f(\eta; \mathcal{P})$ defines the flowfield and the profile parameter is chosen as $\mathcal{P} = H(x)$, the shape factor. An integral kinetic energy equation can be obtained by a suitable manipulation of the integral continuity and momentum equations (see White,[14.5] pp. 266–267) and is given in dimensional and dimensionless forms as

$$\frac{d}{dx}(U_e^3 \theta^*) = \frac{2\Delta}{\rho} \tag{14.100}$$

$$\frac{d\theta^*}{dx} + 3\frac{\theta^*}{U_e}\frac{dU_e}{dx} = 2C_\Delta \tag{14.101}$$

The dissipation integral Δ and associated dissipation coefficient C_Δ are defined as follows:

$$\Delta = \int_0^{z_e} \tau \frac{\partial u}{\partial z} dz = \int_0^{z_e} \mu \left(\frac{\partial u}{\partial z}\right)^2 dz \tag{14.102a}$$

$$C_\Delta = \frac{\Delta}{\frac{1}{2}\rho U_e^3} \tag{14.102b}$$

and the kinetic energy thickness θ^* is defined as

$$\theta^* = \int_0^{z_e} \left(1 - \frac{u^2}{U_e^2}\right)\frac{u}{U_e} dz \tag{14.103}$$

The kinetic energy shape parameter equation is obtained by combining the integral momentum and kinetic energy equations (Eqs. (14.67c) and (14.101)):

$$\frac{\theta}{\theta^*}\left\{\frac{d\theta^*}{dx} + \frac{3\theta^*}{U_e}\frac{dU_e}{dx} - 2C_\Delta\right\} - \left\{\frac{d\theta}{dx} + (H+2)\frac{\theta}{U_e}\frac{dU_e}{dx} - \frac{C_f}{2}\right\} = 0$$

and after rearranging the terms we get

$$\frac{\theta}{H^*}\frac{dH^*}{dx} = \frac{2C_\Delta}{H^*} - \frac{C_f}{2} + (H-1)\frac{\theta}{U_e}\frac{dU_e}{dx} \tag{14.104}$$

where the kinetic energy shape parameter is defined as $H^* = \theta^*/\theta$. Now there is no explicit link between H and the local external velocity.

Three closure relations for H^*, C_f, and C_Δ are now required to integrate the momentum and kinetic energy shape parameter equations simultaneously. They may be written as

$$H^* = f_1(H) \tag{14.105}$$

$$Re_\theta \frac{C_f}{2} = f_2(H) \tag{14.106}$$

$$Re_\theta \frac{2C_\Delta}{H^*} = f_3(H) \tag{14.107}$$

where $Re_\theta = U_e \theta / \nu$ is the momentum thickness Reynolds number. These three functions are numerically computed from the Falkner–Skan profiles and the following wall boundary layer curve fits represent these closure relations (see Drela and Giles[14.16]):

$$H^* = \begin{cases} 1.515 + 0.076\frac{(H-4)^2}{H}, & H < 4 \\ 1.515 + 0.040\frac{(H-4)^2}{H}, & H > 4 \end{cases} \tag{14.108}$$

$$Re_\theta \frac{C_f}{2} = \begin{cases} -0.067 + 0.01977\frac{(7.4-H)^2}{H-1}, & H < 7.4 \\ -0.067 + 0.022\left(1 - \frac{1.4}{H-6}\right)^2, & H > 7.4 \end{cases} \tag{14.109}$$

$$Re_\theta \frac{2C_\Delta}{H^*} = \begin{cases} 0.207 + 0.00205(4-H)^{5.5}, & H < 4 \\ 0.207 - 0.003(H-4)^2, & H > 4 \end{cases} \tag{14.110}$$

The corresponding closure relations for laminar wake profiles are

$$H^* = \begin{cases} 1.50 + 0.025(3.5-H)^3 + 0.001(3.5-H)^5, & H < 3.5 \\ 1.50 + 0.015\frac{(H-3.5)^2}{H}, & H > 3.5 \end{cases} \tag{14.111}$$

$$Re_\theta \frac{C_f}{2} = 0 \tag{14.112}$$

$$Re_\theta \frac{2C_\Delta}{H^*} = 1.52\frac{(H-1)^2}{3+H^3} \tag{14.113}$$

Let us consider the kinetic energy shape parameter equation (Eq. (14.104)). To illustrate the presence of the Goldstein singularity for the direct method (U_e specified) and the absence of the singularity in the inverse method (U_e unspecified), first note that at incipient separation $H = 4.0$ and from the curve fit for H^*, we get $dH^*/dH = 0$ at this point. Note that this equation can be thought of as an equation for $H(x)$ in attached flow and an equation for $U_e(x)$ in separated flow. If the equation is recast as

$$\frac{dH}{dx} = \frac{H^*}{dH^*/dH}\left\{\frac{2C_\Delta}{H^*\theta} - \frac{C_f}{2\theta} + (H-1)\frac{1}{U_e}\frac{dU_e}{dx}\right\} \tag{14.114}$$

and $U_e(x)$ is specified, then the right-hand side is seen to blow up as separation is approached. If, however, the equation is recast as

$$\frac{dU_e}{dx} = U_e\frac{1}{H-1}\left\{\frac{C_f}{2\theta} - \frac{2C_\Delta}{H^*\theta} + \frac{1}{H^*}\frac{dH^*}{dH}\frac{dH}{dx}\right\} \tag{14.115}$$

then it can be integrated for U_e through separation.

14.8 Viscous–Inviscid Interaction Method

To illustrate the development of a viscous–inviscid interaction method, we will consider a steady incompressible laminar flow for simplicity. The viscous flow past an airfoil can be constructed by a combination of the real viscous flow (RVF), which includes the boundary layer and wake, and an equivalent inviscid flow (EIF) over the entire flowfield (see Fig. 14.11). The EIF is fictitious in the boundary layer and wake where it overlaps the

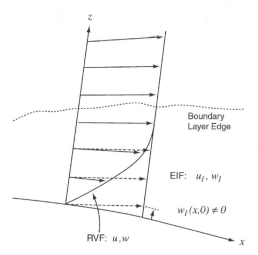

Figure 14.11 Nomenclature for viscous–inviscid interaction. (Courtesy of M. Drela.)

RVF. These two flows are defined with the following properties:

> The RVF is governed by the boundary layer equations.
> The EIF is potential and is governed by Laplace's equation.
> The EIF and RVF match at the outer edge of the RVF region.

The matching requirement is imposed at a location z_e and must be enforced for both velocity components. Since the EIF is irrotational, it can be constructed using a panel method with vortices (or doublets) and sources on the airfoil surface. We have previously shown that the tangential component of velocity at the edge of the boundary layer is

$$u(x, z_e) = U_e(x) = u_I(x, 0) \tag{14.116}$$

where the I subscript stands for inviscid. In addition, the normal component of velocity at the wall for the inviscid flow is (see Eq. (14.27))

$$w_I(x, 0) = \frac{d(U_e \delta^*)}{dx} \tag{14.117}$$

This normal velocity component can be included in the panel method solution by adding an additional source density equal to the transpiration velocity. As an example, consider a panel method using doublets and sources and the Dirichlet boundary condition on the airfoil surface (see Section 11.3.1). Then the source strength of Eq. (9.12) is modified such that

$$\sigma = \mathbf{n} \cdot \mathbf{Q}_\infty - \frac{d(U_e \delta^*)}{dx} \tag{14.118}$$

In addition, we need to add a source distribution along the wake. The wake centerline is taken along the inviscid streamline leaving the airfoil trailing edge, and the displacement thickness and momentum thickness in the wake are simply taken as the sum of these thicknesses above and below the centerline. The displacement body is then continuous.

Numerous methods have been developed for the calculation of flows with strong interaction by iterated coupling between potential flow solvers and traditional boundary layer marching solvers. For simplicity (and without a loss of generality in the discussion of basic principles), we will consider the coupling between an integral boundary layer method and

14.8 Viscous–Inviscid Interaction Method

a potential flow panel method. Three basic approaches have appeared in the literature to tackle a general viscous–inviscid interaction problem with limited separation. They are the quasi-simultaneous method of Veldman,[14.17] the semi-inverse method of Le Balleur,[14.18] and the fully simultaneous method of Drela (see Drela[14.19] and Drela and Giles[14.16]). These approaches are discussed in Wolles and Hoeijmakers[14.20] (with additional details provided in Wolles[14.21]). In the approach of Veldman, an interaction law (which models the behavior of the outer flow) is solved simultaneously with the boundary layer equations. In Le Balleur's approach, the coupling between the inner and outer flows is achieved through a relaxation formula based on the stability analysis of the local flow solution. Even with the coupling (and lack of hierarchy) of these approaches, they are adaptations of the direct and inverse techniques and involve sequential solution of the viscous and inviscid flow equations. The fully simultaneous approach of Drela eliminates this sequential solution of the equations and it is this approach that will be described here. It is also the method used in Wolles and Hoeijmakers.

The coupled system of equations to be solved consists of the two integral boundary layer equations for the RVF, the momentum and kinetic energy shape parameter equations, and the panel method equation for the doublet strength (e.g., Eq. (11.71) from Section 11.3.1) to model the EIF. These are

$$\frac{d\theta}{dx} + (H+2)\frac{\theta}{U_e}\frac{dU_e}{dx} = \frac{C_f}{2} \tag{14.119}$$

$$\frac{1}{H^*}\frac{dH^*}{dx} = \frac{2C_\Delta}{H^*\theta} - \frac{C_f}{2\theta} + (H-1)\frac{1}{U_e}\frac{dU_e}{dx} \tag{14.120}$$

$$\sum_{j=1}^{M} a_{ij} m_j + \sum_{j=1}^{M} c_{ij} \mu_j = \text{RHS}_i \tag{14.121}$$

Note that here the known part of the source distribution (due to the free stream) is moved to the right-hand side as RHS_i and only the additional sources of strength $-dm/dx = -d(U_e\delta^*)/dx$ representing the displacement surface on the airfoil and in the wake remain unknown. Also, M represents the number of nodes on the airfoil and the wake ($M > N$). The new coefficient a_{ij} is related to the original source influence coefficient b_{ij} by

$$a_{ij} = \frac{b_{ij} - b_{i,j-1}}{x_{j-1} - x_j}$$

since the additional source strength is $(m_j - m_{j-1})/(x_{j-1} - x_j)$. In summary, the unknowns are the viscous variables θ, the momentum thickness, and $m = U_e\delta^*$, the mass defect, and the doublet strength μ. The discrete version of the system of equations is written in residual form as

$$\frac{\Delta\theta}{\theta} + (H+2)\frac{\Delta U_e}{U_e} - \frac{C_f}{2}\frac{\Delta x}{\theta} \equiv R_{1_i}(\mu_j, \theta_j, m_j) = 0 \tag{14.122}$$

$$\frac{\Delta H^*}{H^*} + (1-H)\frac{\Delta u_e}{u_e} + \left(\frac{C_f}{2} - \frac{2C_\Delta}{H^*}\right)\frac{\Delta x}{\theta} \equiv R_{2_i}(\mu_j, \theta_j, m_j) = 0 \tag{14.123}$$

$$\sum_{j=1}^{M} a_{ij} m_j + \sum_{j=1}^{M} c_{ij} \mu_j - \text{RHS}_i \equiv Q_i(\mu_j, m_j) = 0 \tag{14.124}$$

The $(\ )_i$ subscript indicates a panel node on the surface and along the wake centerline and the two viscous variables are also obtained at these points. The change $\Delta(\ )$ indicates a finite

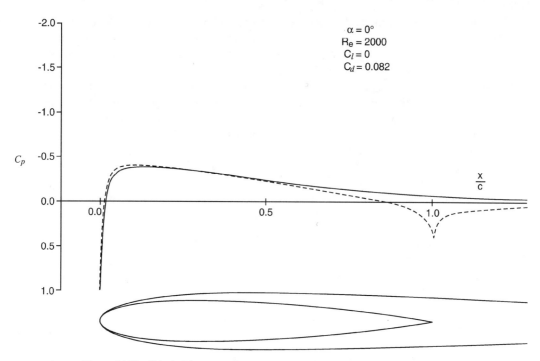

Figure 14.12 Calculated pressure distribution on a NACA 0012 airfoil at zero angle of attack. Dashed line stands for the potential solution and the solid line for the flow with viscous interaction. The airfoil and displacement thickness shapes are depicted at the lower part of the figure.

difference representation of the first derivative, for example,

$$\Delta\theta = \theta_i - \theta_{i-1} \tag{14.125}$$

and variables without a subscript represent a simple average, for example,

$$\theta = \frac{1}{2}(\theta_i + \theta_{i-1}) \tag{14.126}$$

The following secondary variables are computed directly from the primary variables θ_i and m_i using their definitions or closure relations:

$$U_{e_i} = \mathbf{Q}_\infty \cdot \mathbf{t} + \sum_{j=1}^{M} A_{ij} m_j + \sum_{j=1}^{M} C_{ij} \mu_j \tag{14.127}$$

$$\delta_i^* = \frac{m_i}{U_{e_i}} \tag{14.128}$$

$$H_i = \frac{\delta_i^*}{\theta_i} \tag{14.129}$$

$$H_i^* = f_1(H_i) \tag{14.130}$$

$$\left(\frac{C_f}{2}\right)_i = \frac{\nu}{U_{e_i}\theta_i} f_2(H_i) \tag{14.131}$$

$$\left(\frac{2C_\Delta}{H^*}\right)_i = \frac{\nu}{U_{e_i}\theta_i} f_3(H_i) \tag{14.132}$$

The influence matrices A and C in the edge velocity equation are typically related to the particular panel method used. The residuals Q, R_1, and R_2 are driven to zero with a Newton system. The iteration process starts with an initial estimate m, θ, and μ, and the system is solved for dm, $d\theta$, and $d\mu$. The variables are then updated and the iterations continue until convergence.

14.9 Concluding Example: The Flow over a Symmetric Airfoil

As a summary to the approach presented in this chapter, the flow over a NACA 0012 airfoil at zero angle of attack is calculated. The method is as outlined in the previous section, and the results were computed using the computer code XFOIL, based on the method of Drela (Refs. 14.16 and 14.19). The airfoil shape and the resulting pressure distribution are plotted in Fig. 14.12. To present a solution for a case with a laminar boundary layer (as presented in this chapter) a quite low Reynolds number of $Re = 2,000$ was assumed. This may not be a realistic range for an airplane, but some small birds or miniature remotely piloted airplanes may operate in this Reynolds number range. Figure 14.12 also depicts the displacement body shape by adding the boundary layer displacement thickness around the NACA 0012 airfoil. The displacement thickness is quite thick at this low Reynolds number and it continues into the wake. The dashed line in the pressure plot corresponds to the inviscid solution and the formation of a rear stagnation point is clearly visible (note that the NACA 0012 has a finite-angle trailing edge; the full stagnation pressure can be recovered by

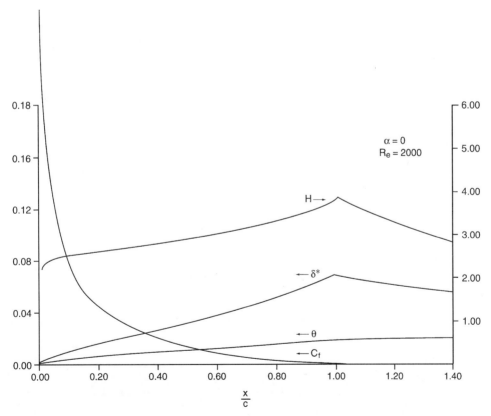

Figure 14.13 Calculated displacement thickness δ^*, momentum thickness θ, shape factor H, and friction coefficient C_f along a NACA 0012 airfoil at zero incidence.

the potential solution when using more panels near the trailing edge). The solid line shows the matched viscous/inviscid solution. Apart from the slight effects of the displacement thickness (on the pressure distribution) the main difference is near the trailing edge, where the rear stagnation point has vanished.

Details on the other boundary layer parameters are presented in Fig. 14.13. Both boundary layer thicknesses (δ^* and θ) grow toward the trailing edge and then slowly decrease in the near wake. The slight discontinuity in shape factor at the trailing edge is a result of the finite trailing-edge thickness added to the wake behind the airfoil. The shape factor begins at a value slightly lower than the Blasius value (of $H = 2.59$) and grows gradually toward the trailing edge. The friction coefficient C_f grows rapidly near the leading edge due to the rapid acceleration of the flow and gradually decreases toward the trailing edge.

The above example demonstrates the principle of matching the viscous and inviscid solutions. The methodology of Section 14.8 was developed here for laminar attached flows but can be extended in a straightforward manner to treat practical airfoil design problems with transition to turbulent flows and even mild flow separation (see Refs. 14.16 and 14.19). For the extension to flows with turbulence, there is no change to the viscous–inviscid coupling strategy or the panel method formulation. An amplification variable is added to the laminar viscous equations (along with an equation for its evolution) to predict transition and an additional equation is added once the flow is turbulent to model the turbulent shear stress transport. Examples including transition and turbulence (for the same NACA 0012 airfoil but at higher Reynolds number) are given in Section 15.2.2.

References

[14.1] Rosenhead, L., *Laminar Boundary Layers*, Oxford University Press, 1963.
[14.2] Van Dyke, M., "Higher-Order Boundary-Layer Theory," *Annual Review of Fluid Mechanics*, Vol. 1, 1969, pp. 265–291.
[14.3] Hildebrand, F. B., *Advanced Calculus for Applications*, Prentice-Hall, 1962, p. 360.
[14.4] Falkner, V. M., and Skan, S. W., "Some Approximate Solutions of the Boundary Layer Equations," *Philosophy Magazine*, Vol. 12, 1931, pp. 865–896.
[14.5] White, F. M., *Viscous Fluid Flow*, McGraw-Hill, 2nd edition, 1991.
[14.6] Thwaites, B., "Approximate Calculation of the Laminar Boundary Layer," *Aeronautical Quarterly*, Vol. 1, 1949, pp. 245–280.
[14.7] Cebeci, T., and Bradshaw, P., *Momentum Transfer in Boundary Layers*, McGraw-Hill, 1977.
[14.8] Goldstein, S., "Concerning Some Solutions of the Boundary Layer Equations in Hydrodynamics," *Proceedings of the Cambridge Philosophical Society*, Vol. 26, 1930, pp. 1–30.
[14.9] Tannehill, J. C., Anderson, D. A., and Pletcher, R. H., *Computational Fluid Mechanics and Heat Transfer*, Taylor and Francis, 2nd edition, 1997, Section 7.4.2.
[14.10] Goldstein, S., "On Laminar Boundary Layer Flow near a Position of Separation," *Quarterly Journal of Mechanics and Applied Mathematics*, Vol. 1, 1948, pp. 43–69.
[14.11] Brown, S. N., and Stewartson, K., "Laminar Separation," *Annual Review of Fluid Mechanics*, Vol. 1, 1969, pp. 45–72.
[14.12] Stewartson, K., "Multistructured Boundary Layers on Flat Plates and Related Bodies," *Advances in Applied Mechanics*, Vol. 14, 1974, pp. 45–239.
[14.13] Catherall, D., and Mangler, K., "The Integration of the Two-Dimensional Laminar Boundary-Layer Equations Past the Point of Vanishing Skin Friction," *Journal of Fluid Mechanics*, Vol. 26, Pt. 1, 1966, pp. 163–182.
[14.14] Stewartson, K., "On the Flow near the Trailing Edge of a Flat Plate II," *Mathematika*, Vol. 16, 1969, pp. 106–121.
[14.15] Messiter, A. F., "Boundary-Layer Flow near the Trailing Edge of a Flat Plate," *S.I.A.M. Journal of Applied Mathematics*, Vol. 18, 1970, pp. 241–257.

[14.16] Drela, M., and Giles, M. B., "Viscous–Inviscid Analysis of Transonic and Low-Reynolds Number Airfoils," *AIAA Journal*, Vol. 25, No. 10, 1987, pp. 1347–1355.

[14.17] Veldman, A. E. P., "New, Quasi-Simultaneous Method to Calculate Interacting Boundary Layers," *AIAA Journal*, Vol. 19, No. 1, 1981, pp. 79–85.

[14.18] Le Balleur, J. C., "Viscid–Inviscid Coupling Calculations for Two and Three Dimensional Flows," in *Computational Fluid Dynamics*, Von Karman Institute for Fluid Dynamics, March 1982. VKI Lecture Series 1982-04.

[14.19] Drela, M., "XFOIL: An Analysis and Design System for Low Reynolds Number Airfoils," in *Low Reynolds Number Aerodynamics*, T. J. Mueller, ed., Springer-Verlag, Lecture Notes in Engineering, No. 54, June 1989.

[14.20] Wolles, B. A., and Hoeijmakers, H. W. M., "On Viscid–Inviscid Interaction Modeling for Airfoils as a Numerical Technique," *AIAA Paper 98-0100*, Jan. 1998.

[14.21] Wolles, B. A. "Computational Viscid–Inviscid Interaction Modeling of the Flow about Airfoils," Ph.D. Thesis, University of Twente, The Netherlands, 1999.

Problems

14.1. For the similarity solution of a boundary layer type problem assume that the stream function has the form
$$\Psi = \nu^{1/2} x^{1/3} f(\eta)$$
where η is
$$\eta = \frac{1}{3\nu^{1/2}} \frac{z}{x^{2/3}}$$
Following the method of Section 14.3, find the differential equation for $f(\eta)$. (This solution describes the free jet flow in a constant pressure.)

14.2. For the similarity solution of a boundary layer type problem assume that the stream function has the form
$$\Psi = 4\nu^{1/2} x^{1/4} f(\eta)$$
where η is
$$\eta = \frac{z}{\nu^{1/2} x^{3/4}}$$
Following the method of Section 14.3, find the differential equation for $f(\eta)$. (This solution describes the flow of a jet parallel to a wall in a constant pressure.)

14.3. Use the von Karman integral method to calculate the flat-plate displacement and momentum thicknesses. Assume a polynomial velocity distibution of
$$\frac{u}{U_e} = a_1 \eta + a_2 \eta^2 + a_3 \eta^3$$
Also, calculate the friction coefficient C_f, and compare with the Blasius solution.

14.4. Calculate the flat-plate displacement and momentum thicknesses and the skin friction coefficient by assuming the velocity profile
$$\frac{u}{U_e} = \sin\left(\frac{\pi}{2} \eta\right)$$
and by using von Karman's integral method.

14.5. As an example for an integral boundary layer method with pressure gradient consider the polynomial velocity profile of Eq. (14.78). By slightly modifying the boundary conditions in Section 14.5.1 such that

$$\text{at} \quad z = 0, \quad \nu \frac{\partial^2 u}{\partial z^2} = -U_e \frac{dU_e}{dx}$$

and defining $\Lambda = (\delta^2/\nu)(dU_e/dx)$, show that

$$a_1 = 2 + \frac{\Lambda}{6}$$

$$a_2 = -\frac{\Lambda}{2}$$

$$a_3 = -2 + \frac{\Lambda}{2}$$

$$a_4 = 1 - \frac{\Lambda}{6}$$

and therefore

$$\frac{u}{U_e} = 2\eta - 2\eta^3 + \eta^4 + \frac{\Lambda}{6}(\eta - 3\eta^2 + 3\eta^3 - \eta^4)$$

Finally, show that

$$\frac{\delta^*}{\delta} = \frac{3}{10} - \frac{\Lambda}{120}$$

$$\frac{\theta}{\delta} = \frac{1}{63}\left(\frac{37}{5} - \frac{\Lambda}{15} - \frac{\Lambda^2}{144}\right)$$

CHAPTER 15

Enhancement of the Potential Flow Model

Toward the end of Chapter 1 (Section 1.8) it is postulated that many flowfields of interest to the low-speed fluid dynamicist lie in the range of high Reynolds number. Consequently, for attached flowfields, the fluid is divided into two regions: (*a*) the thin inner boundary layer and (*b*) the mainly inviscid irrotational outer flow. Chapters 2–13 are entirely devoted to the solution of the inviscid outer flow problem, which indeed is capable of estimating the resulting pressure distribution and lift due to the shape of the given solid boundaries. The laminar boundary layer model was presented in Chapter 14 as an example for modeling the inner part of the complete flowfield. The methodology for obtaining information such as the displacement thickness, the skin friction on the solid surface and resulting drag force (due to surface friction), and the matching process with the outer flow was demonstrated. However, in real high Reynolds number flows over wings the flow is mostly turbulent and the engineering approach to extend the methodology of Chapter 14 to include turbulent or even separated viscous layer models will be discussed briefly in this chapter. The objective of this chapter is to provide a brief survey of some frequently occurring low-speed (wing-related) flowfields and to help the student to place in perspective the relative role of the potential flow methods (presented in this book) and of the viscous effects in order to comprehend the complete real flowfield environment. Additionally, several simplified enhancements to the potential flow model that can help model some nonlinear and viscous effects will be surveyed.

The modifications presented in this chapter will begin with methods of calculating the wake rollup, which from the classical potential-flow solution point of view was denoted as a "slight nonlinear effect." The rest of the presented improvements (or modifications) deal with efforts to include the effects of viscosity and some of them are logical extensions to the potential flow model. Some others (e.g., modeling of two-dimensional flow separation) will clearly fall into the "daring and imaginative" category and their importance is more in providing some explanation of the fluid-mechanical phenomena rather then being in a stage that they can predict unknown flowfields.

In the following discussion, for the sake of simplicity, mainly the lifting characteristics of the experimental observations and the resulting flow models are presented. In a limited number of cases the drag force also is discussed, but important effects such as side forces, moments, and possible crosscoupling of the aerodynamic loads are omitted in favor of brevity. Therefore, the treatment of the various topics in this chapter is by no means complete or comprehensive and the reader is encouraged to further investigate any of the following topics in the referenced literature.

15.1 Wake Rollup

The conditions that the wake will move with the local streamlines (and carry no loads) were introduced as early as Section 4.7 for thin lifting surfaces and later in Section 9.3 when discussing the wake model for panel methods. From the steady-state flow point of view, the shape of the wake is not known, and the process of finding the proper wake shape

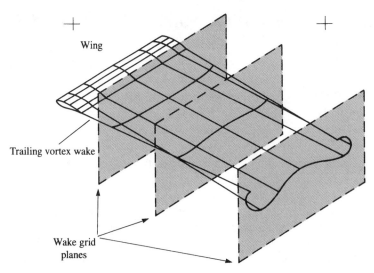

Figure 15.1 Wake grid planes (usually normal to the free stream) used for wake rollup calculations.

(wake rollup) is often denoted as a "slight nonlinearity" in the solution process. Typical remedies to this problem are:

1. Prescribe wake shape: This is done in Chapters 4, 8, and 11, for the lifting line and lifting surface type solution (where the wake is placed on the $z = 0$ plane). A more refined alternative of this option is to prescribe the wake shape based on flow visualizations. This approach is very helpful when analyzing multielement wings where, for example, in the case of a two-element airfoil the wake of the main airfoil is very close to the trailing-edge flap upper surface.
2. Use wake relaxation: This is a process used by several steady-state numerical solutions and to demonstrate the principle of this method let us follow the approach used in the code VSAERO.[9.3,12.11] The initial wake geometry is specified by the programmer (usually as a planar wake extending backward from the trailing edge) and then several wake grid planes (normal to the free stream) are established, as shown in Fig. 15.1. For the first iteration, the flowfield over the wing and the initial wake shape are calculated using the method described in Section 12.5.

For the second iteration the velocity induced by the wing and wake $(u, v, w)_\ell$, at each of the wake points (formed by the intersection between the wake grid planes and the wake lines) is calculated. Next, the wake points are moved with the local induced velocity (see Fig. 15.1) by

$$(\Delta x, \Delta y, \Delta z)_\ell = (u, v, w)_\ell \Delta t \qquad (15.1)$$

(Some methods, for simplicity, will move the wake in the wake grid plane only. If this is done in the free-stream coordinate system, then the wake grid lies in the $x = $ const. plane and only $(\Delta y, \Delta z)_\ell$ are required.) In Eq. (15.1) Δt is an artificial time parameter and its value can be approximated as

$$\Delta t \approx K \frac{\Delta x_\ell}{Q_\infty} \qquad (15.2)$$

where Δx_ℓ is the distance of the wake grid plane from the trailing edge (or between the wake grid planes) and K has values between 0.5 and 5. Once all the wake points have been moved, as a result of the local induced velocity, the flow is computed

15.1 Wake Rollup

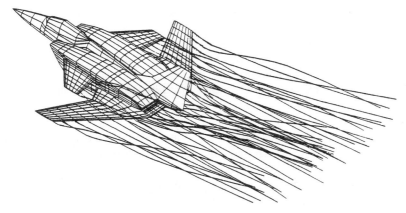

Figure 15.2 Wing and canard wake rollup after two iterations, using the wake relaxation method of Refs. 9.3 and 12.11. From Katz, J., "Evaluation of an Aerodynamic Load Prediction Method on a STOL Fighter Configuration," *AIAA Paper 86-0590*, 1986. Reprinted with permission. Copyright AIAA.

with the new wake geometry and the second iteration cycle (or wake relaxation iteration) has been completed.

These wake relaxation iterations can be continued until convergence is obtained or when sufficient wake rollup has been achieved (with the decision made by the programmer). Since there is always a risk that by too many iterations the wake can reach levels of nonphysical rollup (where the sum of the iteration time steps $\sum \Delta t$ is much larger than $\Delta x_\ell / Q_\infty$) the number of wake rollup iterations is usually limited to less than three. Results of such a procedure (after two wake iterations) are presented in Fig. 15.2. Here the VSAERO[9.3,12.11] code was used and the interaction between a close coupled wing canard configuration was calculated. Figure 15.3

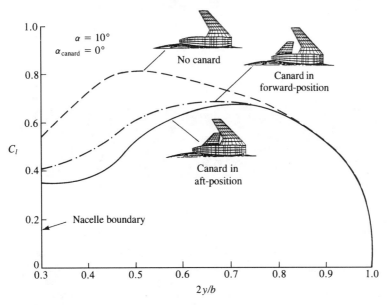

Figure 15.3 Effect of canard position on wing's spanwise loading. From Katz, J., "Evaluation of an Aerodynamic Load Prediction Method on a STOL Fighter Configuration," *AIAA Paper 86-0590*, 1986. Reprinted with permission. Copyright AIAA.

shows the effect of the canard on the wing's spanwise loading. Note the noticeable effect of the canard and its wake, which induces a downwash at the wing root area and thereby reduces its lift there. The proper placement of the wake in such cases of closely spaced lifting surfaces is critical and estimation of the wake motion is important for the solution.

3. Use a time-stepping method: This approach was demonstrated in Chapter 13 (Sections 13.8.2 and 13.10) and in principle is similar to the wake relaxation method, but now the time step is directly related to the motion. (Therefore, the apparent "slight nonlinearity" does not exist.) During the computations, the number of wake points increases with time, and, for example, for N wake lines during K time steps approximately $NK/2$ wake point velocity computations are required. When using the wake relaxation approach, even for the first iteration, all wake grids are used and therefore NK such velocity calculations are required. Thus, even for steady-state flows, this time-stepping wake rollup method may require less computational effort.

As an example for this wake rollup calculation consider the rollup of a single horseshoe vortex. In this case, the wing-bound vortex is modeled by a single vortex line, which sheds two wake line segments of length $Q_\infty \Delta t$ at its tips during each time step. As this shedding process continues the two long trailing vortices are formed, but because of the instability of these two vortex lines a sinusoidal pattern will develop. This instability was first analyzed by Crow,[15.1] who presented the photos appearing in Fig. 15.4, taken from Ref. 15.1. The numerical solution presented in Fig. 15.4a is obtained by using only one panel in the method described

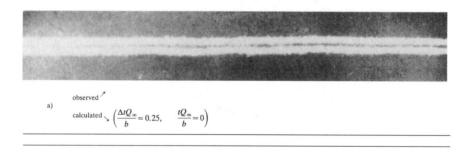

a) observed ↗
calculated ↘ $\left(\dfrac{\Delta t Q_\infty}{b}=0.25, \quad \dfrac{t Q_\infty}{b}\approx 0\right)$

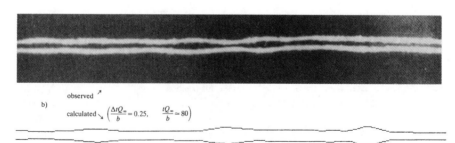

b) observed ↗
calculated ↘ $\left(\dfrac{\Delta t Q_\infty}{b}=0.25, \quad \dfrac{t Q_\infty}{b}\approx 80\right)$

Figure 15.4 Instability of a pair of trailing vortices, and comparison between calculated and observed vortex formations. Panel (a) shows the wake behind the airplane after its passage and (b) depicts the Crow instability, which is shown later at a distance of about 80 wing spans. More details about such calculations can be found in Rossow, V. J., *J. Aircraft*, Vol. 24, No. 7, 1987, pp. 433–440. Photo from Ref. 15.1. Reprinted with permission of AIAA and Meteorology Research Inc. Photo originally appeared in Smith, T. B., and Beemer, K. M., "Contrail Studies of Jet Aircraft," *MRI Report*, Apr. 1959.

in Section 13.12, and the above instability is also visible in the computation. Calculations such as this one and that of Fig. 13.29 indicate that this approach for calculating the rollup of vortex sheets yields satisfactory results (at least when modeling trailing wakes behind wings).

As a closing remark to most of the wake rollup modeling efforts we must emphasize that the velocity induced by a vortex point or line is singular (see, for example, Fig. 3.8a). Therefore, an artificial vortex core (or cutoff distance) must be defined for the purpose of numerical solutions. One possibility is to define the self-induced influence as zero within this radius; however, in some methods a solid body rotation model is used within this core (which is very similar to Fig. 2.11 with ϵ being the core size).

15.2 Coupling between Potential Flow and Boundary Layer Solvers

The concept of coupling between a two-dimensional outer potential flow and an inner laminar boundary layer model was discussed in Section 14.8. This approach can be extended to three-dimensional flows and to turbulent boundary layers that exist in the high Reynolds number case. The information sought from the solution of the attached boundary layer problem is essentially the same as for the laminar boundary layer case:

1. Displacement effects due to the slower velocity inside the boundary layer.
2. Surface skin friction, so that the contribution of friction to the drag force can be estimated.

   However, for engineering applications such as airfoil design, we also want to learn about:
3. Location of boundary layer transition.
4. Indications about the tendency of the flow to detach (or separate). If the boundary layer does separate then the method of Refs. 14.16 and 14.19 can handle cases of "mild" separation. For cases with massive separation additional modeling efforts or the solution of the full Navier–Stokes equations is required.

In this section these topics will be discussed very briefly. More comprehensive descriptions can be found in texts such as Schlichting[1.6] (boundary layers) or AGARD CP-291[15.2] (viscous–inviscid interactions).

15.2.1 *The Laminar/Turbulent Boundary Layer and Transition*

The discussion so far has been limited to laminar flows. For example, steady-state streamlines were described in Section 1.3 by a hypothetical experiment where coloring dye is injected at point 1 (in Fig. 1.3) and then all particles will follow the same path to point 2. However, Osborne Reynolds (in the second half of the 1800s) showed that dye injected into the laminar flow inside a long tube eventually becomes chaotic, a flow we call turbulent. Returning to the example in Fig 1.3, in the case of such *turbulent flow* the particles will be subject to highly unsteady motion and those injected at point 1 will not necessarily pass through point 2. The important conclusion at this point is that in spite of the flow being "seemingly steady" the fluid particles are locally unsteady. However, concepts developed for laminar flow (e.g., streamlines, boundary layer, etc.) can still be used by referring to the time-averaged properties.

Some important features of the turbulent flow can be described by the basic case of the flow over a flat plate. Suppose that the flat plate is submerged in an undisturbed, parallel,

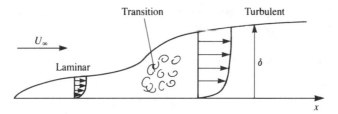

Figure 15.5 Schematic description of the boundary layer on a flat plate and the transition from laminar to turbulent regions.

laminar free stream, as in the example of Fig. 14.3. Initially, a laminar boudary layer will form, but as the distance increases (and Reynolds number increases) a transition to turbulent boundary layer occurs (see Fig. 15.5). The region where this change takes place is called the region of transition. Various models were proposed to explain the origin of turbulence and several models are presented by Panton.[15.3] One possible model (that fits well with the models presented in this book) represents the shear layer near the wall by spanwise vortices (see Fig. 15.6). In the laminar sublayer (which is less than 1% of δ) the shear creates vortex lines, which later become unstable. We have seen vortex instabilities (e.g., Fig. 13.29 or 15.4) and certainly increasing continuously the vorticity in the boundary layer (the length scale) and the wall proximity will create spanwise instability of the lower vortex sheets. Eventually, the higher speed in the upper layer will lift portions of these vortices creating streamwise vortex filaments. As this process continues, the vortex segments reach the outer layers where they break up to form the fully turbulent region. Of course disturbances in the free stream or those generated in the viscous shear layers near the surface (e.g., due to surface roughness) can cause time-dependent fluctuations too, leading to earlier transition and increased turbulence in the boundary layer.

The time-averaged nature of the turbulent flow has some pronounced effects on the physics of the flow. For example, in the case of a turbulent boundary layer over the two-dimensional flat plate of Fig. 15.5, the velocity $u(z)$ at a given x location becomes time dependent and will have the form

$$u(z, t) = \bar{u}(z) + u'(z, t) \tag{15.3a}$$

Similarly, the normal velocity component is

$$w(z, t) = \bar{w}(z) + w'(z, t) \tag{15.3b}$$

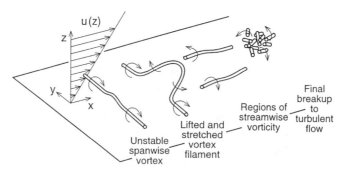

Figure 15.6 Possible scenario for vorticity generation in the laminar sublayer, its liftoff, and breakup, leading to fully developed turbulent boundary layer.

15.2 Coupling between Potential Flow and Boundary Layer Solvers

where $\bar{u}(z)$ and $\bar{w}(z)$ are the mean velocity components and $u'(z,t)$ and $w'(z,t)$ are the time-dependent fluctuating parts. Figure 15.5 schematically indicates that as a result of the fluctuating velocity component (larger momentum transfer) the turbulent boundary layer is thicker. Furthermore, when examining the turbulent boundary layer equations, we see that the shear force near the wall becomes (Schlichting[1.6] p. 562)

$$\tau_{xz} = \mu \left(\frac{\partial u}{\partial z} \right) - \rho \overline{u'w'} \tag{15.4}$$

The second term is the Reynolds stress, which represents additional stress (compared with Eq. (14.47)) due to axial momentum transfer in the vertical direction (time average of the product $u'w'$).

Comparison of experimental laminar and turbulent boundary layer profiles for a flat plate are presented in Fig. 15.7. Clearly, because of the transverse momentum transfer, the velocity increases faster near the wall and the velocity gradient for the turbulent case $(\partial u/\partial z)$ appears to be larger there. Therefore it is expected that the turbulent skin friction is larger than the laminar one. Figure 15.8 shows the trends in skin-friction coefficients for a flat plate. For laminar flow, Eq. (14.49), based on Blasius's solution, seems to be close to experimental observation, but for turbulent boundary layers the skin friction increases dramatically. It also seems that for a wide range of Reynolds number either laminar or turbulent boundary layers can exist. This may be utilized for drag reduction, where by using smooth surfaces and minimizing external disturbances one can maintain a laminar boundary layer.

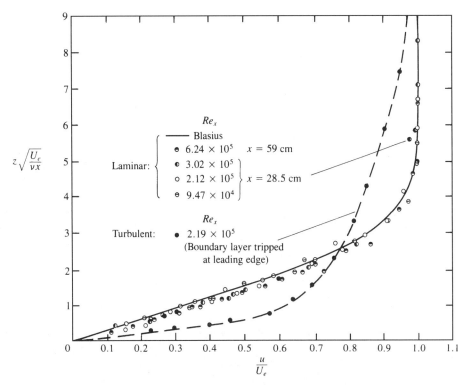

Figure 15.7 Velocity profiles on a flat plate at zero incidence for laminar and turbulent boundary layers. From Dhawan, S., "Direct Measurements of Skin Friction," NACA Report 1121, 1953.

Table 15.1. *Comparison between laminar (Blasius solution) and turbulent (based on the $\frac{1}{7}$-power law) flat-plate, integral boundary layer properties*

	Laminar	Turbulent
$\dfrac{\delta_{99}}{x}$	$\dfrac{5.00}{\sqrt{Re_x}}$	$\dfrac{0.37}{\sqrt[5]{Re_x}}$
$\dfrac{\delta^*}{x}$	$\dfrac{1.721}{\sqrt{Re_x}}$	$\dfrac{0.046}{\sqrt[5]{Re_x}}$
$\dfrac{\theta}{x}$	$\dfrac{0.664}{\sqrt{Re_x}}$	$\dfrac{0.036}{\sqrt[5]{Re_x}}$
C_f	$\dfrac{0.664}{\sqrt{Re_x}}$	$\dfrac{0.0576}{\sqrt[5]{Re_x}}$
H	2.59	1.28

To compare the properties of the two boundary layer profiles, the turbulent velocity distribution must be approximated. A reasonable curve-fit to the experimental profile in Fig. 15.7 ($0.2 \times 10^6 < Re < 10 \times 10^6$), following Schlichting[1.6] (pp. 637–638), is the 1/7 power law:

$$\frac{\bar{u}}{U_e} = \eta^{1/7} \tag{15.5}$$

and here $\eta = z/\delta$ as defined in Section 14.5. When using this velocity profile with the integral boundary layer method of Section 14.5 the properties for the turbulent boundary layer can be calculated and the results are tabulated in Table 15.1. These results are presented next to the results based on Blasius's solution (Eqs. (14.48), (14.50), (14.52), (14.55), and (14.69)).

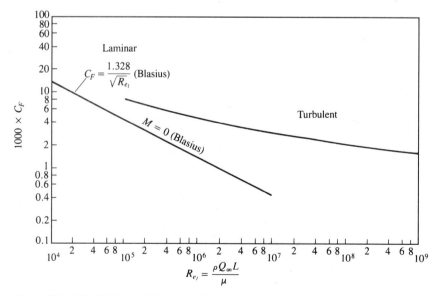

Figure 15.8 Skin friction coefficient on a flat plate at zero incidence for laminar and turbulent boundary layers. From Ref. 1.6, p. 717. Reproduced with permission of McGraw-Hill, Inc.

Now, we can examine the above information and Fig. 15.8 and draw a few conclusions that affect airfoil and wing design:

1. Displacement thickness is larger for turbulent than for laminar boundary layers.
2. The skin-friction coefficient becomes smaller with increased Reynolds number (mainly for laminar flow).
3. Over a certain Reynolds number range (transition) both laminar and turbulent boundary layers are possible. The nature of the actual boundary layer for a particular case depends on flow disturbances, surface roughness, etc.
4. The skin-friction coefficient is considerably larger for the turbulent boundary layer.
5. Because of the vertical momentum transfer in the case of the turbulent boundary layer, flow separations will be delayed somewhat, compared to the laminar boundary layer (see Ref. 15.4, p. 474).

15.2.2 *Viscous–Inviscid Coupling, Including Turbulent Boundary Layer*

The introduction of the turbulent boundary layer concept in the previous section and the results of Table 15.1 indicate that both boundary layer types can be represented by the integral quantities introduced in Chapter 14. Although the relative displacement thickness in the turbulent boundary layer is smaller, the important conclusion is that such a quantity (e.g., δ^*) can be defined for the turbulent flow. Consequently, the coupling process of the viscous inner and the inviscid outer flow can proceed in a manner similar to the description in Section 14.8. This approach can be extended to three-dimensional cases as well (see Ref. 9.3). A possible procedure for solving the coupled potential and boundary layer equations can be established as follows (see Fig. 15.9):

1. Solve for the potential flowfield over the body (solid line in Fig. 15.9) and obtain the surface velocity and pressure distribution.
2. Using this outer velocity distribution obtain the boundary layer solution and generate the displacement thickness and skin friction.
3. Modify the surface boundary condition for the potential flow (e.g., specify it on the displacement surface between the viscous and inviscid regions, as in Fig. 15.9, or add transpiration velocity) and solve for the second iteration.

This iterative process can be repeated several times and as mentioned there are some different approaches for modifying the potential flow boundary conditions. One approach (e.g., Refs. 9.6 and 15.5) is to change the location of the dividing streamline (or the boundary) to account for the displacement thickness. The other approach (which was presented in Section 9.9) is not to change the geometry of the surface but to simulate the displacement by blowing normal to the surface (e.g., Refs. 12.11 or 15.6–15.9). This modification of the boundary condition of Eq. (9.4) is obtained by replacing the zero term on the right-hand

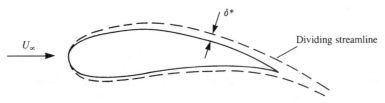

Figure 15.9 Generic shape of an airfoil and the displaced streamline, outside of which the flow can be considered as potential.

side with V_n:

$$\frac{\partial(\Phi + \Phi_\infty)}{\partial n} = V_n \tag{15.6}$$

and then the transpiration (or blowing) velocity V_n is found from the information provided by the boundary layer solution (see Eq. 14.117)):

$$V_n = -\frac{\partial(U_e\delta^*)}{\partial s} \tag{15.7}$$

where s is the line along the surface and the minus sign is a result of **n** pointing into the body. In the case of the Dirichlet boundary condition the source term of Eq. (9.12) (e.g., in the panel code VSAERO[9.3,12.11]) can be modified such that (see Eq. (14.118))

$$\sigma = \mathbf{n} \cdot \mathbf{Q}_\infty - \frac{\partial(U_e\delta^*)}{\partial s} \tag{15.8}$$

This approach is based on the two-dimensional boundary layer model and in practice, when extended to three-dimensional flows, the boundary layer calculations are applied along streamlines or along two-dimensional sections (so in essence this is a quasi-three-dimensional representation and not a true three-dimensional boundary layer model).

To demonstrate these principles let us first consider the case of the symmetric NACA 0012 airfoil, which was investigated for the laminar boundary layer case in Figs. 14.11 and 14.12. For this example (see Fig. 15.10) an angle of attack of 5° was assumed and the Reynolds number is raised to 10^6. The dashed line represents the potential solution and the solid line stands for the coupled viscous solution (using the XFOIL[14.19] code).

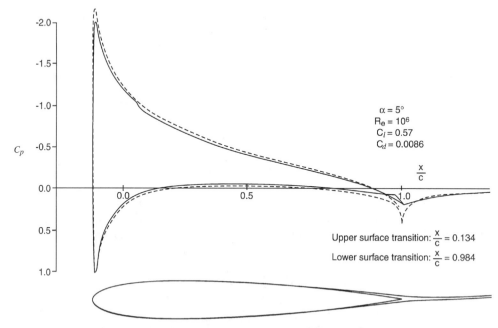

Figure 15.10 Calculated pressure distribution on a NACA 0012 airfoil at 5° angle of attack. Dashed line stands for the potential solution and the solid line for the flow with viscous interaction. The airfoil and displacement thickness shapes are depicted at the lower part of the figure.

15.2 Coupling between Potential Flow and Boundary Layer Solvers

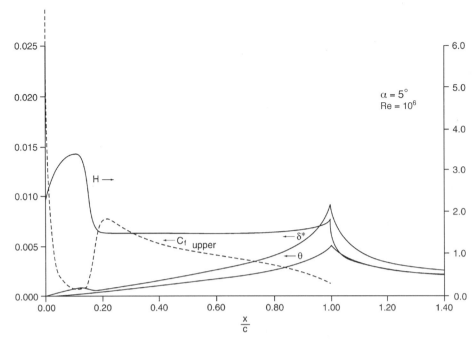

Figure 15.11 Displacement thickness δ^*, momentum thickness θ, shape factor H, and surface friction coefficient C_f along the upper surface of a NACA 0012 airfoil at 5° angle of attack.

Boundary layer transition is estimated at $x/c = 0.134$ on the upper surface and at $x/c = 0.984$ on the lower surface. At the beginning of the transition on the upper surface a small discontinuity in the pressure distribution is visible. This is a result of a local "laminar bubble," which will be described in the next section. The displacement streamline and the airfoil are shown in the lower part of Fig. 15.10 and the displacement δ^* is much smaller than for the low Reynolds number case (Fig. 14.11). The displacement and momentum thickness for the upper surface are presented in Fig. 15.11; note the sharp drop in the displacement thickness for the turbulent flow. This directly affects the shape factor, which drops sharply behind the transition region. The laminar skin friction is initially high owing to the high shear near the leading edge, but, behind the transition point it grows again, sharply demonstrating the large increase in turbulent drag. The above parameters for the airfoil's lower surface were not presented here because of the largely laminar boundary layer there (which, apart from being much thinner, resembles the laminar case presented in the previous chapter).

This approach of coupling between the viscous and inviscid solutions can be extended to treat multiple bodies. As an example, the effect of the displacement thickness on a two-element airfoil (shown in the inset) is presented in Fig. 15.12. Although the treatment of multielement wings seems to be straightforward, in fact, the wake of the leading element must be properly placed relative to the following flap. This problem is even more pronounced for the three-dimensional case and validation against experimental data is recommended. Similarly to the example presented in Fig. 15.10, in this case too, the presence of the thin boundary layer reduces slightly the pressure difference (and hence the lift) obtained by the inviscid solution. This effect increases with the airfoil's angle of attack (see lift coefficient data in Fig. 15.13), as the upper boundary layer becomes thicker and eventually

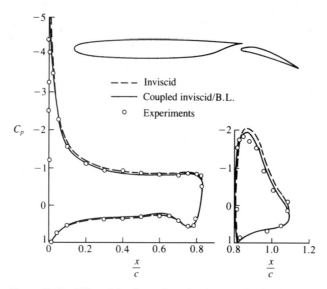

Figure 15.12 Effect of the viscous boundary layer on the chordwise pressure distribution of a two-element airfoil. From Ref. 15.6. Reprinted with permission. Copyright AIAA.

flow separation is initiated near the trailing edge (for α approximately larger than 5°, in Fig. 15.13).

When the flow separates, the streamlines do not follow the surface of the body, as shown schematically in Fig. 15.14. This is a result of an adverse (positive) pressure gradient (which may be caused by high surface curvature), which slows down the fluid inside the boundary

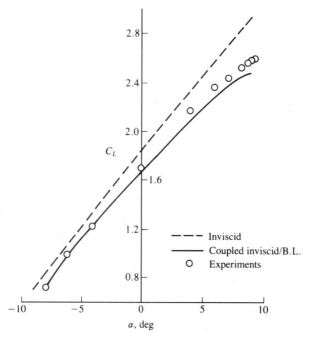

Figure 15.13 Effect of the viscous boundary layer on the lift coefficient of the two-element airfoil of Fig. 15.12. From Ref. 15.6. Reprinted with permission. Copyright AIAA.

15.3 Influence of Viscous Flow Effects on Airfoil Design

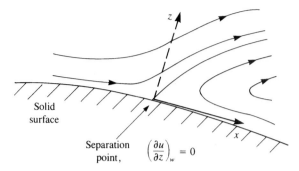

Figure 15.14 Schematic description of the flow in the boundary layer near the point of separation.

layer to a point where the normal velocity gradient at the wall becomes zero. For laminar flows, therefore, at the separation point

$$\left(\frac{\partial u}{\partial z}\right)_w = 0 \tag{15.9}$$

and behind this point reversed flow exists.

15.3   Influence of Viscous Flow Effects on Airfoil Design

One of the earliest applications of panel methods (in their two-dimensional form), when combined with various boundary layer solution methods, was for airfoil shape design. Because of the simplicity of the equations, it was possible to develop inverse methods, where the programmer would specify a modified pressure distribution and then the computer program constructs the airfoil's shape. Figure 15.15 depicts the sensitivity of the chordwise pressure distribution to the airfoil's upper surface shape and emphasizes the importance of such inverse methods. For more details on these airfoil design methods see, for example,

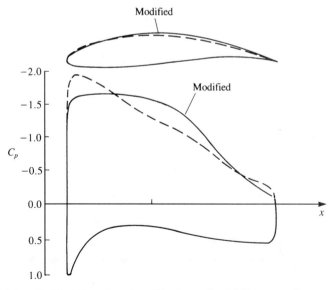

Figure 15.15 Effect of small modifications to the airfoil's upper surface curvature on the chordwise pressure distribution.

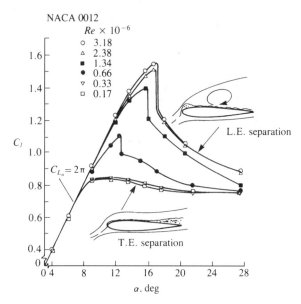

Figure 15.16 Effect of Reynolds number on the lift coefficient of a symmetric NACA 0012 airfoil. From Ericsson, L. E., and Reading, J. P., "Further Considerations of Spilled Leading Edge Vortex Effects on Dynamic Stall," *J. Aircraft*, Vol. 14, No. 6, 1977. Reprinted with permission. Copyright AIAA. (Courtesy of L. E. Ericsson, Lockheed Missiles and Space Company, Inc.)

Refs. 15.9–15.12. Here we will attempt only a brief discussion of some of the more dominant considerations.

To estimate the effects of viscosity on airfoil design let us begin by observing the effect of Reynolds number on the performance of a two-dimensional airfoil. Figure 15.16 shows the lift coefficient versus angle of attack curve of a NACA 0012 airfoil and clearly the angle of attack at which flow separation is initiated depends on the Reynolds number. Note that for the attached flow condition the lift slope is close to 2π, but at a certain angle (e.g., about $\alpha = 8°$ for $Re = 0.17 \times 10^6$) the lift does not increase with an increase in the angle of attack. This is caused by flow separation (see inset in the figure) and the airfoil (or wing) is "stalled."

Let us now have a closer look at the boundary layer on the airfoil's upper surface (that is, the suction side). If the free stream is laminar to begin with then a laminar boundary layer will develop behind the front stagnation point (see Fig. 15.17). At a certain point the laminar flow will not be able to follow the airfoil's upper surface curvature and a "laminar bubble" will form. If the Reynolds number is low (as in the lowest two curves in Fig. 15.16) then the laminar boundary layer will separate at this point. But if the Reynolds number increases then the flow will reattach behind the laminar bubble and a transition to a turbulent boundary layer will take place. The effect of this laminar bubble on the upper surface pressure distribution is shown in Fig. 15.10 and schematically in the upper inset to Fig. 15.17. Because of the modified streamline shape the outer flow will have a higher velocity U_e, resulting in a plateau shape of the pressure distribution (above this bubble). Behind the bubble the velocity is reduced and the pressure increases, thus resulting in the sharp drop of the negative pressure coefficient.

Returning to Fig. 15.16 we can see that for increasing Reynolds numbers, as a result of the momentum transfer from the outer flow into the turbulent boundary layer, the airfoil separation is delayed up to increasingly higher angles of attack (upper curves in Fig. 15.16).

15.3 Influence of Viscous Flow Effects on Airfoil Design

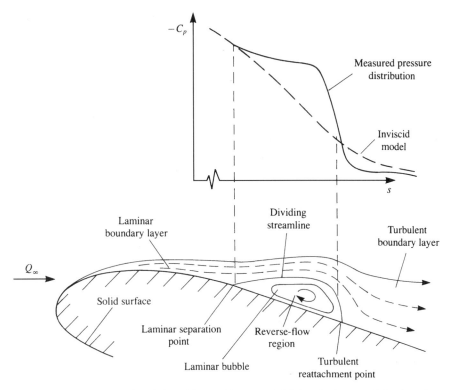

Figure 15.17 Schematic description of the transition on an airfoil from laminar to turbulent boundary layer and the laminar bubble.

This delay in the airfoil's stall angle of attack (caused by increased Reynolds number) results in higher lift coefficients with a maximum lift coefficient of $C_{l_{max}}$, whereas the flow separation now is a "turbulent separation."

Another interesting observation is that, for the low Reynolds number case, the flow starts to separate at the airfoil's trailing edge and gradually moves forward. This is called *trailing-edge separation*, and in this case abrupt changes in the airfoil's lift are avoided. For the high Reynolds number cases the boundary layer becomes turbulent and the flow stays attached for larger angles of attack (e.g., $\alpha = 14°$ for $Re = 3.18 \times 10^6$ in Fig. 15.16). If gradual trailing-edge separation is needed at higher angles of attack (to avoid the abrupt lift loss) then this can be achieved by having a more cambered airfoil section.

Some of the more noticeable considerations, from the airfoil designer's point of view, become clear when observing the effects of the boundary layer with the aid of Figs. 15.8 and 15.16. The first observation is that the drag coefficient of the laminar boundary layer is smaller and for drag reduction purposes larger laminar regions must be maintained on the airfoil. However, when high lift coefficients are sought then an early tripping (causing of transition, e.g., by surface roughness, vortex generators, etc.) of the boundary layer can help to increase the maximum lift coefficient. Also, in many situations the same lifting surface must operate within a wide range of angles of attack and Reynolds number and the final design may be a result of a compromise between some opposing requirements. Consequently, to clarify some of the considerations influencing airfoil design, these two regimes of airfoil performance are discussed briefly in the following sections.

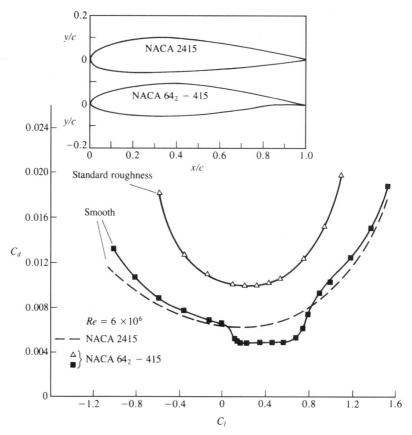

Figure 15.18 Variation of drag coefficient versus lift coefficient for an early NACA airfoil and for a low-drag airfoil, and the effect of rough surface on drag. (Experimental data from Ref. 11.2.)

15.3.1 Low Drag Considerations

When low drag of the lifting surface is sought (e.g., for an airplane cruise configuration) then, as mentioned, large laminar boundary layer regions are desirable. To maintain a laminar boundary layer on the airfoil the surface must be as smooth as possible, and a favorable pressure gradient can also delay the transition to a turbulent boundary layer (Ref. 15.4, Section 17.5). A favorable (negative) pressure gradient occurs when the pressure is decreasing from the leading edge toward the trailing edge (thus adding momentum) and can be achieved by having a gradually increasing thickness distribution of the airfoil. This is demonstrated in the inset to Fig. 15.18 where an earlier NACA 2415 airfoil is compared with a NACA 64_2-415 low-drag airfoil. The inset to the figure clearly shows that the maximum thickness of the low-drag airfoil is moved to the 40% chord area, which is further downstream than the location of the maximum thickness for the NACA 2415 airfoil. The effect of this design on the drag performance is indicated clearly by the comparison between the drag versus lift plots of the two airfoils (at the same Reynolds number). In the case of the low-drag airfoil, a bucket-shaped low-drag area is shown, which is a result of the large laminar flow regions. However, when the angle of attack is increased (resulting in larger C_l) the boundary layer becomes turbulent and this advantage disappears. For comparison the drag of a NACA 64_2-415 airfoil with a standard roughness[11.2] is shown in Fig. 15.18 where the boundary layer is fully turbulent and hence its drag is considerably higher.

15.3 Influence of Viscous Flow Effects on Airfoil Design

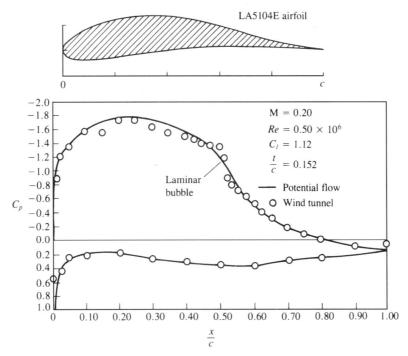

Figure 15.19 Signature of the laminar bubble on the pressure distribution of an airfoil. (Courtesy of Douglas Aircraft Co. and Robert Liebeck.)

A large number of airfoil shapes together with their experimental validation are provided in Ref. 11.2 (e.g., for the 6-series airfoils of Fig. 15.18 on pp. 119–123). Also, the airfoil shape numbering system is explained there in detail and, for example, for the 64_2-415 airfoil the last two digits indicate the airfoil thickness (15%). The first digit (6) is the airfoil series designation and the second digit indicates the chordwise position of minimum low pressure in tenths of chord (or the intention to have about 40% laminar flow). As long as the boundary layer stays laminar in the front of the airfoil, its drag is low (see the bucket shape in Fig. 15.18) and the range of this bucket in terms of ΔC_l is ± 0.2 near the designed C_l of 0.4 (hence the subscript 2 and the digit 4 after the dash).

Most airplane-related airfoils operate with a Reynolds number larger then 10^6, but when the Reynolds number is below this value (as occurs in small-scale testing in wind tunnels or with low-speed gliders and airplanes, etc.) then it is possible to maintain large regions of laminar flow over the airfoil. This condition is more sensitive to stall and usually a larger laminar bubble exists. The effect of such a laminar bubble on the airfoil's pressure distribution is shown in Fig. 15.19, where the plateau caused by the laminar bubble is clearly visible (see also Figs. 15.10 and 15.17). For further details about low Reynolds number airfoils the reader is referred to a review article on this topic by Lissaman.[15.13]

15.3.2 High Lift Considerations

Requirements such as short takeoff and landing can be met by increasing the lift of the lifting surfaces. If this is done by increasing the wing's lift coefficient then a smaller wing surface can be designed (meaning less cruise drag, less weight, etc.). Engineering solutions to this operational requirement within various lift coefficient ranges resulted in

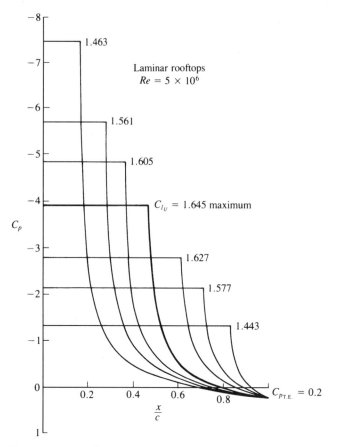

Figure 15.20 Family of possible airfoil upper-surface pressure distributions resulting in an attached flow on the upper surface (for $Re = 5 \times 10^6$). From Liebeck.[15.12] Reprinted with permission. Copyright AIAA. (Courtesy of Douglas Aircraft Co. and Robert Liebeck.)

many ingenious approaches and a comprehensive survey is given by Smith.[15.14] A logical approach is to increase the lift coefficient of a lifting surface by delaying flow separation, but changing the wing area and shape in reaction to the changing flight conditions (e.g., airplane flaps) is also very common. In this section we shall briefly discuss some of the features of single and multielement high-lift airfoils.

One approach is to develop a family of airfoil (upper surface) pressure distributions that will result in the most delayed flow separation. To accomplish this the location of the separation point must be estimated, based on information from the potential flow and the boundary layer solutions. A simplified approach is to use a flow separation criteria such as the Stratford criterion (description of this criterion can be found in several aerodynamic books, e.g., Kuethe and Chow,[15.4] Sections 18.10 and 19.2). Using such a flow separation criterion, Liebeck[15.11,15.12] developed the family of upper surface pressure distributions shown in Fig. 15.20. These curves depend on the Reynolds number, and in the case of Fig. 15.20, for a Reynolds number of 5×10^6, airfoils having any of the described upper pressure distributions will have an attached flow on that surface. Note that the maximum lift coefficient will increase toward the center of the group and the bold curve represents the pressure distribution yielding the highest lift due to the upper surface pressure distribution.

15.3 Influence of Viscous Flow Effects on Airfoil Design

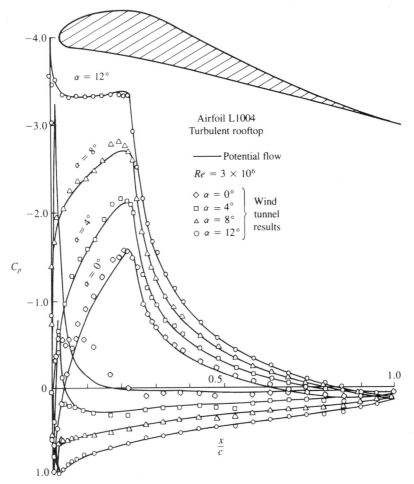

Figure 15.21 Shape of the L1004 airfoil and theoretical and experimental pressure distributions on it at various angles of attack. From Liebeck.[15.11] Reprinted with permission. Copyright AIAA. (Courtesy of Douglas Aircraft Co. and Robert Liebeck.)

At this point it is clear that if, based on the nature of the boundary layer, the shape of the desired pressure distribution can be sketched, then an inverse method is required to find the corresponding (or the closest) practical airfoil shape. Based on this need many inverse, or "design mode," airfoil design methods were developed (e.g., Refs. 14.19 and 15.9–15.12). The airfoil shape based on using one of these pressure distributions is shown in the inset to Fig. 15.21 along with the potential flow based solution and experimental pressure distribution (maximum lift is $C_l \approx 1.8$, at $\alpha = 14°$, and at $Re = 3 \times 10^6$). Note that at the lower angles of attack (at possible cruise conditions) there is a favorable pressure gradient near the front of the airfoil where a laminar boundary layer can be maintained for low drag. (Transition occurs near the maximum thickness section; also, at $\alpha = 0°$ a laminar separation bubble appears on the lower surface near the leading edge, causing the discrepancy between the measured and calculated data).

Another method of obtaining a high lift coefficient is to have a variable wing geometry, where both surface area and airfoil camber can be changed according to the required flight conditions. Mechanically a multielement airfoil can be considered as such a device since by

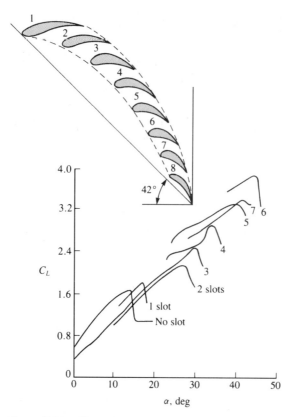

Figure 15.22 Lift coefficient versus angle of attack for the RAF 19 airfoil broken up to different numbers of elements. (Note that a two-element airfoil has 1 slot, a three-element airfoil has 2 slots, etc.). From Ref. 15.14. Reprinted with permission. Copyright AIAA.

changing flap angles the lift coefficient can be altered without changing the wing angle of attack. But the multielement design will inherently possess high lift capabilities. This was realized early in the beginning of this century and Handley Page[15.15] showed experimentally that the greater the number of elements the greater is the maximum lift coefficient. Figure 15.22 shows the results of Ref. 15.15 where the RAF 19 airfoil was broken up into different numbers of elements (note that a two-element airfoil will have one slot, a three-element airfoil two slots, etc.).

The pressure distribution and the lift versus angle of attack for a typical three-element wing section[15.16] is shown in Fig. 15.23; note that lift coefficients of over 3.0 can be obtained (Fig. 15.24). Since the overall effect of a flap is to increase the load on the element ahead of it, the leading-edge slat (if not drooped) is the most likely to separate. Consequently, many airplanes will droop the leading-edge slat at high lift coefficients to delay its flow separation. The effects of these devices is shown schematically in Fig. 15.25 and, in general, extending the slats will extend the range of angle of attack for maximum lift but will not raise the lift curve. Now, recall Example 3 of Section 5.4 about the flapped airfoil, which indicated that a flap at the trailing edge will have a large effect on the airfoil's lift. This is clearly indicated in Fig. 15.25 where bringing the flap down by 50° results in an increase of the lift coefficient by close to 1.0.

The above discussion was mainly aimed at two-dimensional airfoil design but as wing aspect ratio becomes smaller, the pressure distribution will be altered by the three-dimensional

15.3 Influence of Viscous Flow Effects on Airfoil Design

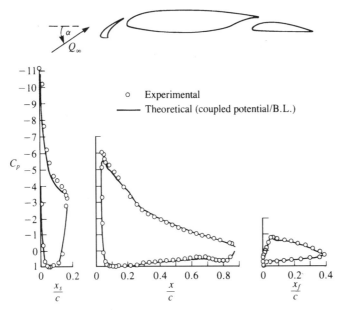

Figure 15.23 Comparison of experimental and computed pressure distribution (based on viscous/inviscid interaction) near stall on a three-element wing. Slat angle is $-42°$, trailing edge flap angle is $10°$, and section lift coefficient is 3.1 at $Re = 3.8 \times 10^6$.

shape of the wing (see Fig. 12.31) and three-dimensional methods (either computational or experimental) must be used. Also, based on Figs. 15.22 and 15.23 it seems that with large aspect ratio wings, section lift coefficients of about 4 are possible, and Smith[15.14] estimates a hypothetical maximum section lift coefficient of 4π and shows a two-element airfoil with an estimated C_l of about 5. For smaller aspect ratio wings a maximum lift coefficient of

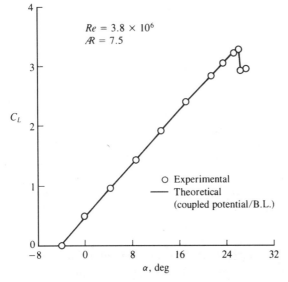

Figure 15.24 Lift coefficient versus angle of attack for a three-element wing shown in the inset to Fig. 15.23. Slat angle is $-40°$, trailing edge flap angle is $10°$, and $Re = 3.8 \times 10^6$.

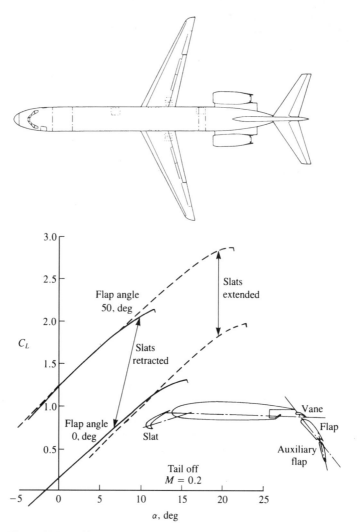

Figure 15.25 Effect of leading-edge slats and trailing-edge flaps on the lift curve of a DC-9-30 airplane (tail off, $M = 0.2$). (Courtesy of Douglas Aircraft Co.)

$C_{L_{max}} = 1.2 \text{Æ}$ is frequently quoted, and this is probably a more conservative version of Hoerner's[15.17] $C_{L_{max}} = 1.94 \text{Æ}$ formula. Hoerner also provides a limit on wing aspect ratio for this formula (p. 4-1) such that $\text{Æ} < 6$.

At this point it is worth mentioning a very simple trailing-edge flap that usually will increase the lift of a wing. This small trailing-edge flap is shown schematically in the inset to Fig. 15.26 and flow visualization indicates that owing to the small vortex created at the pressure side, the trailing-edge upper surface boundary layer will be thinner. This in effect turns the trailing-edge flow downward and increases the wing's circulation. Experimental results[15.12,15.18] usually show a consistent increase in lift due to this device, which in most cases is accompanied by a slight increase in drag. In some limited situations (as in Ref. 15.12 where the trailing-edge flap was attached to a high drag Newman airfoil) a reduction in drag may be observed also. (A sketch of the Newman airfoil's shape can be found in Ref. 15.12.)

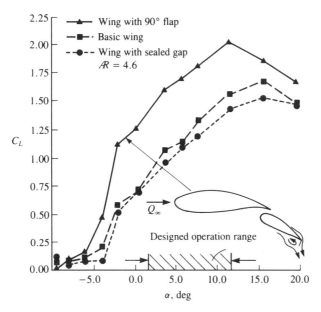

Figure 15.26 Effect of a small 90° flap on the lift of a two-element airfoil ($Re = 0.3 \times 10^6$). From Ref. 15.18. Reprinted with permission of ASME.

15.4 Flow over Wings at High Angles of Attack

Many airplanes and other vehicles that use lifting surfaces face situations where a variable range of lift coefficient is required. For example, the lowest landing speed of a high-speed airplane is dictated by the highest (safe) lift coefficient. This is even more pronounced for supersonic aircraft, which must have swept leading edges (less than the Mach cone) and small wing area for supersonic cruise but, for landing at reasonably low speeds, require very high lift coefficients. Therefore, it is very important to be able to generate high lift coefficients, even if wing stall is approached.

Since the primary function of wings is to generate lift let us observe a typical lift curve of a wing, as shown in Fig. 15.27. At the lower angle of attack range (far from the stall angle) the lift slope versus angle of attack is well defined and predictable (constant C_{L_α} range in Fig. 15.27) and airplane lift can be controlled by changing α. However, when the angle of attack approaches the stall condition the wing lift will not react to α changes with the same intensity as in the so-called linear range. From an airplane point of view the stall should be gradual, as shown by the solid line, and not abrupt as indicated by the dashed line. Even more important is that the stall process not generate strong rolling moments caused by asymmetry (as a result of an earlier stall of either the left or the right wing) such that the airplane will be driven into a stall-spin.[15.19]

A desirable wing stall pattern can be tailored, for example, by having larger section lift coefficients near the wing root. If this is done properly, stall will be initiated there and will gradually spread toward the wing tips, so that the overall stall will be similar to the solid line in Fig. 15.27. Additionally, local wing root flow separations at the beginning of the wing stall yield reasonably controllable rolling moments (assuming that roll control surfaces are located near the wingtips) and such an airplane can safely approach wing stall. This onset of root stall can be obtained by forward wing sweep (see spanwise loading of such wings in Fig. 12.17), or by wing twist, etc. However, the effect of wing taper (see Fig. 12.19) or aft

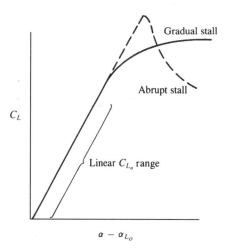

Figure 15.27 Lift curve and stall characteristics of two generic wings.

sweep (Fig. 12.17) results in higher wingtip loading, which (if not corrected by twist, airfoil camber variation, etc.) can cause the high angle of attack flow separations to begin near the wingtips. This may cause a stall of the ailerons and loss of the aircraft lateral control.

Leading-edge sweep also has an important effect on the stalled flowfield over wings. This is illustrated schematically in Fig. 15.28 where typical separated flow patterns as observed by flow visualization are shown for unswept, moderately swept (up to 60°), and highly swept wings. (Note that the discussion in this section is focused mainly on airplane wings where the Reynolds number is considered to be high, e.g., $Re > 10^6$.) The unswept wing, shown in the left-hand side of the figure, at large angles of attack behaves like a flat plate with two shear layers emanating from the leading and trailing edges. In the two-dimensional view of section AA, a time-dependent vortex shedding (sometimes called a "Karman vortex street") is observed. In the case of the highly swept wings, shown in the right-hand side of the figure, the cross-section flowfield is shown schematically in section BB, and the flow separates at the two leading edges, resulting in two strong, concentrated vortices. These leading-edge vortices are located near and above the leading edge, and the low pressure caused by the velocity induced by these vortices increases the wing's lift. In the case of moderate leading-edge sweep, as shown in the center of Fig. 15.28, the leading-edge vortex becomes less visible, and sometimes even two such vortices per side may be observed.[15.20]

At this point we must emphasize that the methods discussed in Chapters 2–13 are based on potential flow and they are applicable (with some viscous corrections, as described in Section 15.2) to attached flows only. The drag force D in a real viscous flow, for example, will have a potential flow component D_i (which was zero in two dimensions) and a viscous flow part D_0:

$$D = D_i + D_0 \quad (15.10)$$

In attached flows most of the viscous drag is due to the skin friction; however, in the case of extensive separations the drag is due to "form drag" or "pressure drag," which is much larger. (As an example, consider the case of a fully stalled airfoil where instead of the attached lifting flow pattern a large separation bubble with a fairly constant negative pressure distribution exists – hence the term "pressure drag"). For further information on the fluid dynamic drag of a particular configuration it is advised first to search through the considerable collection of experimental data provided in *Fluid Dynamic Drag* by Hoerner.[15.21]

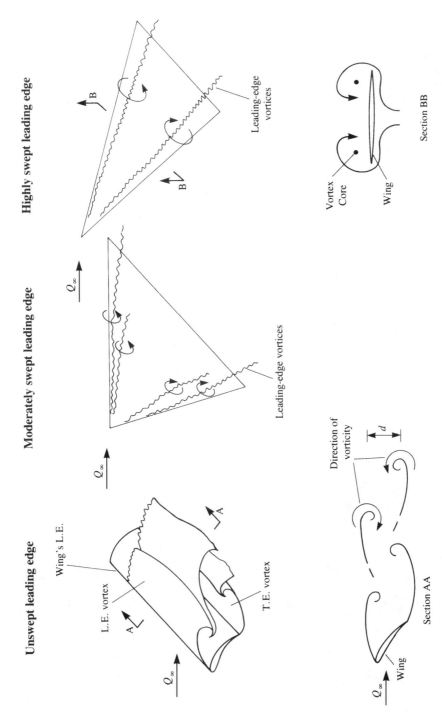

Figure 15.28 Schematic description of the flow patterns observed in the separated flow over unswept, moderately swept, and highly swept leading-edge wings.

The analytical treatment of separated flowfields to determine the resulting pressure distribution is far more difficult and has not yet reached the level of confidence obtained for attached flows. Consequently, in the following four sections, experimental evidence is provided on the problems of unswept and highly swept leading-edge separation along with some simple models for some special cases of flow separation. All of these simple models are in their early state of development and have not reached a level where they can be used as a predictive engineering tool (e.g., similar to some panel methods used within the attached flow domain). However, their importance lies in helping to understand and to explain some fairly complex flow phenomena.

15.4.1 Flow Separation on Wings with Unswept Leading Edge – Experimental Observations

Flow visualization based observations of two-dimensional airfoils indicate that when the angle of attack increases to the point where flow separation is initiated, a shear layer forms near the separation point. The vorticity in this layer seems to have a clockwise value, whereas the shear layer emanating from the trailing edge (wake) mostly has a counterclockwise vorticity (see schematic description in Section AA of Fig. 15.28). These two layers roll up in opposite directions and eventually a periodic wake rollup pattern is observed (Fig. 15.29). This instability of the two shear layers originating from the upper and lower surface boundary layers is present even at zero angle of attack, and results of flow visualizations[15.22] with the hydrogen bubble technique are shown in Fig. 15.30. The first and most important observation is, then, that even for a stationary airfoil, when the flow separates, the problem becomes time dependent. The frequency of this wake oscillation can be related to the wake spacing by the Strouhal number (Eq. (1.52)), which was observed to have values of approximately

$$\frac{f\,d}{Q_\infty} \approx 0.1 - 0.2 \qquad (15.11)$$

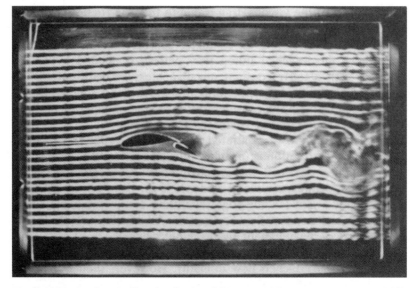

Figure 15.29 Smoke trace flow visualization of the separated flow over a two-element airfoil ($Re = 0.3 \times 10^6$).

15.4 Flow over Wings at High Angles of Attack

Figure 15.30 Hydrogen bubble flow visualization of the wake (created by the upper and lower shear layers of the boundary layer) behind a NACA 0012 airfoil. (Courtesy of K. W. McAlister and L. W. Carr,[15.22] U.S. Army Aeroflightdynamics Directorate, AVSCOM.)

Here d is the spacing between the two shear layer separation points (shown in Section AA in Fig. 15.28) and f is the shedding frequency. Time-dependent chordwise pressure measurements on a stalled airfoil are scarce, but a typical time-averaged pressure distribution is shown in Fig. 15.31. The time-averaged effect of the flow separation is to reduce the lift (reduced circulation) while the pressure stays fairly constant in the region starting behind the separation point and ending at the trailing edge. Note the large difference between the attached, potential flow calculations taken from Ref. 15.23 and the experimental results of the separated flow.

Surface-oil flow visualizations[15.24] with rectangular, finite aspect ratio wings (Fig. 15.32) indicate that in reality there is no "true two-dimensional flow separation." Instead there are three-dimensional cells and at the central plane of each cell a flowfield similar to the "two-dimensional separation" of Fig. 15.28 AA and Fig. 15.29 can be observed. These cells seem

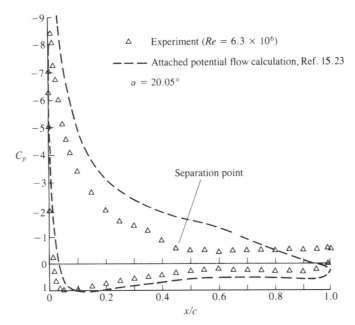

Figure 15.31 Comparison between attached flow (calculation) and partially separated (experimental) pressure distribution on a GAW-1 airfoil. From Ref. 15.23, published by AGARD/NATO.

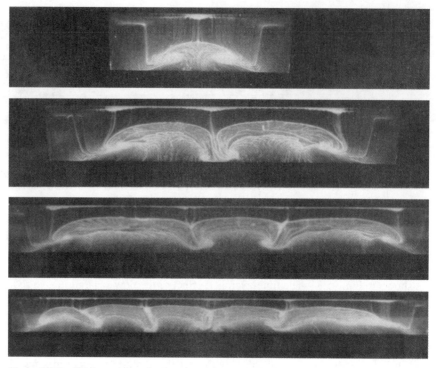

Figure 15.32 Oil flow patterns developed on the upper surface of several stalled rectangular wings. For all wings angle of attack is $\alpha = 18.4°$, the airfoil shape is a 14% Clark Y section, and $Re = 385,000$. (Courtesy of A. Winkelman.[15.24]) Reprinted with permission. Copyright AIAA.

to have some natural aspect ratio, which will adjust itself slightly to the actual wing planform shape. For example, for a rectangular wing with aspect ratio (R) of 3 one cell was visible, whereas for $R = 6$ and 9 two and three cells were visible, respectively. When the wing span was further increased (to $R = 12$) then five cells were observed and the size of some of the cells was somewhat smaller. In conclusion, therefore, the unswept leading-edge problem appears to be always three dimensional and time dependent – and from the experimental point of view still not completely explored.

In the following section some of the inviscid modeling efforts for such flows are presented. It is assumed that if the separated shear layer can be modeled then the rest of the flowfield is still close to being irrotational and therefore the pressures and loads can be estimated.

15.4.2 Flow Separation on Wings with Unswept Leading Edge – Modeling

If we follow the previous assumptions that for high Reynolds number flows the viscous effects are confined within thin shear layers, then the irrotational flow outside these regions can be modeled by inviscid flow methods. The primary objective of this approach is to explain the pressure distribution obtained in separated flow and in some cases also the skin friction can be estimated. It seems clear that the model must be time dependent and numerous such methods can be found in the literature. Two excellent survey papers by Leonard[15.25] and Sarpkaya[10.3] describe and classify a large portion of the available literature on this topic. Some steady-state, two-dimensional models for flow separation[15.5] extend the method of viscous–inviscid interaction of Section 15.2 and model the bubble created by

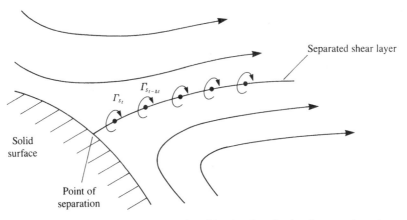

Figure 15.33 Discrete-vortex representation of the shear layer leaving the separation point.

the separated flow with additional sources or simply assume that the pressure is constant there. However, since flow visualization clearly indicates that high Reynolds number flow separation involves time-dependent wake shedding, we shall elaborate a bit more on this approach. Note that the following two-dimensional models are intended to simulate the flowfield at the symmetry plane of a three-dimensional separation cell (as shown in Fig. 15.32).

A typical two-dimensional time-dependent model for the separated flow over an airfoil can be constructed by taking any time-dependent potential flow solver (or any method of Chapter 11 with a time-dependent upgrade) and adding a separated shear layer model. Such a model is depicted schematically in Fig. 15.33, where the flow near a separation point is described. In order to solve the potential flow problem, two additional unknowns characterizing the simplified effects of viscosity must be supplied to the potential flow solver. These are:

1. The location of the separation point or points. (Some information on the time-dependent motion of this point, which may be small, is valuable, too.)
2. The strength of the vortex sheet.

For example, let us use the unsteady thin airfoil method of Section 13.10. It is assumed that the shape of the solid surface and location of the separation point, which is a function of the Reynolds number, are known (e.g., from experiments, flow visualizations, independent viscous calculations, etc.). The separated vortex sheet will be approximated by discrete vortices, as shown in Fig. 15.33, and their strength can be approximated by estimating $d\Gamma_s/dt$ near the separation point (see Fig. 15.34), where the subscript s denotes "separated wake." Using the definition of the circulation as in Eq. (2.36), we write the rate of circulation generation at the separation point as

$$\frac{d\Gamma_s}{dt} = \frac{d}{dt} \oint \mathbf{q} \cdot d\mathbf{l}$$

The integration path can be approximated by the simple rectangle, shown in Fig. 15.34, that is placed near the separation point such that its upper and lower segments are in the potential flow region. If the upper velocity q_u outside the boundary layer and the lower velocity q_l within the separated bubble are known, then the integration on an infinitesimal rectangle

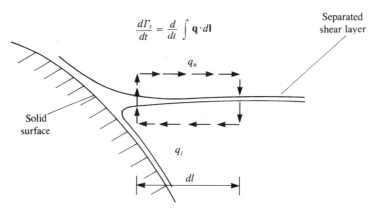

Figure 15.34 Nomenclature used to calculate the vorticity shed from a separation point.

becomes

$$\frac{d\Gamma_s}{dt} \approx \frac{d}{dt}(q_u\,dl - q_l\,dl) \approx (q_u - q_l)\frac{dl}{dt} = (q_u - q_l)\left(\frac{q_u + q_l}{2}\right) = \frac{1}{2}(q_u^2 - q_l^2) \tag{15.12}$$

Here the vertical segments of the rectangle approach zero and their effect on Γ_s is neglected. In practice, the upper velocity is taken as the potential velocity above the separation point and is known (at least from the previous time step), whereas the lower velocity is close to zero. In this case the strength of the latest separated vortex Γ_{s_t} becomes

$$\Gamma_{s_t} = \frac{K}{2}(q_u^2 - q_l^2)\Delta t \tag{15.13}$$

where K is a circulation reduction factor and values of 0.5–0.6 are usually used. This latest separated vortex is placed along the streamline that started at the separation point (say at a distance $(1/2)[(q_u + q_l)/2]\Delta t$ from the separation point).

The momentary solution of the airfoil with the separated wake model is described schematically in Fig. 15.35. If the lifting airfoil is represented by N discrete bound vortices (circles in the figure) with unknown strength, then N equations representing the zero normal flow boundary condition can be specified on the N collocation points. The strength of the latest separated wake vortex Γ_{S_t} is known from Eq. (15.13) whereas the strength of the latest vortex shed at the trailing edge Γ_{W_t} is calculated by using the Kelvin condition (Eq. (13.51)):

$$\frac{d\Gamma}{dt} = \left(\frac{d}{dt}\sum_{i=1}^{N}\Gamma_i\right)_{\text{airfoil}} + \Gamma_{W_t} + \Gamma_{S_t} = 0 \tag{15.14}$$

So at each time step there are $N + 1$ unknown vortices $\Gamma_1, \Gamma_2, \ldots, \Gamma_N, \Gamma_{W_t}$ and $N + 1$ equations (N boundary conditions at the collocation points plus the Kelvin condition, Eq. (15.14)). The problem at each time step is solved exactly as described in Section 13.10, since the addition of the separated wake did not introduce any new additional unknown. Note that in this solution the Kutta condition is specified at the airfoil's trailing edge (as a result of using the lumped-vortex element – see Section 13.10). The wake rollup at each time step can be performed in a manner similar to that described in Section 13.10 (see Eqs. (13.131) and (13.132)) and each vortex of the wake (both trailing edge and separated) will

15.4 Flow over Wings at High Angles of Attack

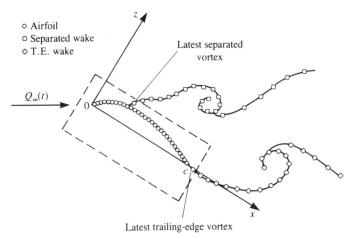

Figure 15.35 Schematic description of the discrete-vortex model for the airfoil, trailing-edge wake, and separated wake. The dashed line rectangle surrounds the vortices whose strength is calculated at each time step.

move with the local velocity $(u, w)_i$ by the amount

$$(\Delta x, \Delta z)_i = (u, w)_i \Delta t \tag{15.15}$$

Here the induced velocity $(u, w)_i$ is the velocity induced by all the vortices (airfoil and wakes) in the field. Owing to the singular nature of the vortices, instabilities can develop during the wake rollup routine calculations, and some methods[10.3] use certain smoothing techniques to improve the wake rollup.

Results of such a calculation without using any wake rollup smoothing technique are presented in Fig. 15.36. The oscillation is obtained either by moving the separation point or by changing the vortex strength. The front separation point in this figure is fixed at a distance of 5% chord from the leading edge.[15.26]

Another approach (e.g., see Ref. 15.25) is to solve the vorticity transport equation (Eq. (2.10)) by using discrete vortices. In this case the flow can be rotational and for thick bodies the Dirichlet boundary condition is applied to the Poisson equation for the stream function ($\nabla^2 \Psi = \zeta = 0$ inside a closed body). For example, the method of modeling the rotational boundary layer is described schematically in Fig. 15.37, where at a certain point in the boundary layer vortex elements are introduced. The strength and initial velocity of the newly introduced vortices can be estimated by assuming a certain effective boundary layer thickness δ_e and outer velocity U_e (U_e is obtained from the potential flow solution, whereas δ_e can be estimated by using existing boundary layer data on flat plates). The initial velocity can then be approximated as $U_e/2$, and the vortex strength is calculated by using Eq. (15.12) with $q_l = 0$ on the surface:

$$\frac{d\Gamma_s}{dt} \approx \frac{U_e^2}{2} \tag{15.16}$$

In some models one set of (bound) discrete vortices is placed around the solid surface in a fixed position and the strengths are calculated at each time step by applying the Dirichlet boundary condition on the surface (see Fig. 15.38). During the second time step these vortices are allowed to translate with the flow and a new set of "bound" vortices is created. Results of such a calculation[15.27] are presented in Fig. 15.39. Note that in this method both

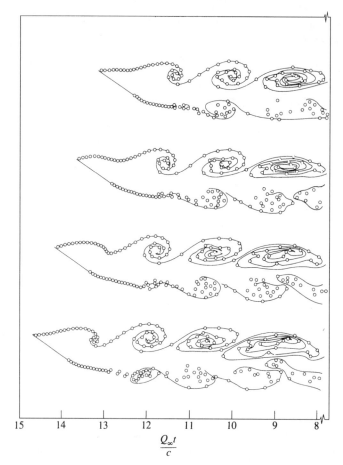

Figure 15.36 Vortex wake rollup behind a separated flat plate. Front separation point was fixed at $x/c = 0.05, \alpha = 30°$, and the time step is $\Delta t Q_\infty / c = 0.1$. From Ref. 15.26. Reprinted with permission from Cambridge University Press.

the tangential and the normal velocity components on the body are zero since the flow is rotational. However, at each time step a large number of new vortices is being created, compared to only two per time step in the method of Ref. 15.23. Consequently, some vortex number reduction schemes (which combine several nearby vortices) and wake rollup reshaping methods can be found in the literature.[10.3,15.25]

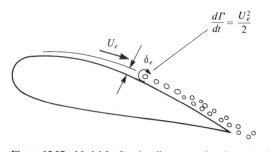

Figure 15.37 Model for forming discrete vortices in an attached boundary layer.

15.4 Flow over Wings at High Angles of Attack

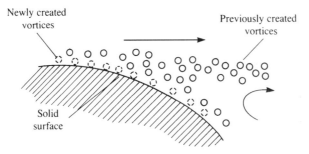

Figure 15.38 Modeling of flow separation by discrete vortices. The dashed circles represent the fixed location of the newly created vortices and the no-slip boundary condition is specified on the solid boundary.

The extension of these two-dimensional models to three dimensions is somewhat more elaborate. For example, methods that are based on the stream function and solve two-dimensional rotational flows cannot be extended automatically to three dimensions (see Chapter 1 or 2). A possible simple extension to the thin lifting surface of Section 13.12 was done in Ref. 15.28 mainly to explain the results of some flow visualizations obtained during high angle of attack testing of unswept wing general aviation airplanes. An imaginary sequence leading to this model is described schematically in Fig. 15.40. The first frame (Fig. 15.40a) shows the time-averaged vortex core positions of the shear layers originating at the leading and trailing edges of a hypothetical two-dimensional flow. The figure shows only the most recent vortices but the complete wake in the two-dimensional section will have a pattern similar to the Karman street shape of Fig. 15.28 (section AA). Also, if the wing geometry is purely two dimensional ($R = \infty$), then those vortex lines will be initially straight (hypothetically). However, a spanwise instability will develop[15.29] between the leading-edge vortex and its image (reflected by the wing upper surface). This pair of vortex lines (the one above the leading edge and its reflection) will develop an instability similar to the one we have seen for the trailing wake vortices (e.g., Fig. 15.4, and also Ref. 15.1). This wave shape disturbance will grow with time and eventually break up into the cellular patterns shown in Fig. 15.40c. Figure 15.40d shows the surface-oil patterns appearing in Fig. 15.32 that can be explained by this simplistic model (note that the two edges of the mushroom shape correspond to the vortex ring "touch down" points).

The addition of large-scale vortex rings[15.28] to the panel model of a thin wing for simulating the large-scale effects of this separated flow are shown in Fig. 15.41. In this model the location and spanwise width of the separated flow cells must be specified (based on flow visualizations). Once this information is supplied to the otherwise potential flow solver, the

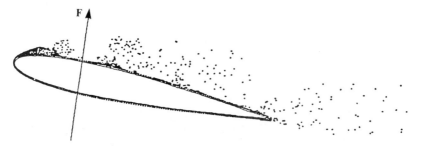

Figure 15.39 Simulation of the separated flow by the vortex tracing method. (Courtesy of P. R. Sparlat,[15.27] NASA Ames Research Center.) Reprinted with permission. Copyright AIAA.

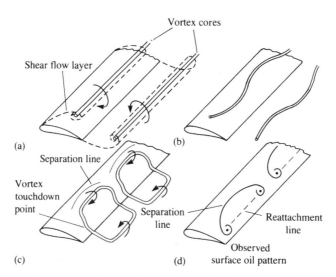

Figure 15.40 Schematic description of the instability sequence of a straight vortex line (representing the vorticity shed from the leading edge) leading to the cellular formations observed on separated rectangular wings.

fluid dynamic loads on the lifting surface can be calculated and, as shown in Ref. 15.28, the lift variation as a result of the separation line movement can be explained.

15.4.3 Flow Separation on Wings with Highly Swept Leading Edge – Experimental Observations

The high angle of attack separated flow pattern, based on numerous flow visualizations (see Refs. 15.30–15.33), over highly swept wings (e.g., a small Æ delta wing) in terms of the cross-flow is depicted in section BB of Fig. 15.28. If the leading-edge radius is small (sharp L.E.) then such L.E. vortices will be present at angles of attack as low as 10°. Because of these vortices, the actual flowfield is entirely different from what would have been expected according to the attached-flow model of slender-wing theory (Section 8.2.2). Also, when the leading edge is sharp, the location of the separation line is fixed along

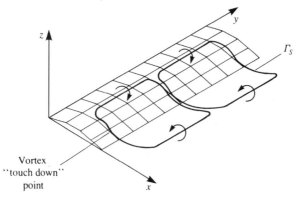

Figure 15.41 Simple vortex ring model for the flow separation over unswept, finite aspect ratio rectangular wings.

15.4 Flow over Wings at High Angles of Attack

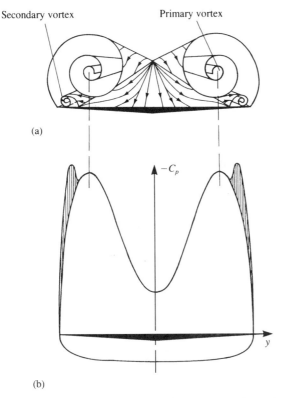

Figure 15.42 Schematic description of the primary and secondary vortex pattern (in the crossflow plane) of the flow over a slender delta wing and the resulting pressure distribution. From Ref. 15.34, published by AGARD/NATO.

the leading edge, and the flowfield appears not to be sensitive to changes in the Reynolds number.

The effect of this vortical flow (due to the leading-edge vortices) on the pressure distribution is shown in Fig. 15.42. The two large suction peaks on the upper surface of the wing are due to the high-speed flow induced by these vortices. This high velocity creates a secondary shear flow near and on the wing's upper surface and results in a secondary (and sometimes even a tertiary) vortex that is much smaller and weaker; its effect is shown in the figure. The above shape of the spanwise pressure distribution (see Ref. 15.34) is maintained along the chord (Fig. 15.43) but the suction force is the strongest near the wing apex. This spanwise pressure distribution is entirely different from the pressure difference data results of the linear theory in Fig. 8.21. Furthermore, the lift coefficient of the wing with leading-edge separation (up to $\alpha \approx 45°$) is considerably larger than predicted by the linear theory (Eq. (8.94)). The difference between the linear value (of $(\pi/2)A\!R\alpha$ in the low angle of attack case) and the actual lift is often referred to as "vortex lift" (and is shown in Fig. 15.44). So in this case of wings with highly swept leading edges, the lift is increased owing to leading-edge separation, unlike in the unswept wing case where the lift is reduced. This fact was realized by many aircraft designers and many modern airplanes have such highly swept lifting surfaces, called strakes (see Fig. 15.45). For example, if such a strake is added in front of a less swept back wing then the vortex originating from the strake will induce low pressures, similar to those in Fig. 15.43, on the upper surface of the main

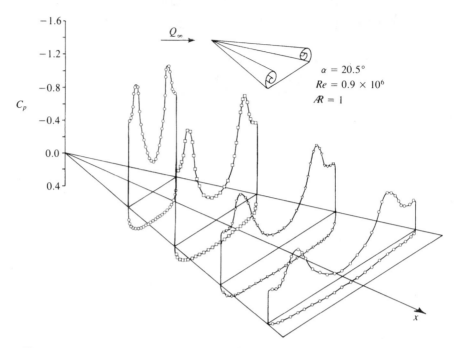

Figure 15.43 Schematic description of upper-surface pressure distribution (based on the results of Ref. 15.34) on an $R = 1$ delta wing at $\alpha = 20°$.

wing. Therefore, the total gain in lift will surpass the lift of the strake alone, as shown in Fig. 15.45.

As mentioned earlier, in contrast to the unswept wing case, the lift of a slender wing is larger when the leading edge is sharper, as shown in Fig. 15.46 (here $\alpha - \alpha_{L_0}$ is used since there is a lift difference due to effective camber between wing A and wing B). In this case

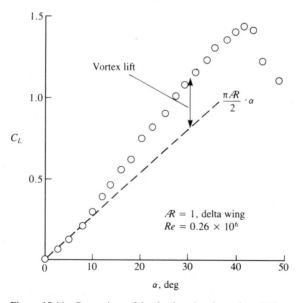

Figure 15.44 Comparison of the slender wing theory based lift curve with experimental results.

15.4 Flow over Wings at High Angles of Attack

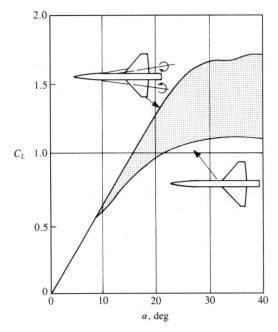

Figure 15.45 Effect of strakes on the lift of a slender wing/body configuration. From Skow, A. M., Titiriga, A., and Moore, W. A., "Forebody/Wing Vortex Interactions and Their Influence on Departure and Spin Resistance," published by AGARD/NATO in CP 247 – High Angle of Attack Aerodynamics, 1978.

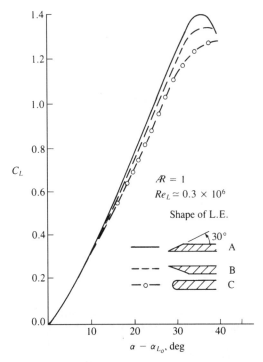

Figure 15.46 Effect of leading-edge shape on the lift of a slender delta wing.

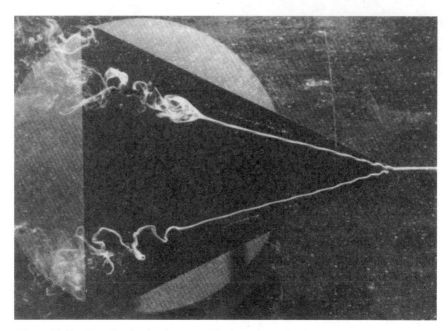

Figure 15.47 Flow visualization in water of leading-edge vortex burst. From Lambourne, N. C., and Bryer, D. W., ARC R&M 3282, 1962. Reproduced with the permission of the Controller of Her Majesty's Stationery Office.

the lift of delta wing A is slightly larger then the lift of the inverted wing, and in both cases the lift is larger than in the case of a rounded leading edge. So in general, when the flow is turned more sharply (when viewed in the two-dimensional cross section as in Fig. 15.28, section BB) the vortex will be stronger, resulting in more suction force. As the leading-edge radius increases, the lift usually decreases and depends more on the Reynolds number. Also, in this case of leading-edge separation the "classical" suction force at the leading edge is lost and therefore the drag force will be larger than for the elliptic case of Section 8.2.2. Consequently, the resultant force due to the pressure difference on the lifting surface will act normal to the surface and therefore the drag can be estimated by

$$C_D = C_L \tan\alpha \qquad (15.17)$$

A more careful examination of Fig. 15.44 reveals that the highly swept wing stalls, too, at a fairly large angle of attack. This stall, though, is somewhat different from the unswept wing stall and is due to "vortex burst" (or breakdown). This condition is shown by the flow visualization of Fig. 15.47, and at a certain point the axial velocity in the vortex core is reduced and the vortex becomes unstable, its core bursts, and the induced suction on the wing disappears. The pressure distribution on the delta wing (from Ref. 15.35) for several angles of attack, shown in Fig. 15.48, shows this effect of vortex lift and vortex burst. So, as a result of the vortex burst the lift of the wing is reduced and a condition similar to stall is observed. Flow visualizations sometimes show the burst as a sudden spiral growth in the vortex core (see Fig. 15.47) and this is called "spiral burst"; in other instances it is seen as a bubble burst (hence it is called "bubble burst" or "bubble instability"). The onset of vortex burst was investigated by many investigators and the results for a delta wing can be summed up best by observing Fig. 15.49 from Polhamus.[15.36] (Incidently, Polhamus developed a method of estimating the vortex lift

15.4 Flow over Wings at High Angles of Attack

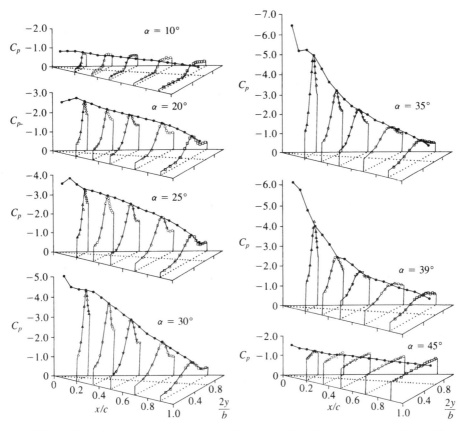

Figure 15.48 Upper-surface pressure distribution on a slender delta wing (semispan, $\mathcal{R} = 1.46$) at various angles of attack and beyond stall. From Ref. 15.35. Reprinted with permission. Copyright AIAA.

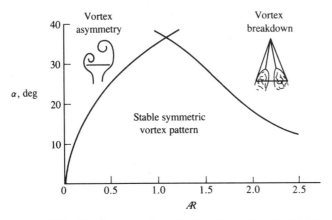

Figure 15.49 Stability boundaries of leading-edge vortices for flat delta wings in incompressible flow. (Adapted from Ref. 15.36.)

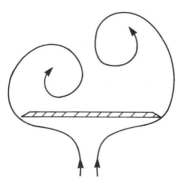

Figure 15.50 Schematic description of the crossflow due to asymmetric leading-edge vortices.

based on the leading-edge suction analogy; for more details the reader is referred to Ref. 15.36 or Section 19.7 of Ref. 15.4.) The abscissa in Fig. 15.49 shows the wing aspect ratio, and the ordinate indicates the angle of attack range. The curve on the right-hand side indicates the boundary at which vortex burst will reach the wing's trailing edge. The method of reading this figure can be demonstrated by taking the delta wing of Fig. 15.44 ($R = 1$) and, for example, gradually increasing its angle of attack. This gradual increase will cause the vortex burst, which is far behind the trailing edge, to move gradually forward. According to this figure, at about $\alpha = 35°$–$40°$ the vortex burst will pass forward of the trailing edge and spoil the lift and initiate the wing stall. Moreover, for larger wing aspect ratios (less L.E. sweep) the burst will occur at lower angles of attack. As the wing becomes very slender the leading-edge vortices become very strong and the burst is delayed. But for these wings another flow phenomenon, called "vortex asymmetry," is observed. This situation occurs when the physical spanwise space is reduced and consequently one vortex raises above the other (Fig. 15.50). Usually any random disturbance can cause this instability to develop and changes in the asymmetry from side to side are also possible. The onset of this condition is depicted by the left-hand curve in Fig. 15.49. For example, if the angle of attack on an $R = 0.5$ delta wing is gradually increased, then over $\alpha \approx 20°$ the vortex asymmetry will develop. If the angle of attack is increased, say up to $\alpha = 40°$, the lift will still grow and probably near $\alpha = 45°$ the vortex burst will advance beyond the trailing edge and wing stall will be initiated. In general the condition of an asymmetric vortex pattern is nondesirable because of the large rolling moments caused by this asymmetry. Furthermore, the pattern of asymmetry is sensitive to disturbances and can arbitrarily flip from side to side. The presence of a vertical fin (e.g., a rudder) between the two vortices or a central body (as in missiles) can have a stabilizing effect and delay the appearance of this vortex asymmetry.

To conclude this discussion on experimental data of slender wings, a set of typical lift coefficient data is presented in Figs. 15.51 and 15.52. Note that in the data of Shanks[15.37] leading-edge sweep angle rather than wing aspect ratio is presented (but for delta wings $R = 4\cot\Lambda$, where Λ is the aft-sweep angle). The lift of slender rectangular wings (Fig. 15.52) is enhanced too by the side edge vortex lift, and the effect of the vortex lift on the wing is similar to the case of the slender delta wings (mainly when wing $R < 1$, and the leading and side edges are sharp). In this case, though, the flowfield is somewhat more complex because of the presence of a leading-edge separation bubble, which is noticeable for sharp leading-edge wings (Fig. 15.53). This bubble is created by the time-dependent leading-edge vortex shedding (as in Fig 15.28) but its effect is small compared to the vortex lift of the side-edge vortices (when $R < 1$).

15.4 Flow over Wings at High Angles of Attack

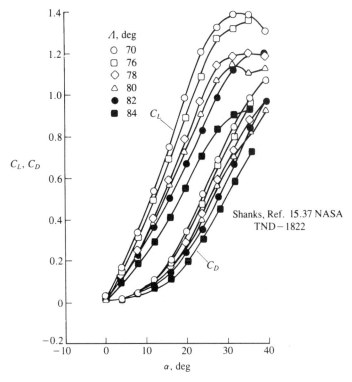

Figure 15.51 Lift and drag coefficient versus angle of attack for several slender delta wings ($Re = 3.2 \times 10^5$) (Adapted from Ref. 15.37.)

15.4.4 Modeling of Highly Swept Leading-Edge Separation

The modeling of leading-edge separation from highly swept wings is somewhat simpler than the modeling of the unswept leading-edge separation. The primary reason is that in this case the vorticity generated at the leading edge is immediately conveyed downstream by the chordwise flow and since there is no vorticity accumulation near the leading edge there is no time-dependent wake shedding (as in the unswept leading-edge case). Modeling is possible if the basic information about the location of the flow separation line and the strength of the separated shear layer is supplied to the potential flow solution. Usually such information is generated by a local viscous solution, experiments, or even by using some parameters obtained from the potential flow solution. For example, when modeling the flow over low aspect ratio delta wings with sharp leading edges, the location of the separation line is fixed along the sharp leading edge (recall that a sharper leading edge results in a stronger L.E. vortex and more vortex lift; Fig. 15.46). In regard to specifying the strength of the leading-edge shear layer (leading-edge wake), two of the more frequently used possibilities are:

1. Estimate vortex strength by using Eq. (15.13), which requires the calculation of the velocity above and below the wake. In this case the effect of leading-edge radius can be included in the K coefficient such that $K = 0.6$ for a sharp leading edge and some smaller values of K may be used as the leading-edge radius increases.
2. Treat the L.E. wake as a regular wake. This second approach is simpler and seems to yield reasonable results in the range of $\alpha = 10°$–$35°$. The strength of the wake

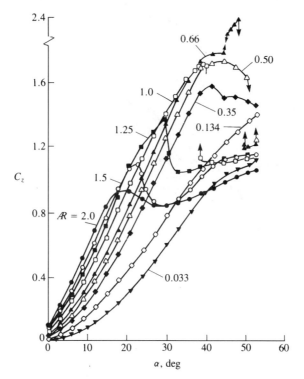

Figure 15.52 Normal force coefficient versus angle of attack for several slender flat rectangular wings ($Re = 0.3\text{–}1.7 \times 10^6$). From Winter, H., "Flow Phenomena on Plates and Airfoils of Short Span," NACA TM 798, 1937.

panel adjacent to the wake (Fig. 15.54) is then calculated by a Kutta condition as in Eq. (9.15):

$$\mu_W = \mu_U - \mu_L \tag{15.18}$$

where μ_U and μ_L are the corresponding upper and lower surface doublet strengths along the separation line (see also Fig. 13.37).

The shape of the separated wake can be calculated by the methods presented in Section 15.1 and in the following examples the calculations are based on the time-stepping

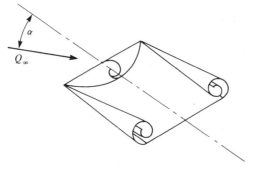

Figure 15.53 Schematic description of the leading- and side-edge vortex rollup on a slender rectangular thin wing.

15.4 Flow over Wings at High Angles of Attack

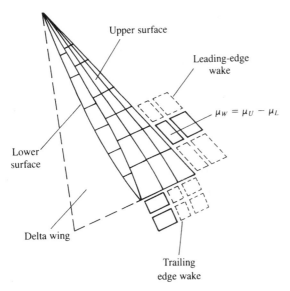

Figure 15.54 Vortex ring (or constant-strength doublet) model for both trailing- and leading-edge wakes shed from slender delta wings.

method[15.38,15.39,12.13] (which is based on the unsteady flow formulation of Chapter 13). The wake rollup is obtained by gradually releasing vortex wake panels from the sharp L.E., similarly to the wake-shedding process at the trailing edge, until the fully developed wake shape is obtained. This is shown schematically in Fig. 15.55 for several time steps and for simplicity only the longitudinal vortex lines are shown (but the wake is constructed by using vortex rings, as shown in Fig. 15.54). The vortex rollup is determined by the momentary velocity induced by the wing and its wakes (as described in Section 13.12, Eqs. (13.153) and (13.154)). Results for the lift curve of this delta wing (based on this model) are presented in Fig. 15.56. At the lower angles of attack (less than $\sim 10°$), the lift curve slope is well predicted by the linear formulation of R.T. Jones (Eq. (8.94)) ($C_L = (\pi \mathcal{R}/2)\alpha$). At higher incidences, however, the leading-edge vortices increase the lift, as indicated by a sample of experimental results (Refs. 15.40–15.42). This vortex lift is not predicted by the basic linear panel method (using only the trailing-edge wake) since the leading-edge wake and its vortex lift is not included. The addition of the separated L.E. vortex model (shown in

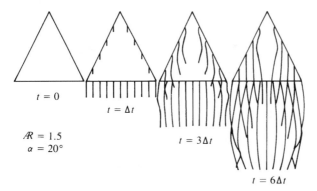

Figure 15.55 Sequence of vortex wake shedding and vortex rollup. From Ref. 15.38. Reprinted with permission. Copyright AIAA.

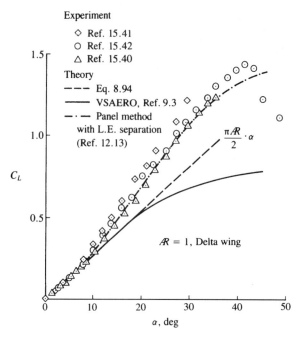

Figure 15.56 Comparison between experimental and calculated lift coefficient for a slender delta wing (using a panel method[12.13] with leading-edge vortex model).

Fig. 15.57) increases the wing's lift and improves the comparison with the experimental data. At very high angles of attack (above 40°), however, vortex breakdown results in the wing's lift loss, a condition that is not modeled here.

The spanwise pressure distribution at the $x/c = 0.5$ station is presented in Fig. 15.58. The wing model consists of 248 panels with 12 spanwise equally spaced panels, and the nondimensional time step is 0.1 chord ($Q_\infty \Delta t/c = 0.1$). Results for a denser computation

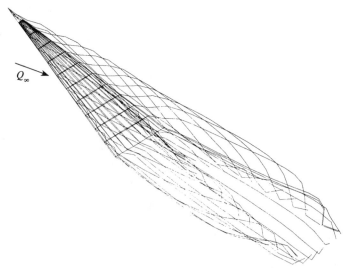

Figure 15.57 Simulation of leading-edge vortex rollup by releasing vortex ring panels from the leading and trailing edges.

15.4 Flow over Wings at High Angles of Attack

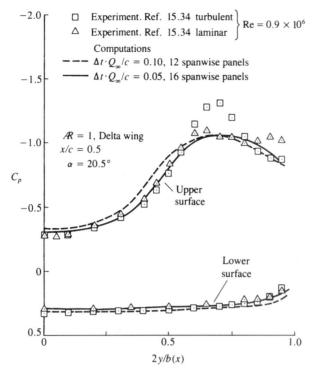

Figure 15.58 Comparison between calculated and experimental spanwise pressure distribution on a slender delta wing.

grid (328 panels, 16 spanwise panels, $Q_\infty \Delta t/c = 0.05$) are shown by the solid line, and both computations show the suction peaks caused by the leading-edge vortices. The experimental results with the turbulent boundary layer of Ref. 15.34 ($Re = 0.9 \times 10^6$ for both experiments) indicate a secondary vortex near the leading edge that was not modeled here. In general, it was found that the lift of the delta wing was less sensitive to a coarser grid and larger time steps than the pressure distribution over the wing's surface. In cases when computer time saving is considered and larger time steps are applied, the spanwise pressure distribution would smear, but the lift will change by only a few percent.

For a demonstration of more complex motions, the wake lines behind a delta wing having an aspect ratio of 0.71, undergoing a coning motion, are presented in Fig. 15.59. The wing angle of attack α was set relative to the x, y, z frame of reference (pitching along $x/c = 0.6$), and then the wing was rotated about the x axis at the rate $\dot\phi$. Computed rolling moments are compared with the experimental data of Ref. 15.43 in Fig. 15.60. The slopes of the computed rolling moment curves, which change rapidly with variations of angle of attack,

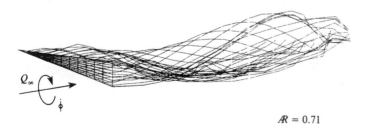

Figure 15.59 Vortex wake lines behind a slender delta wing moving forward in a coning motion.

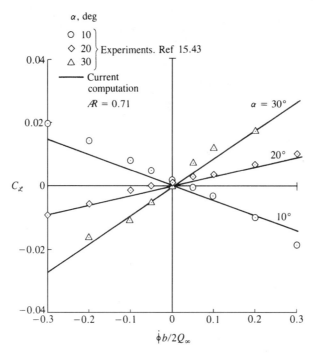

Figure 15.60 Rolling moment versus coning rate for a slender delta wing in a coning motion.

compare reasonably well with the experiments. For angles of attack higher than 30° and for roll rates $\dot{\phi}b/2Q_\infty$ larger than 0.1, vortex breakdown bends the experimental curves and larger differences between the experiment and the computations are detected.

15.5 Possible Additional Features of Panel Codes

a. *Modeling of Propulsion Effects*

Many fluid dynamic problems involve high-energy jets where the jet speed and stagnation pressure are considerably higher than the corresponding free-stream values (as in the case of a jet airplane or a jet-assisted vertical takeoff airplane). Even though the jet flow is compressible and its mixing process with the outer flow is highly viscous, there are several models for simulating the far field effect of confined jets on the otherwise potential outer flow. Such a model is described schematically in Fig. 15.61 and here the outer shear layer of the lifting jet is modeled by linearly varying (along the jet axis) doublet panels[12.11] (which are equal to a constant-strength vortex sheet). Lower speed inlet or auxiliary exit flows can be simulated in many panel codes by simply allowing a transpiration velocity in the normal flow boundary condition, as shown for the inlet flow in Fig. 15.61. The method of including this transpiration velocity in the boundary condition is given by Eq. (15.6):

$$\frac{\partial(\Phi + \Phi_\infty)}{\partial n} = V_n \tag{15.6}$$

but here V_n is a prescribed average jet velocity. When the Dirichlet boundary condition is used then the source term of Eq. (9.12) (e.g., in the panel code VSAERO[12.11]) can be modified such that

$$\sigma = \mathbf{n} \cdot \mathbf{Q}_\infty - V_n \tag{15.19}$$

15.5 Possible Additional Features of Panel Codes

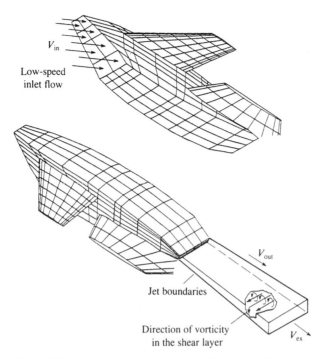

Figure 15.61 Some simple inlet and exhaust jet panel models.

The inlet model of Fig. 15.61 is useful mainly to describe the principle of specifying the inlet inflow but for good computational results the inlet must be modeled with more details as shown in Fig. 15.62. In this case the nacelle inlet is modeled in detail by a large number of panels and the inflow is specified deep inside the nacelle, near an imaginary compressor inlet disk. When the jet is ejected at a large angle relative to the main flow direction then the

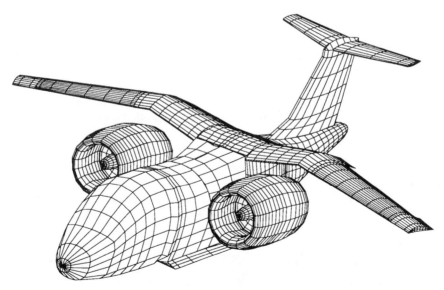

Figure 15.62 Detailed panel model of the tilt-nacelle airplane and its inlets, using 2546 panels per side. (Courtesy of S. Iguchi, M. Dudley, and D. Ashby, 1988, and NASA Ames Research Center.)

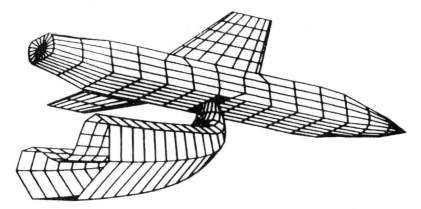

Figure 15.63 Panel model for the plume of the jet ejected normal to the flight direction. (Courtesy of George A. Howell, General Dynamics.)

shape of the jet centerline (Fig. 15.63) is usually calculated with the use of some empirical formula, which relies on the jet initial velocity and the velocity in the potential field into which the jet flows. Consequently, an iterative process is used in some cases in which the jet boundary is treated as a wake (see wake rollup in Section 15.1).

Since a vortex ring model results in a large velocity near the jet outer boundary and a lower velocity at the center of the jet (which is incorrect and limits the use of such models to far field effects only) more refined jet models have been tried. One approach is to model the wake boundary by using constant-strength doublets (similarly to a solid surface), and the jet entrainment that is obtained from empirical data is modeled by a surface source distribution on the jet boundaries (PMARC[9.7]). The jet centerline shape in this case, too, can be calculated again by using empirical formulas or by a time-stepping wake rollup routine (see Section 15.1).

b. *Internal Flows*

In situations when internal flows are modeled as in the case of channel flows, or when studying wind-tunnel/model combined flowfields, then some methods allow the reversal of the direction of the normal **n** to the surface. For example, Fig. 15.64 depicts such a situation, where the turning vane geometry inside a wind tunnel is analyzed. For the basic problem, the free-stream velocity can be specified at the inflow plane as $V_n = Q_\infty$ and some other velocity at the exit plane (usually reduced by the inflow/exit plane cross-sectional area ratio). When the Dirichlet boundary condition is applied to the region outside of the wind tunnel the far field boundary disappears and the influence coefficient matrix becomes singular such that the doublet solution is unique to within an arbitrary constant (also if the free stream is set by prescribed sources at the inlet and exit planes then the other sources will be equal to zero – according to Eq. (9.12)). This difficulty may be overcome by specifying the doublet value on one of the panels (e.g., a value of zero on the wind-tunnel inlet plane). Also, an additional equation is added based on mass conservation of the inlet and exit flows. More details about this approach are provided in Ref. 9.8.

c. *Free-Surface Flows*

The assumptions of inviscid aerodynamics are applicable to many problems in the field of marine hydrodynamics. Most marine vehicles are large, the water flow can be considered as incompressible, and the kinematic viscosity of water is an order of magnitude

15.5 Possible Additional Features of Panel Codes

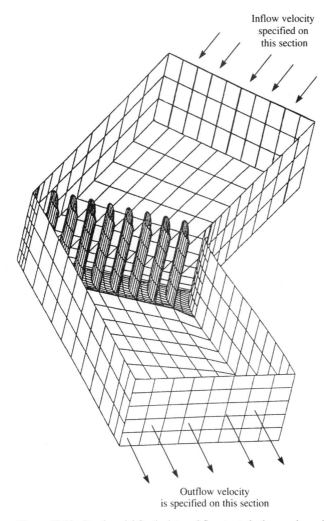

Figure 15.64 Panel model for the internal flow to study the aerodynamics of turning vanes inside a wind tunnel. (Courtesy of D. Ashby, 1986, and NASA Ames Research Center.)

less than for air. Consequently, the Reynolds number is high and the methodology developed here is immediately applicable to deep-water flows (e.g, the flow over a submarine). However, many ships operate near the water surface, which deforms in the presence of the moving vehicle. This deformation of the free surface complicates the boundary conditions when the free-surface shape is not known. For example, Fig. 15.65 depicts the side view of a domain where the flow region of interest V is bounded by the lower surface S_L, by the upper surface S_U, by a floating vessel S_{B1}, and by a submerged but close to the upper surface hydrofoil S_{B2} (the boundary at the left and right sides extends to infinity). For simplicity, let us consider the steady-state problem where the solid surfaces S_{B1} and S_{B2} move to the left at a constant speed U_∞. The shape of the free surface S_U, expressed in an inertial coordinate system (as shown in the figure) attached to the moving vehicle, is

$$z = \eta(x, y,) \tag{15.20}$$

and the $z = 0$ plane coincides with the undisturbed free surface. The kinematic boundary

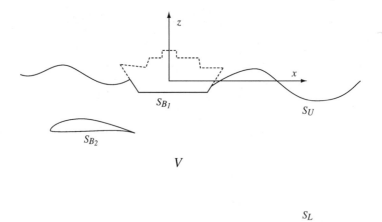

Figure 15.65 Nomenclature used for the free-surface flow model.

condition on this surface (with no flow normal to the boundary) is obtained by using Eq. (2.24). Hence, $F = (z - \eta) = 0$, and

$$\frac{DF}{Dt} = \frac{\partial \Phi^*}{\partial z} - \nabla \Phi^* \cdot \nabla \eta = 0 \tag{15.21}$$

The total velocity potential Φ^* is defined here as $\Phi^* = \Phi + U_\infty x$, where Φ is the perturbation potential (note that $\partial \eta / \partial z = 0$).

A second condition for the free surface emerges because of the non-negligible density of the water and because a deformation of the surface will have an effect on the pressure. Consequently, the Bernoulli equation (Eq. (2.32)) is used for this dynamic boundary condition. Let us write the equation for a point on $z = 0$ far from any disturbance, where p_a is the undisturbed atmospheric pressure, and for a point on the deformed free surface:

$$\frac{p_a}{\rho} + \frac{1}{2}U_\infty^2 = \frac{p}{\rho} + \frac{1}{2}(\nabla \Phi^*)^2 + g\eta \tag{15.22}$$

where g is the gravitational acceleration. Taking the gradient of this equation, and assuming that the pressure variations are negligible, we obtain

$$\frac{1}{2}\nabla(\nabla \Phi^*)^2 + g\nabla \eta = 0$$

Taking the dot product with $\nabla \Phi^*$ and replacing the last term by $\partial \Phi^*/\partial z$ (since based on Eq. (15.21) $\nabla \Phi^* \cdot \nabla \eta = \partial \Phi^*/\partial z$) we get

$$\frac{1}{2}\nabla \Phi^* \cdot \nabla(\nabla \Phi^*)^2 + g\frac{\partial \Phi^*}{\partial z} = 0 \tag{15.23}$$

This is the combined boundary condition for the free surface (S_U). A linearized form can be obtained by neglecting the smaller terms:

$$U_\infty^2 \frac{\partial^2 \Phi^*}{\partial x^2} + g\frac{\partial \Phi^*}{\partial z} = 0 \quad \text{on} \quad z = 0 \tag{15.24}$$

In the light of the derivations in Eq. (15.24), Φ^* can be replaced by the perturbation potential Φ. Several two-dimensional solutions describing surface wave motion, based on a similar

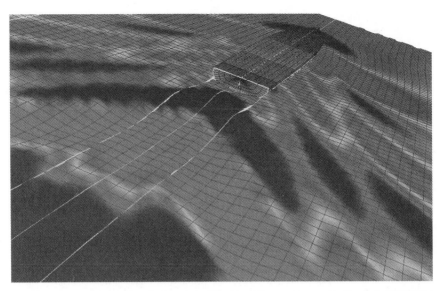

Figure 15.66 Surface waves behind a tanker ship. (Computations with the code VSAERO/FSWAVE.[15.44] Courtesy of M. Hughes.)

formulation, can be found in Chapter 6 of Ref. 7.1. These principles for modeling free-surface flows have been incorporated into three-dimensional panel methods. For example, Ref. 15.44 describes a method where the free surface is divided into rectilinear source panels and those are elevated to a fixed height above the maximum expected wave amplitude (to avoid singularity). These elevated sources account for the normal velocity ("blowing") increments necessary to model the surface normal velocity. During the iterative solution process, the strength of the sources above the free surface are known (from a previous iteration) and the solution of Φ^* in V is obtained by using constant-strength source and doublet panels on the body surfaces (e.g., S_{B1} and S_{B2}) as described in Chapter 12. Once the solution (in terms of the source and doublet distributions on the solid bodies) for this iteration is known, the combined dynamic and kinematic condition (Eq. (15.24), which sometimes is simplified even more) is evaluated at $z = 0$. Reference 15.44 suggests a further simplification of this condition in order to accelerate the iterative process. In following iterations the free-surface source strength is updated, until the condition of Eq. (15.24) converges. Note that in this model the free surface is an imaginary surface since the source panels are not moving with the surface waves, and only their strength changes. Figure 15.66 depicts the results of such a calculation. The surface waves behind a tanker ship are clearly visible, and note the angle of the waves and their spacing, which is similar to experimental observations.

References

[15.1] Crow, S. C., "Stability Theory for a Pair of Trailing Vortices," *AIAA J.*, Vol. 8., 1970, pp. 2172–2179.
[15.2] "Computation of Viscous–Inviscid Interactions," AGARD CP-291, 1980.
[15.3] Panton, R. L., "Self-Sustaining Mechanisms of Wall Turbulence – A Review," *AIAA Paper 99-0552*, Reno NV, Jan. 1999.
[15.4] Kuethe, A. M., and Chow, C. Y., *Foundation of Aerodynamics*, John Wiley & Sons, 5th edition, 1998.
[15.5] Gilmer, B. R., Jasper, D. W., and Bristow, D. R., "Analysis of Stalled Multi-Element Airfoils," *AIAA Paper 84-2196*, Aug. 1984.
[15.6] Cebeci, T., Chang, K. C., Clark, R. W., and Halsey, N. D., "Calculation of Flow over Multielement Airfoils at High Lift," *J. Aircraft*, Vol. 24, No. 8, 1987, pp. 546–550.

[15.7] Lighthill, M. J. "On Displacement Thickness," *J. Fluid Mech.*, Vol. 4, 1958, pp. 383–392.

[15.8] Olson, L. E., and Dvorak, F. A., "Viscous/Potential Flow About Multi-Element Two-Dimensional and Infinite-Span Swept Wings: Theory and Experiment," *AIAA Paper 76-18*, Jan. 1976.

[15.9] Drela, M., and Giles, M. B., "ISES: A Two-Dimensional Viscous Aerodynamic Design Analysis Code," *AIAA Paper 87-0424*, Jan. 1987.

[15.10] Eppler, R., and Sommers, D. M., "A Computer Program for the Design and Analysis of Low-Speed Airfoils," NASA TM 80210, Aug. 1980.

[15.11] Liebeck, R. H., "A Class of Airfoils Designed for High Lift in Incompressible Flow," *J. Aircraft*, Vol. 10, No. 10, 1973, pp. 610–617.

[15.12] Liebeck, R. H., "Design of Subsonic Airfoils for High Lift," *J. Aircraft*, Vol. 15, No. 9, 1978, pp. 547–561.

[15.13] Lissaman, P. B. S., "Low Reynolds Number Airfoils," *Annual Review of Fluid Mechanics*, Vol. 15, 1983, pp. 223–239.

[15.14] Smith, A. M. O., "High Lift Aerodynamics," *J. Aircraft*, Vol. 12, No. 6, 1975, pp. 501–530.

[15.15] Page, F. H., "The Handley Page Wing," *The Aeronautical Journal*, June 1921, p. 263.

[15.16] Olson, L. E., McGowan, P. R., and Guest, C. J., "Leading-Edge Slat Optimization for Maximum Airfoil Lift," NASA TM 78566, July 1979.

[15.17] Hoerner, S. F., *Fluid Dynamic Lift*, Hoerner Fluid Dynamics, 1985.

[15.18] Katz, J., and Largman, R., "Effect of a 90 Degree Flap on the Aerodynamics of a Two-Element Airfoil," *ASME J. Fluids Engineering*, Vol. 111, March 1989, pp. 93–94.

[15.19] Katz, J., and Feistel, T. W., "Propeller Swirl Effect on Single-Engine General-Aviation Aircraft Stall-Spin Tendencies," *J. Aircraft*, Vol. 24, No. 4, pp. 285–287, 1987.

[15.20] Lamar, J. E., and Johnson, T. D., Jr., "Sensitivity of F-106B Leading-Edge Vortex Images to Flight and Vapor-Screen Parameters," NASA TP 2818 June 1988.

[15.21] Hoerner, S. F., *Fluid Dynamic Drag*, Hoerner Fluid Dynamics, 1965.

[15.22] McAlister, K. W., and Carr L. W., "Water-Tunnel Experiments on an Oscillating Airfoil at $Re = 21000$," NASA TM 78446, March 1978.

[15.23] Maskew, B., Rao, B. M., and Dvorak, F. A., "Prediction of Aerodynamic Characteristics for Wings with Extensive Separations," Paper no 31 in "Computation of Viscous-Inviscid Interactions," AGARD CP-291 1980.

[15.24] Winkelmann, A. E., and Barlow, J. B., "Flowfield Model for a Rectangular Planform Wing beyond Stall," *AIAA J.*, Vol. 18, No. 8, 1980, pp. 1006–1008.

[15.25] Leonard, A., "Vortex Methods for Flow Simulation," *J. of Computational Physics*, Vol. 37, No. 3, Oct. 1980, pp. 289–335.

[15.26] Katz, J., "A Discrete Vortex Method for the Non-Steady Separated Flow over an Airfoil," *J. Fluid Mech.*, Vol. 102, 1981, pp. 315–328.

[15.27] Spalart, P. R., and Leonard, A., "Computation of Separated Flows by a Vortex Tracing Algorithm," *AIAA Paper 81-1246*, June 1981.

[15.28] Katz, J., "Large Scale Vortex-Lattice Model for the Locally Separated Flow over Wings," *AIAA J.*, Vol. 20, No. 12, 1982, pp. 1640–1646.

[15.29] Weihs, D., and Katz, J., "Cellular Patterns in Poststall Flow over Unswept Wings," *AIAA J.*, Vol. 21, No. 12, 1983, pp. 1757–1758.

[15.30] "Missile Aerodynamics," AGARD LS-98, Feb. 1979.

[15.31] "High Angle of Attack Aerodynamics," AGARD CP-247, Oct 1978.

[15.32] "Aerodynamics of Vortical Type Flows in Three Dimensions," AGARD CP-342, Apr. 1983.

[15.33] Stahl, W. H., "Aerodynamics of Low Aspect Ratio Wings," Paper No. 3 in "Missile Aerodynamics," AGARD LS-98, Feb. 1979.

[15.34] Hummel, D., "On the Vortex Formation over a Slender Wing at Large Angles of Incidence," Paper No. 15 in "High Angle of Attack Aerodynamics," AGARD CP-247, Oct. 1978.

[15.35] Roos, F. W., and Kegelman, J. T., "An Experimental Investigation of Sweep-Angle Influence on Delta-Wing Flows," *AIAA Paper 90-0383*, Jan. 1990.

[15.36] Polhamus, E. C. "Prediction of Vortex Characteristics by a Leading-Edge Suction Analogy," *J. Aircraft*, Vol. 8, No. 4, 1971, pp. 193–199.

References

[15.37] Shanks, R. E., "Low-Subsonic Measurements of Static and Dynamic Stability Derivatives of Six Flat-Plate Wings Having Leading-Edge Sweep Angles of 70° to 84°," NASA TN D-1822, July 1963.

[15.38] Levin, D., and Katz, J., "Vortex Lattice Method for the Calculation of the Nonsteady Separated Flow over Delta Wings," *J. Aircraft*, Vol. 18, No. 12, 1981, pp. 1032–1037.

[15.39] Katz, J., "Lateral Aerodynamics of Delta Wings with Leading Edge Separation," *AIAA J.*, Vol. 22, No. 3, 1984, pp. 323–328.

[15.40] Peckham, D. H., "Low-Speed Wind-Tunnel Tests on a Series of Uncambered Slender Pointed Wings with Sharp Edges," *R.A.E. Aeronaut. Rep. 2613*, 1958.

[15.41] Davenport, E. E., and Huffman, J. K., "Experimental and Analytical Investigation of Subsonic Longitudinal and Lateral Aerodynamic Characteristics of Slender Sharp-Edge 74° Swept Wings," NASA TN D-6344, 1971.

[15.42] Levin, D., and Katz, J., "Dynamic Load Measurements with Delta Wings Undergoing Self-Induced Roll Oscillations," *J. Aircraft*, Vol. 21, 1984, pp. 30–36.

[15.43] Nguyen, L. T., Yip, L., and Chambers, J. R., "Self-Induced Wing Rock of Slender Delta Wings," *AIAA Paper 81-1883*, Aug. 1981.

[15.44] Hughes, J. M., "Application of CFD to the Prediction of the Wave Height and Energy from High Speed Ferries," Paper No. 4, Int. CFD Conference, May 29–31, 1997, Ulsteinvik, Norway.

APPENDIX A

Airfoil Integrals

The following integrals are reprinted from Ref. 7.2.

1. $\displaystyle\oint_{-1}^{1} \frac{1}{x-\xi}\, d\xi = \ln\frac{1+x}{1-x}$

2. $\displaystyle\oint_{-1}^{1} \frac{\xi}{x-\xi}\, d\xi = x\ln\frac{1+x}{1-x} - 2$

3. $\displaystyle\oint_{-1}^{1} \frac{\xi^2}{x-\xi}\, d\xi = x\left(x\ln\frac{1+x}{1-x} - 2\right)$

4. $\displaystyle\oint_{-1}^{1} \frac{\xi^3}{x-\xi}\, dx = x^2\left(x\ln\frac{1+x}{1-x} - 2\right) - \frac{2}{3}$

5. $\displaystyle\oint_{-1}^{1} \frac{\xi^n}{x-\xi}\, d\xi = x\oint_{-1}^{1} \frac{\xi^{n-1}}{x-\xi}\, d\xi - \frac{1-(-1)^n}{n}$

6. $\displaystyle\oint_{-1}^{1} \frac{1}{\sqrt{1-\xi^2}(x-\xi)}\, d\xi = 0$

7. $\displaystyle\oint_{-1}^{1} \frac{1}{\sqrt{1-\xi^2}(x-\xi)}\, d\xi = -\pi$

8. $\displaystyle\oint_{-1}^{1} \frac{\xi^2}{\sqrt{1-\xi^2}(x-\xi)}\, d\xi = -\pi x$

9. $\displaystyle\oint_{-1}^{1} \frac{\xi^3}{\sqrt{1-\xi^2}(x-\xi)}\, d\xi = -\pi\left(x^2 + \frac{1}{2}\right)$

10. $\displaystyle\oint_{-1}^{1} \frac{\xi^4}{\sqrt{1-\xi^2}(x-\xi)}\, d\xi = -\pi x\left(x^2 + \frac{1}{2}\right)$

11. $\displaystyle\oint_{-1}^{1} \frac{\xi^5}{\sqrt{1-\xi^2}(x-\xi)}\, d\xi = -\pi\left(x^4 + \frac{1}{2}x^2 + \frac{3}{8}\right)$

12. $\displaystyle\oint_{-1}^{1} \frac{\xi^6}{\sqrt{1-\xi^2}(x-\xi)}\, d\xi = -\pi x\left(x^4 + \frac{1}{2}x^2 + \frac{3}{8}\right)$

13. $\displaystyle\oint_{-1}^{1} \frac{\xi^n}{\sqrt{1-\xi^2}(x-\xi)}\, d\xi = x\oint_{-1}^{1} \frac{\xi^{n-1}}{\sqrt{1-\xi^2}(x-\xi)}\, d\xi$
$\displaystyle\qquad\qquad -\frac{\pi}{2}[1-(-1)^n]\frac{1(3)\cdots(n-2)}{2(4)\cdots(n-1)}$

14. $\displaystyle\oint_{-1}^{1}\frac{\sqrt{1-\xi^2}}{x-\xi}\,d\xi = \pi x$

15. $\displaystyle\oint_{-1}^{1}\frac{\xi\sqrt{1-\xi^2}}{x-\xi}\,d\xi = \pi\left(x^2-\frac{1}{2}\right)$

16. $\displaystyle\oint_{-1}^{1}\frac{\xi^2\sqrt{1-\xi^2}}{x-\xi}\,d\xi = \pi x\left(x^2-\frac{1}{2}\right)$

17. $\displaystyle\oint_{-1}^{1}\frac{\xi^3\sqrt{1-\xi^2}}{x-\xi}\,d\xi = \pi\left(x^4-\frac{1}{2}x^2-\frac{1}{8}\right)$

18. $\displaystyle\oint_{-1}^{1}\frac{\sqrt{1+\xi}}{\sqrt{1-\xi}(x-\xi)}\,d\xi = -\pi$

19. $\displaystyle\oint_{-1}^{1}\frac{\ln\frac{1+\xi}{1-\xi}}{x-\xi}\,d\xi = \frac{1}{2}\left(\ln^2\frac{1+x}{1-x}-\pi^2\right)$

20. $\displaystyle\oint_{-1}^{1}\frac{\xi\ln\frac{1+\xi}{1-\xi}}{x-\xi}\,d\xi = \frac{1}{2}x\left(\ln^2\frac{1+x}{1-x}-\pi^2\right)$

21. $\displaystyle\oint_{-1}^{1}\frac{\xi^2\ln\frac{1+\xi}{1-\xi}}{x-\xi}\,d\xi = \frac{1}{2}x^2\left(\ln^2\frac{1+x}{1-x}-\pi^2\right)-2$

22. $\displaystyle\oint_{-1}^{1}\frac{\xi^3\ln\frac{1+\xi}{1-\xi}}{x-\xi}\,d\xi = \frac{1}{2}x^3\left(\ln^2\frac{1+x}{1-x}-\pi^2\right)-2x$

23. $\displaystyle\oint_{-1}^{1}\frac{\xi^4\ln\frac{1+\xi}{1-\xi}}{x-\xi}\,d\xi = \frac{1}{2}x^4\left(\ln^2\frac{1+x}{1-x}-\pi^2\right)-2x^2-\frac{4}{3}$

24. $\displaystyle\oint_{-1}^{1}\frac{\xi^n\ln\frac{1+\xi}{1-\xi}}{x-\xi}\,d\xi = x\oint_{-1}^{1}\frac{\xi^{n-1}\ln\frac{1+\xi}{1-\xi}}{x-\xi}\,d\xi - \frac{2}{n}[1-(-1)^{n-1}]\sum_{v=0}^{(n/2)-1}\frac{1}{n-1-2v}$

25. $\displaystyle\oint_{-1}^{1}\frac{1}{\sqrt{1+\xi}(x-\xi)}\,d\xi = \frac{1}{\sqrt{1+x}}\ln\frac{\sqrt{2}+\sqrt{1+x}}{\sqrt{2}-\sqrt{1+x}}$

26. $\displaystyle\oint_{-1}^{1}\frac{\sqrt{1+\xi}}{x-\xi}\,d\xi = \sqrt{1+x}\ln\frac{\sqrt{2}+\sqrt{1+x}}{\sqrt{2}-\sqrt{1+x}} - 2\sqrt{2}$

27. $\displaystyle\oint_{-1}^{1}\frac{\ln\frac{1+\xi}{1-\xi}}{\sqrt{1-\xi^2}(x-\xi)}\,d\xi = -\frac{\pi^2}{\sqrt{1-x^2}}$

28. $\displaystyle\oint_{-1}^{1}\frac{\xi\ln\frac{1+\xi}{1-\xi}}{\sqrt{1-\xi^2}(x-\xi)}\,d\xi = -\pi^2\frac{x}{\sqrt{1-x^2}}$

29. $\displaystyle\oint_{-1}^{1}\frac{\xi^2\ln\frac{1+\xi}{1-\xi}}{\sqrt{1-\xi^2}(x-\xi)}\,d\xi = -\pi\left(2+\pi\frac{x^2}{\sqrt{1-x^2}}\right)$

30. $\displaystyle\oint_{-1}^{1}\frac{\xi^3\ln\frac{1+\xi}{1-\xi}}{\sqrt{1-\xi^2}(x-\xi)}\,d\xi = -\pi x\left(2+\pi\frac{x^2}{\sqrt{1-x^2}}\right)$

Appendix A / Airfoil Integrals

31. $\displaystyle\oint_{-1}^{1} \frac{\xi^4 \ln \frac{1+\xi}{1-\xi}}{\sqrt{1-\xi^2}(x-\xi)} d\xi = -\pi \left(\frac{5}{3} + 2x^2 + \pi \frac{x^2}{\sqrt{1-x^2}} \right)$

32. $\displaystyle\oint_{-1}^{1} \frac{\xi^n \ln \frac{1+\xi}{1-\xi}}{\sqrt{1-\xi^2}(x-\xi)} d\xi = x \int_{-1}^{1} \frac{\xi^{n-1} \ln \frac{1+\xi}{1-\xi}}{\sqrt{1-\xi^2}(x-\xi)} d\xi - \pi[1-(-1)^{n-1}]\left(\frac{n}{2}-1\right)!$

$\displaystyle \times \sum_{v=0}^{(n/2)-1} \frac{(-1)^v (1)(3)\cdots(v)}{2^v (2v+1)(v!)^2 \left(\frac{n}{2}-1-v\right)!}$

APPENDIX B

Singularity Distribution Integrals

The Integral I_1

Consider the integral

$$I_1 = \int_{x_1}^{x_2} \ln[(x - x_0)^2 + z^2] \, dx_0 \tag{B.1}$$

Let $X = x - x_0$ and $dX = -dx_0$; then

$$I_1 = -\int_{x-x_1}^{x-x_2} \ln(X^2 + z^2) \, dX \tag{B.2}$$

This integral appears in Gradshteyn (Ref. 6.3, p. 205), and the result is

$$I_1 = -\left[X \ln(X^2 + z^2) - 2X + 2z \tan^{-1} \frac{X}{z} \right]\Bigg|_{x-x_1}^{x-x_2} \tag{B.3}$$

Returning to the original variables we obtain

$$I_1 = (x - x_1) \ln[(x - x_1)^2 + z^2] - (x - x_2) \ln[(x - x_2)^2 + z^2]$$
$$- 2(x_2 - x_1) + 2z \left[\tan^{-1} \frac{x - x_1}{z} - \tan^{-1} \frac{x - x_2}{z} \right] \tag{B.4}$$

The derivatives of this integral with respect to the x and z coordinates are

$$\frac{\partial I_1}{\partial x} = \frac{2(x - x_1)^2}{(x - x_1)^2 + z^2} + \ln[(x - x_1)^2 + z^2] - \frac{2(x - x_2)^2}{(x - x_2)^2 + z^2}$$
$$- \ln[(x - x_2)^2 + z^2] + 2z \left[\frac{-z}{(x - x_2)^2 + z^2} + \frac{z}{(x - x_1)^2 + z^2} \right]$$
$$= \frac{\partial I_1}{\partial x} = \ln \frac{(x - x_1)^2 + z^2}{(x - x_2)^2 + z^2} \tag{B.5}$$

and for the derivative in the z direction we have

$$\frac{\partial I_1}{\partial z} = \frac{2(x - x_1)z}{(x - x_1)^2 + z^2} - \frac{2(x - x_2)z}{(x - x_2)^2 + z^2} + 2\left[\tan^{-1} \frac{x - x_1}{z} - \tan^{-1} \frac{x - x_2}{z} \right]$$
$$+ 2z \left[\frac{(x - x_2)}{(x - x_2)^2 + z^2} - \frac{x - x_1}{(x - x_1)^2 + z^2} \right] = \frac{\partial I_1}{\partial z}$$
$$= 2\left[\tan^{-1} \frac{x - x_1}{z} - \tan^{-1} \frac{x - x_2}{z} \right] \tag{B.6}$$

With the use of Eq. (B.1) we can also write

$$\frac{\partial I_1}{\partial z} = 2\int_{x_1}^{x_2} \frac{z}{(x-x_0)^2 + z^2}\, dx_0 \tag{B.7}$$

Let us integrate Eq. (B.7) directly with the use of the polar coordinate transformation (see Fig. 10.6)

$$x - x_0 = r\cos\theta \tag{B.8a}$$
$$z = r\sin\theta \tag{B.8b}$$

Take the differential of Eqs. (B.8) to get

$$-dx_0 = -r\sin\theta\, d\theta + \cos\theta\, dr$$
$$0 = r\cos\theta\, d\theta + \sin\theta\, dr$$

and solve for dx_0:

$$dx_0 = \frac{r\, d\theta}{\sin\theta} \tag{B.8c}$$

Substitute Eqs. (B.8) into Eq. (B.7) to get

$$\frac{\partial I_1}{\partial z} = 2\int_{\theta_1}^{\theta_2} d\theta = 2(\theta_2 - \theta_1)$$

$$= 2\left(\tan^{-1}\frac{z}{x-x_2} - \tan^{-1}\frac{z}{x-x_1}\right) \tag{B.9}$$

We equate Eq. (B.6) to Eq. (B.9) to obtain

$$\tan^{-1}\frac{x-x_1}{z} - \tan^{-1}\frac{x-x_2}{z} = \tan^{-1}\frac{z}{x-x_2} - \tan^{-1}\frac{z}{x-x_1} \tag{B.10}$$

and substitution into Eq. (B.4) yields

$$I_1 = (x-x_1)\ln[(x-x_1)^2 + z^2] - (x-x_2)\ln[(x-x_2)^2 + z^2]$$
$$-2(x_2 - x_1) + 2z\left[\tan^{-1}\frac{z}{x-x_2} - \tan^{-1}\frac{z}{x-x_1}\right] \tag{B.11}$$

The Integral I_2

Consider the integral

$$I_2 = \int_{x_1}^{x_2} \tan^{-1}\frac{z}{x-x_0}\, dx_0 \tag{B.12}$$

Let $X = \frac{z}{x-x_0}$ and $dx_0 = zX^{-2}dX$, and then

$$I_2 = z\int_{\frac{z}{x-x_1}}^{\frac{z}{x-x_2}} X^{-2}\tan^{-1} X\, dX \tag{B.13}$$

Following the results of Ref. 6.3 (p. 210) for this integral, I_2 becomes

$$I_2 = -z \left[X^{-1} \tan^{-1} X + \frac{1}{2} \ln \frac{1+X^2}{X^2} \right] \Big|_{\frac{z}{x-x_1}}^{\frac{z}{x-x_2}}$$

$$= -z \left[\frac{x-x_2}{z} \tan^{-1} \frac{z}{x-x_2} - \frac{x-x_1}{z} \tan^{-1} \frac{z}{x-x_1} \right.$$

$$\left. + \frac{1}{2} \ln \frac{(x-x_2)^2 + z^2}{z^2} - \frac{1}{2} \ln \frac{(x-x_1)^2 + z^2}{z^2} \right]$$

$$I_2 = (x-x_1) \tan^{-1} \frac{z}{x-x_1} - (x-x_2) \tan^{-1} \frac{z}{x-x_2} + \frac{z}{2} \ln \frac{(x-x_1)^2 + z^2}{(x-x_2)^2 + z^2}$$

(B.14)

The Integral I_3

Consider the integral

$$I_3 = \int_{x_1}^{x_2} x_0 \ln[(x-x_0)^2 + z^2] \, dx_0 \tag{B.15}$$

Let $X = x - x_0$ and $dX = -dx_0$; then

$$I_3 = -\int_{x-x_1}^{x-x_2} (x-X) \ln(X^2 + z^2) \, dX$$

$$= -x \int_{x-x_1}^{x-x_2} \ln(X^2 + z^2) \, dX + \int_{x-x_1}^{x-x_2} X \ln(X^2 + z^2) \, dX$$

$$= xI_1 + \frac{1}{2}[(X^2 + z^2) \ln(X^2 + z^2) - X^2]\Big|_{x-x_1}^{x-x_2} \tag{B.16}$$

following p. 205 of Ref. 6.3. This becomes (after inserting the limits)

$$I_3 = xI_1 + \frac{1}{2}\{[(x-x_2)^2 + z^2] \ln[(x-x_2)^2 + z^2]$$
$$- [(x-x_1)^2 + z^2] \ln[(x-x_1)^2 + z^2] + (x-x_1)^2 - (x-x_2)^2\}$$

Substitution of I_1 from (B.11) yields

$$I_3 = \ln[(x-x_1)^2 + z^2] \left[x(x-x_1) - \frac{(x-x_1)^2}{2} - \frac{z^2}{2} \right]$$

$$+ \ln[(x-x_2)^2 + z^2] \left[-x(x-x_2) + \frac{(x-x_2)^2}{2} - \frac{z^2}{2} \right]$$

$$+ 2xz \left[\tan^{-1} \frac{z}{x-x_2} - \tan^{-1} \frac{z}{x-x_1} \right]$$

$$+ \frac{1}{2}[(x-x_1)^2 - (x-x_2)^2] - 2x(x_2 - x_1)$$

and finally

$$I_3 = \frac{x^2 - x_1^2 - z^2}{2} \ln[(x-x_1)^2 + z^2] - \frac{x^2 - x_2^2 - z^2}{2} \ln[(x-x_2)^2 + z^2]$$

$$+ 2xz \left[\tan^{-1} \frac{z}{x-x_2} - \tan^{-1} \frac{z}{x-x_1} \right] + x(x_1 - x_2) + \frac{x_1^2 - x_2^2}{2} \tag{B.17}$$

The derivatives of this integral with respect to the x and z coordinates are

$$\frac{\partial I_3}{\partial x} = x \ln[(x-x_1)^2 + z^2] + (x^2 - x_1^2 - z^2)\frac{(x-x_1)}{(x-x_1)^2 + z^2} - x \ln[(x-x_2)^2 + z^2]$$

$$- (x^2 - x_2^2 - z^2)\frac{(x-x_2)}{(x-x_2)^2 + z^2} + 2z\left[\tan^{-1}\frac{z}{x-x_2} - \tan^{-1}\frac{z}{x-x_1}\right]$$

$$+ 2zx\left[-\frac{z}{(x-x_2)^2 + z^2} + \frac{z}{(x-x_1)^2 + z^2}\right] + (x_1 - x_2)$$

$$= x \ln\frac{(x-x_1)^2 + z^2}{(x-x_2)^2 + z^2} + 2z\left[\tan^{-1}\frac{z}{x-x_2} - \tan^{-1}\frac{z}{x-x_1}\right]$$

$$+ (x_1 - x_2) + (x+x_1)\frac{(x-x_1)^2 + z^2}{(x-x_1)^2 + z^2} - (x+x_2)\frac{(x-x_2)^2 + z^2}{(x-x_2)^2 + z^2}$$

and finally

$$\frac{\partial I_3}{\partial x} = x \ln\frac{(x-x_1)^2 + z^2}{(x-x_2)^2 + z^2} + 2z\left[\tan^{-1}\frac{z}{x-x_2} - \tan^{-1}\frac{z}{x-x_1}\right] + 2(x_1 - x_2) \quad \text{(B.18)}$$

$$\frac{\partial I_3}{\partial z} = -z \ln[(x-x_1)^2 + z^2] + z\frac{(x^2 - x_1^2 - z^2)}{(x-x_1)^2 + z^2} + z \ln[(x-x_2)^2 + z^2]$$

$$- z\frac{(x^2 - x_2^2 - z^2)}{(x-x_2)^2 + z^2} + 2x\left[\tan^{-1}\frac{z}{x-x_2} - \tan^{-1}\frac{z}{x-x_1}\right]$$

$$+ 2zx\left[\frac{x-x_2}{(x-x_2)^2 + z^2} - \frac{x-x_1}{(x-x_1)^2 + z^2}\right] = z \ln\frac{(x-x_2)^2 + z^2}{(x-x_1)^2 + z^2}$$

$$+ 2x\left[\tan^{-1}\frac{z}{x-x_2} - \tan^{-1}\frac{z}{x-x_1}\right] - z\frac{(x-x_1)^2 + z^2}{(x-x_1)^2 + z^2} + z\frac{(x-x_2)^2 + z^2}{(x-x_2)^2 + z^2}$$

$$= z \ln\frac{(x-x_2)^2 + z^2}{(x-x_1)^2 + z^2} + 2x\left[\tan^{-1}\frac{z}{x-x_2} - \tan^{-1}\frac{z}{x-x_1}\right] \quad \text{(B.19)}$$

The Integral I_4

Consider the integral

$$I_4 = \int_{x_1}^{x_2} x_0 \tan^{-1}\frac{z}{x-x_0}\, dx_0 \quad \text{(B.20)}$$

Let $X = \frac{z}{x-x_0}$, $dx_0 = zX^{-2}dX$, and $x_0 = x - zX^{-1}$. Then

$$I_4 = xz\int_{\frac{z}{x-x_1}}^{\frac{z}{x-x_2}} X^{-2} \tan^{-1} X\, dX - z^2 \int_{\frac{z}{x-x_1}}^{\frac{z}{x-x_2}} X^{-3} \tan^{-1} X\, dX = xI_2 - z^2 I_5$$

According to Ref. 6.3 (p. 209)

$$I_5 = \int X^{-3} \tan^{-1} X\, dX = -\frac{X^{-2}}{2}\tan^{-1} X + \frac{1}{2}\int\frac{X^{-2}\, dX}{1+X^2} \quad \text{(B.21)}$$

Substituting the result for the second integral from p. 66 of Ref. 6.3 we obtain

$$I_5 = -\frac{X^{-2}}{2}\tan^{-1} X + \frac{1}{2}[-X^{-1} - \tan^{-1} X] = -\frac{1}{2}\tan^{-1} X(1+X^{-2}) - \frac{1}{2X}$$

With the use of I_2 from (B.14) we get

$$I_4 = x\left[(x-x_1)\tan^{-1}\frac{z}{x-x_1} - (x-x_2)\tan^{-1}\frac{z}{x-x_2} + \frac{z}{2}\ln\frac{(x-x_1)^2+z^2}{(x-x_2)^2+z^2}\right]$$

$$+ \frac{z^2}{2}[(1+X^{-2})\tan^{-1} X + X^{-1}]\Big|_{\frac{z}{x-x_1}}^{\frac{z}{x-x_2}}$$

$$= \frac{xz}{2}\ln\frac{(x-x_1)^2+z^2}{(x-x_2)^2+z^2} + \frac{z}{2}(x_1 - x_2) + \tan^{-1}\frac{z}{x-x_1}\{x(x-x_1)$$

$$- \frac{1}{2}[(x-x_1)^2+z^2]\} - \tan^{-1}\frac{z}{x-x_2}\{x(x-x_2) - \frac{1}{2}[(x-x_2)^2+z^2]\}$$

and finally

$$I_4 = \frac{xz}{2}\ln\frac{(x-x_1)^2+z^2}{(x-x_2)^2+z^2} + \frac{z}{2}(x_1-x_2) + \frac{x^2-x_1^2-z^2}{2}\tan^{-1}\frac{z}{x-x_1}$$

$$- \frac{x^2-x_2^2-z^2}{2}\tan^{-1}\frac{z}{x-x_2} \qquad (B.22)$$

APPENDIX C

Principal Value of the Lifting Surface Integral I_L

Consider the integral appearing in Eq. (8.68). It is of a general singular form

$$I_L = \int_a^b \frac{f(x_0)}{(x-x_0)^2} dx_0 \tag{C.1}$$

The principal value of this integral (also called the finite part) is given by Mangler (see Ashley and Landahl,[4.1] pp. 132–133) as

$$\int_a^b \frac{f(x_0)}{(x-x_0)^2} dx_0 = \lim_{\epsilon \to 0} \left[\int_a^{x-\epsilon} \frac{f(x_0)}{(x-x_0)^2} dx_0 + \int_{x+\epsilon}^b \frac{f(x_0)}{(x-x_0)^2} dx_0 - 2\frac{f(x)}{\epsilon} \right] \tag{C.2}$$

As is the case with the Cauchy principal value in Section 5.1, if the integral can be evaluated in closed form the correct principal value can be obtained by simply ignoring the limit process as long as the arguments of all logarithms are taken as their absolute values.

Brandao[C.1] has provided alternate expressions for the Cauchy (see Eq. (5.18)) and Mangler principal values that do not require a limiting process (which normally must be performed numerically.) These results lead to the following:

$$\int_a^b \frac{f(x_0)}{(x-x_0)} dx_0 = \int_a^b \frac{f(x_0) - f(x)}{(x-x_0)} dx_0 + f(x) \ln \frac{x-a}{b-x} \tag{C.3}$$

$$\int_a^b \frac{f(x_0)}{(x-x_0)^2} dx_0 = \int_a^b \frac{f(x_0) - f(x) + f'(x)(x-x_0)}{(x-x_0)^2} dx_0$$
$$+ f(x) \left[\frac{1}{x-b} - \frac{1}{x-a} \right] + f'(x) \ln \frac{b-x}{x-a}$$

Reference

[C.1] Brandao, M. P., "Improper Integrals in Theoretical Aerodynamics: The Problem Revisited," *AIAA Journal*, Vol. 25, No. 9, 1987, pp. 1258–1260.

APPENDIX D

Sample Computer Programs

This appendix lists several computer programs that are based on the methods presented in the previous chapters. These FORTRAN programs were prepared mainly by students during regular class work and their algorithms were not optimized for clear programming and computational efficiency. Also, an effort was made to list only the simplest versions without interactive and graphic input/output sections owing to the rapid changes and improvements in computer operation systems. In spite of this brevity these computer programs can help the readers to construct their baseline algorithms upon which their customized computer programs may be developed.

D.1   Two-Dimensional Panel Methods

1. Grid generator for van de Vooren airfoil shapes, based on the formulas of Section 6.7. The program also calculates the exact chordwise velocity components and pressure coefficient for the purpose of comparison. All the two-dimensional codes (Programs 3–11) use the input generated by this subroutine.

D.1.1   *Two-Dimensional Panel Methods Based on the Neumann Boundary Condition*

2. Discrete vortex, thin wing method, based on Section 11.1.1.
3. Constant strength source method (based on Section 11.2.1). Note that the matrix solver (SUBROUTINE MATRX) is attached to this program only and is not listed with Programs 4–11, for brevity.
4. Constant strength doublet method, based on Section 11.2.2.
5. Constant strength vortex method, based on Section 11.2.3.
6. Linear strength source method, based on Section 11.4.1.
7. Linear strength vortex method, based on Section 11.4.2.

D.1.2   *Two-Dimensional Panel Methods Based on the Dirichlet Boundary Condition*

8. Constant strength doublet method, based on Section 11.3.2.
9. Constant strength source/doublet method, based on Section 11.3.1.
10. Linear strength doublet method, based on Section 11.5.2.
11. Quadratic strength doublet method, based on Section 11.6.2.

D.2   Three-Dimensional Programs

12. Influence of a three-dimensional, constant strength source/doublet element, based on Sections 10.4.1 and 10.4.2.
13. Three-dimensional vortex lattice method for rectilinear lifting surfaces (with ground effect), based on Section 12.3.

D.3 Time-Dependent Programs

14. Three-dimensional panel method, using constant strength sources and doublets with the Dirichlet boundary condition, based on Section 12.5.

D.3   Time-Dependent Programs

15. Sudden acceleration of a flat plate at angle of attack (using a single lumped vortex element). This program solves numerically the example of Section 13.7, but the strengths of the airfoil vortex and the wake latest vortex are obtained algebraically (and there is no matrix solution phase).
16. Unsteady motion of a thin rectangular lifting surface (solution is based on the model of Section 13.12, which is an upgrade of the vortex lattice method of Program 13).

```
C       PROGRAM No. 1: GRID GENERATOR FOR 2-D AIRFOILS
C       -----------------------------------------------
C       THIS PROGRAM IS AN AUTOMATED COMPLEX AIRFOIL TRANSFORMATION OF THE
C          TYPE PRESENTED BY VAN DE VOOREN AND DE JONG (1970). THE RESULTING
C          AIRFOIL MAY HAVE A NON-ZERO TRAILING EDGE ANGLE. THIS FORMULATION
C          IS FOR NON-CAMBERED AIRFOILS ONLY (PROGRAMMED BY STEVEN YON, 1989).

        OPEN(8,FILE='AFOIL2.DAT',STATUS='NEW')
        OPEN(10,FILE='CP.DAT',STATUS='NEW')

        WRITE(6,*)
     *    'READY TO START VAN DE VOOREN TRANSFORMATION'
        WRITE(6,*) 'ENTER THICKNESS COEFF. E'
        READ(5,*) E
        WRITE(6,*) 'ENTER T.E. ANGLE COEFF. K'
        READ(5,*) AK
        TL=1.0
        A=2*TL*(E+1)**(AK-1)/(2**AK)
        WRITE(6,*) 'ENTER THE ANGLE OF ATTACK IN DEGREES'
        READ(5,*) ALPHA
        AL=ALPHA/57.2958
        WRITE(6,*) 'ENTER NUMBER OF AIRFOIL PANELS, M'
        WRITE(6,*) 'WITH WHICH TO MODEL THE AIRFOIL'
        WRITE(6,*)
     *    '(NOTE THAT M SHOULD BE AN EVEN FACTOR OF 360)'
        READ(5,*) M
        ITHETA=360/M

C       THE DO LOOP WILL RUN THROUGH THE CIRCLE PLANE WITH
C       THE SPECIFIED ANGULAR INTERVAL AND TRANSFORM EACH
C       POINT TO THE AIRFOIL PLANE

        DO I=0,360,ITHETA
        IF(I.EQ.0.OR.I.EQ.360) THEN
          X=1
          Y=0
          CP=1
          WRITE(8,*) X,' ',Y
          IF(AK.EQ.2.AND.I.EQ.0) GOTO 25
          IF(AK.EQ.2.AND.I.EQ.360) GOTO 25
          WRITE(10,*) X,' ',CP
```

```fortran
 25         CONTINUE
               GOTO 100
            ELSE
               GOTO 50
            END IF

 50         CONTINUE
              TH=I/57.2958
              R1=SQRT((A*(COS(TH)-1))**2+(A*SIN(TH))**2)
              R2=SQRT((A*(COS(TH)-E))**2+(A*SIN(TH))**2)
            IF(TH.EQ.0) THEN
            TH1=1.5708
            ELSE
            TH1=(ATAN((A*SIN(TH))/(A*(COS(TH)-1))))+3.1415927
            END IF

            IF(COS(TH)-E.LT.0.AND.SIN(TH).GT.0) THEN
            TH2=(ATAN((A*SIN(TH))/(A*(COS(TH)-E))))+3.1415927
            ELSE IF(COS(TH)-E.LT.0.AND.SIN(TH).LT.0) THEN
            TH2=(ATAN((A*SIN(TH))/(A*(COS(TH)-E))))+3.1415927
            ELSE IF(COS(TH)-E.GT.0.AND.SIN(TH).LT.0) THEN
            TH2=(ATAN((A*SIN(TH))/(A*(COS(TH)-E))))+
     *          2*3.1415927
            ELSE
            TH2=(ATAN((A*SIN(TH))/(A*(COS(TH)-E))))
            END IF

C       THIS PART COMPUTES THE TRANSFORMED POSITIONS

            COM1=((R1**AK)/(R2**(AK-1)))/((COS((AK-1)
     *            *TH2))**2+(SIN((AK-1)*TH2))**2)
            X=COM1*(COS(AK*TH1)*COS((AK-1)*TH2)
     *              +SIN(AK*TH1)*SIN((AK-1)*TH2))+TL
            Y=COM1*(SIN(AK*TH1)*COS((AK-1)*TH2)
     *              -COS(AK*TH1)*SIN((AK-1)*TH2))

            WRITE(8,*) X,' ,',Y

C       THIS PART COMPUTES THE TRANSFORMED PRESSURE
C           DISTRIBUTION

            A1=COS((AK-1)*TH1)*COS(AK*TH2)+SIN((AK-1)*TH1)
     *            *SIN(AK*TH2)
            B1=SIN((AK-1)*TH1)*COS(AK*TH2)-COS((AK-1)*TH1)
     *         *SIN(AK*TH2)
            C1=(COS(AK*TH2))**2+(SIN(AK*TH2))**2
            P=A*(1-AK+AK*E)
            D1=A1*(A*COS(TH)-P)-B1*A*SIN(TH)
            D2=A1*A*SIN(TH)+B1*(A*COS(TH)-P)

            TEMP=2*C1*(SIN(AL)-SIN(AL-TH))/(D1**2+D2**2)
            COM2=TEMP*(R2**AK)/(R1**(AK-1))

            VX=D1*SIN(TH)+D2*COS(TH)
            VY=-(D1*COS(TH)-D2*SIN(TH))
```

```
              CP=1-COM2**2*(VX**2+VY**2)

              WRITE(10,*) X,' ',',CP

  100         CONTINUE
              END DO

              CLOSE(8)
              CLOSE(10)
              STOP
              END

C         PROGRAM No. 2: DISCRETE VORTEX METHOD (THIN WING, ELLIPTIC CAMBER)
C              ------------------------------------------------------------
C
C         DISCRETE VORTEX MODEL FOR THIN AIRFOILS (JOE KATZ, CIRCA 1986)
          DIMENSION GAMMA(52),XC(52),ZC(52),X(52),Z(52)
          DIMENSION ENX(52),ENZ(52),A(52,52),IP(52)
          DIMENSION DL(52),DCP(52),DCP1(52)
C
          N=10
          C=1.0
          EPSILON=0.1*C
          ALFA1=10.0
          PAY=3.141592654
          ALFA=ALFA1*PAY/180.0
          RO=1.
          V=1.
          UINF=COS(ALFA)*V
          WINF=SIN(ALFA)*V
          QUE=0.5*RO*V*V
C
C         GRID GENERATION (N PANELS)
C
          DX=C/N
          DO 1 I=1,N
C         COLLOCATION POINT
          XC(I) = C/N*(I-0.25)
          ZC(I) = 4.*EPSILON*XC(I)/C*(1.-XC(I)/C)
C         VORTEX POINT
          X(I)  = C/N*(I-0.75)
          Z(I)  = 4.*EPSILON*X(I)/C*(1.-X(I)/C)
C         NORMAL AT COLLOCATION POINT; N=(ENX,ENZ)
          DETADX=4.*EPSILON/C*(1.-2.*XC(I)/C)
          SQ=SQRT(1+DETADX**2)
          ENX(I)= -DETADX/SQ
   1      ENZ(I)= 1./SQ
C
C         INFLUENCE COEFFICIENTS
C
          DO 3 I=1,N
          DO 2 J=1,N
          CALL VOR2D(XC(I),ZC(I),X(J),Z(J),1.0,U,W)
          A(I,J)=U*ENX(I)+W*ENZ(I)
   2      CONTINUE
C         THE RHS VECTOR IS PLACED IN THE GAMMA VECTOR
          GAMMA(I)=-UINF*ENX(I)-WINF*ENZ(I)
   3      CONTINUE
```

```
C
C       SOLUTION OF THE PROBLEM: RHS(I)=A(I,J)*GAMMA(I)
C
        CALL DECOMP(N,52,A,IP)
        CALL SOLVER(N,52,A,GAMMA,IP)
C
C       AERODYNAMIC LOADS
C
        BL=0.0
        DO 4 I=1,N
        DL(I)=RO*V*GAMMA(I)
        DCP(I)=DL(I)/DX/QUE
C       DCP1 IS THE ANALYTIC SOLUTION
        DD=32.*EPSILON/C*SQRT(X(I)/C*(1.-X(I)/C))
        DCP1(I)=4.*SQRT((C-X(I))/X(I))*ALFA+DD
   4    BL=BL+DL(I)
        CL=BL/(QUE*C)
        CL1=2.*PAY*(ALFA+2*EPSILON/C)
C       CL1, DCP1 - ARE THE EXACT SOLUTIONS
C
C       OUTPUT
        WRITE(6,14)
        WRITE(6,15) V,CL,CL1,N,ALFA1
        DO 5 I=1,N
   5    WRITE(6,16)I,X(I),DCP(I),DCP1(I)
C
  14    FORMAT( ' THIN AIRFOIL WITH ELLIPTIC CAMBER ')
  15    FORMAT( ' V=',F7.1,3X,'CL=',F7.3,3X,'CL(EXACT)=',F7.3,3X,
       *'N= ',I6,3X,'ALPHA= ',F6.1)
  16    FORMAT( I5,3X,'X=',F8.2,5X,'DCP=',F8.2,3X,'DCP(EXACT)=',5F6.2)
C
C       PLOTTER OUTPUT IS PLACED HERE (e.g. DCP AND DCP1 - VS - X/C)
C
        STOP
        END
C
        SUBROUTINE VOR2D(X,Z,X1,Z1,GAMMA,U,W)
C       CALCULATES INFLUENCE OF VORTEX AT (X1,Z1)
        PAY=3.141592654
        U=0.0
        W=0.0
        RX=X-X1
        RZ=Z-Z1
        R=SQRT(RX**2+RZ**2)
        IF(R.LT.0.001) GOTO 1
        V=0.5/PAY*GAMMA/R
        U=V*(RZ/R)
        W=V*(-RX/R)
   1    CONTINUE
        RETURN
        END
C
C       THE FOLLOWING SUBROUTINES ARE LISTED WITH THE STEADY STATE
C       VORTEX LATTICE SOLVER (PROGRAM No. 13).
C
C       SUBROUTINE DECOMP(N,NDIM,A,IP)
C       SUBROUTINE SOLVER(N,NDIM,A,B,IP)
C
```

D.3 Time-Dependent Programs

```
C       PROGRAM No. 3: CONSTANT STRENGTH SOURCE
C       ---------------------------------------
C       THIS PROGRAM FINDS THE PRESSURE DISTRIBUTION ON AN ARBITRARY AIRFOIL
C         BY REPRESENTING THE SURFACE AS A FINITE NUMBER OF SOURCE PANELS WITH
C         CONST. STRENGTH (ALPHA=0, NEUMANN B.C., PROGRAM BY STEVEN YON, 1989).

        REAL EP(400,2),EPT(400,2),PT1(400,2),PT2(400,2)
        REAL CO(400,2),A(400,400),B(400,400),G(400)
        REAL TH(400)

        OPEN(8,FILE='CPS.DAT',STATUS='NEW')
        OPEN(9,FILE='AFOIL2.DAT',STATUS='OLD')

        WRITE(6,*) 'ENTER NUMBER OF PANELS'
        READ(5,*) M
        N=M+1

        WRITE(6,*)'SKIP THE MATRIX REDUCTION? 1=YES, 2=NO'
        READ(5,*) ANS

C       READ IN THE PANEL END POINTS
        DO I=1,M+1
         READ(9,*) EPT(I,1), EPT(I,2)
        END DO

C       CONVERT PANELING TO CLOCKWISE
        DO I=1,M+1
         EP(I,1)=EPT(N-I+1,1)
         EP(I,2)=EPT(N-I+1,2)
        END DO

C       ESTABLISH COORDINATES OF PANEL END POINTS
        DO I=1,M
         PT1(I,1)=EP(I,1)
         PT2(I,1)=EP(I+1,1)
         PT1(I,2)=EP(I,2)
         PT2(I,2)=EP(I+1,2)
        END DO

C       FIND PANEL ANGLES TH(J)
        DO I=1,M
         DZ=PT2(I,2)-PT1(I,2)
         DX=PT2(I,1)-PT1(I,1)
         TH(I)=ATAN2(DZ,DX)
        END DO

C       ESTABLISH COLLOCATION POINTS
        DO I=1,M
         CO(I,1)=(PT2(I,1)-PT1(I,1))/2+PT1(I,1)
         CO(I,2)=(PT2(I,2)-PT1(I,2))/2+PT1(I,2)
        END DO

C       ESTABLISH INFLUENCE COEFFICIENTS
        DO I=1,M
         DO J=1,M
```

```
C       CONVERT COLLOCATION POINT TO LOCAL PANEL COORDS.
        XT=CO(I,1)-PT1(J,1)
        ZT=CO(I,2)-PT1(J,2)
        X2T=PT2(J,1)-PT1(J,1)
        Z2T=PT2(J,2)-PT1(J,2)

        X=XT*COS(TH(J))+ZT*SIN(TH(J))
        Z=-XT*SIN(TH(J))+ZT*COS(TH(J))
        X2=X2T*COS(TH(J))+Z2T*SIN(TH(J))
        Z2=0

C       FIND R1, R2, TH1, TH2
        R1=SQRT(X**2+Z**2)
        R2=SQRT((X-X2)**2+Z**2)

        TH1=ATAN2(Z,X)

        TH2=ATAN2(Z,X-X2)

C       COMPUTE VELOCITY IN LOCAL REF. FRAME
        IF(I.EQ.J) THEN
          UL=0
          WL=0.5
        ELSE
          UL=1/(2*3.141593)*LOG(R1/R2)
          WL=1/(2*3.141593)*(TH2-TH1)
        END IF

C       RETURN VELOCITY TO GLOBAL REF. FRAME
        U=UL*COS(-TH(J))+WL*SIN(-TH(J))
        W=-UL*SIN(-TH(J))+WL*COS(-TH(J))

C       A(I,J) IS THE INFLUENCE COEFF. DEFINED BY THE
C       TANGENCY CONDITION. B(I,J) IS THE INDUCED LOCAL
C       TANGENTIAL VELOCITY TO BE USED IN CP CALCULATION.
        A(I,J)=-U*SIN(TH(I))+W*COS(TH(I))
        B(I,J)=U*COS(TH(I))+W*SIN(TH(I))

        END DO

        A(I,N)=SIN(TH(I))

        END DO

C       SOLVE FOR THE SOLUTION VECTOR OF SOURCE STRENGTHS

        IF(ANS.EQ.1) GOTO 200
        CALL MATRX(A,N,G)

C       CONVERT SOURCE STRENGTHS INTO TANGENTIAL
C       VELOCITIES ALONG THE AIRFOIL SURFACE AND CP'S
C       ON EACH OF THE PANELS

 200    CONTINUE

        DO I=1,M
        VEL=0
```

```
        DO J=1,M
         VEL=VEL+B(I,J)*G(J)
        END DO
        CP=1-(VEL+COS(TH(I)))**2
        WRITE(8,*) CO(I,1),' ,',CP
       END DO

       WRITE(6,*) ' '
       WRITE(6,*) 'LIFT COEFFICIENT=0'

       STOP
       END

       SUBROUTINE MATRX(A,N,G)

C      MATRX IS A MATRIX REDUCER OF THE GAUSSIAN TYPE
C      A(I,J) IS THE MATRIX, A(I,N) IS THE RHS VECTOR
C         AND G(I) IS THE SOLUTION VECTOR.

       REAL A(400,400),TEMP(400,400),G(400)

C      INITIALIZE THE G VECTOR TO ALL ZEROES
       DO I=1,N-1
        G(I)=0
       END DO

C      CONVERT COEFFICIENT MATRIX TO
C      UPPER TRIANGULAR FORM
       DO I=1,N-1
5      IF(ABS(A(I,I)).LT.0.0000001) GOTO 9

        P=A(I,I)
        DO J=I,N
         A(I,J)=A(I,J)/P
        END DO

        DO K=I+1,N-1
         P2=A(K,I)
         DO L=I,N
          A(K,L)=A(K,L)-P2*A(I,L)
         END DO
        END DO
       END DO

C      BACK SUBSTITUTE TRIANGULARIZED MATRIX TO GET
C      VALUES OF SOLUTION VECTOR
       DO I=N-1,1,-1
        G(I)=A(I,N)
        DO J=1,N-1
         A(I,I)=0
         G(I)=G(I)-A(I,J)*G(J)
        END DO
       END DO

       RETURN

C      ORDER MATRIX SO THAT DIAGONAL COEFFICIENTS ARE
C      NOT =0 AND STOP IS MATRIX IS SINGULAR
```

```
      9         IF(I.NE.N-1) THEN
                  DO J=1,N
                    TEMP(I,J)=A(I,J)
                    A(I,J)=A(I+1,J)
                    A(I+1,J)=TEMP(I,J)
                  END DO
                  GOTO 5
                ELSE
                  GOTO 10
                END IF

     10         WRITE(6,*) 'NO SOLUTION'
                STOP
                END


      C         PROGRAM No. 4: CONSTANT STRENGTH DOUBLET
      C         ----------------------------------------
      C         THIS PROGRAM FINDS THE PRESSURE DISTRIBUTION ON AN ARBITRARY AIRFOIL
      C           BY REPRESENTING THE SURFACE AS A FINITE NUMBER OF DOUBLET PANELS WITH
      C           CONST. STRENGTH (NEUMANN B.C., PROGRAM BY STEVEN YON, 1989).

                REAL EP(400,2),EPT(400,2),PT1(400,2),PT2(400,2)
                REAL CO(400,2),A(400,400),B(400,400),G(400)
                REAL TH(400)

                OPEN(8,FILE='CPD.DAT',STATUS='NEW')
                OPEN(9,FILE='AFOIL2.DAT',STATUS='OLD')

                WRITE(6,*) 'ENTER NUMBER OF PANELS'
                READ(5,*) M
                N=M+1
                WRITE(6,*) 'ENTER ANGLE OF ATTACK IN DEGREES'
                READ(5,*) ALPHA
                AL=ALPHA/57.2958

      C         READ IN THE PANEL END POINTS
                DO I=1,M+1
                  READ(9,*) EPT(I,1), EPT(I,2)
                END DO

      C         CONVERT PANELING TO CLOCKWISE
                DO I=1,M+1
                  EP(I,1)=EPT(N-I+1,1)
                  EP(I,2)=EPT(N-I+1,2)
                END DO

      C         ESTABLISH COORDINATES OF PANEL END POINTS
                DO I=1,M
                  PT1(I,1)=EP(I,1)
                  PT2(I,1)=EP(I+1,1)
                  PT1(I,2)=EP(I,2)
                  PT2(I,2)=EP(I+1,2)
                END DO

      C         FIND PANEL ANGLES TH(J)
                DO I=1,M
                  DZ=PT2(I,2)-PT1(I,2)
                  DX=PT2(I,1)-PT1(I,1)
```

D.3 Time-Dependent Programs

```fortran
              TH(I)=ATAN2(DZ,DX)
            END DO

C       ESTABLISH COLLOCATION POINTS
            DO I=1,M
              CO(I,1)=(PT2(I,1)-PT1(I,1))/2+PT1(I,1)
              CO(I,2)=(PT2(I,2)-PT1(I,2))/2+PT1(I,2)
            END DO

C       ESTABLISH INFLUENCE COEFFICIENTS
            DO I=1,M
              DO J=1,M

C       CONVERT THE COLLOCATION POINT
C       TO LOCAL PANEL COORDS.
              XT=CO(I,1)-PT1(J,1)
              ZT=CO(I,2)-PT1(J,2)
              X2T=PT2(J,1)-PT1(J,1)
              Z2T=PT2(J,2)-PT1(J,2)
              X=XT*COS(TH(J))+ZT*SIN(TH(J))
              Z=-XT*SIN(TH(J))+ZT*COS(TH(J))
              X2=X2T*COS(TH(J))+Z2T*SIN(TH(J))
              Z2=0

              R1=SQRT(X**2+Z**2)
              R2=SQRT((X-X2)**2+Z**2)

C       COMPUTE THE VELOCITY INDUCED AT THE ITH
C       COLLOCATION POINT BY THE JTH PANEL
              IF(I.EQ.J) THEN
                UL=0
                WL=-1/(3.14159265*X)
              ELSE
                UL=0.15916*(Z/(R1**2)-Z/(R2**2))
                WL=-0.15916*(X/(R1**2)-(X-X2)/(R2**2))
              END IF

              U=UL*COS(-TH(J))+WL*SIN(-TH(J))
              W=-UL*SIN(-TH(J))+WL*COS(-TH(J))

C       A(I,J) IS THE COMPONENT OF VELOCITY INDUCED IN THE
C       DIRECTION NORMAL TO PANEL I BY PANEL J AT THE ITH
C       COLLOCATION POINT

              A(I,J)=-U*SIN(TH(I))+W*COS(TH(I))
              B(I,J)=U*COS(TH(I))+W*SIN(TH(I))

              END DO

C       INCLUDE THE INFLUENCE OF THE WAKE PANEL
            R=SQRT((CO(I,1)-PT2(M,1))**2
           *       +(CO(I,2)-PT2(M,2))**2)

            U=0.15916*(CO(I,2)/(R**2))
            W=-0.15916*(CO(I,1)-PT2(M,1))/(R**2)

            A(I,N)=-U*SIN(TH(I))+W*COS(TH(I))
            B(I,N)=U*COS(TH(I))+W*SIN(TH(I))
```

```fortran
      A(I,N+1)=COS(AL)*SIN(TH(I))-SIN(AL)*COS(TH(I))

      END DO

C     PREPARE THE MATRIX FOR SOLUTION BY PROVIDING
C     A KUTTA CONDITION
      DO I=1,N+1
       A(N,I)=0
      END DO
       A(N,1)=-1
       A(N,M)=1
       A(N,N)=-1

C     SOLVE FOR THE SOLUTION VECTOR OF DOUBLET STRENGTHS

      N=N+1

      CALL MATRX(A,N,G)

C     CONVERT DOUBLET STRENGTHS INTO TANGENTIAL
C     VELOCITIES ALONG THE AIRFOIL SURFACE AND CP'S
C     ON EACH OF THE PANELS

 200  CONTINUE

      DO I=1,M
        TEMP=0
       DO J=1,M+1
        TEMP=TEMP+B(I,J)*G(J)
       END DO
        IF(I.NE.1.AND.I.NE.M) THEN
         R=SQRT((CO(I+1,1)-CO(I-1,1))**2
     *          +(CO(I+1,2)-CO(I-1,2))**2)
         VLOC=(G(I+1)-G(I-1))/R
        ELSE IF(I.EQ.1) THEN
         R=SQRT((CO(2,1)-CO(1,1))**2
     *          +(CO(2,2)-CO(1,2))**2)
         VLOC=(G(2)-G(1))/R
        ELSE IF(I.EQ.M) THEN
         R=SQRT((CO(M,1)-CO(M-1,1))**2
     *          +(CO(M,2)-CO(M-1,2))**2)
         VLOC=(G(M)-G(M-1))/R
        END IF

        VEL=COS(AL)*COS(TH(I))+SIN(AL)*SIN(TH(I))
     *       +TEMP+VLOC/2
        CP=1-VEL**2
        WRITE(8,*) CO(I,1),' ,',CP

      END DO

      WRITE(6,*) ' '
      WRITE(6,*) 'LIFT COEFFICIENT=', G(M+1)

      STOP
      END
```

D.3 Time-Dependent Programs

```
C       PROGRAM No. 5: CONSTANT STRENGTH VORTEX
C       ----------------------------------------

C       THIS PROGRAM FINDS THE PRESSURE DISTRIBUTION ON AN ARBITRARY AIRFOIL
C           BY REPRESENTING THE SURFACE AS A FINITE NUMBER OF VORTEX PANELS WITH
C           CONST. STRENGTH (NEUMANN B.C., PROGRAM BY STEVEN YON, 1989).

        REAL EP(400,2),EPT(400,2),PT1(400,2),PT2(400,2)
        REAL CO(400,2),A(400,400),B(400,400),G(400)
        REAL VEL(400),VELT(400),TH(400),DL(400)

        OPEN(8,FILE='CPV.DAT',STATUS='NEW')
        OPEN(9,FILE='AFOIL2.DAT',STATUS='OLD')

        WRITE(6,*) 'ENTER NUMBER OF PANELS'
        READ(5,*) M
        N=M+1
        WRITE(6,*) 'ENTER ANGLE OF ATTACK IN DEGREES'
        READ(5,*) ALPHA
        AL=ALPHA/57.2958

C       READ IN THE PANEL END POINTS
        DO I=1,M+1
         READ(9,*) EPT(I,1), EPT(I,2)
        END DO

C       CONVERT PANELING TO CLOCKWISE
        DO I=1,N
         EP(I,1)=EPT(N-I+1,1)
         EP(I,2)=EPT(N-I+1,2)
        END DO

C       ESTABLISH COORDINATES OF PANEL END POINTS
        DO I=1,M
          PT1(I,1)=EP(I,1)
          PT2(I,1)=EP(I+1,1)
          PT1(I,2)=EP(I,2)
          PT2(I,2)=EP(I+1,2)
        END DO

C       FIND PANEL ANGLES TH(J)
        DO I=1,M
         DZ=PT2(I,2)-PT1(I,2)
         DX=PT2(I,1)-PT1(I,1)
         TH(I)=ATAN2(DZ,DX)
        END DO

C       ESTABLISH COLLOCATION POINTS
        DO I=1,M
          CO(I,1)=(PT2(I,1)-PT1(I,1))/2+PT1(I,1)
          CO(I,2)=(PT2(I,2)-PT1(I,2))/2+PT1(I,2)
        END DO

C       ESTABLISH INFLUENCE COEFFICIENTS
        DO I=1,M
         DO J=1,M

C       CONVERT COLLOCATION POINT INTO LOCAL PANEL COORDS.
        X2T=PT2(J,1)-PT1(J,1)
```

```
              Z2T=PT2(J,2)-PT1(J,2)
              XT=CO(I,1)-PT1(J,1)
              ZT=CO(I,2)-PT1(J,2)

              X2=X2T*COS(TH(J))+Z2T*SIN(TH(J))
              Z2=0
              X=XT*COS(TH(J))+ZT*SIN(TH(J))
              Z=-XT*SIN(TH(J))+ZT*COS(TH(J))

C       SAVE PANEL LENGTHS FOR LATER USE
              IF(I.EQ.1) THEN
               DL(J)=X2
              END IF

              R1=SQRT(X**2+Z**2)
              R2=SQRT((X-X2)**2+Z**2)

              TH1=ATAN2(Z,X)
              TH2=ATAN2(Z,X-X2)

              IF(I.EQ.J) THEN
               UL=0.5
               WL=0
              ELSE
               UL=0.15916*(TH2-TH1)
               WL=0.15916*LOG(R2/R1)
              END IF

               U=UL*COS(-TH(J))+WL*SIN(-TH(J))
               W=-UL*SIN(-TH(J))+WL*COS(-TH(J))

C       A(I,J) IS THE COMPONENT OF VELOCITY NORMAL TO
C       THE AIRFOIL INDUCED BY THE JTH PANEL AT THE
C       ITH COLLOCATION POINT.

              A(I,J)=-U*SIN(TH(I))+W*COS(TH(I))
              B(I,J)=U*COS(TH(I))+W*SIN(TH(I))

                END DO

              A(I,N)=COS(AL)*SIN(TH(I))-SIN(AL)*COS(TH(I))

              END DO

C       REPLACE EQUATION M/4 WITH A KUTTA CONDITION
              DO J=I,M+1
               A(M/4,J)=0
              END DO
               A(M/4,1)=1
               A(M/4,M)=1

C       SOLVE FOR THE SOLUTION VECTOR OF VORTEX STRENGTHS

              CALL MATRX(A,N,G)

C       CONVERT SOURCE STRENGTHS INTO TANGENTIAL
C       VELOCITIES ALONG THE AIRFOIL SURFACE AND CP'S
C       ON EACH OF THE PANELS
```

D.3 Time-Dependent Programs

```
200     CONTINUE
          CL=0
          DO I=1,M
            TEMP=0
            DO J=1,M
              TEMP=TEMP+B(I,J)*G(J)
            END DO
            VEL(I)=TEMP+COS(AL)*COS(TH(I))
     *               +SIN(AL)*SIN(TH(I))
            CL=CL+VEL(I)*DL(I)
          END DO

          WRITE(6,*) 'SMOOTH THE VELOCITY DISTRIBUTION?'
          WRITE(6,*) '1=YES'
          WRITE(6,*) '2=NO'
          READ(5,*) ANS1

          DO I=2,M
            IF(ANS1.EQ.1) THEN
              CP=1-((VEL(I)+VEL(I-1))/2)**2
              WRITE(8,*) PT2(I-1,1),' ,',CP
            ELSE
              CP=1-VEL(I)**2
              WRITE(8,*) CO(I,1),' ,',CP
            END IF
          END DO

          WRITE(6,*) ' '
          WRITE(6,*) 'LIFT COEFFICIENT=', CL

          STOP
          END


C       PROGRAM No. 6: LINEAR STRENGTH SOURCE
C       -------------------------------------
C       THIS PROGRAM FINDS THE PRESSURE DISTRIBUTION ON AN ARBITRARY AIRFOIL
C          BY REPRESENTING THE SURFACE AS A FINITE NUMBER OF SOURCE PANELS WITH
C          LINEAR STRENGTH (ALPHA=0, NEUMANN B.C., PROGRAM BY STEVEN YON, 1989).

        REAL EP(400,2),EPT(400,2),PT1(400,2),PT2(400,2)
        REAL CO(400,2),A(400,400),B(400,400),G(400),V(400)
        REAL TH(400)

        OPEN(8,FILE='CPLS.DAT',STATUS='NEW')
        OPEN(9,FILE='AFOIL2.DAT',STATUS='OLD')

        WRITE(6,*) 'ENTER NUMBER OF PANELS'
        READ(5,*) M
        N=M+1
        AL=0

C       READ IN THE PANEL END POINTS
        DO I=1,M+1
          READ(9,*) EPT(I,1), EPT(I,2)
        END DO

C       CONVERT PANELING TO CLOCKWISE
```

```fortran
      DO I=1,N
       EP(I,1)=EPT(N-I+1,1)
       EP(I,2)=EPT(N-I+1,2)
      END DO

C     ESTABLISH COORDINATES OF PANEL END POINTS
      DO I=1,M
       PT1(I,1)=EP(I,1)
       PT2(I,1)=EP(I+1,1)
       PT1(I,2)=EP(I,2)
       PT2(I,2)=EP(I+1,2)
      END DO

C     FIND PANEL ANGLES TH(J)
      DO I=1,M
       DZ=PT2(I,2)-PT1(I,2)
       DX=PT2(I,1)-PT1(I,1)
       TH(I)=ATAN2(DZ,DX)
      END DO

      TH(M+1)=0

C     ESTABLISH COLLOCATION POINTS
      DO I=1,M
       CO(I,1)=(PT2(I,1)-PT1(I,1))/2+PT1(I,1)
       CO(I,2)=(PT2(I,2)-PT1(I,2))/2+PT1(I,2)
      END DO

      WRITE(6,*) 'ENTER X COORD. OF WAKE POINT'
      READ(5,*) XX
       CO(M+1,1)=XX
       CO(M+1,2)=0

C     ESTABLISH INFLUENCE COEFFICIENTS
      DO I=1,M+1
       DO J=1,M

C     CONVERT COLLOCATION POINT TO LOCAL PANEL COORDS.
      XT=CO(I,1)-PT1(J,1)
      ZT=CO(I,2)-PT1(J,2)
      X2T=PT2(J,1)-PT1(J,1)
      Z2T=PT2(J,2)-PT1(J,2)

      X=XT*COS(TH(J))+ZT*SIN(TH(J))
      Z=-XT*SIN(TH(J))+ZT*COS(TH(J))
      X2=X2T*COS(TH(J))+Z2T*SIN(TH(J))
      Z2=0

C     FIND R1, R2, TH1, TH2
      R1=SQRT(X**2+Z**2)
      R2=SQRT((X-X2)**2+Z**2)

      TH1=ATAN2(Z,X)
      TH2=ATAN2(Z,X-X2)

C     COMPUTE VELOCITY COMPONENTS AS FUNCTIONS OF
C     SIGMA1 AND SIGMA2. THESE VELOCITIES ARE IN
C     THE JTH REFERENCE FRAME.
```

```
            IF(I.EQ.J) THEN
              U1L=0.15916
              U2L=-0.15916
              W1L=-0.5*(X-X2)/X2
              W2L=0.5*(X)/X2
            ELSE
              W1L=-(Z*LOG(R2/R1)+X*(TH2-TH1)-X2*(TH2-TH1))/
     *              (6.28319*X2)
              W2L=(Z*LOG(R2/R1)+X*(TH2-TH1))/(6.28319*X2)
              U1L=((X2-Z*(TH2-TH1))-X*LOG(R1/R2)+
     *              X2*LOG(R1/R2))/(6.28319*X2)
              U2L=-((X2-Z*(TH2-TH1))-X*LOG(R1/R2))/(6.28319*X2)
            END IF

C       TRANSFORM THE LOCAL VELOCITIES INTO THE GLOBAL
C       REFERENCE FRAME.

            U1=U1L*COS(-TH(J))+W1L*SIN(-TH(J))
            U2=U2L*COS(-TH(J))+W2L*SIN(-TH(J))
            W1=-U1L*SIN(-TH(J))+W1L*COS(-TH(J))
            W2=-U2L*SIN(-TH(J))+W2L*COS(-TH(J))

C       COMPUTE THE COEFFICIENTS OF SIGMA IN THE
C       INFLUENCE MATRIX

            IF(J.EQ.1) THEN
              A(I,1)=-U1*SIN(TH(I))+W1*COS(TH(I))
              HOLDA=-U2*SIN(TH(I))+W2*COS(TH(I))
              B(I,1)=U1*COS(TH(I))+W1*SIN(TH(I))
              HOLDB=U2*COS(TH(I))+W2*SIN(TH(I))
            ELSE IF(J.EQ.M) THEN
              A(I,M)=-U1*SIN(TH(I))+W1*COS(TH(I))+HOLDA
              A(I,N)=-U2*SIN(TH(I))+W2*COS(TH(I))
              B(I,M)=U1*COS(TH(I))+W1*SIN(TH(I))+HOLDB
              B(I,N)=U2*COS(TH(I))+W2*SIN(TH(I))
            ELSE
              A(I,J)=-U1*SIN(TH(I))+W1*COS(TH(I))+HOLDA
              HOLDA=-U2*SIN(TH(I))+W2*COS(TH(I))
              B(I,J)=U1*COS(TH(I))+W1*SIN(TH(I))+HOLDB
              HOLDB=U2*COS(TH(I))+W2*SIN(TH(I))
            END IF

          END DO

          A(I,N+1)=SIN(TH(I))

        END DO

        N=M+2

        IF(M.EQ.10) THEN
            DO I=1,11
           WRITE(6,10) A(I,1),A(I,2),A(I,3),A(I,4),A(I,5),A(I,6),
     *          A(I,7),A(I,8),A(I,9),A(I,10),A(I,11)
            END DO
        END IF

C       SOLVE FOR THE SOLUTION VECTOR OF SOURCE STRENGTHS
```

```fortran
            CALL MATRX(A,N,G)

C       CONVERT SOURCE STRENGTHS INTO TANGENTIAL
C       VELOCITIES ALONG THE AIRFOIL SURFACE AND CP'S
C       ON EACH OF THE PANELS.

  200   CONTINUE

            N=M+1

            DO I=1,M
              VEL=0
             DO J=1,N
              VEL=VEL+B(I,J)*G(J)
             END DO
              V(I)=VEL+COS(AL)*COS(TH(I))+SIN(AL)*SIN(TH(I))
            END DO

            WRITE(6,*) ' '
            WRITE(6,*) 'SMOOTH THE VELOCITY DISTRIBUTION?'
            WRITE(6,*) '1=YES'
            WRITE(6,*) '2=NO'
            READ(5,*) ANS1

            DO I=2,M
             IF(ANS1.EQ.1) THEN
              VA=(V(I)+V(I-1))/2
              CP=1-VA**2
              WRITE(8,*) PT1(I,1), ' ,',CP
             ELSE
              CP=1-V(I)**2
              WRITE(8,*) CO(I,1),' ,',CP
             END IF
            END DO

            WRITE(6,*) ' '
            WRITE(6,*) 'LIFT COEFFICIENT=0'

            STOP
   10       FORMAT(/,F6.2,1X,F5.2,1X,F5.2,1X,F5.2,1X,F5.2,1X,F5.2,1X,
          *         F5.2,1X,F5.2,1X,F5.2,1X,F5.2,1X,F5.2)

            END


C       PROGRAM No. 7: LINEAR STRENGTH VORTEX
C       -------------------------------------

C       THIS PROGRAM FINDS THE PRESSURE DISTRIBUTION ON AN ARBITRARY AIRFOIL
C          BY REPRESENTING THE SURFACE AS A FINITE NUMBER OF VORTEX PANELS WITH
C          LINEAR STRENGTH (NEUMANN B.C., PROGRAM BY STEVEN YON, 1989).

            REAL EP(400,2),EPT(400,2),PT1(400,2),PT2(400,2)
            REAL CO(400,2),A(400,400),B(400,400),G(400)
            REAL TH(400),DL(400)

            OPEN(8,FILE='CPLV.DAT',STATUS='NEW')
            OPEN(9,FILE='AFOIL2.DAT',STATUS='OLD')

            WRITE(6,*) 'ENTER NUMBER OF PANELS'
            READ(5,*) M
```

D.3 Time-Dependent Programs

```
            N=M+1
            WRITE(6,*) 'ENTER ANGLE OF ATTACK IN DEGREES'
            READ(5,*) ALPHA
            AL=ALPHA/57.2958

C       READ IN THE PANEL END POINTS
            DO I=1,M+1
              READ(9,*) EPT(I,1), EPT(I,2)
            END DO

            DO I=1,N
              EP(I,1)=EPT(N-I+1,1)
              EP(I,2)=EPT(N-I+1,2)
            END DO

C       ESTABLISH COORDINATES OF PANEL END POINTS
            DO I=1,M
              PT1(I,1)=EP(I,1)
              PT2(I,1)=EP(I+1,1)
              PT1(I,2)=EP(I,2)
              PT2(I,2)=EP(I+1,2)
            END DO

C       FIND PANEL ANGLES TH(J)
            DO I=1,M
              DZ=PT2(I,2)-PT1(I,2)
              DX=PT2(I,1)-PT1(I,1)
              TH(I)=ATAN2(DZ,DX)
            END DO

C       ESTABLISH COLLOCATION POINTS
            DO I=1,M
              CO(I,1)=(PT2(I,1)-PT1(I,1))/2+PT1(I,1)
              CO(I,2)=(PT2(I,2)-PT1(I,2))/2+PT1(I,2)
            END DO

C       ESTABLISH INFLUENCE COEFFICIENTS
            DO I=1,M
              DO J=1,M

C       CONVERT COLLOCATION POINT TO LOCAL PANEL COORDS.
            XT=CO(I,1)-PT1(J,1)
            ZT=CO(I,2)-PT1(J,2)
            X2T=PT2(J,1)-PT1(J,1)
            Z2T=PT2(J,2)-PT1(J,2)

            X=XT*COS(TH(J))+ZT*SIN(TH(J))
            Z=-XT*SIN(TH(J))+ZT*COS(TH(J))
            X2=X2T*COS(TH(J))+Z2T*SIN(TH(J))
            Z2=0

C       SAVE PANEL LENGTHS FOR LIFT COEFF. CALC.
            IF(I.EQ.1) THEN
              DL(J)=X2
            END IF

C       FIND R1, R2, TH1, TH2
            R1=SQRT(X**2+Z**2)
            R2=SQRT((X-X2)**2+Z**2)
```

```
          TH1=ATAN2(Z,X)
          TH2=ATAN2(Z,X-X2)

C         COMPUTE VELOCITY COMPONANTS AS FUNCTIONS OF
C         GAMMA1 AND GAMMA2. THESE VELOCITIES ARE IN
C         THE JTH REFERENCE FRAME.

          IF(I.EQ.J) THEN
           U1L=-0.5*(X-X2)/(X2)
           U2L=0.5*(X)/(X2)
           W1L=-0.15916
           W2L=0.15916
          ELSE
           U1L=-(Z*LOG(R2/R1)+X*(TH2-TH1)-X2*(TH2-TH1))/
     *            (6.28319*X2)
           U2L=(Z*LOG(R2/R1)+X*(TH2-TH1))/(6.28319*X2)
           W1L=-((X2-Z*(TH2-TH1))-X*LOG(R1/R2)
     *            +X2*LOG(R1/R2))/(6.28319*X2)
           W2L=((X2-Z*(TH2-TH1))-X*LOG(R1/R2))/(6.28319*X2)
          END IF

C         TRANSFORM THE LOCAL VELOCITIES INTO THE
C         GLOBAL REFERENCE FRAME.

          U1=U1L*COS(-TH(J))+W1L*SIN(-TH(J))
          U2=U2L*COS(-TH(J))+W2L*SIN(-TH(J))
          W1=-U1L*SIN(-TH(J))+W1L*COS(-TH(J))
          W2=-U2L*SIN(-TH(J))+W2L*COS(-TH(J))

C         COMPUTE THE COEFFICIENTS OF GAMMA IN THE
C         INFLUENCE MATRIX.

          IF(J.EQ.1) THEN
           A(I,1)=-U1*SIN(TH(I))+W1*COS(TH(I))
           HOLDA=-U2*SIN(TH(I))+W2*COS(TH(I))
           B(I,1)=U1*COS(TH(I))+W1*SIN(TH(I))
           HOLDB=U2*COS(TH(I))+W2*SIN(TH(I))
          ELSE IF(J.EQ.M) THEN
           A(I,M)=-U1*SIN(TH(I))+W1*COS(TH(I))+HOLDA
           A(I,N)=-U2*SIN(TH(I))+W2*COS(TH(I))
           B(I,M)=U1*COS(TH(I))+W1*SIN(TH(I))+HOLDB
           B(I,N)=U2*COS(TH(I))+W2*SIN(TH(I))
          ELSE
           A(I,J)=-U1*SIN(TH(I))+W1*COS(TH(I))+HOLDA
           HOLDA=-U2*SIN(TH(I))+W2*COS(TH(I))
           B(I,J)=U1*COS(TH(I))+W1*SIN(TH(I))+HOLDB
           HOLDB=U2*COS(TH(I))+W2*SIN(TH(I))
          END IF

          END DO

          A(I,N+1)=COS(AL)*SIN(TH(I))-SIN(AL)*COS(TH(I))

          END DO

C         ADD THE KUTTA CONDITION
          A(N,1)=1
          A(N,N)=1
```

```
          IF(M.EQ.10) THEN
            DO I=1,11
              WRITE(6,10) A(I,1),A(I,2),A(I,3),A(I,4),A(I,5),A(I,6),A(I,7),
     *                    A(I,8),A(I,9),A(I,10),A(I,11)
            END DO
          END IF

          N=N+1

C         SOLVE FOR THE SOLUTION VECTOR OF VORTEX STRENGTHS

          CALL MATRX(A,N,G)

C         CONVERT VORTEX STRENGTHS INTO TANGENTIAL
C         VELOCITIES ALONG THE AIRFOIL SURFACE AND CP'S
C         ON EACH OF THE PANELS.

 200      CONTINUE

          N=M+1
          CL=0

          DO I=1,M
            VEL=0
            DO J=1,N
              VEL=VEL+B(I,J)*G(J)
            END DO
            V=VEL+COS(AL)*COS(TH(I))+SIN(AL)*SIN(TH(I))
            CL=CL+V*DL(I)
            CP=1-V**2
            WRITE(8,*) CO(I,1),' ',CP
          END DO

          WRITE(6,*) ' '
          WRITE(6,*) 'LIFT COEFFICIENT=',CL

          STOP
 10       FORMAT(/,F6.2,1X,F5.2,1X,F5.2,1X,F5.2,1X,F5.2,1X,F5.2,1X,F5.2,
     *                  1X,F5.2,1X,F5.2,1X,F5.2,1X,F5.2)

          END


C         PROGRAM No. 8: CONSTANT STRENGTH DOUBLET POTENTIAL
C         --------------------------------------------------

C         THIS PROGRAM FINDS THE PRESSURE DISTRIBUTION ON AN ARBITRARY AIRFOIL
C            BY REPRESENTING THE SURFACE AS A FINITE NUMBER OF DOUBLET PANELS WITH
C            CONSTANT STRENGTH (DIRICHLET B.C., PROGRAM BY STEVEN YON, 1989).

          REAL EP(400,2),EPT(400,2),PT1(400,2),PT2(400,2)
          REAL CO(400,2),A(400,400),B(400,400),G(400)
          REAL TH(400),DL(400)

          OPEN(8,FILE='CPDP.DAT',STATUS='NEW')
          OPEN(9,FILE='AFOIL2.DAT',STATUS='OLD')

          WRITE(6,*) 'ENTER NUMBER OF PANELS'
          READ(5,*) M
          N=M+1
```

```
            WRITE(6,*) 'ENTER ANGLE OF ATTACK IN DEGREES'
            READ(5,*) ALPHA
            AL=ALPHA/57.2958

C       READ IN THE PANEL END POINTS
            DO I=1,M+1
             READ(9,*) EPT(I,1), EPT(I,2)
            END DO

C       CONVERT THE PANELING TO CLOCKWISE
            DO I=1,M+1
             EP(I,1)=EPT(N-I+1,1)
             EP(I,2)=EPT(N-I+1,2)
            END DO

C       ESTABLISH COORDINATES OF PANEL END POINTS
            DO I=1,M
             PT1(I,1)=EP(I,1)
             PT2(I,1)=EP(I+1,1)
             PT1(I,2)=EP(I,2)
             PT2(I,2)=EP(I+1,2)
            END DO

C       FIND PANEL ANGLES TH(J)
            DO I=1,M
             DZ=PT2(I,2)-PT1(I,2)
             DX=PT2(I,1)-PT1(I,1)
             TH(I)=ATAN2(DZ,DX)
            END DO

C       ESTABLISH COLLOCATION POINTS
            DO I=1,M
             CO(I,1)=(PT2(I,1)-PT1(I,1))/2+PT1(I,1)
             CO(I,2)=(PT2(I,2)-PT1(I,2))/2+PT1(I,2)
            END DO

C       ESTABLISH INFLUENCE COEFFICIENTS
            DO I=1,M
             DO J=1,M

C       CONVERT COLLOCATION POINTS TO LOCAL
C       PANEL(J) COORDINATES.
            X2T=PT2(J,1)-PT1(J,1)
            Z2T=PT2(J,2)-PT1(J,2)
            XT=CO(I,1)-PT1(J,1)
            ZT=CO(I,2)-PT1(J,2)

            X2=X2T*COS(TH(J))+Z2T*SIN(TH(J))
            Z2=0
            X=XT*COS(TH(J))+ZT*SIN(TH(J))
            Z=-XT*SIN(TH(J))+ZT*COS(TH(J))

C       SAVE PANEL LENGTHS
            IF(I.EQ.1) THEN
             DL(J)=X2
            END IF

C       FIND R AND THETA COMPONENTS
            R1=SQRT(X**2+Z**2)
            R2=SQRT((X-X2)**2+Z**2)
```

D.3 Time-Dependent Programs

```
            TH1=ATAN2(Z,X)
            TH2=ATAN2(Z,X-X2)

C           COMPUTE INFLUENCE COEFS. A(I,J)
            IF(I.EQ.J) THEN
             A(I,J)=0.5
            ELSE
             A(I,J)=-0.15916*(TH2-TH1)
            END IF

            END DO

C           ADD WAKE INFLUENCE
            XW=CO(I,1)-PT2(M,1)
            ZW=CO(I,2)-PT2(M,2)
            DTHW=-ATAN(ZW/XW)

             A(I,N)=-0.15916*(DTHW)
             A(I,N+1)=(CO(I,1)*COS(AL)+CO(I,2)*SIN(AL))

            END DO

C           ADD AN EXPLICIT KUTTA CONDITION
            A(N,1)=-1
            A(N,M)=1
            A(N,N)=-1

C           SOLVE FOR THE SOLUTION VECTOR OF DOUBLET STRENGTHS

            N=N+1

            CALL MATRX(A,N,G)

C           CONVERT DOUBLET STRENGTHS INTO TANGENTIAL
C           VELOCITIES ALONG THE AIRFOIL SURFACE AND
C           CP'S ON EACH OF THE PANELS.

  200       CONTINUE

            DO I=1,M-1
             R=(DL(I)+DL(I+1))/2
             VEL=(G(I+1)-G(I))/R
             CP=1-VEL**2
             WRITE(8,*) PT2(I,1),' ',',CP
            END DO

            WRITE(6,*) ' '
            WRITE(6,*) 'LIFT COEFFICIENT=', G(M+1)

            STOP
            END


C       PROGRAM No. 9: CONSTANT STRENGTH SOURCE/DOUBLET POTENTIAL
C       ---------------------------------------------------------
C       THIS PROGRAM FINDS THE PRESSURE DISTRIBUTION ON AN ARBITRARY AIRFOIL
C          BY REPRESENTING THE SURFACE AS A FINITE NUMBER OF SOURCE/DOUBLET PANELS
C             WITH CONSTANT STRENGTH (DIRICHLET B.C., PROGRAM BY STEVEN YON, 1989).
```

```fortran
      REAL EP(400,2),PT1(400,2),PT2(400,2),TH(400)
      REAL CO(400,2),A(400,400),B(400,400),G(400)
      REAL EPT(400,2),SIG(400),PHI(400),DL(400)

      OPEN(8,FILE='CPSD.DAT',STATUS='NEW')
      OPEN(9,FILE='AFOIL2.DAT',STATUS='OLD')

      WRITE(6,*) 'ENTER NUMBER OF PANELS'
      READ(5,*) M
      N=M+1
      WRITE(6,*) 'ENTER ANGLE OF ATTACK IN DEGREES'
      READ(5,*) ALPHA
      AL=ALPHA/57.2958

C     READ IN THE PANEL END POINTS
      DO I=1,M+1
       READ(9,*) EPT(I,1), EPT(I,2)
      END DO

C     CONVERT PANELING TO CLOCKWISE
      DO I=1,M+1
       EP(I,1)=EPT(N-I+1,1)
       EP(I,2)=EPT(N-I+1,2)
      END DO

C     ESTABLISH COORDINATES OF PANEL END POINTS
      DO I=1,M
       PT1(I,1)=EP(I,1)
       PT2(I,1)=EP(I+1,1)
       PT1(I,2)=EP(I,2)
       PT2(I,2)=EP(I+1,2)
      END DO

C     FIND PANEL ANGLES TH(J)
      DO I=1,M
       DZ=PT2(I,2)-PT1(I,2)
       DX=PT2(I,1)-PT1(I,1)
       TH(I)=ATAN2(DZ,DX)
      END DO

C     ESTABLISH SOURCE STRENGTHS (SIGMA=V DOT N)
      DO I=1,M
       SIG(I)=(COS(AL)*SIN(TH(I))-SIN(AL)*COS(TH(I)))
      END DO

C     ESTABLISH SURFACE POINTS (COLLOCATION POINTS)
      DO I=1,M
       CO(I,1)=(PT2(I,1)-PT1(I,1))/2+PT1(I,1)
       CO(I,2)=(PT2(I,2)-PT1(I,2))/2+PT1(I,2)
      END DO

C     ESTABLISH INFLUENCE COEFFICIENTS
      DO I=1,M
         TEMP=0
       DO J=1,M

C     CONVERT THE COLLOCATION POINT TO LOCAL PANEL COORDS.
       XT=CO(I,1)-PT1(J,1)
       ZT=CO(I,2)-PT1(J,2)
```

D.3 Time-Dependent Programs

```
              X2T=PT2(J,1)-PT1(J,1)
              Z2T=PT2(J,2)-PT1(J,2)

              X=XT*COS(TH(J))+ZT*SIN(TH(J))
              Z=-XT*SIN(TH(J))+ZT*COS(TH(J))
              X2=X2T*COS(TH(J))+Z2T*SIN(TH(J))
              Z2=0

C     SAVE PANEL LENGTHS
              IF(I.EQ.1) THEN
               DL(J)=X2
              END IF

C     COMPUTE R AND THETA VALUES FOR THE COLOC. POINT
              R1=SQRT(X**2+Z**2)
              R2=SQRT((X-X2)**2+Z**2)

              TH1=ATAN2(Z,X)
              TH2=ATAN2(Z,X-X2)

C     COMPUTE THE DOUBLET INFLUENCE COEFFICIENTS

              IF(I.EQ.J) THEN
               A(I,J)=0.5
              ELSE
               A(I,J)=-0.15916*(TH2-TH1)
              END IF

C     COMPUTE THE SOURCE INFLUENCE COEFF'S AND ADD
C     THEM UP     TO GIVE THE RHS
              IF(I.EQ.J) THEN
               TEMP=TEMP+SIG(J)/3.14159265*(X*LOG(R1))
              ELSE
               TEMP=TEMP+SIG(J)/6.28319*(X*LOG(R1)
     *              -(X-X2)*LOG(R2)+Z*(TH2-TH1))
              END IF

              END DO

C     ADD WAKE INFLUENCE COEFF.
              XW=CO(I,1)-PT2(M,1)
              ZW=CO(I,2)-PT2(M,2)
              DTHW=-ATAN(ZW/XW)

              A(I,N)=-0.15916*(DTHW)
              A(I,N+1)=TEMP

              END DO

C     ADD AN EXPLICIT KUTTA CONDITION
              DO I=1,N+1
               A(N,I)=0
              END DO
              A(N,1)=-1
              A(N,M)=1
              A(N,N)=-1

C     SOLVE FOR THE SOLUTION VECTOR OF DOUBLET STRENGTHS
```

```
          N=N+1

          CALL MATRX(A,N,G)

C         CONVERT DOUBLET STRENGTHS INTO TANGENTIAL
C         VELOCITIES ALONG THE AIRFOIL SURFACE AND CP'S
C         ON EACH PANEL.

  200     CONTINUE

          DO I=1,M
            PHI(I)=CO(I,1)*COS(AL)+CO(I,2)*SIN(AL)+G(I)
          END DO

          DO I=1,M-1
           R=(DL(I+1)+DL(I))/2
           VEL=(PHI(I)-PHI(I+1))/R
           CP=1-VEL**2
           WRITE(8,*) PT2(I,1),', ',CP
          END DO

          WRITE(6,*) ' '
          WRITE(6,*) 'LIFT COEFFICIENT=',G(M+1)

          STOP
          END


C     PROGRAM No. 10: LINEAR STRENGTH DOUBLET POTENTIAL
C     ------------------------------------------------

C     THIS PROGRAM FINDS THE PRESSURE DISTRIBUTION ON AN ARBITRARY AIRFOIL
C        BY REPRESENTING THE SURFACE AS A FINITE NUMBER OF DOUBLET PANELS WITH
C        LINEAR STRENGTH (DIRICHLET B.C., PROGRAM BY STEVEN YON, 1989).

      REAL EP(400,2),EPT(400,2),PT1(400,2),PT2(400,2)
      REAL CO(400,2),A(400,400),B(400,400),G(400)
      REAL TH(400),DL(400)

      OPEN(8,FILE='CPLD.DAT',STATUS='NEW')
      OPEN(9,FILE='AFOIL2.DAT',STATUS='OLD')

      WRITE(6,*) 'ENTER NUMBER OF PANELS'
      READ(5,*) M
      N=M+1
      WRITE(6,*) 'ENTER ANGLE OF ATTACK IN DEGREES'
      READ(5,*) ALPHA
      AL=ALPHA/57.2958

C     READ IN THE PANEL END POINTS
      DO I=1,M+1
       READ(9,*) EPT(I,1), EPT(I,2)
      END DO

C     CONVERT PANELING TO CLOCKWISE
      DO I=1,N
       EP(I,1)=EPT(N-I+1,1)
       EP(I,2)=EPT(N-I+1,2)
      END DO
```

D.3 Time-Dependent Programs

```
C       ESTABLISH COORDINATES OF PANEL END POINTS
        DO I=1,M
         PT1(I,1)=EP(I,1)
         PT2(I,1)=EP(I+1,1)
         PT1(I,2)=EP(I,2)
         PT2(I,2)=EP(I+1,2)
        END DO

C       FIND PANEL ANGLES TH(J)
        DO I=1,M
         DZ=PT2(I,2)-PT1(I,2)
         DX=PT2(I,1)-PT1(I,1)
         TH(I)=ATAN2(DZ,DX)
        END DO

C       ESTABLISH COLLOCATION POINTS
        DO I=1,M
         CO(I,1)=(PT2(I,1)-PT1(I,1))/2+PT1(I,1)
         CO(I,2)=(PT2(I,2)-PT1(I,2))/2+PT1(I,2)
        END DO

C       ESTABLISH INFLUENCE COEFFICIENTS
        DO I=1,M
         DO J=1,M

C       CONVERT COLLOCATION POINT TO LOCAL PANEL COORDS.
        XT=CO(I,1)-PT1(J,1)
        ZT=CO(I,2)-PT1(J,2)
        X2T=PT2(J,1)-PT1(J,1)
        Z2T=PT2(J,2)-PT1(J,2)

        X=XT*COS(TH(J))+ZT*SIN(TH(J))
        Z=-XT*SIN(TH(J))+ZT*COS(TH(J))
        X2=X2T*COS(TH(J))+Z2T*SIN(TH(J))
        Z2=0

C       SAVE PANEL LENGTHS
        IF(I.EQ.1) THEN
         DL(J)=X2
        END IF

C       FIND TH1, TH2, AND R1,R2
        R1=SQRT(X**2+Z**2)
        R2=SQRT((X-X2)**2+Z**2)

        TH1=ATAN2(Z,X)
        TH2=ATAN2(Z,X-X2)

C       COMPUTE THE POTENTIAL COMPONENTS AS
C       FUNCTIONS OF R, TH
        IF(I.EQ.J) THEN
         PH1=-0.5*(X/X2-1)
         PH2=0.5*(X/X2)
        ELSE
         PH1=0.15916*(X/X2*(TH2-TH1)+Z/X2*
     *        LOG(R2/R1)-(TH2-TH1))
         PH2=-0.15916*(X/X2*(TH2-TH1)+Z/X2*LOG(R2/R1))
        END IF
```

```
C       COMPUTE THE COEFFICIENTS IN THE INFLUENCE MATRIX

        IF(J.EQ.1) THEN
         A(I,1)=PH1
         HOLDA=PH2
        ELSE IF(J.EQ.M) THEN
         A(I,M)=HOLDA+PH1
         A(I,M+1)=PH2
        ELSE
         A(I,J)=HOLDA+PH1
         HOLDA=PH2
        END IF

        END DO

C       ADD INFLUENCE OF WAKE AS A(M+2)
        XW=CO(I,1)-PT2(M,1)
        ZW=CO(I,2)-PT2(M,2)
        DTHW=-ATAN(ZW/XW)
        A(I,M+2)=-0.15916*DTHW

        A(I,M+3)=(CO(I,1)*COS(AL)+CO(I,2)*SIN(AL))

        END DO

C       ADD THE DOUBLET GRADIENT CONDITION
        A(M+1,1)=-1
        A(M+1,2)=1
        A(M+1,M)=1
        A(M+1,M+1)=-1

C       ADD THE KUTTA CONDITION
         A(M+2,1)=-1
         A(M+2,M+1)=1
         A(M+2,M+2)=-1

        N=M+3

C       SOLVE FOR THE SOLUTION VECTOR OF DOUBLET STRENGTHS

        CALL MATRX(A,N,G)

C       CONVERT DOUBLET STRENGTHS INTO TANGENTIAL
C       VELOCITIES ALONG THE AIRFOIL SURFACE AND
C       CP'S ON EACH OF THE PANELS.

  200   CONTINUE

        DO I=1,M-1
         R=(DL(I)+DL(I+1))/2
         T1=(G(I)+G(I+1))/2
         T2=(G(I+1)+G(I+2))/2
         VEL=(T2-T1)/R
         CP=1-VEL**2
          WRITE(8,*) PT2(I,1),' ,',CP
        END DO

        WRITE(6,*) ' '
        WRITE(6,*) 'LIFT COEFFICIENT=', G(M+2)

        STOP
        END
```

D.3 Time-Dependent Programs

```
C       PROGRAM No. 11: QUADRATIC STRENGTH DOUBLET POTENTIAL
C       -----------------------------------------------------
C       THIS PROGRAM FINDS THE PRESSURE DISTRIBUTION ON AN ARBITRARY AIRFOIL
C         BY REPRESENTING THE SURFACE AS A FINITE NUMBER OF DOUBLET PANELS WITH
C         QUADRATIC STRENGTH (DIRICHLET B.C., PROGRAM BY STEVEN YON, 1989).

        REAL EP(400,2),EPT(400,2),PT1(400,2),PT2(400,2)
        REAL CO(400,2),A(400,400),B(400,400,3),G(400)
        REAL DL(400),U1(400),A1(400),B1(400),TH(400)

        OPEN(8,FILE='CPQD.DAT',STATUS='NEW')
        OPEN(9,FILE='AFOIL2.DAT',STATUS='OLD')

        WRITE(6,*) 'ENTER NUMBER OF PANELS'
        READ(5,*) M
        N=M+1
        WRITE(6,*) 'ENTER ANGLE OF ATTACK IN DEGREES'
        READ(5,*) ALPHA
        AL=ALPHA/57.2958

C       READ IN THE PANEL END POINTS
        DO I=1,M+1
         READ(9,*) EPT(I,1), EPT(I,2)
        END DO

C       CONVERT PANELING TO CLOCKWISE
        DO I=1,M+1
         EP(I,1)=EPT(N-I+1,1)
         EP(I,2)=EPT(N-I+1,2)
        END DO

C       ESTABLISH COORDINATES OF PANEL END POINTS
        DO I=1,M
         PT1(I,1)=EP(I,1)
         PT2(I,1)=EP(I+1,1)
         PT1(I,2)=EP(I,2)
         PT2(I,2)=EP(I+1,2)
        END DO

C       FIND PANEL ANGLES TH(J)
        DO I=1,M
         DZ=PT2(I,2)-PT1(I,2)
         DX=PT2(I,1)-PT1(I,1)
         TH(I)=ATAN2(DZ,DX)
        END DO

C       ESTABLISH COLLOCATION POINTS
        DO I=1,M
         CO(I,1)=(PT2(I,1)-PT1(I,1))/2+PT1(I,1)
         CO(I,2)=(PT2(I,2)-PT1(I,2))/2+PT1(I,2)
        END DO

C       ESTABLISH LOCATION OF ADDITIONAL COLLOCATION POINT
        WRITE(6,*) 'ENTER X COORD. OF INTERNAL POINT'
        READ(5,*) XX
        CO(M+1,1)=XX
        CO(M+1,2)=0.0
```

```
C       ESTABLISH INFLUENCE COEFFICIENTS
        DO I=1,M+1
         DO J=1,M

C       CONVERT COLLOCATION POINT TO LOCAL PANEL COORDS.
        XT=CO(I,1)-PT1(J,1)
        ZT=CO(I,2)-PT1(J,2)
        X2T=PT2(J,1)-PT1(J,1)
        Z2T=PT2(J,2)-PT1(J,2)

        X=XT*COS(TH(J))+ZT*SIN(TH(J))
        Z=-XT*SIN(TH(J))+ZT*COS(TH(J))
        X2=X2T*COS(TH(J))+Z2T*SIN(TH(J))
        Z2=0

C       SAVE PANEL LENGTHS FOR LATER USE
         IF(I.EQ.1) THEN
          DL(J)=X2
         END IF

C       FIND TH1, TH2, AND R1, R2
        R1=SQRT(X**2+Z**2)
        R2=SQRT((X-X2)**2+Z**2)

        TH1=ATAN2(Z,X)
        TH2=ATAN2(Z,X-X2)

C       COMPUTE THE INFLUENCE COEFFICIENTS IN
C       THE 'B' MATRIX (UNREDUCED).
        IF(I.EQ.J) THEN
         B(I,J,1)=0.5
         B(I,J,2)=0.5*X
         B(I,J,3)=0.5*X**2
        ELSE
         B(I,J,1)=-0.15916*(TH2-TH1)
         B(I,J,2)=-0.15916*(X*(TH2-TH1)+Z*LOG(R2/R1))
         B(I,J,3)=0.15916*((X**2-Z**2)*(TH1-TH2)
     *    -2*X*Z*LOG(R2/R1)-Z*X2)
        END IF

         END DO
        END DO

C       ADD DOUBLET GRADIENT CONDITION
         B(M+2,1,1)=0
         B(M+2,1,2)=1
         B(M+2,1,3)=0
         B(M+2,M,1)=0
         B(M+2,M,2)=1
         B(M+2,M,3)=2*DL(M)

C       ADD KUTTA CONDITION
         B(M+3,1,1)=-1
         B(M+3,1,2)=0
         B(M+3,1,3)=0
         B(M+3,M,1)=1
         B(M+3,M,2)=DL(M)
         B(M+3,M,3)=(DL(M))**2
```

D.3 Time-Dependent Programs

```
C       BACK SUBSTITUTE THE 'B' MATRIX WITH THE
C       REGRESSION FORMULA TO GET THE COMPONENTS
C       OF THE 'A' MATRIX.
        DO I=1,M+3
          DO J=M-1,1,-1
            B(I,J,1)=B(I,J,1)+B(I,J+1,1)
            B(I,J,2)=B(I,J,2)+B(I,J+1,1)*DL(J)+B(I,J+1,2)
            B(I,J,3)=B(I,J,3)+B(I,J+1,1)
     *          *(DL(J))**2+2*B(I,J+1,2)*DL(J)
          END DO
            A(I,1)=B(I,1,1)
            A(I,2)=B(I,1,2)
          DO J=1,M
            A(I,J+2)=B(I,J,3)
          END DO
        END DO

C       ADD INFLUENCE OF WAKE AS A(I,M+3)
C       AND RHS AS A(I,M+4)
        DO I=1,M+1
          XW=CO(I,1)-PT2(M,1)
          ZW=CO(I,2)-PT2(M,2)
          DTHW=-ATAN(ZW/XW)

          A(I,M+3)=-0.15916*DTHW
          A(I,M+4)=(CO(I,1)*COS(AL)+CO(I,2)*SIN(AL))

        END DO

C       COMPLETE KUTTA COND. BY ADDING WAKE COEFF AND RHS
C       TO ROWS M+2 AND M+3
        A(M+2,M+3)=0
        A(M+2,M+4)=0

        A(M+3,M+3)=-1
        A(M+3,M+4)=0

        N=M+4

C       SOLVE FOR THE SOLUTION VECTOR OF DOUBLET STRENGTHS

        CALL MATRX(A,N,G)

C       CONVERT DOUBLET STRENGTHS, LINEAR CORRECTION
C       AND QUADRATIC CORRECTION INTO TANGENTIAL
C       VELOCITIES ALONG THE AIRFOIL SURFACE AND CP'S
C       ON EACH OF THE PANELS.

 200    CONTINUE

C       FORWARD SUBSTITUTE USING THE SOLUTION VECTOR TO
C       GET THE DOUBLET STRENGTH PARAMETERS
C       FOR EACH PANEL.
        U1(1)=G(1)
        A1(1)=G(2)
        DO I=3,M+2
          B1(I-2)=G(I)
        END DO
```

```
            DO I=1,M-1
             U1(I+1)=U1(I)+A1(I)*DL(I)+B1(I)*(DL(I))**2
             A1(I+1)=A1(I)+2*B1(I)*DL(I)
            END DO

C           THE DERIVATIVE OF THE DOUBLET STRENGTH IS THE
C           SURFACE SPEED ALONG EACH PANEL.

            DO I=1,M
             VEL=A1(I)+B1(I)*DL(I)
             CP=1-VEL**2
             WRITE(8,*) CO(I,1),' ,',CP
            END DO

            WRITE(6,*) ' '
            WRITE(6,*) 'LIFT COEFFICIENT=', G(M+3)

            STOP
            END


C           PROGRAM No. 12: INFLUENCE COEFF. OF A RECTILINEAR SOURCE/DOUBLET PANEL
C           ----------------------------------------------------------------

C           THIS PROGRAM CALCULATES THE INFLUENCE OF A RECTILINEAR PANEL AT AN
C              ARBITRARY POINT. (PROGRAM BY LINDSEY BROWNE, 1988).
            DIMENSION X(5),Y(5),Z(5)

            PI=3.14159
            TWOPI=2.0*PI
            EPS=1.E-06
            PNDS=1.0
C           INPUT DOUBLET AND SOURCE STRENGTHS
            DUB=1.0/(4.0*PI)
            SIG=1.0/(4.0*PI)

            OPEN(2,FILE='INDIV',STATUS='UNKNOWN')

C           SQUARE/FLAT PANEL
C           INPUT COORDINATES
            X(1)=-.5
            X(2)=.5
            X(3)=.5
            X(4)=-.5
            Y(1)=-.5
            Y(2)=-.5
            Y(3)=.5
            Y(4)=.5
            DO 5 I=1,4
                Z(I)=0.0
5               CONTINUE
            X(5)=X(1)
            Y(5)=Y(1)
            Z(5)=Z(1)

C           MID-POINT AT (0,0,0)
            PXO=0.0
            PYO=0.0
            PZO=0.0
```

D.3 Time-Dependent Programs

```
 100      CONTINUE
         RJ31 = 0.0
         CJK1 = 0.0
         VXS = 0.0
         VYS = 0.0
         VZS = 0.0
         VXD = 0.0
         VYD = 0.0
         VZD = 0.0
         VX = 0.0
         VY = 0.0
         VZ = 0.0
C        INPUT POINT OF INTEREST
         WRITE(*,*) 'ENTER POINT OF INTEREST (X,Y,Z):'
         READ(*,*) PXI, PYI, PZI

                 PX=PXI-PXO
                 PY=PYI-PYO
                 PZ=PZI-PZO

                 RDIST=SQRT(PX*PX+PY*PY+PZ*PZ)

                 PNLX=.25*(X(1)+X(2)+X(3)+X(4))
                 PNLY=.25*(Y(1)+Y(2)+Y(3)+Y(4))
                 PNLZ=.25*(Z(1)+Z(2)+Z(3)+Z(4))
                 PNX=PX-PNLX
                 PNY=PY-PNLY
                 PNZ=PZ-PNLZ
                 PNS=SQRT(PNX*PNX+PNY*PNY+PNZ*PNZ)
                 D1X=X(3)-X(1)
                 D1Y=Y(3)-Y(1)
                 D1Z=Z(3)-Z(1)
                 D2X=X(4)-X(2)
                 D2Y=Y(4)-Y(2)
                 D2Z=Z(4)-Z(2)
                 CRX=D1Y*D2Z-D2Y*D1Z
                 CRY=D2X*D1Z-D1X*D2Z
                 CRZ=D1X*D2Y-D2X*D1Y
                 CRSQ=SQRT(CRX*CRX+CRY*CRY+CRZ*CRZ)
                 AREA=CRSQ/2.
                 CNX=CRX/CRSQ
                 CNY=CRY/CRSQ
                 CNZ=CRZ/CRSQ
                 PNN=CNX*PNX+CNY*PNY+CNZ*PNZ
                 TCMX=(X(3)+X(4))/2. - PNLX
                 TCMY=(Y(3)+Y(4))/2. - PNLY
                 TCMZ=(Z(3)+Z(4))/2. - PNLZ
                 TMS=SQRT(TCMX*TCMX+TCMY*TCMY+TCMZ*TCMZ)
                 CMX=((X(3)+X(4))/2. - PNLX)/TMS
                 CMY=((Y(3)+Y(4))/2. - PNLY)/TMS
                 CMZ=((Z(3)+Z(4))/2. - PNLZ)/TMS
                 CLX=CMY*CNZ-CNY*CMZ
                 CLY=CNX*CMZ-CMX*CNZ
                 CLZ=CMX*CNY-CNX*CMY

                 DO 20 J=1,4
                       K=J+1
                       AX=PX-X(J)
```

```
                        AY=PY-Y(J)
                        AZ=PZ-Z(J)
                        BX=PX-X(K)
                        BY=PY-Y(K)
                        BZ=PZ-Z(K)
                        SX=X(K)-X(J)
                        SY=Y(K)-Y(J)
                        SZ=Z(K)-Z(J)
                        A=SQRT(AX*AX + AY*AY + AZ*AZ)
                        B=SQRT(BX*BX + BY*BY + BZ*BZ)
                        S=SQRT(SX*SX + SY*SY + SZ*SZ)

C       SOURCE CONTRIBUTION
                        SM=SX*CMX+SY*CMY+SZ*CMZ
                        SL=SX*CLX+SY*CLY+SZ*CLZ
                        AM=AX*CMX+AY*CMY+AZ*CMZ
                        AL=AX*CLX+AY*CLY+AZ*CLZ
                        BM=BX*CMX+BY*CMY+BZ*CMZ
                        ALL=AM*SL-AL*SM
                        IF((A+B-S).GT.0.0.AND.S.GT.0.0)THEN
                            RJ3=ALOG((A+B+S)/(A+B-S))/S
                        ELSE
                            RJ3=0.0
                        ENDIF
                        PA=PNZ*PNZ*SL + ALL*AM
                        PB=PA - ALL*SM
                        RNUM=SM*PNZ*(B*PA - A*PB)
                        DNOM=PA*PB + PNZ*PNZ*A*B*SM*SM
                        IF(ABS(PNZ).LT.EPS)THEN
                          DE=0.0
                        ELSE
                          IF(RNUM.NE.0)THEN
                             DE=ATAN2(RNUM,DNOM)
                          ELSE
                             DE=0.0
                          ENDIF
                        ENDIF
                        RJ31 = RJ31 - SIG*ALL*RJ3
                            CJK1 = CJK1 - DUB*DE
                        VXS=VXS+SIG*(RJ3*(SM*CLX-SL*CMX)+DE*CNX)
                        VYS=VYS+SIG*(RJ3*(SM*CLY-SL*CMY)+DE*CNY)
                        VZS=VZS+SIG*(RJ3*(SM*CLZ-SL*CMZ)+DE*CNZ)
C       DOUBLET CONTRIBUTION
                        AVBX=AY*BZ - AZ*BY
                        AVBY=AZ*BX - AX*BZ
                        AVBZ=AX*BY - AY*BX
                        ADB=AX*BX + AY*BY + AZ*BZ
                        VMOD=(A+B)/(A*B*(A*B + ADB))
                        VXD=VXD + DUB*VMOD*AVBX
                        VYD=VYD + DUB*VMOD*AVBY
                        VZD=VZD + DUB*VMOD*AVBZ
20              CONTINUE

C       LIMITING CASES
                        DTT=TWOPI
                        IF(RDIST.GT.0.0)PNDS=PNZ**2/RDIST
                        IF(PNDS.LT.EPS.AND.RDIST.GT.EPS)THEN
                            DTT=PNZ*AREA/SQRT(RDIST)/RDIST
                        ENDIF
```

```
              IF(ABS(DTT).LT.ABS(CJK1))CJK1=DTT
              IF(RDIST.LT.EPS*EPS)CJK1=-TWOPI
C     TOTAL
                    CJK = CJK1
              BJK = RJ31 - PNZ*CJK1
              VX=VXD+VXS
              VY=VYD+VYS
              VZ=VZD+VZS

           TVS=SQRT(VXS*VXS+VYS*VYS+VZS*VZS)
           TVD=SQRT(VXD*VXD+VYD*VYD+VZD*VZD)
           TV=SQRT(VX*VX+VY*VY+VZ*VZ)

       WRITE(*,*)'AREA OF PANEL=',AREA
       WRITE(*,*)'SOURCE (POTENTIAL)=',BJK
      WRITE(*,*)'SOURCE (VELOCITY):'
      WRITE(*,*)'VXS=',VXS
      WRITE(*,*)'VYS=',VYS
      WRITE(*,*)'VZS=',VZS
      WRITE(*,*)'TVS=',TVS
      WRITE(*,*)'DOUBLET (POTENTIAL):', CJK
      WRITE(*,*)'DOUBLET (VELOCITY):'
      WRITE(*,*)'VXD=',VXD
      WRITE(*,*)'VYD=',VYD
      WRITE(*,*)'VZD=',VZD
      WRITE(*,*)'TVD=',TVD
      WRITE(*,*)'TOTAL VELOCITY:'
      WRITE(*,*)'VX=',VX
      WRITE(*,*)'VY=',VY
      WRITE(*,*)'VZ=',VZ
      WRITE(*,*)'TV=',TV
        WRITE(*,*)'DO YOU WANT ANOTHER TRY? 1:YES, 2:NO'
        READ(*,*) NTRY
        IF(NTRY.EQ.1)GO TO 100
      STOP
      END


C     PROGRAM No. 13: RECTANGULAR LIFTING SURFACE (VLM)
C     -------------------------------------------------
C  3D-VLM CODE FOR SIMPLE WING PLANFORMS WITH GROUND EFFECT (BY JOE KATZ, 1974).
      DIMENSION QF(6,14,3),QC(4,13,3),DS(4,13,4)
      DIMENSION GAMA(4,13),DL(4,13),DD(4,13),DP(4,13)
      DIMENSION A(52,52),GAMA1(52),DW(52),IP(52)
      DIMENSION A1(5,13),DLY(13),GAMA1J(5),X(4)
      COMMON/NO1/ DS,X,B,C,S,AR,SN1,CS1
      COMMON/NO2/ IB,JB,CH,SIGN
      COMMON/NO3/ A1
      COMMON/NO4/ QF,QC,DXW
C
C     ==========
C     INPUT DATA
C     ==========
C
      IB=4
      JB=13
      X(1)=0.
      X(2)=0.
      X(3)=4.
```

```
              X(4)=4.
              B=13.
              VT=1.0
              ALPHA1=5.0
              CH=1000.
C       X(1) TO X(4) ARE X-COORDINATES OF THE WING'S FOUR CORNERPOINTS.
C       B - WING SPAN, VT - FREE STREAM SPEED, B - WING SPAN,
C       CH - HEIGHT ABOVE GROUND
C
C       CONSTANTS
              DXW=100.0*B
              DO 1 I=1,IB
              DO 1 J=1,JB
C       GAMA(I,J)=1.0 IS REQUIRED FOR INFLUENCE MATRIX CALCULATIONS.
 1            GAMA(I,J)=1.0
              RO=1.
              PAY=3.141592654
              ALPHA=ALPHA1*PAY/180.
              SN1=SIN(ALPHA)
              CS1=COS(ALPHA)
              IB1=IB+1
              IB2=IB+2
              JB1=JB+1
C
C       ============
C       WING GEOMETRY
C       ============
C
              CALL GRID
              WRITE(6,101)
              WRITE(6,102) ALPHA1,B,C,S,AR,VT,IB,JB,CH
C
C       ========================
C       AERODYNAMIC CALCULATIONS
C       ========================
C
C       INFLUENCE COEFFICIENTS CALCULATION
C
              K=0
              DO 14 I=1,IB
              DO 14 J=1,JB
              SIGN=0.0
              K=K+1
              CALL WING(QC(I,J,1),QC(I,J,2),QC(I,J,3),GAMA,U,V,W,1.0,I,J)
              L=0
              DO 10 I1=1,IB
              DO 10 J1=1,JB
              L=L+1
C       A(K,L) - IS THE NORMAL VELOCITY COMPONENT DUE TO A UNIT VORTEX
C                LATTICE.
 10           A(K,L)=A1(I1,J1)
C       ADD INFLUENCE OF WING'S OTHER HALF
              CALL WING(QC(I,J,1),-QC(I,J,2),QC(I,J,3),GAMA,U,V,W,1.0,I,J)
              L=0
              DO 11 I1=1,IB
              DO 11 J1=1,JB
              L=L+1
 11           A(K,L)=A(K,L)+A1(I1,J1)
              IF(CH.GT.100.0) GOTO 12
```

D.3 Time-Dependent Programs

```
C         ADD INFLUENCE OF MIRROR IMAGE (DUE TO GROUND)
          SIGN=10.0
          CALL WING(QC(I,J,1),QC(I,J,2),-QC(I,J,3),GAMA,U,V,W,1.0,I,J)
          L=0
          DO 8 I1=1,IB
          DO 8 J1=1,JB
          L=L+1
    8     A(K,L)=A(K,L)+A1(I1,J1)
C         ADD MIRROR IMAGE INFLUENCE OF WING'S OTHER HALF.
          CALL WING(QC(I,J,1),-QC(I,J,2),-QC(I,J,3),GAMA,U,V,W,1.0,I,J)
          L=0
          DO 9 I1=1,IB
          DO 9 J1=1,JB
          L=L+1
    9     A(K,L)=A(K,L)+A1(I1,J1)
          SIGN=0.0
   12     CONTINUE
C
   13     CONTINUE
C
C         CALCULATE WING GEOMETRICAL DOWNWASH
C
          UINF=VT
          VINF=0.0
          WINF=0.0
C         THIS IS THE GENERAL FORMULATION FOR RIGHT HAND SIDE.
          DW(K)=-(UINF*DS(I,J,1)+VINF*DS(I,J,2)+WINF*DS(I,J,3))
   14     CONTINUE
C
C         SOLUTION OF THE PROBLEM: DW(I)=A(I,J)*GAMA(I)
C
          K1=IB*JB
          DO 15 K=1,K1
   15     GAMA1(K)=DW(K)
          CALL DECOMP(K1,52,A,IP)
   16     CONTINUE
          CALL SOLVER(K1,52,A,GAMA1,IP)
C         HERE           *     THE SAME ARRAY SIZE IS REQUIRED,
C                              AS SPECIFIED IN THE BEGINNING OF THE CODE
C
C         WING VORTEX LATTICE LISTING
C
          K=0
          DO 17 I=1,IB
          DO 17 J=1,JB
          K=K+1
   17     GAMA(I,J)=GAMA1(K)
C
C         ==================
C         FORCES CALCULATION
C         ==================
C
          FL=0.
          FD=0.
          FM=0.
          QUE=0.5*RO*VT*VT
          DO 20 J=1,JB
          DLY(J)=0.
```

```fortran
            DO 20 I=1,IB
            IF(I.EQ.1) GAMAIJ=GAMA(I,J)
            IF(I.GT.1) GAMAIJ=GAMA(I,J)-GAMA(I-1,J)
            DYM=QF(I,J+1,2)-QF(I,J,2)
            DL(I,J)=RO*VT*GAMAIJ*DYM
C           INDUCED DRAG CALCULATION
            CALL WING(QC(I,J,1),QC(I,J,2),QC(I,J,3),GAMA,U1,V1,W1,0.0,I,J)
            CALL WING(QC(I,J,1),-QC(I,J,2),QC(I,J,3),GAMA,U2,V2,W2,0.0,I,J)
            IF(CH.GT.100.0) GOTO 194
            CALL WING(QC(I,J,1),QC(I,J,2),-QC(I,J,3),GAMA,U3,V3,W3,0.0,I,J)
            CALL WING(QC(I,J,1),-QC(I,J,2),-QC(I,J,3),GAMA,U4,V4,W4,0.0,I,J)
            GOTO 195
 194        W3=0.
            W4=0.
 195        WIND=W1+W2-W3-W4
C           ADD INFLUENCE OF MIRROR IMAGE (GROUND).
            ALFI=-WIND/VT
            DD(I,J)=RO*DYM*VT*GAMAIJ*ALFI
C
            DP(I,J)=DL(I,J)/DS(I,J,4)/QUE
            DLY(J)=DLY(J)+DL(I,J)
            FL=FL+DL(I,J)
            FD=FD+DD(I,J)
            FM=FM+DL(I,J)*(QF(I,J,1)-X(1))
 20         CONTINUE
            CL=FL/(QUE*S)
            CD=FD/(QUE*S)
            CM=FM/(QUE*S*C)
C
C           OUTPUT
C
            WRITE(6,104) CL,FL,CM,CD
            WRITE(6,110)
            DO 21 J=1,JB
            DO 211 I=2,IB
 211        GAMA1J(I)=GAMA(I,J)-GAMA(I-1,J)
            DLYJ=DLY(J)/B*JB
 21         WRITE(6,103) J,DLYJ,DP(1,J),DP(2,J),DP(3,J),DP(4,J),GAMA(1,J),
           1GAMA1J(2),GAMA1J(3),GAMA1J(4)
C
C           END OF PROGRAM
 100        CONTINUE
C
C           FORMATS
C
 101        FORMAT(1H ,/,20X,'WING LIFT DISTRIBUTION CALCULATION (WITH GROUND
           1 EFFECT)',/,20X,56('-'))
 102        FORMAT(1H ,/,10X,'ALFA:',F10.2,8X,'B    :',
           1F10.2,8X,'C    :',F13.2,/,10X,
           2'S    :',F10.2,8X,'AR   :',F10.2,8X,'V(INF) :',F10.2,/,10X,
           3'IB   :',I10,8X,'JB   :',I10,8X,'L.E. HEIGHT:', F6.2,/)
 103        FORMAT(1H ,I3,' I ',F9.3,' II ',4(F9.3,' I '),' I ',4(F9.3,' I '))
 104        FORMAT(/,1H ,'CL=',F10.4,2X,'L=',F10.4,4X,'CM=',F10.4,3X,
           1'CD=',F10.4)
 110        FORMAT(1H ,/,5X,'I      DL',4X,'II',22X,'DCP',22X,'I I',25X,
           1'GAMA',/,118('='),/,5X,'I',15X,'I= 1',11X,'2',11X,'3',11X,
           2'4',5X,'I  I',5X,'1',11X,'2',11X,'3',11X,'4',/,118('='))
 112        FORMAT(1H ,'QF(I=',I2,',J,X.Y.Z)= ',15(F6.1))
 113        FORMAT(1H ,110('='))
```

```
C
          STOP
          END
C
          SUBROUTINE GRID
          DIMENSION QF(6,14,3),QC(4,13,3),DS(4,13,4),X(4)
          COMMON/NO1/ DS,X,B,C,S,AR,SN1,CS1
          COMMON/NO2/ IB,JB,CH,SIGN
          COMMON/NO4/ QF,QC,DXW
C
          PAY=3.141592654
C         X(1) - IS ROOT L.E., X(2) TIP L.E., X(3) TIP T.E., AND X(4) IS ROOT T.E.
C         IB: NO. OF CHORDWISE BOXES, JB: NO. OF SPANWISE BOXES
          IB1=IB+1
          IB2=IB+2
          JB1=JB+1
C
C         WING FIXED VORTICES LOCATION    ( QF(I,J,(X,Y,Z))...)
C
          DY=B/JB
          DO 3 J=1,JB1
          YLE=DY*(J-1)
          XLE=X(1)+(X(2)-X(1))*YLE/B
          XTE=X(4)+(X(3)-X(4))*YLE/B
C         XLE AND XTE ARE L.E. AND T.E. X-COORDINATES
          DX=(XTE-XLE)/IB
          DO 1 I=1,IB1
          QF(I,J,1)=(XLE+DX*(I-0.75))*CS1
          QF(I,J,2)=YLE
          QF(I,J,3)=-QF(I,J,1)*SN1+CH
 1        CONTINUE
C         WAKE FAR FIELD POINTS
          QF(IB2,J,1)=XTE+DXW
          QF(IB2,J,2)=QF(IB1,J,2)
 3        QF(IB2,J,3)=QF(IB1,J,3)
C
C         WING COLLOCATION POINTS
C
          DO 4 J=1,JB
          DO 4 I=1,IB
          QC(I,J,1)=(QF(I,J,1)+QF(I,J+1,1)+QF(I+1,J+1,1)+QF(I+1,J,1))/4
          QC(I,J,2)=(QF(I,J,2)+QF(I,J+1,2)+QF(I+1,J+1,2)+QF(I+1,J,2))/4
          QC(I,J,3)=(QF(I,J,3)+QF(I,J+1,3)+QF(I+1,J+1,3)+QF(I+1,J,3))/4
C
C         COMPUTATION OF NORMAL VECTORS
C
          CALL PANEL(QF(I,J,1),QF(I,J,2),QF(I,J,3),QF(I+1,J,1),QF(I+1,J,2),
         1QF(I+1,J,3),QF(I,J+1,1),QF(I,J+1,2),QF(I,J+1,3),QF(I+1,J+1,1),
         2QF(I+1,J+1,2),QF(I+1,J+1,3),DS(I,J,1),DS(I,J,2),DS(I,J,3),
         3DS(I,J,4))
 4        CONTINUE
C
C         B -IS SEMI SPAN, C -AV. CHORD, S - AREA
          S=0.5*(X(3)-X(2)+X(4)-X(1))*B
          C=S/B
          AR=2.*B*B/S
C
          RETURN
          END
```

```
C
        SUBROUTINE PANEL(X1,Y1,Z1,X2,Y2,Z2,X3,Y3,Z3,X4,Y4,Z4,C1,C2,C3,S)
C       CALCULATION OF PANEL AREA AND NORMAL VECTOR.
        A1=X2-X3
        A2=Y2-Y3
        A3=Z2-Z3
        B1=X4-X1
        B2=Y4-Y1
        B3=Z4-Z1
C       NORMAL VECTOR
        X=A2*B3-A3*B2
        Y=B1*A3-A1*B3
        Z=A1*B2-A2*B1
        A=SQRT(X**2+Y**2+Z**2)
        C1=X/A
        C2=Y/A
        C3=Z/A
C       CALCULATION OF PANEL AREA
        E1=X3-X1
        E2=Y3-Y1
        E3=Z3-Z1
        F1=X2-X1
        F2=Y2-Y1
        F3=Z2-Z1
C       NORMAL AREAS (F*B+B*E)
        S11=F2*B3-F3*B2
        S12=B1*F3-F1*B3
        S13=F1*B2-F2*B1
        S21=B2*E3-B3*E2
        S22=E1*B3-B1*E3
        S23=B1*E2-B2*E1
        S=0.5*(SQRT(S11**2+S12**2+S13**2)+SQRT(S21**2+S22**2+S23**2))
        RETURN
        END
C
        SUBROUTINE VORTEX(X,Y,Z,X1,Y1,Z1,X2,Y2,Z2,GAMA,U,V,W)
C       SUBROUTINE VORTEX CALCULATES THE INDUCED VELOCITY (U,V,W) AT A POI
C       (X,Y,Z) DUE TO A VORTEX ELEMENT VITH STRENGTH GAMA PER UNIT LENGTH
C       POINTING TO THE DIRECTION (X2,Y2,Z2)-(X1,Y1,Z1).
        PAY=3.141592654
        RCUT=1.0E-10
C       CALCULATION OF R1 X R2
        R1R2X=(Y-Y1)*(Z-Z2)-(Z-Z1)*(Y-Y2)
        R1R2Y=-((X-X1)*(Z-Z2)-(Z-Z1)*(X-X2))
        R1R2Z=(X-X1)*(Y-Y2)-(Y-Y1)*(X-X2)
C       CALCULATION OF (R1 X R2 )**2
        SQUARE=R1R2X*R1R2X+R1R2Y*R1R2Y+R1R2Z*R1R2Z
C       CALCULATION OF R0(R1/R(R1)-R2/R(R2))
        R1=SQRT((X-X1)*(X-X1)+(Y-Y1)*(Y-Y1)+(Z-Z1)*(Z-Z1))
        R2=SQRT((X-X2)*(X-X2)+(Y-Y2)*(Y-Y2)+(Z-Z2)*(Z-Z2))
        IF((R1.LT.RCUT).OR.(R2.LT.RCUT).OR.(SQUARE.LT.RCUT)) GOTO 1     GROUND
        R0R1=(X2-X1)*(X-X1)+(Y2-Y1)*(Y-Y1)+(Z2-Z1)*(Z-Z1)
        R0R2=(X2-X1)*(X-X2)+(Y2-Y1)*(Y-Y2)+(Z2-Z1)*(Z-Z2)
        COEF=GAMA/(4.0*PAY*SQUARE)*(R0R1/R1-R0R2/R2)
        U=R1R2X*COEF
        V=R1R2Y*COEF
        W=R1R2Z*COEF
        GOTO 2
C       WHEN POINT (X,Y,Z) LIES ON VORTEX ELEMENT; ITS INDUCED VELOCITY IS
1       U=0.
```

```
              V=0.
              W=0.
    2         CONTINUE
              RETURN
              END
C
              SUBROUTINE WING(X,Y,Z,GAMA,U,V,W,ONOFF,I1,J1)
              DIMENSION GAMA(4,13),QF(6,14,3),A1(5,13)
              DIMENSION DS(4,13,4)
              COMMON/NO1/ DS
              COMMON/NO2/ IB,JB,CH,SIGN
              COMMON/NO3/ A1
              COMMON/NO4/ QF
C
C             CALCULATES INDUCED VELOCITY AT A POINT (X,Y,Z), DUE TO VORTICITY
C             DISTRIBUTION GAMA(I,J), OF SEMI-CONFIGURATION - IN A WING FIXED
C             COORDINATE SYSTEM.
              U=0
              V=0
              W=0
              IB1=IB+1
              DO 1 I=1,IB1
              DO 1 J=1,JB
C                      I3 IS WAKE VORTEX COUNTER
              I3=I
              IF(I.EQ.IB1) I3=IB
              VORTIC=GAMA(I3,J)
              IF(ONOFF.LT.0.1) GOTO 2
              CALL VORTEX(X,Y,Z,QF(I,J,1),QF(I,J,2),QF(I,J,3),QF(I,J+1,1),QF(I,J
             1+1,2),QF(I,J+1,3),VORTIC,U1,V1,W1)
              CALL VORTEX(X,Y,Z,QF(I+1,J+1,1),QF(I+1,J+1,2),QF(I+1,J+1,3),
             3QF(I+1,J,1),QF(I+1,J,2),QF(I+1,J,3),VORTIC,U3,V3,W3)
    2         CALL VORTEX(X,Y,Z,QF(I,J+1,1),QF(I,J+1,2),QF(I,J+1,3),QF(I+1,J+1,1
             2),QF(I+1,J+1,2),QF(I+1,J+1,3),VORTIC,U2,V2,W2)
              CALL VORTEX(X,Y,Z,QF(I+1,J,1),QF(I+1,J,2),QF(I+1,J,3),QF(I,J,1),
             4QF(I,J,2),QF(I,J,3),VORTIC,U4,V4,W4)
C
              U0=U2+U4+(U1+U3)*ONOFF
              V0=V2+V4+(V1+V3)*ONOFF
              W0=W2+W4+(W1+W3)*ONOFF
              A1(I,J)=U0*DS(I1,J1,1)+V0*DS(I1,J1,2)+W0*DS(I1,J1,3)
              IF(SIGN.GE.1.0)
             *              A1(I,J)=U0*DS(I1,J1,1)+V0*DS(I1,J1,2)-W0*DS(I1,J1,3)
              IF(I.EQ.IB1) A1(IB,J)=A1(IB,J)+A1(IB1,J)
              U=U+U0
              V=V+V0
              W=W+W0
C
    1         CONTINUE
              RETURN
              END
C
              SUBROUTINE DECOMP(N,NDIM,A,IP)
              REAL A(NDIM,NDIM),T
              INTEGER IP(NDIM)
C             MATRIX TRIANGULARIZATION BY GAUSSIAN ELIMINATION.
C             N = ORDER OF MATRIX. NDIM = DECLARED DIMENSION OF ARRAY A.
C             A = MATRIX TO BE TRIANGULARIZED.
C             IP(K) , K .LT. N = INDEX OF K-TH PIVOT ROW.
```

```fortran
C
            IP(N) = 1
            DO 6 K = 1, N
            IF(K.EQ.N) GOTO 5
            KP1 = K + 1
            M = K
            DO 1 I = KP1, N
            IF( ABS(A(I,K)).GT.ABS(A(M,K))) M=I
1           CONTINUE
            IP(K) = M
            IF(M.NE.K) IP(N) = -IP(N)
            T = A(M,K)
            A(M,K) = A(K,K)
            A(K,K) = T
            IF(T.EQ.0.E0) GO TO 5
            DO 2 I = KP1, N
2           A(I,K) = -A(I,K)/T
            DO 4 J = KP1, N
            T = A(M,J)
            A(M,J) = A(K,J)
            A(K,J) = T
            IF(T .EQ. 0.E0) GO TO 4
            DO 3 I = KP1, N
3           A(I,J) = A(I,J) + A(I,K)*T
4           CONTINUE
5           IF(A(K,K) .EQ. 0.E0) IP(N) = 0
6           CONTINUE
            RETURN
            END
C
            SUBROUTINE SOLVER(N,NDIM,A,B,IP)
            REAL A(NDIM,NDIM), B(NDIM), T
            INTEGER IP(NDIM)
C           SOLUTION OF LINEAR SYSTEM, A*X = B.
C           N = ORDER OF MATRIX.
C           NDIM = DECLARED DIMENSION OF THE ARRAY A.
C           B = RIGHT HAND SIDE VECTOR.
C           IP = PIVOT VECTOR OBTAINED FROM SUBROUTINE DECOMP.
C           B = SOLUTION VECTOR, X.
C
            IF(N.EQ.1) GOTO 9
            NM1 = N - 1
            DO 7 K = 1, NM1
            KP1 = K + 1
            M = IP(K)
            T = B(M)
            B(M) = B(K)
            B(K) = T
            DO 7 I = KP1, N
7           B(I) = B(I) + A(I,K)*T
            DO 8 KB = 1, NM1
            KM1 = N - KB
            K = KM1 + 1
            B(K) = B(K)/A(K,K)
            T = -B(K)
            DO 8 I = 1, KM1
8           B(I) = B(I) + A(I,K)*T
9           B(1) = B(1)/A(1,1)
            RETURN
            END
```

D.3 Time-Dependent Programs

```
C           PROGRAM No. 14: 3D PANEL METHOD, DIRICHLET B.C. (SOURCE + DOUBLET)
C           ----------------------------------------------------------------

C     3D-PANEL CODE FOR SIMPLE WING PLANFORMS. NO TIP PATCH!!!
      DIMENSION QF(22,14,3),QC(20,13,3),DS(20,13,10),SIGMA(20,13)
      DIMENSION DUB(20,13),DL(20,13),DD(20,13),CP(20,13),DDUBJ(13)
      DIMENSION CR(21,13,12)
      DIMENSION A(260,260),DUB1(260),RHS(260),IP(260)
      COMMON/NO1/ DS,CROOT,CTIP,XTIP,ZTIP,B,S,AR,IB,JB,PAY
      COMMON/NO2/ QF,QC,CR,SIGMA,DXW,UT,WT
C
C     ==========
C     INPUT DATA
C     ==========
C
      ALPHA1=5.0
      CROOT=1.0
      CTIP=1.0
      XTIP=0.0
      ZTIP=0.0
      B=10.
      VT=1.0
      JB=3
C     CROOT, CTIP - ROOT AND TIP CHORD, XTIP - AFT SWEEP OF TIP
C     B - WING SPAN, VT - FREE STREAM SPEED, IB,JP - CHORD, SPANWISE COUNTERS
C         SYMMETRY IS ASSUMED (ONLY THE SEMISPAN IS MODELED)
C     CONSTANTS
      DXW=100.0*B
      RO=1.0
      PAY=3.141592654
      UT=VT*COS(ALPHA1*PAY/180.0)
      WT=VT*SIN(ALPHA1*PAY/180.0)
C
C     =============
C     WING GEOMETRY
C     =============
C
      CALL GRID
      IB1=IB+1
      IB2=IB+2
      JB1=JB+1
      WRITE(6,101)
      WRITE(6,102) ALPHA1,B,CROOT,S,AR,VT,IB,JB
      WRITE(6,111)
      WRITE(6,113)
      DO 8 J=1,JB1
      DO 8 I=1,IB2
   8  WRITE(6,112) I,J,QF(I,J,1),QF(I,J,2),QF(I,J,3)
 111  FORMAT(1H ,' I ',' J   ','QF(,I,J,1)  QF(I,J,2)  QF(I,J,3)')
 112  FORMAT(1H ,I3,3X,I3,3F12.4)
 113  FORMAT(1H ,46('='))
C
C     =======================
C     AERODYNAMIC CALCULATIONS
C     =======================
C
C     INFLUENCE COEFFICIENTS CALCULATION
C
C              COLLOCATION POINT COUNTER
```

```
              K=0
              DO 14 I=1,IB
              DO 14 J=1,JB
              K=K+1
              L=0
              RH=0
C                    INFLUENCING PANEL COUNTER
              DO 10 I1=1,IB
              DO 10 J1=1,JB
              L=L+1
              IF(I1.EQ.1) THEN
C       CALCULATE WAKE CONTRIBUTION
C       FIRST CONVERT COLLOCATION POINT TO PANEL COORDINATES,
C       AND THEN CALCULATE INFLUENCE COEFFICIENTS
              CALL CONVERT(QC(IB1,J1,1),QC(IB1,J1,2),QC(IB1,J1,3),
             1 QC(I,J,1),QC(I,J,2),QC(I,J,3),
             2 DS(IB1,J1,1),DS(IB1,J1,2),DS(IB1,J1,3),
             3 DS(IB1,J1,4),DS(IB1,J1,5),DS(IB1,J1,6),
             4 DS(IB1,J1,7),DS(IB1,J1,8),DS(IB1,J1,9),
             5 XC,YC,ZC )
              CALL INFLUENCE(WDUB,DSIG,XC,YC,ZC,
             1 CR(IB1,J1,1),CR(IB1,J1,2),CR(IB1,J1,3),
             2 CR(IB1,J1,4),CR(IB1,J1,5),CR(IB1,J1,6),
             3 CR(IB1,J1,7),CR(IB1,J1,8),CR(IB1,J1,9),
             4 CR(IB1,J1,10),CR(IB1,J1,11),CR(IB1,J1,12) )
C       ADD WING'S IMAGE (SYMMETRY IS ASSUMED)
              CALL CONVERT(QC(IB1,J1,1),QC(IB1,J1,2),QC(IB1,J1,3),
             1 QC(I,J,1),-QC(I,J,2),QC(I,J,3),
             2 DS(IB1,J1,1),DS(IB1,J1,2),DS(IB1,J1,3),
             3 DS(IB1,J1,4),DS(IB1,J1,5),DS(IB1,J1,6),
             4 DS(IB1,J1,7),DS(IB1,J1,8),DS(IB1,J1,9),
             5 XC,YC,ZC )
              CALL INFLUENCE(WDUB1,DSIG,XC,YC,ZC,
             1 CR(IB1,J1,1),CR(IB1,J1,2),CR(IB1,J1,3),
             2 CR(IB1,J1,4),CR(IB1,J1,5),CR(IB1,J1,6),
             3 CR(IB1,J1,7),CR(IB1,J1,8),CR(IB1,J1,9),
             4 CR(IB1,J1,10),CR(IB1,J1,11),CR(IB1,J1,12) )
              DDUBJ(J1)=WDUB+WDUB1
              DMU2=DDUBJ(J1)
              ELSE
              DMU2=0.0
              ENDIF
              IF(I1.EQ.IB) DMU2=-DDUBJ(J1)
C       END OF WAKE INFLUENCE CALCULATION
C       CONVERT COLLOCATION POINT TO PANEL COORDINATES
              CALL CONVERT(QC(I1,J1,1),QC(I1,J1,2),QC(I1,J1,3),
             1 QC(I,J,1),QC(I,J,2),QC(I,J,3),
             2 DS(I1,J1,1),DS(I1,J1,2),DS(I1,J1,3),
             3 DS(I1,J1,4),DS(I1,J1,5),DS(I1,J1,6),
             4 DS(I1,J1,7),DS(I1,J1,8),DS(I1,J1,9),
             5 XC,YC,ZC )
              CALL INFLUENCE(DMU,DSIG,XC,YC,ZC,
             1 CR(I1,J1,1),CR(I1,J1,2),CR(I1,J1,3),
             2 CR(I1,J1,4),CR(I1,J1,5),CR(I1,J1,6),
             3 CR(I1,J1,7),CR(I1,J1,8),CR(I1,J1,9),
             4 CR(I1,J1,10),CR(I1,J1,11),CR(I1,J1,12) )
                IF((I1.EQ.I).AND.(J1.EQ.J)) DMU=-0.5
C       A PANEL INFLUENCE ON ITSELF IS DMU=1/2
C
```

D.3 Time-Dependent Programs

```
C         ADD INFLUENCE OF WING'S IMAGE (OTHER HALF)

          CALL CONVERT(QC(I1,J1,1),QC(I1,J1,2),QC(I1,J1,3),
         1 QC(I,J,1),-QC(I,J,2),QC(I,J,3),
         2 DS(I1,J1,1),DS(I1,J1,2),DS(I1,J1,3),
         3 DS(I1,J1,4),DS(I1,J1,5),DS(I1,J1,6),
         4 DS(I1,J1,7),DS(I1,J1,8),DS(I1,J1,9),
         5 XC,YC,ZC )
          CALL INFLUENCE(DMU1,DSIG1,XC,YC,ZC,
         1 CR(I1,J1,1),CR(I1,J1,2),CR(I1,J1,3),
         2 CR(I1,J1,4),CR(I1,J1,5),CR(I1,J1,6),
         3 CR(I1,J1,7),CR(I1,J1,8),CR(I1,J1,9),
         4 CR(I1,J1,10),CR(I1,J1,11),CR(I1,J1,12) )
C         A(K,L) - IS THE INFLUENCE MATRIX COEFFICIENT
C
          A(K,L)=DMU+DMU1-DMU2
          RH=RH+(DSIG+DSIG1)*SIGMA(I1,J1)
  10      CONTINUE
C
C         CALCULATE RHS
C
          RHS(K)=RH
  14      CONTINUE
C
C         SOLUTION OF THE PROBLEM: A(K,L)*DUB(K)=RHS(K)
C
          K1=IB*JB
          DO 15 K=1,K1
  15      DUB1(K)=RHS(K)
          CALL DECOMP(K1,260,A,IP)
  16      CONTINUE
          CALL SOLVER(K1,260,A,DUB1,IP)
C         HERE          *     THE SAME ARRAY SIZE IS REQUIRED,
C                             AS SPECIFIED IN THE BEGINNING OF THE CODE
C
C         WING DOUBLET LATTICE LISTING
C
          K=0
          DO 17 I=1,IB
          DO 17 J=1,JB
          K=K+1
  17      DUB(I,J)=DUB1(K)
          DO 18 J=1,JB
  18      DUB(IB1,J)=DUB(1,J)-DUB(IB,J)
C
C         ==================
C         FORCES CALCULATION
C         ==================
C
          FL=0.
          FD=0.
          FM=0.
          QUE=0.5*RO*VT*VT
          DO 20 J=1,JB
          DO 20 I=1,IB
C
          I1=I-1
          I2=I+1
```

```
            J1=J-1
            J2=J+1
            IF(I.EQ.1) I1=1
            IF(I.EQ.IB) I2=IB
            IF(J.EQ.1) J1=1
            IF(J.EQ.JB) J2=JB
C           CHORDWISE VELOCITY
            XF=0.5*(QF(I+1,J,1)+QF(I+1,J+1,1))
            YF=0.5*(QF(I+1,J,2)+QF(I+1,J+1,2))
            ZF=0.5*(QF(I+1,J,3)+QF(I+1,J+1,3))
            XR=0.5*(QF(I,J,1)+QF(I,J+1,1))
            YR=0.5*(QF(I,J,2)+QF(I,J+1,2))
            ZR=0.5*(QF(I,J,3)+QF(I,J+1,3))
            DX2=QC(I2,J,1)-XF
            DY2=QC(I2,J,2)-YF
            DZ2=QC(I2,J,3)-ZF
            DX3=QC(I1,J,1)-XR
            DY3=QC(I1,J,2)-YR
            DZ3=QC(I1,J,3)-ZR
            DL1=SQRT((XF-XR)**2+(YF-YR)**2+(ZF-ZR)**2)
            DL2=SQRT(DX2**2+DY2**2+DZ2**2)
            DL3=SQRT(DX3**2+DY3**2+DZ3**2)
            DLL=DL1+DL2+DL3
            IF(I.EQ.1) DLL=DL1/2.0+DL2
            IF(I.EQ.IB) DLL=DL1/2.0+DL3
            QL=-(DUB(I2,J)-DUB(I1,J))/DLL
C           SPANWISE VELOCITY
            DX=QC(I,J2,1)-QC(I,J1,1)
            DY=QC(I,J2,2)-QC(I,J1,2)
            DZ=QC(I,J2,3)-QC(I,J1,3)
            DR=SQRT(DX**2+DY**2+DZ**2)
            QM=-(DUB(I,J2)-DUB(I,J1))/DR
C           FIRST ORDER CORRECTION FOR PANEL SWEEP
            QL=QL+QM*(DX**2+DZ**2)/DR
            QM=QM*(DY**2+DZ**2)/DR
            QINF=UT*DS(I,J,9)-WT*DS(I,J,7)
            CP(I,J)=1.0-((QINF+QL)**2+QM**2)/(VT**2)
            DL(I,J)=-CP(I,J)*DS(I,J,10)*DS(I,J,9)
            DD(I,J)=CP(I,J)*DS(I,J,10)*DS(I,J,7)
            FL=FL+DL(I,J)
            FD=FD+DD(I,J)
            FM=FM+DL(I,J)*QC(I,J,1)
   20       CONTINUE
            CL=FL/(QUE*S)
            CD=FD/(QUE*S)
            CM=FM/(QUE*S*CROOT)
C
C           OUTPUT
C
            WRITE(6,110)
            DO 21 J=1,JB
            DO 21 I=1,IB1
   21       WRITE(6,105)I,J,QC(I,J,1),CP(I,J),DL(I,J),DD(I,J),DUB(I,J),
          1 SIGMA(I,J)
            WRITE(6,104) CL,FL,CM,CD
C
C           END OF PROGRAM
  100       CONTINUE
```

```
C
C       FORMATS
C
 101    FORMAT(1H ,/,20X,'INTERNAL POTENTIAL BASED PANEL CODE',
       1 /,20X,36('-'))
 102    FORMAT(1H ,/,10X,'ALFA:',F10.2,8X,'B    :',
       1F10.2,8X,'C    :',F13.2,/,10X,
       2'S    :',F10.2,8X,'AR   :',F10.2,8X,'V(INF) :',F10.2,/,10X,
       3'IB   :',I10,8X,'JB   :',I10,8X,/)
 103    FORMAT(1H ,I3,' I ',F9.3,' II ',4(F9.3,' I '),' I ',4(F9.3,' I '))
 104    FORMAT(/,1H ,'CL=',F10.4,2X,'L=',F10.4,4X,'CM=',F10.4,3X,
       1'CD=',F10.4)
 105    FORMAT(2I4,6F10.4)
 110    FORMAT(/,1H ,2X,'I   J',7X,'X',8X,'CP',8X,'DL',8X,'DD',7X,
       1'DUB',6X,'SIGMA',/,68('='))
C
        STOP
        END
C
        SUBROUTINE GRID
        DIMENSION QF(22,14,3),QC(20,13,3),DS(20,13,10),SIGMA(20,13)
        DIMENSION CR(21,13,12)
        COMMON/NO1/ DS,CROOT,CTIP,XTIP,ZTIP,B,S,AR,IB,JB,PAY
        COMMON/NO2/ QF,QC,CR,SIGMA,DXW,UT,WT
C
        WRITE(6,9)
   9    FORMAT( 1X,'AIRFOIL COORDINATES',/,1X,19('='),/,8X,'X       Z')
        READ(5,11)  IB1
        DO 10 I=1,  IB1
        READ(5,12)  QF(I,1,1),QF(I,1,3)
  10    WRITE(6,12) QF(I,1,1),QF(I,1,3)
  11    FORMAT(I3)
  12    FORMAT(3F10.4)
C       IB: NO. OF CHORDWISE PANELS, JB: NO. OF SPANWISE PANELS
        IB=IB1-1
        IB2=IB1+1
        JB1=JB+1
C
C       CALCULATE PANEL CORNERPOINTS;   QF(I,J,(X,Y,Z))
C
        DO 3 J=1,JB1
        Y=B/2.0/JB*(J-1)
        DXLE=XTIP*2.0*Y/B
        DZLE=ZTIP*2.0*Y/B
        CHORD=CROOT-(CROOT-CTIP)*2.0*Y/B
C       B - FULL SPAN, DXLE - LOCAL SWEEP
        DO 1 I=1,IB1
        QF(I,J,1)=QF(I,1,1)*CHORD+DXLE
        QF(I,J,2)=Y
        QF(I,J,3)=QF(I,1,3)*CHORD+DZLE
   1    CONTINUE
C       WAKE FAR FIELD POINTS (QF - IS IN BODY FRAME OF REFERENCE)
        QF(IB2,J,1)=QF(IB1,J,1)+DXW
        QF(IB2,J,2)=QF(IB1,J,2)
        QF(IB2,J,3)=QF(IB1,J,3)
   3    CONTINUE
C
C       WING COLLOCATION POINTS
C
```

```
              DO 4 J=1,JB
              DO 4 I=1,IB1
              QC(I,J,1)=(QF(I,J,1)+QF(I,J+1,1)+QF(I+1,J+1,1)+QF(I+1,J,1))/4
              QC(I,J,2)=(QF(I,J,2)+QF(I,J+1,2)+QF(I+1,J+1,2)+QF(I+1,J,2))/4
              QC(I,J,3)=(QF(I,J,3)+QF(I,J+1,3)+QF(I+1,J+1,3)+QF(I+1,J,3))/4
C
C             COMPUTATION OF CHORDWISE VECTORS DS(IJ,1,2,3),
C             TANGENTIAL AND NORMAL VECTORS DS(IJ,4 TO 9), PANEL AREA DS(IJ,1-10)
C             AND SOURCE STRENGTH (SIGMA)
C
              CALL PANEL(QF(I,J,1),QF(I,J,2),QF(I,J,3),QF(I+1,J,1),QF(I+1,J,2),
             1QF(I+1,J,3),QF(I,J+1,1),QF(I,J+1,2),QF(I,J+1,3),QF(I+1,J+1,1),
             2QF(I+1,J+1,2),QF(I+1,J+1,3),DS(I,J,1),DS(I,J,2),DS(I,J,3),
             3DS(I,J,4),DS(I,J,5),DS(I,J,6),DS(I,J,7),DS(I,J,8),DS(I,J,9),
             4DS(I,J,10))
C
              SIGMA(I,J)=DS(I,J,7)*UT+DS(I,J,9)*WT
    4         CONTINUE
C
C             B -IS FULL SPAN, C -ROOT CHORD, S - AREA
              S=0.5*B*(CROOT+CTIP)
              C=S/B
              AR=B*B/S
C
C             TRANSFORM THE 4 PANEL CORNER POINTS INTO PANEL FRAME OF REF.
C             THIS IS NEEDED LATER TO CALCULATE THE INFLUENCE COEFFICIENTS
              DO 5 J=1,JB
              DO 5 I=1,IB1
              CALL CONVERT(QC(I,J,1),QC(I,J,2),QC(I,J,3),
             1 QF(I,J,1),QF(I,J,2),QF(I,J,3),
             2 DS(I,J,1),DS(I,J,2),DS(I,J,3),
             3 DS(I,J,4),DS(I,J,5),DS(I,J,6),
             4 DS(I,J,7),DS(I,J,8),DS(I,J,9),
             5 CR(I,J,1),CR(I,J,2),CR(I,J,3)  )
              CALL CONVERT(QC(I,J,1),QC(I,J,2),QC(I,J,3),
             1 QF(I+1,J,1),QF(I+1,J,2),QF(I+1,J,3),
             2 DS(I,J,1),DS(I,J,2),DS(I,J,3),
             3 DS(I,J,4),DS(I,J,5),DS(I,J,6),
             4 DS(I,J,7),DS(I,J,8),DS(I,J,9),
             5 CR(I,J,4),CR(I,J,5),CR(I,J,6)  )
              CALL CONVERT(QC(I,J,1),QC(I,J,2),QC(I,J,3),
             1 QF(I+1,J+1,1),QF(I+1,J+1,2),QF(I+1,J+1,3),
             2 DS(I,J,1),DS(I,J,2),DS(I,J,3),
             3 DS(I,J,4),DS(I,J,5),DS(I,J,6),
             4 DS(I,J,7),DS(I,J,8),DS(I,J,9),
             5 CR(I,J,7),CR(I,J,8),CR(I,J,9)  )
              CALL CONVERT(QC(I,J,1),QC(I,J,2),QC(I,J,3),
             1 QF(I,J+1,1),QF(I,J+1,2),QF(I,J+1,3),
             2 DS(I,J,1),DS(I,J,2),DS(I,J,3),
             3 DS(I,J,4),DS(I,J,5),DS(I,J,6),
             4 DS(I,J,7),DS(I,J,8),DS(I,J,9),
             5 CR(I,J,10),CR(I,J,11),CR(I,J,12)  )
    5         CONTINUE
              RETURN
              END
C
              SUBROUTINE PANEL(X1,Y1,Z1,X2,Y2,Z2,X3,Y3,Z3,X4,Y4,Z4,C1,C2,C3,
             1 T1,T2,T3,V1,V2,V3,S)
C             X,Y,Z-PANEL CORNERPOINTS, C,T,V-CHORDWISE, TANGENTIAL, NORMAL VECTORS
```

```
C       FIRST CALCULATE CHORWISE VECTOR
        A1=((X2+X4)-(X1+X3))/2.0
        A2=((Y2+Y4)-(Y1+Y3))/2.0
        A3=((Z2+Z4)-(Z1+Z3))/2.0
        AA=SQRT(A1**2+A2**2+A3**2)
        C1=A1/AA
        C2=A2/AA
        C3=A3/AA
C       NEXT, ANOTHER VECTOR IN THIS PLANE
        B1=X4-X1
        B2=Y4-Y1
        B3=Z4-Z1
C       NORMAL VECTOR
        V1=C2*B3-C3*B2
        V2=B1*C3-C1*B3
        V3=C1*B2-C2*B1
        VV=SQRT(V1**2+V2**2+V3**2)
        V1=V1/VV
        V2=V2/VV
        V3=V3/VV
C       TANGENTIAL VECTOR
        T1=V2*C3-V3*C2
        T2=C1*V3-V1*C3
        T3=V1*C2-V2*C1
C       CALCULATION OF PANEL AREA
        E1=X3-X1
        E2=Y3-Y1
        E3=Z3-Z1
        F1=X2-X1
        F2=Y2-Y1
        F3=Z2-Z1
C       NORMAL AREAS (F*B+B*E)
        S11=F2*B3-F3*B2
        S12=B1*F3-F1*B3
        S13=F1*B2-F2*B1
        S21=B2*E3-B3*E2
        S22=E1*B3-B1*E3
        S23=B1*E2-B2*E1
        S=0.5*(SQRT(S11**2+S12**2+S13**2)+SQRT(S21**2+S22**2+S23**2))
        RETURN
        END
C
        SUBROUTINE CONVERT(XO,YO,ZO,XB,YB,ZB,C1,C2,C3,T1,
       *T2,T3,V1,V2,V3,XP,YP,ZP)
C       TRANSFORMATION OF A FIELD POINT XB,YB,ZB INTO PANEL COORDINATES
C       XO,YO,ZO - PANEL COLLOCATION POINT,  C,T,V - ARE CHORDWISE,
C       TANGENTIAL, AND NORMAL VECTORS
        XP=(XB-XO)*C1+(YB-YO)*C2+(ZB-ZO)*C3
        YP=(XB-XO)*T1+(YB-YO)*T2+(ZB-ZO)*T3
        ZP=(XB-XO)*V1+(YB-YO)*V2+(ZB-ZO)*V3
        RETURN
        END
C
        SUBROUTINE INFLUENCE(A,B,XC,YC,ZC,X1,Y1,Z1,X2,Y2,Z2,X3,Y3,Z3,
       1 X4,Y4,Z4)
C       DOUBLET (A) AND SOURCE (B) INFLUENCE AT POINT (XC,YC,ZC) DUE TO PANEL
C       (X1,Y1,Z1,...X4,Y4,Z4), SEE KATZ & PLOTKIN PP 283-6. BY M. VEST, 1993.
        PI=3.141592653580732
        EP=0.000001
```

```
C           EP, PANEL SIDE CUTOFF DISTANCE
C           PANEL SIDE (D) DISTANCE (R), E, AND H (EQS. 10.90 & 10.92-10.94)
            R1=SQRT((XC-X1)**2+(YC-Y1)**2+ZC**2)
            R2=SQRT((XC-X2)**2+(YC-Y2)**2+ZC**2)
            R3=SQRT((XC-X3)**2+(YC-Y3)**2+ZC**2)
            R4=SQRT((XC-X4)**2+(YC-Y4)**2+ZC**2)
C
            D1=SQRT((X2-X1)**2+(Y2-Y1)**2)
            D2=SQRT((X3-X2)**2+(Y3-Y2)**2)
            D3=SQRT((X4-X3)**2+(Y4-Y3)**2)
            D4=SQRT((X1-X4)**2+(Y1-Y4)**2)
C
            E1=(XC-X1)**2+ZC**2
            E2=(XC-X2)**2+ZC**2
            E3=(XC-X3)**2+ZC**2
            E4=(XC-X4)**2+ZC**2
C
            H1=(XC-X1)*(YC-Y1)
            H2=(XC-X2)*(YC-Y2)
            H3=(XC-X3)*(YC-Y3)
            H4=(XC-X4)*(YC-Y4)
C
C              SOURCE (S, B) AND DOUBLET (Q, A) INFLUENCE IN PANEL COORDINATES
C              FOR TRIANGULAR PANEL THE 4TH SIDE CONTRIBUTION IS ZERO
C
            IF (D1.LT.EP) THEN
              S1=0.
              Q1=0.
            ELSE
              F=(Y2-Y1)*E1-(X2-X1)*H1
              G=(Y2-Y1)*E2-(X2-X1)*H2
              Q1=ATAN2(ZC*(X2-X1)*(F*R2-G*R1),ZC**2*(X2-X1)**2*R1*R2+F*G)
              S1=((XC-X1)*(Y2-Y1)-(YC-Y1)*(X2-X1))/D1*LOG((R1+R2+D1)/
     *           (R1+R2-D1))
            ENDIF
C
            IF (D2.LT.EP) THEN
              S2=0.
              Q2=0.
            ELSE
              F=(Y3-Y2)*E2-(X3-X2)*H2
              G=(Y3-Y2)*E3-(X3-X2)*H3
              Q2=ATAN2(ZC*(X3-X2)*(F*R3-G*R2),ZC**2*(X3-X2)**2*R2*R3+F*G)
              S2=((XC-X2)*(Y3-Y2)-(YC-Y2)*(X3-X2))/D2*LOG((R2+R3+D2)/
     *           (R2+R3-D2))
            ENDIF
C
            IF (D3.LT.EP) THEN
              S3=0.
              Q3=0.
            ELSE
              F=(Y4-Y3)*E3-(X4-X3)*H3
              G=(Y4-Y3)*E4-(X4-X3)*H4
              Q3=ATAN2(ZC*(X4-X3)*(F*R4-G*R3),ZC**2*(X4-X3)**2*R3*R4+F*G)
              S3=((XC-X3)*(Y4-Y3)-(YC-Y3)*(X4-X3))/D3*LOG((R3+R4+D3)/
     *           (R3+R4-D3))
            ENDIF
C
            IF (D4.LT.EP) THEN
```

D.3 Time-Dependent Programs

```fortran
              S4=0.
              Q4=0.
            ELSE
              F=(Y1-Y4)*E4-(X1-X4)*H4
              G=(Y1-Y4)*E1-(X1-X4)*H1
              Q4=ATAN2(ZC*(X1-X4)*(F*R1-G*R4),ZC**2*(X1-X4)**2*R4*R1+F*G)
              S4=((XC-X4)*(Y1-Y4)-(YC-Y4)*(X1-X4))/D4*LOG((R4+R1+D4)/
         *      (R4+R1-D4))
            ENDIF
C
C           ADD CONTRIBUTIONS FROM THE 4 SIDES
C
            A=-(Q1+Q2+Q3+Q4)/4./PI          ! times doublet strength
            IF(ABS(ZC).LT.EP) A=0.
            B=-(S1+S2+S3+S4)/4./PI-ZC*A     ! times source strength
            RETURN
            END
C
C
            SUBROUTINE DECOMP(N,NDIM,A,IP)
            REAL A(NDIM,NDIM),T
            INTEGER IP(NDIM)
C           MATRIX TRIANGULARIZATION BY GAUSSIAN ELIMINATION.
C           N = ORDER OF MATRIX. NDIM = DECLARED DIMENSION OF ARRAY A.
C           A = MATRIX TO BE TRIANGULARIZED.
C           IP(K) , K .LT. N = INDEX OF K-TH PIVOT ROW.
C
            IP(N) = 1
            DO 6 K = 1, N
            IF(K.EQ.N) GOTO 5
            KP1 = K + 1
            M = K
            DO 1 I = KP1, N
            IF( ABS(A(I,K)).GT.ABS(A(M,K))) M=I
  1         CONTINUE
            IP(K) = M
            IF(M.NE.K) IP(N) = -IP(N)
            T = A(M,K)
            A(M,K) = A(K,K)
            A(K,K) = T
            IF(T.EQ.0.E0) GO TO 5
            DO 2 I = KP1, N
  2         A(I,K) = -A(I,K)/T
            DO 4 J = KP1, N
            T = A(M,J)
            A(M,J) = A(K,J)
            A(K,J) = T
            IF(T .EQ. 0.E0) GO TO 4
            DO 3 I = KP1, N
  3         A(I,J) = A(I,J) + A(I,K)*T
  4         CONTINUE
  5         IF(A(K,K) .EQ. 0.E0) IP(N) = 0
  6         CONTINUE
            RETURN
            END
C
            SUBROUTINE SOLVER(N,NDIM,A,B,IP)
            REAL A(NDIM,NDIM), B(NDIM), T
            INTEGER IP(NDIM)
```

```
C       SOLUTION OF LINEAR SYSTEM, A*X = B.
C       N = ORDER OF MATRIX.
C       NDIM = DECLARED DIMENSION OF THE ARRAY A.
C       B = RIGHT HAND SIDE VECTOR.
C       IP = PIVOT VECTOR OBTAINED FROM SUBROUTINE DECOMP.
C       B = SOLUTION VECTOR, X.
C
        IF(N.EQ.1) GOTO 9
        NM1 = N - 1
        DO 7 K = 1, NM1
        KP1 = K + 1
        M = IP(K)
        T = B(M)
        B(M) = B(K)
        B(K) = T
        DO 7 I = KP1, N
7       B(I) = B(I) + A(I,K)*T
        DO 8 KB = 1, NM1
        KM1 = N - KB
        K = KM1 + 1
        B(K) = B(K)/A(K,K)
        T = -B(K)
        DO 8 I = 1, KM1
8       B(I) = B(I) + A(I,K)*T
9       B(1) = B(1)/A(1,1)
        RETURN
        END

C       TYPICAL INPUT FOR SUBROUTINE GRID
19                      NACA 0012 AIRFOIL
        1.000     0.000
        0.905    -0.012
        0.794    -0.026
        0.655    -0.046
        0.500    -0.058
        0.345    -0.060
        0.206    -0.050
        0.095    -0.038
        0.024    -0.021
        0.000     0.000
        0.024     0.021
        0.095     0.038
        0.206     0.050
        0.345     0.060
        0.500     0.058
        0.655     0.046
        0.794     0.026
        0.905     0.012
        1.000     0.000


C       PROGRAM No. 15: SUDDEN ACCELERATION OF A FLAT PLATE (LUMPED VORTEX)
C       ---------------------------------------------------------------
C       TRANSIENT AERODYNAMIC OF A FLAT PLATE REPRESENTED BY A SINGLE LUMPED
C       VORTEX ELEMENT (PREPARED AS A SOLUTION OF HOMEWORK PROBLEMS BY JOE
C       KATZ, 1987).
        DIMENSION VORTIC(50,3),UW(50,2)
C                     VORTIC(IT,X,Z,GAMMA)
        COMMON/NO1/ IT,VORTIC
```

```
C
      NSTEP=20
      PAY=3.141592654
      RO=1.
      UT=50.0
      C=1.0
      ALFA1=5.0
      ALFA=ALFA1*PAY/180.0
      SN=SIN(ALFA)
      CS=COS(ALFA)
      DT=C/UT/4.0
      T=-DT
      DXW=0.3*UT*DT
      WRITE(6,11)
C
C     PROGRAM START
C
      DO 100 IT=1,NSTEP
      T=T+DT
C     PATH OF ORIGIN (SX,SZ)
      SX=-UT*T
      SZ=0.0
C     SHEDDING OF WAKE POINTS
      VORTIC(IT,1)=(C+DXW)*CS+SX
      VORTIC(IT,2)=-(C+DXW)*SN+SZ
C
C     CALCULATE MOMENTARY VORTEX STRENGTH OF WING AND WAKE VORTICES
C
      A=-1/(PAY*C)
      B=1/(2.0*PAY*(C/4.0+DXW))
      RHS2=0.0
      WWAKE=0.0
      IF(IT.EQ.1) GOTO 2
      IT1=IT-1
C     CALCULATE WAKE INFLUENCE
      XX1=0.75*C*CS+SX
      ZZ1=-0.75*C*SN+SZ
      CALL DWASH(XX1,ZZ1,1,IT1,U,W)
      WWAKE=U*SN+W*CS
C     CALCULATION OF RHS
      DO 1 I=1,IT1
    1 RHS2=RHS2-VORTIC(I,3)
    2 CONTINUE
      RHS1=-UT*SN-WWAKE
C     SOLUTION (BASED ON ALGEBRAIC SOLUTION OF TWO EQUATIONS FOR GAMMAT
C     AND THE LATEST WAKE VORTEX STRENGTH VORTIC(IT,3).
      VORTIC(IT,3)=1/(B/A-1.0)*(RHS1/A-RHS2)
      GAMMAT=RHS2-VORTIC(IT,3)
C
C     WAKE ROLLUP
C
      IF(IT.LT.1) GOTO 5
      DO 3 I=1,IT
      XX1=0.25*C*CS+SX
      ZZ1=-0.25*C*SN+SZ
      CALL VOR2D(VORTIC(I,1),VORTIC(I,2),XX1,ZZ1,GAMMAT,U,W)
      CALL DWASH(VORTIC(I,1),VORTIC(I,2),1,IT,U1,W1)
      U=U+U1
```

```
            W=W+W1
            UW(I,1)=VORTIC(I,1)+U*DT
            UW(I,2)=VORTIC(I,2)+W*DT
     3      CONTINUE
            DO 4 I=1,IT
            VORTIC(I,1)=UW(I,1)
     4      VORTIC(I,2)=UW(I,2)
     5      CONTINUE
C
C           AERODYNAMIC LOADS
C
            IF(IT.EQ.1) GAMAT1=0.0
            QUE=0.5*RO*UT*UT
            DGAMDT=(GAMMAT-GAMAT1)/DT
            GAMAT1=GAMMAT
C           CALCULATE WAKE INDUCED DOWNWASH
            XX1=0.75*C*CS+SX
            ZZ1=-0.75*C*SN+SZ
            CALL DWASH(XX1,ZZ1,1,IT,U,W)
            WW=U*SN+W*CS
            L=RO*(UT*GAMMAT+DGAMDT*C)
            D=RO*(-WW*GAMMAT+DGAMDT*C*SN)
            CL=L/QUE/C
            CD=D/QUE/C
C
C           OUTPUT
C
            CLT=CL/(2.0*PAY*SN)
            GAM1=GAMMAT/(PAY*UT*C*SN)
            SX1=SX-UT*DT
            WRITE(6,10) IT,SX1,CD,CL,GAM1,CLT
C
  100       CONTINUE
   10       FORMAT(I3,10F7.3)
   11       FORMAT(' SUDDEN ACCELERATION OF A FLAT PLATE',/,38('='),//,
           *' IT',4X,'SX',5X,'CD',5X,'CL',3X,'GAMMAT',3X,'CLT')
            STOP
            END
C
            SUBROUTINE DWASH(X,Z,I1,I2,U,W)
C           CALCULATES DOWNWASH INDUCED BY IT-1 WAKE VORTICES
            DIMENSION VORTIC(50,3)
            COMMON/NO1/ IT,VORTIC
            U=0.0
            W=0.0
            DO 1 I=I1,I2
            CALL VOR2D(X,Z,VORTIC(I,1),VORTIC(I,2),VORTIC(I,3),U1,W1)
            U=U+U1
            W=W+W1
     1      CONTINUE
            RETURN
            END
C
            SUBROUTINE VOR2D(X,Z,X1,Z1,GAMMA,U,W)
C           CALCULATES INFLUENCE OF VORTEX AT (X1,Z1)
            PAY=3.141592654
            U=0.0
            W=0.0
            RX=X-X1
```

```
          RZ=Z-Z1
          R=SQRT(RX**2+RZ**2)
          IF(R.LT.0.001) GOTO 1
          V=0.5/PAY*GAMMA/R
          U=V*(RZ/R)
          W=V*(-RX/R)
    1     CONTINUE
          RETURN
          END


C         PROGRAM No. 16: UNSTEADY RECTANGULAR LIFTING SURFACE (VLM)
C         ------------------------------------------------------------
C         THIS IS A 3-D LINEAR CODE FOR RECTANGULAR PLANFORMS (WITH GROUND EFFECT)
C         IN UNSTEADY MOTION USING THE VORTEX LATTICE METHOD (BY JOE KATZ, 1975).
          DIMENSION ALF(5),SNO(5),CSO(5),ALAM(4),GAMA1J(5)
          DIMENSION QF(5,16,3),QC(4,13,3),BB(13),DLY(13)
          DIMENSION GAMA(4,13),DL(4,13),DP(4,13),DS(4,13),DLT(4,13),DD(4,13)
          DIMENSION A1(4,13),QW(50,14,3),VORTIC(50,13),UVW(50,14,3)
          DIMENSION QW1(50,14,3),VORT1(50,13),US(13)
          DIMENSION A(52,52),GAMA1(52),WW(52),DW(52),IP(52),ALAMDA(2)
          DIMENSION WTS(4,13),X15(50),Y15(50),Y16(50),Y17(50),Z15(50)
          COMMON VORTIC,QW,VORT1,QW1,QF,A1
          COMMON IT,ALF,SNO,CSO,BB,QC,DS,ALAMDA,DXW
          COMMON/NO1/ SX,SZ,CS1,SN1,GAMA
          COMMON/NO2/ IB,JB,CH,SIGN
          COMMON/NO3/ IW
C
C                          MODES OF OPERATION
C
C         1. STEADY STATE         : SET DT=DX/VT*IB*10, AND NSTEPS=5
C         2. SUDDEN ACCELERATION : SET DT=DX/VT/4. AND NSTEPS= UP TO 50
C         3. HEAVING OSCILLATIONS: BH=HEAVING AMPL. OM= FREQ.
C         4. PITCH OSCILLATIONS   : OMEGA=FREQUENCY, TETA=MOMENTARY ANGLE
C         5. FOR COMPUTATIONAL ECONOMY THE PARAMETER IW MIGHT BE USED
C         NOTE; INDUCED DRAG CALCULATION INCREASES COMPUTATION TIME AND
C         CAN BE DISCONNECTED.
C
C
C         INPUT DATA
C
          IB=4
          JB=13
          NSTEPS=50
          NW=5
          DO 100 IPROG=1,IPROG1
C         NW - IS THE NUMBER OF (TIMEWISE) DEFORMING WAKE ELEMENTS.
C         IB = NUMBER OF CHORDWISE PANELS, JB = NO. OF SPANWISE PANELS
C         NSTEPS = NO. OF TIME STEPS.
          PAY=3.141592654
          RO=1.
          BH=0.0
          OM=0.0
          VT=50.0
          C=1.
          B=6.0
          DX=C/IB
          DY=B/JB
          CH=10000.*C
```

```
C           C=CHORD; C=ROOT CHORD, B=SEMI SPAN,  DX,DY=PANEL DIMENSIONS,
C           CH=GROUND CLEARANCE, VT=FAR FIELD VELOCITY.
            ALFA1=5.0
            ALFAO=0.0
            ALFA=(ALFA1+ALFAO)*PAY/180.0
            DO 2 I=1,IB
   2        ALF(I)=0.
            ALF(IB+1)=ALF(IB)
C           ALF(IB+1) IS REQUIRED ONLY FOR QF(I,J,K) CALCULATION IN GEO.
            ALAMDA(1)=90.*PAY/180.
            ALAMDA(2)=ALAMDA(1)
C           ALAMDA(I) ARE SWEP BACK ANGLES. (ALAMDA < 90, SWEEP BACKWARD).
            DT=DX/VT/4.0
            T=-DT
C           TIME IN SECONDS
            DXW=0.3*VT*DT
            DO 3 J=1,JB
   3        BB(J)=DY
C
C           CONSTANTS
C
            K=0
            DO 1 I=1,IB
            DO 1 J=1,JB
            K=K+1
            WW(K)=0.0
            DLT(I,J)=0.0
            VORTIC(I,J)=0.0
            VORT1(I,J)=0.0
C           GAMA(I,J)=1.  IS REQUIRED FOR INFLUENCE MATRIX CALCULATIONS.
   1        GAMA(I,J)=1.

C           CALCULATION OF COLLOCATION POINTS.
C
            CALL GEO(B,C,S,AR,IB,JB,DX,DY,0.0,ALFA)
C           GEO CALCULATES WING COLLOCATION POINTS QC,AND VORTEX TIPS QF
            WRITE(6,101)
            ALAM1=ALAMDA(1)*180./PAY
            ALAM2=ALAMDA(2)*180./PAY
            WRITE(6,102) ALFA1,ALAM1,B,C,ALAM2,S,AR,IB,JB,CH
            DO 31 I=1,IB
            ALL=ALF(I)*180./PAY
  31        WRITE(6,111) I,ALL
            DO 4 I=1,JB,2
   4        WRITE(6,105) I,BB(I)
            IB1=IB+1
            JB1=JB+1
C
C           ============
C           PROGRAM START
C           ============
C
C
            DO 100 IT=1,NSTEPS
C
            T=T+DT
C
C           PATH INFORMATION
C
```

D.3 Time-Dependent Programs

```
              SX=-VT*T
              DSX=-VT
              CH1=CH
              IF(CH.GT.100.0) CH1=0.0
              SZ=BH*SIN(OM*T)+CH1
              DSZ=BH*OM*COS(OM*T)
C             DSX=DSX/DT    DSZ=DSZ/DT
              TETA=0.0
              OMEGA=0.0
              VT=-COS(TETA)*DSX-SIN(TETA)*DSZ
              SN1=SIN(TETA)
              CS1=COS(TETA)
              WT=SN1*DSX-CS1*DSZ
              DO 6 I=1,IB
              SNO(I)=SIN(ALFA+ALF(I))
    6         CSO(I)=COS(ALFA+ALF(I))
C
C             ==========================
C             VORTEX WAKE SHEDDING POINTS
C             ==========================
C
              DO 7 J=1,JB1
              QW(IT,J,1)=QF(IB1,J,1)*CS1-QF(IB1,J,3)*SN1+SX
              QW(IT,J,2)=QF(IB1,J,2)
              QW(IT,J,3)=QF(IB1,J,1)*SN1+QF(IB1,J,3)*CS1+SZ
    7         CONTINUE
C
C             =======================
C             AERODYNAMIC CALCULATIONS
C             =======================
C
C             INFLUENCE COEFFICIENTS CALCULATION
              K=0
              DO 14 I=1,IB
              DO 14 J=1,JB
              SIGN=0.0
              K=K+1
              IF(IT.GT.1) GOTO 12
C             MATRIX COEFFICIENTS CALCULATION OCCURS ONLY ONCE FOR THE
C             TIME-FIXED-GEOMETRY WING.
              CALL WING(QC(I,J,1),QC(I,J,2),QC(I,J,3),GAMA,U,V,W)
              L=0
              DO 10 I1=1,IB
              DO 10 J1=1,JB
              L=L+1
C             A(K,L) - IS THE NORMAL VELOCITY COMPONENT DUE TO A UNIT VORTEX
C                      LATTICE.
   10         A(K,L)=A1(I1,J1)
C             ADD INFLUENCE OF WING OTHER HALF PART
              CALL WING(QC(I,J,1),-QC(I,J,2),QC(I,J,3),GAMA,U,V,W)
              L=0
              DO 11 I1=1,IB
              DO 11 J1=1,JB
              L=L+1
   11         A(K,L)=A(K,L)+A1(I1,J1)
              IF(CH.GT.100.0) GOTO 12
C             ADD INFLUENCE OF MIRROR IMAGE.
              SIGN=10.0
              XX1=QC(I,J,1)*CS1-QC(I,J,3)*SN1+SX
```

```
      ZZ1=QC(I,J,1)*SN1+QC(I,J,3)*CS1+SZ
      XX2=(XX1-SX)*CS1+(-ZZ1-SZ)*SN1
      ZZ2=-(XX1-SX)*SN1+(-ZZ1-SZ)*CS1
      CALL WING(XX2,QC(I,J,2),ZZ2,GAMA,U,V,W)
      L=0
      DO 8 I1=1,IB
      DO 8 J1=1,JB
      L=L+1
    8 A(K,L)=A(K,L)+A1(I1,J1)
C     ADD MIRROR IMAGE INFLUENCE OF WING'S OTHER HALF.
      CALL WING(XX2,-QC(I,J,2),ZZ2,GAMA,U,V,W)
      L=0
      DO 9 I1=1,IB
      DO 9 J1=1,JB
      L=L+1
    9 A(K,L)=A(K,L)+A1(I1,J1)
      SIGN=0.0
   12 CONTINUE
      IF(IT.EQ.1) GOTO 13
C     CALCULATE WAKE INFLUENCE
      XX1=QC(I,J,1)*CS1-QC(I,J,3)*SN1+SX
      ZZ1=QC(I,J,1)*SN1+QC(I,J,3)*CS1+SZ
      CALL WAKE(XX1,QC(I,J,2),ZZ1,IT,U,V,W)
      CALL WAKE(XX1,-QC(I,J,2),ZZ1,IT,U1,V1,W1)
      IF(CH.GT.100) GOTO 121
      CALL WAKE(XX1,QC(I,J,2),-ZZ1,IT,U2,V2,W2)
      CALL WAKE(XX1,-QC(I,J,2),-ZZ1,IT,U3,V3,W3)
      GOTO 122
  121 U2=0.0
      U3=0.0
      V2=0.0
      V3=0.0
      W2=0.0
      W3=0.0
  122 CONTINUE
C     WAKE INDUCED VELOCITY IS GIVEN IN INERTIAL FRAME
      U=U+U1+U2+U3
      W=W+W1-W2-W3
      U11=U*CS1+W*SN1
      W11=-U*SN1+W*CS1
C     WW(K) IS THE PREPENDICULAR COMPONENT OF WAKE INFLUENCE TO WING.
      WW(K)=U11*SNO(I)+W11*CSO(I)
   13 CONTINUE
C
C     CALCULATE WING GEOMETRICAL DOWNWASH
C
      DW(K)=-VT*SNO(I)+QC(I,J,1)*OMEGA-WT
C     FOR GENERAL MOTION DW(K)=-VT*SIN(ALFA)+OMEGA*X
      WTS(I,J)=W11
C     W11 - IS POSITIVE SINCE THE LATEST UNSTEADY WAKE ELEMENT IS
C     INCLUDED IN SUBROUTINE WING
   14 CONTINUE
C
C     SOLUTION OF THE PROBLEM:   DW(I)=WW(I)+A(I,J)*GAMA(I)
C
      K1=IB*JB
      DO 15 K=1,K1
   15 GAMA1(K)=DW(K)-WW(K)
      IF(IT.GT.1) GOTO 16
```

D.3 Time-Dependent Programs

```
C          FOR NONVARIABLE WING GEOMETRY (WITH TIME), MATRIX INVERSION
C          IS DONE ONLY ONCE.
           CALL DECOMP(K1,52,A,IP)
 16        CONTINUE
           CALL SOLVER(K1,52,A,GAMA1,IP)
C          HERE           *       THE SAME ARRAY SIZE IS REQUIRED,
C                                 AS SPECIFIED IN THE BEGINNING OF THE CODE
C
C          WING VORTEX LATTICE LISTING
C
           K=0
           DO 17 I=1,IB
           DO 17 J=1,JB
           K=K+1
 17        GAMA(I,J)=GAMA1(K)
C
C          WAKE SHEDDING
C
           DO 171 J=1,JB
C          LATEST WAKE ELEMENTS LISTING
 162       VORTIC(IT,J)=GAMA(IB,J)
           VORTIC(IT+1,J)=0.0
 171       CONTINUE
C
C          ===========================
C          WAKE ROLLUP CALCULATION
C          ===========================
C
           IW=1
           IF(IT.EQ.1) GOTO 193
           IF(IT.GE.NW) IW=IT-NW+1
C          NW IS THE NUMBER OF (TIMEWISE) DEFORMING WAKE ELEMENTS.
           I1=IT-1
           JS1=0
           JS2=0
           DO 18 I=IW,I1
           DO 18 J=1,JB1
           CALL VELOCE(QW(I,J,1),QW(I,J,2),QW(I,J,3),U,V,W,IT,JS1,JS2)
           UVW(I,J,1)=U*DT
           UVW(I,J,2)=V*DT
           UVW(I,J,3)=W*DT
 18        CONTINUE
C
           DO 19 I=IW,I1
           DO 19 J=1,JB1
           QW(I,J,1)=QW(I,J,1)+UVW(I,J,1)
           QW(I,J,2)=QW(I,J,2)+UVW(I,J,2)
           QW(I,J,3)=QW(I,J,3)+UVW(I,J,3)
 19        CONTINUE
C
 193       CONTINUE
C
C          ==================
C          FORCES CALCULATION
C          ==================
C
           FL=0.
           FD=0.
           FM=0.
```

```
          FG=0.0
          QUE=0.5*RO*VT*VT
          DO 20 J=1,JB
          SIGMA=0.
          SIGMA1=0.0
          DLY(J)=0.
          DO 20 I=1,IB
          IF(I.EQ.1) GAMAIJ=GAMA(I,J)
          IF(I.GT.1) GAMAIJ=GAMA(I,J)-GAMA(I-1,J)
          DXM=(QF(I,J,1)+QF(I,J+1,1))/2.
C         DXM IS VORTEX DISTANCE FROM LEADING EDGE
          SIGMA1=(0.5*GAMAIJ+SIGMA)*DX
          SIGMA=GAMA(I,J)
          DFDT=(SIGMA1-DLT(I,J))/DT
C         DFDT    IS THE VELOCITY POTENTIAL TIME DERIVATIVE
          DLT(I,J)=SIGMA1
          DL(I,J)=RO*(VT*GAMAIJ+DFDT)*BB(J)*CSO(I)
C         INDUCED DRAG CALCULATION
          CALL WINGL(QC(I,J,1),QC(I,J,2),QC(I,J,3),GAMA,U1,V1,W1)
          CALL WINGL(QC(I,J,1),-QC(I,J,2),QC(I,J,3),GAMA,U2,V2,W2)
          IF(CH.GT.100.0) GOTO 194
          XX1=QC(I,J,1)*CS1-QC(I,J,3)*SN1+SX
          ZZ1=QC(I,J,1)*SN1+QC(I,J,3)*CS1+SZ
          XX2=(XX1-SX)*CS1+(-ZZ1-SZ)*SN1
          ZZ2=-(XX1-SX)*SN1+(-ZZ1-SZ)*CS1
          CALL WINGL(XX2,QC(I,J,2),ZZ2,GAMA,U3,V3,W3)
          CALL WINGL(XX2,-QC(I,J,2),ZZ2,GAMA,U4,V4,W4)
          GOTO 195
  194     W3=0.
          W4=0.
  195     W8=W1+W2-W3-W4
C         ADD INFLUENCE OF MIRROR IMAGE (GROUND).
          CTS=-(WTS(I,J)+W8)/VT
          DD1=RO*BB(J)*DFDT*SNO(I)
          DD2=RO*BB(J)*VT*GAMAIJ*CTS
          DD(I,J)=DD1+DD2
C
          DP(I,J)=DL(I,J)/DS(I,J)/QUE
          DLY(J)=DLY(J)+DL(I,J)
          FL=FL+DL(I,J)
          FD=FD+DD(I,J)
          FM=FM+DL(I,J)*DXM
          FG=FG+GAMAIJ*BB(J)
   20     CONTINUE
          CL=FL/(QUE*S)
          CD=FD/(QUE*S)
          CM=FM/(QUE*S*C)
          CLOO=2.*PAY*ALFA/(1.+2./AR)
          IF(ABS(CLOO).LT.1.E-20) CLOO=CL
          CLT=CL/CLOO
          CFG=FG/(0.5*VT*S)/CLOO
C
C         ======
C         OUTPUT
C         ======
C
C         PLACE PLOTTER OUTPUT HERE (e.g. T,SX,SZ,CL,CD,CM)
C
C         OTHER OUTPUT
C
```

```
              WRITE(6,106) T,SX,SZ,VT,TETA,OMEGA
              WRITE(6,104) CL,FL,CM,CD,CLT,CFG
              I2=5
              IF(IT.NE.I2) GOTO 100
              WRITE(6,110)
              DO 21 J=1,JB
              DO 211 I=2,IB
  211         GAMA1J(I)=GAMA(I,J)-GAMA(I-1,J)
              DLYJ=DLY(J)/BB(J)
   21         WRITE(6,103) J,DLYJ,DP(1,J),DP(2,J),DP(3,J),DP(4,J),GAMA(1,J),
             1GAMA1J(2),GAMA1J(3),GAMA1J(4)
              IF(IT.NE.I2) GOTO 100
              WRITE(6,107)
              DO 23 I=1,IT
              WRITE(6,109) I,(VORTIC(I,K1),K1=1,13)
              DO 23 J=1,3
              WRITE(6,108) J,(QW(I,K,J),K=1,14)
   23         CONTINUE
C
C             END OF PROGRAM
  100         CONTINUE
C
C             FORMATS
C
  101         FORMAT(1H ,/,20X,'WING LIFT DISTRIBUTION CALCULATION (WITH GROUND
             1 EFFECT)',/,20X,56('-'))
  102         FORMAT(1H ,/,10X,'ALFA:',F10.2,8X,'LAMDA(1) :',F10.2,8X,'B  :',
             1F10.2,8X,'C   :',F13.2,/,33X,
             2'LAMDA(2) :',F10.2,8X,'S   :',F10.2,8X,'AR  :',F13.2,/,33X,
             3'IB       :',I10,8X,'JB  :',I10,8X,'L.E. HEIGHT:', F6.2,/)
  103         FORMAT(1H ,I3,' I ',F9.3,' II ',4(F9.3,' I '),' I ',4(F9.3,' I '))
  104         FORMAT(1H ,'CL=',F10.4,2X,'L=',F10.4,4X,'CM=',F10.4,3X,'CD=',F10.4
             1,3X,'L/L(INF)=',F10.4,4X,'GAMA/GAMA(INF)=',F10.4,/)
  105         FORMAT(1H ,9X,'BB(',I3,')=',F10.4)
  106         FORMAT(1H ,/,' T=',F10.2,3X,'SX=',F10.2,3X,'SZ=',F10.2,3X,'VT=',
             1F10.2,3X,'TETA=  ',F10.2,6X,'OMEGA=         ',F10.2)
  107         FORMAT(1H ,//,' WAKE ELEMENTS,',//)
  108         FORMAT(1H ,'QW(',I2,')=',22(F6.2))
  109         FORMAT(1H ,' VORTIC(IT=',I3,')=',17(F6.3))
  110         FORMAT(1H ,/,5X,'I     DL',4X,'II',22X,'DCP',22X,'I I',25X,
             1'GAMA',/,118('='),/,5X,'I',15X,'I= 1',11X,'2',11X,'3',11X,
             2'4',5X,'I  I',5X,'1',11X,'2',11X,'3',11X,'4',/,118('='))
  111         FORMAT(1H ,9X,'ALF(',I2,')=',F10.4)
  112         FORMAT(1H ,'QF(I=',I2,',',J,X.Y.Z)= ',15(F6.1))
  113         FORMAT(1H ,110('='))
C
              STOP
              END
C
C             SUBROUTINE VORTEX(X,Y,Z,X1,Y1,Z1,X2,Y2,Z2,GAMA,U,V,W)
C             USE THIS SUBROUTINE FROM PROGRAM NO. 13.
C
              SUBROUTINE WAKE(X,Y,Z,IT,U,V,W)
              DIMENSION VORTIC(50,13),QW(50,14,3)
              COMMON VORTIC,QW
              COMMON/NO2/ IB,JB,CH,SIGN
              COMMON/NO3/ IW
C             CALCULATES SEMI WAKE INDUCED VELOCITY AT POINT (X,Y,Z) AT T=IT*DT,
C             IN THE INERTIAL FRAME OF REFERENCE
C
```

```
              U=0
              V=0
              W=0
              I1=IT-1
              DO 1 I=1,I1
              DO 1 J=1,JB
              VORTEK=VORTIC(I,J)
              CALL VORTEX(X,Y,Z,QW(I,J,1),QW(I,J,2),QW(I,J,3),QW(I+1,J,1),
             1QW(I+1,J,2),QW(I+1,J,3),VORTEK,U1,V1,W1)
              CALL VORTEX(X,Y,Z,QW(I+1,J,1),QW(I+1,J,2),QW(I+1,J,3),QW(I+1,J+1,1
             2),QW(I+1,J+1,2),QW(I+1,J+1,3),VORTEK,U2,V2,W2)
              CALL VORTEX(X,Y,Z,QW(I+1,J+1,1),QW(I+1,J+1,2),QW(I+1,J+1,3),
             3QW(I,J+1,1),QW(I,J+1,2),QW(I,J+1,3),VORTEK,U3,V3,W3)
              CALL VORTEX(X,Y,Z,QW(I,J+1,1),QW(I,J+1,2),QW(I,J+1,3),QW(I,J,1),
             4QW(I,J,2),QW(I,J,3),VORTEK,U4,V4,W4)
              U=U+U1+U2+U3+U4
              V=V+V1+V2+V3+V4
              W=W+W1+W2+W3+W4
        1     CONTINUE
              RETURN
              END
C
              SUBROUTINE VELOCE(X,Y,Z,U,V,W,IT,JS1,JS2)
              DIMENSION GAMA(4,13)
              COMMON/NO1/ SX,SZ,CS1,SN1,GAMA
              COMMON/NO2/ IB,JB,CH,SIGN
C       SUBROUTINE VELOCE CALCULATES INDUCED VELOCITIES DUE TO THE WING
C       AND ITS WAKES IN A POINT (X,Y,Z) GIVEN IN THE INERTIAL FRAME OF
C       REFERENCE.
C
              X1=(X-SX)*CS1+(Z-SZ)*SN1
              Y1=Y
              Z1=-(X-SX)*SN1+(Z-SZ)*CS1
              CALL WAKE(X,Y,Z,IT,U1,V1,W1)
              CALL WAKE(X,-Y,Z,IT,U2,V2,W2)
              CALL WING(X1,Y1,Z1,GAMA,U3,V3,W3)
              CALL WING(X1,-Y1,Z1,GAMA,U4,V4,W4)
              U33=CS1*(U3+U4)-SN1*(W3+W4)
              W33=SN1*(U3+U4)+CS1*(W3+W4)
C       INFLUENCE OF MIRROR IMAGE
              IF(CH.GT.100.0) GOTO 1
              X2=(X-SX)*CS1+(-Z-SZ)*SN1
              Z2=-(X-SX)*SN1+(-Z-SZ)*CS1
              CALL WAKE(X,Y,-Z,IT,U5,V5,W5)
              CALL WAKE(X,-Y,-Z,IT,U6,V6,W6)
              CALL WING(X2,Y1,Z2,GAMA,U7,V7,W7)
              CALL WING(X2,-Y1,Z2,GAMA,U8,V8,W8)
              U77=CS1*(U7+U8)-SN1*(W7+W8)
              W77=SN1*(U7+U8)+CS1*(W7+W8)
              GOTO 2
        1     CONTINUE
              U5=0.0
              U6=0.0
              U77=0.0
              V5=0.0
              V6=0.0
              V7=0.0
              V8=0.0
              W5=0.0
```

```
              W6=0.0
              W77=0.0
    2         CONTINUE
C             VELOCITIES MEASURED IN INERTIAL FRAME
              U=U1+U2+U33+U5+U6+U77
              V=V1-V2+V3-V4+V5-V6+V7-V8
              W=W1+W2+W33-W5-W6-W77
              RETURN
              END
C
              SUBROUTINE WING(X,Y,Z,GAMA,U,V,W)
              DIMENSION GAMA(4,13),QF(5,16,3),A1(4,13),VORTIC(50,13),QW(50,14,3)
              DIMENSION ALF(5),SNO(5),CSO(5),VORT1(50,13),QW1(50,14,3)
              COMMON VORTIC,QW,VORT1,QW1,QF,A1
              COMMON IT,ALF,SNO,CSO
              COMMON/NO2/ IB,JB,CH,SIGN
C
C             CALCULATES SEMI WING INDUCED VELOCITY AT A POINT (X,Y,Z) DUE TO WI
C             VORTICITY DISTRIBUTION GAMA(I,J) IN A WING FIXED COORDINATE SYSTE
              U=0
              V=0
              W=0
              DO 7 I=1,IB
              DO 7 J=1,JB
C
              CALL VORTEX(X,Y,Z,QF(I,J,1),QF(I,J,2),QF(I,J,3),QF(I,J+1,1),QF(I,J
             1+1,2),QF(I,J+1,3),GAMA(I,J),U1,V1,W1)
              CALL VORTEX(X,Y,Z,QF(I,J+1,1),QF(I,J+1,2),QF(I,J+1,3),QF(I+1,J+1,1
             2),QF(I+1,J+1,2),QF(I+1,J+1,3),GAMA(I,J),U2,V2,W2)
              CALL VORTEX(X,Y,Z,QF(I+1,J+1,1),QF(I+1,J+1,2),QF(I+1,J+1,3),
             3QF(I+1,J,1),QF(I+1,J,2),QF(I+1,J,3),GAMA(I,J),U3,V3,W3)
              CALL VORTEX(X,Y,Z,QF(I+1,J,1),QF(I+1,J,2),QF(I+1,J,3),QF(I,J,1),
             4QF(I,J,2),QF(I,J,3),GAMA(I,J),U4,V4,W4)
C
              U0=U1+U2+U3+U4
              V0=V1+V2+V3+V4
              W0=W1+W2+W3+W4
              A1(I,J)=U0*SNO(I)+W0*CSO(I)
              IF(SIGN.GE.1.0) A1(I,J)=U0*SNO(I)-W0*CSO(I)
              U=U+U0
              V=V+V0
              W=W+W0
C
    7         CONTINUE
              RETURN
              END
C
              SUBROUTINE WINGL(X,Y,Z,GAMA,U,V,W)
              DIMENSION GAMA(4,13),QF(5,16,3),A1(4,13),VORTIC(50,13),QW(50,14,3)
              DIMENSION ALF(5),SNO(5),CSO(5),VORT1(50,13),QW1(50,14,3)
              COMMON VORTIC,QW,VORT1,QW1,QF,A1
              COMMON IT,ALF,SNO,CSO
              COMMON/NO2/ IB,JB,CH,SIGN
C
C             CALCULATES INDUCED VELOCITY AT A POINT (X,Y,Z) DUE TO LONGITUDINAL
C             VORTICITY DISTRIBUTION GAMAX(I,J) ONLY(SEMI-SPAN), IN A WING FIXED
C             COORDINATE SYSTEM + (T.E. UNSTEADY VORTEX).
C             ** SERVES FOR INDUCED DRAG CALCULATION ONLY **
C
```

```
      U=0.
      V=0.
      W=0.
      DO 7 I=1,IB
      DO 7 J=1,JB
C
      CALL VORTEX(X,Y,Z,QF(I,J+1,1),QF(I,J+1,2),QF(I,J+1,3),QF(I+1,J+1,1
     2),QF(I+1,J+1,2),QF(I+1,J+1,3),GAMA(I,J),U2,V2,W2)
      CALL VORTEX(X,Y,Z,QF(I+1,J,1),QF(I+1,J,2),QF(I+1,J,3),QF(I,J,1),
     4QF(I,J,2),QF(I,J,3),GAMA(I,J),U4,V4,W4)
C
      U=U+U2+U4
      V=V+V2+V4
      W=W+W2+W4
   7  CONTINUE
C
C     ADD INFLUENCE OF LATEST UNSTEADY WAKE ELEMENT:
      I=IB
      DO 8 J=1,JB
      CALL VORTEX(X,Y,Z,QF(I+1,J+1,1),QF(I+1,J+1,2),QF(I+1,J+1,3),
     3QF(I+1,J,1),QF(I+1,J,2),QF(I+1,J,3),GAMA(I,J),U3,V3,W3)
      U=U+U3
      V=V+V3
      W=W+W3
   8  CONTINUE
C
      RETURN
      END
C
      SUBROUTINE GEO(B,C,S,AR,IB,JB,DX,DY,DGAP,ALFA)
      DIMENSION BB(13),ALF(5),SN(5),CS(5),SNO(5),CSO(5)
      DIMENSION QF(5,16,3),QC(4,13,3),DS(4,13)
      DIMENSION VORTIC(50,13),QW(50,14,3),A1(4,13),ALAMDA(2)
      DIMENSION VORT1(50,13),QW1(50,14,3)
      COMMON VORTIC,QW,VORT1,QW1,QF,A1
      COMMON IT,ALF,SNO,CSO,BB,QC,DS,ALAMDA,DXW
C
      PAY=3.141592654
C     IB:NO. OF CHORDWISE BOXES,  JB:NO. OF SPANWISE BOXES
      IB1=IB+1
      JB1=JB+1
      DO 2 I=1,IB1
      SN(I)=SIN(ALF(I))
   2  CS(I)=COS(ALF(I))
      CTG1=TAN(PAY/2.-ALAMDA(1))
      CTG2=TAN(PAY/2.-ALAMDA(2))
      CTIP=C+B*(CTG2-CTG1)
      S=B*(C+CTIP)/2.
      AR=2.*B*B/S
C
C     WING FIXED VORTICES LOCATION    ( QF(I,J,(X,Y,Z))...)
C
      BJ=0.
      DO 3 J=1,JB1
      IF(J.GT.1) BJ=BJ+BB(J-1)
      Z1=0.
      DC1=BJ*CTG1
      DC2=BJ*CTG2
      DX1=(C+DC2-DC1)/IB
```

D.3 Time-Dependent Programs

```
C         DC1=LEADING EDGE X,   DC2=TRAILING EDGE X
          DO 1 I=1,IB
          QF(I,J,1)=DC1+DX1*(I-0.75)
          QF(I,J,2)=BJ
          QF(I,J,3)=Z1-0.25*DX1*SN(I)
1         Z1=Z1-DX1*SN(I)
C         THE FOLLOWING LINES ARE DUE TO WAKE DISTANCE FROM TRAILING EDGE
          QF(IB1,J,1)=C+DC2+DXW
          QF(IB1,J,2)=QF(IB,J,2)
3         QF(IB1,J,3)=Z1-DXW*SN(IB)
C
C         WING COLLOCATION POINTS
C
          DO 4 J=1,JB
          Z1=0.
          BJ=QF(1,J,2)+BB(J)/2.
          DC1=BJ*CTG1
          DC2=BJ*CTG2
          DX1=(C+DC2-DC1)/IB
          DO 4 I=1,IB
          QC(I,J,1)=DC1+DX1*(I-0.25)
          QC(I,J,2)=BJ
          QC(I,J,3)=Z1-0.75*DX1*SN(I)
          Z1=Z1-DX1*SN(I)
4         DS(I,J)=DX1*BB(J)
C
C         ROTATION OF WING POINTS DUE TO ALFA
C
          SN1=SIN(-ALFA)
          CS1=COS(-ALFA)
          DO 6 I=1,IB1
          DO 6 J=1,JB1
          QF1=QF(I,J,1)
          QF(I,J,1)=QF1*CS1-QF(I,J,3)*SN1
          QF(I,J,3)=QF1*SN1+QF(I,J,3)*CS1
          IF((I.EQ.IB1).OR.(J.GE.JB1)) GOTO 6
          QC1=QC(I,J,1)
          QC(I,J,1)=QC1*CS1-QC(I,J,3)*SN1
          QC(I,J,3)=QC1*SN1+QC(I,J,3)*CS1
6         CONTINUE
C
          RETURN
          END
C
C         THE FOLLOWING SUBROUTINES ARE LISTED WITH THE STEADY STATE
C         VORTEX LATTICE SOLVER (PROGRAM No. 13).
C
C         SUBROUTINE VORTEX(X,Y,Z,X1,Y1,Z1,X2,Y2,Z2,GAMA,U,V,W)

C         SUBROUTINE DECOMP(N,NDIM,A,IP)

C         SUBROUTINE SOLVER(N,NDIM,A,B,IP)
C
```

Index

Acceleration of fluid particle, 9
Added mass, 192–194, 385–387, 398
Aerodynamic center, 110
Aerodynamic loads, 85–87
Aerodynamic twist, 181
Airfoil
 circular arc, 134–135, 138–139
 Joukowski, 135–137, 139–140
 multielement, 311–312
 NACA nomenclature, 499
 van de Vooren, 137–138, 140–141
Angle of attack
 definition, 75
 effective, 171–172
 induced, 171–172
 zero lift, 109
Angular velocity, 21–22
Aspect ratio, 175

Barrier, 29
Bernoulli's equation, 28–29
Biot-Savart Law, 36–41
Blasius formula, 128
Body forces, 7
Bound vortex, 89–90
Boundary conditions
 Dirichlet, 49, 208–209
 inviscid flow, 18, 27–28
 Neumann, 49, 207–208
 no slip, 11
 small-disturbance, 76–78
 solid surface, 11
 unsteady flow, 372–373
Boundary layer (Laminar)
 Blasius solution, 461–463
 classical equations, 18–19, 448–452
 displacement thickness, 454–455
 far wake solution, 472–473
 friction coefficient, 465
 Goldstein singularity, 471–472
 integral kinetic energy equation, 474
 integral kinetic energy shape factor equation, 474–475
 Karman-Pohlhausen method, 468–469
 momentum thickness, 465
 second-order equations, 452–456
 shape factor, 465
 similar solutions, 457–459
 stagnation point solution, 461–462
 Thwaites method, 469–471
 viscous-inviscid interaction, 475–480
 von Karman integral momentum equation, 463–467

Camber function, 78
Canard, 485–486
Cauchy integral theorum, 123
Cauchy principal value, 98
Cauchy-Riemann conditions, 42, 123
Center of pressure, 109
Circular cylinder flow
 lifting, 65–66
 non-lifting, 62–65
Circulation
 definition, 23
 rate of change, 25–26
Collocation point, 115
Complex plane, 142
Complex variable approach
 circle plane, 128–130
 complex potential, 125–126
 complex velocity, 126
 conformal mapping, 125, 128
 Joukowski transformation, 128–137
 van de Vooren airfoil, 137–138
Composite expansion, 161–162
Compressibility, 19, 90–92, 226–227
Coning motion, 443–445, 527–528
Continuity equation, 7–9, 11–12
Coordinate systems
 cartesian, 1
 cylindrical, 11
 spherical, 12
Corner, flow in, 55–56, 127
Cosine panel spacing, 277–278
Crossflow plane, 185
Crow instability, 486–487
Cusped trailing edge, 137, 211–212, 325–327

d'Alembert's paradox, 107–108
Del (gradient) operator
 cartesian coordinates, 8
 cylindrical coordinates, 11
 definition, 8
 spherical coordinates, 12
Delta wing, 192, 404–407, 516–523
Dihedral, 350–351

Dimensional analysis, 17–19
Dirichlet boundary condition, 49, 208–209
Divergence theorem, 8
Divergence, vector field
  cartesian coordinates, 9
  cylindrical coordinates, 12
  spherical coordinates, 12
Doublet
  distribution, 47–48, 72, 235–236,
    239–241, 242–244
  quadrilateral, 247–250
  three-dimensional, 47–48, 51–54
  two-dimensional, 48, 57–58, 231
Downwash
  definition, 170–171
  lifting-line, 169–172
Drag
  definition, 86
  friction drag, 506
  induced drag, 173–175, 201–204
Drag coefficient, 69, 87

Effective angle of attack, 171–172
Ellipse, flow past, 99–100
Elliptic lift distribution, 173–178
Euler angles, 369
Euler equation, 11
Euler number, 16
Eulerian method, 1

Flap, 113–114, 298, 501–505
Flat plate
  airfoil, 110–112, 130–131
  in normal flow, 133–134
  oscillation of, 396–399
  sudden acceleration of, 381–387
Flow similarity, 19
Fluid element, 1, 5
Force, 4–6
Fourier series, 106
Free stream, 54
Free-surface flows, 530–533
Froude number, 15–16
Fuselage, effect on lift, 360–361

Gap, effect on wing, 364–366
Geometric twist, 181
Glauert integral, 98
Green's theorem, identity, 30, 44–45
Ground effect
  inclusion in computation scheme, 338–340
  lumped vortex, 116–118
  rectangular wings, 350
  unsteady flow, 431, 433

Helmholtz vortex theorems, 34
Horseshoe vortex, 168–171, 256–258

Images, method of
  circle, 144–146
  parallel walls, 142–144
  plane wall, 141–142
Incompressible fluid, 9
Induced angle of attack, 171–172
Induced drag, 173–175, 201–204, 336–338, 346–347
Infinity condition, 28
Influence coefficient definition, 214
Irrotational flow, 23

Jones, R. T., method of, 192–194
Joukowski transformation, 128–137

Kelvin theorem, 25–26
Kinematic viscosity, 16
Kutta condition, 88–89, 209–213, 375–376, 416–419
Kutta-Joukowski theorem, 66–67, 128
  generalized, 146–149

Lagrangian method, 2
Laminar bubble, 496, 499
Laplace's equation
  cylindrical coordinates, 11
  definition, 27
  spherical coordinates, 12
Leading edge separation, 496, 516–528
Leading edge suction, 107–108, 131–133
Lift, 86
Lift coefficient
  definition, 69, 87
  for finite wings, 175
  maximum, 497, 504
  for thin airfoils, 109
Lift slope, 109–110, 175–176
Lifting line, 167–183, 331–338
Lifting surface, 82–85
  numerical, 340–351
  unsteady, 479–491
Local (leading edge) solution, 157–160
Lumped vortex element, 134–135

Mach number, 16
Mapping, 125
Matched asymptotic expansions, 160–163
Material derivative, 9, 11–12
Milne-Thomson circle theorem, 144–146
Moment coefficient, 87
Momentum equation, 7–13
Multielement wing, 363–364, 501–504

Navier-Stokes equations, 10–13
Neumann boundary condition, 30, 49, 207–208
Newton's second law, 7, 10
Normal force, 86
Normal stress, 6
No-slip condition, 11

Index

Panel methods, 206–226, 262–367
Parabolic arc airfoil, 112–113, 268–270
Pathlines, 3
Perturbation methods, 151–166
Poisson's equation, 36
Prandtl-Glauert rule, 91–92
Pressure, 6
Pressure coefficient, 16, 430
Propulsion effects, 528–530

Rankine's oval, 60–62
Reduced frequency, 373, 418–419
Residue theorem, 124
Reynold's number, 16
Rotor, 379, 504–506

Separated flow, 69, 508–516
Separation point, 509, 511
Shear stress, 6
Side force, 86
Similarity of flow, 19
Slat, 534–535
Slender body theory, 195–201
Slender wing theory, 184–195
 in unsteady motion, 400–407
Source
 distribution, 47–48, 70–72, 233–234, 238–239
 quadrilateral, 245–247
 three-dimensional, 47–51
 two-dimensional, 48, 56–57, 230–231
Sphere, flow past, 67–69
Stagnation flow, 55–56
Stagnation point, 56
Stall, 536–537, 542
Starting vortex, 168, 364–365
Stokes theorem, 22
Streak lines, 3
Stream function, 41–43
Streamline, 3–4
Stress vector, 4–6
Strouhal number, 15, 539
Sudden acceleration
 flat plate, 381–387
 rectangular wing, 429–431
Superposition principle, 60
Surface forces, 4

Swept wings, 347–349
Symmetric airfoil, 94–100
Symmetric wing, 79–82

Tandem airfoils, 116
Tangential stress, 6
Taper ratio, 349–350
Theodorsen lift deficiency function, 398–399
Thickness function, 78
Thin-airfoil theory, 94–121
Transition, 523, 526
Transpiration velocity, 227–228, 476, 491–492
Trefftz plane, 202–204
Turbulent boundary layer, 487–495
Twist, 172, 181–183

Uniqueness of solution of Laplace's equation, 30–32

van de Vooren airfoil, 137–138, 140–141
Velocity, 1–2
Velocity potential, 26
 perturbation, 77
 total, 77
Viscosity coefficient, 6
Vortex
 Asymmetry, 520–522
 Burst (breakup), 520–522
 core, 36, 254–255
 distribution, 73, 236–237, 241–242
 filament, 32–34
 horseshoe, 168–171, 296–297
 irrotational, 36, 58–60
 lift, 517–522
 line, 32, 38–41, 251–255
 ring, 250–251, 255–256
 two-dimensional, 34–36, 231–232
Vorticity
 definition, 22
 rate of change, 24–25

Wake, 83–85, 87–90, 364, 508–509
Wake rollup, 483–487, 512–514
Wind tunnel wall interference, 118–119, 163–165, 363–364, 530

Zero-lift angle of attack, 109